Handbook *of* Differential Geometry

VOLUME II

Handbook *of* Differential Geometry

VOLUME II

Editors

Franki J.E. Dillen
Leopold C.A. Verstraelen
Katholieke Universiteit Leuven
Department of Mathematics
Leuven, Belgium

ELSEVIER
Amsterdam • Boston • Heidelberg • London • New York • Oxford
Paris • San Diego • San Francisco • Singapore • Sydney • Tokyo
North-Holland is an imprint of Elsevier
NORTH-HOLLAND

ELSEVIER B.V.
Radarweg 29
P.O. Box 211, 1000 AE Amsterdam
The Netherlands

ELSEVIER Inc.
525 B Street, Suite 1900
San Diego, CA 92101-4495
USA

ELSEVIER Ltd
The Boulevard, Langford Lane
Kidlington, Oxford OX5 1GB
UK

ELSEVIER Ltd
84 Theobalds Road
London WC1X 8RR
UK

First edition 2006

Library of Congress Cataloging in Publication Data
A catalog record is available from the Library of Congress.

British Library Cataloguing in Publication Data
A catalogue record is available from the British Library.

ISBN-13: 978-0-444-52052-4
ISBN-10: 0-444-52052-X

⊗The paper used in this publication meets the requirements of ANSI/NISO Z39.48-1992 (Permanence of Paper).
Printed in The Netherlands.

Dedication

In memory of S.S. Chern and T. Willmore

Preface

"Our goal with the volumes which together will constitute the "Handbook of Differential Geometry" is to give a rather complete survey of the field of differential geometry." Thus reads the opening sentence of the "Handbook of Differential Geometry, Volume I", and only the presence of the word "rather" saves this goal from being an obvious mission impossible. Let us recall the contents of this Volume I: *Differential geometry of webs* (M.A. Akivis and V.V. Goldberg), *Spaces of metrics and curvature functionals* (D.E. Blair), *Riemannian submanifolds* (B.-Y. Chen), *Einstein metrics in dimension four* (A. Derdzinski), *The Atiyah–Singer index theorem* (P.B. Gilkey), *Survey of isospectral manifolds* (C.S. Gordon), *Submanifolds with parallel fundamental form* (Ü. Lumiste), *Sphere theorems* (K. Shiohama), *Affine differential geometry* (U. Simon), *A survey on isoparametric hypersurfaces and their generalizations* (G. Thorbergsson), *Curves* (T. Willmore); with introduction by S.S. Chern.

As in Volume I, we allowed the authors in this Volume II as much freedom as possible concerning style and contents. We are confident that the reader will appreciate this pragmatic point of view. Some contributions will emphasize the basics; some will emphasize the classical results; others the recent developments. Needless to say all authors have spent a lot of time and energy in describing their topic, which we appreciate enormously.

The contributions to this Volume II are: *Some problems on Finsler geometry* (J.C. Álvarez Paiva), *Foliations* (R. Barre and A. El Kacimi), *Symplectic geometry* (A. Cannas da Silva), *Metric Riemannian geometry* (K. Fukaya), *Contact geometry* (H. Geiges), *Complex differential geometry* (I. Mihai), *Compendium on the geometry of Lagrange spaces* (R. Miron), *Certain actual topics on modern Lorentzian geometry* (F.J. Palomo and A. Romero).

Obviously the whole field of differential geometry is not yet covered in the two volumes of this "Handbook of Differential Geometry". Some of the authors explicitly mention topics that should have been covered, but are not for practical reasons; but also other topics are not (yet) treated sufficiently or not treated at all.

Recently Professors Chern and Willmore passed away. Both had a great impact on the development of contemporary geometry and were genuine sources of inspiration, guidance and support for many generations of mathematicians through their books and articles, their fantastic lectures and their warm and truly concerned personal contacts. Together with all authors we gratefully dedicate this book to the memories of Professor S.S. Chern and Professor T.J. Willmore.

Franki Dillen and Leopold Verstraelen

List of Contributors

Álvarez Paiva, J.C., *Polytechnic University, Brooklyn, NY* (Ch. 1).
Barre, R., *Université de Valenciennes, Valenciennes* (Ch. 2).
Cannas da Silva, A., *Instituto Superior Técnico, Lisboa* (Ch. 3).
El Kacimi Alaoui, A., *Université de Valenciennes, Valenciennes* (Ch. 2).
Fukaya, K., *Kyoto University, Kyoto* (Ch. 4).
Geiges, H., *Universität zu Köln, Köln* (Ch. 5).
Mihai, I., *University of Bucharest, Bucharest* (Ch. 6).
Miron, R., *"Al.I. Cuza" University Iasi, Iasi* (Ch. 7).
Palomo, F.J., *Universidad de Málaga, Málaga* (Ch. 8).
Romero, A., *Universidad de Granada, Granada* (Ch. 8).

Contents

Contents of Volume I

CHAPTER 1

Some Problems on Finsler Geometry*

J.C. Álvarez Paiva**

Department of Mathematics, Polytechnic University, Six MetroTech Center, Brooklyn, NY 11201, USA
E-mail: jalvarez@duke.poly.edu

We do like intuitive geometric arguments and uncovering simple geometric reasons underlying seemingly recondite facts.
H. Busemann

Contents

*This work was partially funded by a *crédit aux chercheurs* from the FNRS, Belgium.
**I warmly thank G. Berck, D. Burago, C. Durán, E. Fernandes, P. Foulon, I.M. Gelfand, R. Schneider, Z. Shen and S. Tabachnikov for some very useful conversations and/or criticisms of earlier versions of this work.

HANDBOOK OF DIFFERENTIAL GEOMETRY, VOL. II
Edited by F.J.E. Dillen and L.C.A. Verstraelen

Abstract

This chapter is an unorthodox survey of Finsler geometry presenting both results and open problems. It aims to show that recent progress in convex geometry, the calculus of variations, symplectic geometry, and integral geometry can be powerful tools in the study of Finsler manifolds; and that Finsler geometry can prove useful in solving some of the open problems in these fields.

1. Introduction

Finsler manifolds, manifolds whose tangent spaces carry a norm that varies smoothly with the base point, were born prematurely in 1854 together with their Riemannian counterparts in Riemann's ground-breaking *Habilitationsvortrag*. I say prematurely because in 1854 Minkowski's work on normed spaces and convex bodies (see [69]) was still forty three years away, and thus not even the infinitesimal geometry on which Finsler manifolds are based was understood or appreciated at the time. Apparently, Riemann did not know what to make of these 'more general class' of manifolds whose element of arclength does not originate from a scalar product and, fatefully, put in a bad word for them [44]:

> Investigation of this more general class would actually require no essential different principles, but it would be rather time-consuming and throw relatively little new light on the study of Space, especially since the results cannot be expressed geometrically.

Given the awe with which we rightfully regard Riemann's achievements and uncanny geometrical intuition, it is tempting to take the above quotation out of historical context and to dismiss Finsler geometry altogether. But, if we think of the great advances in convex geometry, the calculus of variations, integral geometry, the theory of metric spaces, and symplectic geometry that have taken place since 1854, then we may be moved to reassess Riemann's statement and to consider applying these new tools to develop the subject in a way that Riemann could not have foreseen.

The paper includes eighteen simply-stated open problems, as well as a survey of the more elementary and geometric chapters of Finsler geometry. It presents a detailed discussion of the Holmes–Thompson volume and its role in integral geometry and geometric inequalities, thus complementing the survey by Álvarez and Thompson [16]. The other highlights of the paper are its presentation of Hilbert's fourth problem and its elementary approach to the differential invariants of Finsler surfaces. These are mostly based on the papers [15,12,13] with I.M. Gelfand, M. Smirnov and E. Fernandes, as well as on the lecture notes [10] written jointly with C. Durán.

In view of the often-made criticisms of Finsler geometry—very few concrete and interesting examples, very few non-Riemannian theorems of real geometric content, and too many subindices—I have tried to include as many concrete examples, simply-stated results, and geometric constructions as possible. In this way, many of the jewels, so to speak, of Finsler geometry find their way into the following pages.

As anyone writing a survey paper, I have had to make some choices. In matters of taste, I have consistently preferred the concrete to the abstract, the elementary to the advanced, the C^∞ to the C^k, and the global to the local. I have stayed clear of Riemann–Finsler geometry and Finsler connections because the book [23] of Bao, Chern and Shen covers the subject in depth as do the lecture notes of Abate and Patrizio [1]. Because of my ignorance of the subject, I have not touched on complex Finsler geometry (see [1] also for this topic) and, despite their undeniable interest and importance, non-reversible Finsler metrics are barely mentioned. Another important topic that is not covered in this survey is Busemann's G-spaces. This approach, which consists in abstracting the properties of geodesics on Finsler manifolds, is one of the most powerful in Finsler geometry, but it is impossible to outdo Busemann's own exposition in [37,38].

The reader wishing to have a broader view of the activity in Finsler geometry should read the short surveys of Chern [44] and Busemann [35]. The book [20] contains survey articles on several topics including non-reversible Finsler metrics, the proceedings [22] contains many open problems and a vista of various approaches to Finsler geometry. The book [84] contains a beautiful exposition of the convex-geometric aspects of the Holmes–Thompson volume as well as most of the convex geometry necessary for the study of Finsler manifolds. The lecture notes [10] are similar in spirit to the present paper. The thesis of Egloff [46] is a good place to learn about the beautiful results of Egloff [47,48] and Foulon [50–52] on the geometry and dynamics of Finsler manifolds with non-positive and negative curvature. Finally, I wholeheartedly recommend looking at the papers [42] and [3] before plunging into other papers where Finsler connections are treated.

The reader can find many of the preprints cited in this paper, along with other works on the interaction between convex, integral, metric, and symplectic geometry, in http://www.math.poly.edu/research/finsler.

2. Preliminaries

If $(V, \|\cdot\|)$ is a real, finite-dimensional normed space, we define the length of a smooth curve $\gamma : [a, b] \to V$ by the formula

$$\text{length of } \gamma := \int_a^b \|\dot{\gamma}(t)\| \, dt.$$

A smooth submanifold $N \subset V$ inherits a metric from the norm: if $\mathbf{x}$ and $\mathbf{y}$ are two points on N, define their distance as the infimum of the lengths of all smooth curves on N joining $\mathbf{x}$ and $\mathbf{y}$. Notice that in order to define the metric on N it suffices to know the restriction of the norm to each tangent space. This motivates the following heuristic definition: A Finsler manifold is a manifold together with the choice of a norm on each tangent space. The precise definition requires us to restrict the class of norms to those where the unit sphere is smooth and *quadratically convex* (i.e., it has positive principal curvatures for some (and therefore any) Euclidean structure on V). The intrinsic definition of these norms is as follows:

Let V be a vector space and let $\varphi : V \to [0, \infty)$ be a norm that is smooth outside the origin. Set $L := \varphi^2/2$ and consider the exterior derivative of L, dL, as a map from V minus the origin to V^* minus the origin. The norm φ is said to be a *Minkowski norm* if dL is a diffeomorphism.

For any non-zero vector $\mathbf{v} \in V$, the differential

$$D(dL)(\mathbf{v}) : T_{\mathbf{v}}V \to T_{dL(\mathbf{v})}V^*$$

is an invertible linear map. In fact, using the natural identification of $T_{\mathbf{v}}V$ with V, and $T_{dL(\mathbf{v})}V^*$ with V^*, we can think of $g_\varphi(\mathbf{v}) := D(dL)(\mathbf{v})$ as a (symmetric) bilinear form on V:

$$g_\varphi(\mathbf{v})(\mathbf{w}_1, \mathbf{w}_2) := \big(D(dL)(\mathbf{v})(\mathbf{w}_1)\big)(\mathbf{w}_2).$$

The norm φ is a Minkowski norm if and only if g_φ is positive definite. When the vector $\mathbf{v}$ belongs to the unit sphere, we will denote $g_\varphi(\mathbf{v})$ as the *osculating Euclidean structure* at $\mathbf{v}$ and the ellipsoid

$$E_{\mathbf{v}} := \{\mathbf{w} \in V\colon\ g_\varphi(\mathbf{v})(\mathbf{w}, \mathbf{w}) = 1\}$$

as the *osculating ellipsoid* at $\mathbf{v}$.

DEFINITION 2.1. A *Finsler metric* on a manifold M is a continuous function defined on its tangent bundle with the property that it is smooth away from the zero section and its restriction to each tangent space is a Minkowski norm.

Some examples of Finsler manifolds are submanifolds of Minkowski spaces and flat tori obtained as quotients of Minkowski spaces.

If $\gamma : [a, b] \to M$ is a smooth curve on a Finsler manifold (M, φ), then the quantity

$$\text{length of } \gamma := \int_a^b \varphi(\dot\gamma(t))\, dt \tag{1}$$

is independent of the parameterization. Using this definition of length we define a metric on M by letting the distance between two points $x, y \in M$ to be the infimum of the lengths of all smooth curves joining x and y. Finsler manifolds are *length spaces*: the length of a curve γ defined by the integral in (1) equals the metric length of the curve given by

$$\sup\left\{\sum_{i=0}^{k-1} \operatorname{dist}(\gamma(t_i), \gamma(t_{i+1}))\colon\ a = t_0 < \cdots < t_k = b \text{ is a partition of } [a, b]\right\}.$$

The condition that the norms in each tangent space be Minkowski norms is necessary for the study of the geodesics. Namely, we want these to be solutions of a second-order differential equation on M.

2.1. *The Hamiltonian point of view*

If (V, φ) is a normed space, then the dual vector space V^* inherits a natural norm defined by the equation

$$\varphi^*(\boldsymbol{\xi}) := \sup\{|\boldsymbol{\xi}(\mathbf{v})|\colon\ \varphi(\mathbf{v}) \leqslant 1\}.$$

A related construction on Minkowski spaces is the *Legendre transform* which assigns to a non-zero vector $\mathbf{v} \in V$ the covector $\varphi(\mathbf{v})\, d\varphi(\mathbf{v}) = g_\varphi(\mathbf{v})(\mathbf{v}, \cdot)$. It is easy to check that if $\mathbf{v}$ belongs to the unit sphere $S \subset (V, \varphi)$, then the Legendre transform of $\mathbf{v}$ is the unique covector $\boldsymbol{\xi}$ such that $\boldsymbol{\xi} = 1$ is the hyperplane tangent to S at the point $\mathbf{v}$. This implies that the image of S under the Legendre transform is the unit sphere in (V^*, φ^*).

Let (M, φ) be a Finsler manifold and for each point $m \in M$ let φ_m denote the Minkowski norm on $T_m M$. If $(T_m^* M, \varphi_m^*)$ is the dual of the normed space $(T_m M, \varphi_m)$, then the function

$$H : T^* M \to \mathbb{R}$$

defined by $H(\mathbf{p}_m) := \varphi_m^*(\mathbf{p}_m)$ is a Hamiltonian whose energy surfaces are fiber-wise convex. Applying the Legendre transform on each tangent space of M defines a diffeomorphism $\mathcal{L} : TM \setminus 0 \to T^*M \setminus 0$ with the property that $H \circ \mathcal{L} = \varphi$.

By passing from a Finsler metric to its associated Hamiltonian, we gain access to the techniques of Hamiltonian mechanics and symplectic geometry. Below, we recall some of the basic definitions and constructions. For more information see [18] and [2].

DEFINITION 2.2. Let $\pi : T^*M \to M$ be the standard projection and let $D\pi : T(T^*M) \to TM$ be its differential. The *canonical 1-form* α on T^*M is defined by the equation $\alpha(\mathbf{v}_{\mathbf{p}_m}) = \mathbf{p}_m(D\pi(\mathbf{v}_{\mathbf{p}_m}))$, where $\mathbf{p}_m \in T_m^*M$ and $\mathbf{v}_{\mathbf{p}_m} \in T_{\mathbf{p}_m}(T^*M)$. The *symplectic 2-form* is defined as $\omega := -d\alpha$.

The form ω is non-degenerate: at each point $\mathbf{p}_m \in T^*M$, the map $\mathbf{v}_{\mathbf{p}_m} \mapsto \omega_{\mathbf{p}_m}(\mathbf{v}_{p_m}, \cdot)$ is an isomorphism from $T_{\mathbf{p}_m}(T^*M)$ to $T^*_{\mathbf{p}_m}(T^*M)$. We can use this isomorphism to pass from 1-forms on T^*M to vector fields on T^*M.

DEFINITION 2.3. Let $H : T^*M \to \mathbb{R}$ be a smooth function. The *Hamiltonian vector field* of H, X_H, is defined by the equality $dH = \omega(X_H, \cdot)$.

As an easy consequence of the definition, we have that *H is constant along the integral curves of the Hamiltonian vector field X_H, and that the symplectic form is invariant under the flow of X_H.*

Because of this result, it is usual to disregard the function H in favor of the *unit co-sphere bundle* $S_H^*M := H^{-1}(1)$. If α is the canonical 1-form on T^*M, then its restriction to the unit co-sphere bundle S_H^*M, which we denote by α_H, is a *contact form* (i.e., the top-order form $\alpha_H \wedge (d\alpha_H)^{n-1}$ never vanishes). Using α_H, we can define the restriction of the Hamiltonian vector field X_H without any reference to the function H:

DEFINITION 2.4. The *Reeb vector field* X_H on S_H^*M is defined by the equations

$$\alpha_H(X_H) = 1, \qquad d\alpha_H(X_H, \cdot) = 0.$$

The projection to M of the integral curves of this vector field are geodesics parameterized with unit speed. Conversely, if γ is a geodesic on M parameterized with unit speed, then the Legendre transform $\mathcal{L}$ maps the velocity curve $\dot{\gamma}$ to an orbit of the Reeb vector field. We remark that if γ is any smooth curve on M parameterized with unit speed, then

$$\text{length of } \gamma = \int_{\mathcal{L} \circ \gamma} \alpha_H. \tag{2}$$

Note that the Finsler manifold (M, φ) is geodesically complete (or metrically complete, since it is easy to verify that the Hopf–Rinow theorem extends to the Finsler setting) if and only if the Reeb vector field defines a flow.

Let us finish this section by remarking that the non-degeneracy of the symplectic form ω on T^*M is equivalent to the fact that ω^n, $n = \dim(M)$, is a volume form. This remark will provide us with a natural way to define the volume of a Finsler manifold.

2.2. *The Riemannian point of view*

Finsler manifolds can also be studied from the point of view of Riemannian manifolds and bundles. Indeed, to every unit vector $\mathbf{v}_m \in T_m M$ we may associate the inner product $g_\varphi(\mathbf{v})$. In this way, we can define a Riemannian structure on the pullback of the tangent bundle of M to the unit tangent bundle of M. This construction underlies many of the definitions of connections associated to Finsler manifolds.

A variation on this theme is to take a nowhere zero vector field X defined on an open subset $\mathcal{O} \subset M$ and to associate to it the Riemannian metric on $\mathcal{O}$ defined by $m \mapsto g_\varphi(X(m))$. This construction has been used by Shen (see [79] and Section 8) to give a simple description of the Finsler curvature.

2.3. *Isometries and isometric embeddings*

The definitions of *isometry* and *isometric embedding* between Finsler manifolds (M, φ_M) and (N, φ_N) are the same as for Riemannian manifolds. Namely, an isometry (respectively isometric embedding) is a diffeomorphism (respectively embedding) $f : M \to N$ such that $f^*\varphi_N = \varphi_M$. Unlike Riemannian manifolds, two Finsler manifolds can fail to be isometric because of what happens at a single tangent space. For example, if at a point $m \in M$ the *indicatrix*

$$S_m M := \{\mathbf{v}_m \in T_m M \colon \varphi_M(\mathbf{v}_m) = 1\}$$

is an ellipsoid while none of the indicatrices of N are ellipsoids, then M and N are not isometric. This remark points at the important role played by the centro-affine geometry of convex hypersurfaces in Finsler geometry.

In [32] Burago and Ivanov showed that *any compact Finsler manifold admits an isometric embedding into a finite-dimensional normed space.* It is likely, but unproved, that the norm can be chosen to be a Minkowski norm. They also give examples of non-compact Finsler manifolds that cannot be isometrically embedded in any finite-dimensional normed space. Examples of Finsler manifolds that cannot be isometrically embedded in any *Minkowski* space have been given by Shen (see [79]) and by Álvarez and Durán (see Section 8.5).

However, new types of embedding problems arise in Finsler geometry. For example, while it is known that every two-dimensional normed space is isometric to a subspace of $L_1([0, 1])$ (see, for example, [29]), it is not clear whether the following problem has an affirmative answer.

PROBLEM 1. Does every two-dimensional Finsler manifold admit an isometric embedding into the Banach space $L_1([0, 1])$?

The relation between intrinsic and extrinsic geometric properties of submanifolds of normed space is not understood. For example, the following problem from [84] is open.

PROBLEM 2 (Thompson). Let X and Y be two normed spaces of dimension n, $n > 2$, such that its unit spheres are isometric as Finsler manifolds. Does it follow that X and Y are isometric as normed spaces?

2.4. *Isometric submersions*

In Riemannian geometry isometric submersions are used to construct examples of Riemannian manifolds while keeping some control on their geodesics and curvature. The Finslerian generalization of this construction is simple, but perhaps not as well known as it should be. What follows is taken from [11].

DEFINITION 2.5. A surjective linear map $\pi : X \to Y$ between two normed spaces is said to be an *isometric submersion* if the image of the closed unit ball on X under the map π equals the closed unit ball on Y.

Clearly, if $\mathbf{x}$ is any vector in X, then $\|\pi(\mathbf{x})\|_Y \leqslant \|\mathbf{x}\|_X$. The vectors for which the equality holds are called *horizontal vectors* and form the *horizontal cone* in X. Notice that if the unit sphere in X is smooth, its intersection with the horizontal cone is the singular set of the restriction of the map π to the unit sphere. This description makes it easy to grasp that, unlike the case where the spaces is Euclidean, the horizontal cone is rarely a subspace.

DEFINITION 2.6. A submersion $\rho : M \to N$ between Finsler manifolds is said to be *isometric* if for every point $m \in M$ the differential

$$D_m \rho : T_m M \to T_{\rho(m)} N$$

is an isometric submersion of normed spaces.

More generally, Berestovskii has defined in [24] isometric submersions, or *submetries* of metric spaces, as maps that send metric balls to metric balls of the same radius. When the metric spaces are Finsler manifolds both notions of isometric submersion agree. However, in the particular case of Finsler manifolds we can also speak of horizontal lifts.

DEFINITION 2.7. Let $\rho : M \to N$ be an isometric submersion. An immersed curve $\gamma : [a, b] \to M$ is said to be *horizontal* if for every $t \in (a, b)$ the velocity vector $\dot{\gamma}(t)$ belongs to the horizontal cone in $T_{\gamma(t)} M$. A curve $\gamma : [a, b] \to M$ is said to be a *horizontal lift* of an immersed curve $\sigma : [a, b] \to N$ if γ is horizontal and $\rho \circ \gamma = \sigma$.

The following result is an easy consequence of the definitions and of the basic properties of geodesics in a Finsler manifold.

THEOREM 2.8 (Álvarez and Durán [11]). *An immersed curve on N is a geodesic if and only if any of its horizontal lifts is a geodesic on M. In particular, the geodesics of N are precisely the projections of horizontal geodesics on M.*

As we shall see in Section 6, this theorem is useful in constructing interesting examples of Finsler metrics on complex and quaternionic projective spaces.

3. Volume and area in Finsler spaces

The theory of volume and area in normed spaces has long been a major driving force in convex geometry. For example, the Blaschke–Santaló inequality, the Mahler conjecture, the Busemann–Petty problems, the Shephard problem, and the numerous works of Busemann, Ewald and Shephard on the notions of convexity on Grassmannians originated from or have applications to the study of volumes and areas in normed spaces (see [16]).

Defining a volume on a finite-dimensional normed space seems easy: a natural volume should be invariant under translations, positive on open sets, and finite on compact sets. By Haar's theorem, such a volume must be a multiple of the Lebesgue measure. However, the choice of this multiple is crucial. To understand this, suppose we have already decided on how to assign those constants to two-dimensional normed spaces. We can now define the area of a two-dimensional polyhedral surface embedded in a 3-dimensional normed space as the sum of the areas of its faces with their induced norms. Making a different choice of constants leads to a completely different way of measuring the area of polyhedral surfaces.

The guiding principle in defining volumes on normed spaces is that the choice of a volume for every k-dimensional normed space leads to the definition of the k-volume integrand in all higher-dimensional normed spaces. Requiring even mild conditions on these area integrands, such as that regions in hyperplanes be area-minimizing, severely restricts our choices for a definition of volume. In fact, in the literature one can only find three reasonable choices of volume on normed spaces: the Busemann definition, the Holmes–Thompson definition, and the Benson definition (also known as Gromov's mass* [56]).

Since the Busemann volume of a Finsler manifold coincides with its Hausdorff measure as a metric space, at first sight it seems the most natural and geometric definition. However, the Holmes–Thompson definition, with its ties to Brunn–Minkowski theory, integral, and symplectic geometry, is rapidly becoming the definition of choice.

DEFINITION 3.1. The *Holmes–Thompson volume* of an n-dimensional Finsler manifold (M,φ), $\mathrm{vol}_n(M,\varphi)$, is the symplectic volume of its unit co-disc bundle divided by the volume of the Euclidean n-dimensional unit ball. The k-volume of a k-dimensional submanifold is the volume of the submanifold with its induced Finsler metric.

Note that, using the notation of the previous section, we have that

$$\mathrm{vol}_n(M) = \frac{1}{\epsilon_n n!} \int_{S_H^* M} \alpha_H \wedge (d\alpha_H)^{n-1}, \tag{3}$$

where ϵ_n is the volume of the Euclidean unit ball of dimension n.

Using the Blaschke–Santaló inequality, Durán has remarked in [45] that *the Hausdorff measure of a Finsler manifold is no less than its Holmes–Thompson volume. Equality holds if and only if the metric is Riemannian.*

One of the basic problems about a given definition of volume on normed and Finsler spaces is determining the convexity or ellipticity properties of the k-volume integrands. Perhaps the most enticing and difficult question of the kind is the following problem of Busemann:

PROBLEM 3. Let P be a compact polyhedron of dimension k in a normed space. Is the k-volume of any given face less than or equal to the sum of the k-volumes of the remaining faces?

In the case of polyhedra of codimension one the answer to this question is affirmative for both the Hausdorff measure (Busemann [34]) and the Holmes–Thompson volume (Holmes and Thompson [62]).

For the Hausdorff measure no other results of this kind are known. For the Holmes–Thompson volume the answer to Busemann's question is known to be affirmative if the normed space is a subspace of $L_1([0, 1])$:

THEOREM 3.2 (Busemann et al. [40]). *Let P be a compact polyhedron of dimension k in the Banach space $L_1([0, 1])$. The (Holmes–Thompson) k-volume of any given face is less than or equal to the sum of the k-volumes of the remaining faces.*

The latest progress on the question of minimality of flats in normed spaces is the following result of Burago and Ivanov:

THEOREM 3.3 (Burago and Ivanov [33]). *Let P be a compact 2-dimensional polyhedron in a normed space. If P is homeomorphic to a sphere, then the (Holmes–Thompson) 2-volume of any given face is less than or equal to the sum of the 2-volumes of the remaining faces.*

In [64], S. Ivanov shows that this result can be extended to a theorem that is new even in the Riemannian case:

THEOREM 3.4 (Ivanov [64]). *Let φ be a Finsler metric on the closed two-dimensional disc D such that every two points on D are joined by a unique geodesic. If ψ is another Finsler metric on D such that the distance induced by ψ on the boundary ∂D of D is greater than or equal to the distance induced by φ on ∂D, then the Holmes–Thompson volume of (D, φ) does not exceed that of (D, ψ).*

It follows from Ivanov's theorem that a two-dimensional totally geodesic submanifold in a Finsler space is minimal—in the usual sense of being an extremal for the 2-volume functional—with respect to the Holmes–Thompson definition of area. This is true in all dimensions:

THEOREM 3.5 (Berck [25]). *A totally geodesic submanifold of a Finsler manifold is minimal with the respect to the Holmes–Thompson definition of area.*

In [9], Álvarez and Berck show that Theorems 3.4 and 3.5 no longer hold if the Holmes–Thompson volume is replaced by the Hausdorff measure.

One of the advantages of working with the Holmes–Thompson definition is that there is a remarkably simple formula for the Holmes–Thompson k-volume densities of a Minkowski space in terms of the Fourier transform of its norm. In a different guise, this formula was first obtained by W. Weil [85]. In the present form it was rediscovered by Álvarez and Fernandes in [13].

FOURIER TRANSFORMS OF NORMS. Let ϕ be a smooth, even homogeneous function of degree one on an n-dimensional vector space V, let $\mathbf{e}_1, \dots, \mathbf{e}_n$ be a basis of V, and let $\boldsymbol{\xi}_1, \dots, \boldsymbol{\xi}_n$ be the dual basis in V^*. These bases allow us to introduce coordinates $(x_1, \dots, x_n)$ in V and $(\xi_1, \dots, \xi_n)$ in V^*, which we can use to compute the standard (distributional) Fourier transform

$$\hat{\phi}(\boldsymbol{\xi}) := \int_{\mathbb{R}^n} e^{i\boldsymbol{\xi}\cdot\mathbf{v}} \phi(\mathbf{v})\, d\mathbf{v}.$$

This transform depends on the choice of basis, or rather on the Lebesgue measure associated to it. However, the form $\hat{\phi}\, d\xi_1 \wedge \cdots \wedge d\xi_n$ does not. Up to a constant factor, the *Fourier transform* of ϕ is the contraction of this n-form with the Euler vector field, $X_E(\boldsymbol{\xi}) = \boldsymbol{\xi}$, in V^*:

$$\check{\phi} := \frac{-1}{4(2\pi)^{n-1}} \hat{\phi}\, d\xi_1 \wedge \cdots \wedge d\xi_n \rfloor X_E.$$

It is known (see [63, pp. 167–168]) that $\hat{\phi}$ is smooth on $V^* \setminus \{0\}$ and homogeneous of degree $-n-1$. It follows that $\check{\phi}$ is a smooth differential form on $V^* \setminus \{0\}$ which is homogeneous of degree -1.

DEFINITION 3.6. Let (V, φ) be an n-dimensional Minkowski space. For each integer k, $1 \leqslant k < n$, define the integrand

$$\varphi_k(\mathbf{v}_1 \wedge \cdots \wedge \mathbf{v}_k) := \int_{(\boldsymbol{\xi}_1, \dots, \boldsymbol{\xi}_k) \in S^{*k}} |\boldsymbol{\xi}_1 \wedge \cdots \wedge \boldsymbol{\xi}_k \cdot \mathbf{v}_1 \wedge \cdots \wedge \mathbf{v}_k| \check{\varphi}^k, \tag{4}$$

where S^* is any closed hypersurface in $V^* \setminus \{0\}$ that is star-shaped with respect to the origin.

THEOREM 3.7 [13]. *Let (V, φ) be an n-dimensional Minkowski space. If $N \subset V$ is an immersed submanifold of dimension k, $1 \leqslant k < n$, then we have the following formula for the Holmes–Thompson k-area of N:*

$$\mathrm{vol}_k(N) = \frac{1}{\epsilon_k} \int_N \varphi_k,$$

where ϵ_k denotes the volume of the Euclidean unit ball of dimension k.

One of the main justifications for adopting the Holmes–Thompson volume comes from its role in integral geometry (see [78,12,13,76,77]). In fact, the formula for the Holmes–Thompson k-area density in terms of the Fourier transform of the norm is equivalent to the following Crofton-type formula for Minkowski spaces:

THEOREM 3.8. *Let $(V, \|\cdot\|)$ be an n-dimensional Minkowski space and let k, $1 \leqslant k \leqslant n-1$, be an integer. There exists a smooth, translation-invariant, and possibly signed measure Φ_{n-k} on the manifold $H_{n,n-k}$ of $(n-k)$-flats of V such that if $N \subset V$ is an immersed k-dimensional submanifold, then*

$$\mathrm{vol}_k(N) = \frac{1}{\epsilon_k} \int_{\lambda \in H_{n,n-k}} \#(N \cap \lambda) \Phi_{n-k}, \tag{5}$$

where ϵ_k is the volume of the Euclidean unit ball of dimension k.

This theorem was first proved for finite-dimensional subspaces of $L_1([0,1])$ by Schneider and Wieacker [78]. In the form given above it is due to Álvarez and Fernandes [12]. It has recently been extended by Schneider [77] to generalized hypermetric spaces (i.e., finite-dimensional normed spaces where the distributional Fourier transform of the norm is a signed measure).

4. Unit spheres in Minkowski spaces

Surprisingly little is known about the Finsler geometry of unit spheres in Minkowski spaces. The classic results in two dimensions are the theorems of Gołąb and Schäffer (see [54] and [73]).

THEOREM 4.1 (Gołąb). *The length of the unit circle of a two-dimensional normed space is greater than or equal to six and less than or equal to eight. Moreover, the lower bound is attained if and only if the unit circle is an affine regular hexagon and the upper bound is attained if and only if the unit circle is a parallelogram.*

THEOREM 4.2 (Schäffer). *The length of the unit circle of a two-dimensional normed space equals the length of the unit circle of its dual space.*

The perimeter of the unit circle is at once its surface area, twice its intrinsic diameter, the length of its shortest closed geodesic, and the length of its shortest closed, symmetric geodesic. Each of these interpretations points to a different, possible, higher-dimensional extension of the theorems of Gołąb and Schäffer. In this section, we shall quickly survey what is known about the higher-dimensional analogues of Gołąb's theorem. The generalizations of Schäffer's theorem have an unexpected relation with symplectic geometry and will be discussed in the next section.

We start with the following upper bounds for the Hausdorff measure and the Holmes–Thompson area of unit spheres in finite-dimensional normed spaces.

THEOREM 4.3 (Busemann and Petty [41]). *The (intrinsic) Hausdorff measure of the unit sphere of an n-dimensional normed space is at most $2n$ times the volume of the Euclidean unit ball of dimension $n-1$. Equality holds if and only if the unit ball is a parallelotope.*

Since the Holmes–Thompson area is always less than or equal to the Hausdorff measure, we have the following corollary:

COROLLARY 4.4 (Thompson). *The Holmes–Thompson area of the unit sphere in an n-dimensional normed space is at most $2n$ times the volume of the Euclidean unit ball of dimension $n-1$. Equality holds if and only if $n=2$ and the unit ball is a parallelogram.*

The quest of the lower bound is much more challenging and interesting. The only result in dimension greater than two is the following (unpublished) sharp lower bound by Álvarez, Ivanov and Thompson.

THEOREM 4.5. *The Holmes–Thompson area of the unit sphere in a normed space of dimension three is at least $36/\pi$. This bound is attained, for example, when the unit sphere of the normed space is the rhombic dodecahedron or the cubo-octahedron.*

As a corollary, we have that the Hausdorff measure of the unit sphere of a three-dimensional normed space is greater than $36/\pi$. This result is, so far, the only contribution to the following problem of Busemann and Petty [41].

PROBLEM 4 (Busemann–Petty Problem 7). Find the sharp lower bound for the Hausdorff measure of the unit sphere of a normed space of dimension n, $n \geqslant 3$.

Of course, we have the analogous problem for the Holmes–Thompson definition of volume.

PROBLEM 5 (Thompson [84]). Find the sharp lower bound for the Holmes–Thompson area of the unit sphere of a normed space of dimension n, $n > 3$.

Schäffer has considered the length of the shortest closed geodesic that is symmetric about the origin. This length, which Schäffer calls the *girth* of the normed space, is twice the length of the shortest non-contractible geodesic, the *systole*, for the induced Finsler metric on the projective space.

THEOREM 4.6 (Schäffer [74]). *The girth of an n-dimensional normed space is at most eight and least $4+2[n/2]^{-1}$, where $[\cdot]$ denotes the greatest-integer function. Moreover if the girth equals eight, then the space is two-dimensional and its unit ball is a parallelogram.*

Here is yet another interesting question of Schäffer about the girth of normed spaces. The problem was posed as a conjecture in [74, p. 97].

PROBLEM 6 (Schäffer [74]). Prove or disprove that girth of a Minkowski space of dimension three is at most 2π and that equality holds if and only the space is Euclidean.

So far, we have only considered the shortest geodesic that is symmetric about the origin. Is it possible that on some unit sphere there is a shorter geodesic that is not symmetric?

PROBLEM 7 (Thompson). Is the shortest closed geodesic on the unit sphere of a Minkowski space symmetric with respect to the origin?

We now shift our attention to a more classic metric invariant, the inner or intrinsic diameter of the unit sphere.

THEOREM 4.7 (Schäffer [74]). *The (intrinsic) diameter of the unit sphere of an n-dimensional normed space is at most four and at least $2+[n/2]^{-1}$, where $[\cdot]$ denotes the greatest-integer function. In particular, the diameter of the unit sphere of a three-dimensional Minkowski space is between three and four.*

PROOF. In order to see that the diameter is at most four, let x and y be any two distinct points on the unit sphere and consider a plane passing through these points and the origin. The intersection of the plane with the sphere is a curve whose length, by Gołąb's theorem, is at most eight. It follows that the distance between x and y is at most four.

To obtain the lower bound, notice that, by Theorem 4.6, the length of the shortest closed, symmetric curve on the unit sphere is greater than or equal to $4+2[n/2]^{-1}$. This implies that there is a pair of antipodal points at a distance greater than or equal to $2+[n/2]^{-1}$ and the inequality follows. □

Schäffer also characterizes those normed spaces for which the diameter of the unit sphere equals four (see Theorem 9G in [74, p. 58]).

PROBLEM 8 (Schäffer [74]). Is the (inner) diameter of the unit sphere of a finite-dimensional normed space attained at a pair of antipodes?

5. Symplectic equivalence of Finsler manifolds

In this section we study several notions of symplectic equivalence between Finsler spaces and consider the higher-dimensional generalizations of Schäffer's Theorem 4.2.

5.1. *Equivalence of unit co-disc bundles*

The unit co-disc bundle of a Finsler manifold is an open subset of the cotangent bundle and, as such, it carries a symplectic structure. A vaguely posed, but possibly fruitful, problem is to relate the symplectic invariants of the unit co-disc bundle to the metric invariants of the Finsler manifold.

A large class of examples of Finsler manifolds with symplectomorphic unit co-disc bundles is furnished by the following result:

THEOREM 5.1 (Álvarez [8]). *Let* $(\mathbb{R}^n, \|\cdot\|_1)$ *and* $(\mathbb{R}^n, \|\cdot\|_2)$ *be two Minkowski spaces and let* S_1 *and* S_2 *denote their unit spheres. The unit co-disc bundle of the Finsler metric on* S_1 *induced by its embedding into* $(\mathbb{R}^n, \|\cdot\|_2)$ *is symplectomorphic to the unit co-disc bundle of the Finsler metric on* S_2^* *induced by its embedding into* $(\mathbb{R}^n, \|\cdot\|_1^*)$.

In particular, the unit sphere of a Minkowski space and its dual have symplectomorphic unit co-disc bundles. The following corollary—the first of our generalizations of Schäffer's theorem—predated the theorem, and was in effect its motivation.

COROLLARY 5.2 (Holmes and Thompson [62]). *The unit sphere of a normed space and that of its dual have the same Holmes–Thompson area.*

Besides examples of non-isometric Finsler manifolds with symplectomorphic co-disc bundles, it is interesting to look for rigidity results such as the following theorem of Benci and Sikorav (see [80] and [68, p. 365]).

THEOREM 5.3. *If the unit co-disc bundles of two flat Finsler tori are symplectomorphic, then the tori are isometric.*

5.2. *Equivalence of unit co-sphere bundles*

DEFINITION 5.4. Let M and N be two Finsler manifolds with unit co-sphere bundles S^*M and S^*N and canonical 1-forms α_M and α_N. The Finsler manifolds M and N will be said to be *exactly contactomorphic* if there exists a diffeomorphism $F : S^*M \to S^*N$ and a function f on S^*M such that $F^*\alpha_N = \alpha_M + df$. If $F^*\alpha_N = \alpha_M$, we shall say that the metrics are *α-equivalent*.

PROPOSITION 5.5. *If M and N are two exactly contactomorphic Finsler manifolds, then their volumes and their length spectra are equal. Moreover, if M and N are α-equivalent, then their geodesic flows are conjugate.*

PROOF. To see that the volume of M equals that of N we simply use Eq. (3) expressing the Holmes–Thompson volume of the manifold in terms of the canonical 1-form.

The equality of the length spectra follows from Eq. (2), which states that the action of a leaf of the geodesic foliation equals the length of the underlying geodesic. Indeed, if $F : S^*M \to S^*N$ is a diffeomorphism satisfying $F^*\alpha_N = \alpha_M + df$, then F maps the geodesic foliation on S^*M to the geodesic foliation on S^*N. Moreover, we see that closed leaves are taken to closed leaves with the same action.

In the case where $F^*\alpha_N = \alpha_M$, the expression for the geodesic spray as the Reeb vector field of the canonical 1-form immediately implies that the geodesic flows are conjugate. □

PROBLEM 9. Is every (reversible) Finsler metric on the two-sphere exactly contactomorphic or α-equivalent to a Riemannian metric?

A large class of examples of exactly contactomorphic Finsler manifolds is provided by the following analogue of Theorem 5.1.

THEOREM 5.6 (Álvarez [8]). *Let* $(\mathbb{R}^n, \|\cdot\|_1)$ *and* $(\mathbb{R}^n, \|\cdot\|_2)$ *be two Minkowski spaces and let* S_1 *and* S_2 *denote their unit spheres. The Finsler metric on* S_1 *induced by its embedding into* $(\mathbb{R}^n, \|\cdot\|_2)$ *and the Finsler metric on* S_2^* *induced by its embedding into* $(\mathbb{R}^n, \|\cdot\|_1^*)$ *are exactly contactomorphic. Moreover, the diffeomorphism* F *can be taken such that it takes centrally symmetric closed geodesics to centrally symmetric closed geodesics.*

As a corollary we obtain our second generalization of Schäffer's theorem:

COROLLARY 5.7. *The length of shortest closed geodesic on the unit sphere of a Minkowski space equals the length of the shortest closed geodesic on the unit sphere its dual.*

While the length of the shortest closed geodesic on the unit sphere of a Minkowski space is an interesting invariant, it seems very hard to prove that it is continuous with respect to any of the natural topologies in the space of convex bodies. In this respect, the girth—the infimum of the lengths of all centrally symmetric, simple, closed curves—is a much better invariant. Indeed, Schäffer proved in [74, p. 91] that the girth of a normed space is continuous with respect to the topology induced by the Banach–Mazur distance.

Schäffer also conjectured that the girth of a normed space equals the girth of its dual and proved that it is enough to consider the case of finite-dimensional normed spaces. Theorem 5.6 together with Schäffer's results settles the conjecture.

THEOREM 5.8 (Álvarez [8]). *The girth of a normed space equals the girth of its dual.*

The preceding theorem is a third generalization of the fact that the perimeter of the unit circle of a normed plane equals the perimeter of the dual circle. Schäffer showed in [74, p. 110] that a fourth possible generalization—that the (intrinsic) diameter of the unit sphere in a normed space equals the diameter of the dual sphere—is false. In particular, *the diameter of a Finsler manifold is not a symplectic invariant of its unit co-disc bundle.*

The notion of α-equivalence first appeared in Weinstein's work on the volume of Riemannian manifolds all of whose geodesics are closed. The symplectic and topological nature of his proofs implies that they extend unchanged for Finsler metrics.

THEOREM 5.9 (Weinstein [86]). *Let $\varphi_t, t \in [0, 1]$, be a smooth family of Finsler metrics on a compact manifold M. If for every t the geodesics of the Finsler metric φ_t are all closed and of fixed length L, independent of t, then (M, φ_0) and (M, φ_1) are α-equivalent. In particular, (M, φ_0) and (M, φ_1) have the same volume.*

Weinstein's theorem follows from the fact that the manifolds of geodesics of the metrics involved are symplectomorphic. In the next paragraph we will review the natural symplectic structure on spaces of geodesics and some of their applications to integral geometry and the minimality of submanifolds in Finsler spaces.

5.3. *Manifolds of geodesics*

Let M be a Finsler manifold such that its space of oriented geodesics is a manifold $G(M)$. Let S^*M denote its unit co-sphere bundle and let $\pi : S^*M \to G(M)$ be the canonical projection which sends a given unit covector to the geodesic that has this covector as initial condition.

PROPOSITION–DEFINITION 5.10 ([18] and [27]). Let M be a Finsler manifold with manifold of geodesics $G(M)$ and let

$$\begin{array}{ccc} S^*M & \xrightarrow{\;i\;} & T^*M \\ {\scriptstyle \pi}\big\downarrow & & \\ G(M) & & \end{array}$$

be the canonical projection onto $G(M)$ and the canonical inclusion into T^*M. If ω_M is the standard symplectic form on T^*M, then there is a unique symplectic form ω on $G(M)$ which satisfies the equation $\pi^*\omega = i^*\omega_M$.

At first sight there seem to be very few examples of Riemannian or Finsler manifolds whose space of geodesics is smooth. The following examples will convince the reader that this is not so.

EXAMPLES.

1. *Strictly convex balls and Hadamard manifolds* [49]. Around any point x in a Finsler manifold there is an open geodesic ball with the property that the function that assigns to every point in the ball its distance from x is strictly convex. The space of geodesics in such a geodesic ball is a smooth manifold. As a result, the space of geodesics of any complete Riemannian metric on $\mathbb{R}^n$ with non-positive sectional curvature (a Hadamard manifold) is a smooth manifold.
2. *Projective Finsler metrics*. These are Finsler metrics on open, convex subsets of $\mathbb{R}P^n$ such that projective line segments are geodesics. We will review their construction in the next section.

3. *Zoll manifolds*. These are Finsler metrics all of whose geodesics are periodic with the same minimal period. A great number of Riemannian examples has been constructed by Weinstein (see [27]).

It is interesting to determine when the manifolds of geodesics of two Finsler manifolds are symplectomorphic. Here are two results in this direction:

THEOREM 5.11 (Ferrand [49]). *The manifold of geodesics of a Hadamard manifold of dimension n is symplectomorphic to the cotangent bundle of the $(n-1)$-dimensional sphere.*

THEOREM 5.12 (Ono [70]). *The manifold of geodesics of a Zoll Finsler metric on S^3 is symplectomorphic to a complex hyperquadric in $\mathbb{C}P^3$.*

It follows from these theorems that all Hadamard manifolds and all Zoll metrics on S^3 are α-equivalent among themselves.

PROBLEM 10. Is the manifold of geodesics of a Zoll Finsler metric on the n-sphere symplectomorphic to a complex hyperquadric in $\mathbb{C}P^n$? Is the space of all Zoll Finsler metrics on the n-sphere connected?

The study of the symplectic geometry of the space of geodesics has interesting applications to the integral geometry of Finsler manifolds. For example, the classical integral-geometric theorem of Cauchy and its extension to finite-dimensional normed spaces is a consequence of the following symplectic equivalence.

THEOREM 5.13 (Álvarez [8]). *Let $(V, \|\cdot\|)$ be a Minkowski space, and let $M \subset V$ be a smooth quadratically convex hypersurface. The unit co-disc bundle for the induced Finsler metric on M and the set of all oriented lines in V which pass through the interior of M are symplectomorphic.*

The Crofton formula for hypersurfaces of Finsler spaces, announced by Chakerian in [43], follows easily from the co-area formula and symplectic reduction (see [9] for a proof).

THEOREM 5.14. *Let M be an n-dimensional Finsler manifold with manifold of geodesics $G(M)$. If $N \subset M$ is an immersed hypersurface and if ω^{n-1} denotes the Liouville volume form on $G(M)$, then*

$$\mathrm{vol}_{n-1}(N) = \frac{1}{2\epsilon_{n-1} n!} \cdot \int_{\gamma \in G(M)} \#(N \cap \gamma) \left|\omega^{n-1}\right|,$$

where ϵ_{n-1} is the volume of the Euclidean unit ball of dimension $n-1$.

6. Around Hilbert's fourth problem

In modern terminology, Hilbert's fourth problem asks to construct and study all Finsler metrics on open convex subsets of $\mathbb{R}P^n$ (including $\mathbb{R}P^n$ itself) such that geodesics lie on projective lines. At the root of this problem are Minkowski's work on normed spaces and the following generalization of the Cayley–Klein model of hyperbolic geometry given by Hilbert himself:

Let $D \subset \mathbb{R}^n$ be an open domain bounded by a convex hypersurface C. If $\mathbf{x}$ and $\mathbf{y}$ are two distinct points on D, denote by $\mathbf{a}$ and $\mathbf{b}$ the points of intersection of C with the line determined by $\mathbf{x}$ and $\mathbf{y}$ (see Figure 1), and define the distance between these points by the equation

$$d(\mathbf{x}, \mathbf{y}) := \frac{1}{2} \ln\left(\frac{\|\mathbf{y} - \mathbf{a}\| \, \|\mathbf{x} - \mathbf{b}\|}{\|\mathbf{x} - \mathbf{a}\| \, \|\mathbf{y} - \mathbf{b}\|}\right). \tag{6}$$

THEOREM 6.1 (Hilbert [59]). *The function d is a distance function on D. Moreover, straight line segments are geodesics.*

The metric space (D, d) is called a *Hilbert geometry*. The following elegant description of the Finsler metrics that gives rise to Hilbert geometries, and which I learned from R. Ambartzumian, is apparently well known.

Let $D \subset \mathbb{R}^n$ be an open domain bounded by a smooth, quadratically convex hypersurface C. Define a Finsler metric φ on D by setting its value at a non-zero vector $\mathbf{v_x} \in T_\mathbf{x}D$ to be $\varphi(\mathbf{v_x}) := (t_1^{-1} + t_2^{-1})/2$, where t_1 and t_2 are the two positive real numbers for which $\mathbf{x} + t_1\mathbf{v}$ and $\mathbf{x} - t_2\mathbf{v}$ belong to C.

PROPOSITION 6.2. *If $\mathbf{x}$ and $\mathbf{y}$ are two points on D and $\overline{\mathbf{xy}}$ is the line segment joining them, then*

$$\int_{\overline{\mathbf{xy}}} \varphi = \frac{1}{2} \ln\left(\frac{\|\mathbf{y} - \mathbf{a}\| \, \|\mathbf{x} - \mathbf{b}\|}{\|\mathbf{x} - \mathbf{a}\| \, \|\mathbf{y} - \mathbf{b}\|}\right).$$

Hamel, a student of Hilbert, was the first to study Hilbert's fourth problem. Among other things he showed that Lagrangians on $\mathbb{R}^n$ whose extremals are straight lines are characterized by a system of linear partial differential equations.

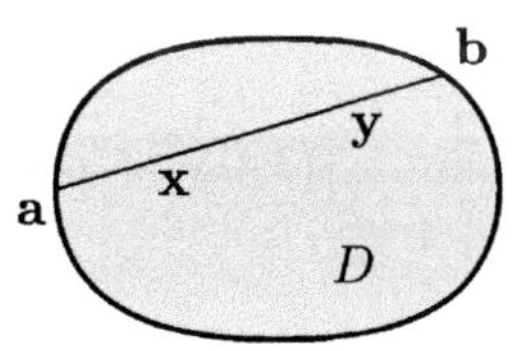

Fig. 1.

THEOREM 6.3 (Hamel [58]). *Let $\varphi : T\mathbb{R}^n \setminus 0 \to \mathbb{R}$ be a smooth Lagrangian which is homogeneous of order one in the velocities. Straight lines are extremals of the functional $\gamma \mapsto \int \varphi(\dot\gamma(t))\,dt$ if and only if φ satisfies the system of equations*

$$\frac{\partial^2 \varphi}{\partial x_i \partial v_j} = \frac{\partial^2 \varphi}{\partial x_j \partial v_i} \quad \text{for } 1 \leqslant i, j \leqslant n. \tag{7}$$

It is mainly through the work of Busemann and Pogorelov (see [37,72,82,15]) that the construction of projective Finsler metrics in terms of a class of (possibly signed) measures is now well understood.

DEFINITION 6.4 [15,82]. Let $D \subset \mathbb{R}P^n$ be an open convex set and let $H_{n-1}(D)$ be the set of all hyperplanes passing through D. A possibly-signed measure on $H_{n-1}(D)$ is said to be *quasi-positive* if for any two line segments $\overline{\mathbf{xy}}$ and $\overline{\mathbf{yz}}$ not on the same line, the measure of the set of hyperplanes intersecting twice the wedge formed by $\overline{\mathbf{xy}}$ and $\overline{\mathbf{yz}}$ is positive.

THEOREM 6.5. *A Finsler metric φ on an open convex set $D \subset \mathbb{R}P^n$ is projective if and only if there exists a smooth quasi-positive measure Φ_{n-1} on the space of hyperplanes passing through D, $H_{n-1}(D)$, such that for any smooth curve γ,*

$$\int_\gamma \varphi = \frac{1}{2} \int_{\zeta \in H_{n-1}(D)} \#(\zeta \cap \gamma)\Phi_{n-1}. \tag{8}$$

Notice that, in particular, the length of a line segment equals half the Φ_{n-1}-measure of the set of all hyperplanes intersecting it.

In Pogorelov's approach to Hilbert's fourth problem, Theorem 6.5 follows from the following integral representation for the solution of Hamel's equations.

THEOREM 6.6. *A Lagrangian $\varphi : T\mathbb{R}^n \setminus 0 \to \mathbb{R}$ which is homogeneous of order one in the velocities satisfies Hamel's equations if and only if there exists a smooth even function $\nu(r, \boldsymbol{\xi})$ on $\mathbb{R} \times S^{n-1}$ such that*

$$\varphi(\mathbf{x}, \mathbf{v}) = \int_{\boldsymbol{\xi} \in S^{n-1}} |\boldsymbol{\xi} \cdot \mathbf{v}| \nu(\boldsymbol{\xi} \cdot \mathbf{x}, \boldsymbol{\xi})\Omega, \tag{9}$$

where Ω is the standard area form on the unit sphere in $\mathbb{R}^n$.

EXAMPLE [12]. Applying formula (9) to the function $\nu : \mathbb{R} \times S^1 \to \mathbb{R}$ defined by $\nu(r, \theta) = 1 + r^2$, we obtain the Finsler metric

$$\varphi(x_1, x_2, v_1, v_2) = \frac{1}{3\sqrt{v_1^2 + v_2^2}} \big[\big(3 + x_1^2 + x_2^2\big)\big(v_1^2 + v_2^2\big) + (x_1 v_1 + x_2 v_2)^2 \big].$$

Theorem 6.5 states that projective Finsler metrics are exactly those Finsler spaces for which there is a Crofton formula for the lengths of curves. Do the Crofton formulas for the areas of submanifolds also hold? Does the first Crofton formula imply all others? The answer is *yes* if by area we mean Holmes–Thompson area:

THEOREM 6.7 (Álvarez and Fernandes [12]). *Let φ be a projective Finsler metric on an open convex domain $D \subset \mathbb{R}P^n$ and let k, $1 \leqslant k \leqslant n-1$, be a natural number. There exists a smooth (possibly signed) measure Φ_{n-k} on the manifold $H_{n-k}(D)$ of $(n-k)$-flats passing through D such that if $N \subset \mathbb{R}^n$ is an immersed submanifold of dimension k, then*

$$\mathrm{vol}_k(N) = \frac{1}{\epsilon_k} \int_{\zeta \in H_{n-k}(D)} \#(N \cap \zeta)\Phi_{n-k}, \tag{10}$$

where ϵ_k is the volume of the Euclidean unit ball of dimension k.

The construction of the measures Φ_{n-k} given in [12] and the formula (4) for the Holmes–Thompson volume imply that all the tangent spaces of a projective Finsler metric are hypermetric if and only if the measures Φ_{n-k}, $k = 1, \ldots, n-1$, are positive. In this case, just like in the case of the standard Riemannian metric on $\mathbb{R}P^n$, projective subspaces are area-minimizing.

THEOREM 6.8 (Álvarez and Fernandes). *If the geodesics of a Finsler metric on $\mathbb{R}P^n$ are projective lines and all its tangent spaces are hypermetric, then the projective subspaces are area-minimizing in their homology class.*

PROOF. If $N \subset \mathbb{R}P^n$ is a k-dimensional submanifold which is homologous to a projective subspace, then the number of points of intersection of N with a projective subspace of complementary dimension is at least one. Using the Crofton formula (10) and positivity of the measure Φ_{n-k}, we have that

$$\begin{aligned} \mathrm{vol}_k(N) &= \frac{1}{\epsilon_k} \int_{\zeta \in H_{n-k}(\mathbb{R}P^n)} \#(N \cap \zeta)\Phi_{n-k} \\ &\geqslant \frac{1}{\epsilon_k} \int_{H_{n-k}(\mathbb{R}P^n)} \Phi_{n-k} = \mathrm{vol}_k\big(\mathbb{R}P^k\big). \end{aligned}$$

□

R. Schneider has recently shown in [77] that the two previous theorems remain valid even if the regularity assumptions on both the Finsler metric and the submanifold are significantly weakened.

The following results also points to the similarity between projective Finsler metrics and the standard Riemannian metric on $\mathbb{R}P^n$.

PROPOSITION 6.9. *If $(\mathbb{R}P^n, \varphi)$ is a projective Finsler space for which the length of the projective lines is equal to π, then the Holmes–Thompson volume of $(\mathbb{R}P^n, \varphi)$ equals the volume of $\mathbb{R}P^n$ with its standard Riemannian metric.*

PROOF. If φ_0 denotes the standard Riemannian metric on $\mathbb{R}P^n$, then, for each number t, $t \in [0, 1]$, the metric $\varphi_t = (1-t)\varphi_0 + t\varphi$ is a projective Finsler metric and the length of its closed geodesics is π. Applying Theorem 5.9, we conclude that the Holmes–Thompson volumes of $(\mathbb{R}P^n, \varphi_0)$ and $(\mathbb{R}P^n, \varphi)$ are equal. □

It seems that the analogues of Hilbert's fourth problem for rank-one symmetric spaces other than $\mathbb{R}P^n$ have never been studied. For example, the following problem is open:

PROBLEM 11. Construct all Finsler metrics on $\mathbb{C}P^2$ such that the geodesics coincide as point sets with those of the standard Riemannian metric on $\mathbb{C}P^2$.

In trying to solve this problem, Álvarez and Durán stumbled upon the following construction of Finsler metrics on complex and quaternionic projective spaces for which the projective lines are totally geodesic and the geodesics are circles:

Use the Busemann–Pogorelov construction of projective metrics on real projective spaces and spheres to construct a projective metric φ on S^{2n+1} (respectively S^{4n+3}) that is invariant under the Hopf action. There is a unique Finsler metric ψ on $\mathbb{C}P^n$ (respectively $\mathbb{H}P^n$) for which the projection map of the Hopf fibration is an isometric submersion from (S^{2n+1}, φ) to $(\mathbb{C}P^n, \psi)$ (respectively from (S^{4n+3}, φ) to $(\mathbb{H}P^n, \psi)$). For the metric ψ, projective lines are totally geodesic and geodesics are (geometric) circles.

This construction suggests yet another inverse problem:

PROBLEM 12. Construct all Finsler metrics on $\mathbb{C}P^n$ such that the geodesics are circles.

For $n = 1$ this problem has been solved by Álvarez and Berck (unpublished) in terms of space-type surfaces (also called *elliptic congruences*) on the Grassmannian of Lagrangian planes in $\mathbb{R}^4$.

There have been many attempts to define the Finsler analogue of Kähler metrics, but none seems to have enjoyed any measure of success. Since the metric properties of Kähler manifolds are not so well understood as to proceed in this direction, it makes sense to use some remarkable geometric property of Kähler manifolds in an attempt to define their Finsler analogues. One such remarkable geometric property is that complex submanifolds are minimal.

PROBLEM 13. Construct and study all the Finsler metrics on $\mathbb{C}P^n$ for which complex submanifolds are minimal. Are (Riemannian) Kähler metrics the only ones with this property?

7. Closed geodesics

In Riemannian geometry, the study of closed geodesics has a long and glorious history with contributors like Poincaré, Birkhoff and Morse. However, there do not seem to be many interesting results about closed geodesics in Finsler manifolds. If an existence result in Riemannian geometry depends only on Morse theory and the method of broken geodesics, then it automatically holds in Finsler geometry.

A Riemannian result that would be interesting to extend to the Finsler setting is a theorem of Bangert and Franks [19,53] stating that any Riemannian metric on the two-dimensional sphere has infinitely many closed geodesics. Here we emphasize that we are considering symmetric or reversible Finsler metrics. Indeed, Katok has constructed non-reversible Finsler metrics on the 2-sphere which have only two closed geodesics (see [65,87]).

PROBLEM 14 (Bangert). Does every Finsler metric on S^2 have infinitely many closed geodesics?

Franks' work on the periodic points of area preserving maps of the annulus and recent work by Hofer, Wysocki and Zehnder (see [61]) suggest the following question:

PROBLEM 15. Is the number of distinct closed geodesics on a non-reversible Finsler metric on the 2-sphere either two or infinity?

8. Differential invariants of Finsler surfaces

In this section we study the differential invariants of Finsler surfaces without the aid of the Cartan connection and then introduce Cartan's structure equations in order to uncover the relations between these invariants.

8.1. *Convex geometry and the invariant I*

A smooth, centrally symmetric, and quadratically convex curve S on a two-dimensional vector space V parameterizes a family of Euclidean structures on V. Indeed, for each point $\mathbf{v} \in S$, there is a unique ellipse $E_{\mathbf{v}}$ which is centered at the origin and osculates S up to second order at this point. We associate to $\mathbf{v}$ the Euclidean structure with $E_{\mathbf{v}}$ as unit circle.

DEFINITION 8.1. Let (V, φ) be a two-dimensional Minkowski space with unit circle S. *An orthonormal basis* of V is an ordered pair of vectors $(\mathbf{v}, \mathbf{v}^{\perp})$, where $\mathbf{v} \in S$ and $\mathbf{v}^{\perp}$ is both of unit length and perpendicular to $\mathbf{v}$ with respect to the Euclidean structure associated to $\mathbf{v}$.

Using the Euclidean structures associated to the curve we may define a *distinguished parameterization of* S: orient the vector space V and parameterize the curve S by a map γ in such a way that $\gamma(t), \dot{\gamma}(t)$ is an oriented orthonormal basis. Tabachnikov has remarked (private communication) that this parameterization is, up to shifts in the parameter, the only one satisfying the equation

$$\det\bigl(\gamma(t), \dot{\gamma}(t)\bigr) = \det\bigl(\dot{\gamma}(t), \ddot{\gamma}(t)\bigr).$$

The period of γ is a linear invariant of S which we shall call the *total angle* of S.

THEOREM 8.2 (Schneider [75]). *The total angle of a smooth, centrally-symmetric, and quadratically convex curve on the plane is less than or equal to 2π. Equality holds if and only if the curve is an ellipse.*

DEFINITION 8.3. Let S be a smooth, centrally symmetric, and quadratically convex curve on a two-dimensional vector space V. If γ is a distinguished parameterization of S and $\mathbf{v} = \gamma(t)$ is a point on S, then $I(\mathbf{v})$ is defined by the equation

$$\ddot{\gamma}(t) := -\gamma(t) + I(\mathbf{v})\dot{\gamma}(t).$$

The quantity $I(\mathbf{v})$ is zero if and only if the osculating ellipse $E(\mathbf{v})$ osculates S up to third order or higher at $\mathbf{v}$. It follows that if I is identically zero, then S is an ellipse. Using a theorem of Ghys of the zeros of the Schwartzian derivative (see [71]), Álvarez has showed in [10] that the invariant I vanishes at least eight times.

If (M, φ) is a Finsler surface, then every tangent space $T_m M$, $m \in M$, is a Minkowski space and the indicatrix

$$S_m M := \{\mathbf{v}_m \in T_m M \colon \varphi(\mathbf{v}_m) = 1\}$$

is smooth, centrally symmetric, and quadratically convex. For each unit tangent vector $\mathbf{v}_m$ we define $I(\mathbf{v}_m)$ as the value of the invariant I of $S_m M$ at the point $\mathbf{v}_m$. The result is a smooth function, which we again denote by I, defined of the unit circle bundle of M. Clearly, this function is identically zero if and only if the Finsler surface is Riemannian.

We can also use the previous geometric constructions to define a vector field on the unit circle bundle of an oriented Finsler surface (M, φ): If $\mathbf{v}_m$ is a unit tangent vector and $\gamma(t)$ is a distinguished parameterization of $S_m M$ with $\gamma(0) = \mathbf{v}_m$, then $X_3(\mathbf{v}_m) := \dot{\gamma}(0)$.

8.2. *The invariant J*

We shall now define a smooth function on the unit bundle of M which measures how the invariant I changes along the geodesics.

DEFINITION 8.4. Let (M, φ) be a Finsler surfaces, let $\mathbf{v}_m \in T_m M$ be a unit tangent vector, and let $\sigma : (-\epsilon, \epsilon) \to M$ be the geodesic with initial condition $\mathbf{v}_m$. Define

$$J(\mathbf{v}_m) = d/dt I(\dot{\sigma}(t))\big|_{t=0}.$$

Note that $J = 0$ means that I is an invariant of motion. The Finsler surfaces for which this occurs are called *Landsberg surfaces*. Originally, they came up in the study of Finslerian analogues of the Gauss–Bonnet theorem (see [66] and Section 8.5). Unfortunately, all the known examples of Landsberg surfaces are either Riemannian or locally isometric to Minkowski planes. For example, the following problem remains unsolved.

PROBLEM 16. Is there any non-Riemannian Landsberg metric on the 2-sphere?

8.3. *Curvature of Finsler surfaces*

The formula of second variation really belongs to variational calculus and not to Riemannian geometry. It should not come then as a surprise that many of the definitions and theorems in Riemannian geometry, including the notion of curvature, extend to Finsler geometry.

Let us start by defining the obvious extensions of Jacobi fields and conjugate points:

Given a geodesic $\gamma : [a,b] \to M$, a *geodesic variation* of γ is a smooth map $\Gamma : (-\epsilon, \epsilon) \times [a,b] \to M$ such that

- $\Gamma(0,t) = \gamma(t)$.
- For each fixed s_0, the curve $t \mapsto \Gamma(s_0, t)$ is a geodesic.

If $\Gamma(s,t)$ is a geodesic variation of the geodesic γ, the vector field Y along γ defined by

$$Y(t) = \left.\frac{\partial \Gamma(s,t)}{\partial s}\right|_{s=0}$$

is called a *Jacobi field.* A Jacobi field is said to be *proper* if, for each t, $\dot\gamma(t)$ and $Y(t)$ form an orthonormal basis of the Minkowski plane $T_m M$ in the sense of Definition 8.1.

DEFINITION 8.5. Let p be a point in a Finsler surface and let γ be a geodesic starting from p. A point $\gamma(s)$ is said to be *conjugate* to p along γ if there exists a non-zero Jacobi field $Y(t)$ along γ such that $Y(0) = Y(s) = 0$.

Just as in Riemannian geometry, geodesics minimize length up to their first conjugate point.

PROPOSITION 8.6. *Let p be a point in a Finsler surface and let γ be a geodesic starting from p. The curve γ restricted to the interval $[0,s]$ minimizes length if and only if there is no conjugate point between $p = \gamma(0)$ and $\gamma(s)$.*

If we are given a geodesic γ let us orient the tangent spaces $T_{\gamma(t)}M$ and let us define $\dot\gamma^{\perp}(t)$ in such a way that $\dot\gamma(t)$ and $\dot\gamma^{\perp}(t)$ form an oriented orthonormal basis on $T_{\gamma(t)}M$. Note that any proper Jacobi field Y along γ can be uniquely expressed as

$$Y(t) = y(t)\dot\gamma^{\perp}(t),$$

where y is a real-valued function.

PROPOSITION–DEFINITION 8.7. There is a unique smooth function K on the unit circle bundle of M such that for any geodesic γ parameterized with unit speed and for any proper Jacobi field $Y(t) = y(t)\dot\gamma^{\perp}(t)$ the *Jacobi equation*

$$y''(t) + K\big(\dot\gamma(t)\big)y(t) = 0$$

holds.

The function K is called the *curvature* of (M, φ). In contrast with the Riemannian case, K depends on both the point $\gamma(t) \in M$ and the direction of $\dot\gamma(t)$.

The Bonnet–Myers theorem follows, as usual, from the Jacobi equation and Sturm's comparison theorem.

THEOREM 8.8. *If the curvature function of a Finsler surface* (M, φ) *is greater than or equal to a positive number* δ, *then the diameter of* M *is less than or equal to* $\pi/\sqrt{\delta}$.

A very helpful way of describing the Finsler curvature in terms of auxiliary Riemannian metrics has been given by Shen [79]: Let $\mathbf{v}_m \in T_m M$ be a unit vector and let X be a *geodesic vector field* (i.e., the integral curves of X are geodesics parameterized with unit speed) defined on a neighborhood $\mathcal{O}$ of m and such that $X(m) = \mathbf{v}_m$. If we define a Riemannian metric on $\mathcal{O}$ by $x \mapsto g_\varphi(X(x))$ as in Section 2, then the Riemannian curvature of this metric at m equals the Finsler curvature of (M, φ) at $\mathbf{v}_m$.

8.4. *Cartan's structure equations*

By now we have defined three geometric invariants of Finsler surfaces: I, J and K. The invariant I is a centro-affine invariant which describes the shape of each unit tangent circle, the invariant K belongs to the calculus of variations and measures the focusing of geodesics, and the invariant J, by measuring how I varies along geodesics, partakes of both convex geometry and variational calculus. All three invariants can be defined, as we have done, by elementary geometric and variational considerations, but there is nothing to suggest the deep and interesting relations between the three.

DEFINITION 8.9. If (M, φ) is a Finsler manifold, the *geodesic spray* of M is the vector field X_1 defined on the unit tangent bundle and whose value at a unit tangent vector $\mathbf{v}_m$ is defined as follows: take $\sigma(t)$ to be the geodesic with initial condition v_m and set $X_1(\mathbf{v}_m) := d/dt\, \dot\sigma(t)|_{t=0}$.

THEOREM 8.10 (Cartan [42]). *Let* (M, φ) *be an oriented Finsler surface, let* X_1 *denote its geodesic spray, and let* X_3 *be the vector field defined at the end of Section* 8.1. *If we define* $X_2 := [X_3, X_1]$, *then we have the following equations*:

$$
\begin{aligned}
&[X_3, X_1] = X_2, \\
&[X_1, X_2] = K X_3, \\
&[X_3, X_2] = -X_1 + I X_2 + J X_3.
\end{aligned}
$$

Of course, Cartan preferred differential forms to vector fields and he wrote the above equations in terms of the dual forms ω_1, ω_2 and ω_3 defined by the equations $\omega_i(X_j) = \delta_{ij}$. Cartan's *structure equations* are:

$$
\begin{aligned}
d\omega_1 &= -\omega_2 \wedge \omega_3, \\
d\omega_2 &= \omega_1 \wedge \omega_3 - I\omega_2 \wedge \omega_3, \\
d\omega_3 &= -K\omega_1 \wedge \omega_2 - J\omega_2 \wedge \omega_3.
\end{aligned}
$$

Note that by differentiating the structure equations we obtain the following *Bianchi identities*:

$$J = I_1,$$

$$K_3 + KI + J_1 = 0.$$

In these identities the subindices represent differentiation with respect to the vector fields X_1, X_2 and X_3. In general if F is a function on the unit bundle we will write

$$dF = F_1\omega_1 + F_2\omega_2 + F_3\omega_3.$$

Cartan shows that the forms ω_1, ω_2, and ω_3 solve the problem of equivalence. From this it follows, at least in theory, that all microlocal invariant properties of Finsler surfaces can be written in terms of the functions I, J, K, and their derivatives with respect to the vector fields X_1, X_2, and X_3. As an example, we have Berwald's characterizations of locally Minkowski and projectively flat Finsler surfaces.

THEOREM 8.11 (Berwald). *A Finsler surface is locally Minkowski if and only if K, I_1, and I_2 are identically zero.*

Let us recall that a Finsler manifold is said to be *projectively flat* if around every point we can find a small neighborhood and a diffeomorphism of this neighborhood to an open subset of Euclidean space such that geodesics are mapped onto straight lines.

THEOREM 8.12 (Berwald [26]). *The (non-parameterized) geodesics of a Finsler surface are locally the geodesics of some affine connection if and only if the following equation holds*:

$$I_{23} + J_{33} + 2I(I_2 + J_3) + 6J = 0.$$

The dual system of curves to the geodesics of a Finsler surface are locally the geodesics of some affine connection if and only if the following equation holds:

$$K_{31} - 3K_2 = 0.$$

Moreover, both of the above equations hold if and only if the surface is projectively flat.

For dual systems of curves and an elementary exposition of path geometry see Arnold's book [17, pp. 42–56]. A very clear exposition of Berwald's theorem and its proof is given in [31].

It is amusing to prove the following classic theorem of Beltrami by using Berwald's result.

COROLLARY 8.13 (Beltrami). *A Riemannian surface is projectively flat if and only if its curvature is constant.*

Berwald was also interested in Finsler metrics whose geodesics coincide as parameterized curves with the geodesics of some affine connection. These Finsler metrics are called *Berwald metrics*. In two dimensions they are characterized by the equations $I_1 = 0$ and $I_2 = 0$.

8.5. *Applications of Cartan's structure equations*

Armed with Cartan's structure equations and their Bianchi identities, we are now in a position to prove some of the deepest results in the theory of Finsler surfaces.

THEOREM 8.14 (Akbar–Zadeh [4]). *A compact Finsler surface of constant negative curvature is Riemannian.*

PROOF. Using the second Bianchi identity, which tells us that $K_3 + KI + J_1 = 0$, we have that if $K \equiv c$ is a constant and $\sigma(t)$ is a geodesic on M, then the function $I(t) := I(\dot{\sigma}(t))$ satisfies the differential equation

$$\frac{d^2}{dt^2} I = -cI.$$

If c is negative, then $I(t)$ must be a linear combination of exponentials and, therefore, if the initial condition is not $I(0) = 0$, $I'(0) = 0$, the function $I(t)$ is unbounded. Since I is bounded whenever the Finsler surface is compact, the only possibility that remains is that I be identically zero, and that the surface be Riemannian. □

The classical non-Riemannian examples of Finsler metrics with constant negative curvature are the Hilbert geometries. Since the existence of an isometric embedding of a Finsler surface on a Minkowski space implies that the invariant I is bounded (this follows immediately from the geometric interpretation of I given in Section 8.1) we have the following remark of Álvarez and Durán [10]: *the Hilbert geometry given by a smooth, quadratically convex curve C does not admit an isometric embedding into a Minkowski space unless C is an ellipse.*

If C is an ellipse, then the Hilbert geometry is the Cayley–Klein model of hyperbolic geometry and, by a theorem of Rosendorn, it admits an explicit isometric embedding into $\mathbb{R}^5$ (see [57, p. 276]).

THEOREM 8.15 (Akbar–Zadeh [4]). *A compact Finsler surface with zero curvature is locally isometric to a Minkowski plane.*

PROOF. By Theorem 8.11, we need to show that a compact Finsler surface of zero curvature also satisfies $I_1 \equiv 0$ and $I_2 \equiv 0$.

Reasoning as in the proof of the previous theorem, if $\sigma(t)$ is a geodesic on M, then the function $I(t) := I(\dot{\sigma}(t))$ satisfies the differential equation

$$\frac{d^2}{dt^2} I = 0.$$

This implies that I is a linear function of t. Since I is bounded, it must be a constant and, therefore, $I_1 = 0$.

The proof that $I_2 = 0$ is just slightly more involved. First note from the structure equations that when $K \equiv 0$ the vector fields X_1 and X_2 commute. From this we gather that $I_{21} = I_{12} = 0$, and so I_2 is constant along geodesics. Using this we have that

$$I_2 = X_2 I = [X_3, X_1] I = -I_{31}.$$

This implies that $0 = I_{21} = -I_{311}$ and, hence, I_3 is a linear function of t. Since I_3 is bounded, I_3 must be constant on geodesics and $I_2 = -I_{31} = 0$. □

This proof also shows that a surface with a complete Finsler metric of zero curvature that is isometrically embedded in a Minkowski space must be locally Minkowski.

PROBLEM 17. Is there any complete Finsler metric on $\mathbb{R}^2$ with zero curvature that is not locally Minkowski?

THEOREM 8.16. *Let (M, φ) be a Landsberg surface. If M is connected, then the total angle of any two of its tangent unit circles is the same.*

The proof is taken from Bryant's beautiful paper [30].

PROOF. If x and y be two points on M, the difference of the total angle of the unit circle S_y over y and the total angle of the unit circle S_x over x is given by

$$\int_{S_y} \omega_3 - \int_{S_x} \omega_3,$$

where the orientation over S_y and S_x is taken so that the integrals are positive.

Let $\gamma : [0, 1] \to M$ be a smooth curve joining them. Let $\pi : UM \to M$ denote the natural projection and set $\mathcal{C}$ be the cylinder $\pi^{-1}(\gamma)$. Note that the oriented boundary of $\mathcal{C}$ is $S_y - S_x$ and that the 2-form $\omega_1 \wedge \omega_2$ vanishes identically on $\mathcal{C}$.

Using Stokes theorem and the structure equations we have that

$$\int_{S_y} \omega_3 - \int_{S_x} \omega_3 = \int_{\mathcal{C}} d\omega_3 = \int_{\mathcal{C}} -K\omega_1 \wedge \omega_2 - J\omega_2 \wedge \omega_3 = 0.$$ □

Another application that uses the full power of Cartan's structure equations is the Finsler version of the Gauss–Bonnet theorem given by Bao and Chern in [21].

Let (M, φ) be a compact, oriented Finsler surface and let X be a vector field on M with a finite number of non-degenerate zeros. Cut out small discs, say of radius r, around the zeros of X and denote the resulting manifold with boundary by M_r. Normalizing the vector field X on M_r we obtain a section $\sigma_r : M_r \to SM_r$ over the unit circle bundle of M_r.

Using the fact that the total angle of the tangent unit circle at a point x is given by the integral of ω_3 over $S_x M$, the equation

$$d\omega_3 = -K\omega_1 \wedge \omega_2 - J\omega_2 \wedge \omega_3, \tag{11}$$

and Stokes' theorem, Chern and Bao arrive at the following result:

THEOREM 8.17 [21]. *Let (M, φ) be a compact, oriented Finsler surface and let X be a vector field on M with zeros $x_1, \ldots, x_n$ which are non-degenerate and with indices $\mathcal{I}(x_1), \ldots, \mathcal{I}(x_n)$. Using the notation above, the limit as r tends to zero of the quantity*

$$\int_{M_r} \sigma_r^*(-K\omega_1 \wedge \omega_2 - J\omega_2 \wedge \omega_3) \tag{12}$$

is well defined and equals $\sum_{i=1}^n \mathcal{I}(x_i)\mathcal{A}(x_i)$, where $\mathcal{A}(x_1), \ldots, \mathcal{A}(x_n)$ are the total angles of the unit tangent circles at the points $x_1, \ldots, x_n$.

If, as in the case of Landsberg surfaces, the total angle of the unit tangent circles does not vary from point to point, then we have that the Euler characteristic of M can be written as an integral in terms of the differential invariants of the Finsler surface.

Unfortunately, the Gauss–Bonnet theorem for Finsler surfaces does not have as many geometric implications as its Riemannian counterpart. For example, it cannot be used to prove that a metric of non-positive, or non-negative, curvature on a two-dimensional Finsler torus must be flat, or that two simple closed geodesics in a positively curved Finsler two-sphere must intersect. The reader is invited to prove these seemingly simple results by him/herself and thereby gain some insight into some of the difficulties of extending Riemannian results to the Finsler setting.

A greater conceptual challenge is that the standard Riemannian technique of comparing arbitrary metrics to metrics of constant curvature does not generalize to the Finsler setting. As we saw earlier in this section, a compact Finsler surface with constant negative curvature is Riemannian. Likewise, it has been recently remarked by Bryant that the main result of LeBrun and Mason, in [67], implies the following important result:

THEOREM 8.18 (Bryant). *A Finsler metric of constant positive curvature on the two-sphere is Riemannian.*

The analogous result in higher dimensions is still open:

PROBLEM 18 (Bryant). Is there any non-Riemannian Finsler metric on S^n, $n > 2$, with constant (positive) curvature?

It is possible that we must instead compare arbitrary Finsler metrics to metrics that have a given, simple, dynamical property. For example, there is evidence that the following generalization of the uniformization theorem is true:

CONJECTURE (Álvarez [6]). If φ is a Finsler metric on $\mathbb{R}P^2$, there exists a smooth function ρ on $\mathbb{R}P^2$ such that all the geodesics of $(\mathbb{R}P^2, e^{\rho}\varphi)$ are closed.

References

[1] M. Abate and G. Patrizio, *Finsler Metrics—A Global Approach*, Lecture Notes in Math., vol. 1591, Springer-Verlag (1994).

[2] R. Abraham and J.E. Marsden, *Foundations of Mechanics*, 2nd ed., Benjamin/Cummings, Reading, MA (1978).

[3] H. Akbar-Zadeh, *Les espaces de Finsler et certaines de leurs généralisations*, Ann. Sci. École Norm. Sup. (3) **80** (1963), 1–79.

[4] H. Akbar-Zadeh, *Sur les espaces de Finsler à courbures sectionelles constantes*, Acad. Roy. Belg. Bull. Cl. Sci. (5) **74** (1988), 281–322.

[5] J.C. Álvarez Paiva, *The symplectic geometry of spaces of geodesics*, Ph.D. thesis, Rutgers University (1995).

[6] J.C. Álvarez Paiva, *Dual mixed volumes and isosystolic inequalities*, preprint (2002).

[7] J.C. Álvarez Paiva, *Hilbert's fourth problem in two dimensions*, Mass Selecta: Teaching and Learning Advanced Undergraduate Mathematics, Amer. Math. Soc., Providence, RI (2003), 165–183.

[8] J.C. Álvarez Paiva, *Dual spheres have the same girth*, Amer. J. Math., to appear.

[9] J.C. Álvarez Paiva and G. Berck, *What is wrong with the Hausdorff measure in Finsler spaces*, Adv. in Math., to appear.

[10] J.C. Álvarez Paiva and C. Durán, *An Introduction to Finsler Geometry*, Notas de la Escuela Venezolana de Matématicas (1998).

[11] J.C. Álvarez Paiva and C. Durán, *Isometric submersions of Finsler manifolds*, Proc. Amer. Math. Soc. **129** (2001), 2409–2417.

[12] J.C. Álvarez Paiva and E. Fernandes, *Crofton formulas in projective Finsler spaces*, Electron. Res. Announc. Amer. Math. Soc. **4** (1998), 91–100.

[13] J.C. Álvarez Paiva and E. Fernandes, *Fourier transforms and the Holmes–Thompson volume of Finsler manifolds*, Int. Math. Res. Notices **19** (1999), 1031–1042.

[14] J.C. Álvarez Paiva and E. Fernandes, *Crofton formulas and Gelfand transforms*, Selecta Math. (NS), to appear.

[15] J.C. Álvarez Paiva, I.M. Gelfand and M. Smirnov, *Crofton densities, symplectic geometry, and Hilbert's fourth problem*, Arnold–Gelfand Mathematical Seminars, Geometry and Singularity Theory, V.I. Arnold, I.M. Gelfand and V.S. Retakh, eds, Birkhäuser, Boston (1997), 77–92.

[16] J.C. Álvarez and A.C. Thompson, *Volumes in normed and Finsler spaces*, A Sampler in Riemann–Finsler Geometry, D. Bao, R. Bryant, S.S. Chern and Z. Shen, eds, Cambridge Univ. Press (2004), 1–49.

[17] V.I. Arnold, *Geometrical Methods in the Theory of Ordinary Differential Equations*, Grundlehren der Mathematischen Wissenschaften, vol. 250, Springer-Verlag, Berlin (1983).

[18] V.I. Arnold and A.B. Givental, *Symplectic geometry*, Dynamical Systems IV, Encyclopaedia of Mathematical Sciences, vol. 4, V.I. Arnold and S.P. Novikov, eds, Springer-Verlag, Berlin (1990).

[19] V. Bangert, *On the existence of closed geodesics on the two-spheres*, Internat. J. Math. **4** (1993), 403–418.

[20] D. Bao, R. Bryant, S.S. Chern and Z. Shen, eds, *Some Perspectives in Finsler Geometry*, MSRI Series, Cambridge Univ. Press, in press.

[21] D. Bao and S.S. Chern, *A note on the Gauss–Bonnet theorem for Finsler spaces*, Ann. of Math. **143** (1996), 233–252.

[22] D. Bao, S.S. Chern and Z. Shen, eds, *Finsler Geometry*, Contemporary Mathematics, vol. 196, Amer. Math. Soc. (1996).

[23] D. Bao, S.S. Chern and Z. Shen, *An Introduction to Riemann–Finsler Geometry*, Graduate Texts in Mathematics, vol. 200, Springer-Verlag, Berlin (2000).

[24] V.N. Berestovskii, *Submetries of space forms of nonnegative curvature*, Siberian Math. J. **16** (1975), 651–662.

[25] G. Berck, *Minimalité des sous-variétés totalement géodésiques*, preprint (2004).

[26] L. Berwald, *On Finsler and Cartan geometries. III. Two-dimensional Finsler spaces with rectilinear extremals*, Ann. of Math. (2) **42** (1941), 84–112.

[27] A. Besse, *Manifolds of all Whose Geodesics are Closed*, Springer-Verlag, Berlin (1978).

[28] G.D. Birkhoff, *Dynamical systems with two degrees of freedom*, Trans. Amer. Math. Soc. **18** (1917), 199–300.

[29] E.D. Bolker, *A class of convex bodies*, Trans. Amer. Math. Soc. **145** (1969), 323–345.

[30] R.L. Bryant, *Finsler structures on the 2-sphere satisfying $K = 1$*, in [22], 27–42.

[31] R.L. Bryant, *Projectively flat Finsler 2-spheres of constant curvature*, Selecta Math. (N.S.) **3** (1997), 161–203.

[32] D. Burago and S. Ivanov, *Isometric embeddings of Finsler manifolds*, St. Petersburg Math. J. **5** (1994), 159–169.

[33] D. Burago and S. Ivanov, *On asymptotic volume of Finsler tori, minimal surfaces in normed spaces, and symplectic filling volume*, Ann. of Math. (2) **156** (3) (2002), 891–914.

[34] H. Busemann, *A theorem on convex bodies of the Brunn–Minkowski type*, Proc. Nat. Acad. Sci. USA **35** (1949), 27–31.

[35] H. Busemann, *The geometry of Finsler spaces*, Bull. Amer. Math. Soc. **56** (1950), 5–16.

[36] H. Busemann, *The Geometry of Geodesics*, Pure and Applied Mathematics, vol. 6, Academic Press, New York (1955).

[37] H. Busemann, *Geometries in which the planes minimize area*, Ann. Mat. Pura Appl. (4) **55** (1961), 171–190.

[38] H. Busemann, *Convexity on Grassmann manifolds*, Enseign. Math. (2) **7** (1961), 139–152.

[39] H. Busemann, *Recent Synthetic Differential Geometry*, Ergebnisse der Mathematik und ihrer Grenzgebiete, vol. 54, Springer-Verlag, Berlin (1970).

[40] H. Busemann, G. Ewald and G.C. Shephard, *Convex bodies and convexity on Grassmann cones: I–IV*, Math. Ann. **151** (1963), 1–41.

[41] H. Busemann and C.M. Petty, *Problems on convex bodies*, Math. Scand. **4** (1956), 88–94.

[42] E. Cartan, *Sur un problème d'équivalence et la théorie des espaces métriques généralisés*, Mathematica **4** (1930), 114–136.

[43] G.D. Chakerian, *Integral geometry in Minkowski spaces*, in [22], 43–50.

[44] S.S. Chern, *Finsler geometry is just Riemannian geometry without the quadratic restriction*, Notices Amer. Math. Soc. **43** (9) (1996), 959–963.

[45] C.E. Durán, *A volume comparison theorem for Finsler manifolds*, Proc. Am. Math. Soc. **126** (1998), 3079–3082.

[46] D. Egloff, *Some new developments in Finsler geometry*, Inaugural-Dissertation, Universität Freiburg (Schweiz) (1995).

[47] D. Egloff, *Uniform Finsler Hadamard manifolds*, Ann. Inst. H. Poincaré Phys. Théor. **66** (1997), 323–357.

[48] D. Egloff, *On the dynamics of uniform Finsler manifolds of negative flag curvature*, Ann. Global Anal. Geom. **15** (1997), 101–116.

[49] E. Ferrand, *Sur la structure symplectique de l'espace des géodésiques d'un espace de Hadamard*, Geom. Dedicata **68** (1997), 79–89.

[50] P. Foulon, *Géométrie des équations différentielles du second ordre*, Ann. Inst. H. Poincaré Phys. Théor. **45** (1986), 1–28.

[51] P. Foulon, *Locally symmetric Finsler spaces in negative curvature*, C. R. Acad. Sci. Paris **324** (1997), 1127–1132.

[52] P. Foulon, *Curvature and global rigidity in Finsler manifolds*, Houston J. Math. **28** (2002), 263–292.

[53] J. Franks, *Geodesics on S^2 and periodic points of annulus homeomorphisms*, Invent. Math. **108** (1992), 403–418.

[54] S. Gołąb, *Quelques problèmes métriques de la géometrie de Minkowski*, Trav. l'Acad. Mines Cracovie **6** (1932), 1–79.

[55] P. Goodey and W. Weil, *Zonoids and generalisations*, Handbook of Convex Geometry, vol. B, P.M. Gruber et al., eds, North-Holland, Amsterdam (1993), 1297–1326.

[56] M. Gromov, *Filling Riemannian manifolds*, J. Differential Geom. **18** (1983), 1–147.

[57] M. Gromov, *Partial Differential Relations*, Ergebnisse der Mathematik und ihrer Grenzgebiete, vol. 9, Springer-Verlag, Berlin (1986).

[58] G. Hamel, *Über die Geometrien, in denen die Geraden die kürzesten sind*, Math. Ann. **57** (1903), 231–264.

[59] D. Hilbert, *Foundations of Geometry*, Open Court Classics, Lasalle, IL (1971).

[60] D. Hilbert, *Mathematical problems*, translation in Mathematical developments arising from Hilbert problems, Proceedings of Symposia in Pure Math., vol. XXVII, Part 1, F. Browder, ed., Amer. Math. Soc., Providence, RI (1974).

[61] H. Hofer, K. Wysocki and E. Zehnder, *The dynamics on three-dimensional strictly convex energy surfaces*, Ann. of Math. **148** (1998), 197–289.

[62] R.D. Holmes and A.C. Thompson, *N-dimensional area and content in Minkowski spaces*, Pacific J. Math. **85** (1979), 77–110.

[63] L. Hörmander, *The Analysis of Linear Partial Differential Operators I*, Grundlehren der Mathematischen Wissenschaften, vol. 256, Springer-Verlag, Berlin (1983).

[64] S. Ivanov, *On two-dimensional minimal fillings*, St. Petersburg Math. J. **13** (2002), 17–25.

[65] A. Katok, *Ergodic perturbations of degenerate integrable Hamiltonian systems* (in English); Russian original in Math. USSR Izv. (1974), 535–571.

[66] G. Landsberg, *Über die Totalkrümmung*, Jahresber. Deutsch. Math.-Verein. **16** (1907), 36–46.

[67] C. LeBrun and L.J. Mason, *Zoll manifolds and complex surfaces*, J. Differential Geom. **61** (2002), 453–535.

[68] D. McDuff and D. Salamon, *Introduction to Symplectic Topology*, 2nd. ed., Oxford Mathematical Monographs, Clarendon Press, Oxford (1998).

[69] H. Minkowski, *Gesammelte Abhandlungen*, Chelsea, New York (1967).

[70] K. Ono, *Space of geodesics on Zoll three-spheres*, preprint (1999).

[71] V. Ovsienko and S. Tabachnikov, *Sturm theory, Ghys theorem on zeros of the Schwartzian derivative, and flattening of Legendrian curves*, Selecta Math. (N.S.) **2** (1996), 297–307.

[72] A.V. Pogorelov, *Hilbert's Fourth Problem*, Scripta Series in Mathematics, Winston and Sons (1979).

[73] J.J. Schäffer, *The self-circumference of polar convex bodies*, Arch. Math. **24** (1973), 87–90.

[74] J.J. Schäffer, *Geometry of Spheres in Normed Spaces*, Lecture Notes in Pure and Applied Mathematics, vol. 20, Dekker, New York (1976).

[75] R. Schneider, *Über die Finslerräume mit $S_{ijkl} = 0$*, Arch. Math. **19** (1968), 656–658.

[76] R. Schneider, *Crofton formulas in hypermetric projective Finsler spaces*, Festschrift: Erich Lamprecht, Arch. Math. (Basel) **77** (1) (2001), 85–97.

[77] R. Schneider, *On integral geometry in projective Finsler spaces*, Izv. Nats. Akad. Nauk Armenii Mat. **37** (2002) 34–51.

[78] R. Schneider and J.A. Wieacker, *Integral geometry in Minkowski spaces*, Adv. in Math. **129** (1997), 222–260.

[79] Z. Shen, *Curvature, distance, and volume in Finsler geometry*, IHES preprint (1997).

[80] J.-C. Sikorav, *Rigidité symplectique dans le cotangent de T^n*, Duke Math. J. **59** (1989), 759–763.

[81] E. Socié-Méthou, *Comportements asymptotiques et rigidités en géométries de Hilbert*, Thèse, Université de Strasbourg (2000).

[82] Z.I. Szabo, *Hilbert's fourth problem, I*, Adv. in Math. **59** (1986), 185–301.

[83] A.C. Thompson, *Applications of various inequalities to Minkowski geometry*, Geom. Dedicata **46** (1993), 215–231.

[84] A.C. Thompson, *Minkowski Geometry*, Encyclopedia of Math. and Its Applications, vol. 63, Cambridge Univ. Press, Cambridge (1996).

[85] W. Weil, *Centrally symmetric convex bodies and distributions II*, Israel J. Math. **32** (1979), 173–182.

[86] A. Weinstein, *Fourier integral operators, quantization, and the spectra of Riemannian manifolds*, Géométrie Symplectique et Physique Mathématique, Editions CNRS, Paris (1975).

[87] W. Ziller, *Geometry of the Katok examples*, Ergodic Theory Dynam. Systems **3** (1982), 135–157.

CHAPTER 2

Foliations*

Raymond Barre and Aziz El Kacimi Alaoui

LAMATH, Université de Valenciennes, Le Mont Houy, 59313 Valenciennes Cedex 9, France
E-mail: {raymond.barre, aziz.elkacimi}@univ-valenciennes.fr

Contents

*We thank Cédric Rousseau for drawing all the pictures in this paper.

HANDBOOK OF DIFFERENTIAL GEOMETRY, VOL. II
Edited by F.J.E. Dillen and L.C.A. Verstraelen

0. Foreword

Foliation Theory is the qualitative study of Differential Equations. It was initiated by the works of H. Poincaré, I. Bendixon and developed later by C. Ehresmann, G. Reeb and many other people. Since then the subject has been a wide field in mathematical research. Actually it is almost impossible to describe all the results and the different steps of its development. So the purpose of this chapter is to give definitions, some examples and the fundamental concepts like holonomy, transverse structures, etc. Some themes in the point of view of Differential Geometry are discussed: characteristic classes, basic Hodge theory, deformations, etc. A complete account on Foliation Theory can be found in the book [135] by C. Godbillon. The bibliography is not complete. It is motivated by two reasons: the first one is to indicate references for the reader who wants to learn much more on foliations; the second one is to mention people who make contributions to the subject; for most of them the selected list is nonexhaustive. All foliations considered are *regular* that is, all leaves have the same dimension. The theory of *singular foliations* and specially *holomorphic singular foliations* is well developed with a plentiful literature. It merits to be presented independently. References on the subject can be found on the paper [56] by D. Cerveau.

Unless otherwise stated, all the objects (manifolds, maps, functions, etc.) are assumed to be of class C^∞. Moreover, for simplicity, we will suppose that all the manifolds are orientable. For any manifold M, we denote by A the algebra of functions on M. If $E \longrightarrow M$ is a vector bundle, $C^\infty(E)$ will denote the space of its global sections; this is an A-module and, equipped with the C^∞-topology, it is a Fréchet space. In case E is the tangent bundle TM of M, we denote $C^\infty(TM)$ simply by $\chi(M)$ (the space of vector fields tangent to M). For $r \in \mathbf{N}$, $\Omega^r(M)$ is the space of differential forms of degree r on M which is by definition $C^\infty(\Lambda^r T^*M)$ where $\Lambda^r T^*M \longrightarrow M$ is the vector bundle with fibre at $x \in M$ the vector space of skew-symmetric forms of degree r on T_xM; $\Omega^0(M)$ is just A. The other notations will be introduced at need.

1. Definitions and examples

Let M be the Euclidean space $\mathbf{R}^{m+n} = \mathbf{R}^m \times \mathbf{R}^n$ with canonical coordinates denoted $(x, y) = (x_1, \ldots, x_m, y_1, \ldots, y_n)$ and consider the family of affine subspaces F_y of M where $y \in \mathbf{R}^n$, defined by the differential system: $dy_1 = \cdots = dy_n = 0$. Then M, considered as a disjoint union of these spaces, is a nonconnected manifold of dimension m. Its topology is the product of the usual topology on $\mathbf{R}^m$ and the discrete one on $\mathbf{R}^n$. We say that M, with this structure, is a *foliated manifold* of *dimension* m and *codimension* n. It constitutes the *local model* of a *foliation* of codimension n on a manifold of dimension $m + n$. Let $\mathcal{O}$ be an open set of $\mathbf{R}^{m+n}$; let us call a *plaque* of $\mathcal{O}$ any intersection of $\mathcal{O}$ with a horizontal space F_y.

DEFINITION 1. Let M be a manifold of dimension $m + n$. A codimension n *foliation* $\mathcal{F}$ on M is given by an open cover $\mathcal{U} = (U_i)_{i \in I}$ and for each i, a diffeomorphism

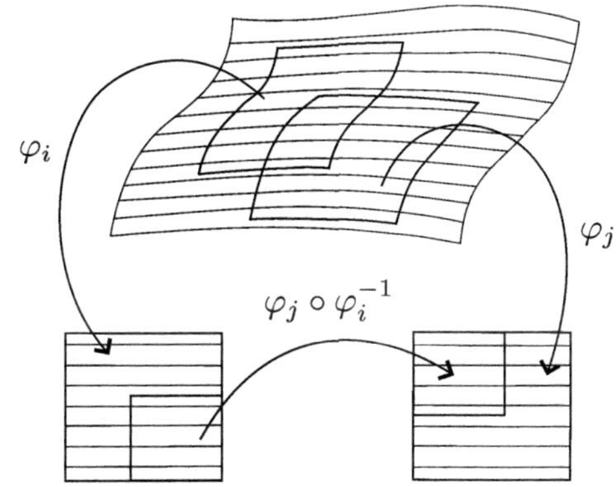

Fig. 1.

$\varphi_i : \mathbf{R}^{m+n} \longrightarrow U_i$ such that, on each nonempty intersection $U_i \cap U_j$, the coordinate change $\varphi_j^{-1} \circ \varphi_i : (x, y) \in \varphi_i^{-1}(U_i \cap U_j) \longrightarrow (x', y') \in \varphi_j^{-1}(U_i \cap U_j)$ has the form:

$$x' = \varphi_{ij}(x, y) \quad \text{and} \quad y' = \gamma_{ij}(y). \tag{1}$$

This means that the diffeomorphism $\varphi_j^{-1} \circ \varphi_i$ sends a plaque of $\varphi_i^{-1}(U_i \cap U_j)$ into a plaque of $\varphi_j^{-1}(U_i \cap U_j)$. The manifold M is decomposed into connected submanifolds of dimension m. Each of these submanifolds is called a *leaf* of $\mathcal{F}$. A subset U of M is *saturated* for $\mathcal{F}$ if it is union of leaves that is, if $x \in U$ then the leaf passing through x is contained in U.

Coordinate patches (U_i, φ_i) satisfying conditions of Definition 1 are said to be *distinguished* for the foliation $\mathcal{F}$.

Let $\mathcal{F}$ be a codimension n foliation on M defined by a maximal atlas $(U_i, \varphi_i)_{i \in I}$ like in Definition 1. Let $\pi : \mathbf{R}^{m+n} = \mathbf{R}^m \times \mathbf{R}^n \longrightarrow \mathbf{R}^n$ be the second projection. Then the map $f_i : U_i \overset{\pi \circ \varphi_i^{-1}}{\longrightarrow} \mathbf{R}^n$ is a submersion. On $U_i \cap U_j \neq \emptyset$ we have $f_j = \gamma_{ij} \circ f_i$. The submersions f_i and the local diffeomorphisms γ_{ij} of $\mathbf{R}^n$ give a complete characterization of $\mathcal{F}$.

DEFINITION 2. A codimension *n foliation* on M is given by an open cover $(U_i)_{i \in I}$, submersions $f_i : U_i \longrightarrow T$ over an n-dimensional transverse manifold T and, for $U_i \cap U_j \neq \emptyset$, a diffeomorphism $\gamma_{ij} : f_i(U_i \cap U_j) \subset T \longrightarrow f_j(U_i \cap U_j) \subset T$ satisfying:

$$f_j(x) = \gamma_{ij} \circ f_i(x) \quad \text{for } x \in U_i \cap U_j. \tag{2}$$

We say that $\{U_i, f_i, T, \gamma_{ij}\}$ is a *foliated cocycle* defining $\mathcal{F}$.

The proof of the equivalence of Definitions 1 and 2 is not difficult; it is left to the reader.

The foliation $\mathcal{F}$ is said to be *transversely orientable* if T can be given an orientation preserved by all the local diffeomorphisms γ_{ij}.

1.1. *Morphisms of foliations*

Let M and M' be two manifolds endowed respectively with two foliations $\mathcal{F}$ and $\mathcal{F}'$. A map $f : M \longrightarrow M'$ will be called *foliated* or a *morphism* between $\mathcal{F}$ and $\mathcal{F}'$ if, for every leaf L of $\mathcal{F}$, $f(L)$ is contained in a leaf of $\mathcal{F}'$; we say that f is an *isomorphism* if, in addition, f is a diffeomorphism; in this case the restriction of f to any leaf $L \in \mathcal{F}$ is a diffeomorphism on the leaf $L' = f(L) \in \mathcal{F}'$.

Suppose now that f is a diffeomorphism of M with a codimension n foliation $\mathcal{F}$. Then for every leaf $L \in \mathcal{F}$, $f(L)$ is a leaf of a codimension n foliation $\mathcal{F}'$ on M; we say that $\mathcal{F}'$ is the *image* of $\mathcal{F}$ by the diffeomorphism f and we write $\mathcal{F} = f^*(\mathcal{F}')$. Two foliations $\mathcal{F}$ and $\mathcal{F}'$ on M are said to be C^r*-conjugated* (*topologically* if $r = 0$, *differentiably* if $r = \infty$ and *analytically* in the case $r = \omega$) if there exists a C^r-homeomorphism $f : M \longrightarrow M$ such that $f^*(\mathcal{F}') = \mathcal{F}$.

1.2. *The concept of holonomy*

This is a very important notion in Foliation Theory. In many situations it determines completely the structure of the foliation. In this subsection, we will introduce this concept and give the statement of the local and global *stability theorems*.

Let $\mathcal{F}$ be a codimension n foliation on M, let L be a leaf of $\mathcal{F}$ and $x \in L$. Let T be a small transversal to $\mathcal{F}$ passing through x. Let $\sigma : [0, 1] \longrightarrow L$ be a continuous path such that $\sigma(0) = \sigma(1) = x$. Then there exist a finite open cover U_i, $i = 0, 1, \ldots, k$, of M with $U_0 = U_k$ and a subdivision $0 = t_0 < t_1 < \cdots < t_k = 1$ of $[0, 1]$ such that:

– $\sigma([t_{i-1}, t_i]) \subset U_i$,

– if $U_i \cap U_j \neq \emptyset$ then $U_i \cup U_j$ is contained in a distinguished chart of $\mathcal{F}$.

We say that U_i is a *subordinated chain* to σ. For $i = 0, 1, \ldots, k$ let T_i be a small transversal to $\mathcal{F}$ passing through $\sigma_i(t)$ with $T_0 = T_k = T$. For every point $z \in T_i$, sufficiently close to $\sigma(t_i)$, the plaque of $\mathcal{F}$ passing through z intersects T_{i+1} in a unique point $f_i(z)$. The domain of f_i contains a transversal T_i' passing through $\sigma(t_i)$ and homeomorphic to an open ball of $\mathbf{R}^n$. Then, it is clear that the map: $f_\sigma = f_{k-1} \circ f_{k-2} \circ \cdots \circ f_0$ is well defined on an open neighborhood of x; it is called the *holonomy map* associated to σ. We can prove (see [39], for instance) that the germ of f_σ:

– does not depend on the chain U_i, $i = 1, \ldots, k$, and in the choice of σ in its homotopy class in the group $\pi_1(L, x)$ of the homotopy classes of loops based at x,

– satisfies $f_\sigma(x) = x$.

So we get a homomorphism $h : [\sigma] \in \pi_1(L, x) \longrightarrow f_\sigma \in G(T, x)$ where $G(T, x)$ is the group of germs of diffeomorphisms of T fixing the point x. This representation h is called the *holonomy* of the leaf L at x. It is trivial if L is simply connected. The foliation $\mathcal{F}$ is said to be *without holonomy* if this representation is trivial for every leaf L of $\mathcal{F}$ and every point $x \in L$.

THEOREM 1 (Local stability). *Suppose that $\mathcal{F}$ admits a compact leaf L with finite fundamental group. Then L admits a saturated neighborhood V such that every leaf contained in V is compact with finite fundamental group.*

THEOREM 2 (Global stability). *Suppose that M is compact, the codimension of $\mathcal{F}$ is one and that $\mathcal{F}$ admits a compact leaf with finite fundamental group. Then all leaves of $\mathcal{F}$ are compact with finite fundamental group.*

The proof can be found in the original paper of G. Reeb [285] or in the book [39] by C. Camacho and A. Lins Neto.

1.3. *Examples of foliations*

(i) *Simple foliations.* On every manifold M we have a foliation by taking points as leaves. Its codimension is equal to the dimension of M. Also M can be equipped with a codimension zero foliation with only one leaf, namely, M itself.

In general, every submersion $M \stackrel{\pi}{\longrightarrow} B$ with connected fibres defines a foliation. The leaves being the fibres $\pi^{-1}(b)$, $b \in B$. In particular, every product $F \times B$ is a foliation with leaves $F \times \{b\}$, $b \in B$. These foliations are transversely orientable if, and only if, the manifold B is orientable.

These are *simple foliations.* We shall give more interesting examples in different situations.

(ii) *One-dimensional foliations.* Let us begin by surfaces. Let $\tilde{M} = \mathbf{R}^2$ and consider the differential equation $dy - \alpha\, dx = 0$ where α is a real number. This equation has $y = \alpha x + c$, $c \in \mathbf{R}$, as general solution. When c varies, we obtain a family of parallel lines which defines a foliation $\tilde{\mathcal{F}}$ in $\tilde{M}$.

The natural action of $\mathbf{Z}^2$ on $\tilde{M}$ preserves the foliation $\tilde{\mathcal{F}}$ (i.e. the image of any leaf of $\tilde{\mathcal{F}}$ by an integer translation is a leaf of $\tilde{\mathcal{F}}$). Then $\tilde{\mathcal{F}}$ induces a foliation $\mathcal{F}$ on the torus $\mathbf{T}^2 = \mathbf{R}^2/\mathbf{Z}^2$. The leaves are all diffeomorphic to the circle $\mathbf{S}^1$ if α is rational and to the real line if α is not rational (Figure 2). In fact, if α is not rational, every leaf of $\mathcal{F}$ is dense; this shows that even if locally a foliation is simple, globally it can be complicated.

Let M be a closed orientable surface. The fact that M admits a one-dimensional foliation depends on the topology of M, which is described by the *Euler–Poincaré number* $\chi(M)$;

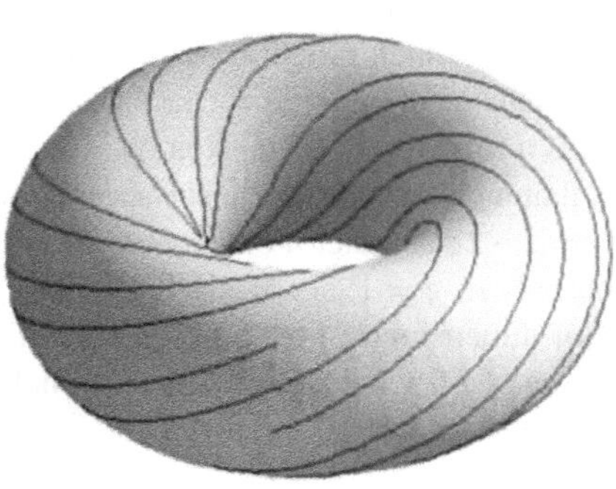

Fig. 2.

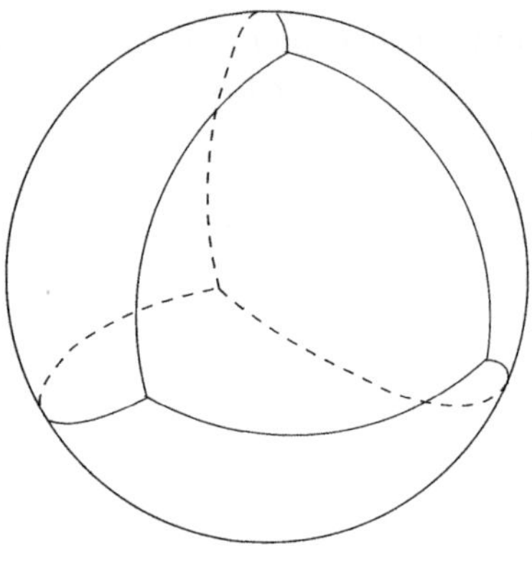

Fig. 3.

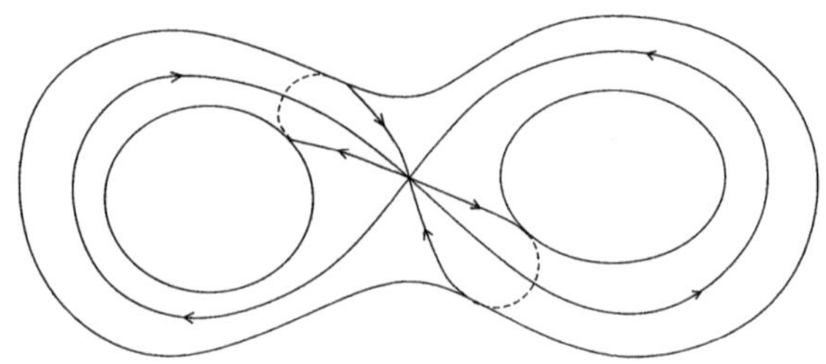

Fig. 4.

this number can be defined as follows: take a triangulation of M, i.e. a decomposition of M into triangles such as shown for the 2-sphere $\mathbf{S}^2$ (Figure 3).

Let b_0, b_1 and b_2 be the numbers respectively of vertices, edges and triangles. Then $\chi(M) = b_0 - b_1 + b_2$ is independent of the triangulation; it is called the *Euler–Poincaré number* of M. (There are many books on Algebraic or Differential Topology where we can find the proof of this fact.) It classifies completely the topology of closed orientable surfaces, i.e. M and M' are homeomorphic if, and only if, $\chi(M) = \chi(M')$. For the triangulation of $\mathbf{S}^2$ in Figure 3 we have $b_0 = 4$, $b_1 = 6$ and $b_2 = 4$. So $\chi(\mathbf{S}^2) = 2$.

The Euler–Poincaré number of M is the only obstruction to the existence of dimension one foliation on M: M admits such foliation if, and only if, $\chi(M) = 0$. For example, $\mathbf{S}^2$ cannot support a one-dimensional foliation. In fact, $\mathbf{T}^2$ is the only one compact orientable surface which admits a foliation of dimension one. The reader can prove, by using an adequate triangulation, that a closed orientable surface M_g of genus g (see in Figure 4 the case $g = 2$) has $\chi(M_g) = 2 - 2g$ as Euler–Poincaré number. Then M_g admits a foliation $\mathcal{F}$ of dimension one if, and only if, $g = 1$, i.e. M_g is $\mathbf{T}^2$.

Suppose M is compact of dimension n. For each $r = 0, 1, \ldots, n$, let $H^r(M, \mathbf{R})$ denote the real rth *cohomology space* of M which is finite dimensional. Then the number

$$\chi(M) = \sum_{r=0}^{n} (-1)^r \dim H^r(M, \mathbf{R})$$

is a *topological invariant* called the *Euler–Poincaré number* of M. For a surface, it is exactly the number defined above by using a triangulation. The manifold M admits a one-dimensional foliation if, and only if, $\chi(M) = 0$.

(iii) *Reeb foliation on the 3-sphere* $\mathbf{S}^3$. Let M be the 3-dimensional sphere $\mathbf{S}^3 = \{(z_1, z_2) \in \mathbf{C}^2\colon |z_1|^2 + |z_2|^2 = 1\}$. Denote by $\mathbf{D}$ the open unit disc in $\mathbf{C}$ and $\bar{\mathbf{D}}$ its closure which is the closed unit disc $\{z \in \mathbf{C}\colon |z| \leqslant 1\}$. The two subsets:

$$M_+ = \left\{(z_1, z_2) \in \mathbf{S}^3\colon |z_1|^2 \leqslant \frac{1}{2}\right\} \quad \text{and} \quad M_- = \left\{(z_1, z_2) \in \mathbf{S}^3\colon |z_2|^2 \leqslant \frac{1}{2}\right\}$$

are diffeomorphic to $\bar{\mathbf{D}} \times \mathbf{S}^1$. They have $\mathbf{T}^2$ as common boundary:

$$\partial M_+ = \partial M_- = \left\{(z_1, z_2) \in \mathbf{S}^3\colon |z_1|^2 = |z_2|^2 = \frac{1}{2}\right\}$$

and their union is equal to $\mathbf{S}^3$. Then $\mathbf{S}^3$ can be obtained by gluing M_+ and M_- along their boundaries by the diffeomorphism $(z_1, z_2) \in \partial M_+ \longrightarrow (z_2, z_1) \in \partial M_-$, i.e. we identify (z_1, z_2) with (z_2, z_1) in the disjoint union $M_+ \coprod M_-$. Let $f: \mathbf{D} \longrightarrow \mathbf{R}$ be the function defined by:

$$f(z) = \exp\left(\frac{1}{1 - |z|^2}\right).$$

Let t denote the second coordinate in $\mathbf{D} \times \mathbf{R}$. The family of surfaces $(S_t)_{t \in \mathbf{R}}$ obtained by translating the graph S of f along the t-axis defines a foliation on $\mathbf{D} \times \mathbf{R}$. If we add the cylinder $\mathbf{S}^1 \times \mathbf{R}$, where $\mathbf{S}^1$ is viewed as the boundary of $\bar{\mathbf{D}}$, we obtain a codimension one foliation $\tilde{\mathcal{F}}$ on $\bar{\mathbf{D}} \times \mathbf{R}$. By construction, $\tilde{\mathcal{F}}$ is invariant by the transformation

$$(z, t) \in \bar{\mathbf{D}} \times \mathbf{R} \longrightarrow (z, t + 1) \in \bar{\mathbf{D}} \times \mathbf{R};$$

so it induces a foliation $\mathcal{F}_0$ on the quotient:

$$\bar{\mathbf{D}} \times \mathbf{R}/(z, t) \sim (z, t + 1) \simeq \bar{\mathbf{D}} \times \mathbf{S}^1.$$

It has the boundary $\mathbf{T}^2 = \mathbf{S}^1 \times \mathbf{S}^1$ as a closed leaf. The others are diffeomorphic to $\mathbf{R}^2$ (see Figure 5).

Because M_+ and M_- are diffeomorphic to $\bar{\mathbf{D}} \times \mathbf{S}^1$, $\mathcal{F}_0$ defines on M_+ and M_- respectively two foliations $\mathcal{F}_+$ and $\mathcal{F}_-$ which give a codimension one foliation $\mathcal{F}$ on $\mathbf{S}^3$ called the *Reeb foliation*.

(iv) *Lie group actions*. Let M be a manifold of dimension $m + n$ and G a connected Lie group of dimension m. An *action* of G on M is a map $G \times M \xrightarrow{\Phi} M$ such that:

- $\Phi(e, x) = x$ for every $x \in M$ (where e is the unit element of G),
- $\Phi(g', \Phi(g, x)) = \Phi(g'g, x)$ for every $x \in M$ and every $g, g' \in G$.

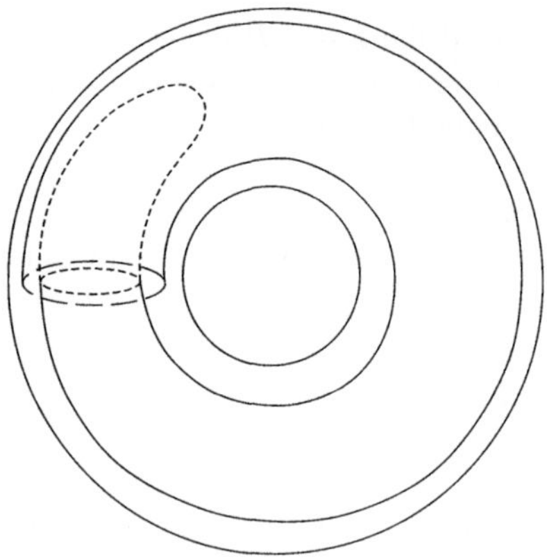

Fig. 5.

Suppose that, for every point $x \in M$, the dimension of the *isotropy subgroup*:

$$G_x = \{g \in G\colon\ \Phi(g,x) = x\}$$

is independent of x. Then the action Φ defines a foliation $\mathcal{F}$ of dimension $= m - \dim G_x$; its leaves are the orbits $\{\Phi(g,x)\colon\ g \in G\}$. In particular, this is the case if Φ is *locally free*, i.e. if, for every $x \in M$, the isotropy subgroup G_x is discrete. An explicit example is given when M is the quotient H/Γ of a Lie group H by a discrete subgroup Γ and G is a connected Lie subgroup of H; the action of G on M being induced by the left action of G on H. We say that $\mathcal{F}$ is a *homogeneous foliation.* Let us give an explicit example (for more details see [93]).

Let $A \in \mathrm{SL}(m+n-1, \mathbf{Z})$, where $m+n \geqslant 3$, be a matrix diagonalizable and having all its eigenvalues $\mu_1, \dots, \mu_{m-1}, \lambda_1, \dots, \lambda_n$ real and positive. We can think of A as a diffeomorphism of the $(m+n-1)$-torus $\mathbf{T}^{m+n-1}$. Let $X_1, \dots, X_{m-1}, Y_1, \dots, Y_n$ be linear vector fields on $\mathbf{T}^{m+n-1}$ such that:

$$A_* X_j = \mu_j X_j, \quad A_* Y_k = \lambda_k Y_k \quad \text{for } j = 1, \dots, m-1 \text{ and } k = 1, \dots, n,$$

and denote by $\mathcal{F}_0$ the foliation on $\mathbf{T}^{m+n-1}$ defined by the vector fields $X_1, \dots, X_{m-1}$. The product of $\mathcal{F}_0$ by $\mathbf{R}$ gives a codimension n foliation on $\mathbf{T}^{m+n-1} \times \mathbf{R}$ which is invariant by the diffeomorphism ϕ of $\mathbf{T}^{m+n-1} \times \mathbf{R}$ sending (z,t) to $(A(z), t+1)$. So, it induces a codimension n foliation $\mathcal{F}$ on the quotient manifold $\mathbf{T}_A^{m+n} = \mathbf{T}^{m+n-1} \times \mathbf{R}/\phi$. Notice that $\mathbf{T}_A^{m+n}$ is a flat bundle over the circle $\mathbf{S}^1$ with fibre $\mathbf{T}^{m+n-1}$. In fact $\mathbf{T}_A^{m+n}$ is the homogeneous space H/Γ where H is the semi-direct product of $\mathbf{R}^{m+n-1}$ by $\mathbf{R}$ given by the action:

$$(t,z) \in \mathbf{R} \times \mathbf{R}^{m+n-1} \longrightarrow A^t z \in \mathbf{R}^{m+n-1}$$

and Γ is the subgroup:

$$\{(\mathbf{m}, k) \in H \mid \mathbf{m} \in \mathbf{Z}^{m+n-1},\ k \in \mathbf{Z}\}.$$

If $v_1, \ldots, v_{m-1} \in \mathbf{R}^{m+n-1}$ are eigenvectors of A corresponding respectively to the eigenvalues $\mu_1, \ldots, \mu_{m-1}$ then the subgroup:

$$G = \left\{ \left(\sum_{i=1}^{m-1} a_i v_i, b \right) \in H \,\middle|\, a_1, \ldots, a_{m-1},\ b \in \mathbf{R} \right\}$$

is isomorphic to the semi-direct product of $\mathbf{R}^{m-1}$ by $\mathbf{R}_+^*$ where $\mathbf{R}_+^*$ acts on $\mathbf{R}^{m-1}$ by homotheties on each factor. The action of G on $\mathbf{T}_A^{m+n}$, induced by this identification, is a locally free action whose orbits define the foliation $\mathcal{F}$.

(v) *Foliations obtained by suspension.* Let B and F be two manifolds, respectively of dimensions m and n. Suppose that the fundamental group $\pi_1(B)$ of B is finitely generated. Let $\rho : \pi_1(B) \longrightarrow \mathrm{Diff}(F)$ be an injective representation, where $\mathrm{Diff}(F)$ is the diffeomorphism group of F. Denote by $\tilde{B}$ the universal covering of B and $\tilde{\mathcal{F}}$ the horizontal foliation on $\tilde{M} = \tilde{B} \times F$, i.e. the foliation whose leaves are the subsets $\tilde{B} \times \{y\}$, $y \in F$. This foliation is invariant by all the transformations $T_\gamma : \tilde{M} \longrightarrow \tilde{M}$ defined by $T_\gamma(\tilde{x}, y) = (\gamma \cdot \tilde{x}, \rho(\gamma)(y))$ where $\gamma \cdot \tilde{x}$ is the natural action of $\gamma \in \pi_1(B)$ on $\tilde{B}$; then $\tilde{\mathcal{F}}$ induces a codimension n foliation $\mathcal{F}_\rho$ on the quotient manifold:

$$M = \tilde{M}/(\tilde{x}, y) \sim (\gamma \cdot \tilde{x}, \rho(\gamma)(y)).$$

We say that $\mathcal{F}_\rho$ is the *suspension* of the diffeomorphism group $\Gamma = \rho(\pi_1(B))$. The leaves of $\mathcal{F}_\rho$ are transverse to the fibres of the natural fibration induced by the first projection $\tilde{B} \times F \longrightarrow \tilde{B}$.

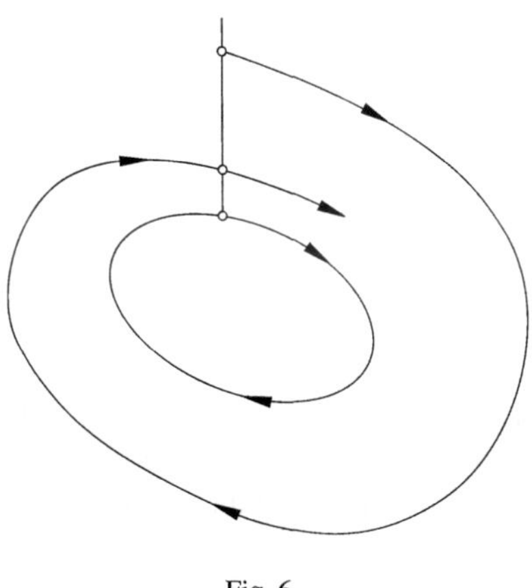

Fig. 6.

Conversely, suppose that $F \longrightarrow M \xrightarrow{\pi} B$ is a fibration with compact fibre F and that $\mathcal{F}$ is a codimension n foliation ($n =$ dimension of F) transverse to the fibres of π. Then there exists a representation $\rho : \pi_1(B) \longrightarrow \mathrm{Diff}(F)$ such that $\mathcal{F} = \mathcal{F}_\rho$.

Concrete example: let B be the circle $\mathbf{S}^1$ and $F = \mathbf{R}_+ = [0, +\infty[$. Let ρ be the representation of $\mathbf{Z} = \pi_1(\mathbf{S}^1)$ in $\mathrm{Diff}([0, +\infty[)$ defined by $\rho(1) = \varphi$ where $\varphi(y) = \lambda y$ with $\lambda \in]0, 1[$. Because φ is isotopic to the identity map of F, the manifold M is diffeomorphic to $\mathbf{S}^1 \times \mathbf{R}_+$ and the foliation $\mathcal{F}_\rho$ has one closed leaf diffeomorphic to the circle $\mathbf{S}^1$, corresponding to the fixed point $\varphi(0) = 0$ (see Figure 6).

1.4. *Foliations and differential systems*

Let M be a manifold of dimension $m + n$. Denote by TM the tangent bundle of M and let E be a subbundle of rank m. Let U be an open set of M such that on U, TM is equivalent to the product $U \times \mathbf{R}^{m+n}$. At each point $x \in U$, the fibre E_x can be considered as the kernel of n differential 1-forms $\omega_1, \ldots, \omega_n$ linearly independent:

$$E_x = \bigcap_{j=1}^{n} \ker \omega_j(x). \tag{S}$$

The subbundle E is called an *m-plane field* on M. We say that E is *involutive*, if for every vector field X and Y tangent to E (i.e. sections of E), the bracket $[X, Y]$ is also tangent to E. We say that E is *completely integrable* if, through each point $x \in M$, there exists a submanifold P_x of dimension m which admits $E_{|P_x}$ (the restriction of E to P_x) as tangent bundle. The maximal connected submanifolds satisfying this property are called the *integral submanifolds* of the differential system (S). They define a partition of M, i.e. a codimension n foliation. We have

THEOREM 3 (Frobenius). *Let E be a subbundle of rank m given locally by a differential system like in* (S). *Then the following assertions are equivalent*:

- *E is involutive,*
- *E is completely integrable,*
- *there exist differential* 1*-forms* (*defined locally*) (β_{ij}), $i, j = 1, \ldots, n$, *such that* $d\omega_i = \sum_{j=1}^{n} \beta_{ij} \wedge \omega_j$, $i = 1, \ldots, n$.

For example, let ω be a nonsingular 1-form. The corresponding subbundle E has fibre $E_x = \ker(\omega_x)$. It defines a codimension one foliation if, and only if, there exists a 1-form β such that $d\omega = \beta \wedge \omega$; this condition is equivalent to $d\omega \wedge \omega = 0$. In particular, if ω is closed it defines a codimension one foliation $\mathcal{F}$. If M is compact, all leaves are diffeomorphic and integration of ω over loops of M gives rise to a morphism $h : \pi_1(M) \longrightarrow \mathbf{R}$. The range $\Gamma = h(\pi_1(M))$ of h is a subgroup of $\mathbf{R}$ called the *holonomy group* of $\mathcal{F}$. Example (ii) is of this type: $M = \mathbf{T}^2$, $\omega = dy - \alpha\, dx$ which is closed. The fundamental group of $\mathbf{T}^2$ is $\mathbf{Z} \oplus \mathbf{Z}$ and it is easy to see that $\Gamma = \{p + q\alpha : p, q \in \mathbf{Z}\}$.

1.5. *Notations*

Let $\mathcal{F}$ be a codimension n foliation on M. We denote by $T\mathcal{F}$ the tangent bundle to $\mathcal{F}$ and $\nu\mathcal{F}$ the quotient $TM/T\mathcal{F}$ which is the *normal bundle* to $\mathcal{F}$; $\chi(\mathcal{F})$ will denote the space of sections of $T\mathcal{F}$ (elements of $\chi(\mathcal{F})$ are vector fields $X \in \chi(M)$ tangent to $\mathcal{F}$).

A differential form $\alpha \in \Omega^r(M)$ is said to be *basic* if it satisfies $i_X\alpha = 0$ and $L_X\alpha = 0$ for every $X \in \chi(\mathcal{F})$. (Here i_X and L_X denote respectively the inner product and the Lie derivative with respect to the vector field X.) For a function $f : M \longrightarrow \mathbf{R}$, these conditions are equivalent to $X \cdot f = 0$ for every $X \in \chi(\mathcal{F})$, i.e. f is constant on the leaves of $\mathcal{F}$; we denote by $\Omega^r(M/\mathcal{F})$ the space of basic forms of degree r on the foliated manifold $(M, \mathcal{F})$; this is a module over the algebra A_b of basic functions. A vector field $Y \in \chi(M)$ is said to be *foliated*, if for every $X \in \chi(\mathcal{F})$, the bracket $[X, Y] \in \chi(\mathcal{F})$. We can easily see that the set $\chi(M, \mathcal{F})$ of foliated vector fields is a Lie algebra and an A_b-module; by definition $\chi(\mathcal{F})$ is an ideal of $\chi(M, \mathcal{F})$ and the quotient

$$\chi(M/\mathcal{F}) = \chi(M, \mathcal{F})/\chi(\mathcal{F})$$

is called the Lie algebra of *transverse* (or *basic*) *vector fields* on the foliated manifold $(M, \mathcal{F})$. Also, it has a module structure over the algebra A_b.

2. Transverse structures

Let M be a manifold of dimension $m + n$ endowed with a codimension n foliation $\mathcal{F}$ defined by a foliated cocycle $\{U_i, f_i, T, \gamma_{ij}\}$ like in Definition 2.

DEFINITION 3. A *transverse structure* to $\mathcal{F}$ is a geometric structure on T invariant by the local diffeomorphisms γ_{ij}.

This is a very important notion in Foliation Theory. To make it clear, let us give the main examples.

2.1. *Measurable foliations*

Let $\mathcal{B}_T$ denote the family of Borel sets on T. A *transverse invariant measure* to $\mathcal{F}$ is a measure μ on $\mathcal{B}_T$ such that, for any $A \in \mathcal{B}_T$ in the domain of definition of γ_{ij}, we have

$$\mu(\gamma_{ij}(A)) = \mu(A).$$

We say that $\mathcal{F}$ is a *measurable foliation* if it admits a transverse measure. The notion of measurable foliation was introduced firstly by J.F. Plante; he obtained many interesting results on the qualitative behavior of codimension one measurable foliations on compact manifolds (cf. [273]).

2.2. *Lie foliations*

We say that $\mathcal{F}$ is a *Lie foliation*, if T is a Lie group G and γ_{ij} are restrictions of left translations on G.

Such foliation can also be defined by a 1-form ω on M with values in the Lie algebra $\mathcal{G}$ such that:

(i) $\omega_x : T_x M \longrightarrow \mathcal{G}$ is surjective for every $x \in M$,

(ii) $d\omega + \frac{1}{2}[\omega, \omega] = 0$.

If $\mathcal{G}$ is Abelian, ω is given by n linearly independent closed scalar 1-forms $\omega_1, \ldots, \omega_n$. In particular, if $n = 1$, an important topological property of compact manifolds supporting such foliation is given by the following theorem due to Tischler [350].

THEOREM 4. *If a compact manifold admits a closed nonsingular* 1*-form, then it is a locally trivial fibration over the circle* $\mathbf{S}^1$.

The hypothesis $\mathcal{G}$ is Abelian is important: D. Lehmann [211] proved that, in general, the result is false even if $\mathcal{G}$ is nilpotent.

Foliations defined by nonsingular closed 1-forms can be considered as topological prototype of codimension one foliations without holonomy as it is illustrated by Sacksteder's theorem [311]:

THEOREM 5. *Let* $\mathcal{F}$ *be a* C^r $(r \geqslant 2)$ *codimension one foliation on a connected compact manifold. If* $\mathcal{F}$ *has no holonomy, then it is topologically conjugated to a foliation defined by a nonsingular closed* 1*-form.*

In the general case, the structure of a Lie foliation on a compact manifold, is given by the following theorem due to E. Fédida [101]:

THEOREM 6. *Let* $\mathcal{F}$ *be a Lie* G*-foliation on a compact manifold* M*. Let* $\tilde{M}$ *be the universal covering of* M *and* $\tilde{\mathcal{F}}$ *the lift of* $\mathcal{F}$ *to* $\tilde{M}$*. Then there exist a homomorphism* $h : \pi_1(M) \longrightarrow G$ *and a locally trivial fibration* $D : \tilde{M} \longrightarrow G$ *whose fibres are the leaves of* $\tilde{\mathcal{F}}$ *and such that, for every* $\gamma \in \pi_1(M)$*, the following diagram is commutative*:

$$\begin{array}{ccc} \tilde{M} & \stackrel{\gamma}{\longrightarrow} & \tilde{M} \\ D\downarrow & & \downarrow D \\ G & \stackrel{h(\gamma)}{\longrightarrow} & G \end{array}$$

where the first line denotes the deck transformation of $\gamma \in \pi_1(M)$ *on* $\tilde{M}$.

The subgroup $\Gamma = h(\pi_1(M)) \subset G$ is called the *holonomy group* of $\mathcal{F}$ although the holonomy of each leaf is trivial. The fibration $D : \tilde{M} \longrightarrow G$ is called the *developing map* of $\mathcal{F}$.

2.3. *Transversely parallelizable foliations*

We say that $\mathcal{F}$ is *transversely parallelizable* if there exist on M, foliated vector fields $Y_1, \ldots, Y_n$, transverse to $\mathcal{F}$ and everywhere linearly independent. This means that the manifold T admits a parallelism $(Y_1, \ldots, Y_n)$ invariant by all the local diffeomorphisms γ_{ij} or, equivalently, that the A_b-module $\chi(M/\mathcal{F})$ is free of rank n. The structure of a transversely parallelizable foliation on a compact manifold is given by the following theorem due to L. Conlon [60] for $n = 2$ and in general to P. Molino [243].

THEOREM 7. *Let $\mathcal{F}$ be a transversely parallelizable foliation of codimension n on a compact manifold M. Then*:

(1) *the closures of the leaves are submanifolds which are fibres of a locally trivial fibration $\pi : M \longrightarrow W$ where W is a compact manifold,*
(2) *there exists a simply connected Lie group G_0 such that the restriction $\mathcal{F}_0$ of $\mathcal{F}$ to any leaf closure F is a G_0-Lie foliation,*
(3) *the cocycle of the fibration $\pi : M \longrightarrow W$ has values in the group of diffeomorphisms of F preserving $\mathcal{F}_0$.*

The fibration $\pi : M \longrightarrow W$ and the manifold W are called respectively the *basic fibration* and the *basic manifold* associated to $\mathcal{F}$. Theorem 7 says that if, in particular, the leaves of $\mathcal{F}$ are closed, then the foliation is just a fibration over W. This is still true even if the leaves are not closed: the manifold M is a fibration over the leaf space $M/\mathcal{F}$ which is, in this case, a *Q-manifold* in the sense of [13]. Theorem 7 is still valid for *transversely complete foliations* on noncompact manifolds (cf. [242]).

It is not difficult to see that any Lie foliation is transversely parallelizable. This is a consequence of the fact that a Lie group is parallelizable and that the parallelism can be chosen invariant by left translations.

2.4. *Riemannian foliations*

The foliation $\mathcal{F}$ is said to be *Riemannian* if there exists on T a Riemannian metric such that the local diffeomorphisms γ_{ij} are isometries. Using the submersions $f_i : U_i \longrightarrow T$ one can construct on M a Riemannian metric which can be written in local coordinates:

$$ds^2 = \sum_{i,j=1}^{m} \theta_i \otimes \theta_j + \sum_{k,\ell=1}^{n} g_{k\ell}(y)\, dy_k \otimes dy_\ell.$$

Equivalently, $\mathcal{F}$ is Riemannian, if any geodesic orthogonal to the leaves at a point is orthogonal to the leaves everywhere [289].

Let $\mathcal{F}$ be Riemannian. Then there exists a Levi-Civita connection, transverse to the leaves which, by unicity argument, coincides on any distinguished open set, with the pull-back of the Levi-Civita connection on the Riemannian manifold T. This connection is said to be *projectable*. Let $\mathrm{O}(n) \longrightarrow M^\# \stackrel{\tau}{\longrightarrow} M$ be the principal bundle of orthonormal frames transverse to $\mathcal{F}$; this is an $\mathcal{F}$-bundle, in the sense of Section 7.1. The following theorem is due to P. Molino [243].

THEOREM 8. *Suppose M is compact. Then, the foliation $\mathcal{F}$ can be lifted to a foliation $\mathcal{F}^{\#}$ on $M^{\#}$ of the same dimension and such that*:

(1) *$\mathcal{F}^{\#}$ is transversely parallelizable,*

(2) *$\mathcal{F}^{\#}$ is invariant under the action of $\mathrm{O}(n)$ on $M^{\#}$ and projects, by τ, on $\mathcal{F}$.*

The basic manifold $W^{\#}$ and the basic fibration $F^{\#} \longrightarrow M^{\#} \xrightarrow{\pi^{\#}} W^{\#}$ are called respectively the *basic manifold* and the *basic fibration* of $\mathcal{F}$.

We have the following properties:

- the restriction of τ to a leaf of $\mathcal{F}^{\#}$ is a covering over a leaf of $\mathcal{F}$. So all leaves of $\mathcal{F}$ have the same universal covering,
- the closure of any leaf of $\mathcal{F}$ is a submanifold of M and the leaf closures define a *singular foliation* (the leaves have different dimensions) on M. (For more details about this notion see [243].)

Another interesting result for Riemannian foliations is the Global Reeb Stability Theorem which is valid even if the codimension is greater than 1.

THEOREM 9. *Let $\mathcal{F}$ be a Riemannian foliation on a compact manifold M. If there exists a compact leaf with finite fundamental group, then all leaves are compact with finite fundamental group.*

The property $\mathcal{F}$ is Riemannian means that the leaf space $Q = M/\mathcal{F}$ is a Riemannian manifold even if Q does not support any differentiable structure!

2.5. *Transversely holomorphic foliations*

The foliation $\mathcal{F}$ is said to be *transversely holomorphic* if T is a complex manifold and the γ_{ij} are local biholomorphisms. Particular case is a *holomorphic foliation*: the manifolds M and T are complex, all the f_i are holomorphic and all γ_{ij} are local biholomorphisms.

If T is Kählerian and γ_{ij} biholomorphisms which preserve the Kähler form on T we say that $\mathcal{F}$ is *transversely Kählerian*. For example, any codimension 2 Riemannian foliation which is transversely orientable is transversely Kählerian.

Let us give concrete examples of such foliations. Let M be the unit sphere in the Hermitian space $\mathbf{C}^{n+1}$:

$$M = \mathbf{S}^{2n+1} = \left\{ (z_1, \ldots, z_{n+1}) \in \mathbf{C}^{n+1} \colon \sum_{k=1}^{n+1} |z_k|^2 = 1 \right\}.$$

Let Z be the holomorphic vector field on $\mathbf{C}^{n+1}$ given by the formula:

$$Z = \sum_{k=1}^{n+1} a_k z_k \frac{\partial}{\partial z_k},$$

where $a_k = \alpha_k + i\beta_k \in \mathbf{C}$. There exists a good choice of the numbers a_k such that the orbits of Z intersect transversely the sphere M; then Z induces on M a real vector field X which defines a foliation $\mathcal{F}$. It is not difficult to see that $\mathcal{F}$ is transversely holomorphic. It is transversely Kählerian if we choose in addition $\alpha_k = 0$ for any $k = 1, \dots, n+1$.

3. Codimension one foliations

Codimension one foliations constitute a rich theme which was studied extensively by many people. The richness comes from the existence, for such foliations, of nonsingular transverse vector fields which give a way to go from a leaf to an other. Most of the results in Foliation Theory were first obtained in the codimension one case; this section is devoted to summarize some of them.

Let $\mathcal{F}$ be a codimension one foliation on a manifold M and ν a transverse vector bundle to $\mathcal{F}$. Because ν is of rank one, it is integrable and defines a foliation $\mathcal{V}$ transverse to $\mathcal{F}$. So we have clearly $\chi(M) = 0$. It is natural to ask if this condition is sufficient for the existence of a codimension one foliation on M; this was conjectured by E. Thomas [342]. The reader can see the paper [207] by B. Lawson about the history of the different steps for solving this conjecture. The final solution was given by W. Thurston [348] who proved

THEOREM 10 (Thurston). *Let M be a compact manifold. Then M admits a codimension one foliation if, and only if, the Euler–Poincaré number $\chi(M)$ of M is zero.*

Recall that two vector bundles $E \longrightarrow M$ and $E' \longrightarrow M$ are said to be *homotopic* if there exists a continuous family $E_t \longrightarrow M$, $t \in [0, 1]$, of vector bundles such that $E_0 = E$ and $E_1 = E'$. So we can formulate the question of existence of codimension one foliations, in general, in the following:

Let M be a compact manifold. Then any codimension one plane field on M is homotopic to an integrable one.

The first results solving (in some particular cases), this conjecture were obtained by J. Wood (see [376]) and also by P. Schweitzer and W. Thurston in the C^0-case. As far as we know this conjecture is still open.

Notice that the compactness of the manifold is a big constraint. Indeed on open manifolds the answer to this conjecture is positive [266].

The regularity property seems to be very important in the existence of foliations on compact manifolds. In particular, there is a big difference in the treatment between the C^∞ case and the real analytic one. In this direction A. Haefliger proved in [143] the following important theorem.

THEOREM 11 (Haefliger). *Let M be a compact manifold with a finite fundamental group. Then M has no real analytic codimension one foliation.*

Let us end this section with one of the most important results obtained in codimension one foliation theory on 3-manifolds [261].

THEOREM 12 (Novikov). *Let M be a compact 3-manifold with a finite fundamental group. Then any codimension one foliation on M has a compact leaf diffeomorphic to the torus $\mathbf{T}^2$.*

4. Γ-structures

This notion was introduced by A. Haefliger and became a key ingredient in studying characteristic classes of foliations.

DEFINITION 4. A groupoid is given by a set Γ, a subset $\Gamma^{(2)}$ of $\Gamma \times \Gamma$ with a law

$$(\gamma, \sigma) \in \Gamma^{(2)} \longrightarrow \gamma\sigma \in \Gamma$$

and an inverse map $\gamma \in \Gamma \longrightarrow \gamma^{-1} \in \Gamma$ satisfying the following properties:

(i) $(\gamma^{-1})^{-1} = \gamma$,

(ii) if $(\gamma, \sigma), (\sigma, \tau) \in \Gamma^{(2)}$, then $(\gamma\sigma, \tau), (\gamma, \sigma\tau) \in \Gamma^{(2)}$ and $(\gamma\sigma)\tau = \gamma(\sigma\tau)$,

(iii) if $(\gamma^{-1}, \gamma) \in \Gamma^{(2)}$ and $(\gamma, \sigma) \in \Gamma^{(2)}$, then $\gamma^{-1}(\gamma\sigma) = \sigma$,

(iv) if $(\gamma, \gamma^{-1}) \in \Gamma^{(2)}$ and $(\tau, \gamma) \in \Gamma^{(2)}$, then $(\tau\gamma)\gamma^{-1} = \tau$.

For $\gamma \in \Gamma$, $s(\gamma) = \gamma^{-1}\gamma$ is called the *source* of γ and $r(\gamma) = \gamma\gamma^{-1}$ the *range* of γ.

Then, there are two projections s, r (or α, β) $: \Gamma \longrightarrow \Gamma^{(0)} = \operatorname{Im} r$. The subset $\Gamma^{(0)}$ of Γ is called the *unit space* of Γ.

A *topological groupoid* is a groupoid with a topology compatible with the composition and inverse maps. As a consequence, the two projections s, r on the unit space are also continuous.

A *differentiable structure* on Γ is given by a manifold structure on Γ and $\Gamma^{(0)}$ compatible with the composition and inverse maps and such that:

- $s : \Gamma \longrightarrow \Gamma^{(0)}$ is a submersion,
- the canonical injection $\Gamma^{(0)} \longrightarrow \Gamma$ is an embedding.

The differentiable (or topological) groupoid Γ is *étale* if s is étale.

Let M be a manifold, Γ a topological groupoid and $\{U_i\}$ an open cover of M; a 1-*cocycle* on M with values in Γ is given as follows: for each pair (i, j), let

$$\gamma_{ij} : U_i \cap U_j \longrightarrow \Gamma$$

be a continuous map such that, if $x \in U_i \cap U_j \cap U_k$, then $(\gamma_{ij}(x), \gamma_{jk}(x)) \in \Gamma^{(2)}$ and

$$\gamma_{ik}(x) = \gamma_{ij}(x)\gamma_{jk}(x).$$

Two 1-cocycles are said to be *cohomologous* if they are restrictions of the same cocyle on the union of their coverings. A Γ-*structure* on M, or an element of $H^1(M, \Gamma)$, is an equivalence class of 1-cocycles.

Let Γ be the groupoid of germs of local diffeomorphisms of $\mathbf{R}^n$; then the unit space $\Gamma^{(0)}$ may be identified to $\mathbf{R}^n$. A codimension n foliation $\mathcal{F}$ on M may be viewed as a particular Γ-structure for which a representative is a 1-cocycle on an open covering $\{U_i\}$ such that the following maps $f_i = \gamma_{ii} : U_i \longrightarrow \Gamma^{(0)} = \mathbf{R}^q$ are submersions.

5. The leaf space

Let $\mathcal{F}$ be a codimension n foliation on M. Let U be a subset of M and denote by $\hat{U}$ the union of the leaves intersecting U. Recall that U is *saturated* if $U = \hat{U}$. It is easy to see that if U is open, so is $\hat{U}$. Then, the equivalence relation on M, $x \sim y$ if, and only if, x and y are in the same leaf, is open. The set of equivalence classes $M/\sim$, endowed with the quotient topology, is called the *leaf space* of $\mathcal{F}$ and usually denoted by $M/\mathcal{F}$.

We can think of $M/\mathcal{F}$ as follows. The foliation $\mathcal{F}$ is the geometric realization of a completely integrable differential system (S) on M. Each integral submanifold is a leaf of $\mathcal{F}$ and corresponds to an initial condition of (S). So we can consider Q as a parameter space of the initial conditions of this differential system. In general Q is not a manifold, but we can define on this space many geometrical objects like functions, differential forms, differential operators, etc. (cf., for instance, Section 7). They correspond to their analogues on M invariant along the leaves (in a sense to be determined following the context).

There were many attempts to give the leaf space of a foliation a differentiable structure, even if its topology is, generally poor.

A first one was from Satake, whose point of view was recovered by W. Thurston. In other domains, let us cite G.W. Mackey [219] who introduced the virtual group notion in *Ergodic Theory* and M. Artin [12] the algebraic space notion. The former corresponds to the measurable version of the S-atlas of W.T. Van Est, the latter suggested the definition of a Q-manifold.

In fact, there is no uniform definition. Each corresponds to a given situation or a particular problem. Nevertheless, the point of view of *Noncommutative Geometry*, by A. Connes, using the C^*-algebra of the groupoid of a foliation, or the crossed-product of the C^*-algebra of a manifold by a group acting on it, is attractive and efficient too. For example, there are Longitudinal and Transversal Index Theorems for foliations; one gets also Godbillon–Vey classes, etc.

5.1. *V-manifolds*

Let Ω be an open set in $\mathbf{R}^n$ and let Σ be a finite group of diffeomorphisms of Ω. Denote by Ω/Σ the orbit space with quotient topology and p the canonical projection $\Omega \longrightarrow \Omega/\Sigma$. If Ω' is another open set of $\mathbf{R}^n$, Σ' a finite group of diffeomorphisms of Ω' and p' the canonical projection $\Omega' \longrightarrow \Omega'/\Sigma'$, then a *morphism* from Ω/Σ to Ω'/Σ' is a continuous map f from Ω/Σ to Ω'/Σ', which admits local coverings by smooth local maps from Ω to Ω'. An *isomorphism* is a bijective morphism, the inverse of which is a morphism.

If V is a second countable Hausdorff space, a *Satake atlas* of dimension n is a family $\mathcal{A} = (U_i, \Phi_i)$ where (U_i) is an open covering of V and $\Phi_i : U_i \longrightarrow \Omega_i/\Sigma_i$ is a homeomorphism of U_i on the quotient of an open subset Ω_i of $\mathbf{R}^n$ by a finite group of diffeomorphisms, with following coherence condition: for all i, j such that $U_i \cap U_j \neq \emptyset$ the map $\Phi_j \circ \Phi_i^{-1} : \Phi_i(U_i \cap U_j) \longrightarrow \Phi_j(U_i \cap U_j)$ is a morphism as previously defined.

A *V-manifold* (or a *Satake manifold* or an *orbifold*) of dimension n is a space V with a maximal Satake atlas of dimension n. The following are simple examples illustrating the notion of a V-manifold.

(i) Let Γ be a finite group of isometries of a Riemannian manifold M of dimension n. Then the quotient space M/Γ is a V-manifold of dimension n.

(ii) A waterdrop obtained by gluing two open discs along their boundaries, one of them being implemented with a rotation of $\frac{2\pi}{3}$ around its center.

It is proved in [243] that *every leaf space of a Riemannian foliation with compact leaves on a compact manifold is a V-manifold.*

Conversely: *every compact V-manifold is the leaf space of a Riemannian foliation with compact leaves on a compact manifold* (cf. [132]).

5.2. *QF-manifolds*

Let (X, p, S) be a triple where X is a manifold, S a set and p a surjective map from X to S; this is an *étale QF-atlas* of S if it satisfies the following conditions:

(H) for every pair (x, y) in X^2 such that $p(x) = p(y)$, there are open neighborhoods U and V respectively of x and y and a diffeomorphism h from U to V such that $h(x) = y$ and $p \circ h(t) = p(t)$ for every $t \in U$,

(QF) every morphism from a manifold Z to X such that $p \circ f$ is constant is locally constant.

As usual two étale QF-atlases (X_1, p_1, S) and (X_2, p_2, S) are *equivalent* if (X, p, S) is an étale QF-atlas where X is the disjoint union of X_1 and X_2 and p is p_1 on X_1 and p_2 on X_2. A *QF-manifold structure* on S is an equivalence class of étale QF-atlases on S. All the leaf spaces of foliated second countable manifolds are in this category.

5.3. *Q-manifolds*

Let (X, p, S) be a triple where X is a manifold, S a set and p a surjective map from X to S; this is a *Q-atlas* of S if it satisfies the following conditions:

(H) is as in the definition of an étale QF-atlas,

(Q) let $f = (f_1, f_2)$ be a morphism from a manifold Z to X^2 such that $p \circ f_1 = p \circ f_2$; then the subset $T = \{z \in Z\colon\ f_1(z) = f_2(z)\}$ is open in Z.

A *Q-manifold structure* on S is an equivalence class of Q-atlases of S. The following are examples of Q-manifolds:

(i) the leaf space of foliated torus with geodesics having irrational slope,

(ii) more generally, the leaf space of a transversely parallelizable foliation on a compact manifold.

It was first tried to generalize to leaf spaces the classical theorems and tools (Gauss–Bonnet, de Rham cohomology, Poincaré duality, Leray–Serre spectral sequence, fundamental group, etc.) to get results on the transverse structure.

The V-manifolds are met in natural way and there exist many examples of them. They appear also with ramified coverings.

The Q-manifolds permitted to restore the third Lie theorem for Banach Lie algebras (cf. [268]); they appear also in the structure theorem of P. Molino. Recently, G. Meigniez got a characterization of Godbillon Homotopy Extension Property for foliations, where they play a role (cf. [231]).

6. Characteristic classes

We follow here the lectures of R. Bott as written by L. Conlon in [23]. We will restrict ourself to only one result: Bott vanishing theorem. The reader can find more material in [23] or in [147].

6.1. *The classifying space and the universal bundle*

Let $\mathcal{H}$ be a separable real Hilbert space with norm $\| \ \|$. If u and v are nonzero vectors in $\mathcal{H}$, (u, v) will be the angle defined by u and v; it is immediate to see that for every positive number λ and μ we have $(\lambda u, \mu v) = (u, v)$. Denote by BGL_n the set of n-dimensional subspaces of $\mathcal{H}$. Let $\tau, \sigma \in \mathrm{BGL}_n$ and set $\delta(\tau, \sigma) = \inf(u, v)$ where the infimum is taken over all the vector $u \in \tau$ and $v \in \sigma$ with $\|u\| = \|v\| = 1$. It is not difficult to see that δ defines a distance on BGL_n. The topological space BGL_n is called the *classifying space* of the group $\mathrm{GL}(n, \mathbf{R})$ of linear transformations of the vector space $\mathbf{R}^n$.

The cohomology $H^*(\mathrm{BGL}_n, \mathbf{R})$ of BGL_n is a polynomial ring $\mathbf{R}[p_1, \ldots, p_{[n/2]}]$ where the $p_i \in H^{4i}(\mathrm{BGL}_n, \mathbf{R})$ are the universal Pontryagin classes (cf. [21]).

On BGL_n we have a canonical real vector bundle $S \longrightarrow \mathrm{BGL}_n$ of rank n whose fibre at each τ is the space τ itself; it is called the *universal bundle* on BGL_n.

6.2. *Classification of real vector bundles*

As M is paracompact, it admits a countable locally finite open cover $\mathcal{U} = \{U_i\}$ which, in addition, can be chosen such that each finite intersection $U_{i_1} \cap \cdots \cap U_{i_\ell}$ is contractible. Such an open cover is called a *good cover*; it always exists: take a Riemannian metric on M and a countable family of geodesically convex open balls which covers M. If $E \stackrel{\pi}{\longrightarrow} M$ is a real vector bundle of rank n, its restriction $E_{|U_i}$ to any U_i is trivial, i.e. there exists a diffeomorphism $\varphi_i : E_{|U_i} \longrightarrow U_i \times \mathbf{R}^n$ which sends the fibre E_x isomorphically on $\{x\} \times \mathbf{R}^n$. Let $(s_i^1, \ldots, s_i^n)$ be a basis of the free module $C^\infty(E_{|U_i})$ over the algebra $A(U_i)$ of real valued C^∞-functions on U_i. Let $\{\rho_i\}$ be a partition of the unity subordinated to $\{U_i\}$ and let V_i be the real vector space spanned by $(\rho_i s_i^1, \ldots, \rho_i s_i^n)$. For each i we set $\psi_i = q_i \circ \varphi_i$ where $q_i : U_i \times \mathbf{R}^n \longrightarrow \mathbf{R}^n$ is the second projection. Now express $\mathcal{H}$ as orthogonal direct sum of the $V_i \simeq \mathbf{R}^n$ and denote by $\zeta_i : V_i \longrightarrow \mathcal{H}$ the inclusion of the ith summand. Define $\Phi : E \longrightarrow \mathcal{H}$ by

$$\Phi(x, \xi) = \sum_{i=1}^{\infty} \rho_i(x) \cdot \zeta_i\big(\psi_i(x, \xi)\big).$$

Then Φ is continuous and sends each fibre $\pi^{-1}(x)$ of E isomorphically on an n-dimensional subspace of $\mathcal{H}$. Thus $f(x) = \Phi(\pi^{-1}(x))$ defines a continuous map $f : M \longrightarrow \mathrm{BGL}_n$ called the *classifying map* for the vector bundle E namely E is the pullback by f of the universal bundle $S \longrightarrow \mathrm{BGL}_n$. In fact there is a natural one–one correspondence between the set $\mathrm{Vect}_n(M)$ of isomorphism classes of real vector bundles of rank n on M and the set $[M, \mathrm{BGL}_n]$ of homotopy classes of maps $M \longrightarrow \mathrm{BGL}_n$. The ith-Pontryagin class of the real vector bundle $E \longrightarrow M$ is by definition $p_i(E) = f^*(p_i)$. The graded subring

$$\mathrm{Pont}^*(E) = f^*\big(H^*(\mathrm{BGL}_n, \mathbf{R})\big) \subset H^*(M, \mathbf{R})$$

is called the Pontryagin ring of E. The first important result obtained in the theory of characteristic classes of foliations is the following

THEOREM 13 (Bott vanishing theorem). *Let $\mathcal{F}$ be a foliation of codimension n with normal bundle $\nu\mathcal{F}$. Then* $\mathrm{Pont}^i(\nu\mathcal{F}) = 0$ *for* $i > 2n$.

As a nontrivial example of characteristic class of a foliation $\mathcal{F}$, we have the *Godbillon–Vey invariant* $GV(\mathcal{F})$ (discovered by C. Godbillon and J. Vey [136]) which is, in general, nonzero as shown by R. Roussarie. An elementary construction of this invariant in the codimension one case is as follows.

Let M be a compact manifold endowed with a codimension one foliation defined by a differential 1-form ω. Then the integrability condition implies the equality $\omega \wedge d\omega = 0$, i.e. $d\omega = \alpha \wedge \omega$. It is easy to see that $\alpha \wedge d\alpha$ is closed and that its cohomology class in $H^3(M, \mathbf{R})$, which is by definition $GV(\mathcal{F})$, is independent of the choice of α.

One of the most important results in the study of the Godbillon–Vey invariant for codimension one foliations was obtained by G. Duminy in [73]. Let us describe it briefly; a complete account is given in [119]. Let $\mathcal{F}$ be a codimension one foliation on a compact manifold M. A leaf L of $\mathcal{F}$ is called *resilient* if there exist a loop $\sigma : [0,1] \longrightarrow L$ and a transversal T to $\mathcal{F}$ passing through $\sigma(0)$ such that the following conditions are satisfied:

(i) there exists a point $x \in L \cap T$ in the domain of holonomy h_σ of σ and different from $\sigma(0)$;

(ii) the sequence $h_\sigma^n(x)$ converges to $\sigma(0)$ as $n \to +\infty$.

G. Duminy proved that *if $\mathcal{F}$ has no resilient leaf then the Godbillon–Vey invariant of $\mathcal{F}$ is zero.*

Recently, A. Connes and H. Moscovici have discovered a universal Hopf algebra with cohomology from which one is able to recover the characteristic classes of a foliation without use of Chern–Weil homomorphism or connections (cf. [63]).

7. Basic global analysis

Let M be a manifold endowed with a foliation $\mathcal{F}$ of codimension n. We suppose for simplicity that $\mathcal{F}$ is transversely orientable.

7.1. *Foliated vector bundles and basic sections*

Let $\mathcal{P} : G \hookrightarrow P \overset{\iota}{\longrightarrow} M$ be a principal bundle with structural group $G \subset \mathrm{GL}(N, \mathbf{C})$. The group G acts on P on the right and on its Lie algebra $\mathcal{G}$ by the adjoint representation.

Denote by $\mathcal{V}$ the vector bundle whose fibre V_z at a point $z \in P$ is the tangent space at z of the fibre of $\mathcal{P}$. A *connection* on $\mathcal{P}$ is a subbundle $\mathcal{H}$ of TP such that:

- for every $z \in \mathcal{P}$, $T_z P = V_z \oplus H_z$,
- for every $g \in G$ and every $z \in P$, $H_{zg} = (R_g)_*(H_z)$ where R_g is the right action of g on P.

As is well known the subbundle $\mathcal{H}$ is also the kernel of an invariant (under the action of G) 1-form ω on P (called the *connection form*) with values in $\mathcal{G}$.

It is easy to see that the restriction of ι_* (the derivative of ι) to H_z is an isomorphism onto $T_{\iota(z)}M$. Let $\tau = \iota_*^{-1}(T\mathcal{F})$. We say that $\mathcal{P}$ is *foliated* if τ is integrable. In this case, τ defines a foliation $\tilde{\mathcal{F}}$ on P such that

- $\dim(\tilde{\mathcal{F}}) = \dim(\mathcal{F})$,
- $\tilde{\mathcal{F}}$ is invariant under the action of G.

We say that the connection $\mathcal{H}$ is *basic*, if the ω is basic (cf. Section 1.5). A foliated bundle $\mathcal{E}$ is said to be an *$\mathcal{F}$-bundle*, if it admits a basic connection.

Let $E \longrightarrow M$ be a complex vector bundle defined by a cocycle $\{U_i, \gamma_{ij}, G\}$ where U_i is an open cover of M and $\gamma_{ij}: U_i \cap U_j \longrightarrow G \subset \mathrm{GL}(N, \mathbf{C})$ are the transition functions. We say that E is an *$\mathcal{F}$-bundle*, if the associated principal bundle $G \longrightarrow P \longrightarrow M$ is an $\mathcal{F}$-bundle. Because $E = P \times_G \mathbf{C}^N$, $\tilde{\mathcal{F}}$ induces a foliation $\mathcal{F}_E$ on E. An *$\mathcal{F}$-morphism* $\varphi: (E, \omega) \longrightarrow (E', \omega')$ between two $\mathcal{F}$-bundles is a morphism of vector bundles which sends leaves of $\mathcal{F}_E$ into leaves of $\mathcal{F}_{E'}$.

(Notice that the collection of $\mathcal{F}$-bundles and $\mathcal{F}$-morphisms is a category. So we can define the group $K(M, \mathcal{F})$ of *foliated K-theory* as in the classical case.)

Let $E \longrightarrow M$ be an $\mathcal{F}$-bundle. Then the dual bundle E^* and all its exterior powers $\Lambda^* E^*$ are $\mathcal{F}$-bundles; also $\mathcal{H}^2 E = \{\text{Hermitian forms on } E\}$ is an $\mathcal{F}$-bundle.

7.2. *Transversely elliptic operators*

Let $E \longrightarrow M$ be a $\mathcal{F}$-foliated vector bundle. Denote by ∇ the covariant derivative $\chi(M) \times C^\infty(E) \stackrel{\nabla}{\longrightarrow} C^\infty(E)$ associated to the connection $\mathcal{H}$. We say that a section $\alpha \in C^\infty(E)$ is *basic*, if it satisfies the condition $\nabla_X \alpha = 0$ for every $X \in \chi(\mathcal{F})$. The space $C^\infty(E/\mathcal{F})$ of basic sections of E is an A_b-module.

Let E and E' two $\mathcal{F}$-bundles (with the same rank N for simplicity). A *basic differential operator of order ℓ* from E to E' is a linear map $C^\infty(E/\mathcal{F}) \stackrel{D}{\longrightarrow} C^\infty(E'/\mathcal{F})$ such that on local coordinates $(x_1, \ldots, x_m, y_1, \ldots, y_n)$ for which $\mathcal{F}$ is defined by the differential equations $dy_1 = \cdots = dy_n = 0$, D has the expression:

$$D = \sum_{|s| \leqslant \ell} a_s(y) \frac{\partial^{|s|}}{\partial y_1^{s_1} \ldots \partial y_n^{s_n}},$$

where $s = (s_1, \ldots, s_n) \in \mathbf{N}^n$, $|s| = s_1 + \cdots + s_n$ and a_s are $(N \times N)$-matrices whose coefficients are basic functions. The *principal symbol* of D at the point $z = (x, y)$ and the basic covector $\xi \in \nu_z^* \mathcal{F}$ is the linear map $\sigma(D)(z,\xi): E_z \longrightarrow E'_z$ defined by

$$\sigma(D)(z,\xi)(\eta) = \sum_{|s|=\ell} \xi_1^{s_1} \cdot \dots \cdot \xi_n^{s_n} a_s(y)(\eta).$$

We say that D is *transversely elliptic* if $\sigma(D)(z,\xi)$ is an isomorphism for every $z \in M$ and every basic covector ξ different from 0. If $\mathcal{F}$ is Riemannian, its conormal bundle $\nu^*\mathcal{F}$ is an $\mathcal{F}$-bundle and is equipped with a foliation $\mathcal{F}^*$. Then the principal symbol $\sigma(D)(z,\xi)$ of a transversely elliptic operator D defines an element $[D]$ in the group $K(\nu^*\mathcal{F}, \mathcal{F}^*)$.

A *Hermitian metric* on E is a positive definite section h of $\mathcal{H}^2 E$. If h is basic we say that E is a *Hermitian $\mathcal{F}$-bundle*.

Let $E \longrightarrow M$ be a Hermitian $\mathcal{F}$-foliated bundle with Hermitian metric h and let D be a basic differential operator of order $\ell = 2\ell'$ on $C^\infty(E/\mathcal{F})$. For every $z \in M$ and every basic covector $\xi \in \nu_z^*\mathcal{F}$ we define a quadratic form $A(D)(z,\xi): E_z \longrightarrow \mathbf{C}$ by

$$A(D)(z,\xi)(\eta) = (-1)^{\ell'} \langle \sigma(D)(z,\xi)(\eta), \eta \rangle.$$

We say that D is *strongly transversely elliptic*, if $A(D)(z,\xi)$ is positive definite for every $z \in M$ and every nonzero ξ. Obviously every strongly transversely elliptic operator is transversely elliptic.

From now on we suppose that M is compact and connected. Assume that the foliation $\mathcal{F}$ is Riemannian transversely oriented. Let $E^\#$ be the pullback of E to the principal bundle $\mathrm{SO}(n) \longrightarrow M^\# \xrightarrow{p} M$ of the orthonormal direct frames transverse to $\mathcal{F}$ (cf. Theorem 8). Then $E^\#$ is a $\mathrm{SO}(n)$-bundle and a Hermitian $\mathcal{F}^\#$-bundle equipped with a Hermitian metric $h^\#$. Let $W^\#$ be the basic manifold associated to the transversely parallelizable foliation $\mathcal{F}^\#$ on $M^\#$. The basic sections of E are canonically identified to basic sections of $E^\#$ which are invariant under the action of $\mathrm{SO}(n)$. In particular, if $f: M \longrightarrow \mathbf{C}$ is a basic function, $f \circ p$ is a basic function on $M^\#$ (with respect to $\mathcal{F}^\#$); moreover $f \circ p$ is invariant by the action of $\mathrm{SO}(n)$. Because $f \circ p$ is continuous, it is constant on the leaf closures of $\mathcal{F}^\#$ so it induces an $\mathrm{SO}(n)$-invariant C^∞ function on the basic manifold $W^\#$. We can prove, by the converse process, that any $\mathrm{SO}(n)$-invariant C^∞ function on the basic manifold $W^\#$ defines a C^∞ basic function on M; in other words, the algebra A_b of basic functions on M is canonically isomorphic to the algebra $A_{\mathrm{SO}(n)}(W^\#)$ of functions on $W^\#$ invariant by $\mathrm{SO}(n)$. The bundle like metric on $M^\#$ induces a Riemannian metric on $W^\#$ for which $\mathrm{SO}(n)$ acts by isometries. Let μ be the measure on $W^\#$ associated to this metric.

On $C^\infty(E/\mathcal{F})$ we define an inner product as follows. Let α and β be two elements of $C^\infty(E/\mathcal{F})$. The function $\Theta(\alpha,\beta): z \in M \longrightarrow h_z(\alpha(z), \beta(z)) \in \mathbf{C}$ is basic; so it defines an $\mathrm{SO}(n)$-invariant function $\Theta^\#(\alpha,\beta)$ on $W^\#$. We set

$$\langle \alpha, \beta \rangle = \int_W \Theta^\#(\alpha,\beta)(w)\, d\mu(w).$$

For any basic differential operator D from a Hermitian $\mathcal{F}$-bundle E to a Hermitian $\mathcal{F}$-bundle E', denote by $N(D)$ the kernel of D and $R(D)$ its range.

THEOREM 14. *Let E and E' be two Hermitian $\mathcal{F}$-bundles on M and let D be a transversely elliptic operator from $C^\infty(E/\mathcal{F})$ to $C^\infty(E'/\mathcal{F})$. Denote by D^* the formal adjoint of D which is also a basic transversely elliptic operator from $C^\infty(E'/\mathcal{F})$ to $C^\infty(E/\mathcal{F})$. Then $N(D)$ and $N(D^*)$ are finite dimensional and we have an orthogonal decomposition*:

$$C^\infty(E/\mathcal{F}) = N(D) \oplus R(D^*).$$

In particular, D has an index: $\operatorname{ind}(D/\mathcal{F}) = \dim N(D) - \dim N(D^*)$.

All the details of the proof of this theorem can be found in [85].

7.3. *Transversely elliptic complexes*

Let $(E^r, D_r)_{r=0,1,\dots,n}$ be a family of Hermitian $\mathcal{F}$-bundles and basic differential operators of order one $D_r : C^\infty(E^r/\mathcal{F}) \longrightarrow C^\infty(E^{r+1}/\mathcal{F})$ (by convention $D_n = 0$) such that the sequence

$$\cdots \xrightarrow{D_{r-1}} C^\infty(E^r/\mathcal{F}) \xrightarrow{D_r} C^\infty\big(E^{r+1}/\mathcal{F}\big) \xrightarrow{D_{r+1}} \cdots \tag{$*$}$$

is a differential complex, that is, $D_{r+1} \circ D_r = 0$ for $r = 0, 1, \dots, n-1$. Let $z \in M$ and $\xi \in \nu_z^*\mathcal{F}$; denote by $\sigma(D_r)(z, \xi)$ the principal symbol of D_r at (z, ξ) which is a linear map $\sigma(D_r)(z, \xi) : E_z^r \longrightarrow E_z^{r+1}$. Set $\sigma_r = \sigma(D_r)(z, \xi)$; we say that the complex $(*)$ is *transversely elliptic* if its symbol sequence

$$\cdots \xrightarrow{\sigma_{r-1}} E_z^r \xrightarrow{\sigma_r} E_z^{r+1} \xrightarrow{\sigma_{r+1}} \cdots \tag{$*'$}$$

is exact for every z and every nonzero ξ. Let $D_r^* : C^\infty(E^{r+1}/\mathcal{F}) \longrightarrow C^\infty(E^r/\mathcal{F})$ be the formal adjoint of D_r (with respect to the inner product defined in Section 7.2). Then it is easy to see that the complex $(*)$ is transversely elliptic if and only if the basic operator of order 2: $L_r : C^\infty(E^r/\mathcal{F}) \longrightarrow C^\infty(E^r/\mathcal{F})$ defined by $L_r = D_r^* D_r + D_{r-1} D_{r-1}^*$ is strongly transversely elliptic.

Let (E^r, D_r), $r = 0, 1, \dots, n$, be a transversely elliptic complex with cohomology $H_b^r(E^*)$. Then applying Theorem 14, we have

THEOREM 15.

(i) *For each $r = 0, 1, \dots, n$, the kernel $\mathcal{H}_b^r(E^*)$ of L_r is equal to the space $N(D_r) \cap N(D_{r-1}^*)$.*

(ii) *The space $\mathcal{H}_b^r(E^*)$ is finite dimensional and we have an orthogonal decomposition*

$$C^\infty(E^r/\mathcal{F}) = \mathcal{H}_b^r(E^*) \oplus R(D_{r-1}) \oplus R(D_r^*).$$

(iii) *The orthogonal projection $C^\infty(E^r/\mathcal{F}) \longrightarrow \mathcal{H}_b^r(E^*)$ induces an isomorphism from $H_b^r(E^*)$ to $\mathcal{H}_b^r(E^*)$.*

We will give two concrete examples to illustrate this result: the basic de Rham complex and the basic Dolbeault complex.

Let $r \in \{0, \ldots, n\}$ and denote by E^r the vector bundle of exterior r-forms on the normal bundle $\nu\mathcal{F}$. As it was pointed out, E^r is a Hermitian $\mathcal{F}$-bundle; its basic sections are exactly basic differential forms $\Omega^r(M/\mathcal{F})$ of degree r on M. The de Rham exterior differential d restricted to $\Omega^r(M/\mathcal{F}) = C^\infty(E^r/\mathcal{F})$ is a basic differential operator $d : \Omega^r(M/\mathcal{F}) \longrightarrow \Omega^{r+1}(M/\mathcal{F})$. Thus we obtain a differential complex

$$\cdots \xrightarrow{d} \Omega^r(M/\mathcal{F}) \xrightarrow{d} \Omega^{r+1}(M/\mathcal{F}) \xrightarrow{d} \cdots \qquad (**)$$

called the *basic de Rham complex* of $\mathcal{F}$; its homology $H^r(M/\mathcal{F})$ is called the *basic cohomology* of $\mathcal{F}$ and depends only on the transverse structure of $\mathcal{F}$.

Let $\delta_b : \Omega^{r+1}(M/\mathcal{F}) \longrightarrow \Omega^r(M/\mathcal{F})$ be the formal adjoint of d; this operator can be described explicitly in terms of coefficients of the transverse metric on $\nu\mathcal{F}$ and the Hermitian metrics on the bundles E^r (cf., for instance, [5,352,351,265,289,288,290,291,89, 195–197,85]). Let $\Delta_b = d\delta_b + \delta_b d$; this is a basic differential operator of order 2 on $\Omega^r(M/\mathcal{F})$ called the *basic Laplacian.* A basic form $\alpha \in \Omega^r(M/\mathcal{F})$ which satisfies the equation $\Delta_b\alpha = 0$, or equivalently $d\alpha = 0$ and $\delta_b\alpha = 0$, is called a *basic harmonic* form; denote by $\mathcal{H}^r(M/\mathcal{F})$ the space of such forms. Applying Theorem 14 we obtain the following

THEOREM 16.

(i) *The space $\mathcal{H}^r(M/\mathcal{F})$ is finite dimensional and we have an orthogonal decomposition*

$$\Omega^r(M/\mathcal{F}) = \mathcal{H}^r(M/\mathcal{F}) \oplus R(d) \oplus R(\delta_b).$$

(ii) *The orthogonal projection $\Omega^r(M/\mathcal{F}) \longrightarrow \mathcal{H}^r(M/\mathcal{F})$ induces an isomorphism from $H^r(M/\mathcal{F})$ to $\mathcal{H}^r(M/\mathcal{F})$.*

(iii) *Suppose that the vector space $H^n(M/\mathcal{F})$ is nonzero; then there exists a natural nondegenerate pairing $\Phi : ([\alpha], [\beta]) \in H^r(M/\mathcal{F}) \times H^{n-r}(M/\mathcal{F}) \longrightarrow \Phi([\alpha], [\beta]) \in \mathbf{C}$. So the basic cohomology satisfies Poincaré duality.*

During the last decades, many people contributed to the proof of this theorem. It was first proved by B.L. Reinhart in [290]. But Y. Carrière [51] discovered a mistake which makes assertion (iii) false: B.L. Reinhart does not suppose $H^n(M/\mathcal{F})$ different from $\{0\}$ to obtain Poincaré duality; he was probably thinking that this hypothesis is automatically satisfied. Later on F.W. Kamber and P. Tondeur [196] have shown that the Reinhart's proof is still valid if we suppose the leaves minimal (cf. Section 9.3). Finally the theorem was proved in full generality (without any assumption on the minimality of the leaves) in [89].

Now suppose that $\mathcal{F}$ is Hermitian. Let ν be the complexified normal bundle $\nu\mathcal{F} \otimes_{\mathbf{R}} \mathbf{C}$ of $\nu\mathcal{F}$. Let J be the automorphism of ν associated to the complex structure; J satisfies the relation $J^2 = -\mathrm{id}$ and then has two eigenvalues i and $-i$ with associated eigensubbundles respectively denoted ν^{10} and ν^{01}. We have a splitting $\nu = \nu^{10} \oplus \nu^{01}$ which gives rise to a decomposition

$$\Lambda^r \nu^* = \bigoplus_{p+q=r} \Lambda^{p,q},$$

where $\Lambda^{p,q} = \Lambda^p \nu^{10*} \otimes \Lambda^q \nu^{01*}$. Basic sections of $\Lambda^{p,q}$ are called *basic forms of type* (p,q). They form a vector space denoted $\Omega^{p,q}(M/\mathcal{F})$. We have

$$\Omega^r(M/\mathcal{F}) = \bigoplus_{p+q=r} \Omega^{p,q}(M/\mathcal{F}).$$

The exterior differential decomposes into a sum of two operators

$$\partial : \Omega^{p,q}(M/\mathcal{F}) \longrightarrow \Omega^{p+1,q}(M/\mathcal{F}) \quad \text{and}$$
$$\bar{\partial} : \Omega^{p,q}(M/\mathcal{F}) \longrightarrow \Omega^{p,q+1}(M/\mathcal{F})$$

as in the classical case of a complex manifold. We have $\bar{\partial}^2 = 0$; so we obtain, for p fixed, a differential complex

$$\cdots \xrightarrow{\bar{\partial}} \Omega^{p,q}(M/\mathcal{F}) \xrightarrow{\bar{\partial}} \Omega^{p,q+1}(M/\mathcal{F}) \xrightarrow{\bar{\partial}} \cdots \qquad (***)$$

called the *basic Dolbeault complex* of $\mathcal{F}$; its homology $H^{p,q}(M/\mathcal{F})$ is the *basic Dolbeault cohomology* of the foliation $\mathcal{F}$: even though the leaf space is topologically bad, it can be considered as a "complex manifold" whose Dolbeault cohomology is $H^{p,*}(M/\mathcal{F})$!

Let δ_b'' denote the formal adjoint of $\bar{\partial}$; this is an operator of type $(0,-1)$. The operator $\Delta_b'' = \delta_b''\bar{\partial} + \bar{\partial}\delta_b''$ is selfadjoint; a simple computation in local coordinates, like for the basic Laplacian, shows that Δ_b'' is strongly transversely elliptic. Therefore the complex $(***)$ is transversely elliptic. Let

$$\mathcal{H}^{p,q}(M/\mathcal{F}) = \operatorname{Ker} \Delta_b'' = \{\alpha \in \Omega^{p,q}(M/\mathcal{F}) \colon \bar{\partial}\alpha = 0 \text{ and } \delta_b''\alpha = 0\}.$$

Applying Theorem 14, we obtain

THEOREM 17.

(i) *The space $\mathcal{H}^{p,q}(M/\mathcal{F})$ is finite dimensional and we have an orthogonal decomposition*

$$\Omega^{p,q}(M/\mathcal{F}) = \mathcal{H}^{p,q}(M/\mathcal{F}) \oplus R(\Delta_b'') = \mathcal{H}^{p,q}(M/\mathcal{F}) \oplus R(\bar{\partial}) \oplus R(\delta_b'').$$

(ii) *The orthogonal projection $\Omega^{p,q}(M/\mathcal{F}) \longrightarrow \mathcal{H}^{p,q}(M/\mathcal{F})$ induces an isomorphism from $H^{p,q}(M/\mathcal{F})$ to $\mathcal{H}^{p,q}(M/\mathcal{F})$.*

(iii) *Suppose that the vector space $H^n(M/\mathcal{F})$ is nonzero; then there exists a natural nondegenerate pairing $\Psi : ([\alpha],[\beta]) \in H^{p,q}(M/\mathcal{F}) \times H^{n-p,n-q}(M/\mathcal{F}) \longrightarrow \Psi([\alpha],[\beta]) \in \mathbf{C}$. So the basic Dolbeault cohomology satisfies Serre duality.*

Suppose now that $\mathcal{F}$ is transversely Kählerian with Kähler form ω (it is a basic differential form of degree 2; it is closed and nondegenerate). In this case, we can prove that $\Delta_b = 2\Delta_b''$. Because of the decomposition

$$\Omega^r(M/\mathcal{F}) = \bigoplus_{p+q=r} \Omega^{p,q}(M/\mathcal{F}),$$

every basic differential r-form can be uniquely written as a sum $\alpha = \sum_{p+q=r} \alpha_{pq}$ where $\alpha_{pq} \in \Omega^{p,q}(M/\mathcal{F})$. Then we have the following assertions.

(iv) *α is Δ_b-harmonic if, and only if, each component α_{pq} is Δ_b''-harmonic. So we have a direct decomposition*

$$H^r(M/\mathcal{F}) = \bigoplus_{p+q=r} H^{p,q}(M/\mathcal{F}).$$

(v) *The complex conjugacy induces an isomorphism* (*of real vector spaces*)

$$H^{p,q}(M/\mathcal{F}) \simeq H^{q,p}(M/\mathcal{F}).$$

(vi) *For every odd $r \in \{0, \ldots, 2n\}$, the dimension of the space $H^r(M/\mathcal{F})$ is even. In particular, if $n = 1$ we have $b_1(M/\mathcal{F}) = 2\dim H^{01}(M/\mathcal{F})$.*

The integer $\dim H^{01}(M/\mathcal{F})$ will be denoted $g(\mathcal{F})$ and called the *genus* of the foliation $\mathcal{F}$. It is similar to the genus of a compact Riemann surface; it counts the number of linearly independent basic holomorphic 1-forms.

(vii) *For every $p \in \{0, \ldots, n\}$ the differential form $\omega^p = \omega \wedge \cdots \wedge \omega$ (wedge product p times) is harmonic. So, the space $H^{p,p}(M/\mathcal{F})$ is nonzero.*

Notice that this theorem is also a particular case of Theorem 15. It can be used to establish more properties: basic Hodge structures for transversely Kählerian foliations, basic Calabi–Yau theorem [85] and deformation of transversely holomorphic foliations with a fixed differentiable type [90].

8. Deformation theory of foliations

We will describe only the real case following Hamilton's paper [159]. Deformation theory of holomorphic or generally transversely holomorphic foliations is more rich. The reader can find a good account of the subject in [132].

Let M be a manifold of dimension $m + n$. For each $x \in M$, let $G(x, m)$ be the Grassmanian manifold of m-planes in T_xM. Then:

$$\mathcal{G}(m) = \bigcup_{x \in M} G(x, m)$$

can be given a structure of a differentiable manifold such that the canonical projection $(x, \tau) \in \mathcal{G}(m) \longrightarrow x \in M$ is a locally trivial fibration, the fibre being the Grassmanian

$G(m)$ of m-planes in the space $\mathbf{R}^{m+n}$. Then a subbundle of rank m of TM is just a section of the bundle $\mathcal{G}(m) \longrightarrow M$. Denote by $C^\infty(\mathcal{G}(m))$ the space of sections of this bundle.

Let $\tau \in C^\infty(\mathcal{G}(m))$. By Frobenius theorem, τ is tangent to a foliation if, and only if, for any pair (U, V) of (global) sections of τ, the Lie bracket $[U, V]$ is also a section of τ. Let $(X_1, \ldots, X_m)$ be a local basis of τ. Then

$$U = \sum_{i=1}^{m} a^i X_i \quad \text{and} \quad V = \sum_{j=1}^{m} b^j X_j.$$

So the bracket $[U, V]$ can be expressed as

$$[U, V] = \sum_{i,j=1}^{m} \{a^i b^j [X_i, X_j] + (a^i X_i(b^j) X_j - b^j X_j(a^i) X_i)\}.$$

Therefore the value of $[U, V]$ in $\nu\tau = TM/\tau$ at a point $x \in M$ depends only on the value of U and V at x. Hence $Q_\tau(U, V) = \pi([U, V])$ is a skew-symmetric bilinear map $Q_\tau : \tau \times \tau \longrightarrow \nu\tau$ where $\pi : TM \longrightarrow \nu\tau$ is the canonical projection. In other words, Q_τ is a global section of the vector bundle $\Lambda^2(\tau, \nu\tau)$ of skew-symmetric bilinear forms on the bundle τ. The integrability condition of τ is equivalent to Q_τ identically equal to 0. So we get a map $Q : C^\infty(\mathcal{G}(m)) \longrightarrow \Sigma$ where Σ is a fibre bundle over $\mathcal{G}(m)$ whose fibre over a point $\sigma \in \mathcal{G}(m)$ is the infinite-dimensional space $\Omega^2(\sigma, \nu\sigma)$ of global sections of the bundle $\Lambda^2(\tau, \nu\tau)$. The space $\mathcal{F}ol(M, m)$ of dimension m foliations on M is exactly the set $\{Q = 0\}$. It will be equipped with the C^∞-topology induced by the topology of the Fréchet manifold $C^\infty(\mathcal{G}(m))$ (cf. [160]). Let $\mathcal{D}$ be the diffeomorphism group of M; then $\mathcal{D}$ acts on $C^\infty(\mathcal{G}(m))$ and the action preserves $\mathcal{F}ol(M, m)$. Two foliations $\mathcal{F}, \mathcal{F}' \in \mathcal{F}ol(M, m)$ are conjugated, if they are in the same orbit of the action of $\mathcal{D}$, that is, there exists $\varphi \in \mathcal{D}$ such that $\mathcal{F}' = \varphi^*(\mathcal{F})$.

Now fix τ in $C^\infty(\mathcal{G}(m))$ and suppose that it is tangent to a foliation $\mathcal{F}$. Then the map $P_\tau : \varphi \in \mathcal{D} \longrightarrow \varphi^*(\mathcal{F}) \in C^\infty(\mathcal{G}(m))$ takes its values in $\mathcal{F}ol(M, m)$. So we get a sequence of Fréchet manifolds and differentiable maps

$$\mathcal{D} \xrightarrow{P_\tau} C^\infty(\mathcal{G}(m)) \xrightarrow{Q} \Sigma.$$

Following R. Hamilton, this sequence is called the *nonlinear deformation complex* of the foliation $\mathcal{F}$ [159].

DEFINITION 5. We say that $\mathcal{F}$ is C^∞-stable if there exist an open neighborhood $\mathcal{O}$ of the identity in $\mathcal{D}$ and an open neighborhood $\mathcal{U}$ of $\mathcal{F}$ in $\mathcal{F}ol(M, m)$ such that the sequence $\mathcal{O} \xrightarrow{P_\tau} \mathcal{U} \xrightarrow{Q} \Sigma$ is exact, that is, every dimension m foliation $\mathcal{F}'$ on M, close enough to $\mathcal{F}$ in the C^∞-topology, is conjugated to $\mathcal{F}$ by an element of $\mathcal{O}$.

An important tool to prove the C^∞-stability of a foliation is Hamilton's criterion (cf. [159, p. 47]) that we shall describe. This criterion is based on the implicit function theorem of Nash–Moser which is nicely explained in Hamilton's paper [160].

Given a foliation $\mathcal{F}$ of dimension m on a compact manifold M, let $A^k_{\mathcal{F}}$ denote the space of differentiable sections of $\Lambda^k T\mathcal{F}^* \otimes \nu\mathcal{F}$. Since $\nu\mathcal{F}$ is a foliated bundle there is a well defined "*exterior derivative along the leaves*" $d_{\mathcal{F}} : A^k_{\mathcal{F}} \longrightarrow A^{k+1}_{\mathcal{F}}$ given by:

$$\begin{aligned} &d_{\mathcal{F}}\eta(X_1, \ldots, X_{k+1}) \\ &\quad = \sum_i (-1)^i X_i \eta(X_1, \ldots, \hat{X}_i, \ldots, X_{k+1}) \\ &\qquad + \sum_{i<j} (-1)^{i+j} \eta\big([X_i, X_j], X_1, \ldots, \hat{X}_i, \ldots, \hat{X}_j, \ldots, X_{k+1}\big). \end{aligned}$$

An easy computation shows that $d^2_{\mathcal{F}} = 0$ and thus we obtain a differential complex

$$0 \longrightarrow A^0_{\mathcal{F}} \xrightarrow{d_{\mathcal{F}}} A^1_{\mathcal{F}} \xrightarrow{d_{\mathcal{F}}} A^2_{\mathcal{F}} \xrightarrow{d_{\mathcal{F}}} \cdots \xrightarrow{d_{\mathcal{F}}} A^m_{\mathcal{F}} \longrightarrow 0$$

which is only elliptic along the leaves. Let $\| \ \|_0 \leqslant \| \ \|_1 \leqslant \cdots \leqslant \| \ \|_s \leqslant \cdots$ be an increasing collection of norms (of Sobolev or Hölder type) on the Fréchet space

$$A^*_{\mathcal{F}} = \bigoplus_{k \geqslant 0} A^k_{\mathcal{F}}.$$

With this notation one has

THEOREM 18 (Hamilton). *Assume that there exist continuous linear operators* $H : A^1_{\mathcal{F}} \longrightarrow A^0_{\mathcal{F}}$ *and* $K : A^2_{\mathcal{F}} \longrightarrow A^1_{\mathcal{F}}$ *fulfilling the following conditions*:

(i) $d_{\mathcal{F}} \circ H + K \circ d_{\mathcal{F}} = \mathrm{id}$,

(ii) *there is a fixed number* $r \in \mathbf{N}$ *for which we have tame estimates for all* s,

$$\|H(\beta)\|_s \leqslant C_s \|\beta\|_{s+r} \quad \text{and} \quad \|K(\gamma)\|_s \leqslant C_s \|\gamma\|_{s+r},$$

where C_s *are positive constants depending only on* s.

Then the foliation $\mathcal{F}$ *is* C^∞*-stable.*

Unfortunately Hamilton's paper is still unpublished. In [93], the authors constructed a class of foliations and, using Hamilton's criterion, they proved that these foliations are C^∞-stable. Example 1.3(iv), with some assumptions on the matrix A, is in this class.

9. Some other themes

As we have pointed out in the foreword, Foliation Theory is a wide field in Mathematics and so huge to discuss completely here. For this reason we have chosen only some of the themes related to Differential Geometry which is the main topic to which this book is devoted. The nonwarned reader may be inclined to believe that the theory is reduced to this part. Fortunately this is not the case. We devote this section to other themes which are no less important than the above ones.

9.1. *Compact leaves*

Let M be a connected compact orientable manifold of dimension $m+n$ and $\mathcal{F}$ a codimension n foliation on M. A *compact leaf* of $\mathcal{F}$ is a leaf L which is compact as a subset of M. If $m=1$ such leaf is a periodic orbit and it describes a stationary state of the dynamical system defined by $\mathcal{F}$. The problem of the existence of compact leaves is highly nontrivial. It was first introduced by H. Poincaré in his studies on *limit cycles* for ordinary differential equations. One of the famous problems was the Seifert conjecture: *Every continuous vector field on the* 3*-dimensional sphere* $\mathbf{S}^3$ *has a periodic orbit.*

In 1974, using Denjoy's example of a vector field with exceptional minimal set on the 2-torus, P. Schweitzer [324] constructed a counterexample in class C^1. In 1988, J. Harrison [162] gave a C^2 counterexample. Finally in 1993, K. Kuperberg [200] solved completely the problem by constructing in any compact 3-manifold a real analytic vector field without periodic orbit. However, M. Brunella [31] proved that the conjecture is true if the flow is transversely holomorphic; in fact, he established a complete classification of these flows on compact 3-manifolds.

The most important result concerning the problem of existence of compact leaves was Novikov's theorem stated above (Theorem 12). Nothing is known in higher dimensions and the following question is still open: *is it true that every codimension one foliation on the odd sphere* $\mathbf{S}^{2p+1}$ *(where* $p\geqslant 2$*) admits a compact leaf?*

We say that $\mathcal{F}$ is a *compact foliation* if all leaves are compact. For example, every foliation defined by a locally free action of a connected compact Lie group is a compact foliation. Compact foliations was a theme which interested many people (R. Edwards, K. Millet, D. Sullivan, D. Epstein, E. Vogt, H. Rummler, etc.).

9.2. *When is a manifold a leaf?*

Let L be a noncompact connected manifold. *Does there exist a compact manifold M endowed with a foliation $\mathcal{F}$ with a leaf diffeomorphic to L?* This question was asked by J. Sondow in [335] where he gave some sufficient conditions on L to be a leaf. J. Cantwell and L. Conlon proved in [46] that every surface is a leaf. Along the same lines, G. Hector and W. Bouma proved in [170] that every noncompact surface can be a leaf of a simple foliation of $\mathbf{R}^3$, i.e. a foliation defined by a submersion $\mathbf{R}^3 \longrightarrow \mathbf{R}$.

In [114] E. Ghys observed that the topology of a leaf of a foliation on a compact manifold has to be, in some sense, "recurrent"; then he constructed, for any positive integer d, a noncompact manifold L of dimension d which can not be homeomorphic to any leaf of any foliation on a compact manifold. In [125] he also studies the topology of the generic leaves of a *lamination* by surfaces on a compact metric space and proved that there exist only six noncompact surfaces which can be realized as leaves:

(a) the plane $\mathbf{R}^2$,
(b) the cylinder $\mathbf{S}^1\times\mathbf{R}$,
(c) the "Loch-Ness monster", i.e. the plane with infinitely many handles attached,
(d) the "Jacob ladder",

(e) the "Cantor tree", i.e. the sphere $\mathbf{S}^2$ with a Cantor set removed,
(f) the "flowered Cantor tree", i.e. the Cantor tree with infinitely many handles attached in all directions.

9.3. *Minimal leaves*

Let M be a Riemannian manifold. Denote by ∇ the covariant derivative associated to the Levi-Civita connection. Let L be a submanifold of M (not necessarily properly embedded). Let $x \in L$ and ν a vector field defined on a neighborhood of x and orthogonal to L. For $X \in T_x L$, we set:

$$W_x^\nu(X) = -p_x(\nabla_X \nu),$$

where $p_x : T_x M \longrightarrow T_x L$ is the orthogonal projection. Then W_x^ν is an endomorphism of the vector space $T_x L$, symmetric with respect to the induced metric on $T_x L$; it is called the *Weingarten map* associated to ν. The trace of W_x^ν describes the variation at x of the volume element when L moves in the direction of ν. We say that L is *minimal*, if the trace of W_x^ν is zero for all vector fields ν orthogonal to L. A foliation $\mathcal{F}$ on M is said to be with *minimal leaves*, if all leaves of $\mathcal{F}$ are minimal submanifolds.

Given an m-dimensional foliation on a compact manifold M, does there exist a Riemannian metric on M for which the leaves are minimal?

This question was discussed by H. Rummler [309] and D. Sullivan [339]. They proved the following criterion: *such a metric exists if, and only if, there exists an m-form χ positive on the leaves and relatively closed, namely $d\chi(X_1, \ldots, X_m, Y) = 0$ whenever the vector fields $X_1, \ldots, X_m$ are tangent to $\mathcal{F}$.*

In [149] A. Haefliger proved that the property for $\mathcal{F}$ to be with minimal leaves depends only on the transverse structure. He also gave a criterion in terms of transverse invariant currents and used it to give many examples of minimal foliations and nonminimal ones.

Suppose now that $\mathcal{F}$ is a Riemannian codimension n foliation and denote by υ the volume basic form associated to the metric. If $\mathcal{F}$ is with minimal leaves then υ defines a nonzero class in the basic cohomology $H^n(M/\mathcal{F})$. Indeed, let χ be the m-form given by the Rummler–Sullivan criterion. Suppose that $\upsilon = d\beta$ where $\beta \in \Omega^{n-1}(M/\mathcal{F})$. Then:

$$\chi \wedge \upsilon = \chi \wedge d\beta = (-1)^m \{ d(\chi \wedge \beta) - d\chi \wedge \beta \}.$$

But $d\chi \wedge \beta = 0$ because χ is relatively closed. So $\chi \wedge \upsilon$ is an exact form. But this is impossible because it is a volume form on the compact orientable manifold M. The converse of this assertion was conjectured by Y. Carrière [51] and proved by X. Masa in [227].

In [112] E. Ghys proved that any Riemannian foliation on a simply connected compact manifold admits a bundle-like metric for which the leaves are minimal.

Now let $\mathcal{F}$ be a foliation on a compact Riemannian manifold M. We say that $\mathcal{F}$ is *totally geodesic* if every geodesic tangent to a leaf L at a point is tangent to L everywhere. This is a special class of foliations with minimal leaves which was studied, for instance, by Y. Carrière, G. Cairns, E. Ghys (see [35,36,33,37,54]). In particular, E. Ghys, in [111], has completely classified all the totally geodesic foliations of codimension one on compact manifolds.

References

[1] C. Albert and P. Dazord, *Théorie des groupoï des symplectiques, II*, Pub. Dep. Math. Nouv. Sér. Univ. Claude Bernard, Lyon (1990), 27–99.

[2] F. Alcade Cuesta and G. Hector, *Feuilletages en surfaces, cycles évaouissants et variétés de Poisson*, Monatsh. Math. **124** (3) (1997), 193–213.

[3] J. Álvarez-López, *Duality in the spectral sequence of a Riemannian foliation*, Amer. J. Math. **111** (6) (1989), 905–926.

[4] J. Álvarez-López, *A finiteness theorem for the spectral sequence of a Riemannian foliation*, Illinois J. Math. **33** (1) (1989), 79–92.

[5] J. Álvarez-López, *The basic component of the mean curvature of Riemannian foliations*, Ann. Global Anal. Geom. **10** (1992), 179–194.

[6] J. Álvarez-López, *Morse inequalities for pseudogroups of local isometries*, J. Differential Geom. **37** (3) (1993), 603–638.

[7] J. Álvarez-López and P. Tondeur, *Hodge decomposition along the leaves of a Riemannian foliation*, J. Funct. Anal. **99** (2) (1991), 433–458.

[8] C. Anantharman-Delaroche, *Purely infinite C^*-algebras arising from dynamical systems*, Bull. Soc. Math. France **125** (1997), 199–226.

[9] J.L. Arraut, *A 2-dimensional foliation on* $\mathbf{R}^7$, Topology **12** (1973), 243–245.

[10] J.L. Arraut and M. Craizer, *Foliations of M^3 defined by $\mathbf{R}^2$-actions*, Ann. Inst. Fourier **45** (4) (1995), 1018–1118.

[11] J.L. Arraut and N.M. Dos Santos, *The characteristic mapping of a foliated $\mathbf{S}^1$-bundle*, Topology **31** (3) (1992), 545–555.

[12] M. Artin, *Algebraic spaces*, Yale Monograph Study 3 (1971).

[13] R. Barre, *De quelques aspects de la théorie des Q-variétés différentielles et analytiques*, Ann. Inst. Fourier **23** (3) (1973), 227–312.

[14] V. Belfi, E. Park and K. Richardson, *A Hopf index theorem for foliations*, Differential Geom. Appl. **18** (2003), 319–341.

[15] M. Belliart and O. Birembaux, *Actions localement libres de groupes de Lie résolubles*, Ann. Inst. Fourier **44** (5) (1994), 1519–1537.

[16] A. Benoist and F. Labourie, *Sur les difféomorphismes d'Anosov affines à feuilletages stable et instable*, Invent. Math. **111** (2) (1993), 285–308.

[17] H. Blas, M. Llabres and A. Reventos, *Transverse structures of Lie foliations*, J. Math. Soc. Japan **48** (4) (1996), 769–795.

[18] R.A. Blumenthal, *Transversely homogeneous foliations*, Ann. Inst. Fourier **29** (4) (1979), 143–158.

[19] C. Bonatti and S. Firmo, *Feuilles compactes d'un feuilletage générique en codimension un*, Ann. Sci. École Norm. Sup. **27** (4) (1994), 407–461.

[20] C. Bonatti and A. Haefliger, *Déformations de feuilletages*, Topology **29** (2) (1990), 205–229.

[21] A. Borel and F. Hirzebruch, *Characteristic classes and homogeneous spaces, II*, Amer. J. Math. **81** (1959), 315–382.

[22] R. Bott, *On the topological obstruction to integrability*, Sympos. Pure Math., vol. 16, Amer. Math. Soc., Providence, RI (1970), 127–131.

[23] R. Bott, *Lectures on characteristic classes and foliations*, Lecture Notes in Math. **279** (1972).

[24] R. Bott, *On some formulas for the characteristic classes of group actions*, Lecture Notes in Math. **652** (1978), 25–62.

[25] R. Bott and A. Haefliger, *On characteristic classes of Γ-foliations*, Bull. Amer. Math. Soc. **78** (1972), 1039–1044.

[26] R. Bott and J. Heitsch, *A remark on the integral cohomology of $B\Gamma_q$*, Topology **11** (1972), 141–146.

[27] E. Bouacida, O. Echi and E. Salhi, *Feuilletages et topologie spectrale*, J. Math. Soc. Japan **522** (2000), 447–464.

[28] L.G. Bouma and W.T. Van Est, *Manifolds schemes and foliations on the* 2*-torus and the Klein bottle, I, II, III*, Indag. Math. **40** (1978), 313–347.

[29] F.G.B. Brito, R. Langevin and H. Rosenberg, *Intégrales de courbure sur les variétés feuilletées*, J. Differential Geom. **16** (1981), 19–50.

[30] M. Brunella, *Remarks on structurally stable proper foliations*, Math. Proc. Cambridge Philos. Soc. **115** (1994), 111–120.

[31] M. Brunella, *On transversely holomorphic flows, I*, Invent. Math. **126** (2) (1996), 265–279.

[32] M. Brunella and E. Ghys, *Umbilical foliations and transversely holomorphic flows*, J. Differential Geom. **41** (1) (1995), 1–19.

[33] G. Cairns, *Aspects cohomologiques des feuilletages totalement géodésiques*, C. R. Acad. Sci. Paris Sér. I **299** (20) (1984), 1017–1019.

[34] G. Cairns, *A general description of a totally geodesic foliation*, Tohoku Math. J. (2) **38** (1986), 37–55.

[35] G. Cairns, *Some properties of a cohomology group associated to a totally geodesic foliation*, Math. Z. **192** (1986), 391–403.

[36] G. Cairns, *Feuilletages totalement géodésiques sur les variétés simplement connexes*, Sém. Sud Rhodanien Géom. VII Trav. Cours **26** (1988), 1–14.

[37] G. Cairns and E. Ghys, *Totally geodesic foliations on* 4*-manifolds*, J. Differential Geom. **23** (1986), 241–254.

[38] C. Camacho, *Structural stability theorems for integrable differential forms on* 3*-manifolds*, Topology **17** (1978), 143–155.

[39] C. Camacho and A. Lins Neto, *Geometric Theory of Foliations*, Birkhäuser (1985).

[40] J. Cantwell and L. Conlon, *Growth of leaves*, Comment. Math. Helv. **53** (1978), 93–111.

[41] J. Cantwell and L. Conlon, *Reeb stability for noncompact leaves in foliated* 3*-manifolds*, Proc. Amer. Math. Soc. **83** (1981), 408–410.

[42] J. Cantwell and L. Conlon, *Poincaré–Bendixon theory for leaves of codimension one*, Trans. Amer. Math. Soc. **265** (1981), 181–209.

[43] J. Cantwell and L. Conlon, *Tischler fibrations of open foliated sets*, Ann. Inst. Fourier **31** (2) (1981), 113–135.

[44] J. Cantwell and L. Conlon, *The endsets of leaves*, Topology **21** (1982), 333–352.

[45] J. Cantwell and L. Conlon, *Analytic foliations and the theory of levels*, Math. Ann. **265** (1983), 253–261.

[46] J. Cantwell and L. Conlon, *Every surface is a leaf*, Topology **26** (1987), 265–285.

[47] J. Cantwell and L. Conlon, *The theory of levels*, Contemp. Math. **70** (1988), 1–10.

[48] J. Cantwell and L. Conlon, *Leafwise hyperbolicity of proper foliations*, Comment. Math. Helv. **64** (1989), 329–337.

[49] J. Cantwell and L. Conlon, *Generic leaves*, Comment. Math. Helv. **73** (1998), 306–336.

[50] P. Caron and Y. Carrière, *Flots transversalement de Lie* $\mathbf{R}^n$*, flots minimaux*, C. R. Acad. Sci. Paris **280** (1980), 477–478.

[51] Y. Carrière, *Flots riemanniens*, Structures transverses des feuilletages, Toulouse 1982, Astérisque **116** (1984), 31–52.

[52] Y. Carrière, *Feuilletages riemanniens à croissance polynomiale*, Comment. Math. Helv. **63** (1984), 1–20.

[53] Y. Carrière, *Les propriétés topologiques des flots riemanniens retrouvées à l'aide du théorème des variétés presque plates*, Math. Z. **186** (1984), 393–400.

[54] Y. Carrière and E. Ghys, *Feuilletages totalement géodésiques*, An. Acad. Brasil. Ciênc. **53** (1981), 427–432.

[55] J. Cerf, 1*-formes fermées non singulières sur les variétés compactes de dimension* 3, Sém. Bourbaki, Exposé 574, Lecture Notes in Math. **901** (1981), 205–219.

[56] D. Cerveau, *Feuilletages holomorphes de codimension un. Réduction des singularités en petites dimensions et applications*, IRM de Rennes, Prépublication 98-01 (1998).

[57] G. Chatelet, *Sur les feuilletages induits par l'action de groupes de Lie nilpotents*, Ann. Inst. Fourier **27** (2) (1977), 161–189.

[58] G. Chatelet and H. Rosenberg, *Un théorème de conjugaison des feuilletages*, Ann. Inst. Fourier **21** (3) (1971), 95–106.

[59] J. Chevalley, *Theory of Lie Groups*, Princeton Mathematical Series, vol. 8 (1946).

[60] L. Conlon, *Transversally parallelizable foliations of codimension two*, Trans. Amer. Math. Soc. **194** (1975), 79–102.

[61] A. Connes, *Sur la théorie non commutative de l'intégration*, Lecture Notes in Math. **725** (1979).

[62] A. Connes, *Noncommutative Geometry*, Academic Press (1994).

[63] A. Connes and H. Moscovici, *Hopf algebras, cyclic cohomology and the transverse index theorem*, Commun. Math. Phys. **198** (1998), 199–246.

[64] A. Connes and H. Moscovici, *Differentiable cyclic cohomology and Hopf algebraic structures in transverse geometry*, Essays on Geometry and Related Topics, vols. I, II, Monogr. Enseign. Math., vol. 38, Geneva (2001), 217–255.

[65] M. Craioveanu and M. Puta, *Cohomology on a Riemannian foliated manifold with coefficients in the sheaf of germs of foliated currents*, Math. Nachr. **99** (1980), 43–53.

[66] C. Cumenge, *Cohomologie de l'espace des feuilles d'un feuilletage*, Thèse 3ème cycle, Toulouse (1981).

[67] C. Curras-Bosch and P. Molino, *Voisinage d'une feuille compacte dans un voisinage lagrangien*, Hokkaido Math. J. **23** (2) (1994), 355–360.

[68] A. Dadarlat, *C^*-Algebras and applications*, Contemp. Math. **145** (1993), 393–421.

[69] C. Danthony, *Feuilletages orientés de surfaces; le problème de la section globale*, Ann. Inst. Fourier **38** (2) (1988), 201–227.

[70] A. Denjoy, *Sur les courbes définies par les équations différentielles à la surface du tore*, J. Math. Pures Appl. **11** (1932), 333–375.

[71] P.R. Dipolito, *Codimension one foliations of closed manifolds*, Comment. Math. Helv. **107** (1978), 403–453.

[72] T. Duchamp and M. Kalka, *Deformation theory for holomorphic foliations*, J. Differential Geom. **14** (1979), 317–337.

[73] G. Duminy, *L'invariant de Godbillon–Vey d'un feuilletage se localise sur les feuilles ressort*, Manuscrit Université de Lille I (1982).

[74] G. Duminy, *Sur les cycles feuilletées de codimension un*, Manuscrit Université de Lille I (1982).

[75] G. Duminy and V. Sergiescu, *Sur la nullité de l'invariant de Godbillon–Vey*, C. R. Acad. Sci. Paris **292** (1981), 821–824.

[76] J.L. Dupont, *Simplicial de Rham cohomology and characteristic classes of flat bundles*, Topology **15** (1976), 233–245.

[77] A.H. Durfee, *Foliations of odd-dimensional spheres*, Ann. of Math. **96** (2) (1972), 407–411.

[78] A.H. Durfee and H.B. Lawson, *Fibered knots and foliations of highly connected manifolds*, Invent. Math. **17** (1972), 203–215.

[79] R. Edwards, K. Millet and D. Sullivan, *Foliations with all leaves compact*, Topology **16** (1977), 13–32.

[80] C. Ehresmann, *Les connexions infinitésimales dans un espace fibré différentiable*, Colloque de Topologie, Bruxelles (1950), 29–53.

[81] C. Ehresmann, *Sur la théorie des variétés feuilletées*, Rend. Mat. Appl. (5) **10** (1951), 64–82.

[82] C. Ehresmann, *Structures feuilletées*, Proc. 5th Canadian Congress, Montréal (1961), 109–172.

[83] C. Ehresmann and G. Reeb, *Sur les champs d'éléments de contact de dimension p complètement intégrables sur une variété continûment différentiable*, C. R. Acad. Sci. Paris **218** (1944), 955–957.

[84] C. Ehresmann and S. Weishu, *Sur les espaces feuilletées et théorème de stabilité*, C. R. Acad. Sci. Paris **243** (1956), 344–346.

[85] A. El Kacimi Alaoui, *Opérateurs transversalement elliptiques sur les feuilletages riemanniens et applications*, Compositio Math. **73** (1990), 57–106.

[86] A. El Kacimi Alaoui and E. Gallego, *Applications harmoniques feuilletées*, Illinois J. Math. **40** (1) (1996), 115–122.

[87] A. El Kacimi Alaoui and B. Gmira, *Stabilité du caractère kählérien transverse*, Israel J. Math. **101** (1997), 323–347.

[88] A. El Kacimi Alaoui, G. Guasp and M. Nicolau, *On deformations of transversely homogeneous foliations*, Topology **40** (2001), 1363–1393.

[89] A. El Kacimi Alaoui and G. Hector, *Décomposition de Hodge basique pour un feuilletage riemannien*, Ann. Inst. Fourier **36** (3) (1986), 207–227.

[90] A. El Kacimi Alaoui and M. Nicolau, *Déformations des feuilletages transversalement holomorphes à type différentiable fixé*, Publ. Mat. Barcelona **33** (1989), 485–500.

[91] A. El Kacimi Alaoui and M. Nicolau, *Structures géométriques invariantes et feuilletages de Lie*, Indag. Math. (N.S.) **1** (3) (1990), 323–334.

[92] A. El Kacimi Alaoui and M. Nicolau, *On the topological invariance of the basic cohomology*, Math. Ann. **295** (1993), 627–634.

[93] A. El Kacimi Alaoui and M. Nicolau, *A class of C^{∞}-stable foliations*, Ergodic Theory Dynam. Systems **13** (1993), 697–704.

[94] A. El Kacimi Alaoui, V. Sergiescu and G. Hector, *La cohomologie basique d'un feuilletage riemannien est de dimension finie*, Math. Z. **188** (1985), 593–599.

[95] D.B.A. Epstein, *Periodic flows on* 3*-manifolds*, Ann. of Math. **95** (1972), 68–82.

[96] D.B.A. Epstein, *Foliations with all leaves compact*, Ann. Inst. Fourier **26** (1) (1976), 265–282.

[97] D.B.A. Epstein, K. Millet and D. Tischler, *Leaves without holonomy*, J. London Math. Soc. (2) **16** (1977), 548–552.

[98] D.B.A. Epstein and E. Vogt, *A counterexemple to the periodic orbit conjecture in codimension* 3, Ann. of Math. (2) **108** (1978), 539–552.

[99] T. Fack and G. Skandalis, *Sur les représentations et idéaux de la C^*-algèbre d'un feuilletage*, J. Operator Theory **8** (1982), 95–129.

[100] F.T. Farrell and L.E. Jones, *Foliated control theory, I, II*, K-Theory **2** (1988), 357–399 and 401–430.

[101] E. Fédida, *Sur l'existence des feuilletages de Lie*, C. R. Acad. Sci. Paris **278** (1974), 835–837.

[102] D. Ferus, *Totally geodesic foliations*, Math. Ann. **188** (1970), 313–316.

[103] H.R. Fisher and F.L. Williams, *Complex foliated structures. I. Cohomology of the Dolbeault–Kostant complexes*, Trans. Amer. Math. Soc. **252** (1979), 163–195.

[104] D.B. Fuks, *Characteristic classes of foliations*, Russian Math. Surveys **28** (2) (1973), 1–16.

[105] E. Gallego and A. Reventos, *Lie flows of codimension* 3, Trans. Amer. Math. Soc. **326** (2) (1991), 529–541.

[106] R.B. Gardner, *The Godbillon–Vey invariant and the Bott–Pasternak vanishing theorem*, Lecture Notes in Math. **652** (1978), 75–94.

[107] L. Garnett, *Foliations, the ergodic theorem and Brownian motion*, J. Funct. Anal. **51** (1983), 285–311.

[108] I.M. Gelfand, B.L. Feigin and D.B. Fuks, *Cohomology of the Lie algebra of formal vector fields with coefficients in its adjoint space and variations of characteristic classes of foliations*, Funct. Anal. Appl. **8** (1974), 99–112.

[109] I.M. Gelfand and D.B. Fuks, *PL-foliations*, Funct. Anal. Appl. **7** (1973), 278–284 and **8** (1974), 197–200.

[110] R. Gerard and A. Sec, *Feuilletages de Painlevé*, Bull. Soc. Math. France **100** (1972), 47–72.

[111] E. Ghys, *Classification des feuilletages totalement géodésiques de codimension un*, Comment. Math. Helv. **58** (1983), 543–572.

[112] E. Ghys, *Feuilletages riemanniens sur les variétés simplement connexes*, Ann. Inst. Fourier **34** (4) (1984), 203–223.

[113] E. Ghys, *Actions localement libres du groupe affine*, Invent. Math. **82** (1985), 479–526.

[114] E. Ghys, *Une variété qui n'est pas une feuille*, Topology **24** (1985), 67–73.

[115] E. Ghys, *Classe d'Euler et minimal exceptionnel*, Topology **24** (1985), 67–73.

[116] E. Ghys, *Flots d'Anosov dont les feuilletages stables sont différentiables*, Ann. Sci. École Norm. Sup. **20** (1987), 251–270.

[117] E. Ghys, *Sur l'invariance topologique de la classe de Godbillon–Vey*, Ann. Inst. Fourier **37** (4) (1987), 59–76.

[118] E. Ghys, *Gauss–Bonnet theorem for* 2*-dimensional foliations*, J. Funct. Anal. **77** (1) (1988), 51–59.

[119] E. Ghys, *L'invariant de Godbillon–Vey*, Sém. Bourbaki, Exposé 706, Astérisque **177–178** (1989), 155–181.

[120] E. Ghys, *Déformations des flots d'Anosov et groupes fuchsiens*, Ann. Inst. Fourier **42** (1–2) (1992), 209–247.

[121] E. Ghys, *Rigidité différentiable des groupes fuchsiens*, Publ. IHES **78** (1993), 163–185.

[122] E. Ghys, *Construction de champs de vecteurs sans orbites périodiques*, Sém. Bourbaki, Exposé 785, Astérisque **227** (1993–1994), 283–307.

[123] E. Ghys, *Holomorphic Anosov systems*, Invent. Math. **199** (3) (1995), 585–614.

[124] E. Ghys, *Groupes d'holonomie des feuilletages de Lie*, Indag. Math. **47** (1995), 173–182.

[125] E. Ghys, *Topologie des feuilles génériques*, Ann. of Math. (2) **141** (2) (1995), 387–422.

[126] E. Ghys, *On transversely holomorphic flows*, Invent. Math. **126** (2) (1996), 281–286.

[127] E. Ghys, R. Langevin and P. Walczak, *Entropie géométrique des feuilletages*, Acta Math. **16** (1–2) (1988), 105–142.

[128] E. Ghys and V. Sergiescu, *Stabilité et conjugaison différentiable pour certains feuilletages*, Topology **26** (1987), 93–105.

[129] E. Ghys and E. Sergiescu, *Sur un groupe remarquable de difféomorphismes du cercle*, Comment. Math. Helv. **62** (1987), 185–239.

[130] E. Ghys and T. Tsuboi, *Différentiabilité des conjugaisons entre systèmes dynamiques de dimension un*, Ann. Inst. Fourier **38** (1) (1988), 215–244.

[131] E. Ghys and A. Verjovsky, *Holomorphic actions of the affine group*, Proceedings of Geometric Study of Foliations, Tokyo, World Scientific (1994), 201–217.

[132] J. Girbau, A. Haefliger and D. Sundararaman, *On deformations of transversely holomorphic foliations*, J. Reine Angew. Math. **345** (1983), 122–147.

[133] J. Girbau and M. Nicolau, *On deformations of holomorphic foliations*, Ann. Inst. Fourier **39** (2) (1989), 417–449.

[134] C. Godbillon, *Feuilletages ayant la propriété du prolongement des homotopies*, Ann. Inst. Fourier **17** (2) (1967), 219–260.

[135] C. Godbillon, *Feuilletages. Etudes géométriques*, Progress in Mathematics, vol. 98, Birkhäuser, Basel (1991).

[136] C. Godbillon and J. Vey, *Un invariant des feuilletages de codimension un*, C. R. Acad. Sci. Paris **273** (1971), 92–95.

[137] X. Gomez Mont, *Transversal holomorphic structures*, J. Differential Geom. **15** (1980), 161–185.

[138] S. Goodman, *On the structure of foliated* 3*-manifolds separated by a compact leaf*, Invent. Math. **39** (1977), 213–221.

[139] S. Goodman, *Vector fields with transverse foliations*, Topology **24** (1985), 333–340.

[140] S. Goodman and J.F. Plante, *Holonomy and averaging in foliated sets*, J. Differential Geom. **14** (1979), 401–407.

[141] M. Gromov, *Stable mappings of foliations into manifolds*, Math. USSR Izv. **3** (1969), 671–694.

[142] M. Gromov, *Transversal mappings of foliations*, Soviet Math. Dokl. **9** (1969), 1126–1129.

[143] A. Haefliger, *Structures feuilletées et cohomologie à valeurs dans un faisceau de groupoï des*, Comment. Math. Helv. **32** (1958), 248–329.

[144] A. Haefliger, *Variétés feuilletées*, Ann. Scuola Nom. Sup. Pisa (3) **16** (1962), 367–397.

[145] A. Haefliger, *Feuilletages sur les variétés ouvertes*, Topology **9** (1970), 183–194.

[146] A. Haefliger, *Homotopy, integrability*, Lecture Notes in Math. **197** (1971), 133–163.

[147] A. Haefliger, *Sur les classes caracteristiques des feuilletages*, Sém. Bourbaki 1971/1972, Exposé 412, Lecture Notes in Math. **317** (1973), 239–260.

[148] A. Haefliger, *Cohomology of Lie algebras and foliations*, Lecture Notes in Math. **652** (1978), 1–12.

[149] A. Haefliger, *Some remarks on foliations with minimal leaves*, J. Differential Geom. **15** (1980), 269–284.

[150] A. Haefliger, *Groupoï des d'holonomie et classifiants*, Astérisque **116** (1984), 98–107.

[151] A. Haefliger, *Pseudo-groups of local isometries*, Differential Geometry, Santiago de Compostela, 1984, L. Cordero, ed., Research Notes, vol. 131, Pitman (1985), 174–197.

[152] A. Haefliger, *Deformations of transversely holomorphic flows on spheres and deformations of Hopf fibrations*, Compositio Math. **55** (1985), 241–251.

[153] A. Haefliger, *Complexification of transversely holomorphic foliations*, Math. Ann. **272** (1985), 23–27.

[154] A. Haefliger, *Feuilletages riemanniens*, Sém. Bourbaki 1998/1999, Exposé 707, Astérisque **177–178** (1989), 183–197.

[155] A. Haefliger, *Travaux de Novikov sur les feuilletages*, Sém. Bourbaki 1967/1968, Exposé 339.

[156] A. Haefliger and Quach Ngoc Du, *Une présentation du groupe fondamental d'une orbifold*, Astérisque **116** (1984), 98–107.

[157] A. Haefliger and E. Salem, *Pseudogroupes d'holonomie des feuilletages riemanniens sur des variétés compactes* 1*-connexes*, Colloq. Geom. Phys. Paris/F (1986), Travaux en cours **33** (1988), 141–160.

[158] A. Haefliger and E. Salem, *Riemannian foliations on simply connected manifolds and actions of tori on orbifolds*, Illinois J. Math. **34** (4) (1990), 706–730.
[159] R.S. Hamilton, *Deformation theory of foliations*, Preprint Cornell University, New York (1978).
[160] R.S. Hamilton, *The inverse function theorem of Nash and Moser*, Bull. Amer. Math. Soc. **7** (1982), 65–222.
[161] Y. Hantout, *Classes caractéristiques de (G, H)-structures et finitude de leur évaluation*, Manuscripta Math. **62** (1988), 383–399.
[162] J. Harrison, *A C^2 counterexample to the Seifert conjecture*, Topology **27** (1988), 249–278.
[163] G. Hector, *Sur les feuilletages presque sans holonomie*, C. R. Acad. Sci. Paris Sér. A **274** (1972), 1703–1706.
[164] G. Hector, *Ouverts incompressibles et structures des feuilletages de codimension un*, C. R. Acad. Sci. Paris Sér. A **274** (1972), 741–744.
[165] G. Hector, *Quelques exemples de feuilletages. Espèces rares*, Ann. Inst. Fourier **26** (1) (1976), 239–264.
[166] G. Hector, *Feuilletages en cylindres*, Lecture Notes in Math. **597** (1976), 252–270.
[167] G. Hector, *Leaves whose growth is neither exponential nor polynomial*, Topology **16** (1977), 449–451.
[168] G. Hector, *Croissance des feuilletages presque sans holonomie*, Lecture Notes in Math. **652** (1978), 141–182.
[169] G. Hector, *Architecture des feuilletages de classe C^2*, Astérisque **107–108** (1983), 243–258.
[170] G. Hector and W. Bouma, *All open surfaces are leaves of simple foliations of $\mathbf{R}^3$*, Indag. Math. **45** (1983), 443–452.
[171] G. Hector, E. Ghys and Y. Moriyama, *On codimension one foliations and a theorem of Malcev*, Topology **28** (2) (1989), 197–210.
[172] G. Hector and U. Hirsch, *Introduction to the Geometry of Foliations*, Parts A and B, Vieweg & Sohn (1981).
[173] G. Hector and E. Macias, *Sur le théorème de de Rham pour les feuilletages de Lie*, C. R. Acad. Sci. Paris Sér. I **311** (10) (1990), 633–636.
[174] J.L. Heitsch, *A cohomology of foliated manifolds*, Comment. Math. Helv. **50** (1975), 197–218.
[175] J.L. Heitsch, *Residues and characteristic classes of foliations*, Bull. Amer. Math. Soc. **83** (1977), 397–399.
[176] J.L. Heitsch, *Independent variation of secondary classes*, Ann. of Math. (2) **108** (3) (1978), 421–460.
[177] J.L. Heitsch, *Flat bundles and residues for foliations*, Invent. Math. **73** (1983), 271–285.
[178] J.L. Heitsch and S. Hurder, *Secondary classes, Weil measures and the geometry of foliations*, J. Differential Geom. **20** (1984), 291–309.
[179] J.L. Heitsch and H.B. Lawson, *Transgressions, Chern–Simons invariants and the classical groups*, J. Differential Geom. **9** (1974), 423–434.
[180] J.L. Heitsch and C. Lazarov, *Lefschetz theorem for foliated manifolds*, Topology **29** (2) (1990), 127–162.
[181] J.L. Heitsch and C. Lazarov, *Homotopy invariance of foliation Betti numbers*, Invent. Math. **104** (2) (1991), 321–347.
[182] J.L. Heitsch and C. Lazarov, *Rigidity theorems by surfaces and spin manifolds*, Michigan Math. J. **38** (2) (1991), 285–297.
[183] J.L. Heitsch and C. Lazarov, *Spectral asymptotics of foliated manifolds*, Illinois J. Math. **38** (4) (1994), 653–678.
[184] M. Herman, *The Godbillon–Vey invariant of foliations by planes of $\mathbf{T}^3$*, Lecture Notes in Math. **597** (1977), 294–307.
[185] M. Herman, *Sur la conjugaison différentiable des difféomorphismes du cercle à des rotations*, Publ. Math. IHES **49** (1979), 5–223.
[186] R. Hermann, *The differential geometry of foliations, I*, Ann. of Math. **72** (2) (1960), 445–457.
[187] R. Hermann, *The differential geometry of foliations, II*, J. Math. Mech. **11** (1962), 305–315.
[188] M. Hirsch and W. Thurston, *Foliated bundles, invariant measures and flat manifolds*, Ann. of Math. **101** (1975), 369–390.
[189] S. Hurder, *Problems on rigidity of group actions and cocycles*, Ergodic Theory Dynam. Systems **5** (1985), 473–484.
[190] S. Hurder, *The Godbillon measure of amenable foliations*, J. Differential Geom. **23** (1986), 347–365.
[191] S. Hurder, *Rigidity for Anosov actions of higher lattices*, Ann. of Math. **135** (1992), 361–410.
[192] S. Hurder and A. Katok, *Ergodic theory and Weil measures for foliations*, Ann. of Math. **126** (1987), 221–275.

[193] S. Hurder and Y. Mitsumatsu, *On the intersection products of transverse invariant measures*, Indiana Univ. Math. J. **40** (1992), 1169–1183.

[194] J.-P. Jouanolou, *Une preuve élémentaire d'un théorème de Thurston*, Topology **17** (1978), 109–110.

[195] F. Kamber and P. Tondeur, *Foliated bundles and characteristic classes*, Lecture Notes in Math. **493** (1975).

[196] F. Kamber and P. Tondeur, *Duality for Riemannian foliations*, Proc. Sympos. Pure Math. **40** (1983), 609–618.

[197] F. Kamber and P. Tondeur, *Hodge de Rham theory for Riemannian foliations*, Math. Ann. **277** (1987), 415–431.

[198] M. Kellum, *Uniformly quasi-isometric foliations*, Ergodic Theory Dynam. Systems **13** (1993), 101–122.

[199] K. Kodaira and D.C. Spencer, *Multifoliate structures*, Ann. of Math. **74** (1961), 52–100.

[200] K. Kuperberg, *A smooth counterexample to the Seifert conjecture*, Ann. of Math. **140** (1994), 723–732.

[201] C. Lamoureux, *Holonomie et feuilles exceptionnelles*, Ann. Inst. Fourier **26** (4) (1976), 273–300.

[202] R. Langevin and H. Rosenberg, *On stability of compact leaves and fibrations*, Topology **16** (1977), 107–111.

[203] F. Laudenbach, *Formes différentielles de degré* 1 *fermées non singulières et classes d'homotopie de leurs noyaux*, Comment. Math. Helv. **51** (1971), 447–464.

[204] F. Laudenbach and R. Roussarie, *Un exemple de feuilletage sur* $\mathbf{S}^3$, Topology **9** (1970), 63–70.

[205] H.B. Lawson, *The Quantitative Theory of Foliations*, Regional Conference Series in Mathematics, vol. 27, Amer. Math. Soc. (1970).

[206] H.B. Lawson, *Codimension one foliations of spheres*, Ann. of Math. **94** (2) (1971), 494–503.

[207] H.B. Lawson, *Foliations*, Bull. Amer. Math. Soc. **80** (1974), 369–418.

[208] C. Lazarov, *A vanishing theorem for certain rigid classes*, Michigan Math. J. **28** (1981), 89–95.

[209] C. Lazarov, *An index theorem for foliations*, Illinois J. Math. **3** (1) (1986), 101–121.

[210] C. Lazarov and J.S. Pasternak, *Residues and characteristic classes for Riemannian foliations*, J. Differential Geom. **11** (1976), 599–612.

[211] D. Lehmann, *Sur l'approximation de certains feuilletages nilpotents par des fibrations*, C. R. Acad. Sci. Paris **286** (1978), 251–254.

[212] J. Lehmann-Lejeune, *Cohomologie sur le fibré transverse à un feuilletage*, C. R. Acad. Sci. Paris **295** (1982), 495–498.

[213] G. Levitt, *Foliations and laminations in hyperbolic surfaces*, Topology **22** (1983), 119–135.

[214] G. Levitt, *Groupe fondamental de l'espace des feuilles dans les feuilletages sans holonomie*, J. Differential Geom. **31** (3) (1990), 711–761.

[215] A. Lichnerowicz, *Algèbres de Lie attachées à un feuilletage*, Ann. Fac. Sci. Toulouse I (1979), 45–76.

[216] W.B.R. Lickorish, *A foliation for* 3*-manifolds*, Ann. of Math. **82** (2) (1965), 414–420.

[217] J.-J. Lœb and M. Nicolau, *Holomorphic flows and complex structures on products of odd dimensional spheres*, Math. Ann. **306** (1996), 781–817.

[218] D. Mac Duff, *On groups of volume preserving diffeomorphisms and foliations with transverse volume form*, Proc. London Math. Soc. **43** (1981), 295–320.

[219] G.W. Mackey, *Ergodic theory and virtual groups*, Math. Ann. **166** (1956), 359–363.

[220] T. Matsuaka and S. Morita, *On characteristic classes of Kähler foliations*, Osaka J. Math. **16** (1979), 539–550.

[221] S. Matsumoto, *Some remarks on foliated* S^1*-bundles*, Invent. Math. **90** (1987), 343–358.

[222] S. Matsumoto, *Perturbation of the Hopf flow and transverse foliations*, Nagoya Math. J. **122** (1991), 75–82.

[223] S. Matsumoto, *Lie affine foliations on* 4*-manifolds*, Invent. Math. **109** (1) (1992), 1–16.

[224] S. Matsumoto, *Codimension one foliations on solvable manifolds*, Comment. Math. Helv. **68** (4) (1993), 633–652.

[225] X. Masa, *Quelques propriétés des feuilletages de codimension un à connexion transverse projetable*, C. R. Acad. Sci. Paris Sér. A **284** (1977), 811–812.

[226] X. Masa, *Characteristic classes of subfoliations*, Ann. Inst. Fourier **31** (2) (1981), 61–86.

[227] X. Masa, *Duality and minimality in Riemannian foliations*, Comment. Math. Helv. **67** (1992), 17–27.

[228] G. Meigniez, *Bouts d'un groupe opérant sur la droite. Applications à la topologie des feuilletages*, Tohoku Math. J. (2) **43** (4) (1991), 473–500.

[229] G. Meigniez, *On completeness of transversely projective foliations*, Topology **31** (2) (1992), 421–432.
[230] G. Meigniez, *Feuilletages de Lie résolubles*, Ann. Fac. Sci. Toulouse (6) **4** (1995), 801–817.
[231] G. Meigniez, *Prolongement des homotopies, Q-variétés et cycles tangents*, Ann. Inst. Fourier **47** (3) (1997), 945–965.
[232] A. Miernowski and W. Mozgawa, *On Molino lifting of Riemannian vector fields*, Acta Math. Hungar. **55** (1990), 185–191.
[233] J. Milnor, *On the existence of a connection with curvature zero*, Comment. Math. Helv. **32** (1958), 215–223.
[234] Y. Mitsumatsu, *A relation between the topological invariance of the Godbillon–Vey invariant and the differentiability of Anosov foliations*, Foliations, Adv. Stud. Pure Math., vol. 5 (1985), 159–167.
[235] Y. Mitsumatsu, *On the self intersection of foliation cycles*, Trans. Amer. Math. Soc. **334** (2) (1992), 1169–1183.
[236] T. Mizutani, S. Morita and T. Tsuboi, *The Godbillon–Vey classes of codimension one foliations which are almost without holonomy*, Ann. of Math. (2) **113** (1981), 515–527.
[237] T. Mizutani, S. Morita and T. Tsuboi, *On the cobordism classes of codimension one foliations which are almost without holonomy*, Topology **22** (1983), 325–343.
[238] I. Moerdijk and D.A. Pronk, *Orbifolds, sheaves and groupoids*, K-Theory **12** (1) (1997), 3–21.
[239] I. Moerdijk and D.A. Pronk, *Simplicial cohomology of orbifolds*, Indag. Math. (N.S.) **10** (2) (1999), 269–293.
[240] P. Molino, *Propriétés cohomologiques et propriétés topologiques des feuilletages à connexion transverse projetable*, Topology **12** (1973), 317–325.
[241] P. Molino, *Sur la géométrie transverse des feuilletages*, Ann. Inst. Fourier **25** (1975), 279–284.
[242] P. Molino, *Etude des feuilletages transversalement complets et applications*, Ann. Sci. École Norm. Sup. **10** (1977), 289–307.
[243] P. Molino, *Géométrie globale des feuilletages riemanniens*, Indag. Math. **44** (1982), 45–76.
[244] P. Molino, *Feuilletages riemanniens*, Astérisque **116** (1984).
[245] P. Molino, *Riemannian Foliations*, Birkhäuser (1988).
[246] P. Molino, *Dualité symplectique, feuilletages et géométrie du moment*, Publ. Mat. Barcelona **33** (3) (1989), 533–541.
[247] S. Morita, *On characteristic classes of Riemannian foliations*, Osaka J. Math. **16** (1979), 161–172.
[248] R. Moussu, *Feuilletages sans holonomie d'une variété fermée*, C. R. Acad. Sci. Paris Sér. A **270** (1970), 1308–1311.
[249] R. Moussu, *Feuilletages presque sans holonomie*, C. R. Acad. Sci. Paris Sér. A **272** (1971), 114–117.
[250] R. Moussu, *Sur les classes exotiques des feuilletages*, Lecture Notes in Math. **392** (1974), 37–42.
[251] R. Moussu and F. Pelletier, *Sur le théorème de Poincaré–Bendixon*, Ann. Inst. Fourier **24** (1) (1974), 131–148.
[252] R. Moussu and R. Roussarie, *Relations de conjugaison et de cobordisme entre certains feuilletages*, Publ. Math. IHES **43** (1973), 143–168.
[253] W. Mozgawa, *Feuilletages de Killing*, Collect. Math. **36** (3) (1985), 285–290.
[254] M.P. Muller, *Sur les composantes de Novikov des feuilletages*, Topology **19** (1980), 199–201.
[255] Ngo Van Que and R. Roussarie, *Sur l'isotopie des formes fermées en dimension* 3, Invent. Math. **64** (1981), 69–87.
[256] M. Nicolau and A. Reventos, *On some geometrical properties of Seifert bundles*, Israel J. Math. **47** (1984), 323–334.
[257] S. Nishikawa, *Residues and secondary characteristic classes for projective foliations*, Proc. Japan Acad. **54** (1978), 79–82.
[258] S. Nishikawa and H. Sato, *On characteristic classes of Riemannian, conformal and projective foliations*, J. Math. Soc. Japan **28** (1976), 223–241.
[259] S. Nishikawa and M. Takeushi, *G-foliations and semisimple flat homogeneous spaces*, Tohoku Math. J. **30** (1978), 307–335.
[260] T. Nishimori, *SRH-decompositions of codimension one foliations and the Godbillon–Vey class*, Tohoku Math. J. **32** (1980), 9–34.
[261] S.P. Novikov, *Foliations of codimension one*, Soviet Math. Dokl. **5** (1964), 540–544 and 1023–1025.
[262] S.P. Novikov, *The topology of foliations*, Trans. Moscow Math. Soc. **14** (1965), 268–304.

[263] J. Palis, *Rigidity of the centralizers and structural stability of suspended foliations*, Lecture Notes in Math. **652** (1978), 114–121.

[264] C.F.B. Palmeira, *Open manifolds foliated by planes*, Ann. of Math. **107** (1978), 109–131.

[265] E. Park and K. Richardson, *The basic Laplacian for Riemannian foliations*, Amer. J. Math. **118** (1996), 1249–1275.

[266] A. Phillips, *Foliations on open manifolds*, Comment. Math. Helv. **43** (1968), 204–211 and **44** (1969), 367–370.

[267] H.P. Pittie, *The secondary characteristic classes of parabolic foliations*, Comment. Math. Helv. **54** (1979), 601–614.

[268] M. Plaisant, *Sur l'intégrabilité des algèbres de Lie banachiques dans le cadre des Q-variétés*, Thèse 3ème cycle, Valenciennes (1980).

[269] J.F. Plante, *Anosov flows*, Amer. J. Math. **94** (1972), 729–754.

[270] J.F. Plante, *Asymptotic properties of foliations*, Comment. Math. Helv. **47** (1972), 449–456.

[271] J.F. Plante, *A generalization of the Poincaré–Bendixon theorem for foliations of codimension one*, Topology **12** (1973), 177–181.

[272] J.F. Plante, *Foliations transverse to fibers of a bundle*, Proc. Amer. Math. Soc. **42** (1974), 631–635.

[273] J.F. Plante, *Foliations with measure preserving holonomy*, Ann. of Math. **102** (1975), 327–361.

[274] J.F. Plante, *Foliations of 3-manifolds with solvable fundamental group*, Invent. Math. **51** (1979), 219–230.

[275] J.F. Plante, *Stability of codimension one foliations by compact leaves*, Topology **22** (1983), 173–177.

[276] J.F. Plante, *Polycyclic groups and transversely affine foliations*, J. Differential Geom. **36** (3) (1992), 521–534.

[277] J.F. Plante and W.P. Thurston, *Polynomial growth in holonomy groups of foliations*, Comment. Math. Helv. **51** (1976), 567–584.

[278] J. Pradines, *Sur une classe remarquable de relations d'équivalence sur les variétés*, Rev. An. de Math. Jaca (1973).

[279] J. Pradines, *Un feuilletage sans holonomie transversale dont le quotient n'est pas une Q-variété*, C. R. Acad. Sci. Paris **288** (1979), 245–248.

[280] J. Pradines and B. Bigonnet, *Graphe d'un feuilletage singulier*, C. R. Acad. Sci. Paris **300** (1985), 439–442.

[281] J. Pradines and J. Wouafo-Kamga, *La catégorie des QF-variétés*, C. R. Acad. Sci. Paris Sér. A **288** (1979), 717–719.

[282] M. Puta, *Cohomology classes and foliated manifolds*, Nonlinear Anal. (1987), 137–159.

[283] A. Ramsay, *Virtual groups and group actions*, Adv. in Math. **6** (1971), 253–322.

[284] O.H. Rasmussen, *Exotic characteristic classes for holomorphic foliations*, Invent. Math. **46** (1978), 153–171.

[285] G. Reeb, *Sur certaines propriétés topologiques des variétés feuilletées*, Hermann, Paris (1950).

[286] G. Reeb, *Sur les structures feuilletées de codimension un et sur un théorème de Denjoy*, Ann. Inst. Fourier **11** (1961), 185–200.

[287] G. Reeb, *Feuilletages résultats anciens et nouveaux*, Painlevé, Hector and Martinet, eds, Presses Univ. de Montréal (1974).

[288] B.L. Reinhart, *Harmonic integrals on almost product manifolds*, Trans. Amer. Math. Soc. **88** (1958), 243–276.

[289] B.L. Reinhart, *Foliated manifolds with bundle-like metrics*, Ann. of Math. **69** (1959), 119–132.

[290] B.L. Reinhart, *Harmonic integrals on foliated manifolds*, Amer. J. Math. **81** (1959), 529–586.

[291] B.L. Reinhart, *Differential Geometry of Foliations*, Ergebnisse der Mathematik, vol. 99, Springer (1983).

[292] B.L. Reinhart and J.W. Wood, *A metric formula for the Godbillon–Vey invariant for foliations*, Proc. Amer. Math. Soc. **38** (1973), 427–430.

[293] J. Renault, *A Groupoid Approach to C^*-Algebras*, Lecture Notes in Math., vol. 793, Springer (1980).

[294] K. Richardson, *Asymptotic of heat kernels on Riemannian foliations*, Geom. Funct. Anal. **8** (1998), 356–401.

[295] C. Roger, *Etude des Γ-structures de codimension un sur la sphère $\mathbf{S}^2$*, Ann. Inst. Fourier **23** (4) (1973), 213–227.

[296] C. Roger, *Méthodes homotopiques et cohomologiques en théorie des feuilletages*, Thèse Orsay (1974).

[297] H. Rosenberg, *Actions of $\mathbf{R}^n$ on manifolds*, Comment. Math. Helv. **41** (1966), 170–178.

[298] H. Rosenberg, *Foliations by planes*, Topology **7** (1968), 143–145.

[299] H. Rosenberg and R. Roussarie, *Topological equivalence of Reeb foliations*, Topology **9** (1970), 231–242.

[300] H. Rosenberg and R. Roussarie, *Les feuilles exceptionnelles ne sont pas exceptionnelles*, Comment. Math. Helv. **45** (1970), 517–523.

[301] H. Rosenberg and R. Roussarie, *Reeb foliations*, Ann. of Math. (2) **91** (1970), 1–24.

[302] H. Rosenberg and R. Roussarie, *Some remarks on stability of foliations*, J. Differential Geom. **10** (1975), 207–219.

[303] H. Rosenberg, R. Roussarie and D. Weil, *A classification of closed orientable* 3*-manifolds of rank two*, Ann. of Math. **91** (1970), 449–464.

[304] H. Rosenberg, R. Roussarie and D. Weil, *A classification of the topological types of* $\mathbf{R}^2$*-actions*, Publ. Math. IHES **43** (1973), 261–272.

[305] J. Rosenberg, *K-theory of group* C^**-algebras, foliation* C^**-algebras and crossed products*, Index Theory of Elliptic Operators, Foliations and Operator Algebras, New Orleans, Contemp. Math. **70** (1988), 251–301.

[306] R. Roussarie, *Sur les feuilletages des variétés de dimension* 3, Ann. Inst. Fourier **21** (3) (1971), 13–81.

[307] R. Roussarie, *Plongement dans les variétés feuilletées et classification des feuilletages sans holonomie*, Publ. Math. IHES **43** (1974), 101–141.

[308] D. Ruelle and D. Sullivan, *Currents, flows and diffeomorphisms*, Topology **14** (1975), 319–327.

[309] H. Rummler, *Quelques notions simples en géométrie riemannienne et leurs applications aux feuilletages compacts*, Comment. Math. Helv. **54** (1979), 224–239.

[310] S. Sacksteder, *Some properties of foliations*, Ann. Inst. Fourier **14** (1) (1964), 31–35.

[311] S. Sacksteder, *Foliations and pseudogroups*, Amer. J. Math. **87** (1965), 79–102.

[312] S. Sacksteder, *A remark on Thurston stability theorem*, Ann. Inst. Fourier **25** (2) (1975), 219–220.

[313] S. Sacksteder and J. Schwartz, *Limit sets for foliations*, Ann. Inst. Fourier **15** (2) (1965), 201–213.

[314] E. Salem, *Une généralisation du théorème de Myers–Steenrod aux pseudo-groupes d'isométries locales*, Ann. Inst. Fourier **35** (1985), 137–158.

[315] E. Salhi, *Sur les ensembles minimaux locaux*, C. R. Acad. Sci. Paris **295** (1982), 691–694.

[316] E. Salhi, *Sur un théorème de structure des feuilletages de codimension* 1, C. R. Acad. Sci. Paris **300** (1985), 635–638.

[317] E. Salhi, *Niveau des feuilles*, C. R. Acad. Sci. Paris **301** (1985), 219–222.

[318] M. Saralegi, *The Euler class for flows of isometries*, Differential Geometry, Santiago de Compostela, 1984, Pitman Research Notes in Math., vol. 131 (1985), 220–227.

[319] K.S. Sarkaria, *The de Rham cohomology of foliated manifolds*, Thesis, Stony Brook (1974).

[320] K.S. Sarkaria, *A finiteness theorem for foliated manifolds*, J. Math. Soc. Japan **30** (4) (1978), 687–696.

[321] I. Satake, *On the generalization of the notion of manifold*, Proc. NAS **42** (1956), 359–363.

[322] G. Schwarz, *On the de Rham cohomology of the leaf space of a foliation*, Topology **13** (1974), 185–188.

[323] S. Schwartzman, *Asymptotic cycles*, Ann. of Math. **66** (1957), 270–284.

[324] P. Schweitzer, *Counterexamples to the Seifert conjecture and opening closed leaves of foliations*, Ann. of Math. **100** (1974), 386–400.

[325] P. Schweitzer, *Stability of compact leaves with trivial linear holonomy*, Topology **27** (1) (1988), 37–56.

[326] P. Schweitzer, *Existence of codimension one foliations with minimal leaves*, Ann. Global Anal. Geom. **9** (1) (1991), 77–81.

[327] P. Schweitzer, *Codimension one foliations without compact leaves*, Comment. Math. Helv. **70** (2) (1995), 171–209.

[328] P. Scott, *The geometry of* 3*-manifolds*, Bull. London Math. Soc. **15** (1983), 401–487.

[329] H. Seifert, *Closed integral curves in* 3*-space and isotopic deformations*, Proc. Amer. Math. Soc. **1** (1950), 287–302.

[330] F. Sergeraert, *Feuilletages et difféomorphismes infiniment tangents à l'identité*, Invent. Math. **39** (1977), 253–275.

[331] V. Sergiescu, *Cohomologie basique et dualité des feuilletages riemanniens*, Ann. Inst. Fourier de Grenoble **53** (3) (1985), 135–158.

[332] C. Series, *Foliations of polynomial growth are hyperfinite*, Israel J. Math. **34** (1979), 245–258.

[333] H. Shulman and J. Stasheff, *De Rham theory for* $B\Gamma$, Lecture Notes in Math. **652** (1978), 62–74.

[334] J.-C. Sikorav, *Formes différentielles fermées sur le n-tore*, Comment. Math. Helv. **57** (1982), 79–106.

[335] J.D. Sondow, *When is a manifold a leaf of some foliation*? Bull. Amer. Math. Soc. **81** (1975), 622–624.

[336] D. Sullivan, *A foliation of geodesics is characterized by having no "tangent homologies"*, J. Pure Appl. Algebra **13** (1973), 101–104.

[337] D. Sullivan, *A counterexample to the periodic orbit conjecture*, Publ. Math. IHES **46** (1976), 5–14.

[338] D. Sullivan, *Cycles for the dynamical study of foliated manifolds and complex manifolds*, Invent. Math. **36** (1976), 225–255.

[339] D. Sullivan, *A homological characterization of foliations consisting of minimal surfaces*, Comment. Math. Helv. **54** (1979), 218–223.

[340] H. Takai, *KK-Theory for the C^*-algebras of Anosov foliations*, Geometric Methods in Operator Algebras, Pitman Research Notes in Math., vol. 123 (1986), 387–399.

[341] I. Tamura, *Every odd dimensional homotopy sphere has a codimension one foliation*, Comment. Math. Helv. **47** (1972), 73–79.

[342] E. Thomas, *Vector fields on manifolds*, Bull. Amer. Math. Soc. **75** (1969), 643–683.

[343] W.P. Thurston, *Foliations of* 3*-manifolds which are circle bundles*, Thesis, Berkeley (1972).

[344] W.P. Thurston, *Noncobordant foliations of* $\mathbf{S}^3$, Bull. Amer. Math. Soc. **78** (1972), 511–514.

[345] W.P. Thurston, *A generalization of the Reeb stability theorem*, Topology **13** (1974), 347–352.

[346] W.P. Thurston, *Foliations and groups of diffeomorphisms*, Bull. Amer. Math. Soc. **80** (1974), 304–307.

[347] W.P. Thurston, *The theory of foliations of codimension greater than one*, Comment. Math. Helv. **49** (1974), 214–231.

[348] W.P. Thurston, *Existence of codimension one foliations*, Ann. of Math. **102** (2) (1976), 249–268.

[349] W.P. Thurston, *The geometry and topology of* 3*-manifolds*, Mimeographed Notes, Princeton (1977).

[350] D. Tischler, *On Fibering certain foliated manifolds over* S^1, Topology **9** (1970), 153–154.

[351] P. Tondeur, *Foliations on Riemannian Manifolds*, Springer (1988).

[352] P. Tondeur, *Geometry of Foliations*, Monographs in Mathematics, vol. 90, Birkhäuser (1997).

[353] T. Tsuboi, *Foliations with trivial $\mathcal{F}$-subgroups*, Topology **18** (1979), 223–233.

[354] T. Tsuboi, *On the cobordism classes of codimension one foliations which are almost without holonomy*, Topology **22** (1983), 325–343.

[355] T. Tsuboi, *Foliated cobordism classes of certain foliated* $\mathbf{S}^1$*-bundles*, Topology **23** (1984), 233–244.

[356] T. Tsuboi, *On the foliated product of class* C^1, Ann. of Math. **130** (1989), 227–271.

[357] T. Tsuboi, *The Godbillon–Vey invariant and the foliated cobordism group*, Proc. Japan Acad. **68** (1992), 85–90.

[358] N. Tsuchiya, *The Nishimori decompositions of codimension one foliations and the Godbillon–Vey classes*, Tohoku Math. J. **34** (1982), 343–365.

[359] H. Uzuki, *Characteristic classes of foliated principal GL_r-bundles*, Hokkaido Math. J. **4** (1975), 159–168.

[360] I. Vaisman, *Variétés riemanniennes feuilletées*, Czechoslovak Math. J. **21** (1971), 46–75.

[361] I. Vaisman, *Conformal foliations*, Kodai Math. J. **2** (1979), 22–37.

[362] W.T. Van Est, *Rapport sur les S-atlas*, Astérisque **116** (1984).

[363] A. Verjovsky, *A note on Anosov flows*, Bol. Soc. Mat. Mexicana (1) **1** (1970), 36–37.

[364] A. Verjovsky, *Codimension one Anosov flows*, Bol. Soc. Mat. Mexicana (2) **19** (1974), 49–77.

[365] A. Verjovsky, *A uniformisation theorem for holomorphic foliations*, The Lefschetz Centenial Conference, Mexico, 1984, Contemp. Math. (3) **58** (1987), 233–253.

[366] E. Vogt, *Stable foliations of* 4*-manifolds by closed surfaces. Local structures and free actions of finite cyclic and dihedral groups on surfaces*, Invent. Math. **22** (1973), 321–348.

[367] E. Vogt, *A foliation of* $\mathbf{R}^3$ *and other punctured* 3*-manifolds by circles*, Publ. Math. IHES **69** (1981), 215–232.

[368] E. Vogt, *The first cohomology group of leaves and local stability of compact foliations*, Manuscripta Math. **37** (1982), 229–267.

[369] E. Vogt, *Existence of foliations of Euclidean spaces with all leaves compact*, Math. Ann. **296** (1) (1993), 159–178.

[370] A. Wadsley, *Geodesic foliations by circles*, J. Differential Geom. **10** (1975), 541–550.

[371] P. Walczak, *Dynamics of the geodesic flow of a foliation*, Ergodic Theory Dynam. Systems **8** (1988), 637–650.

[372] H.E. Winkelnkemper, *The graph of a foliation*, Ann. Global Anal. Geom. **1** (1983), 53–75.

[373] R. Wolak, *Characteristic classes of almost flat structures*, Geom. Dedicata **24** (1987), 207–220.

[374] R. Wolak, *Maximal subalgebras in the algebra of foliated vector fields of a Riemannian foliation*, Comment. Math. Helv. **64** (4) (1989), 536–541.

[375] R. Wolak, *The graph of a totally geodesic foliation*, Ann. Polon. Math. **60** (3) (1995), 241–247.

[376] J. Wood, *Foliations on 3-manifolds*, Ann. of Math. **89** (2) (1969), 336–358.

[377] J. Wood, *Foliations of codimension one*, Bull. Amer. Math. Soc. **75** (1970), 1107–1111.

[378] J. Wood, *Bundles with totally disconnected structure group*, Comment. Math. Helv. **46** (1971), 257–273.

[379] J. Wouafo-Kamga, *Décomposition des G-structures d'ordre supérieur. Structures transverse des feuilletages*, Thèse, Toulouse (1979).

[380] A. Zeggar, *Nombre et classe de Lefschetz pour un feuilletage riemannien*, Ann. Fac. Sci. Toulouse (6) **1** (1) (1992), 105–131.

[381] A. Zeghib, *Laminations et hypersurfaces géodésiques des variétés hyperboliques*, Ann. Sci. École Norm. Sup. (4) **24** (2) (1991), 171–188.

[382] A. Zeghib, *Sur les feuilletages géodésiques continus des variétés hyperboliques*, Invent. Math. **114** (1) (1993), 193–206.

[383] A. Zeghib, *Feuilletages géodésiques appliqués*, Math. Ann. **298** (4) (1994), 729–759.

[384] A. Zeghib, *Feuilletages géodésiques des variétés localement symétriques*, Topology **36** (4) (1997), 805–828.

[385] R. Zimmer, *On the Mostow rigidity theorem and measurable foliations*, Israel J. Math. **43** (1982), 281–290.

CHAPTER 3

Symplectic Geometry

Ana Cannas da Silva[1]

Departamento de Matemática, Instituto Superior Técnico, 1049-001 Lisboa, Portugal
E-mail: acannas@math.ist.utl.pt; acannas@math.princeton.edu

Contents

[1]Current address: Department of Mathematics, Princeton University, Princeton, NJ 08544-1000, USA.

HANDBOOK OF DIFFERENTIAL GEOMETRY, VOL. II
Edited by F.J.E. Dillen and L.C.A. Verstraelen

Introduction

This is an overview of symplectic geometry[1]—the geometry of *symplectic manifolds*. From a language for classical mechanics in the XVIII century, symplectic geometry has matured since the 1960's to a rich and central branch of differential geometry and topology. A current survey can thus only aspire to give a partial flavor on this exciting field. The following six topics have been chosen for this handbook:

1. *Symplectic manifolds* are manifolds equipped with *symplectic forms*. A symplectic form is a closed nondegenerate 2-form. The algebraic condition (nondegeneracy) says that the top exterior power of a symplectic form is a volume form, therefore symplectic manifolds are necessarily even-dimensional and orientable. The analytical condition (closedness) is a natural differential equation that forces all symplectic manifolds to being locally indistinguishable: they all locally look like an even-dimensional Euclidean space equipped with the $\sum dx_i \wedge dy_i$ symplectic form. All cotangent bundles admit canonical symplectic forms, a fact relevant for analysis of differential operators, dynamical systems, classical mechanics, etc. Basic properties, major classical examples, equivalence notions, local normal forms of symplectic manifolds and symplectic submanifolds are discussed in Section 1.

2. *Lagrangian submanifolds*[2] are submanifolds of symplectic manifolds of half dimension and where the restriction of the symplectic form vanishes identically. By the *Lagrangian creed* [137], everything is a Lagrangian submanifold, starting with closed 1-forms, real functions modulo constants and symplectomorphisms (diffeomorphisms that respect the symplectic forms). Section 2 also describes normal neighborhoods of Lagrangian submanifolds with applications.

3. *Complex structures* or almost complex structures abound in symplectic geometry: any symplectic manifold possesses almost complex structures, and even so in a *compatible* sense. This is the point of departure for the modern technique of studying pseudoholomorphic curves, as first proposed by Gromov [64]. Kähler geometry lies at the intersection of complex, Riemannian and symplectic geometries, and plays a central role in these three fields. Section 3 includes the local normal form for Kähler manifolds and a summary of Hodge theory for Kähler manifolds.

4. *Symplectic geography* is concerned with existence and uniqueness of symplectic forms on a given manifold. Important results from Kähler geometry remain true in the more general symplectic category, as shown using pseudoholomorphic methods. This viewpoint was more recently continued with work on the existence of certain symplectic

[1]The word *symplectic* in mathematics was coined in the late 1930's by Weyl [142, p. 165] who substituted the Latin root in *complex* by the corresponding Greek root in order to label the symplectic group (first studied by Abel). An English dictionary is likely to list *symplectic* as the name for a bone in a fish's head.

[2]The name *Lagrangian manifold* was introduced by Maslov [93] in the 1960's, followed by *Lagrangian plane*, etc., introduced by Arnold [2].

submanifolds, in the context of Seiberg–Witten invariants, and with topological descriptions in terms of Lefschetz pencils. Both of these directions are particularly relevant to 4-dimensional topology and to mathematical physics, where symplectic manifolds occur as building blocks or as key examples. Section 4 treats constructions of symplectic manifolds and invariants to distinguish them.

5. *Hamiltonian geometry* is the geometry of symplectic manifolds equipped with a *moment map*, that is, with a collection of quantities conserved by symmetries. With roots in Hamiltonian mechanics, moment maps became a consequential tool in geometry and topology. The notion of a moment map arises from the fact that, to any real function on a symplectic manifold, is associated a vector field whose flow preserves the symplectic form and the given function; this is called the *Hamiltonian vector field* of that (Hamiltonian) function. The Arnold conjecture in the 60's regarding Hamiltonian dynamics was a major driving force up to the establishment of Floer homology in the 80's. Section 5 deals mostly with the geometry of moment maps, including the classical Legendre transform, integrable systems and convexity.

6. *Symplectic reduction* is at the heart of many symplectic arguments. There are infinite-dimensional analogues with amazing consequences for differential geometry, as illustrated in a symplectic approach to Yang–Mills theory. Symplectic toric manifolds provide examples of extremely symmetric symplectic manifolds that arise from symplectic reduction using just the data of a polytope. All properties of a symplectic toric manifold may be read from the corresponding polytope. There are interesting interactions with algebraic geometry, representation theory and geometric combinatorics. The variation of reduced spaces is also addressed in Section 6.

1. Symplectic manifolds

1.1. *Symplectic linear algebra*

Let V be a vector space over $\mathbb{R}$, and let $\Omega : V \times V \to \mathbb{R}$ be a skew-symmetric bilinear map. By a skew-symmetric version of the Gram–Schmidt process,[3] there is a basis $u_1, \ldots, u_k, e_1, \ldots, e_n, f_1, \ldots, f_n$ of V for which $\Omega(u_i, v) = \Omega(e_i, e_j) = \Omega(f_i, f_j) = 0$ and $\Omega(e_i, f_j) = \delta_{ij}$ for all i, j and all $v \in V$. Although such a basis is not unique, it is commonly referred to as a *canonical basis*. The dimension k of the subspace $U = \{u \in V \mid \Omega(u, v) = 0 \text{ for all } v \in V\}$ is an invariant of the pair (V, Ω). Since $k + 2n = \dim V$, the even number $2n$ is also an invariant of (V, Ω), called the *rank* of Ω. We denote by $\tilde{\Omega} : V \to V^*$ the linear map defined by $\tilde{\Omega}(v)(u) := \Omega(v, u)$. We say that Ω is *symplectic* (or *nondegenerate*) if the associated $\tilde{\Omega}$ is bijective (i.e., the kernel U of $\tilde{\Omega}$ is the trivial space $\{0\}$). In that case, the map Ω is called a *linear symplectic structure* on V, and the

[3]Let $u_1, \ldots, u_k$ be a basis of $U := \{u \in V \mid \Omega(u, v) = 0 \text{ for all } v \in V\}$, and W a complementary subspace such that $V = U \oplus W$. Take any nonzero $e_1 \in W$. There is $f_1 \in W$ with $\Omega(e_1, f_1) = 1$. Let W_1 be the span of e_1, f_1 and $W_1^{\Omega} := \{v \in V \mid \Omega(v, u) = 0\ \forall u \in W_1\}$. Then $W = W_1 \oplus W_1^{\Omega}$. Take any nonzero $e_2 \in W_1^{\Omega}$. There is $f_2 \in W_1^{\Omega}$ for which $\Omega(e_2, f_2) = 1$. Let W_2 be the span of e_2, f_2, and so on.

pair (V, Ω) is called a *symplectic vector space*. A linear symplectic structure Ω expresses a *duality* by the bijection $\tilde{\Omega} : V \xrightarrow{\simeq} V^*$, similar to the (symmetric) case of an inner product. By considering a canonical basis, we see that the dimension of a symplectic vector space (V, Ω) *must be even*, $\dim V = 2n$, and that V admits a basis $e_1, \ldots, e_n, f_1, \ldots, f_n$ satisfying $\Omega(e_i, f_j) = \delta_{ij}$ and $\Omega(e_i, e_j) = 0 = \Omega(f_i, f_j)$. Such a basis is then called a *symplectic basis* of (V, Ω), and, in terms of exterior algebra, $\Omega = e_1^* \wedge f_1^* + \cdots + e_n^* \wedge f_n^*$, where $e_1^*, \ldots, e_n^*, f_1^*, \ldots, f_n^*$ is the dual basis. With respect to a symplectic basis, the map Ω is represented by the matrix

$$\begin{bmatrix} 0 & \mathrm{Id} \\ -\mathrm{Id} & 0 \end{bmatrix}.$$

EXAMPLES.

1. The *prototype of a symplectic vector space* is $(\mathbb{R}^{2n}, \Omega_0)$ with Ω_0 such that the canonical basis $e_1 = (1, 0, \ldots, 0), \ldots, e_n, f_1, \ldots, f_n = (0, \ldots, 0, 1)$ is a symplectic basis. Bilinearity then determines Ω_0 on other vectors.
2. For any real vector space E, the direct sum $V = E \oplus E^*$ has a *canonical symplectic structure* determined by the formula $\Omega_0(u \oplus \alpha, v \oplus \beta) = \beta(u) - \alpha(v)$. If $e_1, \ldots, e_n$ is a basis of E, and $f_1, \ldots, f_n$ is the dual basis, then $e_1 \oplus 0, \ldots, e_n \oplus 0, 0 \oplus f_1, \ldots, 0 \oplus f_n$ is a symplectic basis for V.

Given a linear subspace W of a symplectic vector space (V, Ω), its *symplectic orthogonal* is the subspace $W^\Omega := \{v \in V \mid \Omega(v, u) = 0 \text{ for all } u \in W\}$. By nondegeneracy, we have $\dim W + \dim W^\Omega = \dim V$ and $(W^\Omega)^\Omega = W$. For subspaces W and Y, we have $(W \cap Y)^\Omega = W^\Omega + Y^\Omega$, and if $W \subseteq Y$ then $Y^\Omega \subseteq W^\Omega$.

There are special types of linear subspaces of a symplectic vector space (V, Ω). A subspace W is a *symplectic subspace* if the restriction $\Omega|_W$ is nondegenerate, that is, $W \cap W^\Omega = \{0\}$, or equivalently $V = W \oplus W^\Omega$. A subspace W is an *isotropic subspace* if $\Omega|_W \equiv 0$, that is, $W \subseteq W^\Omega$. A subspace W is a *coisotropic subspace* if $W^\Omega \subseteq W$. A subspace W is a *Lagrangian subspace* if it is both isotropic and coisotropic, or equivalently, if it is an isotropic subspace with $\dim W = \frac{1}{2} \dim V$. A basis $e_1, \ldots, e_n$ of a Lagrangian subspace can be extended to a symplectic basis: choose f_1 in the symplectic orthogonal to the linear span of $\{e_2, \ldots, e_n\}$, etc.

EXAMPLES.

1. For a symplectic basis as above, the span of e_1, f_1 is symplectic, that of e_1, e_2 isotropic, that of $e_1, \ldots, e_n, f_1$ coisotropic, and that of $e_1, \ldots, e_n$ Lagrangian.
2. The graph of a linear map $A : E \to E^*$ is a Lagrangian subspace of $E \oplus E^*$ with the canonical symplectic structure if and only if A is symmetric (i.e., $(Au)v = (Av)u$). Therefore, the Grassmannian of all Lagrangian subspaces in a $2n$-dimensional symplectic vector space has dimension $\frac{n(n+1)}{2}$.

A *symplectomorphism* φ between symplectic vector spaces (V, Ω) and (V', Ω') is a linear isomorphism $\varphi : V \xrightarrow{\simeq} V'$ such that $\varphi^* \Omega' = \Omega$.[4] If a symplectomorphism exists,

[4] By definition, $(\varphi^* \Omega')(u, v) = \Omega'(\varphi(u), \varphi(v))$.

(V, Ω) and (V', Ω') are said to be *symplectomorphic*. Being symplectomorphic is clearly an equivalence relation in the set of all even-dimensional vector spaces. The existence of canonical bases shows that every $2n$-dimensional symplectic vector space (V, Ω) is symplectomorphic to the prototype $(\mathbb{R}^{2n}, \Omega_0)$; a choice of a symplectic basis for (V, Ω) yields a symplectomorphism to $(\mathbb{R}^{2n}, \Omega_0)$. Hence, nonnegative even integers classify equivalence classes for the relation of being symplectomorphic.

Let $\Omega(V)$ be the space of all linear symplectic structures on the vector space V. Take a $\Omega \in \Omega(V)$, and let $\mathrm{Sp}(V, \Omega)$ be the *group of symplectomorphisms* of (V, Ω). The group $\mathrm{GL}(V)$ of all isomorphisms of V acts *transitively* on $\Omega(V)$ by pullback (i.e., all symplectic structures are related by a linear isomorphism), and $\mathrm{Sp}(V, \Omega)$ is the stabilizer of the given Ω. Hence, $\Omega(V) \simeq \mathrm{GL}(V)/\mathrm{Sp}(V, \Omega)$.

1.2. *Symplectic forms*

Let ω be a de Rham 2-form on a manifold[5] M. For each point $p \in M$, the map $\omega_p : T_pM \times T_pM \to \mathbb{R}$ is skew-symmetric and bilinear on the tangent space to M at p, and ω_p varies smoothly in p.

DEFINITION 1.1. The 2-form ω is *symplectic* if ω is closed (i.e., its exterior derivative $d\omega$ is zero) and ω_p is symplectic for all $p \in M$. A *symplectic manifold* is a pair (M, ω) where M is a manifold and ω is a symplectic form.

Symplectic manifolds must be *even-dimensional*. Moreover, the nth exterior power ω^n of a symplectic form ω on a $2n$-dimensional manifold is a *volume form*.[6] Hence, any symplectic manifold (M, ω) is *canonically oriented*. The form $\frac{\omega^n}{n!}$ is called the *symplectic volume* or *Liouville volume* of (M, ω). When (M, ω) is a *compact* $2n$-dimensional symplectic manifold, the de Rham cohomology class $[\omega^n] \in H^{2n}(M; \mathbb{R})$ must be nonzero by Stokes theorem. Therefore, the class $[\omega]$ must be nonzero, as well as its powers $[\omega]^k = [\omega^k] \neq 0$. *Exact symplectic forms* can only exist on noncompact manifolds. Compact manifolds with a trivial even cohomology group $H^{2k}(M; \mathbb{R})$, $k = 0, 1, \ldots, n$, such as spheres S^{2n} with $n > 1$, can thus never be symplectic. On a manifold of dimension greater than 2, a function multiple $f\omega$ of a symplectic form ω is symplectic if and only if f is a nonzero locally constant function (this follows from the existence of a symplectic basis).

EXAMPLES.

1. Let $M = \mathbb{R}^{2n}$ with linear coordinates $x_1, \ldots, x_n, y_1, \ldots, y_n$. The form

$$\omega_0 = \sum_{i=1}^{n} dx_i \wedge dy_i$$

[5]Unless otherwise indicated, all vector spaces are real and finite-dimensional, all maps are smooth (i.e., C^∞) and all manifolds are smooth, Hausdorff and second countable.

[6]A *volume form* is a nonvanishing form of top degree. If Ω is a symplectic structure on a vector space V of dimension $2n$, its nth exterior power $\Omega^n = \Omega \wedge \cdots \wedge \Omega$ does not vanish. Actually, a skew-symmetric bilinear map Ω is symplectic if and only if $\Omega^n \neq 0$.

is symplectic, and the vectors $(\frac{\partial}{\partial x_1})_p, \ldots, (\frac{\partial}{\partial x_n}t)_p, (\frac{\partial}{\partial y_1})_p, \ldots, (\frac{\partial}{\partial y_n})_p$ constitute a symplectic basis of T_pM.

2. Let $M = \mathbb{C}^n$ with coordinates $z_1, \ldots, z_n$. The form $\omega_0 = \frac{i}{2} \sum dz_k \wedge d\bar{z}_k$ is symplectic. In fact, this form coincides with that of the previous example under the identification $\mathbb{C}^n \simeq \mathbb{R}^{2n}$, $z_k = x_k + iy_k$.
3. The 2-sphere S^2, regarded as the set of unit vectors in $\mathbb{R}^3$, has tangent vectors at p identified with vectors orthogonal to p. The standard symplectic form on S^2 is induced by the standard inner (dot) and exterior (vector) products: $\omega_p(u, v) := \langle p, u \times v\rangle$, for $u, v \in T_pS^2 = \{p\}^\perp$. This is the standard area form on S^2 with total area 4π. In terms of cylindrical polar coordinates $0 \leqslant \theta < 2\pi$ and $-1 \leqslant z \leqslant 1$ away from the poles, it is written $\omega = d\theta \wedge dz$.
4. On any Riemann surface, regarded as a 2-dimensional oriented manifold, any area form, that is, any never vanishing 2-form, is a symplectic form.
5. Products of symplectic manifolds are naturally symplectic by taking the sum of the pullbacks of the symplectic forms from the factors.
6. If a $(2n+1)$-dimensional manifold X admits a *contact form*, that is, a 1-form α such that $\alpha \wedge (d\alpha)^n$ is never vanishing, then the 2-form $d(e^t\alpha)$ is symplectic on $X \times \mathbb{R}$, and the symplectic manifold $(X \times \mathbb{R}, d(e^t\alpha))$ is called the *symplectization* of the *contact manifold* (X, α). For more on *contact geometry*, see for instance the corresponding contribution in this volume.

DEFINITION 1.2. Let (M_1, ω_1) and (M_2, ω_2) be symplectic manifolds. A (smooth) map $\psi : M_1 \to M_2$ is *symplectic* if $\psi^*\omega_2 = \omega_1$.[7] A symplectic diffeomorphism $\varphi : M_1 \to M_2$ is a *symplectomorphism*. (M_1, ω_1) and (M_2, ω_2) are said to be *symplectomorphic* when there exists a symplectomorphism between them.

The classification of symplectic manifolds up to symplectomorphism is an open problem in symplectic geometry. However, the local classification is taken care of by the *Darboux theorem* (Theorem 1.9): the dimension is the only local invariant of symplectic manifolds up to symplectomorphisms. That is, just as any n-dimensional manifold is locally diffeomorphic to $\mathbb{R}^n$, any symplectic manifold (M^{2n}, ω) is locally symplectomorphic to $(\mathbb{R}^{2n}, \omega_0)$. As a consequence, if we prove for $(\mathbb{R}^{2n}, \omega_0)$ a local assertion that is invariant under symplectomorphisms, then that assertion holds for any symplectic manifold. We will hence refer to $\mathbb{R}^{2n}$, with linear coordinates $(x_1, \ldots, x_n, y_1, \ldots, y_n)$, and with symplectic form $\omega_0 = \sum_{i=1}^n dx_i \wedge dy_i$, as the *prototype of a local piece of a* $2n$*-dimensional symplectic manifold*.

1.3. *Cotangent bundles*

Cotangent bundles are major examples of symplectic manifolds. Let $(\mathcal{U}, x_1, \ldots, x_n)$ be a coordinate chart for a manifold X, with associated cotangent coordinates $(T^*\mathcal{U}, x_1, \ldots, x_n,$

[7] By definition of *pullback*, we have $(\psi^*\omega_2)_p(u, v) = (\omega_2)_{\psi(p)}(d\psi_p(u), d\psi_p(v))$, at tangent vectors $u, v \in T_pM_1$.

$\xi_1, \dots, \xi_n$).[8] Define a symplectic form on $T^*\mathcal{U}$ by

$$\omega = \sum_{i=1}^{n} dx_i \wedge d\xi_i .$$

One can check that this ω is intrinsically defined by considering the 1-form on $T^*\mathcal{U}$,

$$\alpha = \sum_{i=1}^{n} \xi_i \, dx_i ,$$

which satisfies $\omega = -d\alpha$ and is coordinate-independent: in terms of the natural projection $\pi : M \to X$, $p = (x, \xi) \mapsto x$, the form α may be equivalently defined pointwise without coordinates by

$$\alpha_p = (d\pi_p)^* \xi \in T_p^* M,$$

where $(d\pi_p)^* : T_x^* X \to T_p^* M$ is the transpose of $d\pi_p$, that is, $\alpha_p(v) = \xi((d\pi_p)v)$ for $v \in T_p M$. Or yet, the form α is uniquely characterized by the property that $\mu^*\alpha = \mu$ for every 1-form $\mu : X \to T^*X$ (see Proposition 2.2). The 1-form α is the *tautological form* (or the *Liouville* 1*-form*) and the 2-form ω is the *canonical symplectic form* on T^*X. When referring to a cotangent bundle as a symplectic manifold, the symplectic structure is meant to be given by this canonical ω.

Let X_1 and X_2 be n-dimensional manifolds with cotangent bundles $M_1 = T^*X_1$ and $M_2 = T^*X_2$, and tautological 1-forms α_1 and α_2. Suppose that $f : X_1 \to X_2$ is a diffeomorphism. Then there is a natural diffeomorphism $f_\sharp : M_1 \to M_2$ which *lifts* f; namely, for $p_1 = (x_1, \xi_1) \in M_1$ we define

$$f_\sharp(p_1) = p_2 = (x_2, \xi_2), \quad \text{with} \begin{cases} x_2 = f(x_1) \in X_2 \quad \text{and} \\ \xi_1 = (df_{x_1})^* \xi_2 \in T_{x_1}^* X_1, \end{cases}$$

where $(df_{x_1})^* : T_{x_2}^* X_2 \xrightarrow{\simeq} T_{x_1}^* X_1$, so $f_\sharp|_{T_{x_1}^*}$ is the inverse map of $(df_{x_1})^*$.

PROPOSITION 1.3. *The lift $f_\sharp$ of a diffeomorphism $f : X_1 \to X_2$ pulls the tautological form on T^*X_2 back to the tautological form on T^*X_1, i.e., $(f_\sharp)^*\alpha_2 = \alpha_1$.*

---

[8] If an n-dimensional manifold X is described by coordinate charts $(\mathcal{U}, x_1, \dots, x_n)$ with $x_i : \mathcal{U} \to \mathbb{R}$, then, at any $x \in \mathcal{U}$, the differentials $(dx_i)_x$ form a basis of $T_x^* X$, inducing a map

$$\begin{aligned} T^*\mathcal{U} &\longrightarrow \mathbb{R}^{2n}, \\ (x, \xi) &\longmapsto (x_1, \dots, x_n, \xi_1, \dots, \xi_n), \end{aligned}$$

where $\xi_1, \dots, \xi_n \in \mathbb{R}$ are the corresponding coordinates of $\xi \in T_x^* X$: $\xi = \sum_{i=1}^{n} \xi_i (dx_i)_x$. Then $(T^*\mathcal{U}, x_1, \dots, x_n, \xi_1, \dots, \xi_n)$ is a coordinate chart for the cotangent bundle T^*X; the coordinates $x_1, \dots, x_n, \xi_1, \dots, \xi_n$ are called the *cotangent coordinates* associated to the coordinates $x_1, \dots, x_n$ on $\mathcal{U}$. One verifies that the transition functions on the overlaps are smooth, so T^*X is a $2n$-dimensional manifold.

PROOF. At $p_1 = (x_1, \xi_1) \in M_1$, the claimed identity says $(df_\sharp)^*_{p_1}(\alpha_2)_{p_2} = (\alpha_1)_{p_1}$, where $p_2 = f_\sharp(p_1)$, that is, $p_2 = (x_2, \xi_2)$ where $x_2 = f(x_1)$ and $(df_{x_1})^*\xi_2 = \xi_1$. This can be proved as follows:

$$
\begin{aligned}
(df_\sharp)^*_{p_1}(\alpha_2)_{p_2} &= (df_\sharp)^*_{p_1}(d\pi_2)^*_{p_2}\xi_2 && \text{by definition of } \alpha_2\\
&= \big(d(\pi_2 \circ f_\sharp)\big)^*_{p_1}\xi_2 && \text{by the chain rule}\\
&= \big(d(f \circ \pi_1)\big)^*_{p_1}\xi_2 && \text{because } \pi_2 \circ f_\sharp = f \circ \pi_1\\
&= (d\pi_1)^*_{p_1}(df)^*_{x_1}\xi_2 && \text{by the chain rule}\\
&= (d\pi_1)^*_{p_1}\xi_1 && \text{by definition of } f_\sharp\\
&= (\alpha_1)_{p_1} && \text{by definition of } \alpha_1. \qquad \square
\end{aligned}
$$

As a consequence of this naturality for the tautological form, a diffeomorphism of manifolds induces a canonical symplectomorphism of cotangent bundles:

COROLLARY 1.4. *The lift $f_\sharp : T^*X_1 \to T^*X_2$ of a diffeomorphism $f : X_1 \to X_2$ is a symplectomorphism for the canonical symplectic forms, i.e., $(f_\sharp)^*\omega_2 = \omega_1$.*

In terms of the group (under composition) of diffeomorphisms $\mathrm{Diff}(X)$ of a manifold X, and the *group of symplectomorphisms* $\mathrm{Sympl}(T^*X, \omega)$ of its cotangent bundle, we see that the injection $\mathrm{Diff}(X) \to \mathrm{Sympl}(T^*X, \omega)$, $f \mapsto f_\sharp$ is a group homomorphism. Clearly this is not surjective: for instance, consider the symplectomorphism $T^*X \to T^*X$ given by translation along cotangent fibers.

EXAMPLE. Let $X_1 = X_2 = S^1$. Then T^*S^1 is a cylinder $S^1 \times \mathbb{R}$. The canonical form is the area form $\omega = d\theta \wedge d\xi$. If $f : S^1 \to S^1$ is any diffeomorphism, then $f_\sharp : S^1 \times \mathbb{R} \to S^1 \times \mathbb{R}$ is a symplectomorphism, i.e., is an area-preserving diffeomorphism of the cylinder. Translation along the $\mathbb{R}$ direction is area-preserving but is not induced by a diffeomorphism of the base manifold S^1.

There is a criterion for which cotangent symplectomorphisms arise as lifts of diffeomorphisms in terms of the tautological form. First note the following feature of symplectic manifolds with *exact symplectic forms*. Let α be a 1-form on a manifold M such that $\omega = -d\alpha$ is symplectic. There exists a unique vector field v whose interior product with ω is α, i.e., $\imath_v\omega = -\alpha$. If $g : M \to M$ is a symplectomorphism that preserves α (that is, $g^*\alpha = \alpha$), then g commutes with the flow[9] of v, i.e., $(\exp tv) \circ g = g \circ (\exp tv)$. When

[9]For $p \in M$, $(\exp tv)(p)$ is the unique curve in M solving the initial value problem

$$
\begin{aligned}
&\tfrac{d}{dt}\big(\exp tv(p)\big) = v\big(\exp tv(p)\big),\\
&(\exp tv)(p)|_{t=0} = p
\end{aligned}
$$

for t in some neighborhood of 0. The one-parameter group of diffeomorphisms $\exp tv$ is called the *flow* of the vector field v.

$M = T^*X$ is the cotangent bundle of an arbitrary n-dimensional manifold X, and α is the tautological 1-form on M, the vector field v is just $\sum \xi_i \frac{\partial}{\partial \xi_i}$ with respect to a cotangent coordinate chart $(T^*\mathcal{U}, x_1, \ldots, x_n, \xi_1, \ldots, \xi_n)$. The flow $\exp tv$, $-\infty < t < \infty$, satisfies $(\exp tv)(x, \xi) = (x, e^t \xi)$, for every (x, ξ) in M.

THEOREM 1.5. *A symplectomorphism $g : T^*X \to T^*X$ is a lift of a diffeomorphism $f : X \to X$ if and only if it preserves the tautological form*: $g^*\alpha = \alpha$.

PROOF. By Proposition 1.3, a lift $f_\sharp : T^*X \to T^*X$ of a diffeomorphism $f : X \to X$ preserves the tautological form. Conversely, if g is a symplectomorphism of M that preserves α, then g preserves the cotangent fibration: by the observation above, $g(x, \xi) = (y, \eta) \Rightarrow g(x, \lambda\xi) = (y, \lambda\eta)$ for all $(x, \xi) \in M$ and $\lambda > 0$, and this must hold also for $\lambda \leqslant 0$ by the differentiability of g at $(x, 0)$. Therefore, there exists a diffeomorphism $f : X \to X$ such that $\pi \circ g = f \circ \pi$, where $\pi : M \to X$ is the projection map $\pi(x, \xi) = x$, and $g = f_\#$. □

The canonical form is natural also in the following way. Given a smooth function $h : X \to \mathbb{R}$, the diffeomorphism τ_h of $M = T^*X$ defined by $\tau_h(x, \xi) = (x, \xi + dh_x)$ turns out to be always a symplectomorphism. Indeed, if $\pi : M \to X$, $\pi(x, \xi) = x$, is the projection, we have $\tau_h^*\alpha = \alpha + \pi^* dh$, so that $\tau_h^*\omega = \omega$.

1.4. *Moser's trick*

There are other relevant notions of equivalence for symplectic manifolds[10] besides being symplectomorphic. Let M be a manifold with two symplectic forms ω_0, ω_1.

DEFINITION 1.6. The symplectic manifolds (M, ω_0) and (M, ω_1) are *strongly isotopic* if there is an isotopy $\rho_t : M \to M$ such that $\rho_1^*\omega_1 = \omega_0$. (M, ω_0) and (M, ω_1) are *deformation-equivalent* if there is a smooth family ω_t of symplectic forms joining ω_0 to ω_1. (M, ω_0) and (M, ω_1) are *isotopic* if they are deformation-equivalent and the de Rham cohomology class $[\omega_t]$ is independent of t.

Hence, being strongly isotopic implies being symplectomorphic, and being isotopic implies being deformation-equivalent. We also have that being strongly isotopic implies being isotopic, because, if $\rho_t : M \to M$ is an isotopy such that $\rho_1^*\omega_1 = \omega_0$, then $\omega_t := \rho_t^*\omega_1$ is a smooth family of symplectic forms joining ω_1 to ω_0 and $[\omega_t] = [\omega_1]$, $\forall t$, by the homotopy invariance of de Rham cohomology.

Moser [105] proved that, on a compact manifold, being isotopic implies being strongly isotopic (Theorem 1.7). McDuff showed that deformation-equivalence is indeed a necessary hypothesis: even if $[\omega_0] = [\omega_1] \in H^2(M; \mathbb{R})$, there are compact examples where (M, ω_0) and (M, ω_1) are not strongly isotopic; see Example 7.23 in [99]. In other words,

[10]Understanding these notions and the normal forms requires tools, such as isotopies (by *isotopy* we mean a smooth one-parameter family of diffeomorphisms starting at the identity, like the flow of a vector field), Lie derivative, tubular neighborhoods and the homotopy formula in de Rham theory, covered in differential geometry or differential topology texts.

fix $c \in H^2(M)$ and define S_c as the set of symplectic forms ω in M with $[\omega] = c$. On a compact manifold, all symplectic forms in the same path-connected component of S_c are symplectomorphic according to the Moser theorem, though there might be symplectic forms in different components of S_c that are not symplectomorphic.

THEOREM 1.7 (Moser). *Let M be a compact manifold with symplectic forms ω_0 and ω_1. Suppose that ω_t, $0 \leqslant t \leqslant 1$, is a smooth family of symplectic forms joining ω_0 to ω_1 with cohomology class $[\omega_t]$ independent of t. Then there exists an isotopy $\rho : M \times \mathbb{R} \to M$ such that $\rho_t^* \omega_t = \omega_0$, $0 \leqslant t \leqslant 1$.*

Moser applied an extremely useful argument, known as *Moser's trick*, starting with the following observation. If there existed an isotopy $\rho : M \times \mathbb{R} \to M$ such that $\rho_t^* \omega_t = \omega_0$, $0 \leqslant t \leqslant 1$, in terms of the associated time-dependent vector field

$$v_t := \frac{d\rho_t}{dt} \circ \rho_t^{-1}, \quad t \in \mathbb{R},$$

we would then have for all $0 \leqslant t \leqslant 1$ that

$$0 = \frac{d}{dt}(\rho_t^* \omega_t) = \rho_t^* \left(\mathcal{L}_{v_t} \omega_t + \frac{d\omega_t}{dt} \right) \quad \Longleftrightarrow \quad \mathcal{L}_{v_t} \omega_t + \frac{d\omega_t}{dt} = 0.$$

Conversely, the existence of a smooth time-dependent vector field v_t, $t \in \mathbb{R}$, satisfying the last equation is enough to produce by integration (since M is compact) the desired isotopy $\rho : M \times \mathbb{R} \to M$ satisfying $\rho_t^* \omega_t = \rho_0^* \omega_0 = \omega_0$, for all t. So everything boils down to solving the equation $\mathcal{L}_{v_t} \omega_t + \frac{d\omega_t}{dt} = 0$ for v_t.

PROOF. By the cohomology assumption that $[\frac{d}{dt}\omega_t] = 0$, there exists a *smooth* family of 1-forms μ_t such that

$$\frac{d\omega_t}{dt} = d\mu_t, \quad 0 \leqslant t \leqslant 1.$$

The argument involves the Poincaré lemma for compactly-supported forms, together with the Mayer–Vietoris sequence in order to use induction on the number of charts in a good cover of M; for a sketch, see page 95 in [99]. In the simplest case where $\omega_t = (1-t)\omega_0 + t\omega_1$ with $[\omega_0] = [\omega_1]$, we have that $\frac{d\omega_t}{dt} = \omega_1 - \omega_0 = d\mu$ is exact.

The nondegeneracy assumption on ω_t, guarantees that we can pointwise solve the equation, known as *Moser's equation*,

$$\imath_{v_t} \omega_t + \mu_t = 0$$

to obtain a unique smooth family of vector fields v_t, $0 \leqslant t \leqslant 1$. Extend v_t to all $t \in \mathbb{R}$. Thanks to the compactness of M, the vector fields v_t generate an isotopy ρ satisfying

$\frac{d\rho_t}{dt} = v_t \circ \rho_t$. Then we indeed have

$$\frac{d}{dt}(\rho_t^* \omega_t) = \rho_t^* \left(\mathcal{L}_{v_t} \omega_t + \frac{d\omega_t}{dt} \right) = \rho_t^* (d\imath_{v_t} \omega_t + d\mu_t) = \rho_t^* d(\imath_{v_t} \omega_t + \mu_t) = 0,$$

where we used Cartan's magic formula in $\mathcal{L}_{v_t} \omega_t = d\imath_{v_t} \omega_t + \imath_{v_t} d\omega_t$. □

EXAMPLE. On a compact oriented 2-dimensional manifold M, a symplectic form is just an area form. Let ω_0 and ω_1 be two area forms on M. If $[\omega_0] = [\omega_1]$, i.e., ω_0 and ω_1 give the same total area, then any convex combination of them is symplectic (because they induce the same orientation), and there is an isotopy $\varphi_t : M \to M$, $t \in [0, 1]$, such that $\varphi_1^* \omega_0 = \omega_1$. Therefore, up to strong isotopy, there is a unique symplectic representative in each nonzero 2-cohomology class of M.

On a *noncompact* manifold, given v_t, we would need to check the existence for $0 \leqslant t \leqslant 1$ of an isotopy ρ_t solving the differential equation $\frac{d\rho_t}{dt} = v_t \circ \rho_t$.

1.5. *Darboux and Moser theorems*

By a *submanifold* of a manifold M we mean either a manifold X with a *closed embedding*[11] $i : X \hookrightarrow M$, or an *open submanifold* (i.e., an open subset of M).

Given a $2n$-dimensional manifold M, a k-dimensional submanifold X, neighborhoods $\mathcal{U}_0, \mathcal{U}_1$ of X, and symplectic forms ω_0, ω_1 on $\mathcal{U}_0, \mathcal{U}_1$, we would like to know whether there exists a *local symplectomorphism preserving* X, i.e., a diffeomorphism $\varphi : \mathcal{U}_0 \to \mathcal{U}_1$ with $\varphi^* \omega_1 = \omega_0$ and $\varphi(X) = X$. Moser's Theorem 1.7 addresses the case where $X = M$. At the other extreme, when X is just one point, there is the classical Darboux theorem (Theorem 1.9). In general, we have:

THEOREM 1.8 (Moser theorem—relative version). *Let ω_0 and ω_1 be symplectic forms on a manifold M, and X a compact submanifold of M. Suppose that the forms coincide, $\omega_0|_p = \omega_1|_p$, at all points $p \in X$. Then there exist neighborhoods $\mathcal{U}_0$ and $\mathcal{U}_1$ of X in M, and a diffeomorphism $\varphi : \mathcal{U}_0 \to \mathcal{U}_1$ such that $\varphi^* \omega_1 = \omega_0$ and φ restricted to X is the identity map.*

PROOF. Pick a tubular neighborhood $\mathcal{U}_0$ of X. The 2-form $\omega_1 - \omega_0$ is closed on $\mathcal{U}_0$, and satisfies $(\omega_1 - \omega_0)_p = 0$ at all $p \in X$. By the homotopy formula on the tubular neighborhood, there exists a 1-form μ on $\mathcal{U}_0$ such that $\omega_1 - \omega_0 = d\mu$ and $\mu_p = 0$ at all $p \in X$. Consider the family $\omega_t = (1-t)\omega_0 + t\omega_1 = \omega_0 + t\, d\mu$ of closed 2-forms on $\mathcal{U}_0$. Shrinking $\mathcal{U}_0$ if necessary, we can assume that ω_t is symplectic for $t \in [0, 1]$, as nondegeneracy is an open property. Solve Moser's equation, $\imath_{v_t} \omega_t = -\mu$, for v_t By integration, shrinking $\mathcal{U}_0$ again if necessary, there exists a local isotopy $\rho : \mathcal{U}_0 \times [0, 1] \to M$ with $\rho_t^* \omega_t = \omega_0$, for all $t \in [0, 1]$. Since $v_t|_X = 0$, we have $\rho_t|_X = \mathrm{id}_X$. Set $\varphi = \rho_1, \mathcal{U}_1 = \rho_1(\mathcal{U}_0)$. □

[11] A *closed embedding* is a *proper* injective immersion. A map is *proper* when its preimage of a compact set is always compact.

THEOREM 1.9 (Darboux). *Let (M, ω) be a symplectic manifold, and let p be any point in M. Then we can find a chart $(\mathcal{U}, x_1, \ldots, x_n, y_1, \ldots y_n)$ centered at p where*

$$\omega = \sum_{i=1}^{n} dx_i \wedge dy_i.$$

Such a coordinate chart $(\mathcal{U}, x_1, \ldots, x_n, y_1, \ldots, y_n)$ is called a *Darboux chart*, and the corresponding coordinates are called *Darboux coordinates*.

The classical proof of Darboux's theorem is by induction on the dimension of the manifold [2], in the spirit of the argument for a symplectic basis (Section 1.1). The proof below, using Moser's theorem, was first provided by Weinstein [136].

PROOF. Apply Moser's relative theorem to $X = \{p\}$. More precisely, use any symplectic basis for (T_pM, ω_p) to construct coordinates $(x'_1, \ldots, x'_n, y'_1, \ldots y'_n)$ centered at p and valid on some neighborhood $\mathcal{U}'$, so that $\omega_p = \sum dx'_i \wedge dy'_i|_p$. There are two symplectic forms on $\mathcal{U}'$: the given $\omega_0 = \omega$ and $\omega_1 = \sum dx'_i \wedge dy'_i$. By Theorem 1.8, there are neighborhoods $\mathcal{U}_0$ and $\mathcal{U}_1$ of p, and a diffeomorphism $\varphi : \mathcal{U}_0 \to \mathcal{U}_1$ such that $\varphi(p) = p$ and $\varphi^*(\sum dx'_i \wedge dy'_i) = \omega$. Since $\varphi^*(\sum dx'_i \wedge dy'_i) = \sum d(x'_i \circ \varphi) \wedge d(y'_i \circ \varphi)$, we simply set new coordinates $x_i = x'_i \circ \varphi$, $y_i = y'_i \circ \varphi$. □

Darboux's theorem is easy in the 2-dimensional case. Being closed ω is locally exact, $\omega = d\alpha$. Every nonvanishing 1-form on a surface can be written locally as $\alpha = g\,dh$ for suitable functions g, h, where h is a coordinate on the local leaf space of the kernel foliation of α. The form $\omega = dg \wedge dh$ is nondegenerate if and only if (g, h) is a local diffeomorphism. By the way, transversality shows that the normal form for a *generic*[12] 2-form is $x\,dx \wedge dy$ near a point where it is degenerate.

1.6. *Symplectic submanifolds*

Moser's argument permeates many other proofs, including those of the next two results regarding *symplectic submanifolds*. Let (M, ω) be a symplectic manifold.

DEFINITION 1.10. A *symplectic submanifold* of (M, ω) is a submanifold X of M where, at each $p \in X$, the space T_pX is a symplectic subspace of (T_pM, ω_p).

If $i : X \hookrightarrow M$ is the inclusion of a symplectic submanifold X, then the restriction of ω to X is a symplectic form, so that $(X, i^*\omega)$ is itself a symplectic manifold.

Let X be a symplectic submanifold of (M, ω). At each $p \in X$, we have $T_pM = T_pX \oplus (T_pX)^{\omega_p}$ (Section 1.1), so the map $(T_pX)^{\omega_p} \to T_pM/T_pX$ is an isomorphism. This canonical identification of the *normal space* of X at p, $N_pX := T_pM/T_pX$, with the symplectic orthogonal $(T_pX)^{\omega_p}$, yields a canonical identification of the *normal bundle* NX

[12] *Generic* here means that the subset of those 2-forms having this behavior is open, dense and invariant under diffeomorphisms of the manifold.

with the *symplectic vector bundle* $(TX)^\omega$. A *symplectic vector bundle* is a vector bundle $E \to X$ equipped with a smooth[13] field Ω of fiberwise nondegenerate skew-symmetric bilinear maps $\Omega_p : E_p \times E_p \to \mathbb{R}$. The *symplectic normal bundle* is the normal bundle of a symplectic submanifold, with the symplectic structure induced by orthogonals. The next theorem, due to Weinstein [136], states that a neighborhood of a symplectic submanifold X is determined by X and (the isomorphism class of) its symplectic normal bundle.

THEOREM 1.11 (Symplectic neighborhood theorem). *Let (M_0, ω_0), (M_1, ω_1) be symplectic manifolds with diffeomorphic compact symplectic submanifolds X_0, X_1. Let $i_0 : X_0 \hookrightarrow M_0$, $i_1 : X_1 \hookrightarrow M_1$ be their inclusions. Suppose there is an isomorphism $\tilde{\phi} : NX_0 \to NX_1$ of the corresponding symplectic normal bundles covering a symplectomorphism $\phi : (X_0, i_0^*\omega_0) \to (X_1, i_1^*\omega_1)$. Then there exist neighborhoods $\mathcal{U}_0 \subset M_0$, $\mathcal{U}_1 \subset M_1$ of X_0, X_1 and a symplectomorphism $\varphi : \mathcal{U}_0 \to \mathcal{U}_1$ extending ϕ such that the restriction of $d\varphi$ to the normal bundle NX_0 is $\tilde{\phi}$.*

As first noted by Thurston [131], the form $\Omega + \pi^*\omega_X$ is symplectic in some neighborhood of the zero section in NX, where $\pi : NX \to X$ is the bundle projection and ω_X is the restriction of ω to X. Therefore, *a compact symplectic submanifold X always admits a tubular neighborhood in the ambient (M, ω) symplectomorphic to a tubular neighborhood of the zero section in the symplectic normal bundle NX.*

PROOF. By the Whitney extension theorem[14] there exist neighborhoods $\mathcal{U}_0 \subset M_0$ and $\mathcal{U}_1 \subset M_1$ of X_0 and X_1, and a diffeomorphism $h : \mathcal{U}_0 \to \mathcal{U}_1$ such that $h \circ i_0 = i_1 \circ \phi$ and the restriction of dh to the normal bundle NX_0 is the given $\tilde{\phi}$. Hence ω_0 and $h^*\omega_1$ are two symplectic forms on $\mathcal{U}_0$ which coincide at all points $p \in X_0$. The result now follows from Moser's relative theorem (Theorem 1.8). □

Carefully combining Moser's argument with the existence of an ambient isotopy that produces a given deformation of a compact submanifold, we can show:

THEOREM 1.12. *Let X_t, $t \in [0, 1]$, be a (smooth) family of compact symplectic submanifolds of a compact symplectic manifold (M, ω). Then there exists an isotopy $\rho : M \times \mathbb{R} \to M$ such that for all $t \in [0, 1]$ we have $\rho_t^*\omega = \omega$ and $\rho_t(X_0) = X_t$.*

Inspired by complex geometry, Donaldson [32] proved the following theorem on the existence of symplectic submanifolds. A major consequence is the characterization of symplectic manifolds in terms of *Lefschetz pencils*; see Section 4.6.

---

[13] Smoothness means that, for any pair of (smooth) sections u and v of E, the real-valued function $\Omega(u, v) : X \to \mathbb{R}$ given by evaluation at each point is smooth.

[14] **Whitney extension theorem.** *Let M be a manifold and X a submanifold of M. Suppose that at each $p \in X$ we are given a linear isomorphism $L_p : T_pM \xrightarrow{\simeq} T_pM$ such that $L_p|_{T_pX} = \mathrm{Id}_{T_pX}$ and L_p depends smoothly on p. Then there exists an embedding $h : \mathcal{N} \to M$ of some neighborhood $\mathcal{N}$ of X in M such that $h|_X = \mathrm{id}_X$ and $dh_p = L_p$ for all $p \in X$.* A proof relies on a tubular neighborhood model.

THEOREM 1.13 (Donaldson). *Let (M, ω) be a compact symplectic manifold. Assume that the cohomology class $[\omega]$ is integral, i.e., lies in $H^2(M; \mathbb{Z})$. Then, for every sufficiently large integer k, there exists a connected codimension-2 symplectic submanifold X representing the Poincaré dual of the integral cohomology class $k[\omega]$.*

Under the same hypotheses, Auroux extended this result to show that given $\alpha \in H_{2m}(M; \mathbb{Z})$ there exist positive $k, \ell \in \mathbb{Z}$ such that $k\mathrm{PD}[\omega^{n-m}] + \ell\alpha$ is realized by a $2m$-dimensional symplectic submanifold.

2. Lagrangian submanifolds

2.1. *First Lagrangian submanifolds*

Let (M, ω) be a symplectic manifold.

DEFINITION 2.1. A submanifold X of (M, ω) is *Lagrangian* (respectively, *isotropic* and *coisotropic*) if, at each $p \in X$, the space T_pX is a Lagrangian (respectively, isotropic and coisotropic) subspace of (T_pM, ω_p).

If $i : X \hookrightarrow M$ is the inclusion map, then X is a *Lagrangian submanifold* if and only if $i^*\omega = 0$ and $\dim X = \frac{1}{2} \dim M$.

The problem of embedding[15] a compact manifold as a Lagrangian submanifold of a given symplectic manifold is often global. For instance, Gromov [64] proved that there can be no Lagrangian spheres in $(\mathbb{C}^n, \omega_0)$, except for the circle in $\mathbb{C}^2$, and more generally no compact *exact Lagrangian* submanifolds, in the sense that $\alpha_0 = \sum y_j\, dx_j$ restricts to an exact 1-form. The argument uses *pseudoholomorphic curves* (Section 3.6). Yet there are *immersed* Lagrangian spheres (Section 2.7). More recently were found topological and geometrical constraints on manifolds that admit Lagrangian embeddings into *compact* symplectic manifolds; see, for instance, [16,17,115].

EXAMPLES.

1. Any 1-dimensional submanifold of a symplectic surface is Lagrangian (because a 1-dimensional subspace of a symplectic vector space is always isotropic).

   Therefore, any product of n embedded curves arises as a Lagrangian submanifold of (a neighborhood of zero in) the prototype $(\mathbb{R}^{2n}, \omega_0)$. In particular, a *torus* $\mathbb{T}^n = S^1 \times \cdots \times S^1$ can be embedded as a Lagrangian submanifold of any $2n$-dimensional symplectic manifold, by Darboux's theorem (Theorem 1.9).
2. Let $M = T^*X$ be the cotangent bundle of a manifold X. With respect to a cotangent coordinate chart $(T^*U, x_1, \dots, x_n, \xi_1, \dots, \xi_n)$, the tautological form is $\alpha = \sum \xi_i\, dx_i$ and the canonical form is $\omega = -d\alpha = \sum dx_i \wedge d\xi_i$.

   The *zero section* $X_0 := \{(x, \xi) \in T^*X \mid \xi = 0 \text{ in } T_x^*X\}$ is an n-dimensional submanifold of T^*X whose intersection with T^*U is given by the equations $\xi_1 = \cdots =$

[15] An *embedding* is an immersion that is a homeomorphism onto its image.

$\xi_n = 0$. Clearly α vanishes on $X_0 \cap T^*U$. Hence, if $i_0 : X_0 \hookrightarrow T^*X$ is the inclusion map, we have $i_0^*\omega = i_0^* d\alpha = 0$, and so X_0 is Lagrangian.

A *cotangent fiber* $T^*_{x_0}X$ is an n-dimensional submanifold of T^*X given by the equations $x_i = (x_0)_i$, $i = 1, \ldots, n$, on T^*U. Since the x_i's are constant, the form α vanishes identically, and $T^*_{x_0}X$ is a Lagrangian submanifold.

Let X_μ be (the image of) an arbitrary section, that is, an n-dimensional submanifold of T^*X of the form $X_\mu = \{(x, \mu_x) \mid x \in X,\ \mu_x \in T^*_x X\}$, where the covector μ_x depends smoothly on x, so $\mu : X \to T^*X$ is a de Rham 1-form. We will investigate when such an X_μ is Lagrangian. Relative to the inclusion $i : X_\mu \hookrightarrow T^*X$ and the cotangent projection $\pi : T^*X \to X$, these X_μ's are exactly the submanifolds for which $\pi \circ i : X_\mu \to X$ is a diffeomorphism.

PROPOSITION 2.2. *The tautological* 1*-form α on T^*X satisfies $\mu^*\alpha = \mu$, for any* 1*-form $\mu : X \to T^*X$.*

PROOF. Denote by $s_\mu : X \to T^*X$, $x \mapsto (x, \mu_x)$, the 1-form μ regarded exclusively as a map. From the definition, $\alpha_p = (d\pi_p)^*\xi$ at $p = (x, \xi) \in M$. For $p = s_\mu(x) = (x, \mu_x)$, we have $\alpha_p = (d\pi_p)^*\mu_x$. Then, since $\pi \circ s_\mu = \mathrm{id}_X$, we have

$$\left(s_\mu^*\alpha\right)_x = (ds_\mu)_x^*\alpha_p = (ds_\mu)_x^*(d\pi_p)^*\mu_x = \left(d(\pi \circ s_\mu)\right)_x^*\mu_x = \mu_x. \qquad \square$$

The map $s_\mu : X \to T^*X$, $s_\mu(x) = (x, \mu_x)$ is an embedding with image the section X_μ. The diffeomorphism $\tau : X \to X_\mu$, $\tau(x) := (x, \mu_x)$, satisfies $i \circ \tau = s_\mu$.

PROPOSITION 2.3. *The sections of T^*X that are Lagrangian are those corresponding to closed* 1*-forms on X.*

PROOF. Using the previous notation, the condition of X_μ being Lagrangian becomes: $i^*\,d\alpha = 0 \Leftrightarrow \tau^*i^*\,d\alpha = 0 \Leftrightarrow s_\mu^*\,d\alpha = 0 \Leftrightarrow d(s_\mu^*\alpha) = 0 \Leftrightarrow d\mu = 0$. $\square$

When $\mu = dh$ for some $h \in C^\infty(X)$, such a primitive h is called a *generating function* for the Lagrangian submanifold X_μ. Two functions generate the same Lagrangian submanifold if and only if they differ by a locally constant function. When X is simply connected, or at least $H^1_{\mathrm{deRham}}(X) = 0$, every Lagrangian X_μ admits a generating function.

Besides the cotangent fibers, there are lots of Lagrangian submanifolds of T^*X not covered by the description in terms of closed 1-forms. Let S be any submanifold of an n-dimensional manifold X. The *conormal space* of S at $x \in S$ is

$$N_x^*S = \left\{\xi \in T_x^*X \mid \xi(v) = 0 \text{ for all } v \in T_xS\right\}.$$

The *conormal bundle* of S is $N^*S = \{(x, \xi) \in T^*X \mid x \in S,\ \xi \in N_x^*S\}$. This is an n-dimensional submanifold of T^*X. In particular, taking $S = \{x\}$ to be one point, the conormal bundle is the corresponding cotangent fiber T_x^*X. Taking $S = X$, the conormal bundle is the zero section X_0 of T^*X.

PROPOSITION 2.4. *If $i : N^*S \hookrightarrow T^*X$ is the inclusion of the conormal bundle of a submanifold $S \subset X$, and α is the tautological 1-form on T^*X, then $i^*\alpha = 0$.*

PROOF. Let $(\mathcal{U}, x_1, \ldots, x_n)$ be a coordinate chart on X adapted to S, so that $\mathcal{U} \cap S$ is described by $x_{k+1} = \cdots = x_n = 0$. Let $(T^*\mathcal{U}, x_1, \ldots, x_n, \xi_1, \ldots, \xi_n)$ be the associated cotangent coordinate chart. The submanifold $N^*S \cap T^*\mathcal{U}$ is described by $x_{k+1} = \cdots = x_n = 0$ and $\xi_1 = \cdots = \xi_k = 0$. Since $\alpha = \sum \xi_i \, dx_i$ on $T^*\mathcal{U}$, we conclude that, at $p \in N^*S$,

$$(i^*\alpha)_p = \alpha_p|_{T_p(N^*S)} = \sum_{i>k} \xi_i \, dx_i \Big|_{\operatorname{span}\{\frac{\partial}{\partial x_i},\ i \leqslant k\}} = 0. \qquad \square$$

COROLLARY 2.5. *For any submanifold S of X, the conormal bundle N^*S is a Lagrangian submanifold of T^*X.*

2.2. *Lagrangian neighborhood theorem*

Weinstein [136] proved that, if a compact submanifold X is Lagrangian with respect to two symplectic forms ω_0 and ω_1, then the conclusion of the Moser relative theorem (Theorem 1.8) still holds. We need some algebra for the Weinstein theorem.

Suppose that U, W are n-dimensional vector spaces, and $\Omega : U \times W \to \mathbb{R}$ is a bilinear pairing; the map Ω gives rise to a linear map $\tilde{\Omega} : U \to W^*$, $\tilde{\Omega}(u) = \Omega(u, \cdot)$. Then Ω is nondegenerate if and only if $\tilde{\Omega}$ is bijective.

PROPOSITION 2.6. *Let (V, Ω) be a symplectic vector space, U a Lagrangian subspace of (V, Ω), and W any vector space complement to U, not necessarily Lagrangian. Then from W we can canonically build a Lagrangian complement to U.*

PROOF. From Ω we get a nondegenerate pairing $\Omega' : U \times W \to \mathbb{R}$, so $\tilde{\Omega}' : U \to W^*$ is bijective. We look for a Lagrangian complement to U of the form $W' = \{w + Aw \mid w \in W\}$ for some linear map $A : W \to U$. For W' to be Lagrangian we need that $\Omega(w_1, w_2) = \tilde{\Omega}'(Aw_2)(w_1) - \tilde{\Omega}'(Aw_1)(w_2)$. Let $A' = \tilde{\Omega}' \circ A$, and look for A' such that $\Omega(w_1, w_2) = A'(w_2)(w_1) - A'(w_1)(w_2)$ for all $w_1, w_2 \in W$. The canonical choice is $A'(w) = -\frac{1}{2}\Omega(w, \cdot)$. Set $A = (\tilde{\Omega}')^{-1} \circ A'$. $\square$

PROPOSITION 2.7. *Let V be a vector space, let Ω_0 and Ω_1 be symplectic forms on V, let U be a subspace of V Lagrangian for Ω_0 and Ω_1, and let W be any complement to U in V. Then from W we can canonically construct a linear isomorphism $L : V \xrightarrow{\simeq} V$ such that $L|_U = \mathrm{Id}_U$ and $L^*\Omega_1 = \Omega_0$.*

PROOF. By Proposition 2.6, from W we canonically obtain complements W_0 and W_1 to U in V such that W_0 is Lagrangian for Ω_0 and W_1 is Lagrangian for Ω_1. The nondegenerate bilinear pairings $\Omega_i : W_i \times U \to \mathbb{R}$, $i = 0, 1$, give isomorphisms $\tilde{\Omega}_i : W_i \xrightarrow{\simeq} U^*$, $i = 0, 1$, respectively. Let $B : W_0 \to W_1$ be the linear map satisfying $\tilde{\Omega}_1 \circ B = \tilde{\Omega}_0$, i.e., $\Omega_0(w_0, u) =$

$\Omega_1(Bw_0, u)$, $\forall w_0 \in W_0$, $\forall u \in U$. Let $L := \mathrm{Id}_U \oplus B : U \oplus W_0 \to U \oplus W_1$ be the extension of B to the rest of V by setting it to be the identity on U. It satisfies:

$$\begin{aligned}(L^*\Omega_1)(u \oplus w_0, u' \oplus w_0') &= \Omega_1(u \oplus Bw_0, u' \oplus Bw_0') \\ &= \Omega_1(u, Bw_0') + \Omega_1(Bw_0, u') \\ &= \Omega_0(u, w_0') + \Omega_0(w_0, u') \\ &= \Omega_0(u \oplus w_0, u' \oplus w_0'). \qquad \square\end{aligned}$$

THEOREM 2.8 (Weinstein Lagrangian neighborhood theorem). *Let M be a $2n$-dimensional manifold, X a compact n-dimensional submanifold, $i : X \hookrightarrow M$ the inclusion map, and ω_0 and ω_1 symplectic forms on M such that $i^*\omega_0 = i^*\omega_1 = 0$, i.e., X is a Lagrangian submanifold of both (M, ω_0) and (M, ω_1). Then there exist neighborhoods $\mathcal{U}_0$ and $\mathcal{U}_1$ of X in M and a diffeomorphism $\varphi : \mathcal{U}_0 \to \mathcal{U}_1$ such that $\varphi^*\omega_1 = \omega_0$ and φ is the identity on X, i.e., $\varphi(p) = p$, $\forall p \in X$.*

PROOF. Put a Riemannian metric g on M. Fix $p \in X$, and let $V = T_pM$, $U = T_pX$ and $W = U^\perp$, the orthocomplement of U in V relative to the inner product $g_p(\cdot,\cdot)$. Since $i^*\omega_0 = i^*\omega_1 = 0$, the subspace U is Lagrangian for both $(V, \omega_0|_p)$ and $(V, \omega_1|_p)$. By Proposition 2.7, we canonically get from $U^\perp$ a linear isomorphism $L_p : T_pM \to T_pM$ depending smoothly on p, such that $L_p|_{T_pX} = \mathrm{Id}_{T_pX}$ and $L_p^*\omega_1|_p = \omega_0|_p$. By the Whitney extension theorem (Section 1.5), there exist a neighborhood $\mathcal{N}$ of X and an embedding $h : \mathcal{N} \hookrightarrow M$ with $h|_X = \mathrm{id}_X$ and $dh_p = L_p$ for $p \in X$. Hence, at any $p \in X$, we have $(h^*\omega_1)_p = (dh_p)^*\omega_1|_p = L_p^*\omega_1|_p = \omega_0|_p$. Applying the Moser relative theorem (Theorem 1.8) to ω_0 and $h^*\omega_1$, we find a neighborhood $\mathcal{U}_0$ of X and an embedding $f : \mathcal{U}_0 \to \mathcal{N}$ such that $f|_X = \mathrm{id}_X$ and $f^*(h^*\omega_1) = \omega_0$ on $\mathcal{U}_o$. Set $\varphi = h \circ f$ and $\mathcal{U}_1 = \varphi(\mathcal{U}_0)$. $\square$

Theorem 2.8 has the following generalization. For a proof see, for instance, either of [61,70,139].

THEOREM 2.9 (Coisotropic embedding theorem). *Let M be a manifold of dimension $2n$, X a submanifold of dimension $k \geqslant n$, $i : X \hookrightarrow M$ the inclusion, and ω_0 and ω_1 symplectic forms on M, such that $i^*\omega_0 = i^*\omega_1$ and X is coisotropic for both (M, ω_0) and (M, ω_1). Then there exist neighborhoods $\mathcal{U}_0$ and $\mathcal{U}_1$ of X in M and a diffeomorphism $\varphi : \mathcal{U}_0 \to \mathcal{U}_1$ such that $\varphi^*\omega_1 = \omega_0$ and $\varphi|_X = \mathrm{id}_X$.*

2.3. *Weinstein tubular neighborhood theorem*

Let (V, Ω) be a symplectic linear space, and let U be a Lagrangian subspace. Then there is a canonical nondegenerate bilinear pairing $\Omega' : V/U \times U \to \mathbb{R}$ defined by $\Omega'([v], u) = \Omega(v, u)$ where $[v]$ is the equivalence class of v in V/U. Consequently, we get a canonical isomorphism $\tilde{\Omega}' : V/U \to U^*$, $\tilde{\Omega}'([v]) = \Omega'([v], \cdot)$.

In particular, if (M,ω) is a symplectic manifold, and X is a Lagrangian submanifold, then T_pX is a Lagrangian subspace of (T_pM,ω_p) for each $p\in X$ and there is a canonical identification of the *normal space* of X at p, $N_pX := T_pM/T_pX$, with the cotangent fiber T_p^*X. Consequently the normal bundle NX and the cotangent bundle T^*X are canonically identified.

THEOREM 2.10 (Weinstein tubular neighborhood theorem). *Let (M,ω) be a symplectic manifold, X a compact Lagrangian submanifold, ω_0 the canonical symplectic form on T^*X, $i_0:X\hookrightarrow T^*X$ the Lagrangian embedding as the zero section, and $i:X\hookrightarrow M$ the Lagrangian embedding given by inclusion. Then there are neighborhoods $\mathcal{U}_0$ of X in T^*X, $\mathcal{U}$ of X in M, and a diffeomorphism $\varphi:\mathcal{U}_0\to\mathcal{U}$ such that $\varphi^*\omega=\omega_0$ and $\varphi\circ i_0=i$.*

PROOF. By the standard tubular neighborhood theorem[16] and since $NX\simeq T^*X$ are canonically identified, we can find a neighborhood $\mathcal{N}_0$ of X in T^*X, a neighborhood $\mathcal{N}$ of X in M, and a diffeomorphism $\psi:\mathcal{N}_0\to\mathcal{N}$ such that $\psi\circ i_0=i$. Let ω_0 be the canonical form on T^*X and $\omega_1=\psi^*\omega$. The submanifold X is Lagrangian for both of these symplectic forms on $\mathcal{N}_0$. By the Weinstein Lagrangian neighborhood theorem (Theorem 2.8), there exist neighborhoods $\mathcal{U}_0$ and $\mathcal{U}_1$ of X in $\mathcal{N}_0$ and a diffeomorphism $\theta:\mathcal{U}_0\to\mathcal{U}_1$ such that $\theta^*\omega_1=\omega_0$ and $\theta\circ i_0=i_0$. Take $\varphi=\psi\circ\theta$ and $\mathcal{U}=\varphi(\mathcal{U}_0)$. Then $\varphi^*\omega=\theta^*\psi^*\omega=\theta^*\omega_1=\omega_0$. □

Theorem 2.10 classifies compact Lagrangian embeddings: up to local symplectomorphism, the set of Lagrangian embeddings is the set of embeddings of manifolds into their cotangent bundles as zero sections.

The classification of compact *isotropic* embeddings is also due to Weinstein in [137, 139]. An *isotropic embedding* of a manifold X into a symplectic manifold (M,ω) is a closed embedding $i:X\hookrightarrow M$ such that $i^*\omega=0$. Weinstein showed that neighborhood equivalence of isotropic embeddings is in one-to-one correspondence with isomorphism classes of symplectic vector bundles.

The classification of compact *coisotropic* embeddings is due to Gotay [61]. A *coisotropic embedding* of a manifold X carrying a closed 2-form α of constant rank into a symplectic manifold (M,ω) is an embedding $i:X\hookrightarrow M$ such that $i^*\omega=\alpha$ and $i(X)$ is coisotropic as a submanifold of M. Let E be the *characteristic distribution* of a closed form α of constant rank on X, i.e., E_p is the kernel of α_p at $p\in X$. Gotay showed that then the total space E^* carries a symplectic structure in a neighborhood of the zero section, such that X embeds coisotropically onto this zero section and, moreover, every coisotropic embedding is equivalent to this in some neighborhood of the zero section.

---

[16]**Tubular neighborhood theorem.** *Let M be a manifold, X a submanifold, NX the normal bundle of X in M, $i_0:X\hookrightarrow NX$ the zero section, and $i:X\hookrightarrow M$ the inclusion. Then there are neighborhoods $\mathcal{U}_0$ of X in NX, $\mathcal{U}$ of X in M and a diffeomorphism $\psi:\mathcal{U}_0\to\mathcal{U}$ such that $\psi\circ i_0=i$.* This theorem can be proved with the exponential map using a Riemannian metric; see, for instance, [120].

2.4. *Application to symplectomorphisms*

Let (M_1, ω_1) and (M_2, ω_2) be two $2n$-dimensional symplectic manifolds. Given a diffeomorphism $f : M_1 \xrightarrow{\simeq} M_2$, there is a way to express the condition of f being a symplectomorphism in terms of a certain submanifold being Lagrangian. Consider the two projection maps $\mathrm{pr}_i : M_1 \times M_2 \to M_i$, $(p_1, p_2) \mapsto p_i$, $i = 1, 2$. The *twisted product form* on $M_1 \times M_2$ is the symplectic[17] form

$$\tilde{\omega} = (\mathrm{pr}_1)^* \omega_1 - (\mathrm{pr}_2)^* \omega_2.$$

PROPOSITION 2.11. *A diffeomorphism $f : M_1 \xrightarrow{\simeq} M_2$ is a symplectomorphism if and only if the graph of f is a Lagrangian submanifold of $(M_1 \times M_2, \tilde{\omega})$.*

PROOF. The graph of f is the $2n$-dimensional submanifold $\mathrm{Graph}\, f = \{(p, f(p)) \mid p \in M_1\} \subseteq M_1 \times M_2$, which is the image of the embedding $\gamma : M_1 \to M_1 \times M_2$, $p \mapsto (p, f(p))$. We have $\gamma^* \tilde{\omega} = \gamma^* \mathrm{pr}_1^* \omega_1 - \gamma^* \mathrm{pr}_2^* \omega_2 = (\mathrm{pr}_1 \circ \gamma)^* \omega_1 - (\mathrm{pr}_2 \circ \gamma)^* \omega_2$, and $\mathrm{pr}_1 \circ \gamma$ is the identity map on M_1 whereas $\mathrm{pr}_2 \circ \gamma = f$. So $\mathrm{Graph}\, f$ is Lagrangian, i.e., $\gamma^* \tilde{\omega} = 0$, if and only if $f^* \omega_2 = \omega_1$, i.e., f is a symplectomorphism. □

Lagrangian submanifolds of $(M_1 \times M_2, \tilde{\omega})$ are called *canonical relations*, when viewed as morphisms between (M_1, ω_1) and (M_2, ω_2), even if $\dim M_1 \neq \dim M_2$. Under a reasonable assumption, there is a notion of composition [137].

Take $M_1 = M_2 = M$ and suppose that (M, ω) is a *compact* symplectic manifold and $f \in \mathrm{Sympl}(M, \omega)$. The graphs $\mathrm{Graph}\, f$ and Δ, of f and of the identity map $\mathrm{id} : M \to M$, are Lagrangian submanifolds of $M \times M$ with $\tilde{\omega} = \mathrm{pr}_1^* \omega - \mathrm{pr}_2^* \omega$. By the Weinstein tubular neighborhood theorem, there exist a neighborhood $\mathcal{U}$ of Δ in $(M \times M, \tilde{\omega})$ and a neighborhood $\mathcal{U}_0$ of M in (T^*M, ω_0) with a symplectomorphism $\varphi : \mathcal{U} \to \mathcal{U}_0$ satisfying $\varphi(p, p) = (p, 0)$, $\forall p \in M$.

Suppose that f is sufficiently *C^1-close*[18] to id, i.e., f is in some sufficiently small neighborhood of the identity id in the C^1-topology. Hence we can assume that $\mathrm{Graph}\, f \subseteq \mathcal{U}$. Let $j : M \hookrightarrow \mathcal{U}$, $j(p) = (p, f(p))$, be the embedding as $\mathrm{Graph}\, f$, and $i : M \hookrightarrow \mathcal{U}$, $i(p) = (p, p)$, be the embedding as $\Delta = \mathrm{Graph}\, \mathrm{id}$. The map j is sufficiently C^1-close to i. These maps induce embeddings $\varphi \circ j = j_0 : M \hookrightarrow \mathcal{U}_0$ and $\varphi \circ i = i_0 : M \hookrightarrow \mathcal{U}_0$ as 0-section, respectively. Since the map j_0 is sufficiently C^1-close to i_0, the image set $j_0(M)$ intersects each fiber T_p^*M at one point μ_p depending smoothly on p. Therefore, the image of j_0 is the image of a smooth section $\mu : M \to T^*M$, that is, a 1-form $\mu = j_0 \circ (\pi \circ j_0)^{-1}$. We conclude that $\mathrm{Graph}\, f \simeq \{(p, \mu_p) \mid p \in M,\ \mu_p \in T_p^*M\}$. Conversely, if μ is a 1-form sufficiently C^1-close to the zero 1-form, then $\{(p, \mu_p) \mid p \in M,\ \mu_p \in T_p^*M\} \simeq \mathrm{Graph}\, f$, for some diffeomorphism $f : M \to M$.

[17] More generally, $\lambda_1 (\mathrm{pr}_1)^* \omega_1 + \lambda_2 (\mathrm{pr}_2)^* \omega_2$ is symplectic for all $\lambda_1, \lambda_2 \in \mathbb{R} \setminus \{0\}$.

[18] Let X and Y be manifolds. A sequence of maps $f_i : X \to Y$ *converges in the C^0-topology* (a.k.a. the *compact-open topology*) to $f : X \to Y$ if and only if f_i converges uniformly on compact sets. A sequence of C^1 maps $f_i : X \to Y$ *converges in the C^1-topology* to $f : X \to Y$ if and only if it and the sequence of derivatives $df_i : TX \to TY$ converge uniformly on compact sets.

By Proposition 2.3, Graph f is Lagrangian if and only if μ is closed. A small C^1-neighborhood of id in $\mathrm{Sympl}(M,\omega)$ is thus homeomorphic to a C^1-neighborhood of zero in the vector space of closed 1-forms on M. So we obtain the model:

$$T_{\mathrm{id}}\big(\mathrm{Sympl}(M,\omega)\big) \simeq \big\{\mu \in \Omega^1(M) \mid d\mu = 0\big\}.$$

In particular, $T_{\mathrm{id}}(\mathrm{Sympl}(M,\omega))$ contains the space of exact 1-forms that correspond to generating functions, $C^\infty(M)/\{\text{locally constant functions}\}$.

THEOREM 2.12. *Let (M,ω) be a compact symplectic manifold (and not just one point) with $H^1_{\mathrm{deRham}}(M) = 0$. Then any symplectomorphism of M that is sufficiently C^1-close to the identity has at least two fixed points.*

PROOF. If $f \in \mathrm{Sympl}(M,\omega)$ is sufficiently C^1-close to id, then its graph corresponds to a closed 1-form μ on M. As $H^1_{\mathrm{deRham}}(M) = 0$, we have that $\mu = dh$ for some $h \in C^\infty(M)$. But h must have at least two critical points because M is compact. A point p where $\mu_p = dh_p = 0$ corresponds to a point in the intersection of the graph of f with the diagonal, that is, a fixed point of f. $\square$

This result has the following analogue in terms of *Lagrangian intersections*: *if X is a compact Lagrangian submanifold of a symplectic manifold (M,ω) with $H^1_{\mathrm{deRham}}(X) = 0$, then every Lagrangian submanifold of M that is C^1-close*[19] *to X intersects X in at least two points.*

2.5. *Generating functions*

We focus on symplectomorphisms between the cotangent bundles $M_1 = T^*X_1$, $M_2 = T^*X_2$ of two n-dimensional manifolds X_1, X_2. Let α_1, α_2 and ω_1, ω_2 be the corresponding tautological and canonical forms. Under the natural identification

$$M_1 \times M_2 = T^*X_1 \times T^*X_2 \simeq T^*(X_1 \times X_2),$$

the tautological 1-form on $T^*(X_1 \times X_2)$ is $\alpha = \mathrm{pr}_1^*\alpha_1 + \mathrm{pr}_2^*\alpha_2$, the canonical 2-form on $T^*(X_1 \times X_2)$ is $\omega = -d\alpha = \mathrm{pr}_1^*\omega_1 + \mathrm{pr}_2^*\omega_2$, and the twisted product form is $\tilde{\omega} = \mathrm{pr}_1^*\omega_1 - \mathrm{pr}_2^*\omega_2$. We define the involution $\sigma_2 : M_2 \to M_2$, $(x_2, \xi_2) \mapsto (x_2, -\xi_2)$, which yields $\sigma_2^*\alpha_2 = -\alpha_2$. Let $\sigma = \mathrm{id}_{M_1} \times \sigma_2 : M_1 \times M_2 \to M_1 \times M_2$. Then $\sigma^*\tilde{\omega} = \mathrm{pr}_1^*\omega_1 + \mathrm{pr}_2^*\omega_2 = \omega$. If L is a Lagrangian submanifold of $(M_1 \times M_2, \omega)$, then its *twist* $L^\sigma := \sigma(L)$ is a Lagrangian submanifold of $(M_1 \times M_2, \tilde{\omega})$.

For producing a symplectomorphism $M_1 = T^*X_1 \to M_2 = T^*X_2$ we can start with a Lagrangian submanifold L of $(M_1 \times M_2, \omega)$, twist it to obtain a Lagrangian submanifold L^σ of $(M_1 \times M_2, \tilde{\omega})$, and, if L^σ happens to be the graph of some diffeomorphism φ: $M_1 \to M_2$, then φ is a symplectomorphism.

[19] We say that a submanifold Y of M is C^1-*close* to another submanifold X when there is a diffeomorphism $X \to Y$ that is, as a map into M, C^1-close to the inclusion $X \hookrightarrow M$.

A method to obtain Lagrangian submanifolds of $M_1 \times M_2 \simeq T^*(X_1 \times X_2)$ relies on generating functions. For any $f \in C^\infty(X_1 \times X_2)$, df is a closed 1-form on $X_1 \times X_2$. The *Lagrangian submanifold generated by* f is $L_f := \{((x,y),(df)_{(x,y)}) \mid (x,y) \in X_1 \times X_2\}$ (cf. Section 2.1). We adopt the loose notation

$$\begin{aligned} d_x f &:= d_x f(x,y) := (df)_{(x,y)} \text{ projected to } T_x^* X_1 \times \{0\}, \\ d_y f &:= d_y f(x,y) := (df)_{(x,y)} \text{ projected to } \{0\} \times T_y^* X_2, \end{aligned}$$

which enables us to write $L_f = \{(x, y, d_x f, d_y f) \mid (x,y) \in X_1 \times X_2\}$ and

$$L_f^\sigma = \{(x, y, d_x f, -d_y f) \mid (x,y) \in X_1 \times X_2\}.$$

When L_f^σ is in fact the graph of a diffeomorphism $\varphi : M_1 = T^*X_1 \to M_2 = T^*X_2$, we call φ the *symplectomorphism generated by* f, and call f the *generating function* of φ. The issue now is to determine whether a given L_f^σ is the graph of a diffeomorphism $\varphi : M_1 \to M_2$. Let $(\mathcal{U}_1, x_1, \ldots, x_n)$, $(\mathcal{U}_2, y_1, \ldots, y_n)$ be coordinate charts for X_1, X_2, with associated charts $(T^*\mathcal{U}_1, x_1, \ldots, x_n, \xi_1, \ldots, \xi_n)$, $(T^*\mathcal{U}_2, y_1, \ldots, y_n, \eta_1, \ldots, \eta_n)$ for M_1, M_2. The set L_f^σ is the graph of $\varphi : M_1 \to M_2$ exactly when, for any $(x,\xi) \in M_1$ and $(y,\eta) \in M_2$, we have $\varphi(x,\xi) = (y,\eta) \Leftrightarrow \xi = d_x f$ and $\eta = -d_y f$. Therefore, given a point $(x,\xi) \in M_1$, to find its image $(y,\eta) = \varphi(x,\xi)$ we must solve the *Hamilton look-alike equations*

$$\begin{cases} \xi_i = \frac{\partial f}{\partial x_i}(x,y), \\ \eta_i = -\frac{\partial f}{\partial y_i}(x,y). \end{cases}$$

If there is a solution $y = \varphi_1(x,\xi)$ of the first equation, we may feed it to the second thus obtaining $\eta = \varphi_2(x,\xi)$, so that $\varphi(x,\xi) = (\varphi_1(x,\xi), \varphi_2(x,\xi))$. By the implicit function theorem, in order to solve the first equation locally and smoothly for y in terms of x and ξ, we need the condition

$$\det\left[\frac{\partial}{\partial y_j}\left(\frac{\partial f}{\partial x_i}\right)\right]_{i,j=1}^n \neq 0.$$

This is a necessary condition for f to generate a symplectomorphism φ. Locally this is also sufficient, but globally there is the usual bijectivity issue.

EXAMPLE. Let $X_1 = X_2 = \mathbb{R}^n$, and $f(x,y) = -\frac{|x-y|^2}{2}$, the square of Euclidean distance up to a constant. In this case, the Hamilton equations are

$$\begin{cases} \xi_i = \frac{\partial f}{\partial x_i} = y_i - x_i, \\ \eta_i = -\frac{\partial f}{\partial y_i} = y_i - x_i, \end{cases} \iff \begin{cases} y_i = x_i + \xi_i, \\ \eta_i = \xi_i. \end{cases}$$

The symplectomorphism generated by f is $\varphi(x,\xi) = (x+\xi, \xi)$. If we use the Euclidean inner product to identify $T^*\mathbb{R}^n$ with $T\mathbb{R}^n$, and hence regard φ as $\tilde{\varphi} : T\mathbb{R}^n \to T\mathbb{R}^n$

and interpret ξ as the velocity vector, then the symplectomorphism φ corresponds to free translational motion in Euclidean space.

The previous example can be generalized to the *geodesic flow on a Riemannian manifold.*[20] Let (X, g) be a geodesically convex Riemannian manifold, where $d(x, y)$ is the Riemannian distance between points x and y. Consider the function

$$f : X \times X \longrightarrow \mathbb{R}, \qquad f(x, y) = -\frac{d(x, y)^2}{2}.$$

We want to investigate if f generates a symplectomorphism $\varphi : T^*X \to T^*X$. Using the identification $\tilde{g}_x : T_xX \xrightarrow{\simeq} T_x^*X$, $v \mapsto g_x(v, \cdot)$, induced by the metric, we translate φ into a map $\tilde{\varphi} : TX \to TX$. We need to solve

$$\begin{cases} \tilde{g}_x(v) = \xi = d_x f(x, y), \\ \tilde{g}_y(w) = \eta = -d_y f(x, y) \end{cases} \tag{1}$$

for (y, η) in terms of (x, ξ) in order to find φ, or, equivalently, for (y, w) in terms (x, v) in order to find $\tilde{\varphi}$. Assume that (X, g) is *geodesically complete*, that is, every geodesic can be extended indefinitely.

PROPOSITION 2.13. *Under the identification $T_xX \simeq T_x^*X$ given by the metric, the symplectomorphism generated by f corresponds to the map*

$$\tilde{\varphi} : TX \longrightarrow TX,$$
$$(x, v) \longmapsto \left(\gamma(1), \frac{d\gamma}{dt}(1)\right),$$

where γ is the geodesic with initial conditions $\gamma(0) = x$ and $\frac{d\gamma}{dt}(0) = v$.

[20] A *Riemannian metric* on a manifold X is a smooth function g that assigns to each point $x \in X$ an *inner product* g_x on T_xX, that is, a symmetric positive-definite bilinear map $g_x : T_xX \times T_xX \to \mathbb{R}$. Smoothness means that for every (smooth) vector field $v : X \to TX$ the real-valued function $x \mapsto g_x(v_x, v_x)$ is smooth. A *Riemannian manifold* is a pair (X, g) where g is a Riemannian metric on the manifold X. The *arc-length* of a piecewise smooth curve $\gamma : [a, b] \to X$ on a Riemannian (X, g) is $\int_a^b \frac{d\gamma}{dt}\, dt$, where $\frac{d\gamma}{dt}(t) = d\gamma_t(1) \in T_{\gamma(t)}X$ and $\frac{d\gamma}{dt} = \sqrt{g_{\gamma(t)}(\frac{d\gamma}{dt}, \frac{d\gamma}{dt})}$ is the *velocity* of γ. A *reparametrization* of a curve $\gamma : [a, b] \to X$ is a curve of the form $\gamma \circ \tau : [c, d] \to X$ for some $\tau : [c, d] \to [a, b]$. By the change of variable formula for the integral, we see that the arc-length of γ is invariant by reparametrization. The *Riemannian distance* between two points x and y of a connected Riemannian manifold (X, g) is the infimum $d(x, y)$ of the set of all arc-lengths for piecewise smooth curves joining x to y. A *geodesic* is a curve that locally minimizes distance and whose velocity is constant. Given any curve $\gamma : [a, b] \to X$ with $\frac{d\gamma}{dt}$ never vanishing, there is a reparametrization $\gamma \circ \tau : [a, b] \to X$ of constant velocity. A *minimizing geodesic* from x to y is a geodesic joining x to y whose arc-length is the Riemannian distance $d(x, y)$. A Riemannian manifold (X, g) is *geodesically convex* if every point x is joined to every other point y by a unique (up to reparametrization) minimizing geodesic. For instance, $(\mathbb{R}^n, \langle \cdot, \cdot \rangle)$ is a geodesically convex Riemannian manifold (where $g_x(v, w) = \langle v, w \rangle$ is the Euclidean inner product on $T\mathbb{R}^n \simeq \mathbb{R}^n \times \mathbb{R}^n$), for which the Riemannian distance is the usual Euclidean distance $d(x, y) = |x - y|$.

This map $\tilde{\varphi}$ is called the *geodesic flow* on (X, g).

PROOF. Given $(x, v) \in TX$, let $\exp(x, v) : \mathbb{R} \to X$ be the unique geodesic with initial conditions $\exp(x, v)(0) = x$ and $\frac{d\exp(x,v)}{dt}(0) = v$. In this notation, we need to show that the unique solution of the system of equations (1) is $\tilde{\varphi}(x, v) = (\exp(x, v)(1), d\frac{\exp(x,v)}{dt}(1))$.

The Gauss lemma in Riemannian geometry (see, for instance, [120]) asserts that geodesics are orthogonal to the level sets of the distance function. To solve the first equation for $y = \exp(x, u)(1)$ for some $u \in T_x X$, evaluate both sides at v and at vectors $v' \in T_x X$ orthogonal to v,

$$|v|^2 = \frac{d}{dt}\left[\frac{-d(\exp(x, v)(t), y)^2}{2}\right]_{t=0} \quad \text{and} \quad 0 = \frac{d}{dt}\left[\frac{-d(\exp(x, v')(t), y)^2}{2}\right]_{t=0}$$

to conclude that $u = v$, and thus $y = \exp(x, v)(1)$.

We have $-d_y f(x, y)(w') = 0$ at vectors $w' \in T_y X$ orthogonal to $W := \frac{d\exp(x,v)}{dt}(1)$, because $f(x, y)$ is essentially the arc-length of a minimizing geodesic. Hence $w = kW$ must be proportional to W, and $k = 1$ since

$$k|v|^2 = g_y(kW, W) = -\frac{d}{dt}\left[\frac{-d(x, \exp(x, v)(1-t))^2}{2}\right]_{t=0} = |v|^2. \qquad \square$$

2.6. *Fixed points*

Let X be an n-dimensional manifold, and $M = T^*X$ its cotangent bundle equipped with the canonical symplectic form ω. Let $f : X \times X \to \mathbb{R}$ be a smooth function generating a symplectomorphism $\varphi : M \to M$, $\varphi(x, d_x f) = (y, -d_y f)$, with the notation of Section 2.5. To describe the fixed points of φ, we introduce the function $\psi : X \to \mathbb{R}$, $\psi(x) = f(x, x)$.

PROPOSITION 2.14. *There is a one-to-one correspondence between the fixed points of the symplectomorphism φ and the critical points of ψ.*

PROOF. At $x_0 \in X$, $d_{x_0}\psi = (d_x f + d_y f)|_{(x,y)=(x_0,x_0)}$. Let $\xi = d_x f|_{(x,y)=(x_0,x_0)}$. Recalling that L_f^σ is the graph of φ, we have that x_0 is a critical point of ψ, i.e., $d_{x_0}\psi = 0$, if and only if $d_y f|_{(x,y)=(x_0,x_0)} = -\xi$, which happens if and only if the point in L_f^σ corresponding to $(x, y) = (x_0, x_0)$ is (x_0, x_0, ξ, ξ), i.e., $\varphi(x_0, \xi) = (x_0, \xi)$ is a fixed point. $\square$

Consider the iterates $\varphi^N = \varphi \circ \varphi \circ \cdots \circ \varphi$, $N = 1, 2, \ldots$, given by N successive applications of φ. According to the previous proposition, if the symplectomorphism $\varphi^N : M \to M$ is generated by some function $f^{(N)}$, then there is a one-to-one correspondence between the set of fixed points of φ^N and the set of critical points of $\psi^{(N)} : X \to \mathbb{R}$, $\psi^{(N)}(x) = f^{(N)}(x, x)$. It remains to know whether φ^N admits a generating function. We will see that to a certain extent it does.

For each pair $x, y \in X$, define a map $X \to \mathbb{R}$, $z \mapsto f(x, z) + f(z, y)$. Suppose that this map has a unique critical point z_0 and that z_0 is nondegenerate. As z_0 is determined for each (x, y) implicitly by the equation $d_y f(x, z_0) + d_x f(z_0, y) = 0$, by nondegeneracy, the implicit function theorem assures that $z_0 = z_0(x, y)$ is a smooth function. Hence, the function

$$f^{(2)} : X \times X \longrightarrow \mathbb{R}, \quad f^{(2)}(x, y) := f(x, z_0) + f(z_0, y)$$

is smooth. Since φ is generated by f, and z_0 is critical, we have

$$\begin{aligned}\varphi^2\big(x, d_x f^{(2)}(x, y)\big) &= \varphi\big(\varphi(x, d_x f(x, z_0))\big) = \varphi\big(z_0, -d_y f(x, z_0)\big)\\ &= \varphi\big(z_0, d_x f(z_0, y)\big) = \big(y, -d_y f(z_0, y)\big)\\ &= \big(y, -d_y f^{(2)}(x, y)\big).\end{aligned}$$

We conclude that the function $f^{(2)}$ is a generating function for φ^2, as long as, for each $\xi \in T_x^* X$, there is a unique $y \in X$ for which $d_x f^{(2)}(x, y)$ equals ξ.

There are similar partial recipes for generating functions of higher iterates. In the case of φ^3, suppose that the function $X \times X \to \mathbb{R}$, $(z, u) \mapsto f(x, z) + f(z, u) + f(u, y)$, has a unique critical point (z_0, u_0) and that it is a nondegenerate critical point. A generating function would be $f^{(3)}(x, y) = f(x, z_0) + f(z_0, u_0) + f(u_0, y)$.

When the generating functions $f, f^{(2)}, f^{(3)}, \ldots, f^{(N)}$ exist given by these formulas, the *N-periodic points* of φ, i.e., the fixed points of φ^N, are in one-to-one correspondence with the critical points of

$$(x_1, \ldots, x_N) \longmapsto f(x_1, x_2) + f(x_2, x_3) + \cdots + f(x_{N-1}, x_N) + f(x_N, x_1).$$

EXAMPLE. Let $\chi : \mathbb{R} \to \mathbb{R}^2$ be a smooth plane curve that is 1-periodic, i.e., $\chi(s + 1) = \chi(s)$, and parametrized by arc-length, i.e., $|\frac{d\chi}{ds}| = 1$. Assume that the region Y enclosed by the image of χ is *convex*, i.e., for any $s \in \mathbb{R}$, the tangent line $\{\chi(s) + t\frac{d\chi}{ds} \mid t \in \mathbb{R}\}$ intersects the image $X := \partial Y$ of χ only at the point $\chi(s)$.

Suppose that a ball is thrown into a billiard table of shape Y rolling with constant velocity and bouncing off the boundary subject to the usual law of reflection. The map describing successive points on the orbit of the ball is

$$\begin{aligned}\varphi : \mathbb{R}/\mathbb{Z} \times (-1, 1) &\longrightarrow \mathbb{R}/\mathbb{Z} \times (-1, 1),\\ (x, v) &\longmapsto (y, w),\end{aligned}$$

saying that when the ball bounces off $\chi(x)$ with angle $\theta = \arccos v$, it will next collide with $\chi(y)$ and bounce off with angle $\nu = \arccos w$. Then the function $f : \mathbb{R}/\mathbb{Z} \times \mathbb{R}/\mathbb{Z} \to \mathbb{R}$ defined by $f(x, y) = -|\chi(x) - \chi(y)|$ is smooth off the diagonal, and for $\varphi(x, v) = (y, w)$ satisfies

$$\begin{cases}\frac{\partial f}{\partial x}(x, y) = \frac{\chi(y)-\chi(x)}{|\chi(x)-\chi(y)|} \cdot \frac{d\chi}{ds}\Big|_{s=x} = \cos\theta = v,\\ \frac{\partial f}{\partial y}(x, y) = \frac{\chi(x)-\chi(y)}{|\chi(x)-\chi(y)|} \cdot \frac{d\chi}{ds}\Big|_{s=y} = -\cos\nu = -w.\end{cases}$$

We conclude that f is a generating function for φ. Similar approaches work for higher-dimensional billiard problems. Periodic points are obtained by finding critical points of real functions of N variables in X,

$$(x_1,\dots,x_N)\longmapsto \left|\chi(x_1)-\chi(x_2)\right|+\cdots+\left|\chi(x_{N-1})-\chi(x_N)\right| \\ +\left|\chi(x_N)-\chi(x_1)\right|,$$

that is, by finding the N-sided (generalized) polygons inscribed in X of critical perimeter. Notice that $\mathbb{R}/\mathbb{Z}\times(-1,1)\simeq\{(x,v)\mid x\in X,\ v\in T_xX,\ |v|<1\}$ is the open unit tangent ball bundle of a circle X, which is an open annulus A, and the map $\varphi:A\to A$ is area-preserving, as in the next two theorems.

While studying *Poincaré return maps* in dynamical systems, Poincaré arrived at the following results.

THEOREM 2.15 (Poincaré recurrence theorem). *Let $\varphi:A\to A$ be a volume-preserving diffeomorphism of a finite-volume manifold A, and $\mathcal{U}$ a nonempty open set in A. Then there is $q\in\mathcal{U}$ and a positive integer N such that $\varphi^N(q)\in\mathcal{U}$.*

Hence, under iteration, a mechanical system governed by φ will eventually return arbitrarily close to the initial state.

PROOF. Let $\mathcal{U}_0=\mathcal{U}$, $\mathcal{U}_1=\varphi(\mathcal{U})$, $\mathcal{U}_2=\varphi^2(\mathcal{U}),\dots$. If all of these sets were disjoint, then, since $\mathrm{Volume}(\mathcal{U}_i)=\mathrm{Volume}(\mathcal{U})>0$ for all i, the volume of A would be greater or equal to $\sum_i \mathrm{Volume}(\mathcal{U}_i)=\infty$. To avoid this contradiction we must have $\varphi^k(\mathcal{U})\cap\varphi^\ell(\mathcal{U})\neq\emptyset$ for some $k>\ell$, which implies $\varphi^{k-\ell}(\mathcal{U})\cap\mathcal{U}\neq\emptyset$. □

THEOREM 2.16 (Poincaré's last geometric theorem). *Suppose that $\varphi:A\to A$ is an area-preserving diffeomorphism of the closed annulus $A=\mathbb{R}/\mathbb{Z}\times[-1,1]$ that preserves the two components of the boundary and twists them in opposite directions. Then φ has at least two fixed points.*

This theorem was proved in 1913 by Birkhoff [18], and hence is also called the *Poincaré–Birkhoff theorem*. It has important applications to dynamical systems and celestial mechanics. The *Arnold conjecture* on the existence of fixed points for symplectomorphisms of compact manifolds (see Section 5.2) may be regarded as a generalization of the Poincaré–Birkhoff theorem. This conjecture has motivated a significant amount of research involving a more general notion of generating function; see, for instance, [41,55].

2.7. *Lagrangians and special Lagrangians in $\mathbb{C}^n$*

The standard *Hermitian inner product* $h(\cdot,\cdot)$ on $\mathbb{C}^n$ has real and imaginary parts given by the Euclidean inner product $\langle\cdot,\cdot\rangle$ and (minus) the symplectic form ω_0, respectively: for

$v = (x_1 + iy_1, \ldots, x_n + iy_n)$, $u = (a_1 + ib_1, \ldots, a_n + ib_n) \in \mathbb{C}^n$,

$$h(v,u) = \sum_{k=1}^{n}(x_k + iy_k)(a_k - ib_k) = \sum_{k=1}^{n}(x_k a_k + y_k b_k) - i\sum_{k=1}^{n}(x_k b_k - y_k a_k)$$
$$= \langle v, u\rangle - i\omega_0(v,u).$$

LEMMA 2.17. *Let W be a subspace of $(\mathbb{C}^n, \omega_0)$ and $e_1, \ldots, e_n$ vectors in $\mathbb{C}^n$. Then:*
 (a) *W is Lagrangian if and only if $W^\perp = iW$;*
 (b) *$(e_1, \ldots, e_n)$ is an orthonormal basis of a Lagrangian subspace if and only if $(e_1, \ldots, e_n)$ is a unitary basis of $\mathbb{C}^n$.*

PROOF. (a) We always have $\omega_0(v,u) = -\operatorname{im} h(v,u) = \operatorname{re} h(iv,u) = \langle iv, u\rangle$. It follows that, if W is Lagrangian, so that $\omega_0(v,u) = 0$ for all $v, u \in W$, then $iW \subseteq W^\perp$. These spaces must be equal because they have the same dimension. Reciprocally, when $\langle iv, u\rangle = 0$ for all $v, u \in W$, the equality above shows that W must be isotropic. Since $\dim W = \dim iW = \dim W^\perp = 2n - \dim W$, the dimension of W must be n.

(b) If $(e_1, \ldots, e_n)$ is an orthonormal basis of a Lagrangian subspace W, then, by the previous part, $(e_1, \ldots, e_n, ie_1, \ldots, ie_n)$ is an orthonormal basis of $\mathbb{C}^n$ as a real vector space. Hence $(e_1, \ldots, e_n)$ must be a complex basis of $\mathbb{C}^n$ and it is unitary because $h(e_j, e_k) = \langle e_j, e_k\rangle - i\omega_0(e_j, e_k) = \delta_{jk}$. Conversely, if $(e_1, \ldots, e_n)$ is a unitary basis of $\mathbb{C}^n$, then the real span of these vectors is Lagrangian ($\omega_0(e_j, e_k) = -\operatorname{im} h(e_j, e_k) = 0$) and they are orthonormal ($\langle e_j, e_k\rangle = \operatorname{re} h(e_j, e_k) = \delta_{jk}$). □

The *Lagrangian Grassmannian* Λ_n is the set of all Lagrangian subspaces of $\mathbb{C}^n$. It follows from part (b) of Lemma 2.17 that Λ_n is the set of all subspaces of $\mathbb{C}^n$ admitting an orthonormal basis that is a unitary basis of $\mathbb{C}^n$. Therefore, we have

$$\Lambda_n \simeq \mathrm{U}(n)/\mathrm{O}(n).$$

Indeed $\mathrm{U}(n)$ acts transitively on Λ_n: given $W, W' \in \Lambda_n$ with orthonormal bases $(e_1, \ldots, e_n)$, $(e'_1, \ldots, e'_n)$, respectively, there is a unitary transformation of $\mathbb{C}^n$ that maps $(e_1, \ldots, e_n)$ to $(e'_1, \ldots, e'_n)$ as unitary bases of $\mathbb{C}^n$. And the stabilizer of $\mathbb{R}^n \in \Lambda_n$ is the subgroup of those unitary transformations that preserve this Lagrangian subspace, namely $\mathrm{O}(n)$. It follows that Λ_n is a compact connected manifold of dimension $\frac{n(n+1)}{2}$; cf. the last example of Section 1.1.

The Lagrangian Grassmannian comes with a *tautological vector bundle*

$$\tau_n := \{(W, v) \in \Lambda_n \times \mathbb{C}^n \mid v \in W\},$$

whose fiber over $W \in \Lambda_n$ is the n-dimensional real space W. It is a consequence of part (a) of Lemma 2.17 that the following map gives a well-defined global isomorphism of the complexification $\tau_n \otimes_{\mathbb{R}} \mathbb{C}$ with the trivial bundle $\underline{\mathbb{C}^n}$ over Λ_n (i.e., *a global trivialization*): $(W, v \otimes c) \mapsto (W, cv)$, for $W \in \Lambda_n$, $v \in W$, $c \in \mathbb{C}$.

DEFINITION 2.18. A *Lagrangian immersion* of a manifold X is an immersion $f: X \to \mathbb{C}^n$ such that $df_p(T_pX)$ is a Lagrangian subspace of $(\mathbb{C}^n, \omega_0)$, for every $p \in X$.

EXAMPLE. The graph of a map $h: \mathbb{R}^n \to i\mathbb{R}^n$ is an embedded n-dimensional submanifold X of $\mathbb{C}^n$. Its tangent space at $(p, h(p))$ is $\{v + dh_p(v) \mid v \in \mathbb{R}^n\}$. Let $e_1, \ldots, e_n$ be the standard basis of $\mathbb{R}^n$. Since $\omega_0(e_k + dh_p(e_k), e_j + dh_p(e_j)) = \langle e_k, -i\, dh_p(e_j)\rangle + \langle e_j, i\, dh_p(e_k)\rangle$, we see that X is Lagrangian if and only if $\frac{\partial h_k}{\partial x_j} = \frac{\partial h_j}{\partial x_k}$, $\forall j, k$, which in $\mathbb{R}^n$ is if and only if h is the gradient of some $H: \mathbb{R}^n \to i\mathbb{R}$.

If $f: X \to \mathbb{C}^n$ is a Lagrangian immersion, we can define a *Gauss map*

$$\begin{aligned} \lambda_f : X &\longrightarrow \Lambda_n, \\ p &\longmapsto df_p(T_pX). \end{aligned}$$

Since $\lambda_f^* \tau_n = TX$ and $\tau_n \otimes \mathbb{C} \simeq \underline{\mathbb{C}^n}$, we see that a necessary condition for an immersion $X \to \mathbb{C}^n$ to exist is that the complexification of TX be trivializable. Using the h-principle (Section 3.2), Gromov [65] showed that this is also sufficient: *an n-dimensional manifold X admits a Lagrangian immersion into $\mathbb{C}^n$ if and only if the complexification of its tangent bundle is trivializable.*

EXAMPLE. For the unit sphere $S^n = \{(t, x) \in \mathbb{R} \times \mathbb{R}^n \colon t^2 + |x|^2 = 1\}$, the *Whitney sphere immersion* is the map

$$\begin{aligned} f : S^n &\longrightarrow \mathbb{C}^n, \\ (t, x) &\longmapsto x + itx. \end{aligned}$$

The only self-intersection is at the origin where $f(-1, 0, \ldots, 0) = f(1, 0, \ldots, 0)$. Since $T_{(t,x)}S^n = (t, x)^\perp$, the differential $df_{(t,x)} : (u, v) \mapsto v + i(tv + ux)$ is always injective: $v + i(tv + ux) = 0 \Leftrightarrow v = 0$ and $ux = 0$, but when $x = 0$ it is $t = \pm 1$ and $T_{(\pm 1, 0)}S^n = \{0\} \times \mathbb{R}^n$, so it must be $u = 0$. We conclude that f is an immersion. By computing ω_0 at two vectors of the form $v + i(tv + ux)$, we find that the image $df_p(T_pS^n)$ is an n-dimensional isotropic subspace of $\mathbb{C}^n$. Therefore, f is a Lagrangian immersion of S^n, and the complexification $TS^n \otimes \mathbb{C}$ must be always trivializable, though the tangent bundle TS^n is only trivializable in dimensions $n = 0, 1, 3, 7$.

The *special Lagrangian Grassmannian* $S\Lambda_n$ is the set of all *oriented* subspaces of $\mathbb{C}^n$ admitting a *positive* orthonormal basis $(e_1, \ldots, e_n)$ that is a *special* unitary basis of $\mathbb{C}^n$. By the characterization of Lagrangian in the part (b) of Lemma 2.17, it follows that the elements of $S\Lambda_n$ are indeed Lagrangian submanifolds. Similarly to the case of the Lagrangian Grassmannian, we have that

$$S\Lambda_n \simeq \mathrm{SU}(n)/\mathrm{SO}(n)$$

is a compact connected manifold of dimension $\frac{n(n+1)}{2} - 1$.

We can single out the *special* Lagrangian subspaces by expressing the condition on the determinant in terms of the real n-form in $\mathbb{C}^n$,

$$\beta := \operatorname{im}\Omega, \qquad \text{where } \Omega := dz_1 \wedge \cdots \wedge dz_n.$$

Since for $A \in \mathrm{SO}(n)$, we have $\det A = 1$ and $\Omega(e_1, \ldots, e_n) = \Omega(Ae_1, \ldots, Ae_n)$, we see that, for an oriented real n-dimensional subspace $W \subset \mathbb{C}^n$, the number $\Omega(e_1, \ldots, e_n)$ does not depend on the choice of a positive orthonormal basis $(e_1, \ldots, e_n)$ of W, thus can be denoted $\Omega(W)$ and its imaginary part $\beta(W)$.

PROPOSITION 2.19. *A subspace W of $(\mathbb{C}^n, \omega_0)$ has an orientation for which it is a special Lagrangian if and only if W is Lagrangian and $\beta(W) = 0$.*

PROOF. Any orthonormal basis $(e_1, \ldots, e_n)$ of a Lagrangian subspace $W \subset \mathbb{C}^n$ is the image of the canonical basis of $\mathbb{C}^n$ by some $A \in \mathrm{U}(n)$, and $\Omega(W) = \det A \in S^1$. Therefore, W admits an orientation for which such a *positive* $(e_1, \ldots, e_n)$ is a *special* unitary basis of $\mathbb{C}^n$ if and only if $\det A = \pm 1$, i.e., $\beta(W) = 0$. □

DEFINITION 2.20. A *special Lagrangian immersion* of an oriented manifold X is a Lagrangian immersion $f : X \to \mathbb{C}^n$ such that, at each $p \in X$, the space $df_p(T_pX)$ is a special Lagrangian subspace of $(\mathbb{C}^n, \omega_0)$.

For a special Lagrangian immersion f, the Gauss map λ_f takes values in $S\Lambda_n$.

By Proposition 2.19, the immersion f of an n-dimensional manifold X in $(\mathbb{C}^n, \omega_0)$ is *special Lagrangian* if and only if $f^*\omega_0 = 0$ and $f^*\beta = 0$.

EXAMPLE. In $\mathbb{C}^2$, writing $z_k = x_k + iy_k$, we have $\beta = dx_1 \wedge dy_2 + dy_1 \wedge dx_2$. We have seen that the graph of the gradient $i\nabla H$ is Lagrangian, for any function $H : \mathbb{R}^2 \to \mathbb{R}$. So $f(x_1, x_2) = (x_1, x_2, i\frac{\partial H}{\partial x_1}, i\frac{\partial H}{\partial x_2})$ is a Lagrangian immersion. For f to be a *special* Lagrangian immersion, we need the vanish of

$$f^*\beta = dx_1 \wedge d\left(\frac{\partial H}{\partial x_2}\right) + d\left(\frac{\partial H}{\partial x_1}\right) \wedge dx_2 = \left(\frac{\partial^2 H}{\partial x_1^2} + \frac{\partial^2 H}{\partial x_2^2}\right) dx_1 \wedge dx_2.$$

Hence the graph of ∇H is special Lagrangian if and only if H is *harmonic.*

If $f : X \to \mathbb{C}^n$ is a special Lagrangian immersion, then $f^*\Omega$ is an exact (real) volume form: $f^*\Omega = d\,\mathrm{re}(z_1\, dz_2 \wedge \cdots \wedge dz_n)$. We conclude, by Stokes theorem, that there can be no special Lagrangian immersion of a compact manifold in $\mathbb{C}^n$. *Calabi–Yau manifolds*[21] are more general manifolds where a definition of special Lagrangian submanifold makes sense and where the space of special Lagrangian embeddings of a compact manifold is interesting. Special Lagrangian geometry was introduced by Harvey and Lawson [71]. For a treatment of Lagrangian and special Lagrangian submanifolds with many examples; see, for instance, [9].

[21] *Calabi–Yau manifolds* are compact *Kähler manifolds* (Section 3.4) with vanishing first Chern class.

3. Complex structures

3.1. *Compatible linear structures*

A *complex structure* on a vector space V is a linear map $J : V \to V$ such that $J^2 = -\mathrm{Id}$. The pair (V, J) is then called a *complex vector space*.A complex structure J on V is equivalent to a structure of vector space over $\mathbb{C}$, the map J corresponding to multiplication by i. If (V, Ω) is a symplectic vector space, a complex structure J on V is said to be *compatible* (with Ω, or Ω-compatible) if the bilinear map $G_J : V \times V \to \mathbb{R}$ defined by $G_J(u, v) = \Omega(u, Jv)$ is an inner product on V. This condition comprises J being a symplectomorphism (i.e., $\Omega(Ju, Jv) = \Omega(u, v)$, $\forall u, v$) and the so-called *taming*: $\Omega(u, Ju) > 0, \forall u \neq 0$.

EXAMPLE. For the symplectic vector space $(\mathbb{R}^{2n}, \Omega_0)$ with symplectic basis $e_1 = (1, 0, \ldots, 0), \ldots, e_n, f_1, \ldots, f_n = (0, \ldots, 0, 1)$, there is a standard compatible complex structure J_0 determined by $J_0(e_j) = f_j$ and $J_0(f_j) = -e_j$ for all $j = 1, \ldots, n$. This corresponds to a standard identification of $\mathbb{R}^{2n}$ with $\mathbb{C}^n$, and $\Omega_0(u, J_0 v) = \langle u, v\rangle$ is the standard Euclidean inner product. With respect to the symplectic basis $e_1, \ldots, e_n, f_1, \ldots, f_n$, the map J_0 is represented by the matrix

$$\begin{bmatrix} 0 & -\mathrm{Id} \\ \mathrm{Id} & 0 \end{bmatrix}.$$

The *symplectic linear group*, $\mathrm{Sp}(2n) := \{A \in \mathrm{GL}(2n; \mathbb{R}) \mid \Omega_0(Au, Av) = \Omega_0(u, v)$ for all $u, v \in \mathbb{R}^{2n}\}$, is the group of all linear transformations of $\mathbb{R}^{2n}$ that preserve the standard symplectic structure. The *orthogonal group* $\mathrm{O}(2n)$ is the group formed by the linear transformations A that preserve the Euclidean inner product, $\langle Au, Av\rangle = \langle u, v\rangle$, for all $u, v \in \mathbb{R}^{2n}$. The *general complex group* $\mathrm{GL}(n; \mathbb{C})$ is the group of linear transformations $A : \mathbb{R}^{2n} \to \mathbb{R}^{2n}$ commuting with J_0, $A(J_0 v) = J_0(Av)$, for all $v \in \mathbb{R}^{2n}$.[22] The compatibility between the structures Ω_0, $\langle\cdot, \cdot\rangle$ and J_0 implies that the intersection of *any two* of these subgroups of $\mathrm{GL}(2n; \mathbb{R})$ is the same group, namely the *unitary group* $\mathrm{U}(n)$.

As $(\mathbb{R}^{2n}, \Omega_0)$ is the prototype of a $2n$-dimensional symplectic vector space, the preceding example shows that compatible complex structures always exist on symplectic vector spaces.[23] There is yet a way to produce a *canonical* compatible complex structure J after the choice of an inner product G on (V, Ω), though the starting $G(u, v)$ is usually different from $G_J(u, v) := \Omega(u, Jv)$.

PROPOSITION 3.1. *Let (V, Ω) be a symplectic vector space, with an inner product G. Then there is a canonical compatible complex structure J on V.*

---

[22] Identify the complex $n \times n$ matrix $X + iY$ with the real $2n \times 2n$ matrix $\begin{pmatrix} X & -Y \\ Y & X \end{pmatrix}$.

[23] Conversely, given (V, J), there is a symplectic Ω with which J is compatible: take $\Omega(u, v) = G(Ju, v)$ for an inner product G such that $J^t = -J$.

PROOF. By nondegeneracy of Ω and G, the maps $u \mapsto \Omega(u, \cdot)$ and $w \mapsto G(w, \cdot)$ are both isomorphisms between V and V^*. Hence, $\Omega(u, v) = G(Au, v)$ for some linear $A: V \to V$. The map A is skew-symmetric, and the product AA^t is symmetric[24] and positive: $G(AA^t u, u) = G(A^t u, A^t u) > 0$, for $u \neq 0$. By the spectral theorem, these properties imply that AA^t diagonalizes with positive eigenvalues λ_i, say $AA^t = B \operatorname{diag}(\lambda_1, \ldots, \lambda_{2n})\, B^{-1}$. We may hence define an arbitrary real power of AA^t by rescaling the eigenspaces, in particular,

$$\sqrt{AA^t} := B \operatorname{diag}(\sqrt{\lambda_1}, \ldots, \sqrt{\lambda_{2n}}) B^{-1}.$$

The linear transformation $\sqrt{AA^t}$ is symmetric, positive-definite and does not depend on the choice of B nor of the ordering of the eigenvalues. It is completely determined by its effect on each eigenspace of AA^t: on the eigenspace corresponding to the eigenvalue λ_k, the map $\sqrt{AA^t}$ is defined to be multiplication by $\sqrt{\lambda_k}$.

Let $J := (\sqrt{AA^t})^{-1} A$. Since A and $\sqrt{AA^t}$ commute, J is orthogonal ($JJ^t = \mathrm{Id}$), as well as skew-symmetric ($J^t = -J$). It follows that J is a complex structure on V. Compatibility is easily checked:

$$\Omega(Ju, Jv) = G(AJu, Jv) = G(JAu, Jv) = G(Au, v) = \Omega(u, v)$$

and

$$\Omega(u, Ju) = G(Au, Ju) = G(-JAu, u) = G\big(\sqrt{AA^t}u, u\big) > 0, \quad \text{for } u \neq 0. \quad \square$$

The factorization $A = \sqrt{AA^t}\, J$ is called the *polar decomposition* of A.

REMARK. Being *canonical*, this construction may be *smoothly* performed: when (V_t, Ω_t) is a family of symplectic vector spaces with a family G_t of inner products, all depending smoothly on a parameter t, an adaptation of the previous proof shows that there is a smooth family J_t of compatible complex structures on (V_t, Ω_t).

Let (V, Ω) be a symplectic vector space of dimension $2n$, and let J be a complex structure on V. If J is Ω-compatible and L is a Lagrangian subspace of (V, Ω), then JL is also Lagrangian and $JL = L^\perp$, where $\perp$ indicates orthogonality with respect to the inner product $G_J(u, v) = \Omega(u, Jv)$. Therefore, a complex structure J is Ω-compatible *if and only if* there exists a symplectic basis for V of the form

$$e_1, e_2, \ldots, e_n, \quad f_1 = Je_1, \quad f_2 = Je_2, \quad \ldots, \quad f_n = Je_n.$$

Let $\mathcal{J}(V, \Omega)$ be the *set of all compatible complex structures in a symplectic vector space* (V, Ω).

[24] A map $B: V \to V$ is *symmetric*, respectively *skew-symmetric*, when $B^t = B$, respectively $B^t = -B$, where the transpose $B^t: V \to V$ is determined by $G(B^t u, v) = G(u, Bv)$.

PROPOSITION 3.2. *The set $\mathcal{J}(V, \Omega)$ is contractible.*[25]

PROOF. Pick a Lagrangian subspace L_0 of (V, Ω). Let $\mathcal{L}(V, \Omega, L_0)$ be the space of all Lagrangian subspaces of (V, Ω) that intersect L_0 transversally. Let $\mathcal{G}(L_0)$ be the space of all inner products on L_0. The map

$$\Psi : \mathcal{J}(V, \Omega) \longrightarrow \mathcal{L}(V, \Omega, L_0) \times \mathcal{G}(L_0),$$
$$J \longmapsto (JL_0, G_J|_{L_0})$$

is a homeomorphism, with inverse as follows. Take $(L, G) \in \mathcal{L}(V, \Omega, L_0) \times \mathcal{G}(L_0)$. For $v \in L_0$, $v^\perp = \{u \in L_0 \mid G(u, v) = 0\}$ is a $(n-1)$-dimensional space of L_0; its symplectic orthogonal $(v^\perp)^\Omega$ is $(n+1)$-dimensional. Then $(v^\perp)^\Omega \cap L$ is 1-dimensional. Let Jv be the unique vector in this line such that $\Omega(v, Jv) = 1$. If we take v's in some G-orthonormal basis of L_0, this defines an element $J \in \mathcal{J}(V, \Omega)$.

The set $\mathcal{L}(V, \Omega, L_0)$ can be identified with the vector space of all symmetric $n \times n$ matrices. In fact, any n-dimensional subspace L of V that is transverse to L_0 is the graph of a linear map $JL_0 \to L_0$, and the Lagrangian ones correspond to symmetric maps (cf. Section 1.1). Hence, $\mathcal{L}(V, \Omega, L_0)$ is contractible. Since $\mathcal{G}(L_0)$ is contractible (it is even convex), we conclude that $\mathcal{J}(V, \Omega)$ is contractible. □

3.2. *Compatible almost complex structures*

An *almost complex structure* on a manifold M is a smooth[26] field of complex structures on the tangent spaces, $J_p : T_pM \to T_pM$, $p \in M$. The pair (M, J) is then called an *almost complex manifold.*

DEFINITION 3.3. An almost complex structure J on a symplectic manifold (M, ω) is *compatible* (with ω or ω-compatible) if the map that assigns to each point $p \in M$ the bilinear pairing $g_p : T_pM \times T_pM \to \mathbb{R}$, $g_p(u, v) := \omega_p(u, J_pv)$ is a Riemannian metric on M. A triple (ω, g, J) of a symplectic form, a Riemannian metric and an almost complex structure on a manifold M is a *compatible triple* when $g(\cdot, \cdot) = \omega(\cdot, J\cdot)$.

If (ω, J, g) is a compatible triple, each of ω, J or g can be written in terms of the other two.

EXAMPLES.

1. If we identify $\mathbb{R}^{2n}$ with $\mathbb{C}^n$ using coordinates $z_j = x_j + iy_j$, multiplication by i induces a constant linear map J_0 on the tangent spaces such that $J_0^2 = -\mathrm{Id}$, known

---

[25] *Contractibility* of $\mathcal{J}(V, \Omega)$ means that there exists a homotopy $h_t : \mathcal{J}(V, \Omega) \to \mathcal{J}(V, \Omega)$, $0 \leqslant t \leqslant 1$, starting at the identity $h_0 = \mathrm{Id}$, finishing at a trivial map $h_1 : \mathcal{J}(V, \Omega) \to \{J_0\}$, and fixing J_0 (i.e., $h_t(J_0) = J_0$, $\forall t$) for some $J_0 \in \mathcal{J}(V, \Omega)$.

[26] *Smoothness* means that for any vector field v, the image Jv is a (smooth) vector field.

as the *standard almost complex structure* on $\mathbb{R}^{2n}$:

$$J_0\left(\frac{\partial}{\partial x_j}\right)=\frac{\partial}{\partial y_j}, \qquad J_0\left(\frac{\partial}{\partial y_j}\right)=-\frac{\partial}{\partial x_j}.$$

For the standard symplectic form $\omega_0=\sum dx_j\wedge dy_j$ and the Euclidean inner product $g_0=\langle\cdot,\cdot\rangle$, the compatibility relation holds: $\omega_0(u,v)=g_0(J_0(u),v)$.

2. Any oriented hypersurface $\Sigma\subset\mathbb{R}^3$ carries a natural symplectic form and a natural compatible almost complex structure induced by the standard inner (or dot) and exterior (or vector) products. They are given by the formulas $\omega_p(u,v):=\langle\nu_p,u\times v\rangle$ and $J_p(v)=\nu_p\times v$ for $v\in T_p\Sigma$, where ν_p is the outward-pointing unit normal vector at $p\in\Sigma$ (in other words, $\nu:\Sigma\to S^2$ is the *Gauss map*). Cf. Example 3 of Section 1.2. The corresponding Riemannian metric is the restriction to Σ of the standard Euclidean metric $\langle\cdot,\cdot\rangle$.
3. The previous example generalizes to the oriented hypersurfaces $M\subset\mathbb{R}^7$. Regarding $u,v\in\mathbb{R}^7$ as imaginary *octonions* (or *Cayley numbers*), the natural vector product $u\times v$ is the imaginary part of the product of u and v as octonions. This induces a natural almost complex structure on M given by $J_p(v)=\nu_p\times v$, where ν_p is the outward-pointing unit normal vector at $p\in M$. In the case of S^6, at least, this J is not compatible with any symplectic form, as S^6 cannot be a symplectic manifold.

As a consequence of the remark in Section 3.1, we have:

PROPOSITION 3.4. *On any symplectic manifold (M,ω) with a Riemannian metric g, there is a canonical compatible almost complex structure J.*

Since Riemannian metrics always exist, we conclude that *any symplectic manifold has compatible almost complex structures*. The metric $g_J(\cdot,\cdot):=\omega(\cdot,J\cdot)$ tends to be different from the given $g(\cdot,\cdot)$.

PROPOSITION 3.5. *Let (M,J) be an almost complex manifold where J is compatible with two symplectic forms ω_0,ω_1 Then ω_0 and ω_1 are deformation-equivalent.*

PROOF. Simply take the convex combinations $\omega_t=(1-t)\omega_0+t\omega_1$, $0\leqslant t\leqslant 1$. □

A counterexample to the converse of this proposition is provided by the family $\omega_t=\cos\pi t\,dx_1\wedge dy_1+\sin\pi t\,dx_1\wedge dy_2+\sin\pi t\,dy_1\wedge dx_2+\cos\pi t\,dx_2\wedge dy_2$ for $0\leqslant t\leqslant 1$. There is no J in $\mathbb{R}^4$ compatible with both ω_0 and $\omega_1=-\omega_0$.

A submanifold X of an almost complex manifold (M,J) is an *almost complex submanifold* when $J(TX)\subseteq TX$, i.e., we have $J_pv\in T_pX$, $\forall p\in X, v\in T_pX$.

PROPOSITION 3.6. *Let (M,ω) be a symplectic manifold equipped with a compatible almost complex structure J. Then any almost complex submanifold X of (M,J) is a symplectic submanifold of (M,ω).*

PROOF. Let $i : X \hookrightarrow M$ be the inclusion. Then $i^*\omega$ is a closed 2-form on X. Since $\omega_p(u, v) = g_p(J_p u, v)$, $\forall p \in X$, $\forall u, v \in T_pX$, and since $g_p|_{T_pX}$ is nondegenerate, so is $\omega_p|_{T_pX}$, and $i^*\omega$ is nondegenerate. □

It is easy to see that the *set* $\mathcal{J}(M, \omega)$ *of all compatible almost complex structures on a symplectic manifold* (M, ω) is path-connected. From two almost complex structures J_0, J_1 compatible with ω, we get two Riemannian metrics $g_0(\cdot, \cdot) = \omega(\cdot, J_0 \cdot)$, $g_1(\cdot, \cdot) = \omega(\cdot, J_1 \cdot)$. Their convex combinations

$$g_t(\cdot, \cdot) = (1 - t)g_0(\cdot, \cdot) + t g_1(\cdot, \cdot), \quad 0 \leqslant t \leqslant 1,$$

form a smooth family of Riemannian metrics. Applying the polar decomposition to the family (ω, g_t), we obtain a smooth path of compatible almost complex structures J_t joining J_0 to J_1. The set $\mathcal{J}(M, \omega)$ is even *contractible* (this is important for defining invariants). The first ingredient is the contractibility of the set of compatible complex structures on a vector space (Proposition 3.2). Consider the fiber bundle $\mathcal{J} \to M$ with fiber over $p \in M$ being the space $\mathcal{J}_p := \mathcal{J}(T_pM, \omega_p)$ of compatible complex structures on the tangent space at p. A compatible almost complex structure on (M, ω) is a section of $\mathcal{J}$. The space of sections of $\mathcal{J}$ is contractible because the fibers are contractible.[27]

The *first Chern class* $c_1(M, \omega)$ of a symplectic manifold (M, ω) is the first Chern class of (TM, J) for any compatible J. The class $c_1(M, \omega) \in H^2(M; \mathbb{Z})$ is invariant under deformations of ω.

We never used the closedness of ω to obtain compatible almost complex structures. The construction holds for an *almost symplectic manifold* (M, ω), that is, a pair of a manifold M and a nondegenerate 2-form ω, not necessarily closed. We could further work with a *symplectic vector bundle*, that is, a vector bundle $E \to M$ equipped with a smooth field ω of fiberwise nondegenerate skew-symmetric bilinear maps (Section 1.6). The existence of such a field ω is equivalent to being able to reduce the structure group of the bundle from the general linear group to the linear symplectic group. As both $\mathrm{Sp}(2n)$ and $\mathrm{GL}(n; \mathbb{C})$ retract to their common maximal compact subgroup $\mathrm{U}(n)$, a symplectic vector bundle can be always endowed with a structure of complex vector bundle, and vice-versa.

Gromov showed in his thesis [63] that any *open*[28] almost complex manifold admits a symplectic form. The books [42, §10.2] and [99, §7.3] contain proofs of this statement using different techniques.

THEOREM 3.7 (Gromov). *For an open manifold the existence of an almost complex structure J implies that of a symplectic form ω in any given 2-cohomology class and such that J is homotopic to an almost complex structure compatible with ω.*

From an almost complex structure J and a metric g, one builds a nondegenerate 2-form $\omega(u, v) = g(Ju, v)$, which will not be closed in general. Closedness is a *differential re-*

[27] The base being a (second countable and Hausdorff) manifold, a contraction can be produced using a countable cover by trivializing neighborhoods whose closures are compact subsets of larger trivializing neighborhoods, and such that each $p \in M$ belongs to only a finite number of such neighborhoods.

[28] A manifold is *open* if it has no closed connected components, where *closed* means compact and without boundary.

lation, i.e., a condition imposed on the partial derivatives, encoded as a subset of *jet space*. One says that a differential relation satisfies the *h-principle*[29] if any *formal solution* (i.e., a solution for the associated algebraic problem, in the present case a nondegenerate 2-form) is homotopic to a *holonomic solution* (i.e., a genuine solution, in the present case a closed nondegenerate 2-form). Therefore, when the h-principle holds, one may concentrate on a purely topological question (such as the existence of an almost complex structure) in order to prove the existence of a differential solution. Gromov showed that, for an open differential relation on an open manifold, when the relation is invariant under the group of diffeomorphisms of the underlying manifold, the inclusion of the space of holonomic solutions into the space of formal solutions is a weak homotopy equivalence, i.e., induces isomorphisms of all homotopy groups. The previous theorem fits here as an application.

For *closed* manifolds there is no such theorem: as discussed in Section 1.2, the existence of a 2-cohomology class whose top power is nonzero is also necessary for the existence of a symplectic form and there are further restrictions coming from *Gromov–Witten theory* (see Section 4.5).

3.3. *Integrability*

Any *complex manifold*[30] has a canonical almost complex structure J. It is defined locally over the domain $\mathcal{U}$ of a complex chart $\varphi:\mathcal{U}\to\mathcal{V}\subseteq\mathbb{C}^n$, by $J_p(\frac{\partial}{\partial x_j}|_p)=\frac{\partial}{\partial y_j}|_p$ and $J_p(\frac{\partial}{\partial y_j}|_p)=-\frac{\partial}{\partial x_j}|_p$, where these are the tangent vectors induced by the real and imaginary parts of the coordinates of $\varphi=(z_1,\dots,z_n)$, $z_j=x_j+iy_j$. This yields a globally well-defined J, thanks to the *Cauchy–Riemann equations* satisfied by the components of the transition maps.

An almost complex structure J on a manifold M is called *integrable* when J is induced by some underlying structure of complex manifold on M as above. The question arises whether some compatible almost complex structure J on a symplectic manifold (M,ω) is integrable. To understand what is involved, we review Dolbeault theory and the Newlander–Nirenberg theorem.

Let (M,J) be a $2n$-dimensional almost complex manifold. The fibers of the complexified tangent bundle, $TM\otimes\mathbb{C}$, are $2n$-dimensional vector spaces over $\mathbb{C}$. We may extend J linearly to $TM\otimes\mathbb{C}$ by $J(v\otimes c)=Jv\otimes c$, $v\in TM$, $c\in\mathbb{C}$. Since $J^2=-\mathrm{Id}$, on the complex vector space $(TM\otimes\mathbb{C})_p$ the linear map J_p has eigenvalues $\pm i$. The $(\pm i)$-eigenspaces of J are denoted $T_{1,0}$ and $T_{0,1}$, respectively, and called the spaces of *J-holomorphic* and of *J-anti-holomorphic tangent vectors*. We have an isomorphism

$$\begin{aligned}(\pi_{1,0},\pi_{0,1}):TM\otimes\mathbb{C}&\xrightarrow{\simeq}T_{1,0}\oplus T_{0,1},\\ v&\longmapsto\frac{1}{2}(v-iJv,v+iJv),\end{aligned}$$

[29] There are in fact different h-principles depending on the different possible coincidences of homotopy groups for the spaces of formal solutions and of holonomic solutions.

[30] A *complex manifold* of (complex) dimension n is a set M with a complete complex atlas $\{(\mathcal{U}_\alpha,\mathcal{V}_\alpha,\varphi_\alpha),\ \alpha\in$ index set $I\}$ where $M=\bigcup_\alpha\mathcal{U}_\alpha$, the $\mathcal{V}_\alpha$'s are open subsets of $\mathbb{C}^n$, and the maps $\varphi_\alpha:\mathcal{U}_\alpha\to\mathcal{V}_\alpha$ are bijections such that the transition maps $\psi_{\alpha\beta}=\varphi_\beta\circ\varphi_\alpha^{-1}:\mathcal{V}_{\alpha\beta}\to\mathcal{V}_{\beta\alpha}$ are *biholomorphic* (i.e., bijective, holomorphic and with holomorphic inverse) as maps on open subsets of $\mathbb{C}^n$, $\mathcal{V}_{\alpha\beta}=\varphi_\alpha(\mathcal{U}_\alpha\cap\mathcal{U}_\beta)$.

where the maps to each summand satisfy $\pi_{1,0} \circ J = i\pi_{1,0}$ and $\pi_{0,1} \circ J = -i\pi_{0,1}$. Restricting $\pi_{1,0}$ to TM, we see that $(TM, J) \simeq T_{1,0} \simeq \overline{T_{0,1}}$, as complex vector bundles, where the multiplication by i is given by J in (TM, J) and where $\overline{T_{0,1}}$ denotes the complex conjugate bundle of $T_{0,1}$.

Similarly, J^* defined on $T^*M \otimes \mathbb{C}$ by $J^*\xi = \xi \circ J$ has $(\pm i)$-eigenspaces $T^{1,0} = (T_{1,0})^*$ and $T^{0,1} = (T_{0,1})^*$, respectively, called the spaces of *complex-linear* and of *complex-antilinear cotangent vectors*. Under the two natural projections $\pi^{1,0}, \pi^{0,1}$, the complexified cotangent bundle splits as

$$\begin{aligned}(\pi^{1,0}, \pi^{0,1}) : T^*M \otimes \mathbb{C} &\xrightarrow{\simeq} T^{1,0} \oplus T^{0,1},\\ \xi &\longmapsto \frac{1}{2}(\xi - iJ^*\xi, \xi + iJ^*\xi).\end{aligned}$$

Let

$$\Lambda^k(T^*M \otimes \mathbb{C}) := \Lambda^k(T^{1,0} \oplus T^{0,1}) = \bigoplus_{\ell+m=k} \Lambda^{\ell,m},$$

where $\Lambda^{\ell,m} := (\Lambda^\ell T^{1,0}) \wedge (\Lambda^m T^{0,1})$, and let $\Omega^k(M;\mathbb{C})$ be the space of sections of $\Lambda^k(T^*M \otimes \mathbb{C})$, called *complex-valued k-forms on M*. The *differential forms of type* (ℓ, m) on (M, J) are the sections of $\Lambda^{\ell,m}$, and the space of these differential forms is denoted $\Omega^{\ell,m}$. The decomposition of forms by Dolbeault type is $\Omega^k(M;\mathbb{C}) = \bigoplus_{\ell+m=k} \Omega^{\ell,m}$. Let $\pi^{\ell,m} : \Lambda^k(T^*M \otimes \mathbb{C}) \to \Lambda^{\ell,m}$ be the projection map, where $\ell + m = k$. The usual exterior derivative d (extended linearly to smooth complex-valued forms) composed with two of these projections induces the *del* and *del-bar* differential operators, ∂ and $\bar{\partial}$, on forms of type (ℓ, m):

$$\partial := \pi^{\ell+1,m} \circ d : \Omega^{\ell,m} \longrightarrow \Omega^{\ell+1,m}$$

and

$$\bar{\partial} := \pi^{\ell,m+1} \circ d : \Omega^{\ell,m} \longrightarrow \Omega^{\ell,m+1}.$$

If $\beta \in \Omega^{\ell,m}(M)$, with $k = \ell + m$, then $d\beta \in \Omega^{k+1}(M;\mathbb{C})$:

$$d\beta = \sum_{r+s=k+1} \pi^{r,s} d\beta = \pi^{k+1,0} d\beta + \cdots + \partial\beta + \bar{\partial}\beta + \cdots + \pi^{0,k+1} d\beta.$$

In particular, on complex-valued functions we have $df = d(\operatorname{re} f) + i\, d(\operatorname{im} f)$ and $d = \partial + \bar{\partial}$, where $\partial = \pi^{1,0} \circ d$ and $\bar{\partial} = \pi^{0,1} \circ d$. A function $f : M \to \mathbb{C}$ is *J-holomorphic at* $p \in M$ if df_p is complex linear, i.e., $df_p \circ J_p = i\, df_p$ (or $df_p \in T_p^{1,0}$). A function f is *J-holomorphic* if it is holomorphic at all $p \in M$. A function $f : M \to \mathbb{C}$ is *J-anti-holomorphic at* $p \in M$ if df_p is complex antilinear, i.e., $df_p \circ J_p = -i\, df_p$ (or $df_p \in T_p^{0,1}$), that is, when the conjugate function $\bar{f}$ is holomorphic at $p \in M$. In terms of ∂ and $\bar{\partial}$, a function f is J-holomorphic if and only if $\bar{\partial} f = 0$, and f is J-anti-holomorphic if and only if $\partial f = 0$.

When M is a *complex manifold* and J is its canonical almost complex structure, the splitting $\Omega^k(M;\mathbb{C}) = \bigoplus_{\ell+m=k} \Omega^{\ell,m}$ is particularly interesting. Let $\mathcal{U} \subseteq M$ be the domain of a complex coordinate chart $\varphi = (z_1, \dots, z_n)$, where the corresponding real coordinates $x_1, y_1, \dots, x_n, y_n$ satisfy $z_j = x_j + i y_j$. In terms of

$$\frac{\partial}{\partial z_j} := \frac{1}{2}\left(\frac{\partial}{\partial x_j} - i\frac{\partial}{\partial y_j}\right) \quad \text{and} \quad \frac{\partial}{\partial \bar{z}_j} := \frac{1}{2}\left(\frac{\partial}{\partial x_j} + i\frac{\partial}{\partial y_j}\right),$$

the $(\pm i)$-eigenspaces of J_p $(p \in \mathcal{U})$ can be written

$$(T_{1,0})_p = \mathbb{C}\text{-span}\left\{\left.\frac{\partial}{\partial z_j}\right|_p : j = 1, \dots, n\right\} \quad \text{and} \quad (T_{0,1})_p = \mathbb{C}\text{-span}\left\{\left.\frac{\partial}{\partial \bar{z}_j}\right|_p\right\}.$$

Similarly, putting $dz_j = dx_j + i\,dy_j$ and $d\bar{z}_j = dx_j - i\,dy_j$, we obtain simple formulas for the differentials of a $b \in C^\infty(\mathcal{U};\mathbb{C})$, $\partial b = \sum \frac{\partial b}{\partial z_j} dz_j$ and $\bar{\partial} b = \sum \frac{\partial b}{\partial \bar{z}_j} d\bar{z}_j$, and we have $T^{1,0} = \mathbb{C}\text{-span}\{dz_j\colon j = 1, \dots, n\}$ and $T^{0,1} = \mathbb{C}\text{-span}\{d\bar{z}_j\colon j = 1, \dots, n\}$. If we use multi-index notation $J = (j_1, \dots, j_\ell)$ where $1 \leqslant j_1 < \dots < j_\ell \leqslant n$, $|J| = \ell$ and $dz_J = dz_{j_1} \wedge dz_{j_2} \wedge \dots \wedge dz_{j_\ell}$, then the set of (ℓ, m)-forms on $\mathcal{U}$ is

$$\Omega^{\ell,m} = \left\{ \sum_{|J|=\ell,\ |K|=m} b_{J,K}\, dz_J \wedge d\bar{z}_K \mid b_{J,K} \in C^\infty(\mathcal{U};\mathbb{C}) \right\}.$$

A form $\beta \in \Omega^k(M;\mathbb{C})$ may be written over $\mathcal{U}$ as

$$\beta = \sum_{\ell+m=k} \left(\sum_{|J|=\ell,\ |K|=m} b_{J,K}\, dz_J \wedge d\bar{z}_K \right).$$

Since $d = \partial + \bar{\partial}$ on functions, we get

$$\begin{aligned} d\beta &= \sum_{\ell+m=k} \left(\sum_{|J|=\ell,\ |K|=m} db_{J,K} \wedge dz_J \wedge d\bar{z}_K \right) \\ &= \sum_{\ell+m=k} \Bigg(\underbrace{\sum_{|J|=\ell,\ |K|=m} \partial b_{J,K} \wedge dz_J \wedge d\bar{z}_K}_{\in \Omega^{\ell+1,m}} \\ &\qquad + \underbrace{\sum_{|J|=\ell,\ |K|=m} \bar{\partial} b_{J,K} \wedge dz_J \wedge d\bar{z}_K}_{\in \Omega^{\ell,m+1}} \Bigg) \\ &= \partial\beta + \bar{\partial}\beta, \end{aligned}$$

and conclude that, *on a complex manifold*, $d = \partial + \bar{\partial}$ *on forms of any degree*. This cannot be proved for an almost complex manifold, because there are no coordinate functions z_j to give a suitable basis of 1-forms.

When $d = \partial + \bar{\partial}$, for any form $\beta \in \Omega^{\ell,m}$, we have

$$0 = d^2\beta = \underbrace{\partial^2\beta}_{\in\Omega^{\ell+2,m}} + \underbrace{\partial\bar{\partial}\beta + \bar{\partial}\partial\beta}_{\in\Omega^{\ell+1,m+1}} + \underbrace{\bar{\partial}^2\beta}_{\in\Omega^{\ell,m+2}} \quad \Longrightarrow \quad \begin{cases} \bar{\partial}^2 = 0, \\ \partial\bar{\partial} + \bar{\partial}\partial = 0, \\ \partial^2 = 0. \end{cases}$$

Since $\bar{\partial}^2 = 0$, the chain $0 \longrightarrow \Omega^{\ell,0} \xrightarrow{\bar{\partial}} \Omega^{\ell,1} \xrightarrow{\bar{\partial}} \Omega^{\ell,2} \xrightarrow{\bar{\partial}} \cdots$ is a differential complex. Its cohomology groups

$$H^{\ell,m}_{\text{Dolbeault}}(M) := \frac{\ker \bar{\partial} : \Omega^{\ell,m} \to \Omega^{\ell,m+1}}{\operatorname{im} \bar{\partial} : \Omega^{\ell,m-1} \to \Omega^{\ell,m}}$$

are called the *Dolbeault cohomology* groups. The Dolbeault theorem states that for complex manifolds $H^{\ell,m}_{\text{Dolbeault}}(M) \simeq H^m(M; \mathcal{O}(\Omega^{(\ell,0)}))$, where $\mathcal{O}(\Omega^{(\ell,0)})$ is the sheaf of forms of type $(\ell, 0)$ over M.

It is natural to ask whether the identity $d = \partial + \bar{\partial}$ could hold for manifolds other than complex manifolds. Newlander and Nirenberg [106] showed that the answer is no: for an almost complex manifold (M, J), the following are equivalent

$$M \text{ is a complex manifold} \iff \mathcal{N} \equiv 0 \iff d = \partial + \bar{\partial} \iff \bar{\partial}^2 = 0,$$

where $\mathcal{N}$ is the *Nijenhuis tensor*:

$$\mathcal{N}(X, Y) := [JX, JY] - J[JX, Y] - J[X, JY] - [X, Y],$$

for vector fields X and Y on M, $[\cdot, \cdot]$ being the usual bracket.[31] The Nijenhuis tensor can be thought of as a measure of the existence of J-holomorphic functions: if there exist n J-holomorphic functions, $f_1, \ldots, f_n$, on $\mathbb{R}^{2n}$, that are independent at some point p, i.e., the real and imaginary parts of $(df_1)_p, \ldots, (df_n)_p$ form a basis of $T_p^*\mathbb{R}^{2n}$, then $\mathcal{N}$ vanishes identically at p. More material related to Dolbeault theory or to the Newlander–Nirenberg theorem can be found in [23,37,62,76,141].

EXAMPLE. Out of all spheres, only S^2 and S^6 admit almost complex structures [121, §41.20]. As a complex manifold, S^2 if referred to as the *Riemann sphere* $\mathbb{CP}^1$. The almost complex structure on S^6 from Example 3 of Section 3.2 is not integrable, but it is not yet known whether S^6 admits a structure of complex manifold.

[31] The *bracket* of vector fields X and Y is the vector field $[X, Y]$ characterized by the property that $\mathcal{L}_{[X,Y]}f := \mathcal{L}_X(\mathcal{L}_Y f) - \mathcal{L}_Y(\mathcal{L}_X f)$, for $f \in C^\infty(M)$, where $\mathcal{L}_X f = df(X)$.

In the (real) 2-dimensional case $\mathcal{N}$ always vanishes simply because $\mathcal{N}$ is a tensor, i.e., $\mathcal{N}(fX, gY) = fg\mathcal{N}(X, Y)$ for any $f, g \in C^\infty(M)$, and $\mathcal{N}(X, JX) = 0$ for any vector field X. Combining this with the fact that any orientable surface is symplectic, we conclude that any orientable surface is a complex manifold, a result already known to Gauss. However, most almost complex structures on higher-dimensional manifolds are not integrable. In Section 3.5 we see that the existence of a complex structure compatible with a symplectic structure on a compact manifold imposes significant topological constraints.

3.4. *Kähler manifolds*

DEFINITION 3.8. A *Kähler manifold* is a symplectic manifold (M, ω) equipped with an integrable compatible almost complex structure J. The symplectic form ω is then called a *Kähler form.*

As a complex manifold, a Kähler manifold (M, ω, J) has Dolbeault cohomology. As it is also a symplectic manifold, it is interesting to understand where the symplectic form ω sits with respect to the Dolbeault type decomposition.

PROPOSITION 3.9. *A Kähler form ω is a ∂- and $\bar{\partial}$-closed* $(1, 1)$*-form that is given on a local complex chart $(\mathcal{U}, z_1, \dots, z_n)$ by*

$$\omega = \frac{i}{2} \sum_{j,k=1}^{n} h_{jk}\, dz_j \wedge d\bar{z}_k,$$

where, at every point $p \in \mathcal{U}$, $(h_{jk}(p))$ is a positive-definite Hermitian matrix.

In particular, ω defines a Dolbeault $(1, 1)$-cohomology class, $[\omega] \in H^{1,1}_{\mathrm{Dolbeault}}(M)$.

PROOF. Being a form in $\Omega^2(M; \mathbb{C}) = \Omega^{2,0} \oplus \Omega^{1,1} \oplus \Omega^{0,2}$, with respect to a local complex chart, ω can be written

$$\omega = \sum a_{jk}\, dz_j \wedge dz_k + \sum b_{jk}\, dz_j \wedge d\bar{z}_k + \sum c_{jk}\, d\bar{z}_j \wedge d\bar{z}_k$$

for some $a_{jk}, b_{jk}, c_{jk} \in C^\infty(\mathcal{U}; \mathbb{C})$. By the compatibility of ω with the complex structure, J is a symplectomorphism, that is, $J^*\omega = \omega$ where $(J^*\omega)(u, v) := \omega(Ju, Jv)$. Since $J^*\, dz_j = dz_j \circ J = i\, dz_j$ and $J^*\, d\bar{z}_j = d\bar{z}_j \circ J = -i\, d\bar{z}_j$, we have $J^*\omega = \omega$ if and only if the coefficients a_{jk} and c_{jk} all vanish identically, that is, if and only if $\omega \in \Omega^{1,1}$. Since ω is closed, of type $(1, 1)$ and $d\omega = \partial\omega + \bar{\partial}\omega$, we must have $\partial\omega = 0$ and $\bar{\partial}\omega = 0$. Set $b_{jk} = \frac{i}{2} h_{jk}$. As ω is real-valued, i.e., $\omega = \frac{i}{2} \sum h_{jk}\, dz_j \wedge d\bar{z}_k$ and $\bar{\omega} = -\frac{i}{2} \sum \overline{h_{jk}}\, d\bar{z}_j \wedge dz_k$ coincide, we must have $h_{jk} = \overline{h_{kj}}$ for all j and k. In other words, at every point $p \in \mathcal{U}$, the $n \times n$ matrix $(h_{jk}(p))$ is Hermitian. The nondegeneracy amounts to the nonvanishing of

$$\omega^n = n! \left(\frac{i}{2}\right)^n \det(h_{jk})\, dz_1 \wedge d\bar{z}_1 \wedge \cdots \wedge dz_n \wedge d\bar{z}_n.$$

Therefore, at every $p \in M$, the matrix $(h_{jk}(p))$ must be nonsingular. Finally, the positivity condition $\omega(v, Jv) > 0$, $\forall v \neq 0$, from compatibility, implies that, at each $p \in \mathcal{U}$, the matrix $(h_{jk}(p))$ is positive-definite. □

Consequently, if ω_0 and ω_1 are both Kähler forms on a compact manifold M with $[\omega_0] = [\omega_1] \in H^2_{\text{deRham}}(M)$, then (M, ω_0) and (M, ω_1) are strongly isotopic by Moser's Theorem 1.7. Indeed $\omega_t = (1-t)\omega_0 + t\omega_1$ is symplectic for $t \in [0, 1]$, as convex combinations of positive-definite matrices are still positive-definite.

Another consequence is the following recipe for Kähler forms. A smooth real function ρ on a complex manifold M is *strictly plurisubharmonic* (*s.p.s.h.*) if, on each local complex chart $(\mathcal{U}, z_1, \ldots, z_n)$, the matrix $(\frac{\partial^2 \rho}{\partial z_j \partial \bar{z}_k}(p))$ is positive-definite at all $p \in \mathcal{U}$. If $\rho \in C^\infty(M; \mathbb{R})$ is s.p.s.h., then the form

$$\omega = \frac{i}{2} \partial \bar{\partial} \rho$$

is Kähler. The function ρ is then called a (global) *Kähler potential.*

EXAMPLE. Let $M = \mathbb{C}^n \simeq \mathbb{R}^{2n}$, with complex coordinates $(z_1, \ldots, z_n)$ and corresponding real coordinates $(x_1, y_1, \ldots, x_n, y_n)$ via $z_j = x_j + iy_j$. The function

$$\rho(x_1, y_1, \ldots, x_n, y_n) = \sum_{j=1}^n (x_j^2 + y_j^2) = \sum |z_j|^2 = \sum z_j \bar{z}_j$$

is s.p.s.h. and is a Kähler potential for the standard Kähler form:

$$\frac{i}{2} \partial \bar{\partial} \rho = \frac{i}{2} \sum_{j,k} \delta_{jk} \, dz_j \wedge d\bar{z}_k = \frac{i}{2} \sum_j dz_j \wedge d\bar{z}_j = \sum_j dx_j \wedge dy_j = \omega_0.$$

There is a local converse to the previous construction of Kähler forms.

PROPOSITION 3.10. *Let ω be a closed real-valued $(1, 1)$-form on a complex manifold M and let $p \in M$. Then on a neighborhood $\mathcal{U}$ of p we have $\omega = \frac{i}{2} \partial \bar{\partial} \rho$ for some $\rho \in C^\infty(\mathcal{U}; \mathbb{R})$.*

The proof of this theorem requires holomorphic versions of Poincaré's lemma, namely, the local triviality of Dolbeault groups (the fact that any point in a complex manifold admits a neighborhood $\mathcal{U}$ such that $H^{\ell,m}_{\text{Dolbeault}}(\mathcal{U}) = 0$ for all $m > 0$) and the local triviality of the holomorphic de Rham groups; see [62].

For a Kähler ω, such a local function ρ is called a *local Kähler potential.*

PROPOSITION 3.11. *Let M be a complex manifold, $\rho \in C^\infty(M; \mathbb{R})$ s.p.s.h., X a complex submanifold, and $i : X \hookrightarrow M$ the inclusion map. Then $i^* \rho$ is s.p.s.h.*

PROOF. It suffices to verify this locally by considering a complex chart $(z_1, \dots, z_n)$ for M adapted to X so that X is given there by the equations $z_1 = \cdots = z_m = 0$. Being a principal minor of the positive-definite matrix $(\frac{\partial^2}{\partial z_j \partial \bar{z}_k}(0, \dots, 0, z_{m+1}, \dots, z_n))$ the matrix $(\frac{\partial^2 \rho}{\partial z_{m+j} \partial \bar{z}_{m+k}}(0, \dots, 0, z_{m+1}, \dots, z_n))$ is also positive-definite. □

COROLLARY 3.12. *Any complex submanifold of a Kähler manifold is also Kähler.*

DEFINITION 3.13. Let (M, ω) be a Kähler manifold, X a complex submanifold, and $i : X \hookrightarrow M$ the inclusion. Then $(X, i^*\omega)$ is called a *Kähler submanifold*.

EXAMPLES.

1. Complex vector space $(\mathbb{C}^n, \omega_0)$ where $\omega_0 = \frac{i}{2} \sum dz_j \wedge d\bar{z}_j$ is Kähler. According to Corollary 3.12, every complex submanifold of $\mathbb{C}^n$ is Kähler.
2. In particular, *Stein manifolds* are Kähler. *Stein manifolds* are the properly embedded complex submanifolds of $\mathbb{C}^n$. They can be alternatively characterized as being the Kähler manifolds (M, ω) that admit a global proper Kähler potential, i.e., $\omega = \frac{i}{2} \partial \bar{\partial} \rho$ for some proper function $\rho : M \to \mathbb{R}$.
3. The function $z \mapsto \log(|z|^2 + 1)$ on $\mathbb{C}^n$ is strictly plurisubharmonic. Therefore the 2-form

$$\omega_{\text{FS}} = \frac{i}{2} \partial \bar{\partial} \log(|z|^2 + 1)$$

   is another Kähler form on $\mathbb{C}^n$ This is called the *Fubini–Study form* on $\mathbb{C}^n$.
4. Let $\{(\mathcal{U}_j, \mathbb{C}^n, \varphi_j), j = 0, \dots, n\}$ be the usual complex atlas for *complex projective space*.[32] The form ω_{FS} is preserved by the transition maps, hence $\varphi_j^* \omega_{\text{FS}}$ and $\varphi_k^* \omega_{\text{FS}}$ agree on the overlap $\mathcal{U}_j \cap \mathcal{U}_k$. The *Fubini–Study form* on $\mathbb{CP}^n$ is the Kähler form obtained by gluing together the $\varphi_j^* \omega_{\text{FS}}$, $j = 0, \dots, n$.
5. Consequently, all *nonsingular projective varieties* are Kähler submanifolds. Here by nonsingular we mean smooth, and by projective variety we mean the zero locus of some collection of homogeneous polynomials.
6. All *Riemann surfaces* are Kähler, since any compatible almost complex structure is integrable for dimension reasons (Section 3.3).

[32] The *complex projective space* $\mathbb{CP}^n$ is the complex n-dimensional manifold given by the space of complex lines in $\mathbb{C}^{n+1}$. It can be obtained from $\mathbb{C}^{n+1} \setminus \{0\}$ by making the identifications $(z_0, \dots, z_n) \sim (\lambda z_0, \dots, \lambda z_n)$ for all $\lambda \in \mathbb{C} \setminus \{0\}$. One denotes by $[z_0, \dots, z_n]$ the equivalence class of $(z_0, \dots, z_n)$, and calls $z_0, \dots, z_n$ the *homogeneous coordinates* of the point $p = [z_0, \dots, z_n]$. (Homogeneous coordinates are, of course, only determined up to multiplication by a nonzero complex number λ.) Let $\mathcal{U}_j$ be the subset of $\mathbb{CP}^n$ consisting of all points $p = [z_0, \dots, z_n]$ for which $z_j \neq 0$. Let $\varphi_j : \mathcal{U}_j \to \mathbb{C}^n$ be the map defined by

$$\varphi_j([z_0, \dots, z_n]) = \frac{z_0}{z_j}, \dots, \frac{z_{j-1}}{z_j}, \frac{z_{j+1}}{z_j}, \dots, \frac{z_n}{z_j}.$$

The collection $\{(\mathcal{U}_j, \mathbb{C}^n, \varphi_j),\ j = 0, \dots, n\}$ is the *usual complex atlas* for $\mathbb{CP}^n$. For instance, the transition map from $(\mathcal{U}_0, \mathbb{C}^n, \varphi_0)$ to $(\mathcal{U}_1, \mathbb{C}^n, \varphi_1)$ is $\varphi_{0,1}(z_1, \dots, z_n) = (\frac{1}{z_1}, \frac{z_2}{z_1}, \dots, \frac{z_n}{z_1})$ defined from the set $\{(z_1, \dots, z_n) \in \mathbb{C}^n \mid z_1 \neq 0\}$ to itself.

7. The Fubini–Study form on the chart $\mathcal{U}_0 = \{[z_0, z_1] \in \mathbb{CP}^1 \mid z_0 \neq 0\}$ of the *Riemann sphere* $\mathbb{CP}^1$ is given by the formula

$$\omega_{\mathrm{FS}} = \frac{dx \wedge dy}{(x^2 + y^2 + 1)^2},$$

where $\frac{z_1}{z_0} = z = x + iy$ is the usual coordinate on $\mathbb{C}$. The standard area form $\omega_{\mathrm{std}} = d\theta \wedge dh$ is induced by regarding $\mathbb{CP}^1$ as the unit sphere S^2 in $\mathbb{R}^3$ (Example 3 of Section 1.2). Stereographic projection shows that $\omega_{\mathrm{FS}} = \frac{1}{4}\omega_{\mathrm{std}}$.
8. *Complex tori* are Kähler. Complex tori look like quotients $\mathbb{C}^n/\mathbb{Z}^n$ where $\mathbb{Z}^n$ is a lattice in $\mathbb{C}^n$. The form $\omega = \sum dz_j \wedge d\bar{z}_j$ induced by the Euclidean structure is Kähler.
9. Just like products of symplectic manifolds are symplectic, also products of Kähler manifolds are Kähler.

3.5. *Hodge theory*

Hodge [73] identified the spaces of cohomology classes of forms with spaces of actual forms, by picking *the* representative from each class that solves a certain differential equation, namely the *harmonic* representative. We give a sketch of Hodge's idea. The first part makes up ordinary Hodge theory, which works for any compact oriented Riemannian manifold (M, g), not necessarily Kähler.

At a point $p \in M$, let $e_1, \ldots, e_n$ be a positively oriented orthonormal basis of the cotangent space T_p^*M, with respect to the induced inner product and orientation. The *Hodge star operator* is the linear operator on the exterior algebra of T_p^*M defined by

$$\begin{aligned}
&*(1) = e_1 \wedge \cdots \wedge e_n,\\
&*(e_1 \wedge \cdots \wedge e_n) = 1,\\
&*(e_1 \wedge \cdots \wedge e_k) = e_{k+1} \wedge \cdots \wedge e_n.
\end{aligned}$$

We see that $* : \Lambda^k(T_p^*M) \to \Lambda^{n-k}(T_p^*M)$ and satisfies $** = (-1)^{k(n-k)}$. The *codifferential* and the *Laplacian* are the operators defined by

$$\begin{aligned}
&\delta = (-1)^{n(k+1)+1} * d* && : \Omega^k(M) \to \Omega^{k-1}(M),\\
&\Delta = d\delta + \delta d && : \Omega^k(M) \to \Omega^k(M).
\end{aligned}$$

The operator Δ is also called the *Laplace–Beltrami operator* and satisfies $\Delta * = * \Delta$. On $\Omega^0(\mathbb{R}^n) = C^\infty(\mathbb{R}^n)$, it is simply the usual Laplacian $\Delta = -\sum_{i=1}^n \frac{\partial^2}{\partial x_i^2}$. The *inner product on forms* of any degree,

$$\langle \cdot, \cdot \rangle : \Omega^k(M) \times \Omega^k(M) \longrightarrow \mathbb{R}, \quad \langle \alpha, \beta \rangle := \int_M \alpha \wedge *\beta,$$

satisfies $\langle d\alpha, \beta\rangle = \langle \alpha, \delta\beta\rangle$, so the codifferential δ is often denoted by d^* and called the *adjoint*[33] *of* d. Also, Δ is self-adjoint (i.e., $\langle \Delta\alpha, \beta\rangle = \langle \alpha, \Delta\beta\rangle$), and $\langle \Delta\alpha, \alpha\rangle = |d\alpha|^2 + |\delta\alpha|^2 \geqslant 0$, where $|\cdot|$ is the norm with respect to this inner product. The *harmonic* k*-forms* are the elements of $\mathcal{H}^k := \{\alpha \in \Omega^k \mid \Delta\alpha = 0\}$. Note that $\Delta\alpha = 0$ if and only if $d\alpha = \delta\alpha = 0$. Since a harmonic form is d-closed, it defines a de Rham cohomology class.

THEOREM 3.14 (Hodge). *Every de Rham cohomology class on a compact oriented Riemannian manifold* (M, g) *possesses a unique harmonic representative, i.e., there is an isomorphism* $\mathcal{H}^k \simeq H^k_{\text{deRham}}(M; \mathbb{R})$. *In particular, the spaces* $\mathcal{H}^k$ *are finite-dimensional. We also have the following orthogonal decomposition with respect to the inner product on forms:* $\Omega^k \simeq \mathcal{H}^k \oplus \Delta(\Omega^k(M)) \simeq \mathcal{H}^k \oplus d\Omega^{k-1} \oplus \delta\Omega^{k+1}$.

This decomposition is called the *Hodge decomposition on forms*. The proof of this and the next theorem involves functional analysis, elliptic differential operators, pseudodifferential operators and Fourier analysis; see for instance [62,83,141].

Here is where *complex Hodge theory* begins. When M is Kähler, the Laplacian satisfies $\Delta = 2(\bar\partial\bar\partial^* + \bar\partial^*\bar\partial)$ (see, for example, [62]) and preserves the decomposition according to type, $\Delta : \Omega^{\ell,m} \to \Omega^{\ell,m}$. Hence, harmonic forms are also bigraded

$$\mathcal{H}^k = \bigoplus_{\ell+m=k} \mathcal{H}^{\ell,m}$$

and satisfy a Künneth formula $\mathcal{H}^{\ell,m}(M \times N) \simeq \bigoplus_{p+r=\ell,\ q+s=m} \mathcal{H}^{p,q}(M) \otimes \mathcal{H}^{r,s}(N)$.

THEOREM 3.15 (Hodge). *Every Dolbeault cohomology class on a compact Kähler manifold* (M, ω) *possesses a unique harmonic representative, i.e., there is an isomorphism* $\mathcal{H}^{\ell,m} \simeq H^{\ell,m}_{\text{Dolbeault}}(M)$.

Combining the two theorems of Hodge, we find the decomposition of cohomology groups for a compact Kähler manifold

$$H^k_{\text{deRham}}(M; \mathbb{C}) \simeq \bigoplus_{\ell+m=k} H^{\ell,m}_{\text{Dolbeault}}(M),$$

known as the *Hodge decomposition*. In particular, the Dolbeault cohomology groups $H^{\ell,m}_{\text{Dolbeault}}$ are finite-dimensional and $H^{\ell,m} \simeq \overline{H^{m,\ell}}$.

Let $b^k(M) := \dim H^k_{\text{deRham}}(M)$ be the usual *Betti numbers* of M, and let $h^{\ell,m}(M) := \dim H^{\ell,m}_{\text{Dolbeault}}(M)$ be the *Hodge numbers* of M.

For an arbitrary compact symplectic manifold (M, ω), the even Betti numbers must be positive, because ω^k is closed but not exact ($k = 0, 1, \ldots, n$). In fact, if it were $\omega^k = d\alpha$, by Stokes' theorem we would have $\int_M \omega^n = \int_M d(\alpha \wedge \omega^{n-k}) = 0$, which contradicts ω^n being a volume form.

[33]When M is not compact, we still have a *formal adjoint* of d with respect to the nondegenerate bilinear pairing $\langle\cdot,\cdot\rangle : \Omega^k(M) \times \Omega^k_c(M) \to \mathbb{R}$ defined by a similar formula, where $\Omega^k_c(M)$ is the space of compactly supported k-forms.

For a compact Kähler manifold (M, ω), there are finer topological consequences coming from the Hodge theorems, as we must have $b^k = \sum_{\ell+m=k} h^{\ell,m}$ and $h^{\ell,m} = h^{m,\ell}$. The odd Betti numbers must be even because $b^{2k+1} = \sum_{\ell+m=2k+1} h^{\ell,m} = 2\sum_{\ell=0}^{k} h^{\ell,(2k+1-\ell)}$. The number $h^{1,0} = \frac{1}{2}b^1$ must be a topological invariant. The numbers $h^{\ell,\ell}$ are positive, because $0 \neq [\omega^\ell] \in H^{\ell,\ell}_{\text{Dolbeault}}(M)$. First of all, $[\omega^\ell]$ defines an element of $H^{\ell,\ell}_{\text{Dolbeault}}$ as $\omega \in \Omega^{1,1}$ implies that $\omega^\ell \in \Omega^{\ell,\ell}$, and the closedness of ω^ℓ implies that $\bar\partial\omega^\ell = 0$. If it were $\omega^\ell = \bar\partial\beta$ for some $\beta \in \Omega^{\ell-1,\ell}$, then $\omega^n = \omega^\ell \wedge \omega^{n-\ell} = \bar\partial(\beta \wedge \omega^{n-\ell})$ would be $\bar\partial$-exact. But $[\omega^n] \neq 0$ in $H^{2n}_{\text{deRham}}(M;\mathbb{C}) \simeq H^{n,n}_{\text{Dolbeault}}(M)$ since it is a volume form. A popular diagram to describe relations among Hodge numbers is the *Hodge diamond*:

$$
\begin{array}{ccccccc}
 & & & h^{n,n} & & & \\
 & & h^{n,n-1} & & h^{n-1,n} & & \\
 & h^{n,n-2} & & h^{n-1,n-1} & & h^{n-2,n} & \\
\cdots & & & \vdots & & & \cdots \\
 & h^{2,0} & & h^{1,1} & & h^{0,2} & \\
 & & h^{1,0} & & h^{0,1} & & \\
 & & & h^{0,0} & & &
\end{array}
$$

Complex conjugation gives symmetry with respect to the middle vertical, whereas the Hodge star operator induces symmetry about the center of the diamond. The middle vertical axis is all nonzero.

There are further symmetries and ongoing research on how to compute $H^{\ell,m}_{\text{Dolbeault}}$ for a compact Kähler manifold (M, ω). In particular, the *hard Lefschetz theorem* states isomorphisms $L^k : H^{n-k}_{\text{deRham}}(M) \xrightarrow{\simeq} H^{n+k}_{\text{deRham}}(M)$ given by wedging with ω^k at the level of forms and the *Lefschetz decompositions* $H^m_{\text{deRham}}(M) \simeq \bigoplus_k L^k(\ker L^{n-m+2k+1}|_{H^{m-2k}})$. The *Hodge conjecture* claims, for projective manifolds M (i.e., complex submanifolds of complex projective space), that the Poincaré duals of elements in $H^{\ell,\ell}_{\text{Dolbeault}}(M) \cap H^{2\ell}(M;\mathbb{Q})$ are rational linear combinations of classes of complex codimension ℓ subvarieties of M. This has been proved only for the $\ell = 1$ case (it is the Lefschetz theorem on $(1,1)$-classes; see, for instance, [62]).

3.6. *Pseudoholomorphic curves*

Whereas an almost complex manifold (M, J) tends to have no J-holomorphic functions $M \to \mathbb{C}$ at all,[34] it has plenty of *J-holomorphic curves* $\mathbb{C} \to M$. Gromov first realized that *pseudoholomorphic curves* provide a powerful tool in symplectic topology in an extremely influential paper [64]. Fix a closed Riemann surface (Σ, j), that is, a compact complex 1-dimensional manifold Σ without boundary and equipped with the canonical almost complex structure j.

[34] However, the study of *asymptotically J-holomorphic functions* has been recently developed to obtain important results [32,34,13]; see Section 4.6.

DEFINITION 3.16. A parametrized *pseudoholomorphic curve* (or *J-holomorphic curve*) in (M, J) is a (smooth) map $u : \Sigma \to M$ whose differential intertwines j and J, that is, $du_p \circ j_p = J_p \circ du_p, \forall p \in \Sigma$.

In other words, the *Cauchy–Riemann equation* $du + J \circ du \circ j = 0$ holds.

Pseudoholomorphic curves are related to parametrized 2-dimensional symplectic submanifolds. If a pseudoholomorphic curve $u : \Sigma \to M$ is an embedding, then its image $S := u(\Sigma)$ is a 2-dimensional almost complex submanifold, hence a symplectic submanifold. Conversely, the inclusion $i : S \hookrightarrow M$ of a 2-dimensional symplectic submanifold can be seen as a pseudoholomorphic curve. An appropriate compatible almost complex structure J on (M, ω) can be constructed starting from S, such that TS is J-invariant. The restriction j of J to TS is necessarily integrable because S is 2-dimensional.

The group G of complex diffeomorphisms of (Σ, j) acts on (parametrized) pseudoholomorphic curves by reparametrization: $u \mapsto u \circ \gamma$, for $\gamma \in G$. This normally means that each curve u has a noncompact orbit under G. The orbit space $\mathcal{M}_g(A, J)$ is the set of unparametrized pseudoholomorphic curves in (M, J) whose domain Σ has genus g and whose image $u(\Sigma)$ has homology class $A \in H_2(M; \mathbb{Z})$. The space $\mathcal{M}_g(A, J)$ is called the *moduli space of unparametrized pseudoholomorphic curves* of genus g representing the class A. For generic J, Fredholm theory shows that pseudoholomorphic curves occur in finite-dimensional smooth families, so that the moduli spaces $\mathcal{M}_g(A, J)$ can be manifolds, after avoiding singularities given by *multiple coverings*.[35]

EXAMPLE. Usually Σ is the Riemann sphere $\mathbb{CP}^1$, whose complex diffeomorphisms are those given by *fractional linear transformations* (or *Möbius transformations*). So the 6-dimensional noncompact group of projective linear transformations PSL(2; $\mathbb{C}$) acts on *pseudoholomorphic spheres* by reparametrization $u \mapsto u \circ \gamma_A$, where $A = \left[\begin{smallmatrix} a & b \\ c & d \end{smallmatrix}\right] \in$ PSL(2; $\mathbb{C}$) acts by $\gamma_A : \mathbb{CP}^1 \to \mathbb{CP}^1$, $\gamma_A[z, 1] = [\frac{az+b}{cz+d}, 1]$.

When J is an almost complex structure *compatible* with a symplectic form ω, the area of the image of a pseudoholomorphic curve u (with respect to the metric $g_J(\cdot, \cdot) = \omega(\cdot, J\cdot)$) is determined by the class A that it represents. The number

$$E(u) := \omega(A) = \int_\Sigma u^*\omega = \text{area of the image of } u \text{ with respect to } g_J$$

is called the *energy* of the curve u and is a topological invariant: it only depends on $[\omega]$ and on the homotopy class of u. Gromov proved that the constant energy of all the pseudoholomorphic curves representing a homology class A ensured that the space $\mathcal{M}_g(A, J)$, though not necessarily compact, had natural *compactifications* $\overline{\mathcal{M}}_g(A, J)$ by including what he called *cusp-curves*.

THEOREM 3.17 (Gromov's compactness theorem). *If (M, ω) is a compact manifold equipped with a generic compatible almost complex structure J, and if u_j is a sequence*

[35] A curve $u : \Sigma \to M$ is a *multiple covering* if u factors as $u = u' \circ \sigma$ where $\sigma : \Sigma \to \Sigma'$ is a holomorphic map of degree greater than 1.

of pseudoholomorphic curves in $\mathcal{M}_g(A, J)$, then there is a subsequence that weakly converges to a cusp-curve in $\overline{\mathcal{M}}_g(A, J)$.

Hence the cobordism class of the compactified moduli space $\overline{\mathcal{M}}_g(A, J)$ might be a nice symplectic invariant of (M, ω), as long as it is not empty or null-cobordant. Actually a nontrivial regularity criterion for J ensures the existence of pseudoholomorphic curves. And even when $\overline{\mathcal{M}}_g(A, J)$ is null-cobordant, we can define an invariant to be the (signed) number of pseudoholomorphic curves of genus g in class A that intersect a specified set of representatives of homology classes in M [112,128,145]. For more on pseudoholomorphic curves; see, for instance, [100] (for a comprehensive discussion of the genus 0 case) or [11] (for higher genus). Here is a selection of applications of (developments from) pseudoholomorphic curves:

- Proof of the *nonsqueezing theorem* [64]: for $R > r$ there is no symplectic embedding of a ball B_R^{2n} of radius R into a cylinder $B_r^2 \times \mathbb{R}^{2n-2}$ of radius r, both in $(\mathbb{R}^{2n}, \omega_0)$.
- Proof that there are *no Lagrangian spheres* in $(\mathbb{C}^n, \omega_0)$, except for the circle in $\mathbb{C}^2$, and more generally *no compact exact Lagrangian submanifolds*, in the sense that the tautological 1-form α restricts to an exact form [64].
- Proof that if (M, ω) is a connected symplectic 4-manifold symplectomorphic to $(\mathbb{R}^4, \omega_0)$ outside a compact set and containing no symplectic S^2's, then (M, ω) symplectomorphic to $(\mathbb{R}^4, \omega_0)$ [64].
- Study questions of *symplectic packing* [15,98,134] such as: for a given $2n$-dimensional symplectic manifold (M, ω), what is the maximal radius R for which there is a symplectic embedding of N disjoint balls B_R^{2n} into (M, ω)?
- Study *groups of symplectomorphisms* of 4-manifolds (for a review see [97]). Gromov [64] showed that $\mathrm{Sympl}(\mathbb{CP}^2, \omega_{\mathrm{FS}})$ and $\mathrm{Sympl}(S^2 \times S^2, \mathrm{pr}_1^*\sigma \oplus \mathrm{pr}_2^*\sigma)$ deformation retract onto the corresponding groups of standard isometries.
- Development of *Gromov–Witten invariants* allowing to prove, for instance, the nonexistence of symplectic forms on $\mathbb{CP}^2 \# \mathbb{CP}^2 \# \mathbb{CP}^2$ or the classification of symplectic structures on *ruled surfaces* (Section 4.3).
- Development of *Floer homology* to prove the Arnold conjecture on the fixed points of symplectomorphisms of compact symplectic manifolds, or on the intersection of Lagrangian submanifolds (Section 5.2).
- Development of *symplectic field theory* introduced by Eliashberg, Givental and Hofer [40] extending Gromov–Witten theory, exhibiting a rich algebraic structure and also with applications to *contact geometry*.

4. Symplectic geography

4.1. *Existence of symplectic forms*

The utopian goal of symplectic classification addresses the standard questions:

- (*Existence*) Which manifolds carry symplectic forms?
- (*Uniqueness*) What are the distinct symplectic structures on a given manifold?

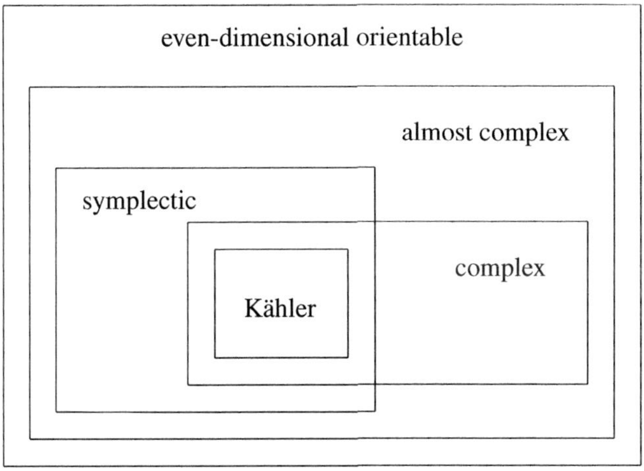

Fig. 1.

Existence is tackled through central examples in this subsection and symplectic constructions in the next two sections. Uniqueness is treated in the remainder of this subsection dealing with invariants that allow to distinguish symplectic manifolds.

A Kähler structure naturally yields both a symplectic form and a complex structure (compatible ones). Either a symplectic or a complex structure on a manifold implies the existence of an almost complex structure. Figure 1 represents the relations among these structures. In dimension 2, orientability trivially guarantees the existence of all other structures, so the picture collapses. In dimension 4, the first interesting dimension, the picture above is faithful—we will see that there are *closed* 4-dimensional examples in each region. *Closed* here means compact and without boundary.

Not all 4-dimensional manifolds are almost complex. A result of Wu [146] gives a necessary and sufficient condition in terms of the signature σ and the Euler characteristic χ of a 4-dimensional closed manifold M for the existence of an almost complex structure: $3\sigma + 2\chi = h^2$ *for some* $h \in H^2(M;\mathbb{Z})$ *congruent with the second Stiefel–Whitney class* $w_2(M)$ *modulo* 2. For example, S^4 and $(S^2 \times S^2) \# (S^2 \times S^2)$ are not almost complex. When an almost complex structure exists, the first Chern class of the tangent bundle (regarded as a complex vector bundle) satisfies the condition for h. The sufficiency of Wu's condition is the remarkable part.[36]

According to Kodaira's classification of closed complex surfaces [82], such a surface admits a Kähler structure if and only if its first Betti number b_1 is even. The necessity of this condition is a Hodge relation on the Betti numbers (Section 3.5). The complex projective plane $\mathbb{CP}^2$ with the Fubini–Study form (Section 3.4) might be called the simplest example of a closed Kähler 4-manifold.

The *Kodaira–Thurston example* [131] first demonstrated that a manifold that admits both a symplectic and a complex structure does not have to admit any Kähler structure.

[36]Moreover, such solutions h are in one-to-one correspondence with *isomorphism* classes of almost complex structures.

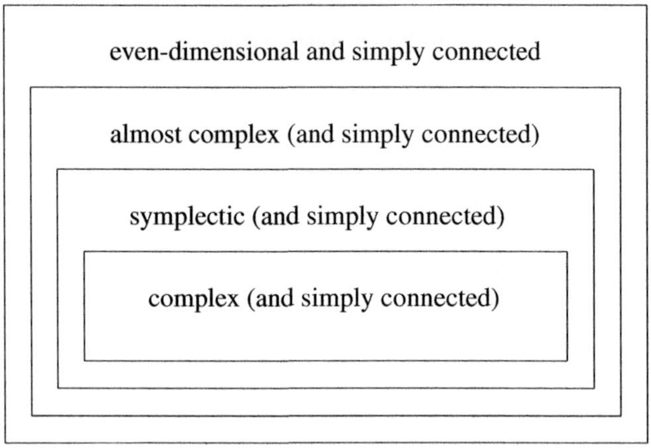

Fig. 2.

Take $\mathbb{R}^4$ with $dx_1 \wedge dy_1 + dx_2 \wedge dy_2$, and Γ the discrete group generated by the four symplectomorphisms:

$$\begin{aligned}
(x_1, x_2, y_1, y_2) &\longmapsto (x_1 + 1, x_2, y_1, y_2), \\
(x_1, x_2, y_1, y_2) &\longmapsto (x_1, x_2 + 1, y_1, y_2), \\
(x_1, x_2, y_1, y_2) &\longmapsto (x_1, x_2 + y_2, y_1 + 1, y_2), \\
(x_1, x_2, y_1, y_2) &\longmapsto (x_1, x_2, y_1, y_2 + 1).
\end{aligned}$$

Then $M = \mathbb{R}^4/\Gamma$ is a symplectic manifold that is a 2-torus bundle over a 2-torus. Kodaira's classification [82] shows that M has a complex structure. However, $\pi_1(M) = \Gamma$, hence $H_1(\mathbb{R}^4/\Gamma; \mathbb{Z}) = \Gamma/[\Gamma, \Gamma]$ has rank 3, so $b_1 = 3$ is *odd*.

Fernández–Gotay–Gray [44] first exhibited symplectic manifolds that do not admit any complex structure at all. Their examples are circle bundles over circle bundles (i.e., a *tower* of circle bundles) over a 2-torus.

The *Hopf surface* is the complex surface diffeomorphic to $S^1 \times S^3$ obtained as the quotient $\mathbb{C}^2 \setminus \{0\}/\Gamma$ where $\Gamma = \{2^n \mathrm{Id} \mid n \in \mathbb{Z}\}$ is a group of *complex* transformations, i.e., we factor $\mathbb{C}^2 \setminus \{0\}$ by the equivalence relation $(z_1, z_2) \sim (2z_1, 2z_2)$. The Hopf surface is not symplectic because $H^2(S^1 \times S^3) = 0$.

The manifold $\mathbb{CP}^2 \# \mathbb{CP}^2 \# \mathbb{CP}^2$ is almost complex but is neither complex (since it does not fit Kodaira's classification [82]), nor symplectic as shown by Taubes [126] using Seiberg–Witten invariants (Section 4.5).

We could go through the previous discussion restricting to closed 4-dimensional examples *with a specific fundamental group*. We will do this restricting to simply connected examples, where Figure 2 holds.

It is a consequence of Wu's result [146] that a simply connected manifold admits an almost complex structure if and only if b_2^+ is odd.[37] In particular, the connected sum

[37] The *intersection form* of an oriented *topological* closed 4-manifold M is the bilinear pairing $Q_M : H^2(M; \mathbb{Z}) \times H^2(M; \mathbb{Z}) \to \mathbb{Z}$, $Q_M(\alpha, \beta) := \langle \alpha \cup \beta, [M] \rangle$, where $\alpha \cup \beta$ is the *cup product* and $[M]$ is the *fundamental class*.

$\#_m \mathbb{CP}^2 \#_n \overline{\mathbb{CP}^2}$ (of m copies of $\mathbb{CP}^2$ with n copies of $\overline{\mathbb{CP}^2}$) has an almost complex structure if and only if m is odd.[38]

By Kodaira's classification [82], a simply connected complex surface always admits a compatible symplectic form (since $b^1 = 0$ is even), i.e., it is always Kähler.

Since they are simply connected, S^4, $\mathbb{CP}^2 \# \mathbb{CP}^2 \# \mathbb{CP}^2$ and $\mathbb{CP}^2$ live in three of the four regions in the picture for simply connected examples. All of $\mathbb{CP}^2 \#_m \overline{\mathbb{CP}^2}$ are also simply connected Kähler manifolds because they are *pointwise blow-ups* $\mathbb{CP}^2$ and the *blow-down map* is holomorphic; see Section 4.3.

There is a family of manifolds obtained from $\mathbb{CP}^2 \#_9 \overline{\mathbb{CP}^2} =: E(1)$ by a *knot surgery* [45] that were shown by Fintushel and Stern to be symplectic and confirmed not to admit a complex structure [109]. The first example of a closed simply connected symplectic manifold that cannot be Kähler, was a 10-dimensional manifold obtained by McDuff [94] as follows. The Kodaira–Thurston example $\mathbb{R}^4/\Gamma$ (not simply connected) embeds symplectically in $(\mathbb{CP}^5, \omega_{\mathrm{FS}})$ [65,132]. McDuff's example is a *blow-up* of $(\mathbb{CP}^5, \omega_{\mathrm{FS}})$ along the image of $\mathbb{R}^4/\Gamma$.

Geography problems are problems on the existence of simply connected closed oriented 4-dimensional manifolds with some additional structure (such as, a symplectic form or a complex structure) for each pair of *topological coordinates*. As a consequence of the work of Freedman [51] and Donaldson [30] in the 80's, it became known that the homeomorphism class of a connected simply connected closed oriented *smooth* 4-manifold is determined by the two integers—the second Betti number and the signature (b_2, σ)—and the *parity*[39] of the intersection form. Forgetting about the parity, the numbers (b_2, σ) can be treated as *topological coordinates*. For each pair (b_2, σ) there could well be infinite different (i.e., nondiffeomorphic) smooth manifolds. Using Riemannian geometry, Cheeger [22] showed that there are at most *countably many* different smooth types for closed 4-manifolds. There are no known finiteness results for the smooth types of a given topological 4-manifold, in contrast to other dimensions.

Traditionally, the numbers used are $(c_1^2, c_2) := (3\sigma + 2\chi, \chi) = (3\sigma + 4 + 2b_2, 2 + b_2)$, and frequently just the *slope* c_1^2/c_2 is considered. If M admits an almost complex structure J, then (TM, J) is a complex vector bundle, hence has Chern classes $c_1 = c_1(M, J)$ and $c_2 = c_2(M, J)$. Both $c_1^2 := c_1 \cup c_1$ and c_2 may be regarded as numbers since $H^4(M; \mathbb{Z}) \simeq \mathbb{Z}$. They satisfy $c_1^2 = 3\sigma + 2\chi$ (by Hirzebruch's signature formula) and $c_2 = \chi$ (because the top Chern class is always the Euler class), justifying the notation for the topological coordinates in this case.

Since Q_M always vanishes on torsion elements, descending to $H^2(M; \mathbb{Z})/\text{torsion}$ it can be represented by a matrix. When M is smooth and simply connected, this pairing is $Q_M(\alpha, \beta) := \int_M \alpha \wedge \beta$ since nontorsion elements are representable by 2-forms. As Q_M is symmetric (in the smooth case, the wedge product of 2-forms is symmetric) and unimodular (the determinant of a matrix representing Q_M is ± 1 by Poincaré duality), it is diagonalizable over $\mathbb{R}$ with eigenvalues ± 1. We denote by b_2^+ (respectively b_2^-) the number of positive (respectively negative) eigenvalues of Q_M counted with multiplicities, i.e., the dimension of a maximal subspace where Q_M is positive-definite (respectively negative-definite). The *signature* of M is the difference $\sigma := b_2^+ - b_2^-$, whereas the second Betti number is the sum $b_2 = b_2^+ + b_2^-$, i.e., the *rank* of Q_M. The *type* of an intersection form is *definite* if it is positive or negative definite (i.e., $|\sigma| = b_2$) and *indefinite* otherwise.

[38] The intersection form of a connected sum $M_0 \# M_1$ is (isomorphic to) $Q_{M_0} \oplus Q_{M_1}$.

[39] We say that the *parity* of an intersection form Q_M is *even* when $Q_M(\alpha, \alpha)$ is even for all $\alpha \in H^2(M; \mathbb{Z})$, and *odd* otherwise.

EXAMPLES. The manifold $\mathbb{CP}^2$ has $(b_2, \sigma) = (1, 1)$, i.e., $(c_1^2, c_2) = (9, 3)$. Reversing the orientation $\overline{\mathbb{CP}^2}$ has $(b_2, \sigma) = (1, -1)$, i.e., $(c_1^2, c_2) = (3, 3)$. Their connected sum $\mathbb{CP}^2 \# \overline{\mathbb{CP}^2}$ has $(b_2, \sigma) = (2, 0)$, i.e., $(c_1^2, c_2) = (8, 0)$. The product $S^2 \times S^2$ also has $(b_2, \sigma) = (2, 0)$, i.e., $(c_1^2, c_2) = (8, 4)$. But $\mathbb{CP}^2 \# \overline{\mathbb{CP}^2}$ has an *odd* intersection form whereas $S^2 \times S^2$ has an *even* intersection form: $\left[\begin{smallmatrix} 1 & 0 \\ 0 & -1 \end{smallmatrix}\right]$ vs. $\left[\begin{smallmatrix} 0 & 1 \\ 1 & 0 \end{smallmatrix}\right]$.

Symplectic geography [60,122] addresses the following question: What is the set of pairs of integers $(m, n) \in \mathbb{Z} \times \mathbb{Z}$ for which there exists a connected simply connected closed *symplectic* 4-manifold M having second Betti number $b_2(M) = m$ and signature $\sigma(M) = n$? This problem includes the usual geography of simply connected complex surfaces, since all such surfaces are Kähler according to Kodaira's classification [82]. Often, instead of the numbers (b_2, σ), the question is equivalently phrased in terms of the Chern numbers (c_1^2, c_2) for a compatible almost complex structure, which satisfy $c_1^2 = 3\sigma + 2\chi$ [146] and $c_2 = \chi$, where $\chi = b_2 + 2$ is the *Euler number*. Usually only *minimal* (Section 4.3) or *irreducible* manifolds are considered to avoid trivial examples. A manifold is *irreducible* when it is not a connected sum of other manifolds, except when one of the summands is a homotopy sphere.

It was speculated that perhaps any simply connected closed smooth 4-manifold other than S^4 is diffeomorphic to a connected sum of symplectic manifolds, where any orientation is allowed on each summand (the so-called *minimal conjecture* for smooth 4-manifolds). Szabó [124,125] provided counterexamples in a family of irreducible simply connected closed nonsymplectic smooth 4-manifolds.

All these problems could be posed for other fundamental groups. Gompf [57] used *symplectic sums* (Section 4.2) to prove the following theorem. He also proved that his surgery construction can be adapted to produce *non*-Kähler examples. Since finitely-presented groups are not classifiable, this shows that compact symplectic 4-manifold are not classifiable.

THEOREM 4.1 (Gompf). *Every finitely-presented group occurs as the fundamental group $\pi_1(M)$ of a compact symplectic 4-manifold (M, ω).*

4.2. *Fibrations and sums*

Products of symplectic manifolds are naturally symplectic. As we will see, special kinds of *twisted products*, i.e., fibrations,[40] are also symplectic.

[40] A *fibration* (or *fiber bundle*) is a manifold M (called the *total space*) with a submersion $\pi : M \to X$ to a manifold X (the *base*) that is locally trivial in the sense that there is an open covering of X, such that, to each set $\mathcal{U}$ in that covering corresponds a diffeomorphism of the form $\varphi_{\mathcal{U}} = (\pi, s_{\mathcal{U}}) : \pi^{-1}(\mathcal{U}) \to \mathcal{U} \times F$ (a *local trivialization*) where F is a fixed manifold (the *model fiber*). A collection of local trivializations such that the sets $\mathcal{U}$ cover X is called a *trivializing cover* for π. Given two local trivializations, the second entry of the composition $\varphi_{\mathcal{V}} \circ \varphi_{\mathcal{U}}^{-1} = (\mathrm{id}, \psi_{\mathcal{U}\mathcal{V}})$ on $(\mathcal{U} \cap \mathcal{V}) \times F$ gives the corresponding *transition function* $\psi_{\mathcal{U}\mathcal{V}}(x) : F \to F$ at each $x \in \mathcal{U} \cap \mathcal{V}$.

DEFINITION 4.2. A *symplectic fibration* is a fibration $\pi : M \to X$ where the model fiber is a symplectic manifold (F, σ) and with a trivializing cover for which all the transition functions are symplectomorphisms $F \to F$.

In a symplectic fibration each *fiber* $\pi^{-1}(x)$ carries a *canonical symplectic form* σ_x defined by the restriction of $s_{\mathcal{U}}^* \sigma$, for any domain $\mathcal{U}$ of a trivialization covering x (i.e., $x \in \mathcal{U}$). A symplectic form ω on the total space M of a symplectic fibration is called *compatible* with the fibration if each fiber $(\pi^{-1}(x), \sigma_x)$ is a symplectic submanifold of (M, ω), i.e., σ_x is the restriction of ω to $\pi^{-1}(x)$.

EXAMPLES.

1. *Every compact oriented*[41] *fibration whose model fiber F is an oriented surface admits a structure of symplectic fibration* for the following reason. Let σ_0 be an area form on F. Each transition function $\psi_{\mathcal{UV}}(x) : F \to F$ pulls σ_0 back to a cohomologous area form σ_1 (depending on $\psi_{\mathcal{UV}}(x)$). Convex combinations $\sigma_t = (1-t)\sigma_0 + t\sigma_1$ give a path of area forms from σ_0 to σ_1 with constant class $[\sigma_t]$. By Moser's argument (Section 1.4), there exists a diffeomorphism $\rho(x) : F \to F$ isotopic to the identity, depending smoothly on $x \in \mathcal{U} \cap \mathcal{V}$, such that $\psi_{\mathcal{UV}}(x) \circ \rho(x)$ is a symplectomorphism of (F, σ_0). By successively adjusting local trivializations for a finite covering of the base, we can make all transition functions into symplectomorphisms.
2. *Every fibration with connected base and compact fibers having a symplectic form ω for which all fibers are symplectic submanifolds admits a structure of symplectic fibration compatible with ω.* Indeed, under trivializations, the restrictions of ω to the fibers give cohomologous symplectic forms in the model fiber F. So by Moser's Theorem 1.7, all fibers are strongly isotopic to (F, σ) where σ is the restriction of ω to a chosen fiber. These isotopies can be used to produce a trivializing cover where each $s_{\mathcal{U}}(x)$ is a symplectomorphism.

In the remainder of this subsection, assume that for a fibration $\pi : M \to X$ the total space is compact and the base is connected. For the existence of a compatible symplectic form on a symplectic fibration, a necessary condition is the existence of a cohomology class in M that restricts to the classes of the fiber symplectic forms. Thurston [131] showed that, when the base admits also a symplectic form, this condition is sufficient. Yet not all symplectic fibrations with a compatible symplectic form have a symplectic base [138].

THEOREM 4.3 (Thurston). *Let $\pi : M \to X$ be a compact symplectic fibration with connected symplectic base (X, α) and model fiber (F, σ). If there is a class $[\nu] \in H^2(M)$ pulling back to $[\sigma]$, then, for sufficiently large $k > 0$, there exists a symplectic form ω_k on M that is compatible with the fibration and is in $[\nu + k\pi^*\alpha]$.*

PROOF. We first find a form τ on M in the class $[\nu]$ that restricts to the canonical symplectic form on each fiber. Pick a trivializing cover $\{\varphi_i = (\pi, s_i) \mid i \in I\}$ with contractible

[41] An *oriented fibration* is a fibration whose model fiber is oriented and there is a trivializing cover for which all transition functions preserve orientation.

domains $\mathcal{U}_i$. Let ρ_i, $i \in I$, be a partition of unity subordinate to this covering and let $\tilde{\rho}_i := \rho_i \circ \pi : M \to \mathbb{R}$. Since $[\nu]$ always restricts to the class of the canonical symplectic form $[\sigma_x]$, and the $\mathcal{U}_i$'s are contractible, on each $\pi_i^{-1}(\mathcal{U}_i)$ the forms $s_i^*\sigma - \nu$ are exact. Choose 1-forms λ_i such that $s_i^*\sigma = \nu + d\lambda_i$, and set

$$\tau := \nu + \sum_{i \in I} d(\tilde{\rho}_i \lambda_i).$$

Since τ is nondegenerate on the (vertical) subbundle given by the kernel of $d\pi$, for $k > 0$ large enough the form $\tau + k\pi^*\alpha$ is nondegenerate on M. □

COROLLARY 4.4. *Let $\pi : M \to X$ be a compact oriented fibration with connected symplectic base (X, α) and model fiber an oriented surface F of genus $g(F) \neq 1$. Then π admits a compatible symplectic form.*

PROOF. By Example 1 above, $\pi : M \to X$ admits a structure of symplectic fibration with model fiber (F, σ). Since the fiber is not a torus ($g(F) \neq 1$), the Euler class of the tangent bundle TF (which coincides with $c_1(F, \sigma)$) is $\lambda[\sigma]$ for some $\lambda \neq 0$. Hence, the first Chern class $[c]$ of the *vertical* subbundle given by the kernel of $d\pi$ (assembling the tangent bundles to the fibers) restricts to $\lambda[\sigma_x]$ on the fiber over $x \in X$. We can apply Theorem 4.3 using the class $[\nu] = \lambda^{-1}[c]$. □

A pointwise connected sum $M_0 \# M_1$ of symplectic manifolds (M_0, ω_0) and (M_1, ω_1) tends to not admit a symplectic form, even if we only require the eventual symplectic form to be isotopic to ω_i on each M_i minus a ball. The reason [7] is that such a symplectic form on $M_0 \# M_1$ would allow to construct an almost complex structure on the sphere formed by the union of the two removed balls, which is known not to exist except on S^2 and S^6. Therefore:

PROPOSITION 4.5. *Let (M_0, ω_0) and (M_1, ω_1) be two compact symplectic manifolds of dimension not 2 nor 6. Then the connected sum $M_0 \# M_1$ does not admit any symplectic structure isotopic to ω_i on M_i minus a ball, $i = 1, 2$.*

For connected sums to work in the symplectic category, they should be done along codimension 2 symplectic submanifolds. The following construction, already mentioned in [65], was dramatically explored and popularized by Gompf [57] (he used it to prove Theorem 4.1). Let (M_0, ω_0) and (M_1, ω_1) be two $2n$-dimensional symplectic manifolds. Suppose that a compact symplectic manifold (X, α) of dimension $2n - 2$ admits symplectic embeddings to both $i_0 : X \hookrightarrow M_0$, $i_1 : X \hookrightarrow M_1$. For simplicity, assume that the corresponding normal bundles are trivial (in general, they need to have symmetric Euler classes). By the symplectic neighborhood theorem (Theorem 1.11), there exist symplectic embeddings $j_0 : X \times B_\varepsilon \to M_0$ and $j_1 : X \times B_\varepsilon \to M_1$ (called *framings*) where B_ε is a ball of radius ε and centered at the origin in $\mathbb{R}^2$ such that $j_k^*\omega_k = \alpha + dx \wedge dy$ and $j_k(p, 0) = i_k(p)$, $\forall p \in X$, $k = 0, 1$. Chose an area- and orientation-preserving diffeomorphism ϕ of the annulus $B_\varepsilon \setminus B_\delta$ for $0 < \delta < \varepsilon$ that interchanges the two boundary components. Let $\mathcal{U}_k = j_k(X \times B_\delta) \subset M_k$, $k = 0, 1$. A *symplectic sum* of M_0 and M_1 along X is defined to be

$$M_0 \#_X M_1 := (M_0 \setminus \mathcal{U}_0) \cup_\phi (M_1 \setminus \mathcal{U}_1),$$

where the symbol $\cup_\phi$ means that we identify $j_1(p,q)$ with $j_0(p, \phi(q))$ for all $p \in X$ and $\delta < |q| < \varepsilon$. As ω_0 and ω_1 agree on the regions under identification, they induce a symplectic form on $M_0 \#_X M_1$. The result depends on j_0, j_1, δ and ϕ.

Rational blow-down is a surgery on 4-manifolds that replaces a neighborhood of a chain of embedded S^2's with boundary a *lens space* $L(n^2, n-1)$ by a manifold with the same rational homology as a ball. This simplifies the homology possibly at the expense of complicating the fundamental group. Symington [123] showed that rational blow-down preserves a symplectic structure if the original spheres are symplectic surfaces in a symplectic 4-manifold.

4.3. *Symplectic blow-up*

Symplectic blow-up is the extension to the symplectic category of the blow-up operation in algebraic geometry. It is due to Gromov according to the first printed exposition of this operation in [94].

Let L be the *tautological line bundle* over $\mathbb{CP}^{n-1}$, that is,

$$L = \{([p], z) \mid p \in \mathbb{C}^n \setminus \{0\},\ z = \lambda p \text{ for some } \lambda \in \mathbb{C}\}$$

with projection to $\mathbb{CP}^{n-1}$ given by $\pi : ([p], z) \mapsto [p]$. The fiber of L over the point $[p] \in \mathbb{CP}^{n-1}$ is the complex line in $\mathbb{C}^n$ represented by that point. The *blow-up of* $\mathbb{C}^n$ *at the origin* is the total space of the bundle L, sometimes denoted $\tilde{\mathbb{C}}^n$. The corresponding *blow-down map* is the map $\beta : L \to \mathbb{C}^n$ defined by $\beta([p], z) = z$. The total space of L may be decomposed as the disjoint union of two sets: the zero section

$$E := \{([p], 0) \mid p \in \mathbb{C}^n \setminus \{0\}\}$$

and

$$S := \{([p], z) \mid p \in \mathbb{C}^n \setminus \{0\},\ z = \lambda p \text{ for some } \lambda \in \mathbb{C}^*\}.$$

The set E is called the *exceptional divisor*; it is diffeomorphic to $\mathbb{CP}^{n-1}$ and gets mapped to the origin by β. On the other hand, the restriction of β to the complementary set S is a diffeomorphism onto $\mathbb{C}^n \setminus \{0\}$. Hence, we may regard L as being obtained from $\mathbb{C}^n$ by smoothly replacing the origin by a copy of $\mathbb{CP}^{n-1}$. Every biholomorphic map $f : \mathbb{C}^n \to \mathbb{C}^n$ with $f(0) = 0$ lifts uniquely to a biholomorphic map $\tilde{f} : L \to L$ with $\tilde{f}(E) = E$. The lift is given by the formula

$$\tilde{f}([p], z) = \begin{cases} ([f(z)], f(z)) & \text{if } z \neq 0, \\ ([p], 0) & \text{if } z = 0. \end{cases}$$

There are actions of the unitary group $U(n)$ on L, E and S induced by the standard linear action on $\mathbb{C}^n$, and the map β is $U(n)$-equivariant. For instance, $\beta^*\omega_0 + \pi^*\omega_{FS}$ is a $U(n)$-invariant Kähler form on L.

DEFINITION 4.6. A *blow-up symplectic form* on the tautological line bundle L is a $U(n)$-invariant symplectic form ω such that the difference $\omega - \beta^*\omega_0$ is compactly supported, where $\omega_0 = \frac{i}{2}\sum_{k=1}^n dz_k \wedge d\bar{z}_k$ is the standard symplectic form on $\mathbb{C}^n$.

Two blow-up symplectic forms are *equivalent* if one is the pullback of the other by a $U(n)$-equivariant diffeomorphism of L. Guillemin and Sternberg [69] showed that two blow-up symplectic forms are equivalent if and only if they have equal restrictions to the exceptional divisor $E \subset L$. Let Ω^ε ($\varepsilon > 0$) be the set of all blow-up symplectic forms on L whose restriction to the exceptional divisor $E \simeq \mathbb{CP}^{n-1}$ is $\varepsilon\omega_{FS}$, where ω_{FS} is the Fubini–Study form (Section 3.4). An *ε-blow-up* of $\mathbb{C}^n$ at the origin is a pair (L, ω) with $\omega \in \Omega^\varepsilon$.

Let (M, ω) be a $2n$-dimensional symplectic manifold. It is a consequence of Darboux's theorem (Theorem 1.9) that, for each point $p \in M$, there exists a complex chart $(\mathcal{U}, z_1, \ldots, z_n)$ centered at p and with image in $\mathbb{C}^n$ where $\omega|_{\mathcal{U}} = \frac{i}{2}\sum_{k=1}^n dz_k \wedge d\bar{z}_k$. It is shown in [69] that, for ε small enough, we can perform an ε-blow-up of M at p modeled on $\mathbb{C}^n$ at the origin, without changing the symplectic structure outside of a small neighborhood of p. The resulting manifold is called an *ε-blow-up of M at p*. As a manifold, the blow-up of M at a point is diffeomorphic to the *connected sum*[42] $M \# \overline{\mathbb{CP}^n}$, where $\overline{\mathbb{CP}^n}$ is the manifold $\mathbb{CP}^n$ equipped with the orientation opposite to the natural complex one.

EXAMPLE. Let $\mathbb{P}(L \oplus \mathbb{C})$ be the $\mathbb{CP}^1$-bundle over $\mathbb{CP}^{n-1}$ obtained by projectivizing the direct sum of the tautological line bundle L with a trivial complex line bundle. Consider the map

$$\begin{aligned} \beta : \mathbb{CP}(L \oplus \mathbb{C}) &\longrightarrow \mathbb{CP}^n, \\ \big([p], [\lambda p : w]\big) &\longmapsto [\lambda p : w], \end{aligned}$$

where $[\lambda p : w]$ on the right represents a line in $\mathbb{C}^{n+1}$, forgetting that, for each $[p] \in \mathbb{CP}^{n-1}$, that line sits in the 2-complex-dimensional subspace $L_{[p]} \oplus \mathbb{C} \subset \mathbb{C}^n \oplus \mathbb{C}$. Notice that β maps the *exceptional divisor*

$$E := \big\{\big([p], [0 : \ldots : 0 : 1]\big) \mid [p] \in \mathbb{CP}^{n-1}\big\} \simeq \mathbb{CP}^{n-1}$$

to the point $[0 : \ldots : 0 : 1] \in \mathbb{CP}^n$, and β is a diffeomorphism on the complement

$$S := \big\{\big([p], [\lambda p : w]\big) \mid [p] \in \mathbb{CP}^{n-1},\ \lambda \in \mathbb{C}^*,\ w \in \mathbb{C}\big\} \simeq \mathbb{CP}^n \setminus \big\{[0 : \ldots : 0 : 1]\big\}.$$

[42]The *connected sum* of two oriented m-dimensional manifolds M_0 and M_1 is the manifold, denoted $M_0 \# M_1$, obtained from the union of those manifolds each with a small ball removed $M_i \setminus B_i$ by identifying the boundaries via a (smooth) map $\phi : \partial B_1 \to \partial B_2$ that extends to an orientation-preserving diffeomorphism of neighborhoods of ∂B_1 and ∂B_2 (interchanging the inner and outer boundaries of the annuli).

Therefore, we may regard $\mathbb{CP}(L \oplus \mathbb{C})$ as being obtained from $\mathbb{CP}^n$ by smoothly replacing the point $[0 : \ldots : 0 : 1]$ by a copy of $\mathbb{CP}^{n-1}$. The space $\mathbb{CP}(L \oplus \mathbb{C})$ is the blow-up of $\mathbb{CP}^n$ at the point $[0 : \ldots : 0 : 1]$, and β is the corresponding blow-down map. The manifold $\mathbb{CP}(L \oplus \mathbb{C})$ for $n = 2$ is a *Hirzebruch surface*.

When $(\mathbb{CP}^{n-1}, \omega_{\mathrm{FS}})$ is symplectically embedded in a symplectic manifold (M, ω) with image X and normal bundle isomorphic to the tautological bundle L, it can be subject to a *blow-down* operation. By the symplectic neighborhood theorem (Theorem 1.11), some neighborhood $\mathcal{U} \subset M$ of the image X is symplectomorphic to a neighborhood $\mathcal{U}_0 \subset L$ of the zero section. It turns out that some neighborhood of $\partial\mathcal{U}_0$ in L is symplectomorphic to a spherical shell in $(\mathbb{C}^n, \omega_0)$. The *blow-down of M along X* is a manifold obtained from the union of $M \setminus \mathcal{U}$ with a ball in $\mathbb{C}^n$. For more details, see [99, §7.1].

Following algebraic geometry, we call *minimal* a $2n$-dimensional symplectic manifold (M, ω) without any symplectically embedded $(\mathbb{CP}^{n-1}, \omega_{\mathrm{FS}})$, so that (M, ω) is not the blow-up at a point of another symplectic manifold. In dimension 4, a manifold is minimal if it does not contain any embedded sphere S^2 with self-intersection -1. Indeed, by the work of Taubes [126,129], if such a sphere S exists, then either the homology class $[S]$ or its symmetric $-[S]$ can be represented by a *symplectically* embedded sphere with self-intersection -1.

For a symplectic manifold (M, ω), let $i : X \hookrightarrow M$ be the inclusion of a symplectic submanifold. The normal bundle NX to X in M admits a structure of complex vector bundle (as it is a symplectic vector bundle). Let $\mathbb{P}(NX) \to X$ be the projectivization of the bundle $NX \to X$, let Z be the zero section of NX, let $L(NX)$ be the corresponding *tautological line bundle* (given by assembling the tautological line bundles over each fiber) and let $\beta : L(NX) \to NX$ be the blow-down map. On the *exceptional divisor*

$$E := \{([p], 0) \in L(NX) \mid p \in NX \setminus Z\} \simeq \mathbb{P}(NX)$$

the map β is just projection to the zero section Z. The restriction of β to the complement $L(NX) \setminus E$ is a diffeomorphism to $NX \setminus Z$. Hence, $L(NX)$ may be viewed as being obtained from NX by smoothly replacing each point of the zero section by the projectivization of its normal space. We symplectically identify some tubular neighborhood $\mathcal{U}$ of X in M with a tubular neighborhood $\mathcal{U}_0$ of the zero section Z in NX. A *blow-up of the symplectic manifold (M, ω) along the symplectic submanifold X* is the manifold obtained from the union of $M \setminus \mathcal{U}$ and $\beta^{-1}(\mathcal{U}_0)$ by identifying neighborhoods of $\partial\mathcal{U}$, and equipped with a symplectic form that restricts to ω on $M \setminus \mathcal{U}$ [94]. When X is one point, this construction reduces to the previous symplectic blow-up at a point.

Often symplectic geography concentrates on minimal examples. McDuff [95] showed that a minimal symplectic 4-manifold with a symplectically embedded S^2 with nonnegative self-intersection is symplectomorphic either to $\mathbb{CP}^2$ or to an S^2-bundle over a surface. Using Seiberg–Witten theory it was proved:

THEOREM 4.7. *Let (M, ω) be a minimal closed symplectic 4-manifold.*

(a) (Taubes [129]) *If $b_2^+ > 1$, then $c_1^2 \geqslant 0$.*

(b) (Liu [89]) *If* $b_2^+ = 1$ *and* $c_1^2 < 0$, *then* M *is the total space of an* S^2*-fibration over a surface of genus* g *where* ω *is nondegenerate on the fibers, and* $(c_1^2, c_2) = (8 - 8g, 4 - 4g)$, *i.e.,* (M, ω) *is a* symplectic ruled surface.

A *symplectic ruled surface*[43] is a symplectic 4-manifold (M, ω) that is the total space of an S^2-fibration where ω is nondegenerate on the fibers.

A *symplectic rational surface* is a symplectic 4-manifold (M, ω) that can be obtained from the standard $(\mathbb{CP}^2, \omega_{\mathrm{FS}})$ by blowing up and blowing down.

With $b_2^+ = 1$ and $c_1^2 = 0$, we have symplectic manifolds $\mathbb{CP}^2 \#_9 \overline{\mathbb{CP}^2} =: E(1)$, the *Dolgachev surfaces* $E(1, p, q)$, the results $E(1)_K$ of surgery on a fibered knot $K \subset S^3$, etc. With $b_2^+ = 1$ and $c_1^2 > 0$, we have symplectic manifolds $\mathbb{CP}^2$, $S^2 \times S^2$, $\mathbb{CP}^2 \#_n \overline{\mathbb{CP}^2}$ for $n \leqslant 8$ and the *Barlow surface*. For $b_2^+ = 1$ and $c_1^2 \geqslant 0$, Park [109] gave a criterion for a symplectic 4-manifold to be rational or ruled in terms of Seiberg–Witten theory.

4.4. *Uniqueness of symplectic forms*

Besides the notions listed in Section 1.4, the following equivalence relation for symplectic manifolds is considered. As it allows the cleanest statements about uniqueness, this relation is simply called *equivalence*.

DEFINITION 4.8. Symplectic manifolds (M, ω_0) and (M, ω_1) are *equivalent* if they are related by a combination of deformation-equivalences and symplectomorphisms.

Recall that (M, ω_0) and (M, ω_1) are *deformation-equivalent* when there is a smooth family ω_t of symplectic forms joining ω_0 to ω_1 (Section 1.4), and they are *symplectomorphic* when there is a diffeomorphism $\varphi : M \to M$ such that $\varphi^* \omega_1 = \omega_0$ (Section 1.2). Hence, equivalence is the relation generated by deformations and diffeomorphisms. The corresponding equivalence classes can be viewed as the connected components of the moduli space of symplectic forms up to diffeomorphism. This is a useful notion when focusing on topological properties.

EXAMPLES.

1. The complex projective plane $\mathbb{CP}^2$ has a unique symplectic structure up to symplectomorphism and scaling. This was shown by Taubes [128] relating Seiberg–Witten invariants (Section 4.5) to pseudoholomorphic curves to prove the existence of a pseudoholomorphic sphere. Previous work of Gromov [64] and McDuff [96] showed that the existence of a pseudoholomorphic sphere implies that the symplectic form is standard.

   Lalonde and McDuff [85] concluded similar classifications for symplectic ruled surfaces and for symplectic rational surfaces (Section 4.3). The symplectic form on

[43] A (rational) *ruled surface* is a complex (Kähler) surface that is the total space of a holomorphic fibration over a Riemann surface with fiber $\mathbb{CP}^1$. When the base is also a sphere, these are the *Hirzebruch surfaces* $\mathbb{P}(L \oplus \mathbb{C})$ where L is a holomorphic line bundle over $\mathbb{CP}^1$.

a symplectic ruled surface is unique up to symplectomorphism in its cohomology class, and is isotopic to a standard Kähler form. In particular, any symplectic form on $S^2 \times S^2$ is symplectomorphic to $a\pi_1^*\sigma + b\pi_2^*\sigma$ for some $a, b > 0$ where σ is the standard area form on S^2.

Li–Liu [88] showed that the symplectic structure on $\mathbb{CP}^2 \#_n \overline{\mathbb{CP}^2}$ for $2 \leqslant n \leqslant 9$ is unique up to equivalence.

2. McMullen and Taubes [101] first exhibited simply connected closed 4-manifolds admitting inequivalent symplectic structures. Their examples were constructed using 3-dimensional topology, and distinguished by analyzing the structure of Seiberg–Witten invariants to show that the first Chern classes (Section 3.2) of the two symplectic structures lie in disjoint orbits of the diffeomorphism group. In higher dimensions there were previously examples of manifolds with inequivalent symplectic forms; see, for instance, [111].

   With symplectic techniques and avoiding gauge theory, Smith [117] showed that, for each $n \geqslant 2$, there is a simply connected closed 4-manifold that admits at least n inequivalent symplectic forms, also distinguished via the first Chern classes. It is not yet known whether there exist inequivalent symplectic forms on a 4-manifold with the same first Chern class.

4.5. *Invariants for* 4*-manifolds*

Very little was known about 4-dimensional manifolds until 1981, when Freedman [51] provided a complete classification of closed simply connected *topological* 4-manifolds, and shortly thereafter Donaldson [30] showed that the panorama for *smooth* 4-manifolds was much wilder.[44] Freedman showed that, modulo homeomorphism, such topological manifolds are essentially classified by their intersection forms (for an *even* intersection form there is exactly one class, whereas for an *odd* intersection form there are exactly two classes distinguished by the *Kirby–Siebenmann invariant KS*, at most one of which admits smooth representatives—smoothness requires $KS = 0$). Donaldson showed that, whereas the existence of a smooth structure imposes strong constraints on the topological type of a manifold, for the same topological manifold there can be infinite different smooth structures.[45] In other words, by far not all intersection forms can occur for smooth 4-manifolds and the same intersection form may correspond to nondiffeomorphic manifolds.

Donaldson's key tool was a set of gauge-theoretic invariants, defined by counting with signs the equivalence classes (modulo gauge equivalence) of connections on SU(2)- (or SO(3)-) bundles over M whose curvature has vanishing self-dual part. For a dozen years there was hard work on the invariants discovered by Donaldson but limited advancement on the understanding of smooth 4-manifolds.

[44] It had been proved by Rokhlin in 1952 that if such a smooth manifold M has even intersection form Q_M (i.e., $w_2 = 0$), then the signature of Q_M must be a multiple of 16. It had been proved by Whitehead and Milnor that two such topological manifolds are homotopy equivalent if and only if they have the same intersection form.

[45] It is known that in dimensions $\leqslant 3$, each topological manifold has exactly one smooth structure, and in dimensions $\geqslant 5$ each topological manifold has at most finitely many smooth structures. For instance, whereas each topological $\mathbb{R}^n$, $n \neq 4$, admits a unique smooth structure, the topological $\mathbb{R}^4$ admits uncountably many smooth structures.

EXAMPLES. Finding *exotic*[46] smooth structures on closed simply connected manifolds with small b_2 has long been an interesting problem, especially in view of the smooth Poincaré conjecture for 4-manifolds. The first exotic smooth structures on a rational surface $\mathbb{CP}^2 \#_n \overline{\mathbb{CP}^2}$ were found in the late 80's for $n = 9$ by Donaldson [31] and for $n = 8$ by Kotschick [84]. There was no progress until the recent work of Park [110] constructing a symplectic exotic $\mathbb{CP}^2 \#_7 \overline{\mathbb{CP}^2}$ and using this to exhibit a third distinct smooth structure $\mathbb{CP}^2 \#_8 \overline{\mathbb{CP}^2}$, thus illustrating how the existence of symplectic forms is tied to the existence of different smooth structures. This stimulated research by Fintushel, Ozsváth, Park, Stern, Stipsicz and Szabó, which together shows that there are infinitely many exotic smooth structures on $\mathbb{CP}^2 \#_n \overline{\mathbb{CP}^2}$ for $n = 5, 6, 7, 8$ (the case $n = 9$ had been shown in the late 80's by Friedman–Morgan and by Okonek–Van de Ven).

In 1994 Witten brought about a revolution in Donaldson theory by introducing a new set of invariants—the *Seiberg–Witten invariants*—which are much simpler to calculate and to apply. This new viewpoint was inspired by developments due to Seiberg and Witten in the understanding of $N = 2$ *supersymmetric Yang–Mills*.

Let M be a smooth oriented closed 4-dimensional manifold with $b_2^+(M) > 1$ (there is a version for $b_2^+(M) = 1$). All such 4-manifolds M (with any $b_2^+(M)$) admit a spin-c structure, i.e., a $\mathrm{Spin}^c(4)$-bundle over M with an isomorphism of the associated SO(4)-bundle to the bundle of oriented frames on the tangent bundle for some chosen Riemannian metric. Let $\mathcal{C}_M = \{a \in H^2(M; \mathbb{Z}) \mid a \equiv w_2(TM)(2)\}$ be the set of characteristic elements, and let $\mathrm{Spin}^c(M)$ be the set of spin-c structures on M. For simplicity, assume that M is simply connected (or at least that $H_1(M; \mathbb{Z})$ has no 2-torsion), so that $\mathrm{Spin}^c(M)$ is isomorphic to $\mathcal{C}_M$ with isomorphism given by the first Chern class of the *determinant line bundle* (the *determinant line bundle* is the line bundle associated by a natural group homomorphism $\mathrm{Spin}^c(4) \to \mathrm{U}(1)$). Fix an orientation of a maximal-dimensional positive-definite subspace $H^2_+(M; \mathbb{R}) \subset H^2(M; \mathbb{R})$. The Seiberg–Witten invariant is the function

$$\mathrm{SW}_M : \mathcal{C}_M \longrightarrow \mathbb{Z}$$

defined as follows. Given a spin-c structure $\alpha \in \mathrm{Spin}^c(M) \simeq \mathcal{C}_M$, the image $\mathrm{SW}_M(\alpha) = [\mathcal{M}] \in H_d(\mathcal{B}^*; \mathbb{Z})$ is the homology class of the moduli space $\mathcal{M}$ of solutions (called *monopoles*) of the Seiberg–Witten (SW) equations modulo gauge equivalence. The SW equations are nonlinear differential equations on a pair of a connection A on the determinant line bundle of α and of a section φ of an associated U(2)-bundle, called the positive (half) spinor bundle:

$$F_A^+ = iq(\varphi) \quad \text{and} \quad D_A\varphi = 0,$$

where F_A^+ is the self-dual part of the (imaginary) curvature of A, q is a squaring operation taking sections of the positive spinor bundle to self-dual 2-forms, and D_A is the corresponding Dirac operator. For a generic perturbation of the equations (replacing the first equation by $F_A^+ = iq(\varphi) + i\nu$, where ν is a self-dual 2-form) and of the Riemannian

[46] A manifold homeomorphic but not diffeomorphic to a smooth manifold M is called an *exotic* M.

metric, a transversality argument shows that the moduli space $\mathcal{M}$ is well-behaved and actually inside the space $\mathcal{B}^*$ of gauge-equivalence classes of irreducible pairs (those (A, φ) for which $\varphi \neq 0$), which is homotopy-equivalent to $\mathbb{CP}^\infty$ and hence has even-degree homology groups $H_d(\mathcal{B}^*; \mathbb{Z}) \simeq \mathbb{Z}$. When the dimension d of $\mathcal{M}$ is odd or when $\mathcal{M}$ is empty, the invariant $\mathrm{SW}_M(\alpha)$ is set to be zero. The *basic classes* are the classes $\alpha \in \mathcal{C}_M$ for which $\mathrm{SW}_M(\alpha) \neq 0$. The set of basic classes is always finite, and if α is a basic class then so is $-\alpha$. The main results are that the Seiberg–Witten invariants are invariants of the diffeomorphism type of the 4-manifold M and satisfy vanishing and nonvanishing theorems, which allowed to answer an array of questions about specific manifolds.

Taubes [128] discovered an equivalence between Seiberg–Witten and Gromov invariants (using pseudoholomorphic curves) for symplectic 4-manifolds, by proving the existence of pseudoholomorphic curves from solutions of the Seiberg–Witten equations and vice-versa. As a consequence, he proved:

THEOREM 4.9 (Taubes). *Let (M, ω) be a compact symplectic 4-manifold.*
If $b_2^+ > 1$, then $c_1(M, \omega)$ admits a smooth pseudoholomorphic representative.
If $M = M_1 \# M_2$, then one of the M_i's has negative definite intersection form.

There are results also for $b_2^+ = 1$, and follow-ups describe the set of basic classes of a connected sum $M \# N$ in terms of the set of basic classes of M when N is a manifold with negative definite intersection form (starting with $\overline{\mathbb{CP}^2}$).

In an attempt to understand other 4-manifolds via Seiberg–Witten and Gromov invariants, some analysis of pseudoholomorphic curves has been extended to nonsymplectic 4-manifolds by equipping these with a *nearly nondegenerate closed 2-form*. In particular, Taubes [130] has related Seiberg–Witten invariants to pseudoholomorphic curves for compact oriented 4-manifolds with $b_2^+ > 0$. Any compact oriented 4-manifold M with $b_2^+ > 0$ admits a closed 2-form that vanishes along a union of circles and is symplectic elsewhere [54,75]. In fact, for a generic metric on M, there is a self-dual harmonic form ω which is transverse to zero as a section of $\Lambda^2 T^* M$. The vanishing locus of ω is the union of a finite number of embedded circles, and ω is symplectic elsewhere.

The generic behavior of closed 2-forms on orientable 4-manifolds is partially understood [3, pp. 23–24]. Here is a summary. Let ω be a generic closed 2-form on a 4-manifold M. At the points of some hypersurface Z, the form ω has rank 2. At a generic point of M, ω is nondegenerate; in particular, has the Darboux normal form $dx_1 \wedge dy_1 + dx_2 \wedge dy_2$. There is a codimension-1 submanifold Z where ω has rank 2, and there are no points where ω vanishes. At a generic point of Z, the kernel of $\tilde{\omega}$ is transverse to Z; the normal form near such a point is $x_1\, dx_1 \wedge dy_1 + dx_2 \wedge dy_2$. There is a curve C where the kernel of $\tilde{\omega}$ is not transverse to Z, hence sits in TZ. At a generic point of C, the kernel of $\tilde{\omega}$ is transverse to C; there are two possible normal forms near such points, called *elliptic* and *hyperbolic*, $d(x - \frac{z^2}{2}) \wedge dy + d(xz \pm ty - \frac{z^3}{3}) \wedge dt$. The hyperbolic and elliptic sections of C are separated by *parabolic* points, where the kernel is tangent to C. It is known that there exists at least one continuous family of inequivalent degeneracies in a parabolic neighborhood [56].

4.6. *Lefschetz pencils*

Lefschetz pencils in symplectic geometry imitate linear systems in complex geometry. Whereas holomorphic functions on a projective surface must be constant, there are interesting functions on the complement of a finite set, and generic such functions have only quadratic singularities. A Lefschetz pencil can be viewed as a complex Morse function or as a very singular fibration, in the sense that, not only some fibers are singular (have ordinary double points) but all *fibers* go through some points.

DEFINITION 4.10. A *Lefschetz pencil* on an oriented 4-manifold M is a map $f : M \setminus \{b_1, \ldots, b_n\} \to \mathbb{CP}^1$ defined on the complement of a finite set in M, called the *base locus*, that is a submersion away from a finite set $\{p_1, \ldots, p_{n+1}\}$, and obeying local models $(z_1, z_2) \mapsto z_1/z_2$ near the b_j's and $(z_1, z_2) \mapsto z_1z_2$ near the p_j's, where (z_1, z_2) are oriented local complex coordinates.

Usually it is also required that each fiber contains at most one singular point. By blowing up M at the b_j's, we obtain a map to $\mathbb{CP}^1$ on the whole manifold, called a *Lefschetz fibration*. Lefschetz pencils and Lefschetz fibrations can be defined on higher-dimensional manifolds where the b_j's are replaced by codimension 4 submanifolds. By working on the Lefschetz fibration, Gompf [59,58] proved that a structure of Lefschetz pencil (with a nontrivial base locus) gives rise to a symplectic form, canonical up to isotopy, such that the fibers are symplectic.

Using asymptotically holomorphic techniques [12,32], Donaldson [34] proved that symplectic 4-manifolds admit Lefschetz pencils. More precisely:

THEOREM 4.11 (Donaldson). *Let J be a compatible almost complex structure on a compact symplectic 4-manifold (M, ω) where the class $[\omega]/2\pi$ is integral. Then J can be deformed through almost complex structures to an almost complex structure J' such that M admits a Lefschetz pencil with J'-holomorphic fibers.*

The closure of a smooth fiber of the Lefschetz pencil is a symplectic submanifold Poincaré dual to $k[\omega]/2\pi$; cf. Theorem 1.13. Other perspectives on Lefschetz pencils have been explored, including in terms of representations of the free group $\pi_1(\mathbb{CP}^1 \setminus \{p_1, \ldots, p_{n+1}\})$ in the mapping class group Γ_g of the generic fiber surface [118].

Similar techniques were used by Auroux [13] to realize symplectic 4-manifolds as *branched covers* of $\mathbb{CP}^2$, and thus reduce the classification of symplectic 4-manifolds to a (hard) algebraic question about factorization in the braid group. Let M and N be compact oriented 4-manifolds, and let ν be a symplectic form on N.

DEFINITION 4.12. A map $f : M \to N$ is a *symplectic branched cover* if for any $p \in M$ there are complex charts centered at p and $f(p)$ such that ν is positive on each complex line and where f is given by: a local diffeomorphism $(x, y) \to (x, y)$, or a simple branching $(x, y) \to (x^2, y)$, or an ordinary cusp $(x, y) \to (x^3 - xy, y)$.

THEOREM 4.13 (Auroux). *Let (M, ω) be a compact symplectic 4-manifold where the class $[\omega]$ is integral, and let k be a sufficiently large integer. Then there is a symplectic* *branched cover* $f_k : (M, k\omega) \to \mathbb{CP}^2$, *that is canonical up to isotopy for* k *large enough. Conversely, given a symplectic branched cover* $f : M \to N$, *the domain* M *inherits a symplectic form canonical up to isotopy in the class* $f^*[\nu]$.

5. Hamiltonian geometry

5.1. *Symplectic and Hamiltonian vector fields*

Let (M, ω) be a symplectic manifold and let $H : M \to \mathbb{R}$ be a smooth function. By nondegeneracy, there is a unique vector field X_H on M such that $\imath_{X_H}\omega = dH$. Supposing that X_H is complete (this is always the case when M is compact), let $\rho_t : M \to M$, $t \in \mathbb{R}$, be its flow (cf. Section 1.3). Each diffeomorphism ρ_t preserves ω, i.e., $\rho_t^*\omega = \omega$, because $\frac{d}{dt}\rho_t^*\omega = \rho_t^*\mathcal{L}_{X_H}\omega = \rho_t^*(d\imath_{X_H}\omega + \imath_{X_H} d\omega) = 0$. Therefore, every function on (M, ω) produces a family of symplectomorphisms. Notice how this feature involves both the *nondegeneracy* and the *closedness* of ω.

DEFINITION 5.1. A vector field X_H such that $\imath_{X_H}\omega = dH$ for some $H \in C^\infty(M)$ is a *Hamiltonian vector field* with *Hamiltonian function* H.

Hamiltonian vector fields preserve their Hamiltonian functions ($\mathcal{L}_{X_H} H = \imath_{X_H} dH = \imath_{X_H}\imath_{X_H}\omega = 0$), so each integral curve $\{\rho_t(x) \mid t \in \mathbb{R}\}$ of a Hamiltonian vector field X_H must be contained in a level set of the Hamiltonian function H. In $(\mathbb{R}^{2n}, \omega_0 = \sum dx_j \wedge dy_j)$, the *symplectic gradient* $X_H = \sum(\frac{\partial H}{\partial y_j}\frac{\partial}{\partial x_j} - \frac{\partial H}{\partial x_j}\frac{\partial}{\partial y_j})$ and the usual (Euclidean) gradient $\nabla H = \sum_j(\frac{\partial H}{\partial x_j}\frac{\partial}{\partial x_j} + \frac{\partial H}{\partial y_j}\frac{\partial}{\partial y_j})$ of a function H are related by $JX_H = \nabla H$, where J is the standard almost complex structure.

EXAMPLES.

1. For the height function $H(\theta, h) = h$ on the sphere $(M, \omega) = (S^2, d\theta \wedge dh)$, from $\imath_{X_H}(d\theta \wedge dh) = dh$ we get $X_H = \frac{\partial}{\partial\theta}$. Thus, $\rho_t(\theta, h) = (\theta + t, h)$, which is rotation about the vertical axis, preserving the height H.
2. Let X be any vector field on a manifold W. There is a unique vector field $X_\sharp$ on the cotangent bundle T^*W whose flow is the lift of the flow of X. Let α be the tautological form and $\omega = -d\alpha$ the canonical symplectic form on T^*W. The vector field $X_\sharp$ is Hamiltonian with Hamiltonian function $H := \imath_{X_\sharp}\alpha$.
3. Consider Euclidean space $\mathbb{R}^{2n}$ with coordinates $(q_1, \ldots, q_n, p_1, \ldots, p_n)$ and $\omega_0 = \sum dq_j \wedge dp_j$. The curve $\rho_t = (q(t), p(t))$ is an integral curve for a Hamiltonian vector field X_H exactly when it satisfies the *Hamilton equations*:

$$\begin{cases} \frac{dq_i}{dt}(t) = \frac{\partial H}{\partial p_i}, \\ \frac{dp_i}{dt}(t) = -\frac{\partial H}{\partial q_i}. \end{cases}$$

4. *Newton's second law* states that a particle of mass m moving in *configuration space* $\mathbb{R}^3$ with coordinates $q = (q_1, q_2, q_3)$ under a potential $V(q)$ moves along a curve $q(t)$ such that

$$m\frac{d^2q}{dt^2} = -\nabla V(q).$$

Introduce the *momenta* $p_i = m\frac{dq_i}{dt}$ for $i = 1, 2, 3$, and *energy* function $H(q, p) = \frac{1}{2m}|p|^2 + V(q)$ on the *phase space*[47] $\mathbb{R}^6 = T^*\mathbb{R}^3$ with coordinates $(q_1, q_2, q_3, p_1, p_2, p_3)$. The energy H is conserved by the motion and Newton's second law in $\mathbb{R}^3$ is then equivalent to the Hamilton equations in $\mathbb{R}^6$:

$$\begin{cases} \frac{dq_i}{dt} = \frac{1}{m}p_i = \frac{\partial H}{\partial p_i}, \\ \frac{dp_i}{dt} = m\frac{d^2q_i}{dt^2} = -\frac{\partial V}{\partial q_i} = -\frac{\partial H}{\partial q_i}. \end{cases}$$

DEFINITION 5.2. A vector field X on M preserving ω (i.e., such that $\mathcal{L}_X\omega = 0$) is a *symplectic vector field*.

Hence, a vector field X on (M, ω) is called *symplectic* when $\imath_X\omega$ is closed, and *Hamiltonian* when $\imath_X\omega$ is exact. In the latter case, a *primitive* H of $\imath_X\omega$ is called a *Hamiltonian function* of X. On a contractible open set every symplectic vector field is Hamiltonian. Globally, the group $H^1_{\text{deRham}}(M)$ measures the obstruction for symplectic vector fields to be Hamiltonian. For instance, the vector field $X_1 = \frac{\partial}{\partial\theta_1}$ on the 2-torus $(M, \omega) = (\mathbb{T}^2, d\theta_1 \wedge d\theta_2)$ is symplectic but not Hamiltonian.

A vector field X is a differential operator on functions: $X \cdot f := \mathcal{L}_X f = df(X)$ for $f \in C^\infty(M)$. As such, the bracket $W = [X, Y]$ is the commutator: $\mathcal{L}_W = [\mathcal{L}_X, \mathcal{L}_Y] = \mathcal{L}_X\mathcal{L}_Y - \mathcal{L}_Y\mathcal{L}_X$ (cf. Section 3.3). This endows the set $\chi(M)$ of vector fields on a manifold M with a structure of *Lie algebra*.[48] For a symplectic manifold (M, ω), using $\imath_{[X,Y]} = [\mathcal{L}_X, \imath_Y]$ and Cartan's magic formula, we find that $\imath_{[X,Y]}\omega = d\imath_X\imath_Y\omega + \imath_X d\imath_Y\omega - \imath_Y d\imath_X\omega - \imath_Y\imath_X\, d\omega = d(\omega(Y, X))$. Therefore:

PROPOSITION 5.3. *If X and Y are symplectic vector fields on a symplectic manifold (M, ω), then $[X, Y]$ is Hamiltonian with Hamiltonian function $\omega(Y, X)$.*

Hence, Hamiltonian vector fields and symplectic vector fields form Lie subalgebras for the Lie bracket $[\cdot, \cdot]$.

DEFINITION 5.4. The *Poisson bracket* of two functions $f, g \in C^\infty(M)$ is the function $\{f, g\} := \omega(X_f, X_g) = \mathcal{L}_{X_g} f$.

---

[47] The *phase space* of a system of n particles is the space parametrizing the position and momenta of the particles. The mathematical model for a phase space is a symplectic manifold.

[48] A (real) *Lie algebra* is a (real) vector space $\mathfrak{g}$ together with a *Lie bracket* $[\cdot, \cdot]$, i.e., a bilinear map $[\cdot, \cdot] : \mathfrak{g} \times \mathfrak{g} \to \mathfrak{g}$ satisfying *antisymmetry*, $[x, y] = -[y, x]$, $\forall x, y \in \mathfrak{g}$, and the *Jacobi identity*, $[x, [y, z]] + [y, [z, x]] + [z, [x, y]] = 0$, $\forall x, y, z \in \mathfrak{g}$.

By Proposition 5.3 we have $X_{\{f,g\}} = -[X_f, X_g]$. Moreover, the bracket $\{\cdot,\cdot\}$ satisfies the *Jacobi identity*, $\{f,\{g,h\}\} + \{g,\{h,f\}\} + \{h,\{f,g\}\} = 0$, and the *Leibniz rule*, $\{f, gh\} = \{f, g\}h + g\{f, h\}$.

DEFINITION 5.5. A *Poisson algebra* $(\mathcal{P}, \{\cdot,\cdot\})$ is a commutative associative algebra $\mathcal{P}$ with a Lie bracket $\{\cdot,\cdot\}$ satisfying the Leibniz rule.

When (M,ω) is a symplectic manifold, $(C^\infty(M), \{\cdot,\cdot\})$ is a Poisson algebra, and the map $C^\infty(M) \to \chi(M)$, $H \mapsto X_H$ is a Lie algebra anti-homomorphism.

EXAMPLES.

1. For the prototype $(\mathbb{R}^{2n}, \sum dx_i \wedge dy_i)$, we have $X_{x_i} = -\frac{\partial}{\partial y_i}$ and $X_{y_i} = \frac{\partial}{\partial x_i}$, so that $\{x_i, x_j\} = \{y_i, y_j\} = 0$ and $\{x_i, y_j\} = \delta_{ij}$ for all i, j. Arbitrary functions $f, g \in C^\infty(\mathbb{R}^{2n})$ have the *classical Poisson bracket*

$$\{f,g\} = \sum_{i=1}^{n} \left(\frac{\partial f}{\partial x_i}\frac{\partial g}{\partial y_i} - \frac{\partial f}{\partial y_i}\frac{\partial g}{\partial x_i} \right).$$

2. Let G be a Lie group,[49] $\mathfrak{g}$ its Lie algebra and $\mathfrak{g}^*$ the dual vector space of $\mathfrak{g}$. The vector field ${}^{\mathfrak{g}}X^\#$ generated by $X \in \mathfrak{g}$ for the adjoint action[50] of G on $\mathfrak{g}$ has value $[X, Y]$ at $Y \in \mathfrak{g}$. The vector field $X^\#$ generated by $X \in \mathfrak{g}$ for the coadjoint action of G on $\mathfrak{g}^*$ is $\langle X^\#_\xi, Y\rangle = \langle \xi, [Y, X]\rangle$, $\forall \xi \in \mathfrak{g}^*$, $Y \in \mathfrak{g}$. The skew-symmetric pairing ω on $\mathfrak{g}$ defined at $\xi \in \mathfrak{g}^*$ by

$$\omega_\xi(X,Y) := \langle \xi, [X,Y]\rangle$$

has kernel at ξ the Lie algebra $\mathfrak{g}_\xi$ of the stabilizer of ξ for the coadjoint action. Therefore, ω restricts to a nondegenerate 2-form on the tangent spaces to the orbits of the coadjoint action. As the tangent spaces to an orbit are generated by the vector fields $X^\#$, the Jacobi identity in $\mathfrak{g}$ implies that this form is closed. It is called the *canonical symplectic form* (or the *Lie–Poisson* or *Kirillov–Kostant–Souriau symplectic structure*) on the *coadjoint orbits*. The corresponding Poisson structure on $\mathfrak{g}^*$ is the canonical one induced by the Lie bracket:

$$\{f,g\}(\xi) = \langle \xi, [df_\xi, dg_\xi]\rangle$$

[49] A *Lie group* is a manifold G equipped with a group structure where the group operation $G \times G \to G$ and inversion $G \to G$ are smooth maps. An *action* of a Lie group G on a manifold M is a group homomorphism $G \to \mathrm{Diff}(M)$, $g \mapsto \psi_g$, where the *evaluation map* $M \times G \to M$, $(p, g) \mapsto \psi_g(p)$ is a smooth map. The *orbit* of G through $p \in M$ is $\{\psi_g(p) \mid g \in G\}$. The *stabilizer* (or *isotropy*) of $p \in M$ is $G_p := \{g \in G \mid \psi_g(p) = p\}$.

[50] Any Lie group G acts on itself by *conjugation*: $g \in G \mapsto \psi_g \in \mathrm{Diff}(G)$, $\psi_g(a) = g \cdot a \cdot g^{-1}$. Let $\mathrm{Ad}_g : \mathfrak{g} \to \mathfrak{g}$ be the derivative at the identity of $\psi_g : G \to G$. We identify the Lie algebra $\mathfrak{g}$ with the tangent space T_eG. For matrix groups, $\mathrm{Ad}_g X = gXg^{-1}$. Letting g vary, we obtain the *adjoint action* of G on its Lie algebra $\mathrm{Ad} : G \to \mathrm{GL}(\mathfrak{g})$. Let $\langle\cdot,\cdot\rangle : \mathfrak{g}^* \times \mathfrak{g} \to \mathbb{R}$ be the natural pairing $\langle \xi, X\rangle = \xi(X)$. Given $\xi \in \mathfrak{g}^*$, we define $\mathrm{Ad}^*_g \xi$ by $\langle \mathrm{Ad}^*_g \xi, X\rangle = \langle \xi, \mathrm{Ad}_{g^{-1}} X\rangle$, for any $X \in \mathfrak{g}$. The collection of maps Ad^*_g forms the *coadjoint action* of G on the dual of its Lie algebra $\mathrm{Ad}^* : G \to \mathrm{GL}(\mathfrak{g}^*)$. These satisfy $\mathrm{Ad}_g \circ \mathrm{Ad}_h = \mathrm{Ad}_{gh}$ and $\mathrm{Ad}^*_g \circ \mathrm{Ad}^*_h = \mathrm{Ad}^*_{gh}$.

for $f, g \in C^\infty(\mathfrak{g}^*)$ and $\xi \in \mathfrak{g}^*$. The differential $df_\xi : T_\xi \mathfrak{g}^* \simeq \mathfrak{g}^* \to \mathbb{R}$ is identified with an element of $\mathfrak{g} \simeq \mathfrak{g}^{**}$.

5.2. *Arnold conjecture and Floer homology*

There is an important generalization of Poincaré's last geometric theorem (Theorem 2.16) conjectured by Arnold starting around 1966. Let (M, ω) be a compact symplectic manifold, and $h_t : M \to \mathbb{R}$ a 1-periodic (i.e., $h_t = h_{t+1}$) smooth family of functions. Let $\rho : M \times \mathbb{R} \to M$ be the isotopy generated by the time-dependent Hamiltonian vector field v_t defined by the equation $\omega(v_t, \cdot) = dh_t$. The symplectomorphism $\varphi = \rho_1$ is then said to be *exactly homotopic to the identity*. In other words, a symplectomorphism exactly homotopic to the identity is the time-1 map of the isotopy generated by some time-dependent 1-periodic Hamiltonian function. There is a one-to-one correspondence between the fixed points of φ and the period-1 orbits of ρ. When all the fixed points of such φ are nondegenerate (generic case), we call φ *nondegenerate*. The *Arnold conjecture* [2, Appendix 9] predicted that

$$\#\{\text{fixed points of a nondegenerate } \varphi\} \geqslant \sum_{i=0}^{2n} \dim H^i(M; \mathbb{R})$$

(or even that the number of fixed points of a nondegenerate φ is at least the minimal number of critical points of a Morse function[51]). When the Hamiltonian $h : M \to \mathbb{R}$ is independent of t, this relation is trivial: a point p is critical for h if and only if $dh_p = 0$, if and only if $v_p = 0$, if and only if $\rho(t, p) = p$, $\forall t \in \mathbb{R}$, which implies that p is a fixed point of $\rho_1 = \varphi$, so the Arnold conjecture reduces to a Morse inequality. Notice that, according to the Lefschetz fixed point theorem, the Euler characteristic of M, i.e., the *alternating* sum of the Betti numbers, $\sum(-1)^i \dim H^i(M; \mathbb{R})$, is a (weaker) lower bound for the number of fixed points of φ.

The Arnold conjecture was gradually proved from the late 70's to the late 90's by Eliashberg [39], Conley–Zehnder [24], Floer [49], Sikorav [116], Weinstein [140], Hofer–Salamon [74], Ono [108], culminating with independent proofs by Fukaya–Ono [52] and Liu–Tian [90]. There are open conjectures for sharper bounds on the number of fixed points. The breakthrough tool for establishing the Arnold conjecture was *Floer homology*—an ∞-dimensional analogue of Morse theory. Floer homology was defined by Floer [46–50] and developed through the work of numerous people after Floer's death. It combines the variational approach of Conley and Zehnder [25], with Witten's Morse–Smale complex [144], and with Gromov's compactness theorem for pseudoholomorphic curves [64].

Floer theory starts from a symplectic action functional on the space of loops $\mathcal{L}M$ of a symplectic manifold (M, ω) whose zeros of the differential $dF : T(\mathcal{L}M) \to \mathbb{R}$ are the period-1 orbits of the isotopy ρ above. The tangent bundle $T(\mathcal{L}M)$ is the space of loops with vector fields over them: pairs (ℓ, v), where $\ell : S^1 \to M$ and $v : S^1 \to \ell^*(TM)$ is a

[51] A *Morse function* is a smooth function $f : M \to \mathbb{R}$ all of whose critical points are nondegenerate, i.e., at any critical point the Hessian matrix is nondegenerate.

section. Then $df(\ell, v) = \int_0^1 \omega(\dot{\ell}(t) - X_{h_t}(\ell(t), v(t))\,dt$. The *Floer complex*[52] is the chain complex freely generated by the critical points of F (corresponding to the fixed points of φ), with *relative grading* index(x, y) given by the difference in the number of positive eigenvalues from the spectral flow. The Floer differential is given by counting the number $n(x, y)$ of pseudoholomorphic surfaces (the *gradient flow lines* joining two fixed points):

$$C_* = \bigoplus_{x \in \mathrm{Crit}(F)} \mathbb{Z}\langle x\rangle \quad \text{and} \quad \partial\langle x\rangle = \sum_{\substack{y \in \mathrm{Crit}(F) \\ \mathrm{index}(x,y)=1}} n(x, y)\langle y\rangle.$$

Pondering transversality, compactness and orientation, Floer's theorem states that the homology of (C_*, ∂) is isomorphic to the ordinary homology of M. In particular, the sum of the Betti numbers is a lower bound for the number of fixed points of φ.

From the above *symplectic Floer homology*, Floer theory has branched out to tackle other differential geometric problems in symplectic geometry and 3- and 4-dimensional topology. It provides a rigorous definition of invariants viewed as homology groups of infinite-dimensional Morse-type theories, with relations to gauge theory and quantum field theory. There is *Lagrangian Floer homology* (for the case of Lagrangian intersections, i.e., intersection of a Lagrangian submanifold with a Hamiltonian deformation of itself), *instanton Floer homology* (for invariants of 3-manifolds), *Seiberg–Witten Floer homology*, *Heegaard Floer homology* and *knot Floer homology*. For more on Floer homology; see, for instance, [35,113].

5.3. *Euler–Lagrange equations*

The equations of motion in classical mechanics arise from *variational principles*. The physical path of a general mechanical system of n particles is the path that *minimizes* a quantity called the *action*. When dealing with systems with constraints, such as the simple

[52] The *Morse complex* for a Morse function on a compact manifold, $f: M \to \mathbb{R}$, is the chain complex freely generated by the critical points of f, graded by the *Morse index* ι and with differential given by counting the number $n(x, y)$ of flow lines of the negative gradient $-\nabla f$ (for a metric on X) from the point x to the point y whose indices differ by 1:

$$C_* = \bigoplus_{x \in \mathrm{Crit}(f)} \mathbb{Z}\langle x\rangle \quad \text{and} \quad \partial\langle x\rangle = \sum_{\substack{y \in \mathrm{Crit}(f) \\ \iota(y)=\iota(x)-1}} n(x, y)\langle y\rangle.$$

The coefficient $n(x, y)$ is thus the number of solutions (modulo $\mathbb{R}$-reparametrization) $u: \mathbb{R} \to X$ of the ordinary differential equation $\frac{d}{dt}u(t) = -\nabla f(u(t))$ with conditions $\lim_{t\to-\infty} u(t) = x$, $\lim_{t\to+\infty} u(t) = y$. The *Morse index* of a critical point of f is the dimension of its unstable manifold, i.e., the number of negative eigenvalues of the Hessian of f at that point. For a generic metric, the unstable manifold of a critical point $W^u(x)$ intersects transversally with the stable manifold of another critical point $W^s(y)$. When $\iota(x) - \iota(y) = 1$, the intersection $W^u(x) \cap W^s(y)$ has dimension 1, so when we quotient out by the $\mathbb{R}$-reparametrization (to count actual image curves) we get a discrete set, which is finite by compactness. That (C_*, ∂) is indeed a complex, i.e., $\partial^2 = 0$, follows from counting broken flow lines between points whose indices differ by 2. Morse's theorem states that the homology of the Morse complex coincides with the ordinary homology of M. In particular, the sum of all the Betti numbers $\sum \dim H^i(M; \mathbb{R})$ is a lower bound for the number of critical points of a Morse function.

pendulum, or two point masses attached by a rigid rod, or a rigid body, the language of variational principles becomes more appropriate than the explicit analogues of Newton's second laws. Variational principles are due mostly to D'Alembert, Maupertius, Euler and Lagrange.

Let M be an n-dimensional manifold, and let $F : TM \to \mathbb{R}$ be a function on its tangent bundle. If $\gamma : [a,b] \to M$ is a curve on M, the *lift of γ to TM* is the curve on TM given by $\tilde{\gamma} : [a,b] \to TM$, $t \mapsto (\gamma(t), \frac{d\gamma}{dt}(t))$. The *action* of γ is

$$\mathcal{A}_\gamma := \int_a^b (\tilde{\gamma}^* F)(t)\, dt = \int_a^b F\left(\gamma(t), \frac{d\gamma}{dt}(t)\right) dt.$$

For fixed p, q, let $\mathcal{P}(a,b,p,q) = \{\gamma : [a,b] \to M \text{ smooth} \mid \gamma(a) = p,\ \gamma(b) = q\}$. The goal is to find, among all $\gamma \in \mathcal{P}(a,b,p,q)$, the curve that *locally minimizes* $\mathcal{A}_\gamma$. (Minimizing curves are always locally minimizing.) Assume that p, q and the image of γ lie in a coordinate neighborhood $(\mathcal{U}, x_1, \dots, x_n)$. On $T\mathcal{U}$ we have coordinates $(x_1, \dots, x_n, v_1, \dots, v_n)$ associated with a trivialization of $T\mathcal{U}$ by $\frac{\partial}{\partial x_1}, \dots, \frac{\partial}{\partial x_n}$. Using this trivialization, a curve $\gamma : [a,b] \to \mathcal{U}$, $\gamma(t) = (\gamma_1(t), \dots, \gamma_n(t))$ lifts to

$$\tilde{\gamma} : [a,b] \longrightarrow T\mathcal{U}, \quad \tilde{\gamma}(t) = \left(\gamma_1(t), \dots, \gamma_n(t), \frac{d\gamma_1}{dt}(t), \dots, \frac{d\gamma_n}{dt}(t)\right).$$

Consider infinitesimal variations of γ. Let $c_1, \dots, c_n \in C^\infty([a,b])$ be such that $c_k(a) = c_k(b) = 0$. For ε small, let $\gamma_\varepsilon : [a,b] \to \mathcal{U}$ be the curve $\gamma_\varepsilon(t) = (\gamma_1(t) + \varepsilon c_1(t), \dots, \gamma_n(t) + \varepsilon c_n(t))$. Let $\mathcal{A}_\varepsilon := \mathcal{A}_{\gamma_\varepsilon}$. A necessary condition for $\gamma = \gamma_0 \in \mathcal{P}(a,b,p,q)$ to minimize the action is that $\varepsilon = 0$ be a critical point of $\mathcal{A}_\varepsilon$. By the Leibniz rule and integration by parts, we have that

$$\begin{aligned}
\frac{d\mathcal{A}_\varepsilon}{d\varepsilon}(0) &= \int_a^b \sum_k \left[\frac{\partial F}{\partial x_k}\left(\gamma_0(t), \frac{d\gamma_0}{dt}(t)\right) c_k(t) + \frac{\partial F}{\partial v_k}\left(\gamma_0, \frac{d\gamma_0}{dt}\right) \frac{dc_k}{dt}(t)\right] dt \\
&= \int_a^b \sum_k \left[\frac{\partial F}{\partial x_k}(\dots) - \frac{d}{dt}\frac{\partial F}{\partial v_k}(\dots)\right] c_k(t)\, dt.
\end{aligned}$$

For $\frac{d\mathcal{A}_\varepsilon}{d\varepsilon}(0)$ to vanish for all c_k's satisfying boundary conditions $c_k(a) = c_k(b) = 0$, the path γ_0 must satisfy the *Euler–Lagrange equations*:

$$\frac{\partial F}{\partial x_k}\left(\gamma_0(t), \frac{d\gamma_0}{dt}(t)\right) = \frac{d}{dt}\frac{\partial F}{\partial v_k}\left(\gamma_0(t), \frac{d\gamma_0}{dt}(t)\right), \quad k = 1, \dots, n.$$

EXAMPLES.

1. Let (M, g) be a Riemannian manifold. Let $F : TM \to \mathbb{R}$ be the function whose restriction to each tangent space is the quadratic form defined by the Riemannian

metric. On a coordinate chart $F(x,v) = |v|^2 = \sum g_{ij}(x)v^i v^j$. Let $p,q \in M$ and $\gamma : [a,b] \to M$ a curve joining p to q. The *action* of γ is

$$\mathcal{A}_\gamma = \int_a^b \left| \frac{d\gamma}{dt} \right|^2 dt.$$

The Euler–Lagrange equations become the *Christoffel equations* for a geodesic

$$\frac{d^2\gamma^k}{dt^2} + \sum (\Gamma_{ij}^k \circ \gamma) \frac{d\gamma^i}{dt} \frac{d\gamma^j}{dt} = 0,$$

where the *Christoffel symbols* Γ_{ij}^k's are defined in terms of the coefficients of the Riemannian metric (g^{ij} is the matrix inverse to g_{ij}) by

$$\Gamma_{ij}^k = \frac{1}{2} \sum_\ell g^{\ell k} \left(\frac{\partial g_{\ell i}}{\partial x_j} + \frac{\partial g_{\ell j}}{\partial x_i} - \frac{\partial g_{ij}}{\partial x_\ell} \right).$$

2. Consider a point-particle of mass m moving in $\mathbb{R}^3$ under a *force field* G. The *work* of G on a path $\gamma : [a,b] \to \mathbb{R}^3$ is $W_\gamma := \int_a^b G(\gamma(t)) \cdot \frac{d\gamma}{dt}(t)\,dt$. Suppose that G is *conservative*, i.e., W_γ depends only on the initial and final points, $p = \gamma(a)$ and $q = \gamma(b)$. We can define the *potential energy* as $V : \mathbb{R}^3 \to \mathbb{R}$, $V(q) := W_\gamma$, where γ is a path joining a fixed base point $p_0 \in \mathbb{R}^3$ to q. Let $\mathcal{P}$ be the set of all paths going from p to q over time $t \in [a,b]$. By the *principle of least action*, the physical path is the path $\gamma \in \mathcal{P}$ that minimizes a kind of mean value of kinetic minus potential energy, known as the *action*:

$$\mathcal{A}_\gamma := \int_a^b \left(\frac{m}{2} \left| \frac{d\gamma}{dt}(t) \right|^2 - V(\gamma(t)) \right) dt.$$

The Euler–Lagrange equations are then equivalent to *Newton's second law*:

$$m\frac{d^2x}{dt^2}(t) - \frac{\partial V}{\partial x}(x(t)) = 0 \quad \Longleftrightarrow \quad m\frac{d^2x}{dt^2}(t) = G(x(t)).$$

In the case of the earth moving about the sun, both regarded as point-masses and assuming that the sun to be stationary at the origin, the *gravitational potential* $V(x) = \frac{\text{const}}{|x|}$ yields the *inverse square law* for the motion.

3. Consider now n point-particles of masses $m_1, \dots, m_n$ moving in $\mathbb{R}^3$ under a conservative force corresponding to a potential energy $V \in C^\infty(\mathbb{R}^{3n})$. At any instant t, the configuration of this system is described by a vector $x = (x_1, \dots, x_n)$ in configuration space $\mathbb{R}^{3n}$, where $x_k \in \mathbb{R}^3$ is the position of the kth particle. For fixed $p, q \in \mathbb{R}^{3n}$, let

$\mathcal{P}$ be the set of all paths $\gamma = (\gamma_1, \dots, \gamma_n) : [a, b] \to \mathbb{R}^{3n}$ from p to q. The *action* of a path $\gamma \in \mathcal{P}$ is

$$\mathcal{A}_\gamma := \int_a^b \left(\sum_{k=1}^n \frac{m_k}{2} \left| \frac{d\gamma_k}{dt}(t) \right|^2 - V(\gamma(t)) \right) dt.$$

The Euler–Lagrange equations reduce to Newton's law for each particle. Suppose that the particles are restricted to move on a submanifold M of $\mathbb{R}^{3n}$ called the *constraint set*. By the *principle of least action for a constrained system*, the physical path has minimal action among all paths satisfying the rigid constraints. I.e., we single out the actual physical path as the one that minimizes $\mathcal{A}_\gamma$ among all $\gamma : [a, b] \to M$ with $\gamma(a) = p$ and $\gamma(b) = q$.

In the case where $F(x, v)$ does not depend on v, the Euler–Lagrange equations are simply $\frac{\partial F}{\partial x_i}(\gamma_0(t), \frac{d\gamma_0}{dt}(t)) = 0$. These are satisfied if and only if the curve γ_0 sits on the critical set of F. For generic F, the critical points are isolated, hence $\gamma_0(t)$ must be a constant curve. In the case where $F(x, v)$ depends affinely on v, $F(x, v) = F_0(x) + \sum_{j=1}^n F_j(x) v_j$, the Euler–Lagrange equations become

$$\frac{\partial F_0}{\partial x_i}(\gamma(t)) = \sum_{j=1}^n \left(\frac{\partial F_i}{\partial x_j} - \frac{\partial F_j}{\partial x_i} \right) (\gamma(t)) \frac{d\gamma_j}{dt}(t).$$

If the $n \times n$ matrix $(\frac{\partial F_i}{\partial x_j} - \frac{\partial F_j}{\partial x_i})$ has an inverse $G_{ij}(x)$, we obtain the system of first order ordinary differential equations $\frac{d\gamma_j}{dt}(t) = \sum G_{ji}(\gamma(t)) \frac{\partial F_0}{\partial x_i}(\gamma(t))$. Locally it has a unique solution through each point p. If q is not on this curve, there is no solution at all to the Euler–Lagrange equations belonging to $\mathcal{P}(a, b, p, q)$.

Therefore, we need nonlinear dependence of F on the v variables in order to have appropriate solutions. From now on, assume the *Legendre condition*:

$$\det\left(\frac{\partial^2 F}{\partial v_i \partial v_j} \right) \neq 0.$$

Letting $G_{ij}(x, v) = (\frac{\partial^2 F}{\partial v_i \partial v_j}(x, v))^{-1}$, the Euler–Lagrange equations become

$$\frac{d^2\gamma_j}{dt^2} = \sum_i G_{ji} \frac{\partial F}{\partial x_i}\left(\gamma, \frac{d\gamma}{dt} \right) - \sum_{i,k} G_{ji} \frac{\partial^2 F}{\partial v_i \partial x_k}\left(\gamma, \frac{d\gamma}{dt} \right) \frac{d\gamma_k}{dt}.$$

This second order ordinary differential equation has a unique solution given initial conditions $\gamma(a) = p$ and $\frac{d\gamma}{dt}(a) = v$. Assume that $(\frac{\partial^2 F}{\partial v_i \partial v_j}(x, v)) \gg 0$, $\forall (x, v)$, i.e., with the x variable frozen, the function $v \mapsto F(x, v)$ is *strictly convex*. Then the path $\gamma_0 \in \mathcal{P}(a, b, p, q)$ satisfying the above Euler–Lagrange equations does indeed locally minimize $\mathcal{A}_\gamma$ (globally it is only critical):

PROPOSITION 5.6. *For every sufficiently small subinterval* $[a_1, b_1]$ *of* $[a, b]$, $\gamma_0|_{[a_1,b_1]}$ *is locally minimizing in* $\mathcal{P}(a_1, b_1, p_1, q_1)$ *where* $p_1 = \gamma_0(a_1)$, $q_1 = \gamma_0(b_1)$.

PROOF. Take $c = (c_1, \dots, c_n)$ with $c_i \in C^\infty([a, b])$, $c_i(a) = c_i(b) = 0$. Let $\gamma_\varepsilon = \gamma_0 + \varepsilon c \in \mathcal{P}(a, b, p, q)$, and let $\mathcal{A}_\varepsilon = \mathcal{A}_{\gamma_\varepsilon}$. Suppose that $\gamma_0 : [a, b] \to \mathcal{U}$ satisfies the Euler–Lagrange equations, i.e., $\frac{d\mathcal{A}_\varepsilon}{d\varepsilon}(0) = 0$. Then

$$\frac{d^2\mathcal{A}_\varepsilon}{d\varepsilon^2}(0) = \int_a^b \sum_{i,j} \frac{\partial^2 F}{\partial x_i \partial x_j}\left(\gamma_0, \frac{d\gamma_0}{dt}\right) c_i c_j \, dt \tag{A}$$

$$+ 2\int_a^b \sum_{i,j} \frac{\partial^2 F}{\partial x_i \partial v_j}\left(\gamma_0, \frac{d\gamma_0}{dt}\right) c_i \frac{dc_j}{dt} \, dt \tag{B}$$

$$+ \int_a^b \sum_{i,j} \frac{\partial^2 F}{\partial v_i \partial v_j}\left(\gamma_0, \frac{d\gamma_0}{dt}\right) \frac{dc_i}{dt}\frac{dc_j}{dt} \, dt. \tag{C}$$

Since $(\frac{\partial^2 F}{\partial v_i \partial v_j}(x, v)) \gg 0$ at all x, v, we have

$$|(\mathrm{A})| \leqslant K_{\mathrm{A}} |c|^2_{L^2[a,b]}, \qquad |(\mathrm{B})| \leqslant K_{\mathrm{B}} |c|_{L^2[a,b]} \left|\frac{dc}{dt}\right|_{L^2[a,b]}$$

and

$$(\mathrm{C}) \geqslant K_{\mathrm{C}} \left|\frac{dc}{dt}\right|^2_{L^2[a,b]},$$

where $K_{\mathrm{A}}, K_{\mathrm{B}}, K_{\mathrm{C}}$ are positive constants. By the Wirtinger inequality[53], if $b - a$ is very small, then $(\mathrm{C}) > |(\mathrm{A})| + |(\mathrm{B})|$ when $c \not\equiv 0$. Hence, γ_0 is a local minimum. □

In Section 5.1 we saw that solving Newton's second law in *configuration space* $\mathbb{R}^3$ is equivalent to solving in *phase space* for the integral curve in $T^*\mathbb{R}^3 = \mathbb{R}^6$ of the Hamiltonian vector field with Hamiltonian function H. In the next subsection we will see how this correspondence extends to more general Euler–Lagrange equations.

5.4. *Legendre transform*

The Legendre transform gives the relation between the variational (Euler–Lagrange) and the symplectic (Hamilton–Jacobi) formulations of the equations of motion.

[53]The *Wirtinger inequality* states that, for $f \in C^1([a, b])$ with $f(a) = f(b) = 0$, we have

$$\int_a^b \left|\frac{df}{dt}\right|^2 dt \geqslant \frac{\pi^2}{(b-a)^2} \int_a^b |f|^2 dt.$$

This can be proved with Fourier series.

Let V be an n-dimensional vector space, with $e_1, \ldots, e_n$ a basis of V and $v_1, \ldots, v_n$ the associated coordinates. Let $F: V \to \mathbb{R}$, $F = F(v_1, \ldots, v_n)$, be a smooth function. The function F is *strictly convex* if and only if for every pair of elements $p, v \in V$, $v \neq 0$, the restriction of F to the line $\{p + xv \mid x \in \mathbb{R}\}$ is strictly convex.[54] It follows from the case of real functions on $\mathbb{R}$ that, for a strictly convex function F on V, the following are equivalent:[55]

(a) F has a critical point, i.e., a point where $dF_p = 0$;
(b) F has a local minimum at some point;
(c) F has a unique critical point (global minimum); and
(d) F is proper, that is, $F(p) \to +\infty$ as $p \to \infty$ in V.

A strictly convex function F is *stable* when it satisfies conditions (a)–(d) above.

DEFINITION 5.7. The *Legendre transform* associated to $F \in C^\infty(V)$ is the map

$$\begin{aligned} L_F : V &\longrightarrow V^*, \\ p &\longmapsto dF_p \in T_p^* V \simeq V^*, \end{aligned}$$

where $T_p^* V \simeq V^*$ is the canonical identification for a vector space V.

From now on, assume that F is a strictly convex function on V. Then, for every point $p \in V$, L_F maps a neighborhood of p diffeomorphically onto a neighborhood of $L_F(p)$. Given $\ell \in V^*$, let

$$F_\ell : V \longrightarrow \mathbb{R}, \quad F_\ell(v) = F(v) - \ell(v).$$

Since $(d^2 F)_p = (d^2 F_\ell)_p$, F is strictly convex if and only if F_ℓ is strictly convex. The *stability set* of F is

$$S_F = \{\ell \in V^* \mid F_\ell \text{ is stable}\}.$$

The set S_F is open and convex, and L_F maps V diffeomorphically onto S_F. (A way to ensure that $S_F = V^*$ and hence that L_F maps V diffeomorphically onto V^*, is to assume that a strictly convex function F has *quadratic growth at infinity*, i.e., there exists a positive-definite quadratic form Q on V and a constant K such that $F(p) \geqslant Q(p) - K$, for all p.) The inverse to L_F is the map $L_F^{-1} : S_F \to V$ described as follows: for $\ell \in S_F$,

[54] A function $F: V \to \mathbb{R}$ is *strictly convex* if at every $p \in V$ the *Hessian* $d^2 F_p$ is positive definite. Let $u = \sum_{i=1}^n u_i e_i \in V$. The *Hessian* of F at p is the quadratic function on V,

$$(d^2 F)_p(u) := \sum_{i,j} \frac{\partial^2 F}{\partial v_i \partial v_j}(p) u_i u_j = \frac{d^2}{dt^2} F(p + tu)\bigg|_{t=0}.$$

[55] A smooth function $f: \mathbb{R} \to \mathbb{R}$ is *strictly convex* if $f''(x) > 0$ for all $x \in \mathbb{R}$. Assuming that f is strictly convex, the following four conditions are equivalent: $f'(x) = 0$ at some point, f has a local minimum, f has a unique (global) minimum, and $f(x) \to +\infty$ as $x \to \pm\infty$. The function f is *stable* if it satisfies one (and hence all) of these conditions. For instance, $e^x + ax$ is strictly convex for any $a \in \mathbb{R}$, but it is stable only for $a < 0$. The function $x^2 + ax$ is strictly convex and stable for any $a \in \mathbb{R}$.

the value $L_F^{-1}(\ell)$ is the unique minimum point $p_\ell \in V$ of F_ℓ. Indeed p is the minimum of $F(v) - dF_p(v)$.

DEFINITION 5.8. The *dual function* F^* to F is

$$F^* : S_F \longrightarrow \mathbb{R}, \quad F^*(\ell) = -\min_{p \in V} F_\ell(p).$$

The dual function F^* is smooth and, for all $p \in V$ and all $\ell \in S_F$, satisfies the *Young inequality* $F(p) + F^*(\ell) \geqslant \ell(p)$.

On one hand we have $V \times V^* \simeq T^*V$, and on the other hand, since $V = V^{**}$, we have $V \times V^* \simeq V^* \times V \simeq T^*V^*$. Let α_1 be the tautological 1-form on T^*V and α_2 be the tautological 1-form on T^*V^*. Via the identifications above, we can think of both of these forms as living on $V \times V^*$. Since $\alpha_1 = d\beta - \alpha_2$, where $\beta : V \times V^* \to \mathbb{R}$ is the function $\beta(p, \ell) = \ell(p)$, we conclude that the forms $\omega_1 = -d\alpha_1$ and $\omega_2 = -d\alpha_2$ satisfy $\omega_1 = -\omega_2$.

THEOREM 5.9. *For a strictly convex function F we have that $L_F^{-1} = L_{F^*}$.*

PROOF. The graph Λ_F of the Legendre transform L_F is a Lagrangian submanifold of $V \times V^*$ with respect to the symplectic form ω_1. Hence, Λ_F is also Lagrangian for ω_2. Let $\mathrm{pr}_1 : \Lambda_F \to V$ and $\mathrm{pr}_2 : \Lambda_F \to V^*$ be the restrictions of the projection maps $V \times V^* \to V$ and $V \times V^* \to V^*$, and let $i : \Lambda_F \hookrightarrow V \times V^*$ be the inclusion map. Then $i^*\alpha_1 = d(\mathrm{pr}_1)^*F$ as both sides have value dF_p at $(p, dF_p) \in \Lambda_F$. It follows that $i^*\alpha_2 = d(i^*\beta - (\mathrm{pr}_1)^*F) = d(\mathrm{pr}_2)^*F^*$, which shows that Λ_F is the graph of the inverse of L_{F^*}. From this we conclude that the inverse of the Legendre transform associated with F is the Legendre transform associated with F^*. □

Let M be a manifold and $F : TM \to \mathbb{R}$. We return to the Euler–Lagrange equations for minimizing the action $\mathcal{A}_\gamma = \int \tilde{\gamma}^* F$. At $p \in M$, let $F_p := F|_{T_pM} : T_pM \to \mathbb{R}$. Assume that F_p is strictly convex for all $p \in M$. To simplify notation, assume also that $S_{F_p} = T_p^*M$. The Legendre transform on each tangent space $L_{F_p} : T_pM \xrightarrow{\simeq} T_p^*M$ is essentially given by the first derivatives of F in the v directions. Collect these and the dual functions $F_p^* : T_p^*M \to \mathbb{R}$ into maps

$$\mathcal{L} : TM \longrightarrow T^*M, \quad \mathcal{L}|_{T_pM} = L_{F_p} \quad \text{and} \quad H : T^*M \longrightarrow \mathbb{R}, \quad H|_{T_p^*M} = F_p^*.$$

The maps H and $\mathcal{L}$ are smooth, and $\mathcal{L}$ is a diffeomorphism.

THEOREM 5.10. *Let $\gamma : [a, b] \to M$ be a curve, and $\tilde{\gamma} : [a, b] \to TM$ its lift. Then γ satisfies the Euler–Lagrange equations on every coordinate chart if and only if $\mathcal{L} \circ \tilde{\gamma} : [a, b] \to T^*M$ is an integral curve of the Hamiltonian vector field X_H.*

PROOF. Let $(\mathcal{U}, x_1, \dots, x_n)$ be a coordinate chart in M, with associated tangent $(T\mathcal{U}, x_1, \dots, x_n, v_1, \dots, v_n)$ and cotangent $(T^*\mathcal{U}, x_1, \dots, x_n, \xi_1, \dots, \xi_n)$ coordinates. On $T\mathcal{U}$ we have $F = F(x, v)$, on $T^*\mathcal{U}$ we have $H = H(x, \xi)$, and

$$\begin{array}{ll} \mathcal{L}: T\mathcal{U} \longrightarrow T^*\mathcal{U}, & H: T^*\mathcal{U} \longrightarrow \mathbb{R}, \\ (x, v) \longmapsto (x, \xi), & (x, \xi) \longmapsto F_x^*(\xi) = \xi \cdot v - F(x, v), \end{array}$$

where $\xi := L_{F_x}(v) = \frac{\partial F}{\partial v}(x, v)$ is called the *momentum*. Integral curves $(x(t), \xi(t))$ of X_H satisfy the Hamilton equations:

$$\text{(H)} \quad \begin{cases} \frac{dx}{dt} = \frac{\partial H}{\partial \xi}(x, \xi), \\ \frac{d\xi}{dt} = -\frac{\partial H}{\partial x}(x, \xi), \end{cases}$$

whereas the physical path $x(t)$ satisfies the Euler–Lagrange equations:

$$\text{(E–L)} \quad \frac{\partial F}{\partial x}\left(x, \frac{dx}{dt}\right) = \frac{d}{dt}\frac{\partial F}{\partial v}\left(x, \frac{dx}{dt}\right).$$

Let $(x(t), \xi(t)) = \mathcal{L}(x(t), \frac{dx}{dt}(t))$. For an arbitrary curve $x(t)$, we want to prove that $t \mapsto (x(t), \xi(t))$ satisfies (H) if and only if $t \mapsto (x(t), \frac{dx}{dt}(t))$ satisfies (E–L). The first line of (H) comes automatically from the definition of ξ:

$$\xi = L_{F_x}\left(\frac{dx}{dt}\right) \iff \frac{dx}{dt} = L_{F_x}^{-1}(\xi) = L_{F_x^*}(\xi) = \frac{\partial H}{\partial \xi}(x, \xi).$$

If $(x, \xi) = \mathcal{L}(x, v)$, by differentiating both sides of $H(x, \xi) = \xi \cdot v - F(x, v)$ with respect to x, where $\xi = L_{F_x}(v) = \xi(x, v)$ and $v = \frac{\partial H}{\partial \xi}$, we obtain

$$\frac{\partial H}{\partial x} + \frac{\partial H}{\partial \xi}\frac{\partial \xi}{\partial x} = \frac{\partial \xi}{\partial x} \cdot v - \frac{\partial F}{\partial x} \iff \frac{\partial F}{\partial x}(x, v) = -\frac{\partial H}{\partial x}(x, \xi).$$

Using the last equation and the definition of ξ, the second line of (H) becomes (E–L):

$$\frac{d\xi}{dt} = -\frac{\partial H}{\partial x}(x, \xi) \iff \frac{d}{dt}\frac{\partial F}{\partial v}(x, v) = \frac{\partial F}{\partial x}(x, v).$$

□

5.5. *Integrable systems*

DEFINITION 5.11. A *Hamiltonian system* is a triple (M, ω, H), where (M, ω) is a symplectic manifold and $H \in C^\infty(M)$ is the *Hamiltonian function*.

PROPOSITION 5.12. *For a function f on a symplectic manifold (M, ω) we have that $\{f, H\} = 0$ if and only if f is constant along integral curves of X_H.*

PROOF. Let ρ_t be the flow of X_H. Then

$$\frac{d}{dt}(f \circ \rho_t) = \rho_t^* \mathcal{L}_{X_H} f = \rho_t^* \imath_{X_H} df = \rho_t^* \imath_{X_H} \imath_{X_f} \omega = \rho_t^* \omega(X_f, X_H)$$
$$= \rho_t^* \{f, H\}. \qquad \square$$

A function f as in Proposition 5.12 is called an *integral of motion* (or a *first integral* or a *constant of motion*). In general, Hamiltonian systems do not admit integrals of motion that are *independent* of the Hamiltonian function. Functions $f_1, \ldots, f_n$ are said to be *independent* if their differentials $(df_1)_p, \ldots, (df_n)_p$ are linearly independent at all points p in some dense subset of M. Loosely speaking, a Hamiltonian system is (*completely*) *integrable* if it has as many *commuting* integrals of motion as possible. *Commutativity* is with respect to the Poisson bracket. If $f_1, \ldots, f_n$ are commuting integrals of motion for a Hamiltonian system (M, ω, H), then $\omega(X_{f_i}, X_{f_j}) = \{f_i, f_j\} = 0$, so at each $p \in M$ the Hamiltonian vector fields generate an isotropic subspace of T_pM. When $f_1, \ldots, f_n$ are independent, by symplectic linear algebra n can be at most half the dimension of M.

DEFINITION 5.13. A Hamiltonian system (M, ω, H) where M is a $2n$-dimensional manifold is (*completely*) *integrable* if it possesses n independent commuting integrals of motion, $f_1 = H, f_2, \ldots, f_n$.

Any 2-dimensional Hamiltonian system (where the set of nonfixed points is dense) is trivially integrable. Basic examples are the simple pendulum and the harmonic oscillator. A Hamiltonian system (M, ω, H) where M is 4-dimensional is integrable if there is an integral of motion independent of H (the commutativity condition is automatically satisfied). A basic example is the spherical pendulum. Sophisticated examples of integrable systems can be found in [8,72].

EXAMPLES.

1. The *simple pendulum* is a mechanical system consisting of a massless rigid rod of length ℓ, fixed at one end, whereas the other end has a bob of mass m, which may oscillate in the vertical plane. We assume that the force of gravity is constant pointing vertically downwards and the only external force acting on this system. Let θ be the oriented angle between the rod and the vertical direction. Let ξ be the coordinate along the fibers of T^*S^1 induced by the standard angle coordinate on S^1. The energy function $H: T^*S^1 \to \mathbb{R}$, $H(\theta, \xi) = \frac{\xi^2}{2m\ell^2} + m\ell(1 - \cos\theta)$, is an appropriate Hamiltonian function to describe the simple pendulum. Gravity is responsible for the potential energy $V(\theta) = m\ell(1 - \cos\theta)$, and the kinetic energy is given by $K(\theta, \xi) = \frac{1}{2m\ell^2}\xi^2$.
2. The *spherical pendulum* consists of a massless rigid rod of length ℓ, fixed at one end, whereas the other end has a bob of mass m, which may oscillate *freely in all directions*. For simplicity let $m = \ell = 1$. Again assume that gravity is the only external force. Let φ, θ $(0 < \varphi < \pi,\ 0 < \theta < 2\pi)$ be spherical coordinates for the bob, inducing coordinates η, ξ along the fibers of T^*S^2. An appropriate Hamiltonian function for this system is the energy function $H: T^*S^2 \to \mathbb{R}$, $H(\varphi, \theta, \eta, \xi) =$

$\frac{1}{2}(\eta^2 + \frac{\xi^2}{(\sin\varphi)^2}) + \cos\varphi$. The function $J(\varphi, \theta, \eta, \xi) = \xi$ is an independent integral of motion corresponding to the group of symmetries given by rotations about the vertical axis (Section 5.6). The points $p \in T^*S^2$ where dH_p and dJ_p are linearly dependent are:

- the two critical points of H (where both dH and dJ vanish);
- if $x \in S^2$ is in the southern hemisphere ($x_3 < 0$), then there exist exactly two points, $p_+ = (x, \eta, \xi)$ and $p_- = (x, -\eta, -\xi)$, in the cotangent fiber above x where dH_p and dJ_p are linearly dependent;
- since dH_p and dJ_p are linearly dependent along the trajectory of the Hamiltonian vector field of H through p_+, this trajectory is also a trajectory of the Hamiltonian vector field of J and hence its projection onto S^2 is a latitudinal (or horizontal) circle. The projection of the trajectory through p_- is the same latitudinal circle traced in the opposite direction.

Let (M, ω, H) be an integrable system of dimension $2n$ with integrals of motion $f_1 = H, f_2, \ldots, f_n$. Let $c \in \mathbb{R}^n$ be a regular value of $f := (f_1, \ldots, f_n)$. The corresponding level set $f^{-1}(c)$ is a Lagrangian submanifold, as it is n-dimensional and its tangent bundle is isotropic. If the flows are complete on $f^{-1}(c)$, by following them we obtain global coordinates. Any compact component of $f^{-1}(c)$ must hence be a torus. These components, when they exist, are called *Liouville tori*. A way to ensure that compact components exist is to have one of the f_i's proper.

THEOREM 5.14 (Arnold–Liouville [2]). *Let (M, ω, H) be an integrable system of dimension $2n$ with integrals of motion $f_1 = H, f_2, \ldots, f_n$. Let $c \in \mathbb{R}^n$ be a regular value of $f := (f_1, \ldots, f_n)$. The level $f^{-1}(c)$ is a Lagrangian submanifold of M.*

(a) *If the flows of the Hamiltonian vector fields $X_{f_1}, \ldots, X_{f_n}$ starting at a point $p \in f^{-1}(c)$ are complete, then the connected component of $f^{-1}(c)$ containing p is a homogeneous space for $\mathbb{R}^n$, i.e., is of the form $\mathbb{R}^{n-k} \times \mathbb{T}^k$ for some k, $0 \leqslant k \leqslant n$, where $\mathbb{T}^k$ is a k-dimensional torus.. With respect to this affine structure, that component has coordinates $\varphi_1, \ldots, \varphi_n$, known as* angle coordinates, *in which the flows of $X_{f_1}, \ldots, X_{f_n}$ are linear.*

(b) *There are coordinates $\psi_1, \ldots, \psi_n$, known as* action coordinates, *complementary to the angle coordinates, such that the ψ_i's are integrals of motion and $\varphi_1, \ldots, \varphi_n$, $\psi_1, \ldots, \psi_n$ form a Darboux chart.*

Therefore, the dynamics of an integrable system has a simple explicit solution in action-angle coordinates. The proof of part (a)—the easy part of the theorem—is sketched above. For the proof of part (b) see, for instance, [2,36]. Geometrically, regular levels being Lagrangian submanifolds implies that, in a neighborhood of a regular value, the map $f : M \to \mathbb{R}^n$ collecting the given integrals of motion is a *Lagrangian fibration*, i.e., it is locally trivial and its fibers are Lagrangian submanifolds. Part (a) states that there are coordinates along the fibers, the angle coordinates,[56] in which the flows of $X_{f_1}, \ldots, X_{f_n}$ are linear. Part (b) guarantees the existence of coordinates on $\mathbb{R}^n$, the action coordinates,

[56] The name *angle coordinates* is used even if the fibers are not tori.

$\psi_1, \ldots, \psi_n$, complementary to the angle coordinates, that (Poisson) commute among themselves and satisfy $\{\varphi_i, \psi_j\} = \delta_{ij}$. The action coordinates are generally not the given integrals of motion because $\varphi_1, \ldots, \varphi_n, f_1, \ldots, f_n$ do not form a Darboux chart.

5.6. *Symplectic and Hamiltonian actions*

Let (M, ω) be a symplectic manifold, and G a Lie group.

DEFINITION 5.15. An action[57] $\psi : G \to \mathrm{Diff}(M)$, $g \mapsto \psi_g$, is a *symplectic action* if each ψ_g is a symplectomorphism, i.e., $\psi : G \to \mathrm{Sympl}(M, \omega) \subset \mathrm{Diff}(M)$.

In particular, symplectic actions of $\mathbb{R}$ on (M, ω) are in one-to-one correspondence with complete symplectic vector fields on M:

$$\psi = \exp tX \quad \longleftrightarrow \quad X_p = \left.\frac{d\psi_t(p)}{dt}\right|_{t=0}, \qquad p \in M.$$

We may define a symplectic action ψ of S^1 or $\mathbb{R}$ on (M, ω) to be *Hamiltonian* if the vector field X generated by ψ is Hamiltonian, that is, when there is $H : M \to \mathbb{R}$ with $dH = \imath_X \omega$. An action of S^1 may be viewed as a periodic action of $\mathbb{R}$.

EXAMPLES.

1. On $(\mathbb{R}^{2n}, \omega_0)$, the orbits of the action generated by $X = -\frac{\partial}{\partial y_1}$ are lines parallel to the y_1-axis, $\{(x_1, y_1 - t, x_2, y_2, \ldots, x_n, y_n) \mid t \in \mathbb{R}\}$. Since X is Hamiltonian with Hamiltonian function x_1, this is a *Hamiltonian action* of $\mathbb{R}$.
2. On the 2-sphere $(S^2, d\theta \wedge dh)$ in cylindrical coordinates, the one-parameter group of diffeomorphisms given by rotation around the vertical axis, $\psi_t(\theta, h) = (\theta + t, h)$ ($t \in \mathbb{R}$) is a symplectic action of the group $S^1 \simeq \mathbb{R}/\langle 2\pi \rangle$, as it preserves the area form $d\theta \wedge dh$. Since the vector field corresponding to ψ is Hamiltonian with Hamiltonian function h, this is a *Hamiltonian action* of S^1.

When G is a product of S^1's or $\mathbb{R}$'s, an action $\psi : G \to \mathrm{Sympl}(M, \omega)$ is called *Hamiltonian* when the restriction to each 1-dimensional factor is Hamiltonian in the previous sense *with Hamiltonian function preserved by the action of the rest of G*.

For an arbitrary Lie group G, we use an upgraded Hamiltonian function μ, known as a *moment map*, determined up to an additive local constant by coordinate functions μ_i indexed by a basis of the Lie algebra of G. We require that the constant be such that μ is *equivariant*, i.e., μ intertwines the action of G on M and the coadjoint action of G on the dual of its Lie algebra. (If M is compact, equivariance can be achieved by adjusting the constant so that $\int_M \mu\omega^n = 0$. Similarly when there is a fixed point p (on each component of M) by imposing $\mu(p) = 0$.)

Let G be a Lie group, $\mathfrak{g}$ the Lie algebra of G, and $\mathfrak{g}^*$ the dual vector space of $\mathfrak{g}$.

[57] A (smooth) *action* of G on M is a group homomorphism $G \to \mathrm{Diff}(M)$, $g \mapsto \psi_g$, whose evaluation map $M \times G \to M$, $(p, g) \mapsto \psi_g(p)$, is smooth.

DEFINITION 5.16. An action $\psi : G \to \mathrm{Diff}(M)$ on a symplectic manifold (M, ω) is a *Hamiltonian action* if there exists a map $\mu : M \to \mathfrak{g}^*$ satisfying:

- For each $X \in \mathfrak{g}$, we have $d\mu^X = \imath_{X^\#}\omega$, i.e., μ^X is a Hamiltonian function for the vector field $X^\#$, where
  - $\mu^X : M \to \mathbb{R}$, $\mu^X(p) := \langle \mu(p), X \rangle$, is the component of μ along X,
  - $X^\#$ is the vector field on M generated by the one-parameter subgroup $\{\exp tX \mid t \in \mathbb{R}\} \subseteq G$.
- The map μ is *equivariant* with respect to the given action ψ on M and the coadjoint action: $\mu \circ \psi_g = \mathrm{Ad}_g^* \circ \mu$, for all $g \in G$.

Then (M, ω, G, μ) is a *Hamiltonian G-space* and μ is a *moment map*.

This definition matches the previous one when G is an Abelian group $\mathbb{R}$, S^1 or $\mathbb{T}^n$, for which equivariance becomes invariance since the coadjoint action is trivial.

EXAMPLES.

1. Let $\mathbb{T}^n = \{(t_1, \ldots, t_n) \in \mathbb{C}^n\colon\ |t_j| = 1,$ for all $j\}$ be a torus acting on $\mathbb{C}^n$ by $(t_1, \ldots, t_n) \cdot (z_1, \ldots, z_n) = (t_1^{k_1} z_1, \ldots, t_n^{k_n} z_n)$, where $k_1, \ldots, k_n \in \mathbb{Z}$ are fixed. This action is Hamiltonian with a moment map $\mu : \mathbb{C}^n \to (\mathfrak{t}^n)^* \simeq \mathbb{R}^n$, $\mu(z_1, \ldots, z_n) = -\frac{1}{2}(k_1|z_1|^2, \ldots, k_n|z_n|^2)$.
2. When a Lie group G acts on two symplectic manifolds (M_j, ω_j), $j = 1, 2$, with moment maps $\mu_j : M_j \to \mathfrak{g}^*$, the diagonal action of G on $M_1 \times M_2$ has moment map $\mu : M_1 \times M_2 \to \mathfrak{g}^*$, $\mu(p_1, p_2) = \mu_1(p_1) + \mu_2(p_2)$.
3. Equip the coadjoint orbits of a Lie group G with the canonical symplectic form (Section 5.1). Then, for each $\xi \in \mathfrak{g}^*$, the coadjoint action on the orbit $G \cdot \xi$ is Hamiltonian with moment map simply the inclusion map $\mu : G \cdot \xi \hookrightarrow \mathfrak{g}^*$.
4. Identify the Lie algebra of the unitary group $\mathrm{U}(n)$ with its dual via the inner product $\langle A, B \rangle = \mathrm{trace}(A^* B)$. The natural action of $\mathrm{U}(n)$ on $(\mathbb{C}^n, \omega_0)$ is Hamiltonian with moment map $\mu : \mathbb{C}^n \to \mathfrak{u}(n)$ given by $\mu(z) = \frac{i}{2} z z^*$. Similarly, a moment map for the natural action of $\mathrm{U}(k)$ on the space $(\mathbb{C}^{k \times n}, \omega_0)$ of complex $(k \times n)$-matrices is given by $\mu(A) = \frac{i}{2} A A^*$ for $A \in \mathbb{C}^{k \times n}$. Thus the $\mathrm{U}(n)$-action by conjugation on the space $(\mathbb{C}^{n^2}, \omega_0)$ of complex $(n \times n)$-matrices is Hamiltonian, with moment map given by $\mu(A) = \frac{i}{2}[A, A^*]$.
5. For the spherical pendulum (Section 5.5), the *energy-momentum map* $(H, J) : T^* S^2 \to \mathbb{R}^2$ is a moment map for the $\mathbb{R} \times S^1$ action given by time flow and rotation about the vertical axis.
6. Suppose that a compact Lie group acts on a symplectic manifold (M, ω) in a Hamiltonian way, and that $q \in M$ is a fixed point for the G-action. Then, by an equivariant version of Darboux's theorem,[58] there exists a Darboux chart $(\mathcal{U}, z_1, \ldots, z_n)$ centered at q that is G-equivariant with respect to a linear action of G on $\mathbb{C}^n$. Consider an ε-blow-up of M relative to this chart, for ε sufficiently small. Then G acts on the blow-up in a Hamiltonian way.

[58]**Equivariant Darboux theorem** [136]. *Let (M, ω) be a $2n$-dimensional symplectic manifold equipped with a symplectic action of a compact Lie group G, and let q be a fixed point. Then there exists a G-invariant chart*

The concept of a moment map was introduced by Souriau [119] under the French name *application moment*; besides the more standard English translation to *moment map*, the alternative *momentum map* is also used, and recently James Stasheff has proposed the short unifying new word *momap*. The name comes from being the generalization of *linear and angular momenta* in classical mechanics.

Let $\mathbb{R}^3$ act on $(\mathbb{R}^6 \simeq T^*\mathbb{R}^3, \omega_0 = \sum dx_i \wedge dy_i)$ by *translations*:

$$a \in \mathbb{R}^3 \longmapsto \psi_a \in \mathrm{Sympl}(\mathbb{R}^6, \omega_0), \quad \psi_a(x, y) = (x + a, y).$$

The vector field generated by $X = a = (a_1, a_2, a_3)$ is $X^\# = a_1 \frac{\partial}{\partial x_1} + a_2 \frac{\partial}{\partial x_2} + a_3 \frac{\partial}{\partial x_3}$, and the *linear momentum* map

$$\mu : \mathbb{R}^6 \longrightarrow \mathbb{R}^3, \quad \mu(x, y) = y$$

is a moment map, with $\mu^a(x, y) = \langle \mu(x, y), a\rangle = y \cdot a$. Classically, y is called the *momentum vector* corresponding to the *position vector* x.

The SO(3)-action on $\mathbb{R}^3$ by *rotations* lifts to a symplectic action ψ on the cotangent bundle $\mathbb{R}^6$. The infinitesimal version of this action is[59]

$$a \in \mathbb{R}^3 \longmapsto d\psi(a) \in \chi^{\mathrm{sympl}}(\mathbb{R}^6), \quad d\psi(a)(x, y) = (a \times x, a \times y).$$

Then the *angular momentum* map

$$\mu : \mathbb{R}^6 \longrightarrow \mathbb{R}^3, \quad \mu(x, y) = x \times y$$

is a moment map, with $\mu^a(x, y) = \langle \mu(x, y), a\rangle = (x \times y) \cdot a$.

The notion of a moment map associated to a group action on a symplectic manifold formalizes the *Noether principle*, which asserts that there is a one-to-one correspondence between *symmetries* (or one-parameter group actions) and *integrals of motion* (or conserved quantities) for a mechanical system.

$(\mathcal{U}, x_1, \ldots, x_n, y_1, \ldots, y_n)$ centered at q and G-equivariant with respect to a linear action of G on $\mathbb{R}^{2n}$ such that

$$\omega|_{\mathcal{U}} = \sum_{k=1}^{n} dx_k \wedge dy_k.$$

A suitable linear action on $\mathbb{R}^{2n}$ is equivalent to the induced action of G on T_qM. The proof relies on an equivariant version of the Moser trick and may be found in [70].

[59] The Lie group $\mathrm{SO}(3) = \{A \in \mathrm{GL}(3; \mathbb{R}) \mid A^t A = \mathrm{Id}$ and $\det A = 1\}$, has Lie algebra, $\mathfrak{g} = \{A \in \mathfrak{gl}(3; \mathbb{R}) \mid A + A^t = 0\}$, the space of 3×3 skew-symmetric matrices. The standard identification of $\mathfrak{g}$ with $\mathbb{R}^3$ carries the Lie bracket to the exterior product:

$$A = \begin{bmatrix} 0 & -a_3 & a_2 \\ a_3 & 0 & -a_1 \\ -a_2 & a_1 & 0 \end{bmatrix} \longmapsto a = (a_1, a_2, a_3),$$

$$[A, B] = AB - BA \longmapsto a \times b.$$

DEFINITION 5.17. An *integral of motion* of a Hamiltonian G-space (M, ω, G, μ) is a G-invariant function $f : M \to \mathbb{R}$. When μ is constant on the trajectories of a Hamiltonian vector field X_f, the corresponding flow $\{\exp tX_f \mid t \in \mathbb{R}\}$ (regarded as an $\mathbb{R}$-action) is a *symmetry* of the Hamiltonian G-space (M, ω, G, μ).

THEOREM 5.18 (Noether). *Let (M, ω, G, μ) be a Hamiltonian G-space where G is connected. If f is an integral of motion, the flow of its Hamiltonian vector field X_f is a symmetry. If the flow of some Hamiltonian vector field X_f is a symmetry, then a corresponding Hamiltonian function f is an integral of motion.*

PROOF. Let $\mu^X = \langle \mu, X\rangle : M \to \mathbb{R}$ for $X \in \mathfrak{g}$. We have $\mathcal{L}_{X_f}\mu^X = \imath_{X_f} d\mu^X = \imath_{X_f}\imath_{X^\#}\omega = -\imath_{X^\#}\imath_{X_f}\omega = -\imath_{X^\#} df = -\mathcal{L}_{X^\#} f$. So μ is invariant over the flow of X_f if and only if f is invariant under the infinitesimal G-action. □

We now turn to the questions of existence and uniqueness of moment maps.

Let $\mathfrak{g}$ be a Lie algebra, and let $C^k := \Lambda^k\mathfrak{g}^*$ be the set of *k-cochains* on $\mathfrak{g}$, that is, of alternating k-linear maps $\mathfrak{g} \times \cdots \times \mathfrak{g} \to \mathbb{R}$. The linear operator $\delta : C^k \to C^{k+1}$ defined by $\delta c(X_0, \ldots, X_k) = \sum_{i<j} (-1)^{i+j} c([X_i, X_j], X_0, \ldots, \hat{X}_i, \ldots, \hat{X}_j, \ldots, X_k)$ satisfies $\delta^2 = 0$. The *Lie algebra cohomology groups* (or *Chevalley cohomology groups*) of $\mathfrak{g}$ are the cohomology groups of the complex $0 \xrightarrow{\delta} C^0 \xrightarrow{\delta} C^1 \xrightarrow{\delta} \cdots$:

$$H^k(\mathfrak{g}; \mathbb{R}) := \frac{\ker \delta : C^k \to C^{k+1}}{\operatorname{im} \delta : C^{k-1} \to C^k}.$$

It is always $H^0(\mathfrak{g}; \mathbb{R}) = \mathbb{R}$. If $c \in C^1 = \mathfrak{g}^*$, then $\delta c(X, Y) = -c([X, Y])$. The *commutator ideal* $[\mathfrak{g}, \mathfrak{g}]$ is the subspace of $\mathfrak{g}$ spanned by $\{[X, Y] \mid X, Y \in \mathfrak{g}\}$. Since $\delta c = 0$ if and only if c vanishes on $[\mathfrak{g}, \mathfrak{g}]$, we conclude that $H^1(\mathfrak{g}; \mathbb{R}) = [\mathfrak{g}, \mathfrak{g}]^0$, where $[\mathfrak{g}, \mathfrak{g}]^0 \subseteq \mathfrak{g}^*$ is the *annihilator* of $[\mathfrak{g}, \mathfrak{g}]$. An element of C^2 is an alternating bilinear map $c : \mathfrak{g} \times \mathfrak{g} \to \mathbb{R}$, and $\delta c(X, Y, Z) = -c([X, Y], Z) + c([X, Z], Y) - c([Y, Z], X)$. If $c = \delta b$ for some $b \in C^1$, then $c(X, Y) = (\delta b)(X, Y) = -b([X, Y])$.

If $\mathfrak{g}$ is the Lie algebra of a compact connected Lie group G, then by averaging one can show that the de Rham cohomology may be computed from the subcomplex of G-invariant forms, and hence $H^k(\mathfrak{g}; \mathbb{R}) = H^k_{\text{deRham}}(G)$.

PROPOSITION 5.19. *If $H^1(\mathfrak{g}; \mathbb{R}) = H^2(\mathfrak{g}, \mathbb{R}) = 0$, then any symplectic G-action is Hamiltonian.*

PROOF. Let $\psi : G \to \operatorname{Sympl}(M, \omega)$ be a symplectic action of G on a symplectic manifold (M, ω). Since $H^1(\mathfrak{g}; \mathbb{R}) = 0$ means that $[\mathfrak{g}, \mathfrak{g}] = \mathfrak{g}$, and since commutators of symplectic vector fields are Hamiltonian, we have $d\psi : \mathfrak{g} = [\mathfrak{g}, \mathfrak{g}] \to \chi^{\text{ham}}(M)$. The action ψ is Hamiltonian if and only if there is a Lie algebra homomorphism $\mu^* : \mathfrak{g} \to C^\infty(M)$ such that the Hamiltonian vector field of $\mu^*(\xi)$ is $d\psi(\xi)$. We first take an arbitrary vector space lift $\tau : \mathfrak{g} \to C^\infty(M)$ with this property, i.e., for each basis vector $X \in \mathfrak{g}$, we choose $\tau(X) = \tau^X \in C^\infty(M)$ such that $v_{(\tau^X)} = d\psi(X)$. The map $X \mapsto \tau^X$ may not be a Lie algebra homomorphism. By construction, $\tau^{[X,Y]}$ is a Hamiltonian function for $[X, Y]^\#$, and (as computed in Section 5.5) $\{\tau^X, \tau^Y\}$ is a Hamiltonian function for $-[X^\#, Y^\#]$. Since $[X, Y]^\# = -[X^\#, Y^\#]$, the corresponding Hamiltonian functions must differ by a constant:

$$\tau^{[X,Y]} - \{\tau^X, \tau^Y\} = c(X, Y) \in \mathbb{R}.$$

By the Jacobi identity, $\delta c = 0$. Since $H^2(\mathfrak{g}; \mathbb{R}) = 0$, there is $b \in \mathfrak{g}^*$ satisfying $c = \delta b$, $c(X, Y) = -b([X, Y])$. We define

$$\begin{aligned} \mu^* : \mathfrak{g} &\longrightarrow C^\infty(M), \\ X &\longmapsto \mu^*(X) = \tau^X + b(X) = \mu^X. \end{aligned}$$

Now μ^* is a Lie algebra homomorphism: $\mu^*([X, Y]) = \{\tau^X, \tau^Y\} = \{\mu^X, \mu^Y\}$. □

By the Whitehead lemmas (see, for instance, [77, pp. 93–95]) a semisimple Lie group G has $H^1(\mathfrak{g}; \mathbb{R}) = H^2(\mathfrak{g}; \mathbb{R}) = 0$. As a corollary, *when G is semisimple, any symplectic G-action is Hamiltonian.*[60]

PROPOSITION 5.20. *For a connected Lie group G, if $H^1(\mathfrak{g}; \mathbb{R}) = 0$, then moment maps for Hamiltonian G-actions are unique.*

PROOF. Suppose that μ_1 and μ_2 are two moment maps for an action ψ. For each $X \in \mathfrak{g}$, μ_1^X and μ_2^X are both Hamiltonian functions for $X^\#$, thus $\mu_1^X - \mu_2^X = c(X)$ is locally constant. This defines $c \in \mathfrak{g}^*$, $X \mapsto c(X)$. Since the corresponding $\mu_i^* : \mathfrak{g} \to C^\infty(M)$ are Lie algebra homomorphisms, we have $c([X, Y]) = 0$, $\forall X, Y \in \mathfrak{g}$, i.e., $c \in [\mathfrak{g}, \mathfrak{g}]^0 = \{0\}$. Hence, $\mu_1 = \mu_2$. □

In general, if $\mu : M \to \mathfrak{g}^*$ is a moment map, then given any $c \in [\mathfrak{g}, \mathfrak{g}]^0$, $\mu_1 = \mu + c$ is another moment map. In other words, moment maps are unique up to elements of the dual of the Lie algebra that annihilate the commutator ideal.

The two extreme cases are when

- G is semisimple: any symplectic action is Hamiltonian, moment maps are unique;
- G is Abelian: symplectic actions may not be Hamiltonian, moment maps are unique up to a constant $c \in \mathfrak{g}^*$.

[60] A compact Lie group G has $H^1(\mathfrak{g}; \mathbb{R}) = H^2(\mathfrak{g}; \mathbb{R}) = 0$ if and only if it is semisimple. In fact, a compact Lie group G is semisimple when $\mathfrak{g} = [\mathfrak{g}, \mathfrak{g}]$. The unitary group $U(n)$ is not semisimple because the multiples of the identity, $S^1 \cdot \mathrm{Id}$, form a nontrivial center; at the level of the Lie algebra, this corresponds to the subspace $\mathbb{R} \cdot \mathrm{Id}$ of scalar matrices, which are not commutators since they are not traceless. Any Abelian Lie group is *not* semisimple. Any direct product of the other compact classical groups $SU(n)$, $SO(n)$ and $Sp(n)$ is semisimple. An arbitrary compact Lie group admits a finite cover by a direct product of tori and semisimple Lie groups.

5.7. *Convexity*

Atiyah, Guillemin and Sternberg [4,68] showed that the image of the moment map for a Hamiltonian torus action on a compact connected symplectic manifold is always a polytope.[61] A proof of this theorem can also be found in [99].

THEOREM 5.21 (Atiyah, Guillemin–Sternberg). *Let (M,ω) be a compact connected symplectic manifold. Suppose that $\psi:\mathbb{T}^m \to \mathrm{Sympl}(M,\omega)$ is a Hamiltonian action of an m-torus with moment map $\mu: M \to \mathbb{R}^m$. Then:*

(a) *the levels $\mu^{-1}(c)$ are connected* $(c \in \mathbb{R}^m)$;

(b) *the image $\mu(M)$ is convex*;

(c) *$\mu(M)$ is the convex hull of the images of the fixed points of the action.*

The image $\mu(M)$ of the moment map is called the *moment polytope*.

EXAMPLES.

1. Suppose that $\mathbb{T}^m$ acts linearly on $(\mathbb{C}^n,\omega_0)$. Let $\lambda^{(1)},\dots,\lambda^{(n)} \in \mathbb{Z}^m$ be the *weights* appearing in the corresponding weight space decomposition, that is,

$$\mathbb{C}^n \simeq \bigoplus_{k=1}^{n} V_{\lambda^{(k)}},$$

where, for $\lambda^{(k)} = (\lambda_1^{(k)},\dots,\lambda_m^{(k)})$, the torus $\mathbb{T}^m$ acts on the complex line $V_{\lambda^{(k)}}$ by $(e^{it_1},\dots,e^{it_m})\cdot v = e^{i\sum_j \lambda_j^{(k)} t_j} v$. If the action is effective[62], then $m \leqslant n$ and the weights $\lambda^{(1)},\dots,\lambda^{(n)}$ are part of a $\mathbb{Z}$-basis of $\mathbb{Z}^m$. If the action is symplectic (hence Hamiltonian in this case), then the weight spaces $V_{\lambda^{(k)}}$ are symplectic subspaces. In this case, a moment map is given by

$$\mu(v) = -\frac{1}{2}\sum_{k=1}^{n} \lambda^{(k)} |v_{\lambda^{(k)}}|^2,$$

where $|\cdot|$ is the standard norm[63] and $v = v_{\lambda^{(1)}} + \cdots + v_{\lambda^{(n)}}$ is the weight space decomposition of v. We conclude that, if $\mathbb{T}^n$ acts on $\mathbb{C}^n$ in a linear, effective and Hamiltonian way, then any moment map μ is a submersion, i.e., each differential $d\mu_v : \mathbb{C}^n \to \mathbb{R}^n$ $(v \in \mathbb{C}^n)$ is surjective.

---

[61] A *polytope* in $\mathbb{R}^n$ is the convex hull of a finite number of points in $\mathbb{R}^n$. A *convex polyhedron* is a subset of $\mathbb{R}^n$ that is the intersection of a finite number of affine half-spaces. Hence, polytopes coincide with bounded convex polyhedra.

[62] An action of a group G on a manifold M is called *effective* if each group element $g \neq e$ moves at least one point $p \in M$, that is, $\bigcap_{p\in M} G_p = \{e\}$, where $G_p = \{g \in G \mid g\cdot p = p\}$ is the stabilizer of p.

[63] The standard inner product satisfies $\langle v, w\rangle = \omega_0(v, Jv)$ where $J\frac{\partial}{\partial z} = i\frac{\partial}{\partial z}$ and $J\frac{\partial}{\partial \bar z} = -i\frac{\partial}{\partial \bar z}$. In particular, the standard norm is invariant for a symplectic complex-linear action.

2. Consider a coadjoint orbit $\mathcal{O}_\lambda$ for the unitary group $\mathrm{U}(n)$. Multiplying by i, the orbit $\mathcal{O}_\lambda$ can be viewed as the set of Hermitian matrices with a given eigenvalue spectrum $\lambda = (\lambda_1 \geqslant \cdots \geqslant \lambda_n)$. The restriction of the coadjoint action to the maximal torus $\mathbb{T}^n$ of diagonal unitary matrices is Hamiltonian with moment map $\mu : \mathcal{O}_\lambda \to \mathbb{R}^n$ taking a matrix to the vector of its diagonal entries. Then the moment polytope $\mu(\mathcal{O}_\lambda)$ is the convex hull C of the points given by all the permutations of $(\lambda_1, \dots, \lambda_n)$. This is a rephrasing of the classical theorem of Schur ($\mu(\mathcal{O}_\lambda) \subseteq C$) and Horn ($C \subseteq \mu(\mathcal{O}_\lambda)$).

Example 1 is related to the universal local picture for a moment map near a fixed point of a Hamiltonian torus action:

THEOREM 5.22. *Let $(M^{2n}, \omega, \mathbb{T}^m, \mu)$ be a Hamiltonian $\mathbb{T}^m$-space, where q is a fixed point. Then there exists a chart $(\mathcal{U}, x_1, \dots, x_n, y_1, \dots, y_n)$ centered at q and weights $\lambda^{(1)}, \dots, \lambda^{(n)} \in \mathbb{Z}^m$ such that*

$$\omega|_{\mathcal{U}} = \sum_{k=1}^{n} dx_k \wedge dy_k \quad \textit{and} \quad \mu|_{\mathcal{U}} = \mu(q) - \frac{1}{2}\sum_{k=1}^{n} \lambda^{(k)}\left(x_k^2 + y_k^2\right).$$

The following two results use the crucial fact that any effective action of an m-torus on a manifold has orbits of dimension m; a proof may be found in [19].

COROLLARY 5.23. *Under the conditions of the convexity theorem, if the $\mathbb{T}^m$-action is effective, then there must be at least $m + 1$ fixed points.*

PROOF. At a point p of an m-dimensional orbit the moment map is a submersion, i.e., $(d\mu_1)_p, \dots, (d\mu_m)_p$ are linearly independent. Hence, $\mu(p)$ is an interior point of $\mu(M)$, and $\mu(M)$ is a nondegenerate polytope. A nondegenerate polytope in $\mathbb{R}^m$ has at least $m+1$ vertices. The vertices of $\mu(M)$ are images of fixed points. □

PROPOSITION 5.24. *Let $(M, \omega, \mathbb{T}^m, \mu)$ be a Hamiltonian $\mathbb{T}^m$-space. If the $\mathbb{T}^m$-action is effective, then* $\dim M \geqslant 2m$.

PROOF. Since the moment map is constant on an orbit $\mathcal{O}$, for $p \in \mathcal{O}$ the differential $d\mu_p : T_pM \to \mathfrak{g}^*$ maps $T_p\mathcal{O}$ to 0. Thus $T_p\mathcal{O} \subseteq \ker d\mu_p = (T_p\mathcal{O})^\omega$, where $(T_p\mathcal{O})^\omega$ is the symplectic orthogonal of $T_p\mathcal{O}$. This shows that orbits $\mathcal{O}$ of a Hamiltonian torus action are isotropic submanifolds of M. In particular, by symplectic linear algebra we have that $\dim \mathcal{O} \leqslant \frac{1}{2} \dim M$. Now consider an m-dimensional orbit. □

For a Hamiltonian action of an arbitrary compact Lie group G on a compact symplectic manifold (M, ω), the following *non-Abelian* convexity theorem was proved by Kirwan [81]: if $\mu : M \to \mathfrak{g}^*$ is a moment map, then the intersection $\mu(M) \cap \mathfrak{t}_+^*$ of the image of μ with a Weyl chamber for a Cartan subalgebra $\mathfrak{t} \subseteq \mathfrak{g}$ is a convex polytope. This had been conjectured by Guillemin and Sternberg and proved by them in particular cases.

6. Symplectic reduction

6.1. *Marsden–Weinstein–Meyer theorem*

Classical physicists realized that, whenever there is a symmetry group of dimension k acting on a mechanical system, the number of degrees of freedom for the position and momenta of the particles may be reduced by $2k$. Symplectic reduction formulates this process mathematically.

THEOREM 6.1 (Marsden–Weinstein, Meyer [92,102]). *Let (M, ω, G, μ) be a Hamiltonian G-space (Section 5.6) for a compact Lie group G. Let $i : \mu^{-1}(0) \hookrightarrow M$ be the inclusion map. Assume that G acts freely on $\mu^{-1}(0)$. Then*

(a) *the orbit space $M_{\text{red}} = \mu^{-1}(0)/G$ is a manifold,*

(b) *$\pi : \mu^{-1}(0) \to M_{\text{red}}$ is a principal G-bundle, and*

(c) *there is a symplectic form ω_{red} on M_{red} satisfying $i^*\omega = \pi^*\omega_{\text{red}}$.*

DEFINITION 6.2. The symplectic manifold $(M_{\text{red}}, \omega_{\text{red}})$ is the *reduction* (or *reduced space*, or *symplectic quotient*) of (M, ω) with respect to G, μ.

When M is Kähler and the action of G preserves the complex structure, we can show that the symplectic reduction has a natural Kähler structure.

Let (M, ω, G, μ) be a Hamiltonian G-space for a compact Lie group G. To reduce at a level $\xi \in \mathfrak{g}^*$ of μ, we need $\mu^{-1}(\xi)$ to be preserved by G, or else take the G-orbit of $\mu^{-1}(\xi)$, or else take the quotient by the maximal subgroup of G that preserves $\mu^{-1}(\xi)$. Since μ is equivariant, G preserves $\mu^{-1}(\xi)$ if and only if $\mathrm{Ad}^*_g \xi = \xi$, $\forall g \in G$. Of course, the level 0 is always preserved. Also, when G is a torus, any level is preserved and *reduction at ξ* for the moment map μ, is equivalent to reduction at 0 for a shifted moment map $\phi : M \to \mathfrak{g}^*$, $\phi(p) := \mu(p) - \xi$. In general, let $\mathcal{O}$ be a coadjoint orbit in $\mathfrak{g}^*$ equipped with the *canonical symplectic form*$\omega_{\mathcal{O}}$ (defined in Section 5.1). Let $\mathcal{O}^-$ be the orbit $\mathcal{O}$ equipped with $-\omega_{\mathcal{O}}$. The natural product action of G on $M \times \mathcal{O}^-$ is Hamiltonian with moment map $\mu_{\mathcal{O}}(p, \xi) = \mu(p) - \xi$. If the hypothesis of Theorem 6.1 is satisfied for $M \times \mathcal{O}^-$, then one obtains a *reduced space with respect to the coadjoint orbit $\mathcal{O}$*.

EXAMPLES.

1. The standard symplectic form on $\mathbb{C}^n$ is $\omega_0 = \frac{i}{2} \sum dz_i \wedge d\bar{z}_i = \sum dx_i \wedge dy_i = \sum r_i \, dr_i \wedge d\theta_i$ in polar coordinates. The S^1-action on $(\mathbb{C}^n, \omega_0)$ where $e^{it} \in S^1$ acts as multiplication by e^{it} has vector field $X^\# = \frac{\partial}{\partial \theta_1} + \frac{\partial}{\partial \theta_2} + \cdots + \frac{\partial}{\partial \theta_n}$. This action is Hamiltonian with moment map $\mu : \mathbb{C}^n \to \mathbb{R}$, $\mu(z) = -\frac{|z|^2}{2}$, since $\imath_{X^\#}\omega = \sum r_i dr_i = -\frac{1}{2} \sum dr_i^2 = d\mu$. The level $\mu^{-1}(-\frac{1}{2})$ is the unit sphere S^{2n-1}, whose orbit space is the projective space,

$$\mu^{-1}\left(-\frac{1}{2}\right)/S^1 = S^{2n-1}/S^1 = \mathbb{CP}^{n-1}.$$

The reduced symplectic form at level $-\frac{1}{2}$ is $\omega_{\rm red} = \omega_{\rm FS}$ the Fubini–Study symplectic form. Indeed, if $\mathrm{pr}: \mathbb{C}^{n+1} \setminus \{0\} \to \mathbb{CP}^n$ is the standard projection, the forms $\mathrm{pr}^*\omega_{\rm FS} = \frac{i}{2}\partial\bar\partial \log(|z|^2)$ and ω_0 have the same restriction to S^{2n+1}.

2. Consider the natural action of $\mathrm{U}(k)$ on $\mathbb{C}^{k\times n}$ with moment map $\mu(A) = \frac{i}{2}AA^* + \frac{\mathrm{Id}}{2i}$ for $A \in \mathbb{C}^{k\times n}$ (Section 5.6). Since $\mu^{-1}(0) = \{A \in \mathbb{C}^{k\times n} \mid AA^* = \mathrm{Id}\}$, the reduced manifold is the Grassmannian of k-planes in $\mathbb{C}^n$:

$$\mu^{-1}(0)/\mathrm{U}(k) = \mathbb{G}(k, n).$$

For the case where $G = S^1$ and $\dim M = 4$, here is a glimpse of reduction. Let $\mu : M \to \mathbb{R}$ be the moment map and $p \in \mu^{-1}(0)$. Choose local coordinates near p: θ along the orbit through p, μ given by the moment map, and η_1, η_2 the pullback of coordinates on $M_{\rm red} = \mu^{-1}(0)/S^1$. Then the symplectic form can be written

$$\omega = A\, d\theta \wedge d\mu + \sum B_j\, d\theta \wedge d\eta_j + \sum C_j\, d\mu \wedge d\eta_j + D\, d\eta_1 \wedge d\eta_2.$$

As $d\mu = \imath(\frac{\partial}{\partial\theta})\omega$, we must have $A = 1$, $B_j = 0$. Since ω is symplectic, it must be $D \neq 0$. Hence, $i^*\omega = D\, d\eta_1 \wedge d\eta_2$ is the pullback of a symplectic form on $M_{\rm red}$.

The actual proof of Theorem 6.1 requires some preliminary ingredients.

Let $\mu : M \to \mathfrak{g}^*$ be the moment map for an (Hamiltonian) action of a Lie group G on a symplectic manifold (M, ω). Let $\mathfrak{g}_p$ be the Lie algebra of the stabilizer of a point $p \in M$, let $\mathfrak{g}_p^0 = \{\xi \in \mathfrak{g}^* \mid \langle \xi, X\rangle = 0,\ \forall X \in \mathfrak{g}_p\}$ be the annihilator of $\mathfrak{g}_p$, and let $\mathcal{O}_p$ be the G-orbit through p. Since $\omega_p(X_p^\#, v) = \langle d\mu_p(v), X\rangle$, for all $v \in T_pM$ and all $X \in \mathfrak{g}$, the differential $d\mu_p : T_pM \to \mathfrak{g}^*$ has

$$\ker d\mu_p = (T_p\mathcal{O}_p)^{\omega_p} \quad \text{and} \quad \operatorname{im} d\mu_p = \mathfrak{g}_p^0.$$

Consequently, the action is locally free[64] at p if and only if p is a regular point of μ (i.e., $d\mu_p$ is surjective), and we obtain:

LEMMA 6.3. *If G acts freely on $\mu^{-1}(0)$, then 0 is a regular value of μ, the level $\mu^{-1}(0)$ is a submanifold of M of codimension* $\dim G$, *and, for $p \in \mu^{-1}(0)$, the tangent space $T_p\mu^{-1}(0) = \ker d\mu_p$ is the symplectic orthogonal to $T_p\mathcal{O}_p$ in T_pM.*

In particular, *orbits in $\mu^{-1}(0)$ are isotropic*. Since any tangent vector to the orbit is the value of a vector field generated by the group, we can show this directly by computing, for any $X, Y \in \mathfrak{g}$ and $p \in \mu^{-1}(0)$, the Hamiltonian function for $[Y^\#, X^\#] = [Y, X]^\#$ at that point: $\omega_p(X_p^\#, Y_p^\#) = \mu^{[Y,X]}(p) = 0$.

LEMMA 6.4. *Let (V, Ω) be a symplectic vector space, and I an isotropic subspace. Then Ω induces a canonical symplectic structure $\Omega_{\rm red}$ on I^Ω/I.*

---

[64] The action is *locally free* at p when $\mathfrak{g}_p = \{0\}$, i.e., the stabilizer of p is a discrete group. The action is *free* at p when the stabilizer of p is trivial, i.e., $G_p = \{e\}$.

PROOF. Let $[u], [v]$ be the classes in I^{Ω}/I of $u, v \in I^{\Omega}$. We have $\Omega(u+i, v+j) = \Omega(u, v)$, $\forall i, j \in I$, because $\Omega(u, j) = \Omega(i, v) = \Omega(i, j) = 0$. Hence, we can define $\Omega_{\text{red}}([u],[v]) := \Omega(u, v)$. This is nondegenerate: if $u \in I^{\Omega}$ has $\Omega(u, v) = 0$, for all $v \in I^{\Omega}$, then $u \in (I^{\Omega})^{\Omega} = I$, i.e., $[u] = 0$. □

PROPOSITION 6.5. *If a compact Lie group G acts freely on a manifold M, then M/G is a manifold and the map $\pi : M \to M/G$ is a principal G-bundle.*

PROOF. We first show that, for any $p \in M$, the G-orbit through p is a compact submanifold of M diffeomorphic to G.[65] The G-orbit through p is the image of the smooth injective map $\text{ev}_p : G \to M$, $\text{ev}_p(g) = g \cdot p$. The map ev_p is proper because, if A is a compact, hence closed, subset of M, then its inverse image $(\text{ev}_p)^{-1}(A)$, being a closed subset of the compact Lie group G, is also compact. The differential $d(\text{ev}_p)_e$ is injective because $d(\text{ev}_p)_e(X) = 0 \Leftrightarrow X_p^{\#} = 0 \Leftrightarrow X = 0$, $\forall X \in T_eG$, as the action is free. At any other point $g \in G$, for $X \in T_gG$ we have $d(\text{ev}_p)_g(X) = 0 \Leftrightarrow d(\text{ev}_p \circ R_g)_e \circ (dR_{g^{-1}})_g(X) = 0$, where $R_g : G \to G$, $h \mapsto hg$, is right multiplication by g. But $\text{ev}_p \circ R_g = \text{ev}_{g \cdot p}$ has an injective differential at e, and $(dR_{g^{-1}})_g$ is an isomorphism. It follows that $d(\text{ev}_p)_g$ is always injective, so ev_p is an immersion. We conclude that ev_p is a closed embedding.

We now apply the slice theorem[66] which is an equivariant tubular neighborhood theorem. For $p \in M$, let $q = \pi(p) \in M/G$. Choose a G-invariant neighborhood $\mathcal{U}$ of p as in the slice theorem, so that $\mathcal{U} \simeq G \times S$ where S is an appropriate slice. Then $\pi(\mathcal{U}) = \mathcal{U}/G =: \mathcal{V}$ is a neighborhood of q in M/G homeomorphic[67] to S. Such neighborhoods $\mathcal{V}$ are used as charts on M/G. To show that the associated transition maps are smooth, consider two G-invariant open sets $\mathcal{U}_1, \mathcal{U}_2$ in M and corresponding slices S_1, S_2. Then $S_{12} = S_1 \cap \mathcal{U}_2$, $S_{21} = S_2 \cap \mathcal{U}_1$ are both slices for the G-action on $\mathcal{U}_1 \cap \mathcal{U}_2$. To compute the transition map $S_{12} \to S_{21}$, consider the sequence $S_{12} \xrightarrow{\simeq} \{e\} \times S_{12} \hookrightarrow G \times S_{12} \xrightarrow{\simeq} \mathcal{U}_1 \cap \mathcal{U}_2$ and similarly for S_{21}. The composition $S_{12} \hookrightarrow \mathcal{U}_1 \cap \mathcal{U}_2 \xrightarrow{\simeq} G \times S_{21} \xrightarrow{\text{pr}} S_{21}$ is smooth.

Finally, we show that $\pi : M \to M/G$ is a principal G-bundle. For $p \in M$, $q = \pi(p)$, choose a G-invariant neighborhood $\mathcal{U}$ of p of the form $\eta : G \times S \xrightarrow{\simeq} \mathcal{U}$. Then $\mathcal{V} = \mathcal{U}/G \simeq S$ is the corresponding neighborhood of q in M/G:

$$\begin{array}{ccccc} M \supseteq \mathcal{U} & \overset{\eta}{\simeq} & G \times S & \simeq G \times \mathcal{V} \\ \downarrow \pi & & & \downarrow \\ M/G \supseteq \mathcal{V} & & = & \mathcal{V} \end{array}$$

[65] Even if the action is not free, the orbit through p is a compact submanifold of M. In that case, the orbit of a point p is diffeomorphic to the quotient G/G_p of G by the stabilizer of p.

[66] **Slice theorem.** *Let G be a compact Lie group acting on a manifold M such that G acts freely at $p \in M$. Let S be a transverse section to $\mathcal{O}_p$ at p (this is called a* slice*). Choose a coordinate chart $x_1, \ldots, x_n$ centered at p such that $\mathcal{O}_p \simeq G$ is given by $x_1 = \cdots = x_k = 0$ and S by $x_{k+1} = \cdots = x_n = 0$. Let $S_\varepsilon = S \cap B_\varepsilon$ where B_ε is the ball of radius ε centered at 0 with respect to these coordinates. Let $\eta : G \times S \to M$, $\eta(g, s) = g \cdot s$. Then, for sufficiently small ε, the map $\eta : G \times S_\varepsilon \to M$ takes $G \times S_\varepsilon$ diffeomorphically onto a G-invariant neighborhood $\mathcal{U}$ of the G-orbit through p.* In particular, if the action of G is free at p, then the action is free on $\mathcal{U}$, so the set of points where G acts freely is open.

[67] We equip the orbit space M/G with the *quotient topology*, i.e., $\mathcal{V} \subseteq M/G$ is open if and only if $\pi^{-1}(\mathcal{V})$ is open in M.

Since the projection on the right is smooth, π is smooth. By considering the overlap of two trivializations $\phi_1 : \mathcal{U}_1 \to G \times \mathcal{V}_1$ and $\phi_2 : \mathcal{U}_2 \to G \times \mathcal{V}_2$, we check that the transition map $\phi_2 \circ \phi_1^{-1} = (\sigma_{12}, \mathrm{id}) : G \times (\mathcal{V}_1 \cap \mathcal{V}_2) \to G \times (\mathcal{V}_1 \cap \mathcal{V}_2)$ is smooth. □

PROOF OF THEOREM 6.1. Since G acts freely on $\mu^{-1}(0)$, by Lemma 6.3 the level $\mu^{-1}(0)$ is a submanifold. Applying Proposition 6.5 to the free action of G on the manifold $\mu^{-1}(0)$, we conclude the assertions (a) and (b).

At $p \in \mu^{-1}(0)$ the tangent space to the orbit $T_p\mathcal{O}_p$ is an isotropic subspace of the symplectic vector space (T_pM, ω_p). By Lemma 6.4 there is a canonical symplectic structure on the quotient $T_p\mu^{-1}(0)/T_p\mathcal{O}_p$. The point $[p] \in M_{\mathrm{red}} = \mu^{-1}(0)/G$ has tangent space $T_{[p]}M_{\mathrm{red}} \simeq T_p\mu^{-1}(0)/T_p\mathcal{O}_p$. This gives a well-defined nondegenerate 2-form ω_{red} on M_{red} because ω is G-invariant. By construction $i^*\omega = \pi^*\omega_{\mathrm{red}}$ where

$$\begin{array}{ccc} \mu^{-1}(0) & \stackrel{i}{\hookrightarrow} & M \\ \downarrow \pi & & \\ M_{\mathrm{red}} & & \end{array}$$

The injectivity of π^* yields closedness: $\pi^* \, d\omega_{\mathrm{red}} = d\pi^*\omega_{\mathrm{red}} = d\imath^*\omega = \imath^* \, d\omega = 0$. □

6.2. *Applications and generalizations*

Let (M, ω, G, μ) be a Hamiltonian G-space for a compact Lie group G. Suppose that another Lie group H acts on (M, ω) in a Hamiltonian way with moment map $\phi : M \to \mathfrak{h}^*$. Suppose that the H-action commutes with the G-action, that ϕ is G-invariant and that μ is H-invariant. Assuming that G acts freely on $\mu^{-1}(0)$, let $(M_{\mathrm{red}}, \omega_{\mathrm{red}})$ be the corresponding reduced space. Since the action of H preserves $\mu^{-1}(0)$ and ω and commutes with the G-action, the reduced space $(M_{\mathrm{red}}, \omega_{\mathrm{red}})$ inherits a symplectic action of H. Since ϕ is preserved by the G-action, the restriction of this moment map to $\mu^{-1}(0)$ descends to a moment map $\phi_{\mathrm{red}} : M_{\mathrm{red}} \to \mathfrak{h}^*$ satisfying $\phi_{\mathrm{red}} \circ \pi = \phi \circ i$, where $\pi : \mu^{-1}(0) \to M_{\mathrm{red}}$ and $i : \mu^{-1}(0) \hookrightarrow M$. Therefore, $(M_{\mathrm{red}}, \omega_{\mathrm{red}}, H, \phi_{\mathrm{red}})$ is a Hamiltonian H-space.

Consider now the action of a *product group* $G = G_1 \times G_2$, where G_1 and G_2 are compact connected Lie groups. We have $\mathfrak{g} = \mathfrak{g}_1 \oplus \mathfrak{g}_2$ and $\mathfrak{g}^* = \mathfrak{g}_1^* \oplus \mathfrak{g}_2^*$. Suppose that (M, ω, G, ψ) is a Hamiltonian G-space with moment map

$$\psi = (\psi_1, \psi_2) : M \longrightarrow \mathfrak{g}_1^* \oplus \mathfrak{g}_2^*,$$

where $\psi_i : M \to \mathfrak{g}_i^*$ for $i = 1, 2$. The fact that ψ is equivariant implies that ψ_1 is invariant under G_2 and ψ_2 is invariant under G_1. Assume that G_1 acts freely on $Z_1 := \psi_1^{-1}(0)$. Let $(M_1 = Z_1/G_1, \omega_1)$ be the reduction of (M, ω) with respect to G_1, ψ_1. From the observation above, (M_1, ω_1) inherits a Hamiltonian G_2-action with moment map $\mu_2 : M_1 \to \mathfrak{g}_2^*$ such that $\mu_2 \circ \pi = \psi_2 \circ i$, where $\pi : Z_1 \to M_1$ and $i : Z_1 \hookrightarrow M$. If G acts freely on $\psi^{-1}(0, 0)$, then G_2 acts freely on $\mu_2^{-1}(0)$, and there is a natural symplectomorphism

$$\mu_2^{-1}(0)/G_2 \simeq \psi^{-1}(0, 0)/G.$$

This technique of performing reduction with respect to one factor of a product group at a time is called *reduction in stages*. It may be extended to reduction by a normal subgroup $H \subset G$ and by the corresponding quotient group G/H.

EXAMPLE. Finding symmetries for a mechanical problem may reduce degrees of freedom by two at a time: an integral of motion f for a $2n$-dimensional Hamiltonian system (M, ω, H) may allow to understand the trajectories of this system in terms of the trajectories of a $(2n-2)$-dimensional Hamiltonian system $(M_{\mathrm{red}}, \omega_{\mathrm{red}}, H_{\mathrm{red}})$. Locally this process goes as follows. Let $(\mathcal{U}, x_1, \dots, x_n, \xi_1, \dots, \xi_n)$ be a Darboux chart for M such that $f = \xi_n$.[68] Since ξ_n is an integral of motion, $0 = \{\xi_n, H\} = -\frac{\partial H}{\partial x_n}$, the trajectories of the Hamiltonian vector field X_H lie on a constant level $\xi_n = c$ (Proposition 5.12), and H does not depend on x_n. The *reduced space* is $\mathcal{U}_{\mathrm{red}} = \{(x_1, \dots, x_{n-1}, \xi_1, \dots, \xi_{n-1}) \mid \exists a\colon (x_1, \dots, x_{n-1}, a, \xi_1, \dots, \xi_{n-1}, c) \in \mathcal{U}\}$ and the *reduced Hamiltonian* is $H_{\mathrm{red}}: \mathcal{U}_{\mathrm{red}} \to \mathbb{R}$, $H_{\mathrm{red}}(x_1, \dots, x_{n-1}, \xi_1, \dots, \xi_{n-1}) = H(x_1, \dots, x_{n-1}, a, \xi_1, \dots, \xi_{n-1}, c)$ for some a. In order to find the trajectories of the original system on the hypersurface $\xi_n = c$, we look for the trajectories $(x_1(t), \dots, x_{n-1}(t), \xi_1(t), \dots, \xi_{n-1}(t))$ of the reduced system on $\mathcal{U}_{\mathrm{red}}$, and integrate the equation $\frac{dx_n}{dt}(t) = \frac{\partial H}{\partial \xi_n}$ to obtain the original trajectories where

$$\begin{cases} x_n(t) = x_n(0) + \int_0^t \frac{\partial H}{\partial \xi_n}(x_1(t), \dots, x_{n-1}(t), \xi_1(t), \dots, \xi_{n-1}(t), c)\, dt, \\ \xi_n(t) = c. \end{cases}$$

By Sard's theorem, the singular values of a moment map $\mu: M \to \mathfrak{g}^*$ form a set of measure zero. So, perturbing if necessary, we may assume that a level of μ is regular hence, when G is compact, that any point p of that level has finite stabilizer G_p. Let $\mathcal{O}_p$ be the orbit of p. By the slice theorem for the case of orbifolds, near $\mathcal{O}_p$ the orbit space of the level is modeled by S/G_p, where S is a G_p-invariant disk in the level and transverse to $\mathcal{O}_p$ (a *slice*). Thus, the orbit space is an *orbifold*.[69] This implies that, when $G = \mathbb{T}^n$ is an n-torus, for most levels reduction goes through, however the quotient space is not necessarily a manifold but an orbifold. Roughly speaking, orbifolds are singular manifolds where each singularity is locally modeled on $\mathbb{R}^m/\Gamma$, for some finite group $\Gamma \subset \mathrm{GL}(m; \mathbb{R})$. The differential-geometric notions of vector fields, differential forms, exterior

[68] To obtain such a chart, in the proof of Darboux's Theorem 1.9 start with coordinates $(x'_1, \dots, x'_n, y'_1, \dots y'_n)$ such that $y'_n = f$ and $\frac{\partial}{\partial x'_n} = X_f$.

[69] Let $|M|$ be a Hausdorff topological space satisfying the second axiom of countability. An *orbifold chart* on $|M|$ is a triple $(\mathcal{V}, \Gamma, \varphi)$, where $\mathcal{V}$ is a connected open subset of some Euclidean space $\mathbb{R}^m$, Γ is a finite group that acts linearly on $\mathcal{V}$ so that the set of points where the action is not free has codimension at least two, and $\varphi: \mathcal{V} \to |M|$ is a Γ-invariant map inducing a homeomorphism from $\mathcal{V}/\Gamma$ onto its image $\mathcal{U} \subset |M|$. An *orbifold atlas* $\mathcal{A}$ for $|M|$ is a collection of orbifold charts on $|M|$ such that: the collection of images $\mathcal{U}$ forms a basis of open sets in $|M|$, and the charts are compatible in the sense that, whenever two charts $(\mathcal{V}_1, \Gamma_1, \varphi_1)$ and $(\mathcal{V}_2, \Gamma_2, \varphi_2)$ satisfy $\mathcal{U}_1 \subseteq \mathcal{U}_2$, there exists an injective homomorphism $\lambda: \Gamma_1 \to \Gamma_2$ and a λ-equivariant open embedding $\psi: \mathcal{V}_1 \to \mathcal{V}_2$ such that $\varphi_2 \circ \psi = \varphi_1$. Two orbifold atlases are *equivalent* if their union is still an atlas. An m-dimensional *orbifold* M is a Hausdorff topological space $|M|$ satisfying the second axiom of countability, plus an equivalence class of orbifold atlases on $|M|$. We do not require the action of each group Γ to be effective. Given a point p on an orbifold M, let $(\mathcal{V}, \Gamma, \varphi)$ be an orbifold chart for a neighborhood $\mathcal{U}$ of p. The *orbifold structure group* of p, Γ_p, is (the isomorphism class of) the stabilizer of a preimage of p under ϕ. Orbifolds were introduced by Satake in [114].

differentiation, group actions, etc., extend naturally to orbifolds by gluing corresponding local Γ-invariant or Γ-equivariant objects. In particular, a *symplectic orbifold* is a pair (M,ω) where M is an orbifold and ω is a closed 2-form on M that is nondegenerate at every point.

EXAMPLES. The S^1-action on $\mathbb{C}^2$ given by $e^{i\theta}\cdot(z_1,z_2)=(e^{ik\theta}z_1,e^{i\ell\theta}z_2)$, for some integers k and ℓ, has moment map $\mu:\mathbb{C}^2\to\mathbb{R}$, $(z_1,z_2)\mapsto -\frac{1}{2}(k|z_1|^2+\ell|z_2|^2)$. Any $\xi<0$ is a regular value and $\mu^{-1}(\xi)$ is a 3-dimensional ellipsoid.

When $\ell=1$ and $k\geqslant 2$, the stabilizer of (z_1,z_2) is $\{1\}$ if $z_2\neq 0$ and is $\mathbb{Z}_k=\{e^{i\frac{2\pi m}{k}}\mid m=0,1,\dots,k-1\}$ if $z_2=0$. The reduced space $\mu^{-1}(\xi)/S^1$ is then called a *teardrop* orbifold or *conehead*; it has one *cone* (or *dunce cap*) singularity with cone angle $\frac{2\pi}{k}$, that is, a point with orbifold structure group $\mathbb{Z}_k$.

When $k,\ell\geqslant 2$ are relatively prime, for $z_1,z_2\neq 0$ the stabilizer of $(z_1,0)$ is $\mathbb{Z}_k$, of $(0,z_2)$ is $\mathbb{Z}_\ell$ and of (z_1,z_2) is $\{1\}$. The quotient $\mu^{-1}(\xi)/S^1$ is called a *football* orbifold: it has two cone singularities, with angles $\frac{2\pi}{k}$ and $\frac{2\pi}{\ell}$.

For S^1 acting on $\mathbb{C}^n$ by $e^{i\theta}\cdot(z_1,\dots,z_n)=(e^{ik_1\theta}z_1,\dots,e^{ik_n\theta}z_n)$ the reduced spaces are orbifolds called *weighted* (or *twisted*) *projective spaces*.

Let (M,ω) be a symplectic manifold where S^1 acts in a Hamiltonian way, $\rho:S^1\to\mathrm{Diff}(M)$, with moment map $\mu:M\to\mathbb{R}$. Suppose that:

- M has a unique nondegenerate minimum at q where $\mu(q)=0$, and
- for ε sufficiently small, S^1 acts freely on the level set $\mu^{-1}(\varepsilon)$.

Let $\mathbb{C}$ be equipped with the symplectic form $-i\,dz\wedge d\bar{z}$. Then the action of S^1 on the product $\psi:S^1\to\mathrm{Diff}(M\times\mathbb{C})$, $\psi_t(p,z)=(\rho_t(p),t\cdot z)$, is Hamiltonian with moment map

$$\phi:M\times\mathbb{C}\longrightarrow\mathbb{R},\qquad \phi(p,z)=\mu(p)-|z|^2.$$

Observe that S^1 acts freely on the ε-level of ϕ for ε small enough:

$$\begin{aligned}\phi^{-1}(\varepsilon)&=\bigl\{(p,z)\in M\times\mathbb{C}\mid \mu(p)-|z|^2=\varepsilon\bigr\}\\&=\bigl\{(p,0)\in M\times\mathbb{C}\mid \mu(p)=\varepsilon\bigr\}\\&\quad\cup\bigl\{(p,z)\in M\times\mathbb{C}\mid |z|^2=\mu(p)-\varepsilon>0\bigr\}.\end{aligned}$$

The reduced space is hence

$$\phi^{-1}(\varepsilon)/S^1\simeq\mu^{-1}(\varepsilon)/S^1\cup\bigl\{p\in M\mid \mu(p)>\varepsilon\bigr\}.$$

The open submanifold of M given by $\{p\in M\mid\mu(p)>\varepsilon\}$ embeds as an open dense submanifold into $\phi^{-1}(\varepsilon)/S^1$. The reduced space $\phi^{-1}(\varepsilon)/S^1$ is the ε-blow-up of M at q (Section 5.6). This global description of blow-up for Hamiltonian S^1-spaces is due to Lerman [86], as a particular instance of his *cutting* technique. *Symplectic cutting* is the application of symplectic reduction to the product of a Hamiltonian S^1-space with the standard

$\mathbb{C}$ as above, in a way that the reduced space for the original Hamiltonian S^1-space embeds symplectically as a codimension 2 submanifold in a symplectic manifold. As it is a local construction, the cutting operation may be more generally performed at a local minimum (or maximum) of the moment map μ. There is a remaining S^1-action on the cut space $M_{\text{cut}}^{\geqslant\varepsilon} := \phi^{-1}(\varepsilon)/S^1$ induced by

$$\tau : S^1 \longrightarrow \text{Diff}(M \times \mathbb{C}), \quad \tau_t(p, z) = \big(\rho_t(p), z\big).$$

In fact, τ is a Hamiltonian S^1-action on $M \times \mathbb{C}$ that commutes with ψ, thus descends to an action $\tilde{\tau} : S^1 \to \text{Diff}(M_{\text{cut}}^{\geqslant\varepsilon})$, which is also Hamiltonian.

Loosely speaking, the cutting technique provides a Hamiltonian way to close the open manifold $\{p \in M \mid \mu(p) > \varepsilon\}$, by using the reduced space at level ε, $\mu^{-1}(\varepsilon)/S^1$. We may similarly close $\{p \in M \mid \mu(p) < \varepsilon\}$. The resulting Hamiltonian S^1-spaces are called *cut spaces*, and denoted $M_{\text{cut}}^{\geqslant\varepsilon}$ and $M_{\text{cut}}^{\leqslant\varepsilon}$. If another group G acts on M in a Hamiltonian way that commutes with the S^1-action, then the cut spaces are also Hamiltonian G-spaces.

6.3. *Moment map in gauge theory*

Let G be a Lie group and P a principal G-bundle over B.[70] If A is a connection (form)[71] on P, and if $a \in \Omega^1_{\text{horiz}} \otimes \mathfrak{g}$ is G-invariant for the product action, then $A + a$ is also a connection on P. Reciprocally, any two connections on P differ by an $a \in (\Omega^1_{\text{horiz}} \otimes \mathfrak{g})^G$.

[70] Let G be a Lie group and B a manifold. A *principal G-bundle over B* is a fibration $\pi : P \to B$ (Section 4.2) with a free action of G (the *structure group*) on the total space P, such that the base B is the orbit space, the map π is the point-orbit projection and the local trivializations are of the form $\varphi_{\mathcal{U}} = (\pi, s_{\mathcal{U}}) : \pi^{-1}(\mathcal{U}) \to \mathcal{U} \times G$ with $s_{\mathcal{U}}(g \cdot p) = g \cdot s_{\mathcal{U}}(p)$ for all $g \in G$ and all $p \in \pi^{-1}(\mathcal{U})$. A principal G-bundle is represented by a diagram

$$\begin{array}{ccc} G & \hookrightarrow & P \\ & & \downarrow \pi \\ & & B \end{array}$$

For instance, the *Hopf fibration* is a principal S^1-bundle over $S^2 (= \mathbb{CP}^1)$ with total space S^3 regarded as unit vectors in $\mathbb{C}^2$ where circle elements act by complex multiplication.

[71] An action $\psi : G \to \text{Diff}(P)$ induces an infinitesimal action $d\psi : \mathfrak{g} \to \chi(P)$ mapping $X \in \mathfrak{g}$ to the vector field $X^\#$ generated by the one-parameter group $\{\exp tX(e) \mid t \in \mathbb{R}\} \subseteq G$. Fix a basis $X_1, \ldots, X_k$ of $\mathfrak{g}$. Let P be a principal G-bundle over B. Since the G-action is free, the vector fields $X_1^\#, \ldots, X_k^\#$ are linearly independent at each $p \in P$. The *vertical bundle* V is the rank k subbundle of TP generated by $X_1^\#, \ldots, X_k^\#$. Alternatively, V is the set of vectors tangent to P that lie in the kernel of the derivative of the bundle projection π, so V is indeed independent of the choice of basis for $\mathfrak{g}$. An (*Ehresmann*) *connection* on P is a choice of a splitting $TP = V \oplus H$, where H (called the *horizontal bundle*) is a G-invariant subbundle of TP complementary to the vertical bundle V. A *connection form* on P is a Lie-algebra-valued 1-form $A = \sum_{i=1}^k A_i \otimes X_i \in \Omega^1(P) \otimes \mathfrak{g}$ such that A is G-invariant, with respect to the product action of G on $\Omega^1(P)$ (induced by the action on P) and on $\mathfrak{g}$ (the adjoint action), and A is vertical, in the sense that $\imath_{X^\#} A = X$ for any $X \in \mathfrak{g}$. A connection $TP = V \oplus H$ determines a connection (form) A and vice-versa by the formula $H = \ker A = \{v \in TP \mid \imath_v A = 0\}$. Given a connection on P, the splitting $TP = V \oplus H$ induces splittings for bundles $T^*P = V^* \oplus H^*$, $\wedge^2 T^*P = (\wedge^2 V^*) \oplus (V^* \wedge H^*) \oplus (\wedge^2 H^*)$, etc., and for their sections: $\Omega^1(P) = \Omega^1_{\text{vert}} \oplus \Omega^1_{\text{horiz}}$, $\Omega^2(P) = \Omega^2_{\text{vert}} \oplus \Omega^2_{\text{mix}} \oplus \Omega^2_{\text{horiz}}$, etc. The corresponding connection form A is in $\Omega^1_{\text{vert}} \otimes \mathfrak{g}$.

We conclude that the *set $\mathcal{A}$ of all connections* on the principal G-bundle P is an affine space modeled on the linear space $\mathfrak{a} = (\Omega^1_{\text{horiz}} \otimes \mathfrak{g})^G$.

Now let P be a principal G-bundle over a compact Riemann surface. Suppose that the group G is compact or semisimple. Atiyah and Bott [6] noticed that the corresponding space $\mathcal{A}$ of all connections may be treated as an *infinite-dimensional symplectic manifold*. This requires choosing a G-invariant inner product $\langle\cdot,\cdot\rangle$ on $\mathfrak{g}$, which always exists, either by averaging any inner product when G is compact, or by using the *Killing form* on semisimple groups.

Since $\mathcal{A}$ is an affine space, its tangent space at any point A is identified with the model linear space $\mathfrak{a}$. With respect to a basis $X_1, \dots, X_k$ for the Lie algebra $\mathfrak{g}$, elements $a, b \in \mathfrak{a}$ are written

$$a = \sum a_i \otimes X_i \quad \text{and} \quad b = \sum b_i \otimes X_i .$$

If we wedge a and b, and then integrate over B, we obtain a real number:

$$\begin{aligned} \omega : \mathfrak{a} \times \mathfrak{a} &\longrightarrow \left(\Omega^2_{\text{horiz}}(P)\right)^G \simeq \Omega^2(B) \longrightarrow \mathbb{R}, \\ (a,b) &\longmapsto \sum_{i,j} a_i \wedge b_j \langle X_i, X_j\rangle \qquad \longmapsto \int_B \sum_{i,j} a_i \wedge b_j \langle X_i, X_j\rangle . \end{aligned}$$

We used that the pullback $\pi^* : \Omega^2(B) \to \Omega^2(P)$ is an isomorphism onto its image $(\Omega^2_{\text{horiz}}(P))^G$. When $\omega(a,b) = 0$ for all $b \in \mathfrak{a}$, then a must be zero. The map ω is nondegenerate, skew-symmetric, bilinear and constant in the sense that it does not depend on the base point A. Therefore, it has the right to be called a symplectic form on $\mathcal{A}$, so the pair $(\mathcal{A}, \omega)$ is an *infinite-dimensional symplectic manifold*.

A diffeomorphism $f : P \to P$ commuting with the G-action determines a diffeomorphism $f_{\text{basic}} : B \to B$ by projection. Such a diffeomorphism f is called a *gauge transformation* if the induced f_{basic} is the identity. The *gauge group* of P is the group $\mathcal{G}$ of all gauge transformations of P.

The derivative of an $f \in \mathcal{G}$ takes an *Ehresmann connection* $TP = V \oplus H$ to another connection $TP = V \oplus H_f$, and thus induces an action of $\mathcal{G}$ in the space $\mathcal{A}$ of all connections. Atiyah and Bott [6] noticed that the action of $\mathcal{G}$ on $(\mathcal{A}, \omega)$ is Hamiltonian, where the moment map (appropriately interpreted) is

$$\begin{aligned} \mu : \mathcal{A} &\longrightarrow \left(\Omega^2(P) \otimes \mathfrak{g}\right)^G, \\ A &\longmapsto \operatorname{curv} A, \end{aligned}$$

i.e., the moment map *is* the curvature.[72] The reduced space $\mathcal{M} = \mu^{-1}(0)/\mathcal{G}$ is the space of *flat connections* modulo gauge equivalence, known as the *moduli space of flat connections*, which is a finite-dimensional symplectic orbifold.

[72] The exterior derivative of a connection A decomposes into three components,

$$dA = (dA)_{\text{vert}} + (dA)_{\text{mix}} + (dA)_{\text{horiz}} \in \left(\Omega^2_{\text{vert}} \oplus \Omega^2_{\text{mix}} \oplus \Omega^2_{\text{horiz}}\right) \otimes \mathfrak{g}$$

EXAMPLE. We describe the Atiyah–Bott construction for the case of a circle bundle

$$\begin{array}{ccc} S^1 & \hookrightarrow & P \\ & & \downarrow \pi \\ & & B \end{array}$$

Let v be the generator of the S^1-action on P, corresponding to the basis 1 of $\mathfrak{g} \simeq \mathbb{R}$. A connection form on P is an ordinary 1-form $A \in \Omega^1(P)$ such that $\mathcal{L}_v A = 0$ and $\imath_v A = 1$. If we fix one particular connection A_0, then any other connection is of the form $A = A_0 + a$ for some $a \in \mathfrak{a} = (\Omega^1_{\text{horiz}}(P))^G = \Omega^1(B)$. The symplectic form on $\mathfrak{a} = \Omega^1(B)$ is simply

$$\begin{aligned} \omega : \mathfrak{a} \times \mathfrak{a} &\longrightarrow \Omega^2(B) \longrightarrow \mathbb{R}, \\ (a, b) &\longmapsto \ a \wedge b \ \longmapsto \int_B a \wedge b. \end{aligned}$$

The gauge group is $\mathcal{G} = \text{Maps}(B, S^1)$, because a gauge transformation is multiplication by some element of S^1 over each point in B encoded in a map $h : B \to S^1$. The action $\phi : \mathcal{G} \to \text{Diff}(P)$ takes $h \in \mathcal{G}$ to the diffeomorphism

$$\phi_h : p \longmapsto h\big(\pi(p)\big) \cdot p.$$

The Lie algebra of $\mathcal{G}$ is $\text{Lie}\,\mathcal{G} = \text{Maps}(B, \mathbb{R}) = C^\infty(B)$ with dual $(\text{Lie}\,\mathcal{G})^* = \Omega^2(B)$, where the (smooth) duality is provided by integration $C^\infty(B) \times \Omega^2(B) \to \mathbb{R}$, $(h, \beta) \mapsto \int_B h\beta$. The gauge group acts on the space of all connections by

$$\begin{aligned} \psi : \mathcal{G} &\longrightarrow \text{Diff}(\mathcal{A}), \\ \big(h : x \mapsto e^{i\theta(x)}\big) &\longmapsto \big(\psi_h : A \mapsto A - \pi^* \, d\theta\big). \end{aligned}$$

(In the case where $P = S^1 \times B$ is a trivial bundle, every connection can be written $A = dt + \beta$, with $\beta \in \Omega^1(B)$. A gauge transformation $h \in \mathcal{G}$ acts on P by $\phi_h : (t, x) \mapsto (t + \theta(x), x)$ and on $\mathcal{A}$ by $A \mapsto \phi^*_{h^{-1}}(A)$.) The infinitesimal action is

$$\begin{aligned} d\psi : \text{Lie}\,\mathcal{G} &\longrightarrow \chi(\mathcal{A}), \\ X &\longmapsto X^\# = \text{vector field described by } (A \mapsto A - dX), \end{aligned}$$

so that $X^\# = -dX$. It remains to check that

$$\begin{aligned} \mu : \mathcal{A} &\longrightarrow (\text{Lie}\,\mathcal{G})^* = \Omega^2(B), \\ A &\longmapsto \text{curv}\, A \end{aligned}$$

satisfying $(dA)_{\text{mix}} = 0$ and $(dA)_{\text{vert}}(X, Y) = [X, Y]$, i.e., $(dA)_{\text{vert}} = \frac{1}{2} \sum_{i,\ell,m} c^i_{\ell m} A_\ell \wedge A_m \otimes X_i$, where the $c^i_{\ell m}$'s are the *structure constants* of the Lie algebra with respect to the chosen basis, and defined by $[X_\ell, X_m] = \sum_{i,\ell,m} c^i_{\ell m} X_i$. So the relevance of dA may come only from its horizontal component, called the *curvature form* of the connection A, and denoted $\text{curv}\, A = (dA)_{\text{horiz}} \in \Omega^2_{\text{horiz}} \otimes \mathfrak{g}$. A connection is called *flat* if its curvature is zero.

is indeed a moment map for the action of the gauge group on $\mathcal{A}$. Since in this case $\operatorname{curv} A = dA \in (\Omega^2_{\text{horiz}}(P))^G = \Omega^2(B)$, the action of $\mathcal{G}$ on $\Omega^2(B)$ is trivial and μ is $\mathcal{G}$-invariant, the equivariance condition is satisfied. Take any $X \in \operatorname{Lie}\mathcal{G} = C^\infty(B)$. Since the map $\mu^X : A \mapsto \langle X, dA\rangle = \int_B X \cdot dA$ is linear in A, its differential is

$$\begin{aligned} d\mu^X : \mathfrak{a} &\longrightarrow \mathbb{R}, \\ a &\longmapsto \int_B X\, da. \end{aligned}$$

By definition of ω and the Stokes theorem, we have that

$$\omega(X^\#, a) = \int_B X^\# \cdot a = -\int_B dX \cdot a = \int_B X \cdot da = d\mu^X(a), \quad \forall a \in \Omega^1(B),$$

so we are done in proving that μ is the moment map.

The function $\|\mu\|^2 : \mathcal{A} \to \mathbb{R}$ giving the square of the L^2 norm of the curvature is the *Yang–Mills functional*, whose Euler–Lagrange equations are the *Yang–Mills equations*. Atiyah and Bott [6] studied the topology of $\mathcal{A}$ by regarding $\|\mu\|^2$ as an equivariant Morse function. In general, it is a good idea to apply Morse theory to the norm square of a moment map [80].

6.4. *Symplectic toric manifolds*

Toric manifolds are smooth *toric varieties*.[73] When studying the symplectic features of these spaces, we refer to them as *symplectic toric manifolds*. Relations between the algebraic and symplectic viewpoints on toric manifolds are discussed in [21].

DEFINITION 6.6. A *symplectic toric manifold* is a compact connected symplectic manifold (M, ω) equipped with an effective Hamiltonian action of a torus $\mathbb{T}$ of dimension equal to half the dimension of the manifold, $\dim \mathbb{T} = \frac{1}{2}\dim M$, and with a choice of a corresponding moment map μ. Two symplectic toric manifolds, $(M_i, \omega_i, \mathbb{T}_i, \mu_i)$, $i = 1, 2$, are *equivalent* if there exists an isomorphism $\lambda : \mathbb{T}_1 \to \mathbb{T}_2$ and a λ-equivariant symplectomorphism $\varphi : M_1 \to M_2$ such that $\mu_1 = \mu_2 \circ \varphi$.

EXAMPLES.

1. The circle S^1 acts on the 2-sphere $(S^2, \omega_{\text{standard}} = d\theta \wedge dh)$ by rotations, $e^{i\nu} \cdot (\theta, h) = (\theta + \nu, h)$. with moment map $\mu = h$ equal to the height function and moment polytope $[-1, 1]$ (see Figure 3).

[73] Toric varieties were introduced by Demazure in [29]. There are many nice surveys of the theory of toric varieties in algebraic geometry; see, for instance, [27,53,79,107]. Toric geometry has recently become an important tool in physics in connection with mirror symmetry [26].

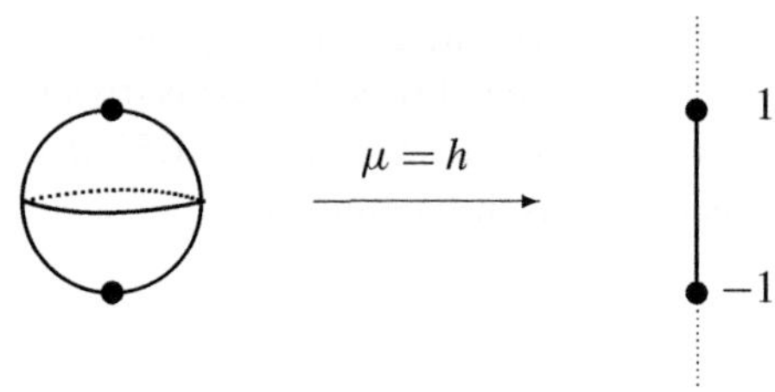

Fig. 3.

Analogously, S^1 acts on the Riemann sphere $\mathbb{CP}^1$ with the Fubini–Study form $\omega_{\text{FS}} = \frac{1}{4}\omega_{\text{standard}}$, by $e^{i\theta} \cdot [z_0, z_1] = [z_0, e^{i\theta} z_1]$. This is Hamiltonian with moment map $\mu[z_0, z_1] = -\frac{1}{2} \cdot \frac{|z_1|^2}{|z_0|^2+|z_1|^2}$, and moment polytope $[-\frac{1}{2}, 0]$.

2. For the $\mathbb{T}^n$-action on the product of n Riemann spheres $\mathbb{CP}^1 \times \cdots \times \mathbb{CP}^1$ by

$$\left(e^{i\theta_1}, \ldots, e^{i\theta_n}\right) \cdot \left([z_1, w_1], \ldots, [z_n, w_n]\right) = \left(\left[z_1, e^{i\theta_1} w_1\right], \ldots, \left[w_0, e^{i\theta_n} w_1\right]\right),$$

the moment polytope is an n-dimensional cube.

3. Let $(\mathbb{CP}^2, \omega_{\text{FS}})$ be 2-(complex-)dimensional complex projective space equipped with the Fubini–Study form defined in Section 3.4. The $\mathbb{T}^2$-action on $\mathbb{CP}^2$ by $(e^{i\theta_1}, e^{i\theta_2}) \cdot [z_0, z_1, z_2] = [z_0, e^{i\theta_1} z_1, e^{i\theta_2} z_2]$ has moment map

$$\mu[z_0, z_1, z_2] = -\frac{1}{2}\left(\frac{|z_1|^2}{|z_0|^2 + |z_1|^2 + |z_2|^2}, \frac{|z_2|^2}{|z_0|^2 + |z_1|^2 + |z_2|^2}\right).$$

The image is the isosceles triangle with vertices $(0, 0)$, $(-\frac{1}{2}, 0)$ and $(0, -\frac{1}{2})$.

4. For the $\mathbb{T}^n$-action on $(\mathbb{CP}^n, \omega_{\text{FS}})$ by

$$\left(e^{i\theta_1}, \ldots, e^{i\theta_n}\right) \cdot [z_0, z_1, \ldots, z_n] = \left[z_0, e^{i\theta_1} z_1, \ldots, e^{i\theta_n} z_n\right]$$

the moment polytope is an n-dimensional simplex.

Since the coordinates of the moment map are commuting integrals of motion, a symplectic toric manifold gives rise to a completely integrable system. By Proposition 5.24, symplectic toric manifolds are optimal Hamiltonian torus-spaces. By Theorem 5.21, they have an associated polytope. It turns out that the moment polytope contains enough information to sort all symplectic toric manifolds. We now define the class of polytopes that arise in the classification. For a symplectic toric manifold the weights $\lambda^{(1)}, \ldots, \lambda^{(n)}$ in Theorem 5.22 form a $\mathbb{Z}$-basis of $\mathbb{Z}^m$, hence the moment polytope is a *Delzant polytope*:

DEFINITION 6.7. A *Delzant polytope* in $\mathbb{R}^n$ is a polytope satisfying:

- *simplicity*, i.e., there are n edges meeting at each vertex;
- *rationality*, i.e., the edges meeting at the vertex p are rational in the sense that each edge is of the form $p + tu_i$, $t \geqslant 0$, where $u_i \in \mathbb{Z}^n$;

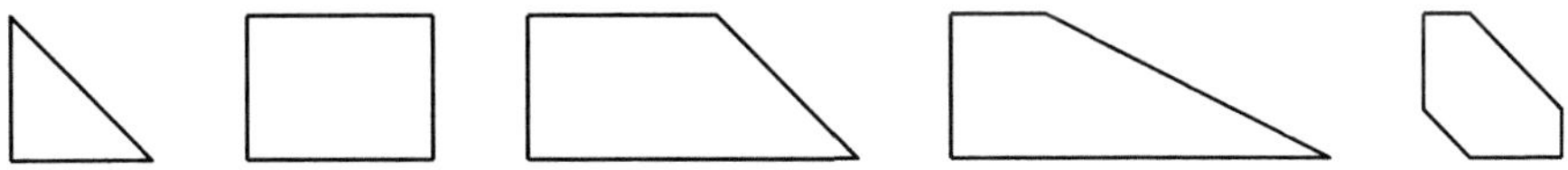

Fig. 4.

- *smoothness*, i.e., for each vertex, the corresponding $u_1, \ldots, u_n$ can be chosen to be a $\mathbb{Z}$-basis of $\mathbb{Z}^n$.

In $\mathbb{R}^2$ the simplicity condition is always satisfied (by nondegenerate polytopes). In $\mathbb{R}^3$, for instance, a square pyramid fails the simplicity condition.

EXAMPLES. Figure 4 represents Delzant polytopes in $\mathbb{R}^2$.

The following theorem classifies (equivalence classes of) symplectic toric manifolds in terms of the combinatorial data encoded by a Delzant polytope.

THEOREM 6.8 (Delzant [28]). *Toric manifolds are classified by Delzant polytopes, and their bijective correspondence is given by the moment map*:

$$\{\textit{toric manifolds}\} \longleftrightarrow \{\textit{Delzant polytopes}\},$$
$$\left(M^{2n}, \omega, \mathbb{T}^n, \mu\right) \longmapsto \mu(M).$$

Delzant's construction (Section 6.5) shows that for a toric manifold the moment map takes the fixed points bijectively to the vertices of the moment polytope and takes points with a k-dimensional stabilizer to the codimension k faces of the polytope. The moment polytope is exactly the orbit space, i.e., the preimage under μ of each point in the polytope is exactly one orbit. For instance, consider $(S^2, \omega = d\theta \wedge dh, S^1, \mu = h)$, where S^1 acts by rotation. The image of μ is the line segment $I = [-1, 1]$. The product $S^1 \times I$ is an open-ended cylinder. We can recover the 2-sphere by collapsing each end of the cylinder to a point. Similarly, we can build $\mathbb{CP}^2$ from $\mathbb{T}^2 \times \Delta$ where Δ is a rectangular isosceles triangle, and so on.

EXAMPLES.

1. By a linear transformation in $\mathrm{SL}(2; \mathbb{Z})$, we can make one of the angles in a Delzant triangle into a right angle. Out of the rectangular triangles, only the isosceles one satisfies the smoothness condition. Therefore, up to translation, change of scale and the action of $\mathrm{SL}(2; \mathbb{Z})$, there is just one 2-dimensional Delzant polytope with three vertices, namely an *isosceles triangle*. We conclude that the projective space $\mathbb{CP}^2$ is the only 4-dimensional toric manifold with three fixed points, up to choices of a constant in the moment map, of a multiple of ω_{FS} and of a lattice basis in the Lie algebra of $\mathbb{T}^2$.
2. Up to translation, change of scale and the action of $\mathrm{SL}(n; \mathbb{Z})$, the *standard n-simplex* Δ in $\mathbb{R}^n$ (spanned by the origin and the standard basis vectors $(1, 0, \ldots, 0), \ldots,$

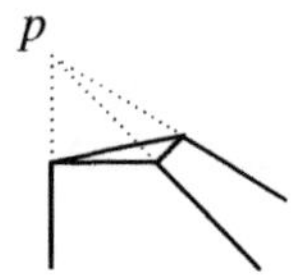

Fig. 5.

$(0,\dots,0,1))$ is the only n-dimensional Delzant polytope with $n+1$ vertices. Hence, $M_\Delta = \mathbb{CP}^n$ is the only $2n$-dimensional toric manifold with $n+1$ fixed points, up to choices of a constant in the moment map, of a multiple of ω_{FS} and of a lattice basis in the Lie algebra of $\mathbb{T}^N$.

3. A transformation in $\mathrm{SL}(2;\mathbb{Z})$ makes one of the angles in a Delzant quadrilateral into a right angle. Automatically an adjacent angle also becomes 90°. Smoothness imposes that the slope of the skew side be integral. Thus, up to translation, change of scale and $\mathrm{SL}(2;\mathbb{Z})$-action, the 2-dimensional Delzant polytopes with four vertices are trapezoids with vertices $(0,0)$, $(0,1)$, $(\ell,1)$ and $(\ell+n,0)$, for n a nonnegative integer and $\ell > 0$. Under Delzant's construction (that is, under symplectic reduction of $\mathbb{C}^4$ with respect to an action of $(S^1)^2$), these correspond to the so-called *Hirzebruch surfaces*—the only 4-dimensional symplectic toric manifolds that have four fixed points up to equivalence as before. Topologically, they are S^2-bundles over S^2, either the trivial bundle $S^2 \times S^2$ when n is even or the nontrivial bundle (given by the blow-up of $\mathbb{CP}^2$ at a point; see Section 4.3) when n is odd.

Let Δ be an n-dimensional Delzant polytope, and let $(M_\Delta, \omega_\Delta, \mathbb{T}^n, \mu_\Delta)$ be the associated symplectic toric manifold. The ε-blow-up of $(M_\Delta, \omega_\Delta)$ at a fixed point of the $\mathbb{T}^n$-action is a new symplectic toric manifold (Sections 4.3 and 5.6). Let q be a fixed point of the $\mathbb{T}^n$-action on $(M_\Delta, \omega_\Delta)$, and let $p = \mu_\Delta(q)$ be the corresponding vertex of Δ. Let $u_1,\dots,u_n$ be the primitive (inward-pointing) edge vectors at p, so that the rays $p + tu_i$, $t \geqslant 0$, form the edges of Δ at p.

PROPOSITION 6.9. *The ε-blow-up of $(M_\Delta, \omega_\Delta)$ at a fixed point q is the symplectic toric manifold associated to the polytope Δ_ε obtained from Δ by replacing the vertex p by the n vertices $p + \varepsilon u_i$, $i = 1,\dots,n$.*

In other words, the moment polytope for the blow-up of $(M_\Delta, \omega_\Delta)$ at q is obtained from Δ by chopping off the corner corresponding to q, thus substituting the original set of vertices by the same set with the vertex corresponding to q replaced by exactly n new vertices. The truncated polytope is Delzant. We may view the ε-blow-up of $(M_\Delta, \omega_\Delta)$ as being obtained from M_Δ by smoothly replacing q by $(\mathbb{CP}^{n-1}, \varepsilon\omega_{\mathrm{FS}})$ (whose moment polytope is an $(n-1)$-dimensional simplex). (See Figure 5.)

EXAMPLE. The moment polytope for the standard $\mathbb{T}^2$-action on $(\mathbb{CP}^2, \omega_{\mathrm{FS}})$ is a right isosceles triangle Δ. If we blow up $\mathbb{CP}^2$ at $[0:0:1]$ we obtain a symplectic toric manifold associated to the trapezoid below: a *Hirzebruch surface* (see Figure 6).

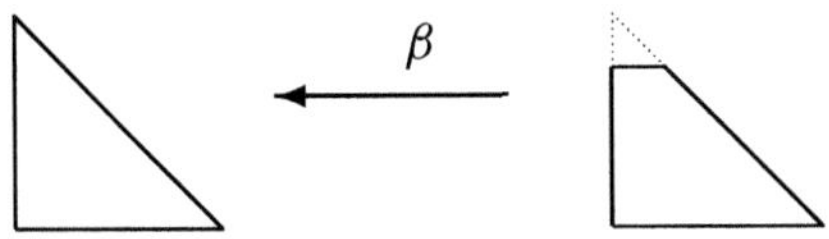

Fig. 6.

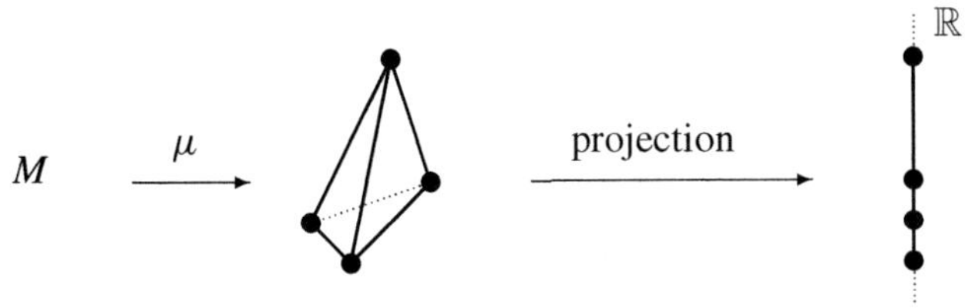

Fig. 7.

Let $(M, \omega, \mathbb{T}^n, \mu)$ be a $2n$-dimensional symplectic toric manifold. Choose a suitably generic direction in $\mathbb{R}^n$ by picking a vector X whose components are independent over $\mathbb{Q}$. This condition ensures that:

- the one-dimensional subgroup $\mathbb{T}^X$ generated by the vector X is dense in $\mathbb{T}^n$,
- X is not parallel to the facets of the moment polytope $\Delta := \mu(M)$, and
- the vertices of Δ have different projections along X.

Then the fixed points for the $\mathbb{T}^n$-action are exactly the fixed points of the action restricted to $\mathbb{T}^X$, that is, are the zeros of the vector field, $X^\#$ on M generated by X. The projection of μ along X, $\mu^X := \langle \mu, X \rangle : M \to \mathbb{R}$, is a Hamiltonian function for the vector field $X^\#$ generated by X. We conclude that the critical points of μ^X are precisely the fixed points of the $\mathbb{T}^n$-action (see Figure 7).

By Theorem 5.22, if q is a fixed point for the $\mathbb{T}^n$-action, then there exists a chart $(\mathcal{U}, x_1, \dots, x_n, y_1, \dots, y_n)$ centered at q and weights $\lambda^{(1)}, \dots, \lambda^{(n)} \in \mathbb{Z}^n$ such that

$$\mu^X|_{\mathcal{U}} = \langle \mu, X \rangle|_{\mathcal{U}} = \mu^X(q) - \frac{1}{2}\sum_{k=1}^{n} \langle \lambda^{(k)}, X \rangle \left(x_k^2 + y_k^2\right).$$

Since the components of X are independent over $\mathbb{Q}$, all coefficients $\langle \lambda^{(k)}, X \rangle$ are nonzero, so q is a *nondegenerate* critical point of μ^X. Moreover, the *index*[74] of q is twice the number of labels k such that $-\langle \lambda^{(k)}, X \rangle < 0$. But the $-\lambda^{(k)}$'s are precisely the edge vectors u_i which satisfy Delzant's conditions. Therefore, geometrically, the index of q can be read from the moment polytope Δ, by taking twice the number of edges whose inward-pointing

[74] A *Morse function* on an m-dimensional manifold M is a smooth function $f : M \to \mathbb{R}$ all of whose critical points (where df vanishes) are nondegenerate (i.e., the *Hessian matrix* is nonsingular). Let q be a nondegenerate critical point for $f : M \to \mathbb{R}$. The *index of f at q* is the index of the Hessian $H_q : \mathbb{R}^m \times \mathbb{R}^m \to \mathbb{R}$ regarded as a symmetric bilinear function, that is, the maximal dimension of a subspace of $\mathbb{R}$ where H is negative definite.

edge vectors at $\mu(q)$ *point up relative to* X, that is, whose inner product with X is positive. In particular, μ^X is a *perfect Morse function*[75] and we have

PROPOSITION 6.10. *Let $X \in \mathbb{R}^n$ have components independent over $\mathbb{Q}$. The degree-2k homology group of the symplectic toric manifold $(M, \omega, \mathbb{T}, \mu)$ has dimension equal to the number of vertices of the moment polytope where there are exactly k (primitive inward-pointing) edge vectors that point up relative to the projection along the X. All odd-degree homology groups of M are zero.*

By Poincaré duality (or by taking $-X$ instead of X), the words *point up* may be replaced by *point down*. The Euler characteristic of a symplectic toric manifold is simply the number of vertices of the corresponding polytope. There is a combinatorial way of understanding the cohomology ring [53].

A *symplectic toric orbifold* is a compact connected symplectic orbifold (M, ω) equipped with an effective Hamiltonian action of a torus of dimension equal to half the dimension of the orbifold, and with a choice of a corresponding moment map. Symplectic toric orbifolds were classified by Lerman and Tolman [87] in a theorem that generalizes Delzant's: a symplectic toric orbifold is determined by its moment polytope plus a positive integer label attached to each of the polytope facets. The polytopes that occur are more general than the Delzant polytopes in the sense that only simplicity and rationality are required; the edge vectors $u_1, \ldots, u_n$ need only form a rational basis of $\mathbb{Z}^n$. When the integer labels are all equal to 1, the failure of the polytope smoothness accounts for all orbifold singularities.

6.5. *Delzant's construction*

Following [28,66], we prove the existence part (or surjectivity) in Delzant's theorem, by using symplectic reduction to associate to an n-dimensional Delzant polytope Δ a symplectic toric manifold $(M_\Delta, \omega_\Delta, \mathbb{T}^n, \mu_\Delta)$.

Let Δ be a Delzant polytope in $(\mathbb{R}^n)^*$[76] and with d facets.[77] We can algebraically describe Δ as an intersection of d halfspaces. Let $v_i \in \mathbb{Z}^n$, $i = 1, \ldots, d$, be the primitive[78] outward-pointing normal vectors to the facets of Δ. Then, for some $\lambda_i \in \mathbb{R}$, we can write $\Delta = \{x \in (\mathbb{R}^n)^* \mid \langle x, v_i \rangle \leqslant \lambda_i,\ i = 1, \ldots, d\}$.

[75] A *perfect Morse function* is a Morse function f for which the *Morse inequalities* [103,104] are equalities, i.e., $b_\lambda(M) = C_\lambda$ and $b_\lambda(M) - b_{\lambda-1}(M) + \cdots \pm b_0(M) = C_\lambda - C_{\lambda-1} + \cdots \pm C_0$ where $b_\lambda(M) = \dim H_\lambda(M)$ and C_λ be the number of critical points of f with index λ. If all critical points of a Morse function f have even index, then f is a perfect Morse function.

[76] Although we identify $\mathbb{R}^n$ with its dual via the Euclidean inner product, it may be more clear to see Δ in $(\mathbb{R}^n)^*$ for Delzant's construction.

[77] A *face* of a polytope Δ is a set of the form $F = P \cap \{x \in \mathbb{R}^n \mid f(x) = c\}$ where $c \in \mathbb{R}$ and $f \in (\mathbb{R}^n)^*$ satisfies $f(x) \geqslant c$, $\forall x \in P$. A *facet* of an n-dimensional polytope is an $(n-1)$-dimensional face.

[78] A lattice vector $v \in \mathbb{Z}^n$ is *primitive* if it cannot be written as $v = ku$ with $u \in \mathbb{Z}^n$, $k \in \mathbb{Z}$ and $|k| > 1$; for instance, $(1, 1)$, $(4, 3)$, $(1, 0)$ are primitive, but $(2, 2)$, $(3, 6)$ are not.

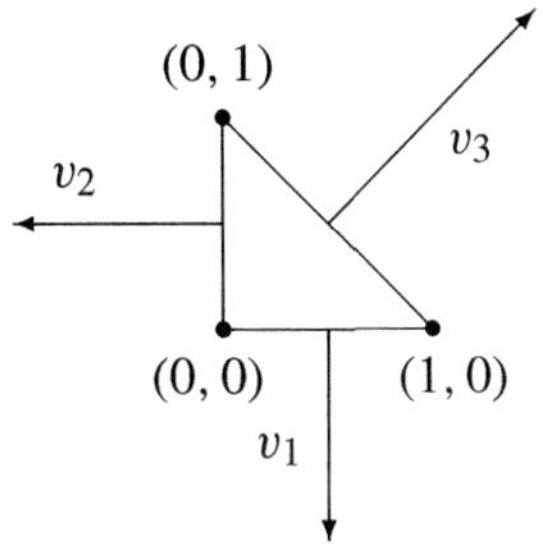

Fig. 8.

EXAMPLE. When Δ is the triangle shown in Figure 8, we have

$$\Delta = \{x \in (\mathbb{R}^2)^* \mid \langle x, (-1,0)\rangle \leqslant 0,\ \langle x, (0,-1)\rangle \leqslant 0,\ \langle x, (1,1)\rangle \leqslant 1\}.$$

For the standard basis $e_1 = (1, 0, \dots, 0), \dots, e_d = (0, \dots, 0, 1)$ of $\mathbb{R}^d$, consider

$$\pi : \mathbb{R}^d \longrightarrow \mathbb{R}^n,$$
$$e_i \longmapsto v_i.$$

LEMMA 6.11. *The map π is onto and maps $\mathbb{Z}^d$ onto $\mathbb{Z}^n$.*

PROOF. We need to show that the set $\{v_1, \dots, v_d\}$ spans $\mathbb{Z}^n$. At a vertex p, the edge vectors $u_1, \dots, u_n \in (\mathbb{R}^n)^*$ form a basis for $(\mathbb{Z}^n)^*$ which, by a change of basis if necessary, we may assume is the standard basis. Then the corresponding primitive normal vectors to the facets meeting at p are $-u_1, \dots, -u_n$. □

We still call π the induced surjective map $\mathbb{T}^d = \mathbb{R}^d/(2\pi\mathbb{Z}^d) \xrightarrow{\pi} \mathbb{T}^n = \mathbb{R}^n/(2\pi\mathbb{Z}^n)$. The kernel N of π is a $(d-n)$-dimensional Lie subgroup of $\mathbb{T}^d$ with inclusion $i : N \hookrightarrow \mathbb{T}^d$. Let $\mathfrak{n}$ be the Lie algebra of N. The exact sequence of tori

$$1 \longrightarrow N \xrightarrow{i} \mathbb{T}^d \xrightarrow{\pi} \mathbb{T}^n \longrightarrow 1$$

induces an exact sequence of Lie algebras

$$0 \longrightarrow \mathfrak{n} \xrightarrow{i} \mathbb{R}^d \xrightarrow{\pi} \mathbb{R}^n \longrightarrow 0$$

with dual exact sequence

$$0 \longrightarrow (\mathbb{R}^n)^* \xrightarrow{\pi^*} (\mathbb{R}^d)^* \xrightarrow{i^*} \mathfrak{n}^* \longrightarrow 0.$$

Consider $\mathbb{C}^d$ with symplectic form $\omega_0 = \frac{i}{2}\sum dz_k \wedge d\bar{z}_k$, and standard Hamiltonian action of $\mathbb{T}^d$ given by $(e^{it_1}, \ldots, e^{it_d}) \cdot (z_1, \ldots, z_d) = (e^{it_1}z_1, \ldots, e^{it_d}z_d)$. A moment map is $\phi : \mathbb{C}^d \to (\mathbb{R}^d)^*$ defined by

$$\phi(z_1, \ldots, z_d) = -\frac{1}{2}\left(|z_1|^2, \ldots, |z_d|^2\right) + (\lambda_1, \ldots, \lambda_d),$$

where the constant is chosen for later convenience. The subtorus N acts on $\mathbb{C}^d$ in a Hamiltonian way with moment map $i^* \circ \phi : \mathbb{C}^d \to \mathfrak{n}^*$. Let $Z = (i^* \circ \phi)^{-1}(0)$.

In order to show that Z (a closed set) is compact it suffices (by the Heine–Borel theorem) to show that Z is bounded. Let Δ' be the image of Δ by π^*. First we show that $\phi(Z) = \Delta'$. A value $y \in (\mathbb{R}^d)^*$ is in the image of Z by ϕ if and only if

$$\text{(a)}\quad y \text{ is in the image of } \phi \quad \text{and} \quad \text{(b)}\quad i^* y = 0$$

if and only if (using the expression for ϕ and the third exact sequence)

$$\begin{aligned} &\text{(a)}\quad \langle y, e_i\rangle \leqslant \lambda_i \quad \text{for } i = 1, \ldots, d \quad \text{and}\\ &\text{(b)}\quad y = \pi^*(x) \quad \text{for some } x \in (\mathbb{R}^n)^*. \end{aligned}$$

Suppose that $y = \pi^*(x)$. Then

$$\begin{aligned} \langle y, e_i\rangle \leqslant \lambda_i, \quad \forall i \quad &\Longleftrightarrow \quad \left\langle x, \pi(e_i)\right\rangle \leqslant \lambda_i, \quad \forall i\\ &\Longleftrightarrow \quad \langle x, v_i\rangle \leqslant \lambda_i, \quad \forall i \quad \Longleftrightarrow \quad x \in \Delta. \end{aligned}$$

Thus, $y \in \phi(Z) \Leftrightarrow y \in \pi^*(\Delta) = \Delta'$. Since Δ' is compact, ϕ is proper and $\phi(Z) = \Delta'$, we conclude that Z must be bounded, and hence compact.

In order to show that N acts freely on Z, pick a vertex p of Δ, and let $I = \{i_1, \ldots, i_n\}$ be the set of indices for the n facets meeting at p. Pick $z \in Z$ such that $\phi(z) = \pi^*(p)$. Then p is characterized by n equations $\langle p, v_i\rangle = \lambda_i$ where $i \in I$:

$$\begin{aligned} \langle p, v_i\rangle = \lambda_i \quad &\Longleftrightarrow \quad \left\langle p, \pi(e_i)\right\rangle = \lambda_i\\ &\Longleftrightarrow \quad \left\langle \pi^*(p), e_i\right\rangle = \lambda_i\\ &\Longleftrightarrow \quad \left\langle \phi(z), e_i\right\rangle = \lambda_i\\ &\Longleftrightarrow \quad i\text{th coordinate of } \phi(z) \text{ is equal to } \lambda_i\\ &\Longleftrightarrow \quad -\frac{1}{2}|z_i|^2 + \lambda_i = \lambda_i\\ &\Longleftrightarrow \quad z_i = 0. \end{aligned}$$

Hence, those z's are points whose coordinates in the set I are zero, and whose other coordinates are nonzero. Without loss of generality, we may assume that $I = \{1, \ldots, n\}$. The stabilizer of z is

$$\left(\mathbb{T}^d\right)_z = \left\{(t_1, \ldots, t_n, 1, \ldots, 1) \in \mathbb{T}^d\right\}.$$

As the restriction $\pi : (\mathbb{R}^d)_z \to \mathbb{R}^n$ maps the vectors $e_1, \ldots, e_n$ to a $\mathbb{Z}$-basis $v_1, \ldots, v_n$ of $\mathbb{Z}^n$ (respectively), at the level of groups $\pi : (\mathbb{T}^d)_z \to \mathbb{T}^n$ must be bijective. Since $N = \ker(\pi : \mathbb{T}^d \to \mathbb{T}^n)$, we conclude that $N \cap (\mathbb{T}^d)_z = \{e\}$, i.e., $N_z = \{e\}$. Hence all N-stabilizers at points mapping to vertices are trivial. But this was the worst case, since other stabilizers $N_{z'}$ ($z' \in Z$) are contained in stabilizers for points z that map to vertices. We conclude that N acts freely on Z.

We now apply reduction. Since i^* is surjective, $0 \in \mathfrak{n}^*$ is a regular value of $i^* \circ \phi$. Hence, Z is a compact submanifold of $\mathbb{C}^d$ of (real) dimension $2d - (d-n) = d+n$. The orbit space $M_\Delta = Z/N$ is a compact manifold of (real) dimension $\dim Z - \dim N = (d+n) - (d-n) = 2n$. The point-orbit map $p : Z \to M_\Delta$ is a principal N-bundle over M_Δ. Consider the diagram

$$\begin{array}{ccc} Z & \stackrel{j}{\hookrightarrow} & \mathbb{C}^d \\ p \downarrow & & \\ M_\Delta & & \end{array}$$

where $j : Z \hookrightarrow \mathbb{C}^d$ is inclusion. The Marsden–Weinstein–Meyer theorem (Theorem 6.1) guarantees the existence of a symplectic form ω_Δ on M_Δ satisfying

$$p^*\omega_\Delta = j^*\omega_0.$$

Since Z is connected, the symplectic manifold $(M_\Delta, \omega_\Delta)$ is also connected.

It remains to show that $(M_\Delta, \omega_\Delta)$ is a Hamiltonian $\mathbb{T}^n$-space with a moment map μ_Δ having image $\mu_\Delta(M_\Delta) = \Delta$. Let z be such that $\phi(z) = \pi^*(p)$ where p is a vertex of Δ. Let $\sigma : \mathbb{T}^n \to (\mathbb{T}^d)_z$ be the inverse for the earlier bijection $\pi : (\mathbb{T}^d)_z \to \mathbb{T}^n$. This is a *section*, i.e., a right inverse for π, in the sequence

$$1 \longrightarrow N \stackrel{i}{\longrightarrow} \mathbb{T}^d \underset{\stackrel{\longleftarrow}{\sigma}}{\stackrel{\pi}{\longrightarrow}} \mathbb{T}^n \longrightarrow 1,$$

so it *splits*, i.e., becomes like a sequence for a product, as we obtain an isomorphism $(i, \sigma) : N \times \mathbb{T}^n \stackrel{\simeq}{\longrightarrow} \mathbb{T}^d$. The action of the $\mathbb{T}^n$ factor (or, more rigorously, $\sigma(\mathbb{T}^n) \subset \mathbb{T}^d$) descends to the quotient $M_\Delta = Z/N$. Consider the diagram

$$\begin{array}{l} Z \stackrel{j}{\hookrightarrow} \mathbb{C}^d \stackrel{\phi}{\longrightarrow} (\mathbb{R}^d)^* \simeq \eta^* \oplus (\mathbb{R}^n)^* \stackrel{\sigma^*}{\longrightarrow} (\mathbb{R}^n)^* \\ p \downarrow \\ M_\Delta \end{array}$$

where the last horizontal map is projection onto the second factor. Since the composition of the horizontal maps is constant along N-orbits, it descends to a map

$$\mu_\Delta : M_\Delta \longrightarrow (\mathbb{R}^n)^*$$

which satisfies $\mu_\Delta \circ p = \sigma^* \circ \phi \circ j$. By reduction for product groups (Section 6.2), this is a moment map for the action of $\mathbb{T}^n$ on $(M_\Delta, \omega_\Delta)$. The image of μ_Δ is

$$\mu_\Delta(M_\Delta) = (\mu_\Delta \circ p)(Z) = (\sigma^* \circ \phi \circ j)(Z) = (\sigma^* \circ \pi^*)(\Delta) = \Delta,$$

because $\phi(Z) = \pi^*(\Delta)$ and $\pi \circ \sigma = \mathrm{id}$. We conclude that $(M_\Delta, \omega_\Delta, \mathbb{T}^n, \mu_\Delta)$ is the required toric manifold corresponding to Δ. This construction via reduction also shows that symplectic toric manifolds are in fact Kähler.

EXAMPLE. Here are the details of Delzant's construction for the case of a segment $\Delta = [0, a] \subset \mathbb{R}^*$ $(n = 1, d = 2)$. Let $v (= 1)$ be the standard basis vector in $\mathbb{R}$. Then Δ is described by $\langle x, -v\rangle \leqslant 0$ and $\langle x, v\rangle \leqslant a$, where $v_1 = -v$, $v_2 = v$, $\lambda_1 = 0$ and $\lambda_2 = a$. The projection $\mathbb{R}^2 \xrightarrow{\pi} \mathbb{R}$, $e_1 \mapsto -v$, $e_2 \mapsto v$, has kernel equal to the span of $(e_1 + e_2)$, so that N is the diagonal subgroup of $\mathbb{T}^2 = S^1 \times S^1$. The exact sequences become

$$\begin{array}{ccccccc}
1 \longrightarrow N & \xrightarrow{i} & \mathbb{T}^2 & \xrightarrow{\pi} & S^1 & \longrightarrow 1, \\
t & \longmapsto & (t,t), & & & \\
 & & (t_1, t_2) & \longmapsto & t_1^{-1} t_2, &
\end{array}$$

$$\begin{array}{ccccccc}
0 \longrightarrow \mathfrak{n} & \xrightarrow{i} & \mathbb{R}^2 & \xrightarrow{\pi} & \mathbb{R} & \longrightarrow 0, \\
x & \longmapsto & (x,x), & & & \\
 & & (x_1, x_2) & \longmapsto & x_2 - x_1, &
\end{array}$$

$$\begin{array}{ccccccc}
0 \longrightarrow \mathbb{R}^* & \xrightarrow{\pi^*} & (\mathbb{R}^2)^* & \xrightarrow{i^*} & \mathfrak{n}^* & \longrightarrow 0, \\
x & \longmapsto & (-x,x), & & & \\
 & & (x_1, x_2) & \longmapsto & x_1 + x_2. &
\end{array}$$

The action of the diagonal subgroup $N = \{(e^{it}, e^{it}) \in S^1 \times S^1\}$ on $\mathbb{C}^2$ by

$$(e^{it}, e^{it}) \cdot (z_1, z_2) = (e^{it} z_1, e^{it} z_2)$$

has moment map $(i^* \circ \phi)(z_1, z_2) = -\frac{1}{2}(|z_1|^2 + |z_2|^2) + a$, with zero-level set

$$(i^* \circ \phi)^{-1}(0) = \{(z_1, z_2) \in \mathbb{C}^2\colon |z_1|^2 + |z_2|^2 = 2a\}.$$

Hence, the reduced space is a projective space, $(i^* \circ \phi)^{-1}(0)/N = \mathbb{CP}^1$.

6.6. *Duistermaat–Heckman theorems*

Throughout this subsection, let (M, ω, G, μ) be a Hamiltonian G-space, where G is an n-torus[79] and the moment map μ is proper.

[79] The discussion in this subsection may be extended to Hamiltonian actions of other compact Lie groups, not necessarily tori; see [66, Exercises 2.1–2.10].

If G acts freely on $\mu^{-1}(0)$, it also acts freely on nearby levels $\mu^{-1}(t)$, $t \in \mathfrak{g}^*$ and $t \approx 0$. (Otherwise, assume only that 0 is a regular value of μ and work with orbifolds.) We study the variation of the reduced spaces by relating

$$\big(M_{\text{red}} = \mu^{-1}(0)/G, \omega_{\text{red}}\big) \quad \text{and} \quad \big(M_t = \mu^{-1}(t)/G, \omega_t\big).$$

For simplicity, assume G to be the circle S^1. Let $Z = \mu^{-1}(0)$ and let $i : Z \hookrightarrow M$ be the inclusion map. Fix a connection form $\alpha \in \Omega^1(Z)$ for the principal bundle

$$\begin{array}{ccc} S^1 & \hookrightarrow & Z \\ & & \downarrow \pi \\ & & M_{\text{red}} \end{array}$$

that is, $\mathcal{L}_{X^\#}\alpha = 0$ and $\imath_{X^\#}\alpha = 1$, where $X^\#$ is the infinitesimal generator for the S^1-action. Construct a 2-form on the product manifold $Z \times (-\varepsilon, \varepsilon)$ by the recipe

$$\sigma = \pi^*\omega_{\text{red}} - d(x\alpha),$$

where x is a linear coordinate on the interval $(-\varepsilon, \varepsilon) \subset \mathbb{R} \simeq \mathfrak{g}^*$. (By abuse of notation, we shorten the symbols for forms on $Z \times (-\varepsilon, \varepsilon)$ that arise by pullback via projection onto each factor.)

LEMMA 6.12. *The 2-form σ is symplectic for ε small enough.*

PROOF. At points where $x = 0$, the form $\sigma|_{x=0} = \pi^*\omega_{\text{red}} + \alpha \wedge dx$ satisfies $\sigma|_{x=0}(X^\#, \frac{\partial}{\partial x}) = 1$, so σ is nondegenerate along $Z \times \{0\}$. Since nondegeneracy is an open condition, we conclude that σ is nondegenerate for x in a sufficiently small neighborhood of 0. Closedness is clear. □

Notice that σ is invariant with respect to the S^1-action on the first factor of $Z \times (-\varepsilon, \varepsilon)$. This action is Hamiltonian with moment map $x : Z \times (-\varepsilon, \varepsilon) \to (-\varepsilon, \varepsilon)$ given by projection onto the second factor (since $\mathcal{L}_{X^\#}\alpha = 0$ and $\imath_{X^\#}\alpha = 1$):

$$\imath_{X^\#}\sigma = -\imath_{X^\#}d(x\alpha) = -\mathcal{L}_{X^\#}(x\alpha) + d\imath_{X^\#}(x\alpha) = dx.$$

LEMMA 6.13. *There exists an equivariant symplectomorphism between a neighborhood of Z in M and a neighborhood of $Z \times \{0\}$ in $Z \times (-\varepsilon, \varepsilon)$, intertwining the two moment maps, for ε small enough.*

PROOF. The inclusion $i_0 : Z \hookrightarrow Z \times (-\varepsilon, \varepsilon)$ as $Z \times \{0\}$ and the natural inclusion $i : Z \hookrightarrow M$ are S^1-equivariant coisotropic embeddings. Moreover, they satisfy $i_0^*\sigma = i^*\omega$ since both sides are equal to $\pi^*\omega_{\text{red}}$, and the moment maps coincide on Z because $i_0^*x = 0 = i^*\mu$. Replacing ε by a smaller positive number if necessary, the result follows from the equivariant version of the coisotropic embedding theorem (Theorem 2.9).[80] □

[80]**Equivariant coisotropic embedding theorem:** *Let (M_0, ω_0), (M_1, ω_1) be symplectic manifolds of dimension $2n$, G a compact Lie group acting on (M_i, ω_i), $i = 0, 1$, in a Hamiltonian way with moment maps μ_0*

Therefore, in order to compare the reduced spaces $M_t = \mu^{-1}(t)/S^1$ for $t \approx 0$, we can work in $Z \times (-\varepsilon, \varepsilon)$ and compare instead the reduced spaces $x^{-1}(t)/S^1$.

PROPOSITION 6.14. *The space* (M_t, ω_t) *is symplectomorphic to* $(M_{\text{red}}, \omega_{\text{red}} - t\beta)$ *where* β *is the curvature form of the connection* α.

PROOF. By Lemma 6.13, (M_t, ω_t) is symplectomorphic to the reduced space at level t for the Hamiltonian space $(Z \times (-\varepsilon, \varepsilon), \sigma, S^1, x)$. Since $x^{-1}(t) = Z \times \{t\}$, where S^1 acts on the first factor, all the manifolds $x^{-1}(t)/S^1$ are diffeomorphic to $Z/S^1 = M_{\text{red}}$. As for the symplectic forms, let $\iota_t : Z \times \{t\} \hookrightarrow Z \times (-\varepsilon, \varepsilon)$ be the inclusion map. The restriction of σ to $Z \times \{t\}$ is

$$\iota_t^* \sigma = \pi^* \omega_{\text{red}} - t\, d\alpha.$$

By definition of curvature, $d\alpha = \pi^*\beta$. Hence, the reduced symplectic form on $x^{-1}(t)/S^1$ is $\omega_{\text{red}} - t\beta$. □

In loose terms, Proposition 6.14 says that the reduced forms ω_t vary linearly in t, for t close enough to 0. However, the identification of M_t with M_{red} as abstract manifolds is not natural. Nonetheless, any two such identifications are isotopic. By the homotopy invariance of de Rham classes, we obtain:

THEOREM 6.15 (Duistermaat–Heckman [38]). *Under the hypotheses and notation before, the cohomology class of the reduced symplectic form* $[\omega_t]$ *varies linearly in* t. *More specifically, if* $c = [-\beta] \in H^2_{\text{deRham}}(M_{\text{red}})$ *is the first Chern class*[81] *of the* S^1*-bundle* $Z \to M_{\text{red}}$, *we have*

$$[\omega_t] = [\omega_{\text{red}}] + tc.$$

---

and μ_1, *respectively,* Z *a manifold of dimension* $k \geqslant n$ *with a* G*-action, and* $\iota_i : Z \hookrightarrow M_i$, $i = 0, 1$, G*-equivariant coisotropic embeddings. Suppose that* $\iota_0^*\omega_0 = \iota_1^*\omega_1$ *and* $\iota_0^*\mu_0 = \iota_1^*\mu_1$. *Then there exist* G*-invariant neighborhoods* $\mathcal{U}_0$ *and* $\mathcal{U}_1$ *of* $\iota_0(Z)$ *and* $\iota_1(Z)$ *in* M_0 *and* M_1, *respectively, and a* G*-equivariant symplectomorphism* $\varphi : \mathcal{U}_0 \to \mathcal{U}_1$ *such that* $\varphi \circ \iota_0 = \iota_1$ *and* $\mu_0 = \varphi^*\mu_1$.

[81] Often the Lie algebra of S^1 is identified with $2\pi i\mathbb{R}$ under the exponential map $\exp : \mathfrak{g} \simeq 2\pi i\mathbb{R} \to S^1$, $\xi \mapsto e^{\xi}$. Given a principal S^1-bundle, by this identification the infinitesimal action maps the generator $2\pi i$ of $2\pi i\mathbb{R}$ to the generating vector field $X^{\#}$. A connection form A is then an imaginary-valued 1-form on the total space satisfying $\mathcal{L}_{X^{\#}} A = 0$ and $\imath_{X^{\#}} A = 2\pi i$. Its curvature form B is an imaginary-valued 2-form on the base satisfying $\pi^* B = dA$. By the Chern–Weil isomorphism, the *first Chern class* of the principal S^1-bundle is $c = [\frac{i}{2\pi} B]$.

Here we identify the Lie algebra of S^1 with $\mathbb{R}$ and implicitly use the exponential map $\exp : \mathfrak{g} \simeq \mathbb{R} \to S^1$, $t \mapsto e^{2\pi i t}$. Hence, given a principal S^1-bundle, the infinitesimal action maps the generator 1 of $\mathbb{R}$ to $X^{\#}$, and here a connection form α is an ordinary 1-form on the total space satisfying $\mathcal{L}_{X^{\#}}\alpha = 0$ and $\imath_{X^{\#}}\alpha = 1$. The curvature form β is an ordinary 2-form on the base satisfying $\pi^*\beta = d\alpha$. Consequently, we have $A = 2\pi i\alpha$, $B = 2\pi i\beta$ and the first Chern class is given by $c = [-\beta]$.

DEFINITION 6.16. The *Duistermaat–Heckman measure*, m_{DH}, on $\mathfrak{g}^*$ is the push-forward of the Liouville measure[82] by $\mu : M \to \mathfrak{g}^*$, that is, for any Borel subset $\mathcal{U}$ of $\mathfrak{g}^*$, we have

$$m_{\mathrm{DH}}(\mathcal{U}) = \int_{\mu^{-1}(\mathcal{U})} \frac{\omega^n}{n!}.$$

The integral with respect to the Duistermaat–Heckman measure of a compactly-supported function $h \in C^\infty(\mathfrak{g}^*)$ is

$$\int_{\mathfrak{g}^*} h \, dm_{\mathrm{DH}} := \int_M (h \circ \mu) \frac{\omega^n}{n!}.$$

On $\mathfrak{g}^*$ regarded as a vector space, say $\mathbb{R}^n$, there is also the Lebesgue (or Euclidean) measure, m_0. The relation between m_{DH} and m_0 is governed by the *Radon–Nikodym derivative*, denoted by $\frac{dm_{\mathrm{DH}}}{dm_0}$, which is a *generalized function* satisfying

$$\int_{\mathfrak{g}^*} h \, dm_{\mathrm{DH}} = \int_{\mathfrak{g}^*} h \frac{dm_{\mathrm{DH}}}{dm_0} \, dm_0.$$

THEOREM 6.17 (Duistermaat–Heckman [38]). *Under the hypotheses and notation before, the Duistermaat–Heckman measure is a piecewise polynomial multiple of Lebesgue measure on $\mathfrak{g}^* \simeq \mathbb{R}^n$, that is, the Radon–Nikodym derivative $f = \frac{dm_{\mathrm{DH}}}{dm_0}$ is piecewise polynomial. More specifically, for any Borel subset $\mathcal{U}$ of $\mathfrak{g}^*$, we have $m_{\mathrm{DH}}(\mathcal{U}) = \int_{\mathcal{U}} f(x)\,dx$, where $dx = dm_0$ is the Lebesgue volume form on $\mathcal{U}$ and $f : \mathfrak{g}^* \simeq \mathbb{R}^n \to \mathbb{R}$ is polynomial on any region consisting of regular values of μ.*

This Radon–Nikodym derivative f is called the *Duistermaat–Heckman polynomial.* In the case of a toric manifold, the Duistermaat–Heckman polynomial is a universal constant equal to $(2\pi)^n$ when Δ is n-dimensional. Thus the symplectic volume of $(M_\Delta, \omega_\Delta)$ is $(2\pi)^n$ times the Euclidean volume of Δ.

EXAMPLE. For the standard spinning of a sphere $(S^2, \omega = d\theta \wedge dh, S^1, \mu = h)$, the image of μ is the interval $[-1, 1]$. The Lebesgue measure of $[a, b] \subseteq [-1, 1]$ is $m_0([a, b]) = b - a$. The Duistermaat–Heckman measure of $[a, b]$ is

$$m_{\mathrm{DH}}\big([a, b]\big) = \int_{\{(\theta,h)\in S^2 \mid a \leqslant h \leqslant b\}} d\theta \, dh = 2\pi(b - a),$$

[82] On an arbitrary symplectic manifold (M^{2n}, ω), with symplectic volume $\frac{\omega^n}{n!}$, the *Liouville measure* (or *symplectic measure*) of a Borel subset $\mathcal{U}$ of M is

$$m_\omega(\mathcal{U}) = \int_{\mathcal{U}} \frac{\omega^n}{n!}.$$

The set $\mathcal{B}$ of *Borel subsets* is the *σ-ring* generated by the set of compact subsets, i.e., if $A, B \in \mathcal{B}$, then $A \setminus B \in \mathcal{B}$, and if $A_i \in \mathcal{B}$, $i = 1, 2, \ldots$, then $\bigcup_{i=1}^\infty A_i \in \mathcal{B}$.

i.e., $m_{\mathrm{DH}} = 2\pi m_0$. Consequently, *the area of the spherical region between two parallel planes depends only on the distance between the planes*, a result that was known to Archimedes around 230 BC.

PROOF. We sketch the proof of Theorem 6.17 for the case $G = S^1$. The proof for the general case, which follows along similar lines, can be found in, for instance, [66], besides the original articles.

Let (M, ω, S^1, μ) be a Hamiltonian S^1-space of dimension $2n$ and let (M_x, ω_x) be its reduced space at level x. Proposition 6.14 or Theorem 6.15 imply that, for x in a sufficiently narrow neighborhood of 0, the symplectic volume of M_x,

$$\mathrm{vol}(M_x) = \int_{M_x} \frac{\omega_x^{n-1}}{(n-1)!} = \int_{M_{\mathrm{red}}} \frac{(\omega_{\mathrm{red}} - x\beta)^{n-1}}{(n-1)!},$$

is a polynomial in x of degree $n-1$. This volume can be also expressed as

$$\mathrm{vol}(M_x) = \int_Z \frac{\pi^*(\omega_{\mathrm{red}} - x\beta)^{n-1}}{(n-1)!} \wedge \alpha,$$

where α is a connection form for the S^1-bundle $Z \to M_{\mathrm{red}}$ and β is its curvature form. Now we go back to the computation of the Duistermaat–Heckman measure. For a Borel subset $\mathcal{U}$ of $(-\varepsilon, \varepsilon)$, the Duistermaat–Heckman measure is, by definition,

$$m_{\mathrm{DH}}(\mathcal{U}) = \int_{\mu^{-1}(\mathcal{U})} \frac{\omega^n}{n!}.$$

Using the fact that $(\mu^{-1}(-\varepsilon, \varepsilon), \omega)$ is symplectomorphic to $(Z \times (-\varepsilon, \varepsilon), \sigma)$ and, moreover, they are isomorphic as Hamiltonian S^1-spaces, we obtain

$$m_{\mathrm{DH}}(\mathcal{U}) = \int_{Z \times \mathcal{U}} \frac{\sigma^n}{n!}.$$

Since $\sigma = \pi^*\omega_{\mathrm{red}} - d(x\alpha)$, its power is $\sigma^n = n(\pi^*\omega_{\mathrm{red}} - x\, d\alpha)^{n-1} \wedge \alpha \wedge dx$. By the Fubini theorem, we then have

$$m_{\mathrm{DH}}(\mathcal{U}) = \int_{\mathcal{U}} \left[\int_Z \frac{\pi^*(\omega_{\mathrm{red}} - x\beta)^{n-1}}{(n-1)!} \wedge \alpha \right] \wedge dx.$$

Therefore, the Radon–Nikodym derivative of m_{DH} with respect to the Lebesgue measure, dx, is

$$f(x) = \int_Z \frac{\pi^*(\omega_{\mathrm{red}} - x\beta)^{n-1}}{(n-1)!} \wedge \alpha = \mathrm{vol}(M_x).$$

The previous discussion proves that, for $x \approx 0$, $f(x)$ is a polynomial in x. The same holds for a neighborhood of any other regular value of μ, because we may change the moment map μ by an arbitrary additive constant. □

Duistermaat and Heckman [38] also applied these results when M is compact to provide a formula for the oscillatory integral $\int_M e^{i\mu^X}\frac{\omega^n}{n!}$ for $X \in \mathfrak{g}$ as a sum of contributions of the fixed points of the action of the one-parameter subgroup generated by X. They hence showed that the *stationary phase approximation*[83] is exact in the case of the moment map. When G is a maximal torus of a compact connected simple Lie group acting on a coadjoint orbit, the Duistermaat–Heckman formula reduces to the Harish–Chandra formula. It was observed by Berline and Vergne [14] and by Atiyah and Bott [5] that the Duistermaat–Heckman formula can be derived by *localization in equivariant cohomology*. This is an instance of *Abelian localization*, i.e., a formula for an integral (in equivariant cohomology) in terms of data at the fixed points of the action, and typically is used for the case of Abelian groups (or of maximal tori). Later *non-Abelian localization* formulas were found, where integrals (in equivariant cohomology) are expressed in terms of data at the zeros of the moment map, normally used for the case of non-Abelian groups. Both localizations gave rise to computations of the cohomology ring structure of reduced spaces [80].

References

[1] V. Arnold, *On a characteristic class entering into conditions of quantization*, Funkcional Anal. i Priložen **1** (1967), 1–14.

[2] V. Arnold, *Mathematical Methods of Classical Mechanics*, Graduate Texts in Math., vol. 60, Springer-Verlag, New York (1978).

[3] V. Arnold and A. Givental, *Symplectic geometry*, Dynamical Systems IV, Symplectic Geometry and Its Applications, V. Arnold and S. Novikov, eds, Encyclopaedia of Math. Sciences, vol. 4, Springer-Verlag, Berlin (1990).

[4] M. Atiyah, *Convexity and commuting Hamiltonians*, Bull. London Math. Soc. **14** (1982), 1–15.

[5] M. Atiyah and R. Bott, *The moment map and equivariant cohomology*, Topology **23** (1984), 1–28.

[6] M. Atiyah and R. Bott, *The Yang–Mills equations over Riemann surfaces*, Topology **23** (1984), 1–28; Philos. Trans. Roy. Soc. London **308** (1983), 523–615.

[7] M. Audin, *Exemples de variétés presque complexes*, Enseign. Math. **37** (1991), 175–190.

[8] M. Audin, *Spinning Tops, A Course on Integrable Systems*, Cambridge Studies in Advanced Mathematics, vol. 51, Cambridge Univ. Press, Cambridge (1996).

[9] M. Audin, *Lagrangian submanifolds*, Symplectic Geometry of Integrable Hamiltonian Systems (Barcelona, 2001), Adv. Courses Math. CRM Barcelona, Birkhäuser, Basel (2003), 1–83.

[10] M. Audin, *Torus Actions on Symplectic Manifolds*, Progress in Mathematics, vol. 93, Birkhäuser, Basel (2004).

[11] M. Audin and J. Lafontaine, eds, *Holomorphic Curves in Symplectic Geometry*, Progress in Mathematics, vol. 117, Birkhäuser, Basel (1994).

[12] D. Auroux, *Asymptotically holomorphic families of symplectic submanifolds*, Geom. Funct. Anal. **7** (1997), 971–995.

[13] D. Auroux, *Symplectic* 4-*manifolds as branched coverings of* $\mathbb{CP}^2$, Invent. Math. **139** (2000), 551–602.

[14] N. Berline and M. Vergne, *Zéros d'un champ de vecteurs et classes caractéristiques équivariantes*, Duke Math. J. **50** (1983), 539–549.

[83] The *stationary phase lemma* gives the asymptotic behavior (for large N) of integrals $(\frac{N}{2\pi})^n \int_M f e^{ig}$ vol, where f and g are real functions and vol is a volume form on a $2n$-dimensional manifold M.

[15] P. Biran, *A stability property of symplectic packing*, Invent. Math. **136** (1999), 123–155.

[16] P. Biran, *Geometry of symplectic intersections*, Proceedings of the I.C.M., vol. II (Beijing, 2002), Higher Ed. Press, Beijing (2002), 241–255.

[17] P. Biran and K. Cieliebak, *Symplectic topology on subcritical manifolds*, Comment. Math. Helv. **76** (2001), 712–753.

[18] G. Birkhoff, *Dynamical Systems*, Amer. Math. Soc. Colloquium Publications, vol. IX, Amer. Math. Soc., Providence (1966), reprinting of the original 1927 edition, with an addendum by J. Moser.

[19] G. Bredon, *Introduction to Compact Transformation Groups*, Pure and Applied Mathematics, vol. 46, Academic Press, New York (1972).

[20] R. Bryant, *An introduction to Lie groups and symplectic geometry*, Geometry and Quantum Field Theory (Park City, UT, 1991), IAS/Park City Math. Ser., vol. 1, Amer. Math. Soc., Providence (1995), 5–181.

[21] A. Cannas da Silva, *Symplectic toric manifolds*, Symplectic Geometry of Integrable Hamiltonian Systems (Barcelona, 2001), Adv. Courses Math. CRM Barcelona, Birkhäuser, Basel (2003), 1–83.

[22] J. Cheeger, *Finiteness theorems for Riemannian manifolds*, Amer. J. Math. **92** (1970), 61–74.

[23] S.S. Chern, *Complex Manifolds Without Potential Theory*, 2nd ed., Universitext, Springer-Verlag, New York (1979), with an appendix on the geometry of characteristic classes.

[24] C. Conley and E. Zehnder, *The Birkhoff–Lewis fixed point theorem and a conjecture of V.I. Arnold*, Invent. Math. **73** (1983), 33–49.

[25] C. Conley and E. Zehnder, *Morse-type index theory for flows and periodic solutions for Hamiltonian equations*, Comm. Pure Appl. Math. **37** (1984), 207–253.

[26] D. Cox, *Recent developments in toric geometry*, Algebraic Geometry—Santa Cruz 1995, Proc. Sympos. Pure Math., vol. 62, part 2, Amer. Math. Soc., Providence (1997), 389–436.

[27] V. Danilov, *The geometry of toric varieties*, Uspekhi Mat. Nauk **33** (2) (1978), (200), 85–134, 247; English translation: Russian Math. Surveys **33** (2) (1978), 97–154.

[28] T. Delzant, *Hamiltoniens périodiques et images convexes de l'application moment*, Bull. Soc. Math. France **116** (1988), 315–339.

[29] M. Demazure, *Sous-groupes algébriques de rang maximum du groupe de Cremona*, Ann. Sci. École Norm. Sup. (4) **3** (1970), 507–588.

[30] S. Donaldson, *An application of gauge theory to four-dimensional topology*, J. Differential Geom. **18** (1983), 279–315.

[31] S. Donaldson, *Irrationality and the h-cobordism conjecture*, J. Differential Geom. **26** (1987), 141–168.

[32] S. Donaldson, *Symplectic submanifolds and almost-complex geometry*, J. Differential Geom. **44** (1996), 666–705.

[33] S. Donaldson, *Lefschetz fibrations in symplectic geometry*, Proceedings of the I.C.M., vol. II (Berlin, 1998), Doc. Math., extra vol. II (1998), 309–314.

[34] S. Donaldson, *Lefschetz pencils on symplectic manifolds*, J. Differential Geom. **53** (1999), 205–236.

[35] S. Donaldson, *Floer Homology Groups in Yang–Mills Theory*, Cambridge Tracts in Mathematics, vol. 147, Cambridge Univ. Press, Cambridge (2002), with the assistance of M. Furuta and D. Kotschick.

[36] J.J. Duistermaat, *On global action-angle coordinates*, Comm. Pure Appl. Math. **33** (1980), 687–706.

[37] J.J. Duistermaat, *The Heat Kernel Lefschetz Fixed Point Formula for the Spin-c Dirac Operator*, Progress in Nonlinear Differential Equations and their Applications, vol. 18, Birkhäuser, Boston (1996).

[38] J.J. Duistermaat and G. Heckman, *On the variation in the cohomology of the symplectic form of the reduced phase space*, Invent. Math. **69** (1982), 259–268; *Addendum*, Invent. Math. **72** (1983), 153–158.

[39] Y. Eliashberg, *Estimates on the number of fixed points of area-preserving transformations*, Syktyvkar University, preprint (1979).

[40] Y. Eliashberg, A. Givental and H. Hofer, *Introduction to symplectic field theory*, GAFA 2000 (Tel Aviv, 1999), Geom. Funct. Anal., special volume, part II (2000), 560–673.

[41] Y. Eliashberg and M. Gromov, *Lagrangian intersection theory: finite-dimensional approach*, Geometry of Differential Equations, Amer. Math. Soc. Transl. Ser. 2, vol. 186, Amer. Math. Soc., Providence (1998), 27–118.

[42] Y. Eliashberg and N. Mishachev, *Introduction to the h-Principle*, Graduate Studies in Mathematics, vol. 48, Amer. Math. Soc., Providence (2002).

[43] Y. Eliashberg and L. Traynor, eds, *Symplectic Geometry and Topology*, Lectures from the Graduate Summer School held in Park City, June 29–July 19, 1997, IAS/Park City Mathematics Series, vol. 7, Amer. Math. Soc., Providence (1999).

[44] M. Fernández, M. Gotay and A. Gray, *Compact parallelizable four-dimensional symplectic and complex manifolds*, Proc. Amer. Math. Soc. **103** (1988), 1209–1212.

[45] R. Fintushel and R. Stern, *Knots, links, and 4-manifolds*, Invent. Math. **134** (1998), 363–400.

[46] A. Floer, *A relative Morse index for the symplectic action*, Comm. Pure Appl. Math. **41** (1988), 393–407.

[47] A. Floer, *The unregularized gradient flow of the symplectic action*, Comm. Pure Appl. Math. **41** (1988), 775–813.

[48] A. Floer, *Morse theory for Lagrangian intersections*, J. Differential Geom. **28** (1988), 513–547.

[49] A. Floer, *Symplectic fixed points and holomorphic spheres*, Comm. Math. Phys. **120** (1989), 575–611.

[50] A. Floer, *Witten's complex and infinite-dimensional Morse theory*, J. Differential Geom. **30** (1989), 207–221.

[51] M. Freedman, *The topology of four-dimensional manifolds*, J. Differential Geom. **17** (1982), 357–453.

[52] K. Fukaya and K. Ono, *Arnold conjecture and Gromov–Witten invariant*, Topology **38** (1999), 933–1048.

[53] W. Fulton, *Introduction to Toric Varieties*, Annals of Mathematics Studies, vol. 131, Princeton Univ. Press, Princeton (1993).

[54] D. Gay and R. Kirby, *Constructing symplectic forms on 4-manifolds which vanish on circles*, Geom. Topol. **8** (2004), 743–777.

[55] A. Givental, *Periodic mappings in symplectic topology*, Funktsional. Anal. i Prilozhen **23** (1989), 37–52 (in Russian); translation in Funct. Anal. Appl. **23** (1989), 287–300.

[56] M. Golubitsky and D. Tischler, *An example of moduli for singular symplectic forms*, Invent. Math. **38** (1976/1977), 219–225.

[57] R. Gompf, *A new construction of symplectic manifolds*, Ann. of Math. **142** (1995), 527–595.

[58] R. Gompf, *Symplectic structures from Lefschetz pencils in high dimensions*, Geom. Topol. Monogr. **7** (2004), 267–290.

[59] R. Gompf, *Toward a topological characterization of symplectic manifolds*, J. Symp. Geom., in press.

[60] R. Gompf and A. Stipsicz, *4-Manifolds and Kirby Calculus*, Graduate Studies in Mathematics, vol. 20, Amer. Math. Soc., Providence (1999).

[61] M. Gotay, *On coisotropic imbeddings of presymplectic manifolds*, Proc. Amer. Math. Soc. **84** (1982), 111–114.

[62] P. Griffiths and J. Harris, *Principles of Algebraic Geometry*, Wiley Classics Library, Wiley, New York (1994), reprint of the 1978 original.

[63] M. Gromov, *Stable mappings of foliations into manifolds*, Izv. Akad. Nauk SSSR Ser. Mat. **33** (1969), 707–734.

[64] M. Gromov, *Pseudoholomorphic curves in symplectic manifolds*, Invent. Math. **82** (1985), 307–347.

[65] M. Gromov, *Partial Differential Relations*, Ergebnisse der Mathematik und ihrer Grenzgebiete, vol. 9, Springer-Verlag, Berlin (1986).

[66] V. Guillemin, *Moment Maps and Combinatorial Invariants of Hamiltonian T^n-spaces*, Progress in Mathematics, vol. 122, Birkhäuser, Boston (1994).

[67] V. Guillemin, V. Ginzburg and Y. Karshon, *Moment Maps, Cobordisms, and Hamiltonian Group Actions*, Mathematical Surveys and Monographs, vol. 98, Amer. Math. Soc., Providence (2002), with appendix J by M. Braverman.

[68] V. Guillemin and S. Sternberg, *Convexity properties of the moment mapping*, Invent. Math. **67** (1982), 491–513.

[69] V. Guillemin and S. Sternberg, *Birational equivalence in the symplectic category*, Invent. Math. **97** (1989), 485–522.

[70] V. Guillemin and S. Sternberg, *Symplectic Techniques in Physics*, 2nd ed., Cambridge Univ. Press, Cambridge (1990).

[71] R. Harvey and H.B. Lawson, *Calibrated geometries*, Acta Math. **148** (1982), 47–157.

[72] N. Hitchin, G. Segal and R. Ward, *Integrable Systems. Twistors, Loop groups, and Riemann Surfaces*, Oxford Graduate Texts in Mathematics, vol. 4, Clarendon, Oxford Univ. Press, New York (1999).

[73] W. Hodge, *The Theory and Applications of Harmonic Integrals*, 2nd ed., Cambridge Univ. Press, Cambridge (1952).

[74] H. Hofer and D. Salamon, *Floer homology and Novikov rings*, The Floer Memorial Volume, Progress in Mathematics, vol. 133, Birkhäuser, Basel (1995), 483–524.

[75] K. Honda, *Transversality theorems for harmonic forms*, Rocky Mountain J. Math. **34** (2004), 629–664.

[76] L. Hörmander, *An Introduction to Complex Analysis in Several Variables*, 3rd ed., North-Holland Mathematical Library, vol. 7, North-Holland, Amsterdam (1990).

[77] N. Jacobson, *Lie Algebras*, Dover, New York (1979), republication of the 1962 original.

[78] L. Jeffrey and F. Kirwan, *Localization for non-Abelian group actions*, Topology **34** (1995), 291–327.

[79] G. Kempf, F. Knudsen, D. Mumford and B. Saint-Donat, *Toroidal Embeddings, I*, Lecture Notes in Mathematics, vol. 339, Springer-Verlag, Berlin (1973).

[80] F. Kirwan, *Cohomology of Quotients in Symplectic and Algebraic Geometry*, Mathematical Notes, vol. 31, Princeton Univ. Press, Princeton (1984).

[81] F. Kirwan, *Convexity properties of the moment mapping, III*, Invent. Math. **77** (1984), 547–552.

[82] K. Kodaira, *On the structure of compact complex analytic surfaces, I*, Amer. J. Math. **86** (1964), 751–798.

[83] J. Kohn, *Harmonic integrals on strongly pseudo-convex manifolds, I*, Ann. of Math. **78** (1963) 112–148.

[84] D. Kotschick, *On manifolds homeomorphic to* $\mathbb{CP}^2 \#_8 \overline{\mathbb{CP}^2}$, Invent. Math. **95** (1989), 591–600.

[85] F. Lalonde and D. McDuff, *J-curves and the classification of rational and ruled symplectic* 4*-manifolds*, Contact and Symplectic Geometry (Cambridge, 1994), Publ. Newton Inst., vol. 8, Cambridge Univ. Press, Cambridge (1996), 3–42.

[86] E. Lerman, *Symplectic cuts*, Math. Res. Lett. **2** (1995), 247–258.

[87] E. Lerman and S. Tolman, *Hamiltonian torus actions on symplectic orbifolds and toric varieties*, Trans. Amer. Math. Soc. **349** (1997), 4201–4230.

[88] T. Li and A. Liu, *Symplectic structure on ruled surfaces and a generalized adjunction formula*, Math. Res. Lett. **2** (1995), 453–471.

[89] A.-K. Liu, *Some new applications of general wall crossing formula, Gompf's conjecture and its applications*, Math. Res. Lett. **3** (1996), 569–585.

[90] G. Liu and G. Tian, *Floer homology and Arnold conjecture*, J. Differential Geom. **49** (1998), 1–74.

[91] J. Marsden and T. Ratiu, *Introduction to Mechanics and Symmetry, A Basic Exposition of Classical Mechanical Systems*, Texts in Applied Mathematics, vol. 17, Springer-Verlag, New York (1994).

[92] J. Marsden and A. Weinstein, *Reduction of symplectic manifolds with symmetry*, Rep. Math. Phys. **5** (1974), 121–130.

[93] V. Maslov, *Perturbation Theory and Asymptotic Methods*, Izdat. Moskov Univ., Moscow (1965) (in Russian).

[94] D. McDuff, *Examples of simply-connected symplectic non-Kählerian manifolds*, J. Differential Geom. **20** (1984), 267–277.

[95] D. McDuff, *Rational and ruled symplectic* 4*-manifolds*, Geometry of low-dimensional manifolds 2 (Durham, 1989), London Math. Soc. Lecture Note Ser., vol. 151, Cambridge Univ. Press, Cambridge (1990), 7–14.

[96] D. McDuff, *The structure of rational and ruled symplectic* 4*-manifolds*, J. Amer. Math. Soc. **3** (1990), 679–712.

[97] D. McDuff, *Lectures on groups of symplectomorphisms*, Rend. Circ. Mat. Palermo (2) **72** (2004), 43–78.

[98] D. McDuff and L. Polterovich, *Symplectic packings and algebraic geometry*, Invent. Math. **115** (1994), 405–434, with an appendix by Yael Karshon.

[99] D. McDuff and D. Salamon, *Introduction to Symplectic Topology*, Oxford Mathematical Monographs, Oxford Univ. Press, New York (1995).

[100] D. McDuff and D. Salamon, *J-holomorphic Curves and Symplectic Topology*, Amer. Math. Soc. Colloquium Publications, vol. 52, Amer. Math. Soc., Providence (2004).

[101] C. McMullen and C. Taubes, 4*-manifolds with inequivalent symplectic forms and* 3*-manifolds with inequivalent fibrations*, Math. Res. Lett. **6** (1999), 681–696.

[102] K. Meyer, *Symmetries and integrals in mechanics*, Dynamical Systems (Proc. Sympos., Univ. Bahia, Salvador, 1971), Academic Press, New York (1973), 259–272.

[103] J. Milnor, *Morse Theory*, Annals of Mathematics Studies, vol. 51, Princeton Univ. Press, Princeton (1963), based on lecture notes by M. Spivak and R. Wells.

[104] M. Morse, *The foundations of a theory in the calculus of variations in the large*, Trans. Amer. Math. Soc. **30** (1928), 213–274.

[105] J. Moser, *On the volume elements on a manifold*, Trans. Amer. Math. Soc. **120** (1965), 286–294.

[106] A. Newlander and L. Nirenberg, *Complex analytic coordinates in almost complex manifolds*, Ann. of Math. **65** (1957), 391–404.

[107] T. Oda, *Convex Bodies and Algebraic Geometry—An Introduction to the Theory of Toric Varieties*, Ergebnisse der Mathematik und ihrer Grenzgebiete (3), vol. 15, Springer-Verlag, Berlin (1988).

[108] K. Ono, *On the Arnold conjecture for weakly monotone symplectic manifolds*, Invent. Math. **119** (1995), 519–537.

[109] J. Park, *Non-complex symplectic* 4*-manifolds with* $b_2^+ = 1$, Bull. London Math. Soc. **36** (2004), 231–240.

[110] J. Park, *Simply connected symplectic* 4*-manifolds with* $b_2^+ = 1$ *and* $c_1^2 = 2$, Invent. Math., in press.

[111] Y. Ruan, *Symplectic topology on algebraic* 3*-folds*, J. Differential Geom. **39** (1994), 215–227.

[112] Y. Ruan, *Topological sigma model and Donaldson-type invariants in Gromov theory*, Duke Math. J. **83** (1996), 461–500.

[113] D. Salamon, *Lectures on Floer homology*, Symplectic Geometry and Topology, Y. Eliashberg and L. Traynor, eds, IAS/Park City Math. Ser., vol. 7, Amer. Math. Soc., Providence (1999), 143–229.

[114] I. Satake, *On a generalization of the notion of manifold*, Proc. Nat. Acad. Sci. U.S.A. **42** (1956), 359–363.

[115] P. Seidel, *Graded Lagrangian submanifolds*, Bull. Soc. Math. France **128** (2000), 103–149.

[116] J.-C. Sikorav, *Points fixes d'un symplectomorphisme homologue à l'identité*, C. R. Acad. Sci. Paris Sér. I Math. **299** (1984), 343–346.

[117] I. Smith, *On moduli spaces of symplectic forms*, Math. Res. Lett. **7** (2000), 779–788.

[118] I. Smith, *Geometric monodromy and the hyperbolic disc*, Quart. J. Math. **52** (2001), 217–228.

[119] J.-M. Souriau, *Structure des Systèmes Dynamiques*, Maî trises de Mathématiques, Dunod, Paris (1970).

[120] M. Spivak, *A Comprehensive Introduction to Differential Geometry*, vol. I, 2nd ed., Publish or Perish, Wilmington (1979).

[121] N. Steenrod, *The Topology of Fibre Bundles*, Princeton Mathematical Series, vol. 14, Princeton Univ. Press, Princeton (1951).

[122] A. Stipsicz, *The geography problem of* 4*-manifolds with various structures*, Acta Math. Hungar. **7** (2000), 267–278.

[123] M. Symington, *Symplectic rational blowdowns*, J. Differential Geom. **50** (1998), 505–518.

[124] Z. Szabó, *Exotic* 4*-manifolds with* $b_2^+ = 1$, Math. Res. Lett. **3** (1996), 731–741.

[125] Z. Szabó, *Simply-connected irreducible* 4*-manifolds with no symplectic structures*, Invent. Math. **132** (1998), 457–466.

[126] C. Taubes, *The Seiberg–Witten invariants and symplectic forms*, Math. Res. Lett. **1** (1994), 809–822.

[127] C. Taubes, *More constraints on symplectic forms from Seiberg–Witten invariants*, Math. Res. Lett. **2** (1995), 9–13.

[128] C. Taubes, *The Seiberg–Witten and Gromov invariants*, Math. Res. Lett. **2** (1995), 221–238.

[129] C. Taubes, SW $\Rightarrow$ Gr*: from the Seiberg–Witten equations to pseudoholomorphic curves*, J. Amer. Math. Soc. **9** (1996), 845–918.

[130] C. Taubes, *Seiberg–Witten invariants and pseudo-holomorphic subvarieties for self-dual, harmonic 2-forms*, Geom. Topol. **3** (1999), 167–210.

[131] W. Thurston, *Some simple examples of symplectic manifolds*, Proc. Amer. Math. Soc. **55** (1976), 467–468.

[132] D. Tischler, *Closed* 2*-forms and an embedding theorem for symplectic manifolds*, J. Differential Geom. **12** (1977), 229–235.

[133] S. Tolman and J. Weitsman, *The cohomology rings of symplectic quotients*, Comm. Anal. Geom. **11** (2003), 751–773.

[134] L. Traynor, *Symplectic packing constructions*, J. Differential Geom. **42** (1995), 411–429.

[135] C. Viterbo, *Symplectic topology as the geometry of generating functions*, Math. Ann. **292** (1992), 685–710.

[136] A. Weinstein, *Symplectic manifolds and their Lagrangian submanifolds*, Adv. in Math. **6** (1971), 329–346.

[137] A. Weinstein, *Lectures on Symplectic Manifolds*, Regional Conference Series in Mathematics, vol. 29, Amer. Math. Soc., Providence (1977).

[138] A. Weinstein, *Fat bundles and symplectic manifolds*, Adv. in Math. **37** (1980), 239–250.

[139] A. Weinstein, *Neighborhood classification of isotropic embeddings*, J. Differential Geom. **16** (1981), 125–128.

[140] A. Weinstein, *On extending the Conley–Zehnder fixed point theorem to other manifolds*, Nonlinear Functional Analysis and its Applications, Part 2 (Berkeley, 1983), Proc. Sympos. Pure Math., vol. 45, Part 2, Amer. Math. Soc., Providence (1986), 541–544.

[141] R.O. Wells, *Differential Analysis on Complex Manifolds*, 2nd ed., Graduate Texts in Mathematics, vol. 65, Springer-Verlag, New York (1980).

[142] H. Weyl, *The Classical Groups. Their Invariants and Representations*, Princeton Landmarks in Mathematics, Princeton Univ. Press, Princeton (1997).

[143] H. Whitney, *Analytic extensions of differentiable functions defined in closed sets*, Trans. Amer. Math. Soc. **36** (1934), 63–89.

[144] E. Witten, *Supersymmetry and Morse theory*, J. Differential Geom. **17** (1982), 661–692.

[145] E. Witten, *Topological sigma models*, Comm. Math. Phys. **118** (1988), 411–449.

[146] W.-T. Wu, *Sur les classes caractíristiques des structures fibrées sphériques*, Actualités Sci. Ind. **1183** (1952).

CHAPTER 4

Metric Riemannian Geometry

Kenji Fukaya*

Department of Mathematics, Kyoto University, Kitashirakawa Oiwake Cho, Sakyo-ku, Kyoto 606-8502, Japan
E-mail: fukaya@math.kyoto-u.ac.jp

Contents

*The author thanks Professors A. Kasue, Y. Otsu, T. Sakai, T. Shioya and T. Yamaguchi (the authors of [92]) who provided detailed information on the topics covered in this article (those which are written or not written in [92]). This article grew from the (English translation of) the introduction of [92]. The author also thanks Professor F. Dillen for his patience and encouragement.

HANDBOOK OF DIFFERENTIAL GEOMETRY, VOL. II
Edited by F.J.E. Dillen and L.C.A. Verstraelen
Published by Elsevier B.V.

1. Introduction

This article is a survey of (a part of) Riemannian geometry. Riemannian geometry is a huge area which occupies, I believe, at least $1/3$ of the whole of differential geometry. So obviously we need to restrict our attention to some part of it to write an article in this handbook. (M. Berger's books [20,19] deals with wider topics.) Let me mention first what is *not* included in this article but should have been included in a survey of Riemannian geometry.

(1) We do not include an elementary or introductory part of Riemannian geometry. For example, topics covered in [103, Sections II, III] or [97] are not in this article. We assume the reader to have some knowledge about it.

(2) We focus our attention to global results, and results of local nature are rarely discussed.

(3) One powerful tool to study global Riemannian geometry is partial differential equations, especially nonlinear one. We do not discuss it.[1] The theory of geodesics (which is a theory of nonlinear *ordinary* differential equations) is one of the main tools used in this article. Linear partial differential equations, especially the Laplacian, is mentioned only when it is closely related to the other topics included in this article.

(4) We do not discuss manifolds of nonpositive curvature.

(5) We do not discuss scalar curvature.

After removing so many important and interesting topics there are still many things missing in this article. For example, results such as filling volume [73] is not discussed. The study of closed geodesics is not included either.

So what is included in this article?

We focus the part of Riemannian geometry which describes relations of curvature (sectional or Ricci curvature) to topology of the underlying manifold. Since we do not discuss nonpositively curved manifolds, the main target is manifolds of (almost) nonnegative curvature and more generally the class of manifolds with curvature bounded from below. The study of such Riemannian manifolds started with sphere theorems in the 50's where comparison theorems are introduced by Rauch as an important tool of study.

At the beginning of the 70's Cheeger (and Weinstein) proved finiteness theorems which provide another kind of statements to be established other than sphere theorems. Soon after that, M. Gromov introduced many new ideas, results and tools, such as Gromov–Hausdorff convergence, almost flat manifold theorem, Betti number estimate, etc., and gave tremendous influence to the area. These present the first turning point of the development of metric Riemannian geometry.

In the 1980's global Riemannian geometry was a very rapidly developing area. Especially the class of Riemannian manifolds with sectional curvature bounded from below and above are studied extensively. An important progress in the 1980's is the theory of collapsing Riemannian manifolds.

Those topics are discussed in Sections 2–13. After a brief review of sphere theorems in Section 2, we describe finiteness theorems in Section 3. In Section 4, while explaining a rough sketch of proofs of sphere theorems, we review several basic facts on global Rie-

[1] So, for example, famous result by Hamilton on the 3 manifold of positive Ricci curvature is not discussed.

mannian geometry, such as Rauch's comparison theorem, cut points, conjugate points, injectivity radius, etc. One of the main tools of global Riemannian geometry is the Gromov–Hausdorff distance, which we define in Section 5, and we will prove Gromov's precompactness theorem. The proofs of finiteness theorems are discussed in Sections 6–9. We try to sketch various (different) techniques used to prove finiteness theorems etc. there, rather than to concentrate on one method and to give its full details. Collapsing Riemannian manifolds (under the bound of the absolute value of sectional curvature) is discussed in Sections 10–13.

In Sections 14–18, we discuss the class of Riemannian manifolds with sectional curvature bounded from below (but not above). The basic tool to study it is Morse theory of the distance function, which was initiated by Grove–Shiohama. We discuss it and its application to sphere theorems in Section 14. We explain applications of the same method to finiteness theorems in Section 15. The theme of Section 16 is noncompact manifolds of nonnegative curvature. Besides its own interest, it is used in many places to study compact Riemannian manifold. Our focus in this article is on the compact case, so we restrict our discussion on noncompact manifolds to ones which have a direct application to compact manifolds.

New turning points of the development of metric Riemannian geometry came at some point in the 1990's when several mathematicians belonging to the new generation (such as Perelman and Colding) began to work in this field. In Sections 17 and 18 we discuss Alexandrov spaces. They are metric spaces which have curvature $> -\infty$ in some generalized sense. The notion of curvature on a metric space which is not a manifold was introduced by Alexandrov a long time ago. Recently various applications of it to Riemannian geometry (study of *smooth* Riemannian manifolds) were discovered. It makes this topic more popular among Riemannian geometers. An important structure theorem of Alexandrov spaces is obtained by Perelman and his collaborators, which we review in Sections 17 and 18.

In Sections 19–23 we discuss the class of Riemannian manifolds with Ricci curvature bounded from below. The first Betti number and the fundamental group are topics studied extensively under this curvature assumption. We review some of such studies in Section 19. The theme of Section 20 is (mainly) a special case, that is the case of Einstein manifolds. Our discussion of Einstein manifolds is restricted to those related to the other parts of this article. We discuss Einstein manifolds here since they provide rich examples of a new phenomenon which appears when we replace the assumption sectional curvature $\geqslant$ const, by the Ricci curvature $\geqslant$ const. Also it is an area where results we discuss in Sections 21–23 provide (and will provide) a powerful tool. Sections 21–23 are reviews of results obtained recently by Colding and Cheeger–Colding on the class of manifolds whose Ricci curvature is bounded from below. Here we emphasize the geometric part of the story and omit most of the analytic parts of the proofs, though analytic parts are as important as geometric parts.

It is of course impossible to write down all details of the proofs in this article. However, rather than stating as many results as possible without proof, the author tried to survey as many ideas, tools, techniques, methods of proofs, etc. as possible. In that sense, the emphasis of this article is on methods of proofs and not on their outcome. (Of course important applications of various techniques are explained.) Since this is a survey article there are no new results in it.

1.1. *Notations used in this article*

$T_pM =$ the tangent space, $\mathrm{Exp}_p : T_pM \to M$, the exponential map,

$B_p(R, X) = \{x \in X \mid d(x, p) < R\}$, for a metric space (X, d) and $p \in X$,

$K_M =$ the sectional curvature of M, $\mathrm{Vol}(M) =$ the volume of M,

$\mathrm{Ricci}_M =$ the Ricci curvature of M, $\mathrm{Diam}(M) =$ the diameter of M,

$i_M(p) =$ the injectivity radius of M at p (Definition 4.1),

$\overline{xy} =$ a minimal geodesic joining x and y,

$\angle xyz =$ the angle between $\overline{xy}$ and $\overline{yz}$ at y,

$\mathfrak{S}_n(D) = \{M \mid \mathrm{Ricci}_M \geqslant -(n-1),\ \dim = n,\ \mathrm{Diam}(M) \leqslant D\}$,

$\mathfrak{S}_n(D, v) = \{M \in \mathfrak{S}_n(D) \mid \mathrm{Vol}(M) \geqslant v\}$,

$\mathfrak{S}_n(D, i > \rho) = \{M \in \mathfrak{S}_n(D) \mid \forall p,\ i_M(p) \geqslant \rho\}$,

$\mathfrak{M}_n(D) = \{M \mid |K_M| \leqslant 1,\ \dim = n,\ \mathrm{Diam}(M) \leqslant D\}$,

$\mathfrak{M}_n(D, v) = \{M \in \mathfrak{M}_n(D) \mid \mathrm{Vol}(M) \geqslant v\}$,

$\mathfrak{M}'_n(D, v) = \{M \mid K_M \geqslant -1,\ \mathrm{Diam}(M) \leqslant D,\ \mathrm{Vol}(M) \geqslant v\}$,

$d_{GH}(X_1, X_2) =$ the Gromov–Hausdorff distance (Definition 3.2),

$\mathbb{S}^n(\kappa) =$ simply connected Riemannian manifold with $K_M \equiv \kappa$,

$A_p(a, b; M) = \{x \in M \mid a \leqslant d(p, x) \leqslant b\}$,

$S_p(a; M) = \{x \in M \mid d(p, x) = a\}$.

$\lim^{GH}_{i\to\infty} X_i = X$ means $\lim_{i\to\infty} d_{GH}(X_i, X) = 0$.

The symbol $\doteqdot$ means almost equal. The argument using this symbol is not rigorous. We use it only when we sketch the proof.

The symbol $\tau(\epsilon_1, \ldots, \epsilon_k | a_1, \ldots, a_m)$ stand for the positive number depending only on $\epsilon_1, \ldots, \epsilon_k, a_1, \ldots, a_m$ and satisfying

$$\lim_{\epsilon_1, \ldots, \epsilon_k \to 0} \tau(\epsilon_1, \ldots, \epsilon_k | a_1, \ldots, a_m) = 0,$$

for each fixed $a_1, \ldots, a_m$. In other words

$$f(\epsilon_1, \ldots, \epsilon_k | a_1, \ldots, a_m) < \tau(\epsilon_1, \ldots, \epsilon_k | a_1, \ldots, a_m)$$

is equivalent to the following statement.

For each $\delta, a_1, \ldots, a_m$ there exists ϵ such that if $\epsilon_1 < \epsilon, \ldots, \epsilon_k < \epsilon$ then

$$f(\epsilon_1, \ldots, \epsilon_k | a_1, \ldots, a_m) < \delta.$$

2. Sphere theorems

There are several pioneering works in metric Riemannian geometry (such as Myers' theorem (Theorem 5.4), Hadamard–Cartan's theorem (Theorem 4.6), study of convex surfaces in $\mathbb{R}^3$, etc.). But let me set the beginning of metric Riemannian geometry at the time when the following theorem was proved. From now on, we denote by K_M the sectional curvature of a Riemannian manifold M. We assume all Riemannian manifolds are complete unless otherwise stated.

THEOREM 2.1 (Rauch's sphere theorem [129]). *There exists a positive constant ϵ_n depending only on the dimension n such that, if a simply connected Riemannian manifold M satisfies $1 \geqslant K_M \geqslant 1 - \epsilon_n$, then M is homeomorphic to a sphere.*

This theorem is one of the first theorems which are called "sphere theorems". In this section, we mention some of the most important sphere theorems.[2]

THEOREM 2.2 (Klingenberg [94], Berger [18]). *If a simply connected Riemannian manifold M satisfies $1 \geqslant K_M > 1/4$, then it is homeomorphic to a sphere.*

If M satisfies $1 \geqslant K_M \geqslant 1/4$, then M is either homeomorphic to a sphere or is isometric to a symmetric space of compact type.[3]

Theorem 2.2 is a generalization of Rauch's theorem, and is an optimal result among those characterizing spheres under an assumption of the sectional curvature bounded from above or below.[4] (We remark that the sectional curvature of a complex, or quaternionic projective space, or Cayley plane is between 1 and $1/4$.)

THEOREM 2.3 (Bochner [157]). *If the curvature tensor R of a simply connected Riemannian manifold M satisfies*

$$\frac{C}{2} \leqslant \frac{-R_{ijk\ell}\xi^{ij}\xi^{k\ell}}{\|\xi\|} \leqslant C$$

for any antisymmetric 2 tensor ξ (where C is a positive constant), then the homology group over $\mathbb{R}$ of M is isomorphic to the homology group of the sphere.

The assumption of Theorem 2.3 is on the curvature operator and is more restrictive than the one on sectional curvature. Hence Theorem 2.3 follows from Theorem 2.2. (Theorem 2.3 was proved earlier.) We mention Theorem 2.3 since the idea of its proof is quite different from the proof of Theorem 2.2. We mention them later in Section 19.

THEOREM 2.4. *If M is simply connected and if $1 \geqslant K_M \geqslant 1 - \epsilon$, then M is diffeomorphic to a sphere.*

The difference between Theorems 2.4 and 2.2 is that the conclusion of Theorem 2.4 is one on the diffeomorphism type and is sharper. The constant $1 - \epsilon$ in Theorem 2.4

---

[2] In this article we mention only a part of many sphere theorems. The reader may find more in [139].

[3] More precisely, one of the complex or quaternionic projective space or the Cayley plane.

[4] Several results which relax the condition of Theorem 2.2 to $1 \geqslant K_M \geqslant 1/4 - \epsilon$ are known. See [3].

was $1 - \epsilon_n$ where ϵ_n is a positive number depending only on dimension n and was not explicit, at the time when it was first proved by Gromoll and Shikata in [65,136]. Later it was improved to a constant $1 - \epsilon$ which is independent of the dimension. It was further improved and an explicit bound ($1 - \epsilon = 0.87$) was found [143]. The explicit bound is improved several times.[5] The possibility that "$1 \geqslant K_M > 0.25$ and $\pi_1(M) = \{1\}$ implies that M is diffeomorphic to S^n" was not yet eliminated. The best constant is not yet found.

Remark 2.1. Hitchin [85] proved that there are some exotic spheres which do not admit a metric of positive scalar curvature, by using the *KO* index theorem of the Dirac operator. Gromoll–Myer [66] (and Grove–Ziller [83]) found examples of exotic spheres which have a metric of nonnegative curvature. So far no example of an exotic sphere which has a metric of (strictly) positive sectional curvature is found.

THEOREM 2.5 (Berger [18], Grove–Shiohama [82]). *If $K_M \geqslant 1/4$ and if the diameter of M is greater than π, then M is homeomorphic to a sphere.*

Berger proved that M is homotopy equivalent to a sphere under the assumption of Theorem 2.5 and Grove–Shiohama proved that M is homeomorphic to a sphere. By the generalized Poincaré conjecture (proved by Smale and Freedman) the latter follows from the former (in case dimension is not 3). But the proof by Grove–Shiohama (which is different from Berger's) uses Morse theory of functions which are not differentiable. This technique turns out to be very useful to study Riemannian manifold under lower (but not upper) curvature bounds. (See Section 14.)

The next theorem is a final form of a series of results due to Shiohama [137], Otsu–Shiohama–Yamaguchi [111], Perelman [114]. We will discuss it in Section 22.

THEOREM 2.6 (Cheeger–Colding [29]). *There exists $\epsilon_n > 0$ such that if M satisfies* $\mathrm{Ricci}_M \geqslant (n-1)$, $\mathrm{Vol}(M) \geqslant \mathrm{Vol}(S^n) - \epsilon_n$ *then M is diffeomorphic to a sphere.*

A sphere theorem is a characterization of a sphere, which is the most basic example of Riemannian manifold.

Let us recall the classification of surfaces (two-manifolds). It was first proved that a "simply connected compact 2-dimensional manifold is a sphere", then the classification in the general case was performed by simplifying a general surface by, say, surgery.

In a similar sense, sphere theorems play an important role in metric Riemannian geometry. Especially the techniques used to prove the sphere theorems we mentioned above play an important role to study more general Riemannian manifolds.

3. Finiteness theorems and Gromov–Hausdorff distance

Another type of important result in metric Riemannian geometry are finiteness theorems. First examples of that kind are proved by Cheeger and by Weinstein, which appeared at the beginning of the 1970's. Cheeger's finiteness theorem is as follows.

[5] The best estimate known at the time of writing this article is about $1 - \epsilon = 0.68$ [86,144].

THEOREM 3.1 (Cheeger [25]). *For all positive numbers D, v, n, the number of diffeomorphism classes of Riemannian manifolds M with* $\mathrm{Diam}(M) \leqslant D$, $\mathrm{Vol}(M) \geqslant v$, *and* $|K_M| \leqslant 1$ *is finite.*

The method of proof of Theorem 3.1 is closely related to the proofs of Rauch's sphere theorem and of Theorems 2.2, 2.4. We will explain it later.

Theorems 2.4 and 3.1 (and their proof) use an idea that if two Riemannian manifolds are "close" to each other then they are diffeomorphic to each other.

One way to formulate precisely what we mean by two Riemannian manifolds to be close, is by using the notion Gromov–Hausdorff distance.[6] Let us first review the definition of (usual or classical) Hausdorff distance. Let (X, d) be a metric space and Y_1, Y_2 be subspaces. We put for any subspace Y of X,

$$N_\epsilon Y = \{x \in X \mid d(x, Y) < \epsilon\},$$

where $d(x, Y) = \inf\{d(x, y) \mid y \in Y\}$.

DEFINITION 3.1. The *Hausdorff distance* $d_X(Y_1, Y_2)$ between Y_1 and Y_2 is the infimum of $\epsilon > 0$ such that $Y_2 \subset N_\epsilon Y_1$, $Y_1 \subset N_\epsilon Y_2$.

The Hausdorff distance defines a complete metric on the set of all compact subsets of a fixed complete metric space (X, d).

The Gromov–Hausdorff distance is an "absolute analogue" of the Hausdorff distance. Namely it defines a distance between two metric spaces (which we do not assume to be embedded somewhere a priori).

DEFINITION 3.2. The *Gromov–Hausdorff distance* $d_{GH}((X_1, d), (X_2, d))$ between two metric spaces (X_1, d) and (X_2, d) is an infimum of the Hausdorff distance $d_Z(X_1, X_2)$, where Z is a metric space such that X_1, X_2 are embedded to Z by isometries.

Hereafter we write $\lim^{GH}_{i\to\infty} X_i = X$ if $\lim_{i\to\infty} d_{GH}(X_i, X) = 0$.

Gromov–Hausdorff distance defines a complete metric on the set of all isometry classes of compact metric spaces.

The following version is sometimes convenient.

DEFINITION 3.3 [52]. A map $\varphi : X_1 \to X_2$ is called an *ϵ-Hausdorff approximation*, if $|d_{X_1}(\varphi(x), \varphi(y)) - d_{X_2}(x, y)| \leqslant \epsilon$ for all $x, y \in X_1$ and if the ϵ-neighborhood of the image $\varphi(X_1)$ is X_2.

If $d_{GH}(X_1, X_2) \leqslant \epsilon$ then there exists a 3ϵ-Hausdorff approximation $X_1 \to X_2$. If there exists an ϵ-Hausdorff approximation $X_1 \to X_2$ then $d_{GH}(X_1, X_2) \leqslant 3\epsilon$.

There are two types of important results on the Gromov–Hausdorff distance which are applied to finiteness theorems. In this section, we explain results which were developed mainly in the 1980's.

[6] See [69,75,57] for more detailed account on it.

We first state Gromov's precompactness theorem on manifolds with Ricci curvature bound. Let n, D be a positive integer and a positive number. We denote by $\mathfrak{S}_n(D)$ the set of all isometry classes of Riemannian manifolds M such that Ricci $\geqslant -(n-1)$ and diameter $\leqslant D$. Here and hereafter the diameter $\mathrm{Diam}(X)$ of a metric space (X, d) is the supremum of $d(x, y)$ where $x, y \in X$.

THEOREM 3.2 (Gromov [69]). *$(\mathfrak{S}_n(D), d_{GH})$ is relatively compact in the space of all isometry classes of compact metric spaces.*

The method of the proof of Theorem 3.2 is related to the proofs of Rauch's sphere theorem and of Theorem 2.2. We will explain it in Section 5.

We next mention a rigidity theorem. Gromov's precompactness theorem assumes bounds from below of the Ricci curvature, which is a rather weak assumption. We need the stronger assumption for the rigidity theorem. We first discuss the case that Gromov studied in [69]. For n, D, v, we denote by $\mathfrak{M}_n(D, v)$ the set of all isometry classes of n-dimensional Riemannian manifolds M such that $|K_M| \leqslant 1$, $\mathrm{Diam}(M) \leqslant D$, and $\mathrm{Vol}(M) \geqslant v$.

THEOREM 3.3 [69,93]. *There exists $\epsilon_n(D, v) > 0$ such that if $M_1, M_2 \in \mathfrak{M}_n(D, v)$ and if $d_{GH}(M_1, M_2) \leqslant \epsilon_n(D, v)$, then M_1 is diffeomorphic to M_2.*

Attempts to prove a similar conclusion as Theorem 3.3 under an assumption milder than $M_1, M_2 \in \mathfrak{M}_n(D, v)$, played a very important role in the development of metric Riemannian geometry. Perelman proved that M_1 is homeomorphic to M_2 if $d_{GH}(M_1, M_2) \leqslant \epsilon_n(D, v)$ under the assumption $K_M \geqslant -1$, which replaces $|K_M| \leqslant 1$ in the definition of $\mathfrak{M}_n(D, v)$. (Theorem 18.2.) Further study is done when we assume Ricci curvature bounds. (See Theorem 22.3.)

Theorem 3.1 follows from Theorems 3.2 and 3.3. (We leave its proof as an exercise to the reader.)

Theorem 3.2 asserts relative compactness. Namely it implies that, for any sequence M_i of elements of $\mathfrak{S}_n(D)$, there exists a converging subsequence. Its limit M_∞ may be regarded as a "weak solution" of various problems of metric Riemannian geometry (when we regard it as an analogy of functional analysis). Then it is natural and important to study the "regularity" of M_∞. It is closely related to the proof of Theorem 3.3. The next result is related to the "regularity" question.

THEOREM 3.4 [69,64,121]. *Each element of $\mathfrak{M}_n(D, v)$ is a Riemannian manifold of $C^{1,\alpha}$-class.*[7]

Here α is any positive number with $\alpha < 1$ and a Riemannian manifold of $C^{1,\alpha}$-class is a manifold with metric tensor g whose first derivative is C^α-Hölder continuous.

The assumption of Theorem 3.4 is rather strong. There are two kinds of study to relax this condition $X \in \mathfrak{M}_n(D, v)$.

[7] The proof of this theorem is completed in [64,121] based on the idea of Gromov [69]. There seems to be various independent research in Russia. (See, for example, [107,108,17].)

One is to remove the assumption $\mathrm{Vol}(M) \geqslant v$. It means that we study the limit of a sequence of Riemannian manifolds which will become degenerate. This is called the study of collapsing Riemannian manifolds. We discuss it in Sections 10–13. (See also [57].)

The other direction is to relax the assumption $|K_M| \leqslant 1$. Theorem 3.1 is generalized as follows towards this direction.

For n, D, v, we denote by $\mathfrak{M}'_n(D, v)$ the set of all isometry classes of n-dimensional Riemannian manifolds M such that $K_M \geqslant -1$, $\mathrm{Diam}(M) \leqslant D$, $\mathrm{Vol}(M) \geqslant v$.

THEOREM 3.5 (Grove–Petersen–Wu [78,81]). *For each n, D, v, the number of homeomorphism classes of elements of $\mathfrak{M}'_n(D, v)$ is finite.*[8]

We explain the proof of Theorem 3.5 in Section 15.

The limit of a sequence of manifolds M satisfying $K_M \geqslant -1$ is an Alexandrov space. We will discuss it in Sections 17 and 18.

Remark 3.1. (1) If M_i is a sequence of Riemannian manifolds such that $N = \lim^{GH}_{i\to\infty} M_i$ and N is a Riemannian manifold. Then $K_{M_i} \geqslant \kappa$ implies $K_N \geqslant \kappa$. Moreover we have $\dim N \leqslant \dim M_i$.

(2) On the other hand, in the case when $\Lambda \geqslant K_{M_i} \geqslant \kappa$, $\Lambda \geqslant K_N$ is, in general, false for $\lim^{GH}_{i\to\infty} M_i = N$. A counterexample can be constructed as follows. Let Rot_θ be the rotation by angle θ of $S^2 = \{(x, y, z) \in \mathbb{R}^3 \mid x^2 + y^2 + z^2 = 1\}$ around the z axis. We consider the quotient of $S^2 \times \mathbb{R}$ by the $\mathbb{Z}$ action generated by $(p, t) \to (\mathrm{Rot}_{\alpha\epsilon}(p), t + \epsilon)$. Let $M_{\epsilon,\alpha}$ be the quotient space with quotient metric. ($M_{\epsilon,\alpha}$ is diffeomorphic to $S^2 \times S^1$.) We have $1 \geqslant K_{M_{\epsilon,\alpha}} \geqslant 0$ since $M_{\epsilon,\alpha}$ is locally isometric to $S^2 \times \mathbb{R}$. The limit of $M_{\epsilon,\alpha}$ as $\epsilon \to 0$ is S^2 with some Riemannian metric g_α. $1 \geqslant (S^2, g_\alpha)$ does not hold unless $\alpha = 0$.

4. Geodesic coordinate, injectivity radius, comparison theorems and sphere theorem

The following theorem in differential topology is used in the proof of Theorem 2.2.

THEOREM 4.1. *If a compact n-dimensional manifold M is a union of two open sets both of which are diffeomorphic to $\mathbb{R}^n$, then M is homeomorphic to a sphere.*

In order to apply Theorem 4.1 to the proofs of sphere theorems, we want to cover M by two coordinate neighborhoods. To estimate the size of the coordinate charts plays an important role in the study of other problems. Let us begin with the following

PROPOSITION 4.2. *For each compact Riemannian manifold M, there exists a positive number ϵ_M with the following properties. If the distance between $p, q \in M$ is smaller than ϵ_M, then there exists a unique geodesic of length $< \epsilon_M$ joining p, q.*

[8] In case the dimension is 3, [78,81] proved only finiteness of homotopy type. Now, Perelman's stability theorem (Theorem 18.2) implies the finiteness of homeomorphism classes in general.

The proof of Proposition 4.2 is given in many standard text books of Riemannian geometry. (For example, in [97,33].)

The uniqueness of such geodesic is essential for our purpose. Let us explain this point. Let M be a complete Riemannian manifold. For each $p \in M$ we define the exponential map, $\mathrm{Exp}_p : T_pM \to M$ as follows. Let $V \in T_p(M)$. There exists a geodesic $\ell : \mathbb{R} \to M$, such that $\frac{d\ell}{dt}(0) = V$. We then put $\ell(1) = \mathrm{Exp}_p V$.

Proposition 4.2 implies that $\mathrm{Exp}_p : T_pM \to M$ is a diffeomorphism on the ball of radius ϵ_M.

DEFINITION 4.1. The *injectivity radius* of a Riemannian manifold M is a function $i_M : M \to \mathbb{R}$ which associates to $p \in M$ the positive number:

$$i_M(p) = \sup\{\epsilon \mid \mathrm{Exp}_p : T_pM \to M \text{ is injective on } \{V \in T_pM \mid \|V\| < \epsilon\}\}.$$

Proposition 4.2 implies $i_M \geqslant \epsilon_M$ for a compact Riemannian manifold M. (It is easy to see that i_M is continuous. Hence $i_M \geqslant \epsilon_M > 0$ follows easily from the implicit function theorem. Proposition 4.2 is a bit more involved.)

If $R < i_M(p)$, then the restriction of the exponential map $\mathrm{Exp}_p : T_pM \to M$ to the metric ball of radius R centered at origin, defines a coordinate of a neighborhood of p. We call it the *geodesic coordinate*.

To prove Theorem 2.2, it is important to estimate the injectivity radius i_M from below. The next result[9] provides such an estimate.

THEOREM 4.3. *Suppose that* $\dim M$ *is even. If* $K_M > 0$, *then* $i_M \geqslant \pi$ *and* M *is simply connected.*[10]

Suppose $\dim M$ *is odd. If* $1 \geqslant K_M \geqslant 1/4$ *and if* M *is simply connected then,* $i_M \geqslant \pi$.

In particular, if M *satisfies the assumption of Theorem* 2.2, *then we have* $i_M \geqslant \pi$.

(There are several results in the nonsimply connected case. We omit it.) Another result we use is the following

PROPOSITION 4.4 (Berger). *Let us assume that* $K_M \geqslant 1/4$ *and* $\mathrm{Diam}(M) \geqslant \pi$. *We take* $p, q \in M$ *such that* $d(p,q) = \mathrm{Diam}(M)$. *Then we have*

$$\mathrm{Int}\, B_p(\pi, M) \cup \mathrm{Int}\, B_q(\pi, M) = M.$$

(*Here* Int *denotes the interior.*)

The proof is in Section 14.

Using Theorem 4.3 and Proposition 4.4, the proof of Theorem 2.2 goes roughly as follows. By Theorem 4.3, the injectivity radius of M is not smaller than π. Especially the diameter of M is not smaller than π.

---

[9] This theorem is due to [18] in even dimension, and to [95,37] in odd dimension.

[10] The second assertion is a classical result due to Synge.

Let us first assume $1 \geqslant K_M > 1/4$. We replace the metric g_M of M by $(1+\delta)g_M$, where δ is a positive number sufficiently close to 0. The assumption $1 \geqslant K_M > 1/4$ is still satisfied. Hence M satisfies the assumption of Proposition 4.4. Hence $\operatorname{Int} B_p(\pi, M) \cup \operatorname{Int} B_q(\pi, M) = M$. Moreover $\operatorname{Int} B_p(\pi, M)$ and $\operatorname{Int} B_q(\pi, M)$ are diffeomorphic to the ball by Theorem 4.3. Therefore, by Theorem 4.1, M is homeomorphic to a sphere.

We next consider the case when $1 \geqslant K_M \geqslant 1/4$. If the diameter of M is strictly greater than π, then again Proposition 4.4 and Theorems 4.1 and 4.3 imply that M is homeomorphic to a sphere.

Finally we consider the case when the diameter of M is π. In this case, we consider the restriction of the exponential map $\operatorname{Exp}_p : T_pM \to M$ to the ball $D^n(\pi)$ of radius π. Then it is a diffeomorphism at the interior. So M is obtained from $D^n(\pi)$ by identifying boundary points only. We examine this situation carefully and conclude that M is a symmetric space of compact type. We omit the details. (See, for example, [33, Chapter 7].) □

We explain the outline of the proof of Theorem 4.3 later in this section. We first explain some basic facts. Let us begin with the following theorem. Let κ be a constant. We put

$$s_\kappa(t) = \begin{cases} \frac{\sin t\sqrt{\kappa}}{\sqrt{\kappa}}, & \kappa > 0, \\ t, & \kappa = 0, \\ \frac{\sinh t\sqrt{-\kappa}}{\sqrt{-\kappa}}, & \kappa < 0. \end{cases} \tag{4.1}$$

THEOREM 4.5 (Rauch). *If $K_M \leqslant \kappa$, then the derivative $d_x \operatorname{Exp}_p$ of the exponential map Exp_p satisfies*

$$\left\| d_x \operatorname{Exp}_p(V) \right\| \geqslant \|V\| s_\kappa(r).$$

Here $x \in T_p(M)$, $\|x\| = r$, $V \in T_x T_p(M) \cong T_p(M)$ and we assume $r \leqslant \pi/\sqrt{\kappa}$ in case $\kappa > 0$.

Let $K_M \geqslant \kappa$. In case $\kappa > 0$, we assume $d_{tx} \operatorname{Exp}_p$ is invertible for $t \in [0, 1]$. Then we have

$$\left\| d_x \operatorname{Exp}_p(V) \right\| \leqslant \|V\| s_\kappa(r).$$

Theorem 4.5 implies that if $K_M \leqslant 1$, then the restriction of $\operatorname{Exp}_p : T_pM \to M$ to the ball of radius π is an immersion. (Namely its Jacobi matrix is invertible.)

We remark that the equality in Theorem 4.5 holds in the case when M is of constant curvature κ.

Theorem 4.5 is used by Rauch to prove his sphere theorem. We use the Jacobi field in the proof of Theorem 4.5 as follows. Let x, V be as in Theorem 4.5, and define a geodesic ℓ_s by

$$\ell_s(t) = \operatorname{Exp}_p\big(t(x + sV)\big).$$

For each s, ℓ_s is a geodesic. Its derivative

$$J(t) = \left.\frac{\partial \ell_s(t)}{\partial s}\right|_{s=0} \in T_{\ell_0(t)}M$$

with respect to s, by definition, is a Jacobi field. Note that $d_x \operatorname{Exp}_p(V) = J(1)$. Therefore, to prove Theorem 4.5, it suffices to estimate the Jacobi field. We use the following equation (which the Jacobi field satisfies) for this purpose.

$$\frac{D^2}{dt^2} J(t) + R\left(\frac{d\ell_0}{dt}(t), J(t)\right)\frac{d\ell_0}{dt}(t) = 0. \tag{4.2}$$

Here $\frac{D}{dt}$ is a covariant derivative with respect to the tangent vector $\frac{d\ell_0}{dt}(t)$ and R is a curvature tensor.

If e_1, e_2 is an orthonormal frame of a plane π in the tangent space, then $g(R(e_1, e_2)e_2, e_1)$ is the sectional curvature of the plane π. (Here g is the metric tensor.) Therefore, the second term of Eq. (4.2) can be written in terms of the sectional curvature. Using it we can compare Eq. (4.2) to the one in case our manifold is of constant curvature. Namely if $K_M \equiv \kappa$ then (4.2) will be

$$\frac{D^2}{dt^2} J(t) + \kappa J(t) = 0. \tag{4.3}$$

Its solution is $J(t) = s_\kappa(t)V(t)$ where $\nabla_{\dot{\ell}(t)} V = 0$. Namely $\|J(t)\| = s_\kappa(t)$ if $K_M \equiv \kappa$. This implies Theorem 4.5. □

Theorem 4.5 implies the following

THEOREM 4.6 (Hadamard–Cartan). *If a complete Riemannian manifold M satisfies $K_M \leqslant 0$, then $\operatorname{Exp}_p : T_pM \to M$ is a covering map. In particular the universal covering space of M is diffeomorphic to $\mathbb{R}^n$.*

In fact, Theorem 4.5 implies that the Jacobi matrix of $\operatorname{Exp}_p : T_pM \to M$ is of maximal rank everywhere. To prove that it is a covering map we need a bit more. We use completeness of metric for this last step. We omit it. □

By integrating the conclusion of Theorem 4.5, we can compare the distance between two points $\operatorname{Exp}_p(x)$, $\operatorname{Exp}_p(y)$ (which are close to p) to the corresponding distance in the space with constant curvature. Actually we can do it more globally and obtain the Toponogov comparison theorem.

To state it we need some notation. Let $\mathbb{S}^n(\kappa)$ be the complete simply connected Riemannian manifold with constant curvature κ. Let $x', y', z' \in \mathbb{S}^n(\kappa)$. We denote by $\overline{x'y'}$, etc. the minimal geodesic joining x' and y', etc. Let $\theta = \angle y'x'z'$ be the angle between $\overline{x'y'}$ and

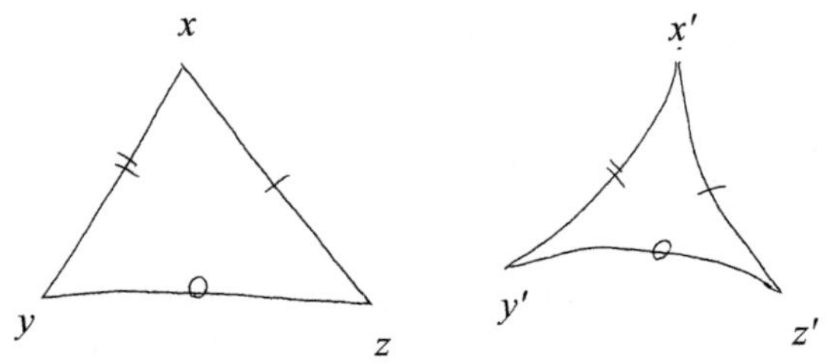

Fig. 4.1.

$\overline{x'z'}$ at x'. We put $a = d(x', y')$, $b = d(x', z')$. It is easy to see that $d(y', z')$ depends only on a, b, θ, κ. We define

$$s(a, b, \theta, \kappa) = d(y', z'). \tag{4.4}$$

We remark that in case $\kappa > 0$, the number $s(a, b, \theta, \kappa)$ is defined only for $a, b < \pi/\sqrt{\kappa}$.

Let M be a Riemannian manifold and $x, y, z \in M$. We denote by $\overline{xy}$ a minimal geodesic joining x and y. (In case there are several minimal geodesics we take any one of them.) Let $\angle yxz$ be the angle between $\overline{xy}$ and $\overline{xz}$ at x (see Figure 4.1).

THEOREM 4.7 (Alexandrov–Toponogov). *If $K_M \geqslant \kappa$ then we have*

$$d(y, z) \leqslant s\bigl(d(x, y), d(x, z), \angle yxz, \kappa\bigr).$$

If $K_M \leqslant \kappa$ and if $d(x, y), d(x, z) \leqslant i_M(x)$ then

$$d(y, z) \geqslant s\bigl(d(x, y), d(x, z), \angle yxz, \kappa\bigr).$$

We remark that in the first inequality we do not need to assume that the triangle x, y, z is small. Actually we only need to assume one of the geodesics joining x to y and to z are minimal and the other may be any geodesic of length $\leqslant \pi/\sqrt{\kappa}$. Theorem 4.7 is proved in many text books (see, for example, [33]).

As we already mentioned, Theorem 4.5 implies that, if $K_M \leqslant 1$, then the exponential map is an immersion on the metric ball of radius π. Especially it is locally an injection there. To prove Theorem 4.1 we need global injectivity. We here introduce several terminology.

DEFINITION 4.2. $q \in M$ is said to be a *conjugate point* of $p \in M$ if there exists x such that $q = \mathrm{Exp}_p(x)$ and that $d_x \mathrm{Exp}_p$ is not of maximal rank.

q is said to be a *cut point* of $p \in M$ if there exists $x \neq y \in T_p(M)$ such that $\mathrm{Exp}_p x = \mathrm{Exp}_p y = q$.

EXAMPLE 4.1. We consider the sphere S^2 of constant curvature 1. Every geodesic which starts at the north pole *np* meets again at the south pole *sp*. Hence the south pole is a conjugate point of the north pole.

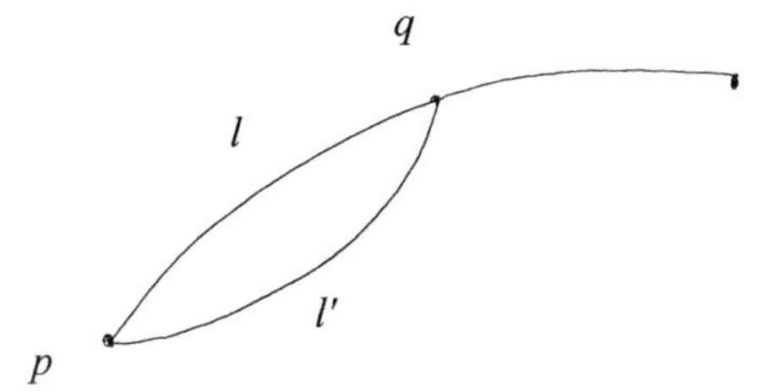

Fig. 4.2.

We next divide S^2 by the involution and obtain the real projective space $\mathbb{R}P^2$. Then np and sp determine the same point $x = [np] = [sp] \in \mathbb{R}P^2$. If $c \in S^2$ is on the equator then there are minimal geodesics ℓ_1, ℓ_2 joining c to np, sp, respectively. ℓ_1, ℓ_2 induce two minimal geodesics $\bar{\ell}_1, \bar{\ell}_2$ in $\mathbb{R}P^2$ joining x to $y = [c]$. Thus y is a cut point of x.

Note that $i_M(p) > r$ holds if there exists neither a cut point nor a conjugate point q of p such that $d(p,q) \leqslant r$. We can use Theorem 4.5 to estimate the distance to the conjugate point. However the problem to estimate the distance to the cut point is a more global one.

We remark the following fact.

LEMMA 4.8. *If $\ell:[a,b] \to M$ is the minimal geodesic, then for $t \in (a,b)$, $q = \ell(t)$ is neither a cut point nor a conjugate point of $p = \ell(a)$.*

In fact if q is a cut point then there is a geodesic ℓ' joining p to q with $|\ell'| = |\ell_{[a,t]}|$ (see Figure 4.2). Then the union $\ell' \cup \ell|_{[t,b]}$ of two geodesics is not smooth and has the same length as the minimal geodesic ℓ. This is a contradiction. If q is a conjugate point then by the Morse index theorem (see [103,97,33]), $\ell_{[a,t+\epsilon]}$ is not minimal. This contradicts the assumption.

Here we state the following basic result about cut points. (See, for example, [33, p. 96] for its proof.)

THEOREM 4.9 (Klingenberg). *Let M be a Riemannian manifold. We assume that q is not a conjugate point of p, for each $p, q \in M$ with $d(p,q) < r$. If there exists $p \in M$ with $i_M(p) < r$ then there exists a closed geodesic of length $< 2r$ in M.*

In view of Theorems 4.5 and 4.9, to prove Theorem 4.3, it suffices to show that the length of a nontrivial closed geodesic of M is greater than 2π. We explain the brief outline of its proof below. (See [33, p. 100] for its details.)

We first consider the case $\dim M$ is even. Let M be a simply connected Riemannian manifold with $1 \geqslant K_M > 0$. Let $\ell : S^1 \to M$ be a nontrivial geodesic of minimal length. We regard $S^1 \cong \mathbb{R}/\mathbb{Z}$. Put $p = \ell(0)$. By the parallel transport along ℓ we have a holonomy homomorphism $\mathrm{hol}_\ell : T_pM \to T_pM$. The tangent vector $\frac{d\ell}{dt}(0)$ is an invariant of the holonomy. Since hol_ℓ is an orthogonal transformation, and $\dim M$ is even, it follows that there exists a nonzero vector $V \in T_pM$ orthogonal to $\frac{d\ell}{dt}(0)$ such that $\mathrm{hol}_\ell(V) = V$. The parallel transport of V defines a vector field $V(t) \in T_{\ell(t)}M$, which is a parallel vector field. We put

$$\ell_s(t) = \mathrm{Exp}_{\ell(t)}\big(sV(t)\big).$$

Using $\nabla V(t) = 0$ and first variation formula (see, for example, [33, Section 1], [97, Vol. II, Theorem 5.1], [103, Theorem 12.2]), we find that $\frac{d\ell_s}{ds}(0) = 0$. Using moreover the second variation formula (see, for example, [33, Section 6], [97, Vol. II, Theorem 5.4], [103, Theorem 13.1]) and the positivity of curvature, we find $\frac{d^2\ell_s}{ds^2}(0) < 0$, which contradicts to the minimality of the length of ℓ.

The proof of the odd-dimensional case is more involved. We remark that the quotient of S^3 by a cyclic group $\mathbb{Z}/p\mathbb{Z}$ has constant positive curvature one (and is not simply connected). Its injectivity radius converges to 0 as $p \to \infty$. This shows that, to prove Theorem 4.3 in odd-dimensional case, we need to use the assumption that M is simply connected.

The proof of the odd-dimensional case is roughly as follows. We assume that there exists a closed geodesic ℓ of length $< 2\pi$. Since M is simply connected, ℓ is null homotopic. Let ℓ_s be a homotopy such that $\ell_0 = \ell$, $\ell_1 = \text{const}$. We may assume that the length of ℓ is minimal among all nontrivial closed geodesics. By using the assumption that $K_M > 1/4$ we can prove that the length of ℓ_s is always smaller than 2π. (This is the essential point of the proof. To prove this we use the fact that the Morse index (with respect to the length) of the closed geodesic of length $> 2\pi$ is not smaller than 2.[11])

Now we consider the exponential map Exp_p at the tangent space of $p = \ell(0)$. Exp_p is a submersion on the ball of radius π. Hence it has a similar property to the covering map up to radius π. Especially it has the homotopy lifting property there. Since the length of ℓ_s is not greater than 2π, its image is of distance $\leqslant \pi$ from p. Therefore we can lift ℓ_s to T_pM. (Note we can lift ℓ_1 since it is a constant map.) Hence we obtain a lift $\tilde{\ell}_0 : S^1 \to T_pM$. But this is a contradiction since $\ell_0 = \ell$ is a geodesic.[12] □

5. Packing and precompactness theorem

A similar argument as in the last section is used in the proof of the finiteness theorem (Theorem 3.1) and of Theorem 3.2. We explain this point here. We first discuss Theorem 3.2. The basic fact we use for its proof is the following

[11] Let us consider the round sphere of radius 2 (that is the round sphere of curvature $1/4$). The geodesic segment of length 2π, that is the geodesic segment joining north pole with south pole, has Morse index $n-1$. (Here we consider the set of all arcs joining north pole with south pole and consider the length as a Morse function on it. $n-1$ is the Morse index with respect to this Morse function.) We compare our closed geodesic with this geodesic segment to obtain the conclusion about the Morse index.

[12] This argument is not enough to handle the case $1 \geqslant K_M \geqslant 1/4$ of Theorem 2.2 (since then we can only show that π is a submersion at the *interior* of the ball of radius π). In that case we need an additional argument. We omit it.

PROPOSITION 5.1. *Let $D > 0$ and $N:(0,1) \to \mathbb{N}$. We denote by $\mathfrak{Met}(D, N)$, the set of all isometry classes of complete metric spaces satisfying* (1), (2) *below. Then* $\mathfrak{Met}(D, N)$ *is compact with respect to the Gromov–Hausdorff distance.*

(1) *The diameter of $M \leqslant D$.*
(2) *For each $\epsilon \in (0,1)$ there exists a finite subset Z of M with the following properties*:
  (2.a) $\sharp Z \leqslant N(\epsilon)$.
  (2.b) *For each $x \in M$, there exists $x_0 \in Z$ satisfying $d(x, x_0) < \epsilon$.*

The proof of Proposition 5.1 is given, for example, in [57, Section 2].

Here we introduce a notation.

DEFINITION 5.1. We call the subset Z an *ϵ-net* if it satisfies (2.b).

To deduce Theorem 3.2 from Proposition 5.1, we use the following Theorem 5.2. Let $\mathbb{S}^n(\kappa)$ be the complete simply connected Riemannian manifold with constant curvature κ. Let $B_p(R, M)$ be the metric ball in M of radius R centered at p.

THEOREM 5.2 (Bishop–Gromov). *If* Ricci $\geqslant (n-1)\kappa$ *then the volume* $\mathrm{Vol}(B_p(R, M))$ *of the metric ball satisfies the following inequality for $r < R$*:

$$\frac{\mathrm{Vol}(B_p(R, M))}{\mathrm{Vol}(B_p(r, M))} \leqslant \frac{\mathrm{Vol}(B_{p_0}(R, \mathbb{S}^n(\kappa)))}{\mathrm{Vol}(B_{p_0}(r, \mathbb{S}^n(\kappa)))}. \tag{5.1}$$

(5.1) is called the Bishop–Gromov inequality. It plays a key role to study the class of Riemannian manifolds with Ricci curvature bounded from below. The equality holds if M is of constant curvature κ.

Let us sketch a proof of Theorem 5.2. We put

$$A(t) = \frac{\mathrm{Vol}(B_p(t, M))}{\mathrm{Vol}(B_{p_0}(t, \mathbb{S}^n(\kappa)))}. \tag{5.2}$$

It suffices to show that A is nonincreasing. (In case $\kappa > 0$, Theorem 5.4 implies that we need to consider $t \leqslant \pi$ only.)

Let $\ell:[0,a) \to M$ be a minimal geodesic with $\ell(0) = p$ parameterized by arc length. Let $v = (d\ell/dt)(0)$.

We take a vector $v_* \in T_{p_0}\mathbb{S}^n(\kappa)$ with unit length. We put

$$a(v, t) = \frac{\det d_{tv} \mathrm{Exp}_p}{\det d_{tv_*} \mathrm{Exp}_{p_*}}. \tag{5.3}$$

Here $\det d_{tv} \mathrm{Exp}_p$ is the determinant of the derivative of the exponential map. We first prove that $a(v, t)$ is a nonincreasing function of t for each fixed v.

We can prove it in a way similar as the proof of Theorem 4.5. One difference however is that our assumption in Theorem 5.2 is only on the Ricci curvature while in Theorem 4.5 the assumption is on the sectional curvature. However since we only need to estimate the determinant of the Jacobi matrix of the exponential map, the assumption on the Ricci curvature, which is the trace of the curvature tensor, is enough. This is half of the idea of the proof of Proposition 5.2. Let us fix p and move $q \in M$, and consider the set

$$V = \left\{ \frac{d\ell_{p,q}}{dt}(0) \in T_pM \,\middle|\, q \in M \right\}, \tag{5.4}$$

where $\ell_{p,q}$ is the minimal geodesics joining p and q. (If there are several we take all of them.) (We take a parametrization of $\ell_{p,q}$ so that the length of $\frac{d\ell_{p,q}}{dt}(0)$ is $d(p,q)$.)

We have

$$\operatorname{Vol}\bigl(B_p(R,M)\bigr) = \int_{V \cap B_0(R,R_pM)} \| \det d_x \operatorname{Exp}_p \| \, dx. \tag{5.5}$$

(Here $\det d_x \operatorname{Exp}$ is the determinant of Jacobi matrix.) (5.5) and the fact that $a(v,t)$ is a nonincreasing function of t implies (5.1) for R, r smaller than the injectivity radius.

To prove Theorem 5.2 beyond injectivity radius, we proceed as follows. We remark that V is star shaped (that is if $x \in V$ $t \in [0,1]$, then $tx \in V$). We then modify a to a' so that $a'(t,v) = a(t,v)$ if $tv \in V$ and $a'(t,v) = 0$ if $tv \notin V$. Then a' is a nonincreasing function of t. Theorem 5.2 follows. □

COROLLARY 5.3. *If* $\operatorname{Ricci}_M \geqslant \kappa$ *and* $p \in M$ *then*

$$\operatorname{Vol}\bigl(B_p(R,M)\bigr) \leqslant \operatorname{Vol}\bigl(B_{p_0}\bigl(R, \mathbb{S}^n(\kappa)\bigr)\bigr).$$

This corollary follows from the fact that the function A in (5.2) is nonincreasing and $\lim_{t\to 0} A(t) = 1$.

Theorem 5.2 and Proposition 5.1 imply Theorem 3.2 as follows. Let us assume that M satisfies the assumption of Theorem 3.2. It suffices to show that M satisfies the assumption of Proposition 5.1. Let $\epsilon > 0$. We take $Z \subset M$ which is maximal (with respect to inclusion) among the subsets of M satisfying "$z_1, z_2 \in Z$, $z_1 \neq z_2$, implies $d(z_1,z_2) > \epsilon$". The maximality implies (2.6). On the other hand, since $B_z(\epsilon/2, M)$, $z \in Z$, are disjoint to each other, it follows that

$$\sum_{z \in Z} \operatorname{Vol}\bigl(B_z(\epsilon/2, M)\bigr) < \operatorname{Vol} M.$$

Since $B_z(D,M) = M$, Proposition 5.1 implies

$$\sharp Z \leqslant \frac{\operatorname{Vol}(M)}{\sup \operatorname{Vol}(B_p(\epsilon/2, M))} \leqslant \frac{\operatorname{Vol}(B_{p_0}(D, \mathbb{S}^n(\kappa)))}{\operatorname{Vol}(B_{p_0}(\epsilon/2, \mathbb{S}^n(\kappa)))}.$$

If we let $N(\epsilon)$ be the right-hand side, then the assumption of Proposition 5.1 is satisfied. Theorem 3.2 follows. □

We remark that the following classical result is actually proved during the proof of Theorem 5.2.

THEOREM 5.4 (Myers). *If M is an n-dimensional complete Riemannian manifold with* Ricci $\geqslant (n-1)\kappa > 0$, *then M is compact and its diameter is not greater than $\pi/\sqrt{\kappa}$.*

In fact during the proof of Theorem 5.2 we proved the following under the assumption $p \in M$, $\mathrm{Ricci}_M \geqslant \kappa$.

"If $t \mapsto \mathrm{Exp}_p(tv)$ is a minimal geodesic for $t \in [0,1]$, then $\det d_v \mathrm{Exp}_p$ is not greater than $\det d_{v_0} \mathrm{Exp}_{p_0}$, where $p_0 \in \mathbb{S}^n(\kappa)$, $v_0 \in T_{p_0}\mathbb{S}^n(\kappa)$ and $|v_0| = |v|$."

We remark that $\det d_{v_0} \mathrm{Exp}_{p_0} = 0$ if $\|v_0\| = \pi/\sqrt{\kappa}$. Therefore there exists no minimal geodesic of length $> \pi/\sqrt{\kappa}$ if $\mathrm{Ricci}_M \geqslant \kappa$. Theorem 5.4 follows immediately. □

In the above argument, $B_z(\epsilon, M)$, $z \in Z$, covers M. Namely we estimate the number of metric balls (geodesic coordinate) to show Theorem 3.2. If ϵ is smaller than the injectivity radius of M, then $B_z(\epsilon, M)$ is diffeomorphic to D^n. The proof of Theorem 3.2 is related to the proof of sphere theorems in this way. Theorem 4.1 deals with the case when two balls cover M and conclude that M is a sphere. If we can replace Theorem 4.1 by a statement such as "if M is covered by the balls whose number is estimated by C, then the number of diffeomorphism classes of such M is estimated by C" then finiteness theorems would follow. Unfortunately the statement in the parenthesis above does not hold. So we need to include information how the balls are glued. Theorem 3.1 can be proved in that way. (See Sections 6–8.) Here we prove a weaker version (Weinstein [150]).

PROPOSITION 5.5. *For each D, ϵ the number of homotopy equivalence classes of n-dimensional Riemannian manifolds satisfying* (1), (2) *below is finite.*

(1) $M \in \mathfrak{M}_n(D)$,

(2) *The injectivity radius of M is greater than ϵ.*

To prove Proposition 5.5 we use the set Z above. We then obtain an open covering $B_z(\epsilon, M)$, $z \in Z$, of M. It is a simple covering. Namely for each $z_1, \dots, z_k \in Z$ the intersection $\bigcap_{i=1}^k B_{z_i}(\epsilon, M)$ is either empty or contractible. It implies that the simplicial complex $K(Z)$ defined below is homotopy equivalent to M.

(1) A vertex of $K(Z)$ corresponds to an element of Z.

(2) $z_0, \dots, z_k \in Z$ is the set of vertices of a k simplex of $K(Z)$ if and only if $\bigcap_{i=0}^k B_{z_i}(\epsilon, M) \neq \emptyset$.

Since the order of Z is estimated by a number depending only on D and ϵ, it follows that there exists only finitely many possibilities for the homotopy type of $K(Z)$. Proposition 5.5 follows. □

In Theorem 3.1, there is no assumption on the injectivity radius but only a bound of volume from below is assumed. An assumption on the volume is more natural and geometric than one on the injectivity radius. However, in case the absolute value of the sectional curvature is bounded, these two assumptions are equivalent.

PROPOSITION 5.6 (Cheeger [25]). *There exists a positive number $c(n, D, v)$ depending only on n, D, v such that if $M \in \mathfrak{M}_n(D, v)$, then $i_M \geqslant c(n, D, v)$.*

The proof of Proposition 5.6 is closely related to the proof of Theorem 3.5. We will explain it in Section 15.

6. Construction of homeomorphism by isotopy theory

In Section 5, we discussed an estimate of the number of open sets which cover M and which are diffeomorphic to D^n, and we showed how it is used to estimate the number of homotopy types (Proposition 5.5). However as we mentioned there, we need more arguments to estimate the number of diffeomorphism classes (or homeomorphism classes). We will explain some of them in the four sections beginning from this section.

We again begin with a sphere theorem, the differentiable sphere theorem (Theorem 2.4) this time.

Let M satisfy the assumptions of Theorem 2.4. Namely we assume that M is simply connected and $1 \geqslant K_M \geqslant 1 - \epsilon$. Then by Proposition 4.4 and Theorem 4.3, M is a union of two balls V_1, V_2 such that $V_i \cong D^n$. We may assume $\partial V_i \cong S^{n-1}$. Moreover we may assume $V_1 \cap V_2 = \partial V_1 = \partial V_2$. So we obtain a diffeomorphism

$$I : S^{n-1} \cong \partial V_1 \to \partial V_2 \cong S^{n-1}. \tag{6.1}$$

It is easy to see that if I is diffeotopic to the identity map (namely if there exists a smooth family I_t of diffeomorphisms such that $I_0 = I$, $I_1 = \mathrm{id}$), then $M = V_1 \cup V_2$ is *diffeomorphic* to S^n.

Now we use the following

PROPOSITION 6.1. *For each compact Riemannian manifold N there exists $\epsilon_N > 0$ such that if the C^1-distance between $F : N \to N$ and the identity is smaller than ϵ_N, then F is diffeotopic to the identity.*

Here we recall

DEFINITION 6.1. Two diffeomorphisms $F_1, F_2 : N \to N'$ are said to be *diffeotopic* to each other if there exists a smooth map $F : [1,2] \times N \to N'$ such that $F(1,x) = F_1(x)$, $F(2,x) = F_2(x)$ and that $x \mapsto F(t,x)$ is a diffeomorphism for each t.

The proof is elementary. To apply Proposition 6.1 to the proof of Theorem 2.5, we use the following lemma.

LEMMA 6.2. *For each $\epsilon > 0$ there exists $\delta_n(\epsilon) > 0$ with the following properties. Let M be an n-dimensional simply connected Riemannian manifold with $1 > K > 1 - \delta_n(\epsilon)$. Then we may choose the gluing map (6.1) so that its C^1-distance from the identity is smaller than ϵ.*

We omit the proof. See, for example, [33, Chapter 7].

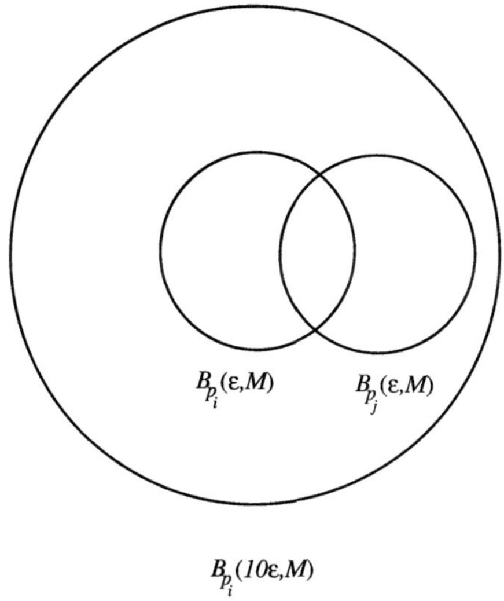

Fig. 6.1.

We are going to explain how we use the idea above for the proof of Theorems 3.3 and 3.1. Cheeger's original proof of Theorem 3.1 [25] is similar to the idea explained in this section.

Let M, N be Riemannian manifolds. We assume that they are covered by the same number of metric balls. Namely we assume $M = \bigcup_{i=1}^{k} B_{p_i}(\epsilon, M)$, $N = \bigcup_{i=1}^{k} B_{q_i}(\epsilon, N)$. We assume also that 10ϵ is smaller than the injectivity radius of M and of N. (We put 10 for a technical reason.) We assume also that intersection pattern of the balls are the same. Namely, for each i, j, $B_{p_i}(\epsilon, M) \cap B_{p_j}(\epsilon, M) \neq \emptyset$ if and only if $B_{q_i}(\epsilon, N) \cap B_{q_j}(\epsilon, N) \neq \emptyset$.

We want to find a sufficient condition for M to be diffeomorphic to N. For this purpose we compare the chart $\bigcup_{i=1}^{k} B_{p_i}(\epsilon, M)$ of M, with the chart $N = \bigcup_{i=1}^{k} B_{q_i}(\epsilon, N)$ of N. To compare, we want to take the same domain for coordinate transformations. For this purpose we proceed as follows. Let $B_{p_i}(\epsilon, M) \cap B_{p_j}(\epsilon, M) \neq \emptyset$ then $B_{p_i}(\epsilon, M) \subset B_{p_j}(10\epsilon, M)$ (see Figure 6.1). For each p_i, q_j, we fix a linear isometry $T_{p_i}M \cong \mathbb{R}^n$, $T_{q_j}N \cong \mathbb{R}^n$ and use it to identify tangent spaces with $\mathbb{R}^n$. (There are various choices of identification. We take one and fix it.)

We consider the composition

$$\varphi_{ji}^{M} = \mathrm{Exp}_{p_j}^{-1} \circ \mathrm{Exp}_{p_i} : B^n(\epsilon) \to B^n(10\epsilon).$$

Here $B^n(\epsilon)$ is a metric ball of radius ϵ in $\mathbb{R}^n$ centered at origin, and $\mathrm{Exp}_{p_j}^{-1}$ is an inverse of the exponential map $\mathrm{Exp}_{p_j} : B^n(10\epsilon) \to N$. We define φ_{ji}^{N} in a similar way.

In the next proposition we *assume* that the C^2 norm (or $C^{1,\alpha}$ norm) of φ_{ji}^{M}, φ_{ji}^{N} is smaller than a constant C.

PROPOSITION 6.3. *There exists $\epsilon_{n,k}(C) > 0$ such that if the C^1 distance between φ_{ji}^{M} and φ_{ji}^{N} is smaller than $\epsilon_{n,k}(C)$, then M is diffeomorphic to N.*

Cheeger proved Proposition 6.3 in the following way. First we use Proposition 6.1 to prove that the coordinate transformation φ_{ji}^M is diffeotopic to φ_{ji}^N. We then use it to construct a diffeomorphism $\bigcup_{i=1}^K U_i \to N$ (to its image) by induction in K. For details see [25]. We prove Proposition 6.3 in a slightly different way in Section 7.

Proposition 6.3 is used to prove Theorem 3.1. For this purpose, we first observe that there is a constant C such that a Riemannian manifold satisfying the assumption of Theorem 3.1 is covered by metric balls whose number is not greater than C. Since the number of metric balls is bounded, the number of possible intersection patterns among them is also bounded. Let us fix intersection pattern of the metric balls we use. We use Proposition 6.3 and find that, if the coordinate transformations φ_{ji}^M are C^1 close to φ_{ji}^N, then M is diffeomorphic to N. If the coordinate transformations φ_{ji}^M are uniformly bounded in the C^2 norm then Ascoli–Alzera's theorem implies that they are precompact in C^1-topology. Theorem 3.1 will follow.

We need however to estimate second derivative of the coordinate transformation uniformly. Our assumption in Theorem 3.1 is on curvature, which is a second derivative of the metric tensor. So one may imagine that it implies the estimate of the second derivative of coordinate transformation. However when we use geodesic coordinates, the assumption of (sectional) curvature is not enough to do so. (Cheeger [25] proved it under the additional assumption that a covariant derivative of the curvature tensor is also bounded.) To go around this trouble, Cheeger in [25] proceeds as follows. Instead of using a statement such as "two diffeomorphism is diffeotopic to each other if they are C^1-close to each other" we can use a statement such as "two homeomorphism are isotopic to each other if they are C^0-close to each other" [48]. And we can use the isotopy extension theorem[13] to construct an homomorphism $\bigcup_{i=1}^k U_i \to N$ by induction on k. This argument implies finiteness of homeomorphism classes and is not enough to prove Theorem 3.1 in four dimensions.[14] (In higher dimensions, one can use surgery etc. to deduce finiteness of diffeomorphism classes from finiteness of homeomorphism classes by purely topological argument.)

We can use harmonic coordinates (which we discuss in the next section) to find a coordinate chart such that the $C^{2,\alpha}$-norm of its coordinate transformation can be estimated uniformly.

7. Harmonic coordinate and its application

As we mentioned in the last section, in order to obtain an estimate of the Hölder norm of the coordinate transformation, taking geodesic coordinates does not give an optimal result. Harmonic coordinates are the best choice for this purpose.[15] There are various other

[13] Which is much less elementary than Proposition 6.1 and is based on highly nontrivial results such as Kirby–Siebenmann's result on the Hauptvermutung. See [48].

[14] [120] added some technical argument and proved Theorem 3.1 in four dimension as well.

[15] In mathematical study of gauge theory, we need to take representative of gauge equivalence class in order to kill freedom of gauge transformation. This is an important point to study moduli space of connections. Here we are studying "gravity" and coordinate transformation plays a role of gauge transformation. The process to find a good coordinate is called gauge fixing in Physics. Harmonic coordinates are used in Riemannian geometry around

applications of harmonic coordinates.[16] It also plays an important role to prove that the limit metric in Theorem 3.4 is of $C^{1,\alpha}$-class.

Let M be a Riemannian manifold. We assume that the injectivity radius of M is much greater than r. Let $p \in M$ and $e_i(p)$, $i = 1, \dots, n$ be an orthonormal frame of T_pM. We put $v_i(p) = \mathrm{Exp}_p(re_i(p))$, $w_i(p) = \mathrm{Exp}_p(-re_i(p))$ and define

$$h_{p,i}(x) = \frac{d(x, w_i(p))^2 - d(x, v_i(p))^2}{4r^2}. \tag{7.1}$$

We call $h_{p,i}$ an *almost linear function*. (We remark that $h_{p,i}$ is a linear function if $M = \mathbb{R}^n$.)

$h_p = (h_{p,1}, \dots, h_{p,n})$ defines a coordinate system in a neighborhood of p. However since h_p is in principle a distance function, this coordinate does not provide optimal results for the estimate of the Hölder norm of coordinate transformation. We will replace it by a harmonic function. We consider a boundary value problem of the Laplace equation $\Delta\varphi = 0$ as follows. Let us take δ such that $r \ll \delta \ll i_M(p)$, and consider $\varphi_{p,i} : B_p(\delta, M) \to \mathbb{R}$ with the following properties:

(1) $\Delta\varphi_{p,i} = 0$.

(2) If $q \in S_p(\delta, M)$, then $\varphi_{p,i}(q) = h_{p,i}(q)$.

DEFINITION 7.1. We call $\varphi_p = (\varphi_{p,1}, \dots, \varphi_{p,n})$ a *harmonic coordinate*.

Using the fact that φ_i^p is C^1-close to h_i^p we can prove that φ_p defines a coordinate in a neighborhood of p.

Now we can prove an estimate of $C^{2,\alpha}$ norm of the coordinate transformation of the harmonic coordinate as follows. We put $D^n(\epsilon) = \{x \in \mathbb{R}^n \mid \|x\| < \epsilon\}$. We take ϵ with $10\epsilon < r$. Let $p, q \in M$ with $d(p,q) < \epsilon$. We consider the inverse φ_p^{-1} of φ_p. Then the image of $\varphi_p^{-1} : D^n(\epsilon) \to M$ is contained in the domain of $\varphi_q : B_q(r, M) \to \mathbb{R}^n$. Therefore we can define

$$\varphi_{q,p}^M = \varphi_q \circ \varphi_p^{-1} : D^n(\epsilon) \to \mathbb{R}^n. \tag{7.2}$$

THEOREM 7.1. *There exists a positive constant $C(r, \epsilon, \alpha, n)$ depending only on r, ϵ, α and the dimension n, such that the $C^{2,\alpha}$-norm of $\varphi_{q,p}^M$ is not greater than $C(r, \epsilon, \alpha, n)$.*

Also the $C^{1,\alpha}$-norm of the metric tensor in harmonic coordinates is estimated by $C(r, \epsilon, \alpha, n)$.

The proof is based on a priori estimate of harmonic functions. See [87,88,64], where the second half is proved. The first half follows easily from the second half. Theorem 7.1 is generalized to Theorem 20.7.

---

the same time when Uhlenbeck etc. used Coulomb gauge in the study of moduli space of connections. The proof of Theorem 3.4 we present in this section is very similar to the proof by Uhlenbeck etc. of the compactification of the moduli space of self dual connections on 4 manifolds.

[16]We can use it to study Gromov–Hausdorff convergence under weaker assumption also. See Section 20.

Let us prove Theorem 3.4 as a typical application of Theorem 7.1.[17] Let us take a sequence M_k of elements of $\mathfrak{M}_n(D, v)$. We denote its limit in Gromov–Hausdorff distance by X. By Theorem 4.3, the injectivity radius of M_k is greater than r, a number independent of k. We take ϵ such that $10\epsilon < r$. In the same way as Section 2, we can take a finite subset $\{p_{i,k} \mid i = 1, \dots, I_k\} \in M_k$ with the following properties:

(1) I_k is smaller than a number independent of k.

(2) $\bigcup_i \varphi_{p_{i,k}}^{-1}(D^n(\epsilon)) = M_k$.

By (1) we may assume that I_k is independent of k by taking a subsequence if necessary. Set $I = I_k$. Then the intersection pattern of the coordinates $\varphi_{p_{i,k}}^{-1}(D^n(\epsilon))$ has only a finite number of possibilities. Hence by taking a subsequence we may assume that the intersection pattern is independent of k. Namely we may assume that for each $i, j \leqslant I$,

$$\varphi_{p_{i,k}}^{-1}\big(D^n(\epsilon)\big) \cap \varphi_{p_{j,k}}^{-1}\big(D^n(\epsilon)\big) \tag{7.3}$$

is empty or not does not depend on k.

Now for any i, j such that (7.3) is not empty, we consider $\varphi_{p_{j,k}, p_{i,k}}^{M_k}$ defined by (7.2). We fix $\alpha < 1$, and apply Theorem 7.1 to α' with $1 > \alpha' > \alpha$. We then find that the $C^{2,\alpha'}$-norm of $\varphi_{p_{j,k}, p_{i,k}}^{M_k}$ is estimated by a number independent of k. Hence we may take a subsequence and assume that $\varphi_{p_{j,k}, p_{i,k}}^{M_k}$ converges in $C^{2,\alpha}$-topology. Let us denotes its limit by

$$\varphi_{p_{j,\infty}, p_{i,\infty}} : D^n(\epsilon) \to \mathbb{R}^n.$$

We use them as a coordinate transformation to obtain a smooth manifold M_∞ of $C^{2,\alpha}$-class. Moreover by the uniform $C^{1,\alpha'}$-boundedness of metric tensor, we find a Riemannian metric g_∞ on M_∞ of $C^{1,\alpha}$-class which is a limit of metrics on M_k. We can prove that M_k converges to (M_∞, g_∞) in Gromov–Hausdorff distance. Hence (M_∞, g_∞) is isometric to X. Theorem 3.4 follows. □

We next prove Theorem 3.3. We assume that the theorem is false. Then there exist $M_{1,k}, M_{2,k} \in \mathfrak{M}_n(D, v)$ such that $d_H(M_{1,k}, M_{2,k}) < 1/k$ but $M_{1,k}$ is not diffeomorphic to $M_{2,k}$. We use Theorem 3.3 to show that, after taking a subsequence, $M_{1,k}, M_{2,k}$ converges to X_1, X_2, respectively. By Theorem 3.4, X_1, X_2 are Riemannian manifolds of $C^{1,\alpha}$-class. By using the center of mass technique we will explain in the next section, we can prove that $M_{1,k}$ is diffeomorphic to X_1 and $M_{2,k}$ is diffeomorphic to X_2 for large k. On the other hand, since the Gromov–Hausdorff distance between X_1 and X_2 is zero, it follows that X_1 is isometric to X_2. Hence X_1 is diffeomorphic to X_2. This is a contradiction. □

8. Center of mass technique

In Section 6 we explained how the isotopy extension theorem can be used to construct a homeomorphism. In fact the isotopy extension theorem is very difficult to prove. We can use a method called the center of mass technique which simplifies those points. The center

[17] The author follows the argument of [90] here.

of mass technique can be applied to various other problems, for example, to group actions. In this section we explain it.

Let us start the explanation of the center of mass technique by beginning a proof of (a modified version of) Proposition 6.3.

In Proposition 6.3, the assumption is about the exponential map Exp_p or coordinate transformation of geodesic coordinates. We actually use the case of harmonic coordinates. So we consider the following situation.

(a) $M = \bigcup_i \varphi_{p_i}(D^n(\epsilon))$, $N = \bigcup_i \psi_{q_i}(D^n(\epsilon))$ are open coverings.

(b) The intersection pattern of coordinate neighborhoods coincide to each other. Namely $\varphi_{p_i}(D^n(\epsilon)) \cap \varphi_{p_j}(D^n(\epsilon)) \neq \emptyset$ if and only if $\psi_{q_i}(D^n(\epsilon)) \cap \psi_{q_j}(D^n(\epsilon)) \neq \emptyset$.

(c) If $\varphi_{p_i}(D^n(\epsilon)) \cap \varphi_{p_j}(D^n(\epsilon)) \neq \emptyset$, then $\varphi_{p_i}(D^n(\epsilon)) \subseteq \varphi_{p_j}(D^n(r))$.

(d) The $C^{2,\alpha}$-norm of the coordinate transformation

$$\Phi_{ij} = \varphi_{p_i}^{-1} \circ \varphi_{p_i} : D^n(\epsilon) \to \mathbb{R}^n$$

is bounded uniformly above by C. The same holds for

$$\Psi_{ij} = \psi_{q_i}^{-1} \circ \psi_{q_i} : D^n(\epsilon) \to \mathbb{R}^n.$$

(e) Φ_{ij} is close to Ψ_{ij} in C^1-norm.

Our purpose is to construct a diffeomorphism $F: M \to N$ under these assumptions.

For each $x \in \varphi_{p_i}(D^n(\epsilon))$, we put

$$F_i(x) = \psi_{q_i} \circ \varphi_{p_i}^{-1}(x) \in N. \tag{8.1}$$

This corresponds to what we defined F on each coordinate chart $\varphi_{p_i}(D^n(\epsilon))$. The main point is whether we can glue them to obtain F globally. Namely in case $x \in \varphi_{p_i}(D^n(\epsilon)) \cap \varphi_{p_j}(D^n(\epsilon))$ we need to know whether

$$\psi_{q_i} \circ \varphi_{p_i}^{-1}(x) \stackrel{?}{=} \psi_{q_j} \circ \varphi_{p_j}^{-1}(x) \tag{8.2}$$

or not. It is easy to see that (8.2) does not hold. What follows from our assumption (assumption of Proposition 6.3 or the assumption (e) above) is

$$d\big(\psi_{q_i} \circ \varphi_{p_i}^{-1}(x), \psi_{q_j} \circ \varphi_{p_j}^{-1}(x)\big) < \epsilon \tag{8.3}$$

(where ϵ is a sufficiently small positive number). (More precisely, (8.3) is in C^0-norm, but assumption (e) is in C^1-norm.)

The basic idea of the center of mass technique is to take the average of $F_i(x)$ over i with $x \in \varphi_{p_i}(D^n(\epsilon))$. Before we continue the proof of Proposition 6.3, we explain the center of mass technique in general here.

Let $\mathfrak{m}$ a Borel probability measure on M (namely a measure on M with $\mathfrak{m}(M) = 1$). Let us denote the support of $\mathfrak{m}$ by $\mathrm{Supp}(\mathfrak{m})$. We define a function $d_{\mathfrak{m}}$ on M by

$$d_{\mathfrak{m}}(x) = \int d(x, p)\, d\mathfrak{m}(p). \tag{8.4}$$

PROPOSITION 8.1. *We assume the injectivity radius of M is larger than 10ϵ. We also assume $K_M \leqslant \kappa$ and $20\epsilon < \pi/\sqrt{\kappa}$.*[18]

If the diameter of Supp($\mathfrak{m}$) *is smaller than ϵ, then on*

$$B_{3\epsilon}\big(\mathrm{Supp}(\mathfrak{m}), M\big) = \big\{x \in M \mid d\big(x, \mathrm{Supp}(\mathfrak{m})\big) < 3\epsilon\big\},$$

the function $d_{\mathfrak{m}}$ is convex.

Here a function on a Riemannian manifold said to be convex if its restriction to each geodesic is convex.

We can prove Proposition 8.1 by using the convexity of the distance function d_p on $B_p(\pi/\sqrt{\kappa}, M)$.[19]

Now we assume that the diameter of Supp($\mathfrak{m}$) is smaller than ϵ. Then outside $B_{3\epsilon}(\mathrm{Supp}(\mathfrak{m}), M)$ the value of the function $d_{\mathfrak{m}}$ is greater than 3ϵ, and on Supp($\mathfrak{m}$) the value of the function $d_{\mathfrak{m}}$ is smaller than ϵ. Therefore Supp($\mathfrak{m}$) attains its minimum on the interior of $B_{3\epsilon}(\mathrm{Supp}(\mathfrak{m}), M)$. Since $d_{\mathfrak{m}}$ is convex there, the minimum is attained at unique point.

DEFINITION 8.1. The *center of mass* is the point where $d_{\mathfrak{m}}$ attains its minimum. We write center of mass by $\mathfrak{CM}(\mathfrak{m})$.

We remark that if $M = \mathbb{R}^n$, then

$$\mathfrak{CM}(\mathfrak{m}) = \int_{\mathbb{R}^n} x \, d\mathfrak{m}(x).$$

We go back to the proof of Proposition 6.3. We take a partition of unity χ_i associated to the covering $M = \bigcup_i B_{p_i}(\epsilon, M)$. We define a measure $\mathfrak{F}(x)$ on N by

$$\mathfrak{F}(x) = \sum_i \chi_i(x)\delta_{F_i(x)}.$$

Here $\delta_{F_i(x)}$ is the delta measure supported at $F_i(x)$ and the summation is taken over all i with $x \in B_{p_i}(\epsilon, M)$.

By (8.3) we have Diam(Supp($\mathfrak{F}(x)$)) $< \epsilon$. Let $F(x)$ be the center of mass of $\mathfrak{F}(x)$. Namely,

$$F(x) = \mathfrak{CM}\big(\mathfrak{F}(x)\big) = \mathfrak{CM}\bigg(\sum_i \chi_i(x)\delta_{F_i(x)}\bigg). \tag{8.5}$$

It is easy to see that $F(x)$ is a continuous function of x. Actually it is smooth. (We can prove it by using implicit function theorem.) We can prove that it is a diffeomorphism by using the following lemma.

[18] In case $\kappa \leqslant 0$ the second condition is void.

[19] This fact is a consequence of Toponogov's comparison Theorem 4.7.

LEMMA 8.2. *If F_i, $i = 1, 2, \dots$, are C^1-close to each other, then F, determined by* (8.5), *is C^1-close to F_i.*

The proof is elementary.

Then, to prove Proposition 6.3, we only need to show that F is injective. Suppose $F(x) = F(y)$, $x \neq y$. By using the fact that the Jacobi matrix of F is invertible, we can show that x cannot be close to y. On the other hand, since F is close of F_i and since F_i is injective, we can prove that x cannot be far from y. This is a contradiction. This is an outline of the proof of Proposition 6.3. □

There are various other applications of the center of mass technique. Let us mention another application of it, that is an application to group actions. Let M be a Riemannian manifold on which G acts. For simplicity we assume G is a finite group. We assume G has two different actions on M and write them as $\psi_1 : G \to \mathrm{Diff}(M)$ and $\psi_2 : G \to \mathrm{Diff}(M)$. We assume that there exists C such that for each element $g \in G$, the C^2 norm of $\psi_1(g)$, $\psi_2(g)$ are smaller than C.

PROPOSITION 8.3 (Grove–Karcher). *There exists a constant ϵ depending only on C, the dimension n, the injectivity radius of M, and the maximum of the absolute value of the sectional curvature of M, with the following property.*

If $d(\psi_1(g)(x), \psi_2(g)(x)) < \epsilon$ for each $g \in G$, $x \in M$, then there exists a diffeomorphism $\phi : M \to M$ such that $\phi(\psi_1(g)(x)) = \psi_2(g)(\phi(x))$.

See [77] for its proof. ([77] is the paper where the center of mass technique first appeared.)

Proposition 8.3 is applied to study Riemannian manifold whose sectional curvature is close to 1 but whose fundamental group is not necessary trivial.

9. Embedding Riemannian manifolds by distance function

In the last section we explained the center of mass technique which we can use to construct a diffeomorphism. In Section 6 we mentioned another way, that is to use isotopy theory. In this section, we discuss the third method which was introduced and used by Gromov [68, 69]. In [53] the author remarked that this method can be used to construct a smooth map (projection of a fiber bundle) in collapsing situation (Theorem 11.2). It was further generalized by Yamaguchi [153] (Theorem 11.3) to the case when we assume a bound of sectional curvature from below (but not above).

We here explain an alternative proof of Theorem 3.3. This proof is completed by Katsuda [93] based on an idea of Gromov [69]. We assume $M, N \in \mathfrak{M}_n(D, v)$, $d_H(M, N) < \epsilon(n, v, D)$. (We choose $\epsilon = \epsilon(n, v, D) > 0$ later.) Let $\psi : M \to N$ be a 3ϵ Hausdorff approximation. We take a 20ϵ-net X of M. We can take X such that

$$\text{if } x, x' \in X,\ x \neq x', \quad \text{then } d(x, x') > 10\epsilon, \tag{$*$}$$

in addition. It is easy to see that $\psi(X)$ is an 30ϵ-net of N. It is also easy to see that

$$\text{if } x, x' \in X,\ x \neq x', \quad \text{then } d\big(\psi(x), \psi(x')\big) > \epsilon. \tag{**}$$

We denote by $[0,1]^X$ the set of all maps $X \to [0,1]$. It is a finite-dimensional Euclidean space. The idea is to embed M (respectively N) in $[0,1]^X$ using the distance function from X (respectively $\psi(X)$). In order to go around the trouble that the distance function is not differentiable, we proceed as follows. We take ϵ so that it is much smaller than the injectivity radius of M and N. We next take $\chi : \mathbb{R}_{>0} \to [0,1]$ such that

$$\chi(t) = \begin{cases} 0, & t < C\epsilon, \\ t, & t \in [C^2\epsilon, C^3\epsilon], \\ \text{const}, & t > C^4\epsilon. \end{cases}$$

Here C is a sufficiently large positive number which will be determined later. We may assume that $C^5\epsilon$ is smaller than the injectivity radius of M and N. (Precisely we first choose C and then choose ϵ so that this condition is satisfied.) Then we define $I_M : M \to [0,1]^X$ by $I_M(p)(x) = \chi(d(p,x))$ and $I_N : N \to \mathbb{R}^X$ by $I_N(p)(x) = \chi(d(p,\psi(x)))$. Note $\chi(t)$ is a constant where t is larger than the injectivity radius. Hence I_M, I_N are smooth. We can prove the following

LEMMA 9.1.

(1) *I_M, I_N are smooth embeddings.*

(2) *$I_M(M)$ is contained in a tubular neighborhood $U(N)$ of $I_N(N)$.*

(3) *We identify $U(N)$ with a normal bundle and let $\pi : U(N) \to N$ be the projection of the normal bundle.*

Then the restriction of π to $I_M(M)$ is a diffeomorphism.

We omit the details of the proof (see [93]), but explain briefly its idea. The reason that (1) holds is that, for each p, there are sufficiently many points $q \in X$ with $d(q,p) \in [C^2\epsilon, C^3\epsilon]$. Namely using the distance function from such q we can show the Jacobi matrix of I_M, I_N are invertible in a neighborhood of p.

To prove (2) we observe that, if $x \in X \subset M$, then the distance between $I_M(x)$ and $I_N(\psi(x))$ is small. (Namely it is something like $\text{const}\, d_H(M,N) = \text{const}\,\epsilon$.) Moreover, since $X, \psi(X)$ are enough dense in M, N, it follows that $I_M(M)$ are sufficiently close to $I_N(N)$. We next need an estimate of the size of the tubular neighborhoods of $I_M(M)$, $I_N(N)$. This follows from the estimate of the second derivative of I_M and I_N, which turn out to be a consequence of the assumption on curvature of M, N. To carry out the actual proof we need to estimate the size of the tubular neighborhood and the distance between $I_M(x)$ and $I_N(\psi(x))$ more precisely.

To prove (3) we need to see that the Jacobi matrix of the restriction of $\pi : U(N) \to N$ to $I_M(M)$ is invertible. This follows from the fact that $I_M(M)$ is C^1-close to $I_N(N)$, namely they are close to each other together with their tangent spaces. Since the derivative of the distance function is written in terms of the angle between edges of geodesic triangles, we can prove this fact by using comparison theorems. □

Theorem 3.3 follows immediately from Lemma 9.1.

Remark 9.1. (A) We used distance functions in the discussion above. We can use eigenfunctions of the Laplace operator (or Green kernel) instead. Then the estimate about the derivatives of the diffeomorphism we get becomes better. (See, for example, [16,56,91].) This approach is closely related to harmonic coordinates.

(B) We took a net and embed Riemannian manifolds to a finite-dimensional Euclidean space in the above argument. We can use distance functions from all the points and can embed Riemannian manifolds to a Hilbert or Banach space. This argument is useful for a generalization of Theorem 3.3 to an equivariant version. (Namely in the situation when a Lie group acts on M, N.) If we use the eigenfunction of the Laplace operator as we mentioned in (A), embedding to finite-dimensional Euclidean space is good enough to show the equivariant version also.

10. Almost flat manifold

In this section we start discussing the case when the injectivity radius goes to zero. In the earlier sections, we began with sphere theorems and applied the method appeared there to finiteness theorems, etc. In sphere theorems, we study manifolds of positive curvature. We here consider another typical Riemannian manifold that is a flat manifold. We first recall the following famous

THEOREM 10.1 (Bieberbach). *If M is a compact Riemannian manifold with $K_M \equiv 0$, then there exists a finite covering $\tilde{M}$ of M such that $\tilde{M}$ is isometric to a flat torus.*

We want to study a Riemannian manifold (M, g) whose curvature is close to zero. To obtain a nontrivial result, we need some normalization. (In fact, the curvature of (M, kg) tends to 0 as $k \to \infty$ for any (M, g).) To normalize volume is not good enough either. (For example, $M \times S^1$ for any M carries a metric with volume 1 and curvature arbitrary small.) So let us normalize the diameter to 1. In other words, we assume $|K_M|\,\mathrm{Diam}(M)^2$ is small. We call such manifold *almost flat manifold.* However the assumption $|K_M|\,\mathrm{Diam}(M)^2$ small does *not* imply that M is diffeomorphic to a flat manifold.

EXAMPLE 10.1. We consider the group N of all 3×3 matrix of the form

$$\begin{pmatrix} 1 & x & z \\ 0 & 1 & y \\ 0 & 0 & 1 \end{pmatrix}.$$

We consider a left invariant metric g_ϵ on N such that $g_\epsilon = \epsilon^2\,dx^2 + \epsilon^2\,dy^2 + \epsilon^4\,dz^2$ at the unit matrix I. Let E_1, E_2, E_3 be left invariant vectors such that $E_1 = \partial/\partial x$, $E_2 = \partial/\partial y$, $E_3 = \partial/\partial z$ at I. It is well known that the curvature of a Lie group with left invariant metric is calculated as follows. If E, F are left invariant orthonormal vectors then the sectional curvature of the plane spanned by them is not greater $6\|[E, F]\|$. (See [24].)

Hence the sectional curvature $K_{(N,g_\epsilon)}$ is bounded as $\epsilon \to 0$. On the other hand, we consider the subgroup $N_{\mathbb{Z}}$ consisting of matrix in N such that $x, y, z \in \mathbb{Z}$. $N_{\mathbb{Z}}$ is a discrete subgroup of N and the quotient space $M = N_{\mathbb{Z}} \setminus N$ is known to be compact. We consider the metric on M induced by g_ϵ and denote it by $\bar{g}_\epsilon$. It is easy to see that the diameter $\mathrm{Diam}(M, \bar{g}_\epsilon)$ goes to zero. Hence $\mathrm{Diam}(M, \bar{g}_\epsilon)^2 K_{g_\epsilon}$ goes to zero. However no finite cover of M is diffeomorphic to T^3.

This example shows that we need to include not only Abelian but also nilpotent Lie groups to characterize almost flat manifolds.

THEOREM 10.2 (Gromov [68]). *There exists $\epsilon_n > 0$ such that if an n-dimensional compact Riemannian manifold M satisfies $|K_M| \mathrm{Diam}(M)^2 < \epsilon_n$, then M has a finite cover $\tilde{M}$ which is diffeomorphic to $\Gamma \setminus N$, where N is a nilpotent Lie group and Γ is a discrete subgroup.*

There is an improvement of Theorem 10.2 due to Ruh [133]. Let N be an nilpotent Lie group. There exists a connection ∇_{can} of TN which is invariant to both left and right actions of N. Let Γ be a discrete subgroup of N. ∇_{can} induces a connection on $\Gamma \setminus N$ which we denote by the same symbol. (We remark that ∇_{can} is not equal to the Levi-Civita connection.) Let Λ be a finite subgroup of $\mathrm{Aut}(\Gamma \setminus N, \nabla_{\mathrm{can}})$. We call $\Lambda \setminus (\Gamma \setminus N)$ an *infranilmanifold*.

THEOREM 10.3 (Ruh). *Under the assumption of Theorem 10.2, M is diffeomorphic to an infranilmanifold.*

Let us sketch some of the essential ideas behind the proof of Theorem 10.2. One important origin is Margulis' lemma. Margulis' lemma first appeared in the study of discrete subgroup of Lie group.

THEOREM 10.4 (Zassenhaus, see [69, 8.44]). *For each Lie group G there exists a neighborhood U of the unit, such that if $\Gamma \subset G$ is a discrete subgroup then $U \cap \Gamma$ generates a nilpotent subgroup.*

The proof is based on the following fact. Let $g_1, g_2 \in G$ be in a neighborhood of the unit 1, then

$$d(1, \{g_1, g_2\}) \leqslant C d(1, g_1) d(1, g_2). \tag{10.1}$$

Here $\{g_1, g_2\}$ is the commutator. This formula (10.1) is a consequence of the fact that the derivative of $(g_1, g_2) \mapsto \{g_1, g_2\}$ at 1 is zero. Once we have (10.1) we can prove Theorem 10.4 as follows. We choose U small enough such that if $g \in U$ then $d(1, g) < 1/(2C)$. Then (10.1) implies that if $g_i \in U$, then $d(1, \{g_1, g_2\})$ is strictly smaller than $d(1, g_i)/C$. We repeat this and find that N hold commutator between elements of U is in the $1/C^N$ neighborhood of 1. Since Γ is discrete, it implies the existence of N such that any N hold commutators between elements of $U \cap \Gamma$ are trivial. It follows that $U \cap \Gamma$ generates a nilpotent group. □

There are various Riemannian geometry versions of Theorem 10.4. The following, which is proved by Cheeger–Colding [31] (improving [59]) is one of the strongest versions.

THEOREM 10.5. *There exists ϵ_n with the following properties. Let M be an n-dimensional complete Riemannian manifold with $\mathrm{Ricci}_M \geqslant -(n-1)$ and $p \in M$. Then the image of $\pi_1(B_p(\epsilon_n, M)) \to \pi_1(B_p(1, M))$ has nilpotent subgroup of finite index.*

If we apply it to the situation of Theorem 10.2 we find that the fundamental group of M has nilpotent subgroup of finite index. (See Section 19 for more discussions on the fundamental group.)

Another idea applied by Gromov to prove Theorem 10.2 is to use local fundamental pseudogroup, which we discuss briefly here. (See [57, Section 7] and [24] for its precise definition.) Let M be a complete Riemannian manifold. We assume $K_M \leqslant 1$. Let $p \in M$. Then by Theorem 4.5 the exponential map $\mathrm{Exp}_p : T_pM \to M$ is an immersion on the ball $B_0(\pi; T_pM)$. Since $B_0(\pi; T_pM)$ has a boundary, $\mathrm{Exp}_p : B_0(\pi; T_pM) \to M$ is not a covering map. So we cannot consider its deck transformation group in the usual sense. But we can define a "pseudogroup" in the following way. Let $\epsilon < \pi/10$. We consider the set of all loops $\ell : S^1 \to B_p(\epsilon, M)$ with $\ell(0) = p$ and $|\ell| < \epsilon$. We say $\ell \sim \ell'$ for such ℓ, ℓ' if there exists a based homotopy ℓ_t between them such that $|\ell_t| < \epsilon$ for each t. Let us denote the set of equivalence classes by $\pi_1(M, p; \epsilon)$. The loop sum $*$ on $\pi_1(M, p; \epsilon)$ is not necessary defined. But when it is defined, its $\sim$ equivalence class is well defined. (We need to use the fact that $\mathrm{Exp}_p : B_0(\pi; T_pM) \to M$ is an immersion to show this.) When loop sum is well defined it is associative. (Here the reader may find some flavor of Klingenberg's argument we mentioned at the end of Section 4.) Thus $(\pi_1(M, p; \epsilon), *)$ is something similar to a group. We call it a *fundamental pseudogroup*. The following pseudogroup version of Margulis' lemma is used in the proof of Theorem 10.2.

LEMMA 10.6. *If $|K_M| \leqslant 1$ and if $\mathrm{Diam}(M) < \epsilon_n$, then there exists a subpseudogroup $(\pi_1^0(M, p; \epsilon), *)$ of $(\pi_1(M, p; \epsilon), *)$ such that $(\pi_1^0(M, p; \epsilon), *)$ is embedded (preserving $*$) into a nilpotent Lie group N, its image generates a discrete subgroup Γ and that the index $[\pi_1(M, p; \epsilon) : \pi_1^0(M, p; \epsilon)]$ is estimated by a number depending only on n. Here $\epsilon_n \ll \epsilon \ll 1$.*

Lemma 10.6 is the main part of the proof of Theorem 10.2. (Actually we need a bit more. Namely we have to show that the action of $(\pi_1(M, p; \epsilon), *)$ to $B_0(\pi; T_p(M))$ is diffeomorphic to an action to $U \subset N$ of some subpseudogroup $\Gamma \cap U$, where N is a nilpotent Lie group and Γ is its discrete subgroup.) □

For the details of the proof, we refer to [24,57].

11. Collapsing Riemannian manifolds—I

Using Theorems 3.3, 3.4, we can describe a sequence of n-dimensional Riemannian manifolds M_i with $|K_{M_i}| \leqslant 1$ and $\mathrm{Vol}(M_i) \geqslant v > 0$ where v is independent of i. Namely the limit X (which exists after taking a subsequence) is a Riemannian manifold of $C^{1,\alpha}$-class and X is diffeomorphic to M_i for sufficiently large i.

In Section 10, we considered a sequence of Riemannian manifolds M_i with $|K_{M_i}| \leqslant 1$ and $\mathrm{Diam}(M_i) \to 0$. (The second condition is equivalent to saying that M_i converges to a point.) Theorem 10.2 implies that M_i is an infranilmanifold for large i.

These are two extremal cases. We now discuss the intermediate case. Namely we consider the case when a sequence of Riemannian manifolds M_i converges to a metric space X (with respect to the Gromov–Hausdorff distance) such that $n > \dim X > 0$. We say that such sequence *collapses to* X. Here we discuss results under the assumption $|K_{M_i}| \leqslant 1$. (The study under weaker assumption is discussed in later sections.)

We first explain some examples of collapsing Riemannian manifolds. The first example is due to Berger and is called the Berger sphere.

EXAMPLE 11.1. We consider the Hopf fibration $\pi : S^3 \to S^2$. (Namely we regard $S^3 = \{(z_1, z_2) \in \mathbb{C}^2 \mid |z_1|^2 + |z_2|^2 = 1\}$, and we associate to (z_1, z_2) the complex one-dimensional space spanned by it, which is an element of $\mathbb{C}P^1 = S^2$.) We put the standard metric on S^3 and regard S^2 as a sphere of radius $1/2$. It is easy to see that π is a Riemannian submersion. (Namely if $V_h \in T_pS^3$ and V is perpendicular to the fiber of π containing p, then $g_{S^3}(V_h, V_h) = g_{S^2}(\pi_* V_h, \pi_* V_h)$.) We define a metric g_ϵ on S^3 as follows. Let $V, W \in T_pS^3$. We write

$$V = V_h + V_v, \qquad W = W_h + W_v,$$

where V_h, W_h are perpendicular to the fiber (with respect to g_{S^3}) and V_v, W_v are tangent to the fiber. We set

$$g_\epsilon(V, W) = g_{S^3}(V_h, W_h) + \epsilon^2 g_{S^3}(V_v, W_v).$$

It is easy to see that $\lim^{GH}_{\epsilon\to 0}(S^3, g_\epsilon) = (S^2, g_{S^2})$. We can check that the sectional curvature of (S^3, g_ϵ) is between 0 and 1 if $\epsilon \in (0, 1]$.

We can generalize this construction and prove the following

PROPOSITION 11.1. *Let M be a compact manifold on which a torus T^m acts. We assume that there is no point p on M such that p is fixed by all the elements of T^m. Then there exists a family of metrics g_ϵ on M such that K_{g_ϵ} is bounded from below and above and that* $\lim^{GH}_{\epsilon\to 0}(M, g_\epsilon) = M/T^m$.

To find such a sequence of metrics, we first take a T^m invariant Riemannian metric g_M on M. We next take X an element of the Lie algebra of T^m such that the subgroup $\cong \mathbb{R}$

generated by X is dense in T^m. We regard X as a (Killing) vector field on M. We remark that X never vanishes on M. For $V, W \in T_pM$ we put

$$V = V_h + c(V)X(p), \qquad W = W_h + c(W)X(p),$$

where $g_M(V_h, X_p) = g_M(W_h, X_p) = 0$. We then define

$$g_\epsilon(V, W) = g_M(V_h, W_h) + \epsilon^2 c(V)c(W)g_M\big(X(p), X(p)\big).$$

We can prove that the limit of (M, g_ϵ) as $\epsilon \to 0$ is M/T^m with quotient metric and the sectional curvature of (M, g_ϵ) is bounded for $\epsilon \in (0, 1]$. □

Let us take, for example, $M = S^3$. We can find an action of T^2 on S^3 satisfying the assumption of Proposition 11.1. Hence there exists a sequence of metrics on S^3 such that the limit is $S^3/T^2 = [0, 1]$, the interval. In particular the limit space is not a manifold.

This construction is further generalized in [38] (Theorem 12.1).

There are two approaches to study collapsing Riemannian manifolds under the assumption $|K_{M_i}| \leqslant 1$. One is due to Cheeger–Gromov [39,38], the other is due to the author [53, 55,56]. These two approaches are unified in [34]. In this section, we discuss the second approach and in the next section we discuss the first (and the joined) approach.

Here we discuss the following two problems. For n, D, we denote by $\mathfrak{M}_n(D)$ the set of all isometry classes of n-dimensional Riemannian manifolds M such that $|K_M| \leqslant 1$, and $\operatorname{Diam}(M) \leqslant D$.

PROBLEM 11.1. Let $M_i \in \mathfrak{M}_n(D)$ and $X = \lim^{GH}_{i\to\infty} M_i = X$.

(1) What kind of singularity can X have?

(2) Describe the relations between X and M_i.

We remark that if we replace $\mathfrak{M}_n(D)$ by $\mathfrak{M}_n(D, v)$, the answers are Theorems 3.3, 3.4. Problem 11.1 will be studied also under milder assumptions on curvature later.

We first discuss Problem 3.4(2) in the special case when X is a smooth manifold.

THEOREM 11.2 (Fukaya [53,56]). *Let $M_i \in \mathfrak{M}_n(D)$. Suppose $B = \lim^{GH}_{i\to\infty} M_i$ is a smooth Riemannian manifold. Then, for large i, there exists a fiber bundle $\pi_i : M_i \to B$ with the following properties*:

(1) *The fiber is diffeomorphic to an infranilmanifold F.*

(2) *The structure group is the group of affine transformations* $\operatorname{Aff}(F, \nabla_{\text{can}})$, *where we define the affine connection ∇_{can} on F as in the last section.*

(3) *π_i is an almost Riemannian submersion in the following sense. If $V \in T_p(M_i)$ is perpendicular to the fiber then*

$$1 - \epsilon_i < \frac{g_{M_i}(V, V)}{g_N(\pi_{i*}V, \pi_{i*}V)} < 1 + \epsilon_i,$$

where $\epsilon_i \to 0$.

Yamaguchi [153] generalized Theorem 11.2 as follows.

THEOREM 11.3 (Yamaguchi). *If M_i is a sequence of n-dimensional Riemannian manifold with $K_{M_i} \geqslant -1$. We assume $B = \lim^{GH}_{i\to\infty} M_i$ is a smooth Riemannian manifold. Then for large i there exists a fiber bundle $\pi_i : M_i \to B$. It satisfies* (3) *above.*

See Section 19 for more results on the fiber of $\pi_i : M_i \to B$ in Theorem 11.3.

The idea of the proof of Theorems 11.2, 11.3 is similar to the discussion in Section 9. Namely we embed the limit space B to $\mathbb{R}^X$ using the distance function ($I_B : B \to \mathbb{R}^X$). (Here X is a net in B.) We then map M_i to the same space ($I_{M_i} : M_i \to \mathbb{R}^X$). We cannot prove that I_{M_i} is an embedding since there is no bound of injectivity radius of M_i. However I_B is an embedding and $I_{M_i}(M_i)$ is contained in a tubular neighborhood $U(I_B(B))$ of $I_B(B)$ for large i. Hence we have a composition of three maps, I_{M_i}, the projection of the normal bundle of $I_B(B)$, and I_B^{-1}. This map is our $\pi_i : M_i \to B$. To check that it satisfies (1), (2) we use a parameterized version of the proof of Theorems 10.2, 10.3. □

In general, the limit space as in Problem 11.1 has singularities. Hence Theorem 11.2 does not apply in the general case. However we can use its equivariant version and a trick (which we explain below) so that we can apply it to the general situation.

Let M be an n-dimensional Riemannian manifold. We define its frame bundle by

$$FM = \left\{ (p; e_1, \dots, e_n) \;\middle|\; \begin{array}{l} p \in M, \\ (e_1, \dots, e_n) \text{ is an orthonormal basis of } T_pM \end{array} \right\}.$$

There exists an $O(n)$ action on FM such that $FM/O(n) = M$. In other words $FM \to M$ is a principal $O(n)$ bundle. The Riemannian metric determines a connection of this principal bundle (that is the Levi-Civita connection). Using it we can canonically define an $O(n)$ invariant Riemannian metric on FM such that $FM \to M$ is a Riemannian submersion and the fiber $\cong O(n)$ has given the standard metric on $O(n)$. From now on we use this metric on FM.

THEOREM 11.4 [55]. *If $M_i \in \mathfrak{M}_n(D)$ and if $Y = \lim^{GH}_{i\to\infty} FM_i$. Then we have the following:*

(1) *Y is a smooth manifold.*
(2) *$O(n)$ acts by isometries on Y such that $\lim^{GH}_{i\to\infty} M_i = Y/O(n)$.*
(3) *There exists a sequence of $O(n)$-equivariant Riemannian metric on g_i on Y and $\epsilon_i \to 0$ such that*

$$1 - \epsilon_i < \frac{d_Y(x, y)}{d_{g_i}(x, y)} < 1 + \epsilon_i$$

for any $x, y \in Y$, where d_Y is the limit metric.

(4) *For each $p \in Y$ the connected component of the isotropy group $\{g \in O(n) \mid gp = p\}$ is Abelian.*

To prove Theorem 11.4, we use the notion of fundamental pseudogroup we explained in the last section as follows. (The idea to use pseudofundamental group to study collapsing is initiated by Gromov in [69, Chapter 8].) Let $p \in X$. We take $p_i \in M_i$ which converges to p. We fix small ϵ and consider $\pi_1(M_i, p_i, \epsilon)$ which acts on $B_0(\epsilon, T_{p_i}M_i)$ such that the quotient space is isometric to an ϵ-neighborhood of p_i in M_i. (We can define a notion of action of a pseudogroup to a space and of the quotient space, in a reasonable way.) We can define the convergence of a pseudogroup action and can find a limit of $\pi_1(M_i, p_i, \epsilon)$, which we denote by N. The group N acts by isometries on the limit $\tilde{B}(p)$ of $B_0(\epsilon, T_{p_i}M_i)$. Here we put a Riemannian metric on $B_\epsilon(0, T_{p_i}M_i)$ which is induced on M_i by the exponential map. Since the injectivity radius of $B_0(\epsilon, T_{p_i}M_i)$ is bounded away from 0, it follows that we can apply Theorem 3.4 to find that $\tilde{B}(p)$ is a Riemannian manifold of $C^{1,\alpha}$-class. The point here is that N is in general not discrete and collapsing occurs exactly when N has positive dimension. We can show that the group germ of the origin of N is a Lie group germ. Note that this is easy in case when the metric on Y is smooth. To avoid using a metric which is not smooth, we approximate g_{M_i} by $g_{M_i,\epsilon}$ such that

$$\left|\nabla^k R_{g_{M_i,\epsilon}}\right| < C_k(\epsilon), \tag{11.1a}$$

$$e^{-\epsilon} g_{M_i} < g_{M_i,\epsilon} < e^{\epsilon} g_{M_i}. \tag{11.1b}$$

Here the left-hand side of (11.1b) is the norm of the kth derivative of the curvature tensor and the right-hand side is a constant depending on k and ϵ but is independent of i. The existence of such approximation is proved in [15] (and generalized in [1] for complete manifolds).

Then the limit of the ball $(\tilde{B}(p), \tilde{g}_{M_i,\epsilon})$ is smooth. Replacing G by its quotient we may assume that the action of G on $\tilde{B}(p)$ is effective.

We now consider the frame bundle $F\tilde{B}(p)$. Using the fact that G is effective and isometric on $\tilde{B}(p)$, it follows that the action of N on $F\tilde{B}(p)$ is *free*. Therefore $F\tilde{B}(p)/N$ is a manifold. We can easily see that $F\tilde{B}(p)/N$ is an open set of the limit Y of FM_i and by changing $p \in X$ it covers Y. Thus Y is a manifold as required. Using Margulis' lemma we find that the connected component of N is nilpotent. Since the isotropy group of $O(n)$ action on Y can be identified to the isotropy group of N action on $\tilde{B}(p)$, it follows that the connected component of the isotropy group is both compact and nilpotent. Hence it is Abelian. □

Using Theorem 11.4, we can improve Theorem 11.2 as follows.

THEOREM 11.5 [55]. *Let M_i, Y be as in Theorem* 11.4. *Then there exists $\tilde{\pi}_i : FM_i \to Y$ for large i with the following properties*:

(1) *$\tilde{\pi}$ is a fiber bundle satisfying* (1)–(3) *of Theorem* 11.2.

(2) *$\tilde{\pi}$ is $O(n)$-equivariant and hence induces a map $\pi : M_i \to Y/O(n)$.*

The proof is an equivariant version of the proof of Theorem 11.4. (Compare Remark 9.1(B).)

12. Collapsing Riemannian manifolds—II

As we mentioned before there are two approaches to study collapsing Riemannian manifolds and we discuss another approach [39,38] in this section. One advantage of this approach (compared with one we discussed in the last section) is that we do not need to assume a diameter bound. Let us first give an example to illustrate a new phenomenon which occurs when we do not assume a diameter bound.

EXAMPLE 12.1 (See [134]). Let Σ_k be a Riemann surface of genus $k > 0$. For each ϵ there exists a Riemannian metric $g_{k,\epsilon}$ on $\Sigma_{k,0} = \Sigma_k \setminus \operatorname{Int} D^2$ with the following properties:

(1) $0 \geqslant K_{g_{k,\epsilon}} \geqslant -1$.

(2) A neighborhood of the boundary of $\Sigma_{k,0}$ is isometric to $S^1(\epsilon) \times [0,1)$. (Here $S^1(\epsilon)$ is a circle with radius ϵ.) (See Figure 12.1.)

Now we consider $(\Sigma_{k_1,0}, g_{k_1,\epsilon}) \times S^1(\epsilon)$ and $(\Sigma_{k_2,0}, g_{k_2,\epsilon}) \times S^1(\epsilon)$ and glue them at their boundaries by the isometry $(s,t) \mapsto (t,s)$, $S^1(\epsilon) \times S^1(\epsilon) \to S^1(\epsilon) \times S^1(\epsilon)$. We thus obtain a family of 3-dimensional Riemannian manifolds $M_{k_1,k_2,\epsilon}$, which satisfy the curvature condition $0 \geqslant K \geqslant -1$. The injectivity radius of it goes to zero everywhere as $\epsilon \to 0$. It is however easy to see that $M_{k_1,k_2,\epsilon}$ is not a S^1-bundle over a surface.

See [4] for a more sophisticated construction.

We remark that the diameter of $\Sigma_{k,\epsilon}$ goes to infinity as ϵ goes to zero. The point of this example is that in each piece $\Sigma_{k_i} \times S^1(\epsilon)$ there is one direction (the direction of second factor) which collapses. But in the domain we glue metrics, there are two factors which collapse. Theorem 11.2 implies that such a phenomenon does not occur. Namely the dimension of the collapsing direction is constant in the case when the limit space is compact. Thus to describe collapsing Riemannian manifolds without diameter bound, we need a language to describe the situation where the dimension of the collapsing direction changes. Cheeger–Gromov [38] used a notion of the local action of a group for this purpose. They call it an F-structure.

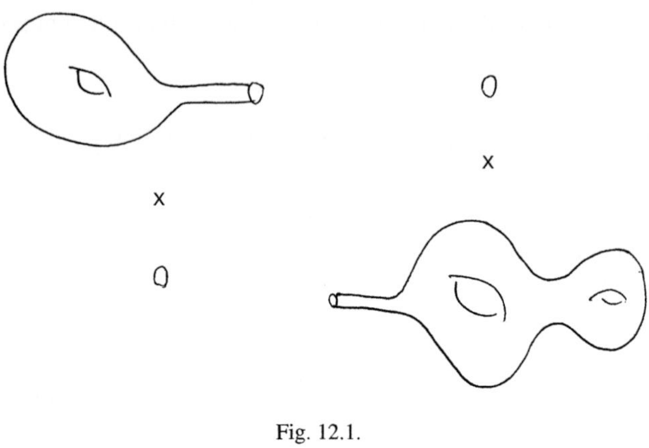

Fig. 12.1.

DEFINITION 12.1. An *F-structure* on M is an open cover $M = \bigcup U_i$ together with an action of T^{n_i} on $\tilde{U}_i$, which is a finite cover of U_i, with the following properties:

(1) There exists no point $x \in \tilde{U}_i$ which is fixed by all the elements of T^{n_i}.
(2) If $U_i \cap U_j \neq \emptyset$ then there exists a covering space $\pi_{ij} : \tilde{U}_{ij} \to U_i \cap U_j$, maps $\pi_{ij,i} : \tilde{U}_{ij} \to \tilde{U}_i$, $\pi_{ij,j} : \tilde{U}_{ij} \to \tilde{U}_j$ such that
(3) $\pi_i \circ \pi_{ij,i} = \pi_j \circ \pi_{ij,j} = \pi_{i,j}$.
(4) There exists an action of $T^{n_{ij}}$ on $\tilde{U}_{ij}$ with the property (1).
(5) There exists an n_i-dimensional subtorus $T^{n_i}_{ij} \subset T^{n_{ij}}$ and a locally isomorphic group homomorphism $T^{n_i}_{ij} \to T^{n_i}$, such that $\pi_{ij,i}$ is equivariant. The same holds when we replace i by j.

Let us consider Example 12.1. We may split M into two pieces $U_i \cong \Sigma_{k_i} \times S^1$. On U_i we have an S^1-action. These two actions do not coincide on the overlapped part $U_1 \cap U_2 \cong S^1 \times S^1 \times (-C, C)$. Namely the S^1-action on U_1 is an action on the first factor while the S^1-action on U_2 is an action on the second factor. However we have a T^2-action which contains both actions. This is a typical situation of an F-structure.

The main theorem in [38] is as follows.

THEOREM 12.1 (Cheeger–Gromov). *If M has an F-structure then there exists a sequence of metrics g_i on M such that $|K_{g_i}| \leqslant 1$ and the injectivity radius of (M, g_i) converges to zero everywhere as $i \to \infty$.*

The proof is a kind of generalization of the proof of Proposition 11.1. The new point which appears in the proof of Theorem 12.1 is that we need to control the curvature at the points where the dimension of the torus acting there changes. Roughly speaking, to keep the curvature bounded from above and below, we expand the direction normal to the action. □

The converse of Theorem 12.1 is the main theorem of [39]. Namely

THEOREM 12.2 (Cheeger–Gromov). *There exists a positive constant ϵ_n such that if M is an n-dimensional complete Riemannian manifold such that $|K_M| \leqslant 1$ and the injectivity radius is everywhere smaller than ϵ_n, then there exists an F-structure on M.*

Remark 12.1. We can modify Theorem 12.2 so that we do not need to assume that the injectivity radius is small everywhere. Namely we can consider any M with $|K_M| \leqslant 1$, and construct the F-structure on $\{p \in M \mid i_M(p) < \epsilon\}$.

Let us sketch the proof of Theorem 12.2 very briefly. We assume $|K_{M_i}| \leqslant 1$ and $\sup i_{M_i} \to 0$ where i_{M_i} is an injectivity radius. We need to construct an F-structure on M_i for large i. There are two steps to do so. One is to construct a torus action on the finite cover locally. The other is to glue them. We explain the first step only. The following is the basis of this step.

LEMMA 12.3. *If a Riemannian manifold X is complete and flat, then there exists a compact flat submanifold S without boundary in X such that X is diffeomorphic to the normal bundle of S.*

This lemma is a special case of the soul Theorem 16.7 which we will discuss in Section 16. By using Theorem 10.1, we find that S has a finite cover which is a flat torus. So we can find a torus action on the finite cover of X in Lemma 12.3. To use Lemma 12.3 in our situation, we proceed as follows. Let $p_i \in M_i$ and $\epsilon_i = i_{M_i}(p_i)$. We consider the metric $g_i' = g_{M_i}/\epsilon_i$ and consider the limit (M_i, g_i'). (The limit is taken with respect to the pointed Hausdorff distance which we define in Section 16.) Since the curvature of (M_i, g_i') goes to zero and since the injectivity radius of (M_i, g_i') at p_i is 1 we have a flat manifold X as a limit. Also a neighborhood of p_i is diffeomorphic to a compact subset of X for large i.

This is a very rough sketch. Actually the gluing part (which we do not discuss here) is harder. □

In the case when we do not assume a diameter bound, there are several possible ways to define collapsing. One definition is that the injectivity radius becomes small everywhere. The other one is that the volume becomes small. (Note Theorem 5.6 implies that they are equivalent in the case when the diameter and the absolute value of the sectional curvature are bounded.) We call the first one (injectivity radius is small) the collapse and the second one (volume is small) the volume collapse. There is an example of a manifold which admits an F-structure but does not admit a volume collapsed metric. Actually $\mathbb{C}P^2$ admits an F-structure but we can use the fact that its Euler number is nonzero to prove the nonexistence of a volume collapsed metric. (This example is due to Januszkiewicz. See [57, p. 229] or [39].) Cheeger–Gromov defined a notion of polarized F-structure which implies the existence of a volume collapsed metric. However we do not know whether a volume collapsed manifold has a polarized F-structure. So the following problem is still open.

DEFINITION 12.2 [72]. A *minimal volume* $\mathrm{MinVol}(M)$ of a compact manifold M without boundary is the infimum of the volume (M, g) where g is a Riemannian metric on M such that $|K_g| \leqslant 1$.

PROBLEM 12.1. Does there exists a positive number ϵ_n depending only on n with the following properties? If n-dimensional compact manifold M satisfies $\mathrm{MinVol}(M) < \epsilon_n$, then $\mathrm{MinVol}(M) = 0$.

There are several partial results toward Problem 12.1.

THEOREM 12.4 (Rong [131,130]). *In case the dimension of M is* 3 *or* 4, *Problem* 12.1 *is affirmative.*

There is a very sharp result in the case when M admits a metric of constant negative curvature.

THEOREM 12.5 (Besson–Courtrois–Gallot [21]). *Let an n-dimensional manifold M admit a metric of constant curvature g_0. Then if g is any metric on M with* Ricci $\geqslant -(n-1)$ *we have* $\mathrm{Vol}(M, g) \geqslant \mathrm{Vol}(M, g_0)$.

Theorem 12.5 in particular implies that $\mathrm{MinVol}(M) = \mathrm{Vol}(M, g_0)$.

The answer to Problem 12.1 is affirmative under an additional assumption on the diameter.

THEOREM 12.6 (Cheeger–Rong [40]). *There exists a positive number $\epsilon(n, D)$ depending only on n and D with the following properties. If an n-dimensional compact manifold M has a Riemannian metric g such that $|K_g| \leqslant 1$,* Diam $\leqslant D$ *and* $\mathrm{Vol}(M, g) < \epsilon_n$, *then for any ϵ there exists a Riemannian metric g_ϵ on M such that $|K_{g_\epsilon}| \leqslant 1$ and* $\mathrm{Vol}(M, g_\epsilon) < \epsilon$.

We next describe the result of [34]. We remark that the results in the last section do not give enough description in the case when the diameter is not bounded. On the other hand, if we consider the case of an almost flat manifold, for example, the F-structure corresponds to the action of the center of the nilpotent group, and hence only a part of the collapsed direction is described by the F-structure. So we need a local action of a nilpotent group to describe collapsing Riemannian manifolds in the general case. Such a structure may be called an N-structure. One trouble to define it rigorously is that the noncommutativity of the group makes it harder to describe a compatibility condition. To have a simplified description we remark the following fact. In the situation of Theorem 12.2, we can approximate the metric by one invariant of the F-structure. (Actually the original metric is "almost invariant" by the action and we can take the average so that it is strictly invariant.) So in place of writing compatibility of actions, we may state that the actions are isometric with respect to the metric nearby (which is independent of the chart).

Note that the fact that we can approximate the metric by an invariant one, is also true in a modified sense for the almost flat manifold and in the situation of Theorem 11.2. Namely we can make the metric "invariant" of the action of a nilpotent group. We need to remark however the following. In case of $\Gamma \setminus N$ (where N is a nilpotent group and Γ is a discrete subgroup), for example, the almost flat metric is *not* an invariant of the *right* action of N. Since the induced metric on $\Gamma \setminus N$ is well defined only if we start with the *left* invariant metric on N, it means that the group acting on $\Gamma \setminus N$ (equipped with an almost flat metric) by isometries is only the center of N. In other words, we can find an isometric action of N only after taking the infinite (universal) cover. This point is different from the case of Abelian group (torus).

Now we are going to state the main result of [34]. Let M be a manifold and $p \in M$. Let $U_p \subset M$ be an open neighborhood of p. We denote by ∇^g the Levi-Civita connection of g.

THEOREM 12.7 (Cheeger–Fukaya–Gromov [34]). *For each $\epsilon > 0$ and $n \in \mathbb{Z}_+$, there exists $\rho = \rho(\epsilon, n) > 0$ with the following properties. Let (M, g) be a complete n-dimensional Riemannian manifold with $|K_g| \leqslant 1$. Then there exists a metric g_ϵ and $U_p, \tilde{U}_p, \Gamma_p, N_p$ for each $p \in M$ such that*:

(1) *N_p is nilpotent. $\Gamma_p \subset N_p$ is a discrete subgroup such that $\pi_0(N_p)$ is finite and N_p is generated by its connected component $N_{p,0}$ and Γ_p.*

(2) *U_p is a neighborhood of p and $U_p \supseteq B_p(\rho, M)$.*
(3) *N_p acts on $(\tilde{U}_p, \tilde{g}_p)$ by isometry. Here $\tilde{U}_p$ is a covering space of U_p and $\tilde{g}_p$ is the metric induced by g_ϵ.*
(4) *If $\tilde{p} \in \tilde{U}_p$ and $[\tilde{p}] = p$ then $i_{\tilde{U}_p}(\tilde{p}) > \rho$.*
(5) $[\Gamma_p : \Gamma_p \cap N_{p,0}] < k$.
(6) *For any $x \in \tilde{U}_p$ $\mathrm{Diam}(\Gamma_p \setminus N_p x) < \epsilon$. Here $N_p x$ is an N_p-orbit.*
Moreover we have
(7) $e^{-\epsilon} g < g_\epsilon < e^{\epsilon} g$.
(8) $|\nabla^g - \nabla^{g_\epsilon}| < \epsilon$.
(9) *$|\nabla^{g_\epsilon} R_{g_\epsilon}| < c(n, i, \epsilon)$, where R_{g_ϵ} is the curvature tensor of g_ϵ and $c(n, i, \epsilon)$ depends only on n, i, ϵ.*

Remark 12.2. The existence of g_ϵ satisfying (7)–(9) is proved by [15,1].

Remark 12.3. The metric satisfying (1)–(6) is called a (ρ, k)-round metric in [34].

We remark that at the point where $i_M(p) > \rho$ we may take $N_p = 1$ and $\tilde{U}_p = U_p$. Hence the statement above is obvious.

On the other hand, condition (4) implies that at the point p where injectivity radius is small, the group N_p is nontrivial. Hence, together with (1), we obtain a local action of a torus by restricting the action of N_p to the center. Using (6) and the fact that the local action of the torus is compatible with the metric g_ϵ, we can prove that these actions are compatible in the sense of Theorem 12.2.

Moreover, in the case when the diameter of M is smaller than a constant depending only on ϵ and n, we can prove that the group N_p is independent of the choice of p. Hence its orbits defines a foliation on the frame bundle of M. It implies Theorem 11.5. Thus Theorem 12.7 unifies two approaches for collapsing Riemannian manifolds.

The proof of Theorem 12.7 is a combination of the proofs of Theorems 12.2 and 11.5. We use Theorem 11.5 and its proof (together with some improvement) to find N_p locally. We then glue them in a way similar to the proof of Theorem 12.2. Finally we take the average and obtain the required metric g_ϵ. □

EXAMPLE 12.2. Let Γ be a lattice of a semisimple Lie group G of noncompact type and G/K be a symmetric space. We assume $\Gamma \setminus G/K$ is noncompact. Then for each $p \in G/K$ the group $\Gamma_p = \{g \in \Gamma \mid d(p, gp) < \epsilon\}$ has nilpotent subgroup $\Gamma_{p,0}$ of finite index $[\Gamma_p : \Gamma_{p,0}] < k$ by Theorem 10.4. (We remark that Γ_p may not be contained in a small neighborhood of the unit in G. But its subgroup of finite index is in a small neighborhood of the unit.) The Zariski closure $N_p \subset G$ of $\Gamma_{p,0}$ is a nilpotent group. This is our N_p. The original metric (the metric of symmetric space) is an invariant of left the N_p action.

Hattori [84] found the following. Let $M = \Gamma \setminus G/K$ be a locally symmetric space of noncompact type. We assume that it is noncompact and of finite volume. Then the limit $(M, g_M/R)$ as R goes to infinity is a cone of a simplicial complex T which is called the Tits building. (Here the limit is taken with respect to the pointed Gromov–Hausdorff distance (Definition 16.3).) Now if we take a simplex Δ of T then a "neighborhood" of it in M is diffeomorphic to $\Delta \times [0, \infty) \times \Gamma(\Delta) \setminus N(\Delta)$. The dimension of the nilmanifold

$\Gamma(\Delta) \setminus N(\Delta)$ depends on Δ. They are glued appropriately, which gives a structure as in Theorem 12.7.

The following addendum to Theorem 12.7 is useful for various applications.

PROPOSITION 12.8 [132]. *If $a \geqslant K_M \geqslant b$ in Theorem 12.7 then we may choose g_ϵ so that $a + \epsilon \geqslant K_M \geqslant b - \epsilon$.*

13. Collapsing Riemannian manifolds—III

In this section, we review some of the applications of collapsing Riemannian manifolds. We recall that $\mathfrak{M}_n(D)$ is the set of isometry classes of n-dimensional Riemannian manifold M with $\mathrm{Diam}(M) \leqslant D$, $|K_M| \leqslant 1$.

THEOREM 13.1 (Fang–Rong [51], Petrunin–Tuschmann [126]). *For each n, D the number of diffeomorphism classes of simply connected manifolds M in $\mathfrak{M}_n(D)$ with finite $\pi_2(M)$ is finite.*

THEOREM 13.2 (Fang–Rong [51], Petrunin–Tuschmann [126]). *There exists $i(n,\delta) > 0$ such that if M is simply connected, $\pi_2(M)$ is finite and if $1 \geqslant K_M \geqslant \delta > 0$, then the injectivity radius of M is larger than $i(n,\delta)$.*

We remark that, in case the dimension is even, Theorem 13.2 follows from Theorem 4.3 without assumption on π_2.

EXAMPLE 13.1. We first consider the Lens space $S^3/\mathbb{Z}_p$ where $\mathbb{Z}_p \subset S^1$ is a cyclic group of order p. Its curvature is 1 and its limit is $S^2 = S^3/S^1$. This example shows the assumption on $\pi_1(M)$ is necessary both in Theorems 13.1, 13.2.

The three examples below show that the assumption on $\pi_2(M)$ is also necessary in Theorems 13.1, 13.2.

We consider the Lie group $SU(3)$. It has a metric with positive sectional curvature. We consider its maximal torus $T^2 \subseteq SU(3)$. Let p_i, q_i be coprime integers such that $\lim p_i/q_i = \alpha \in \mathbb{R} \setminus \mathbb{Q}$. We identify $T^2 = \mathbb{R}^2/\mathbb{Z}^2$ and let x, y be coordinates of $\mathbb{R}^2$. We consider $S_i^1 = \{[x,y] \in T^2 \mid y = p_i x/q_i\}$. We equip $M_i = S_i^1 \setminus SU(3)$ with quotient Riemannian metric. M_i is a sequence of 7-dimensional manifolds of positive curvature. Using the fact $\lim p_i/q_i$ is irrational, we can easily find that the limit of M_i with respect to the Gromov–Hausdorff distance is $T^2 \setminus SU(3)$. We can also prove that the sectional curvature of M_i is uniformly positive. Namely $C \geqslant K_{M_i} \geqslant \delta > 0$ for some δ, C independent of i. (This is a consequence of the fact that p_i/q_i converges. We remark that $\pi_2(M_i) \cong \pi_1(S^1) = \mathbb{Z}$.)

In a similar way we can use the T^2-action to $S^3 \times S^3$ to get a sequence of metrics g_i on $S^2 \times S^3$ with $C \geqslant K_{g_i} \geqslant \delta > 0$ such that $(S^2 \times S^3, g_i)$ converges to $S^2 \times S^2$.

We next consider an action of $T^2 \times T^2$ on $SU(3)$ where the first factor acts by left multiplication and the second factor acts by right multiplication. Using an appropriate family of

$S_i^1 \cong S^1 \subseteq T^2 \times T^2$, Petrunin–Tuschmann [126] (using Eschenburg [50]) found an example of $M_i = S_i^1 \setminus SU(3)$ with $C \geqslant K_{M_i} > \delta > 0$ such that M_i converges to $T^2 \setminus SU(3)/T^2$.

Remark 13.1. A similar π_2-assumption as in Theorem 13.2 was proposed by the author in [57, Remark 15.10]. However [57, Conjecture 15.7] (by the author) turns out to be false. A counterexample (due to Petrunin–Tuschmann) is the last example in Example 13.1.

We now sketch the proof of Theorem 13.1. We start with the following

LEMMA 13.3 (Rong [132]). *If we assume that $\pi_1(M)$ is finite in the situation of Theorem* 11.5 *in addition, then the fiber of $\pi : FM_i \to Y$ in Theorem* 11.5 *is diffeomorphic to a flat manifold.*

Using the fact that the fundamental group of M_i is finite (here we assume $\dim M_i > 2$), it follows easily that the fundamental group of the fiber has an index finite Abelian subgroup. Since the fiber is an infranilmanifold the lemma follows immediately. □

Lemma 13.3 implies that we have an F-structure whose orbits are fibers. (Here our F-structure is one called pure F-structure by Cheeger–Gromov [38]. A pure F-structure is an F-structure such that all the orbits of the local action have the same dimension.) We next apply the averaging process in the proof of Theorem 12.7 to the situation of Theorem 11.5 and of Lemma 13.3. Then we have

LEMMA 13.4. *In the situation of Lemma* 13.3, *we can approximate the Riemannian metric on FM_i by g_ϵ in the same sense as Theorem* 12.7(7)–(9) *so that g_ϵ is an invariant of the local T^k action and of the $O(n)$ action.*

Now we start the proof of Theorem 13.1. We assume that Theorem 13.1 is false. Then there exists a sequence $M_i \in \mathfrak{M}_n(D)$ such that M_i is simply connected, $\pi_2(M_i)$ is finite, and M_i is not diffeomorphic to M_j for $i \neq j$. We take FM_i and may assume that it converges to Y. Since we approximate the metric by one satisfying Theorem 12.7(7)–(9), it follows that Y is a smooth Riemannian manifold. We may replace FM_i by its finite cover $\tilde{F}M_i$ so that it has global a $T^k \times G$-action, where G is a compact group[20] and a T^k-orbits are the fibers of the fibration $\tilde{F}M_i \to Y$. We modify the metric of $\tilde{F}M_i$ so that it is $T^k \times G$-equivariant. The next lemma is the place where we use the key assumption that $\pi_2(M_i)$ is finite.

LEMMA 13.5. *If $\tilde{F}M_i/T^k$ is G-diffeomorphic to $\tilde{F}M_j/T^k$ then $\tilde{F}M_i$ is $T^k \times G$-diffeomorphic to $\tilde{F}M_j$.*

In fact the torus bundle $T^k \to E \to B$ is determined by the (T^k-analogue of) Euler class $\in \operatorname{Hom}(H_2(B), \pi_1(T^k))$ (which is well defined up to $\operatorname{Aut}(\pi_1(T^k))$). In our case where $\pi_2(\tilde{F}M_j)$ is finite and $\pi_1(\tilde{F}M_j)$ is trivial, the Euler class is an isomorphism

[20] Actually it is finite covering group of $O(n)$. (It may be disconnected.)

$H_2(B)/\mathrm{Tor} \to \pi_1(T^k)$, hence it is unique up to $\mathrm{Aut}(\pi_1(T^k))$. To obtain the T^k-equivariant diffeomorphism $\tilde{F}M_i \to \tilde{F}M_j$ which is G-equivariant also, we use the center of mass technique (Proposition 8.3). □

We remark that $\tilde{F}M_j/T^k$ has the same dimension as Y and $\tilde{F}M_j/T^k$ converges to Y with respect to the G-Gromov–Hausdorff topology (which was introduced in [52]). Estimate (9) of Theorem 12.7 implies that Y is a smooth manifold. On the other hand, the sectional curvature of $\tilde{F}M_j/T^k$ is bounded from below. Hence Theorem 11.3 implies that $\tilde{F}M_j/T^k$ is diffeomorphic to Y for large i. We can use the G-equivariant version of Theorem 11.3 (which can be proved in the same way as Theorem 11.3 using an embedding to Hilbert space as in [55][21]), $\tilde{F}M_i/T^k$ is G diffeomorphic to Y for large i. Hence Lemma 13.5 implies that $\tilde{F}M_j$ is G diffeomorphic to $\tilde{F}M_j$ for i, j large. Namely M_i is diffeomorphic to M_j. This is a contradiction. □

To prove Theorem 13.2 we need another result by Petrunin–Rong–Tuschmann.

THEOREM 13.6 [125]. *Let M be a compact manifold. We assume that M admits a sequence of metrics g_i. We assume that $\Lambda \geqslant K_{g_i} \geqslant \lambda$ and that the metric space $X = \lim^{GH}_{i\to\infty}(M, g_i)$ is of dimension strictly smaller than M. We also assume that the distance function $d_i : M \times M \to \mathbb{R}$ induced by g_i converges to a function d which determines a pseudometric[22] on M.*

Then $\lambda \leqslant 0$.

Remark 13.2. Klingenberg and Sakai conjectured a similar statement, but their conjecture does not assume the additional assumption that d_i converges to a pseudometric.

To prove Theorem 13.2 using Theorem 13.6 we proceed as follows. We assume that there exists M_i with $1 \geqslant K_{K_i} \geqslant \delta > 0$ and that the injectivity radius goes to 0. We can discuss in the same way as in the proof of Theorem 11.3 to show that M_i is diffeomorphic to M_j.[23] By looking at the proof carefully we may assume that the diffeomorphism almost preserves distance function. Namely if we identify M_i with M_j then the sequence $M = M_i = M_j$ satisfies the assumption of Theorem 13.6. The conclusion of Theorem 13.6 contradicts to $K_{M_i} \geqslant \delta > 0$. □

One of the ideas of the proof of Theorem 13.6 is the following observation. If the collapsing occurs in the same way as in the proof of Proposition 11.1, then the sectional curvature of the plain spanned by X and other vector is always converges to zero. To make this simple idea works we need a lot of delicate work which is not described here. □

We next discuss some other applications of collapsing theory.

[21] See Remark 9.1.

[22] Namely it satisfies axioms of metric except "$d(x, y)$ implies $= y$".

[23] We use Proposition 12.8 to show $\Lambda + \epsilon \geqslant K_{M_i} \geqslant \lambda - \epsilon$.

THEOREM 13.7 (Rong [132]). *There exists $w(n,\delta)$ such that if a compact n-dimensional Riemannian manifold M satisfies $1 \geqslant K_M \geqslant \delta$ then there exists a cyclic subgroup C of the fundamental group $\pi_1(M)$ such that $[\pi_1(M):C] < w(n,\delta)$.*

Remark 13.3. If we assume $1 \geqslant K_M \geqslant 0$, $\mathrm{Diam}(M) < D$, then there exists an Abelian subgroup C of $\pi_1(M)$ such that $[\pi_1(M):C] < w(n,D)$ [132]. There are results under milder assumption that is the case when M is of almost of nonnegative curvature. See Section 19.

The following is another application of collapsing theory. This time we apply to manifolds of almost nonpositive curvature.

THEOREM 13.8 (Fukaya–Yamaguchi [58]). *There exists $\epsilon(D,n)$ such that if a compact n-dimensional Riemannian manifold M satisfies* $\mathrm{Diam}(M) \leqslant D$, $\epsilon(D,n) \geqslant K_M \geqslant -1$ *then the universal covering space of M is diffeomorphic to $\mathbb{R}^n$.*

This is a generalization of Hadamard–Cartan's theorem (Theorem 4.6), which is the case when $K_M \leqslant 0$.

14. Morse theory of distance function

So far we mainly discussed results assuming the curvature to be bounded from above and below. From this section on, we consider the case when the curvature is bounded from below only.

The next theorem is a corollary of Theorem 4.1.

THEOREM 14.1 (Rauch). *Let M be a compact manifold without boundary. If there exists a Morse function on M with two critical points, then M is homeomorphic to a sphere.*

In Section 4, we started with Theorem 4.1 and showed the way to prove sphere theorems, finiteness theorems and compactness theorems by estimating the number of balls we need to cover a manifold. The number of contractible open subsets one needs to cover the space (plus one), is called the Lusternik–Shnirel'man category and is important in Morse theory. In this section we will try to apply Morse theory directly.

For a given Riemannian manifold M, a function which is determined automatically from the metric is a distance function $d_p(x) = d(p,x)$ from a point. (Note we can use the fact that $p = x$ is the unique critical point of $x \mapsto d_p(x)$ with $d(x,p) < i_M(p)$ to prove that $B_p(r,M)$ is diffeomorphic to a sphere if $r < i_M(p)$.)

The difficulty to apply Morse theory to the distance function is that $x \mapsto d_p(x)$ is not differentiable for $d(x,p) > i_M(p)$. (d_p is not differentiable at p either. But this does not cause serious trouble. We may consider d_p^2 instead, for example.) During the proof of Theorem 2.5, Grove–Shiohama applied Morse theory away from the ball with radius $= i_M(p)$. After that, their method is used in many other places. The main idea of them is the following definition.

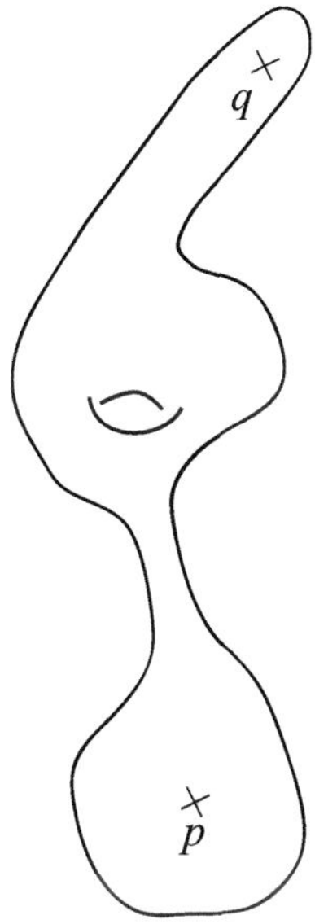

Fig. 14.1.

DEFINITION 14.1. We say q is a *regular point* of d_p if there exists a nonzero vector $\vec{V} \in T_qM$ such that for any minimal geodesic $\ell : [0, d(p,q)] \to M$ joining p and q, the angle between $\overline{\frac{d\ell}{dt}(0)}$ and $\vec{V}$ is not greater than $\pi/2$.

For example, let p, q be as in Figure 14.1. It is not clear how many minimal geodesics are there joining p with q. But it is easy to see that the direction of any of them is downwards at q. Hence q is a regular point of d_p.

Remark 14.1. We may consider various situations similar to Definition 14.1. For example, let us consider a continuous function f which is an infimum of finitely many differentiable functions f_α locally (namely $f = \inf f_\alpha$).[24] In this case we say q is a regular point of f if there exists a vector $\vec{V} \in T_qM$ such that, for each α with $f(q) = f_\alpha(q)$, we have $\vec{V}(f_\alpha) > 0$. We can apply a similar argument to a linear combination of finitely many d_p's or the infimum of them also. Proposition 14.2 holds for such cases.

Based on Definition 14.1, we can prove the following analogue of the Morse lemma for d_p.

PROPOSITION 14.2. *If q with $a \leqslant d_p(q) \leqslant b$ is an arbitrary regular point of d_p, and if $B_p(b, M)$ is compact, then $B_p(b, M) \setminus B_p(a, M) = \{q \in M \mid a \leqslant d(p,q) \leqslant b\}$ is homeomorphic to a direct product of $\partial B_p(b, M) = \{q \in M \mid d(p,q) = b\}$ and* $[0, 1]$.

The proof is similar to the proof of the following famous

[24] We remark that d_p may not satisfy this condition in general.

THEOREM 14.3 (Morse lemma). *We assume that $f : M \to \mathbb{R}$ is differentiable, and arbitrary q with $f(q) \in [a,b]$ is a regular point of f, and that $f^{-1}([a,b])$ is compact. Then $f^{-1}([a,b])$ is diffeomorphic to $f^{-1}(\{a\}) \times [0,1]$.*

The proof of Morse lemma uses an integral curve of $\operatorname{grad} f$. (See [103].) Since d_p is not differentiable, the vector field $\operatorname{grad} d_p$ does not make sense. Instead, we will use the vector field V constructed below.

For $q \in B_p(b,M) \setminus B_p(a,M)$ let $V_q = V$ be the vector $\in T_qM$ as in Definition 14.1. If we can take V_q depending smoothly on q, then we can take the vector field $V(q) = V_q$ in place of $-\operatorname{grad} f$. (The condition in Definition 14.1 implies that d_p decreases along the integral curve of V.)

To find V_q depending smoothly on q, we proceed as follows. We first take $\tilde{V}_q$ which may not depend smoothly on q. We extend it to its neighborhood and denote it by the same symbol $\tilde{V}_q$. Then if q' is in a small neighborhood $U(q)$ of q, then the vector $\tilde{V}_q(q') \in T_{q'}M$ satisfies the condition of Definition 14.1. We cover $B_p(b,M) \setminus B_p(a,M)$ by finitely many $U(q_i)$'s. We then take a partition of unity χ_i and put

$$V(q) = \sum \chi_i(q)\tilde{V}_{q_i}(q).$$

It is easy to see that this V has the required properties.

Using this vector field V, the proof of Proposition 14.2 goes in the same way as the proof of Morse lemma. □

To apply Morse theory of d_p to the proof of Theorem 2.5 we need the following lemma.

LEMMA 14.4. *We assume that M satisfies the assumption of Theorem 2.5. Let $p, q \in M$ with $d(p,q) = \operatorname{Diam}(M)$, and $x \in M$ be a point different from p, q. Let $\ell_p : [0, d(p,x)] \to M$, $\ell_q : [0, d(q,x)] \to M$ be minimal geodesics joining x to p, and x to q, respectively. (In case there are several of them, we assume any of them have the property below.)*

Then the angle between two tangent vectors $\frac{d\ell_p}{dt}(0)$ and $\frac{d\ell_q}{dt}(0) \in T_xM$ is greater than $\pi/2$ (see Figure 14.2).

The proof of Lemma 14.4 uses Toponogov's comparison theorem (Theorem 4.7). Under the assumption of Lemma 14.4 (that is $K_M \geqslant 1/4$), Theorem 4.7 implies the following Sublemma 14.5. Let $x, y, z \in M$. We consider the geodesic triangle whose vertices are those three points. We denote the length of its edges by $|xy|$ etc. and angles by $\angle xyz$ etc. We put $X = |yz|$, $Y = |zx|$, $Z = |xy|$.

SUBLEMMA 14.5. *If $\angle zxy \leqslant \pi/2$, then $\cos\frac{X}{2} \geqslant \cos\frac{Y}{2}\cos\frac{Z}{2}$.*

Note that we have

$$s(Y/2, Z/2, \theta, 1) \leqslant s(Y/2, Z/2, \pi/2, 1) = \cos^{-1}(\cos Y/2 \cos Z/2),$$

where $s(\cdot,\cdot,\cdot,\cdot)$ is as in Theorem 4.7.

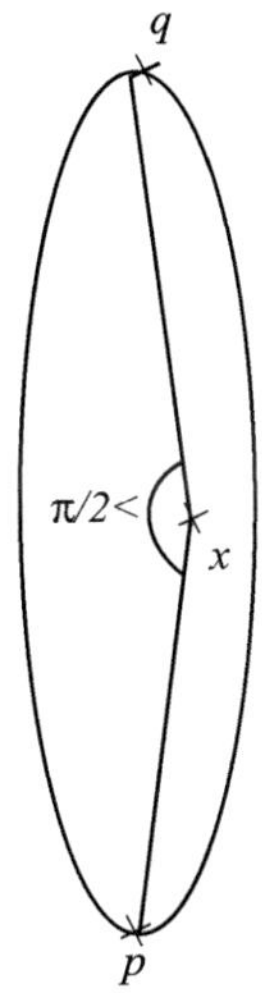

Fig. 14.2.

We start the proof of Lemma 14.4. We put $|\ell_p| = t$, $|\ell_q| = s$, $d(p,q) = D$. Since d_p attains its maximum at q it follows that q is not a regular point of d_p. Hence there exists a geodesic ℓ joining p and q such that the angle between ℓ and ℓ_q is not greater than $\pi/2$. We apply Sublemma 14.5 to the geodesic triangle consisting of ℓ, ℓ_p, ℓ_q and obtain

$$\cos\frac{t}{2} \geqslant \cos\frac{s}{2}\cos\frac{D}{2}. \tag{14.1}$$

Since $D/2 > \pi/2$ we have $\cos\frac{D}{2} < 0$. Therefore one of $\cos\frac{s}{2}$, $\cos\frac{t}{2}$ is positive. We may assume $\cos\frac{s}{2} > 0$.

If the angle between ℓ_p and ℓ_q is not greater than $\pi/2$, then we can again apply Sublemma 14.5 and obtain

$$\cos\frac{D}{2} \geqslant \cos\frac{s}{2}\cos\frac{t}{2}. \tag{14.2}$$

Since $\cos\frac{s}{2} > 0$, (14.1), (14.2) imply

$$\cos\frac{D}{2} \geqslant \cos^2\frac{s}{2}\cos\frac{D}{2}.$$

We remark that $0 < D/2, s < \pi$.[25] This is then a contradiction. □

Now Lemma 14.4 implies that if $x \neq p, q$ then x is a regular point of d_p, d_q. In fact, let V be the tangent vector of ℓ_q at x. It follows from Lemma 14.4 that the vector field

[25] This is a consequence of Myers' theorem (Theorem 5.4).

V satisfies the condition in Definition 14.1. Namely x is a regular point of d_p. Now we can use Proposition 14.2 to prove that M is homeomorphic to sphere. Hence we proved Theorem 2.5. □

We remark that we proved Proposition 4.4 during the proof of Lemma 14.4. In fact, we proved $\cos t/2 > 0$ or $\cos s/2 > 0$ there. It implies $t < \pi$ or $s < \pi$. □

The method we explained above is very useful to study Riemannian manifolds under the bounds of sectional curvature from below. It is also useful to study Alexandrov spaces (see Sections 17, 18).

Theorem 2.5 is a sphere theorem. There are several finiteness theorems corresponding to it. The first one is the following, which is called the Gromov's Betti number estimate.

THEOREM 14.6 (Gromov [70]). *There exists $C(n)$ such that if an n-dimensional compact Riemannian manifold M satisfies $K_M \geqslant -\kappa$ $(\kappa \geqslant 0)$ and if its diameter is D then*

$$\sum_k \operatorname{rank} H_k(M; F) \leqslant C(n)^{1+\kappa D}.$$

Here F is an arbitrary field.

Note in the case when $\kappa = 0$, the right-hand side is independent of D.

It follows from Theorem 14.6 that the connected sum of sufficiently many copies of $\mathbb{C}P^2$ does not carry a metric of nonnegative sectional curvature.

The proof of Theorem 14.6 is based on Morse theory of a kind of distance function. Namely we use an idea similar to the Morse inequality to estimate the Betti number in terms of the number of critical points. However the proof is more involved since Morse theory of the distance function itself does not work. The actual proof requires more complicated argument, which we omit here.

There are many other applications of Morse theory of distance functions to metric Riemannian geometry. For example, Gromov used it to show that a complete manifold M such that $0 > -a^2 \geqslant K_M \geqslant -b^2$ and has finite volume is diffeomorphic to the interior of a compact manifold with boundary [67].

Let us add a few more remarks to Theorem 2.5. If we assume $1 \geqslant K_M \geqslant 1 - \epsilon$ in addition in Theorem 2.2, then we can show that M is not only diffeomorphic but is also close to a sphere as a Riemannian manifold. Namely if a sequence of n-dimensional simply connected Riemannian manifolds M_i satisfies $1 \geqslant K_{M_i} \geqslant 1 - 1/i$, then M_i converges to S^n with standard metric with respect to the Gromov–Hausdorff distance.

On the contrary, the corresponding statement in the situation of Theorem 2.5 does not hold. Namely, let us consider a sequence of Riemannian manifolds M_i such that $K_{M_i} \geqslant 1$ and that the diameter of M_i converges to π as i goes to infinity. Then Theorem 2.5 implies that M_i is homeomorphic to a sphere. However it is *not* true that the limit of M_i with respect to the Gromov–Hausdorff distance is isometric to the sphere with standard metric. We remark however a Riemannian *manifold* with diameter $= \pi$ and $K_M \geqslant 1$ (actually weaker assumption Ricci $\geqslant n - 1$ is enough) is isometric to the sphere. (Theorem 21.11 [148,41].)

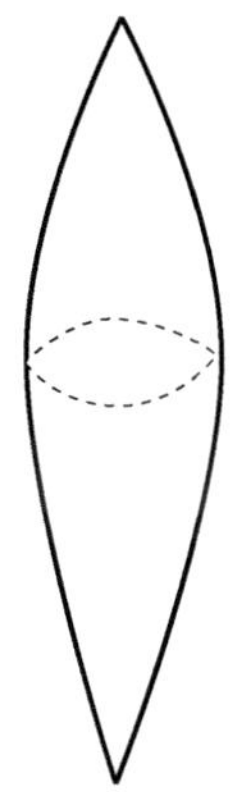

Fig. 14.3.

In fact, let us consider the quotient of S^2 by the action of $S^2/\mathbb{Z}_p$ generated by the rotation of angle $2\pi/p$ around the fixed axis. The quotient is a Riemannian manifold with constant curvature 1, except two points where the axis intersects with S^2. We approximate the quotient space by a Riemannian manifold with curvature $\geqslant 1$ and obtain a sequence of Riemannian manifolds M_i whose diameter converges to π and $K_{M_i} \geqslant 1$. The limit is $S^2/\mathbb{Z}_p$ and is not isometric to the sphere with standard metric. The essential point here is that the Alexandrov space X with diameter $= \pi$ and $K_M \geqslant 1$ is not necessary isometric to a sphere with standard metric. (Compare Theorem 23.11.) (See Figure 14.3.)

This is related to the fact that the limit of Riemannian manifolds M_i with $K_{M_i} \geqslant$ const is rather different from a Riemannian manifold even in the case when the limit has the same dimension. For example, we consider a boundary S of a convex set in $\mathbb{R}^3$. There is a point of S that has no tangent plane. In the situation when the absolute value of the sectional curvature is bounded, the Gromov–Hausdorff convergence is equivalent to the $C^{1,\alpha}$-convergence of the metric tensor (in the situation when the limit has the same dimension), by Theorem 3.4. Therefore the limit space has a tangent space everywhere.

By the reason we explained above the following question is yet open.

PROBLEM 14.1. Is there any $\epsilon_n > 0$ such that if M is an n-dimensional complete Riemannian manifold with $K_M \geqslant 1$ and $\mathrm{Diam}(M) \geqslant \pi - \epsilon_n$, then M is *diffeomorphic* to a sphere?

We remark that in the proof of Theorem 2.5 we consider the distance functions d_p, d_q simultaneously where p, q lie in the different sides from x. This is similar to the notion strainer used in Alexandrov space. (See Section 17.)

15. Finiteness theorem by Morse theory

In this section, we explain the idea of the proof of Theorem 3.5. The first half of it, which was proved in [78], asserts that the number of homotopy classes represented by an element of $\mathfrak{M}'_n(D, v)$ is finite. (We recall that $M \in \mathfrak{M}'_n(D, v)$ if $K_M \geqslant 1$, $\mathrm{Diam}(M) \leqslant D$, and $\mathrm{Vol}(M) \geqslant v$, $\dim M = n$.) In this section we mainly explain this part. The key of the proof is the following proposition.

PROPOSITION 15.1. *There exists $\epsilon = \epsilon(n, D, v) > 0$ such that the following holds for each $M \in \mathfrak{M}'_n(D, v)$. Let $p, q \in M$ with $d(p, q) < \epsilon$, $p \neq q$. Then q is a regular point of d_p.*

Moreover we have the following. We put $\Delta = \{(x, x) \in M \times M \mid x \in M\}$, $\Delta(\epsilon) = \{(x, y) \mid d(x, y) < \epsilon\}$. Then Δ is a deformation retract of $\Delta(\epsilon)$. The deformation retraction $H : \Delta(\epsilon) \times [0, 1] \to \Delta(\epsilon)$ can be chosen so that the length of the curve $t \mapsto H(p, q, t)$ is not greater than $Cd(p, q)$. Here C depends only on n, D, v.

Using Proposition 15.1, the proof of Theorem 3.5 goes in a way similar to the proof of Proposition 5.5. Namely, from the first half of the Proposition 15.1, we find that the metric balls $B_p(\epsilon, M)$ of radius ϵ are contractible in M. On the other hand, the number of metric balls $B_p(\epsilon, M)$ we need to cover M is estimated in the same way as in Section 5 by using Proposition 5.2. However since it is not clear whether the intersection of finitely many metric balls $B_p(\epsilon, M)$ is contractible or not in our case, so we need to modify the proof of Proposition 5.5 a bit. The second half of Proposition 15.1 is used for this purpose. We omit this part of the proof. □

The proof of Proposition 15.1 is closely related to the proof of Proposition 5.6. So we first sketch the proof of Proposition 5.6. By Theorem 4.9 we only need to estimate the length ϵ of closed geodesic of minimal length from below for $M \in \mathfrak{M}_n(D, v)$. Let $\ell : S^1 \to M$ be the closed geodesic of length ϵ. We take an arbitrary point $x \in M$, and let $\ell(t) \in \ell(S^1)$ be the point of smallest distance from x. Then ℓ is orthogonal to $\overline{x\ell(t)}$ at $\ell(t)$.[26] (Here $\overline{x\ell(t)}$ is a minimal geodesic joining x and $\ell(t)$.) We put $\ell(0) = p$. Since $d(p, \ell(t)) \leqslant \epsilon$, it follows that if $d(x, p)$ is sufficiently larger than ϵ, then the angle between ℓ and $\overline{xp}$ is close to $\pi/2$. We thus have proved the following lemma.

LEMMA 15.2. *Let $\delta, \rho > 0$. Then there exists ϵ depending only on n, D, v, δ, such that if ℓ is a closed geodesic with length $< \epsilon$ and if $\ell(0) = p$ then M is contained in the image of the exponential map of the domain $\subset T_pM$ in Figure* 15.1.

We can choose δ sufficiently small compared to the diameter D, so that the volume of the image of the domain II in Figure 15.1 is smaller than $v/2$. By choosing ρ small we may assume the volume of the image of the domain I in Figure 15.1 is smaller than $v/2$ also. Therefore if there exists a closed geodesic of length $< \epsilon$, then the volume of M is smaller than v. □

We turn to the proof of Proposition 15.1. It suffices to show the following lemma.

[26] More precisely in case x is a cut point with respect to the geodesic ℓ. (The notion of the cut point with respect to a submanifold is defined in a similar way to the notion of cut point from a point. See, for example, [33].) ℓ may not be orthogonal to $\overline{x\ell(t)}$. However this does not cause a trouble for the proof of Lemma 15.2 since the measure of the set of cut points is zero.

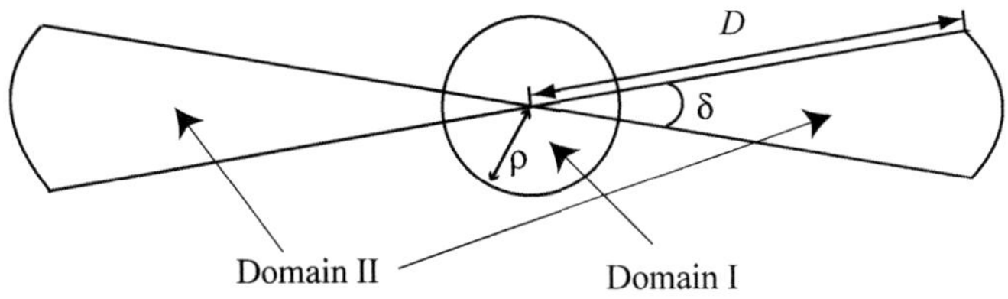

Fig. 15.1.

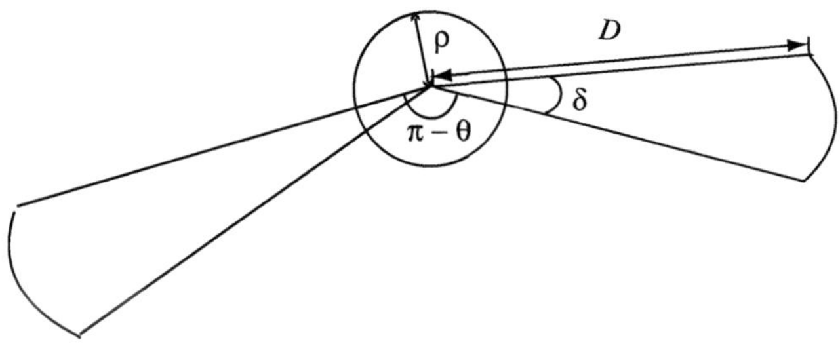

Fig. 15.2.

LEMMA 15.3. *There exist $\theta = \theta(n, v, D) > 0$ and $\epsilon = \epsilon(n, v, D) > 0$ with the following properties. Let $M \in \mathfrak{M}'_n(D, v)$, $p, q \in M$, $d(p, q) < \epsilon$. Let ℓ_1 and ℓ_2 are minimal geodesics joining p and q. Then the angle between ℓ_1 and ℓ_2 at p or q is smaller than $\pi - \theta$.*

Lemma 15.3 implies that q in the lemma is a regular point of d_p. The first half of Proposition 15.1 follows from Proposition 14.2. The second half can also be proved in the same way by examining the proof of Proposition 14.2 carefully. □

The proof of Lemma 15.3 is similar to the proof of Proposition 5.6. Namely we replace Figure 15.1 by Figure 15.2.

We thus explained an outline of the first half of the proof of Theorem 3.5. The other half is the finiteness of the number of homeomorphism classes and requires another deep argument. The main new technique required is the idea from controlled surgery. □

16. Soul theorem and splitting theorem

Typical results on noncompact complete Riemannian manifolds of nonnegative curvature are the soul theorem and the splitting theorem. They also are very useful to study the local structure of the Gromov–Hausdorff limit of Riemannian manifolds or its limit.

We first explain why the study of noncompact manifolds is useful to study local structure of the limit space. Let us begin with the introducing some notations. Let X be a metric space and $\ell : [a, b] \to X$ be a continuous map (that is a curve). The length $|\ell|$ of ℓ is by definition a supremum of the sum

$$\sum d(\ell(t_i), \ell(t_{i+1})),$$

where $a = t_0 < t_1 < \cdots < t_N = b$ runs over all partitions (N moves also).

DEFINITION 16.1. We say that X is a *length space* if for each $p, q \in X$ there exists a curve joining p, q of length $d(p, q)$.

A complete Riemannian manifold is a length space. The Gromov–Hausdorff limit of length spaces is also a length space.

DEFINITION 16.2. A complete metric space is said to be *compactly generated* if all of its metric balls are compact.

The set of all isometry classes of compact metric spaces is complete with respect to the Gromov–Hausdorff distance. A natural metric to put on the set of all isometry classes of complete compactly generated spaces is pointed the Gromov–Hausdorff distance, which we define below.

DEFINITION 16.3. Let X, Y be metric spaces and $x \in X$, $y \in Y$. We say that the *pointed Gromov–Hausdorff distance* $d_{pGH}((X, x), (Y, y))$ between (X, x) and (Y, y) is not greater than ϵ, if the Gromov–Hausdorff distance between the metric balls $B_{1/\epsilon}(x, X)$ and $B_{1/\epsilon}(y, Y)$ is not greater than ϵ. We write $\lim^{pGH}_{i\to\infty}(X_i, x_i) = (X, x)$ if $\lim_{i\to\infty} d_{pGH}((X_i, x_i), (X, x)) = 0$.

The following can be proved in the same way as Theorem 3.2.

THEOREM 16.1. *The set of all isometry classes of a pair (M, p) of an n-dimensional Riemannian manifold M with* $\mathrm{Ricci}_M \geqslant -(n-1)$ *and a point p on it is relatively compact with respect to the pointed Gromov–Hausdorff distance.*

Now we can define the tangent cone. Let (X, d_X) be a length space and $x \in X$.

DEFINITION 16.4. If the limit $\lim^{pGH}_{c\to\infty}((X, c d_X), x)$ exists, we call it the *tangent cone* (at $x \in X$) and write it as $T_x X$.

If X is an n-dimensional Riemannian manifold then the tangent cone of X is isometric to $\mathbb{R}^n$ at each point.

EXAMPLE 16.1. Let $\Omega \subset \mathbb{R}^n$ be a compact convex set. We put $X = \partial\Omega$ and define a length metric on it. (Namely the distance between $x, y \in X$ is the infimum of the length of all curves joining x and y in X.)

Then tangent cone T_xX is described as follows. We consider every ray (half of the straight line) $\ell:[0,\infty)\to\mathbb{R}^n$ such that $\ell(t)\in\Omega$ for small $t>0$. The set of such ℓ is an open subset of $\mathbb{R}^n$. Its boundary in $\mathbb{R}^n$ is the tangent cone T_xX.

If the space X is not so wild then we may expect the tangent cone T_xX exists and a neighborhood of x in X is homeomorphic to a neighborhood of the origin (base point) in T_xX. (This holds for Alexandrov spaces, for example. See Theorem 18.1.) Namely we can study the local structure of X by studying the tangent cone T_xX.

If X is a Gromov–Hausdorff limit of a sequence of Riemannian manifolds M_i and if the sectional curvature of M_i is bounded from below by a constant independent of i, then we may regard the limit X as the space with "curvature bounded from below". Then the infimum of the "curvature" of family of length spaces (X, cd_X) as c goes to infinity will become nonnegative. (Note if we multiply the metric by c then the curvature is multiplied by c^{-2}.) This means that if tangent cone of X exists, then it is of "nonnegative curvature". (The discussion here is informal and heuristic. So for a moment the curvature may either to the Ricci or the sectional curvature.) This is one of the reasons why the study of noncompact spaces with nonnegative curvature is important in the local theory of spaces which are a limit of Riemannian manifolds.

By using Gromov's precompactness theorem (Theorem 16.1) we have the following

PROPOSITION 16.2. *Let M_i be a sequence of Riemannian manifolds with* $\mathrm{Ricci}_{M_i} > -(n-1)$. *Let $X=\lim^{GH}_{i\to\infty} M_i$. Let $x\in X$ and c_k be a sequence of positive numbers with* $\lim c_k=+\infty$. *Then there exists a subsequence of $((X, c_k d_X), x)$ which converges in pointed Gromov–Hausdorff distance.*

In general $((X, c_k d_X), x)$ itself may not converge. (Namely we need to take a subsequence.) Hence X may not have a tangent cone. This is one of the difficulties to study the family of Riemannian manifolds with Ricci curvature bounded from below.

In case X is a limit of Riemannian manifolds with *sectional* curvature bounded from below (or more generally if X is an Alexandrov space), $\lim_{c\to\infty}((X, cd_X), x)$ converges without taking a subsequence (Theorem 17.14).

Let us now state the soul theorem and the splitting theorem. We first define the notions line and ray. Let X be a length space. A curve $\ell:(a,b)\to X$ is called a geodesic if it is length minimizing locally. Namely ℓ is a geodesic if, for each $t\in(a,b)$, there exists ϵ such that $d(\ell(t-\epsilon),\ell(t+\epsilon))$ is equal to the length of the restriction of ℓ to $(t-\epsilon,t+\epsilon)$. We use arc length as a parameter in the next definition.

DEFINITION 16.5. Let X be a length space. A geodesic $\ell:[0,\infty)\to X$ is called a *ray* if $d(\ell(t),\ell(s))=|t-s|$ for any t,s. A geodesic $\ell:(-\infty,\infty)\to X$ is called a *line* if $d(\ell(t),\ell(s))=|t-s|$ for any t,s.

(The difference between line and ray is the domain of its definition.)

If there exists a tangent cone $T_xX=\lim_{c\to\infty}((X, cd_X), x)$ then it is a union of its rays ℓ such that $\ell(0)$ is the base point. We also have the following

LEMMA 16.3. *Let X be a length space and $\ell:(-\epsilon,\epsilon)\to X$ be a minimal geodesic with $\ell(0)=x$. If the tangent cone T_xX exists, then it contains a line.*

In fact, since in (X, cd_X) there exists a minimal geodesic of length $c\epsilon$ containing the origin, its limit in T_xX will be a line.

We assume that a complete metric space X is a length space and satisfies one of the following conditions.

CONDITIONS 16.1.

(a) X is a Riemannian manifold of nonnegative sectional curvature.
(b) $X=\lim^{pGH}_{i\to\infty} M_i$ such that $K_{M_i}\geqslant -\epsilon_i$, $\lim_{i\to\infty}\epsilon_i=0$ and $\dim X=\dim M_i$.
(c) X is a Riemannian manifold with nonnegative Ricci curvature.
(d) $X=\lim^{pGH}_{i\to\infty} M_i$ such that $\mathrm{Ricci}_{M_i}\geqslant -\epsilon_i$, $\lim_{i\to\infty}\epsilon_i=0$ and $\mathrm{Vol}(M_i)\geqslant v>0$.

The next theorem is called the splitting theorem.

THEOREM 16.4. *If X satisfies one of the Conditions* 16.1 *and contains a ray, then X is isometric to a direct product $\mathbb{R}\times X_0$.*

Theorem 16.4 is due to Toponogov [149] in case (a), to Cheeger–Gromoll [35] in case (c), Grove–Petersen [80] and Yamaguchi [153] in case (b) and Cheeger–Colding [28] in case (d).

We will explain an idea of the proof of the cases (a), (c) later in this section. (Case (b) is similar to case (a). Case (d) is discussed in Section 23.)

We explain more how to apply it to study the local structure of the limit space. Note that we can use Theorem 16.4 repeatedly. Namely if X_0 contains a line then we can again apply the theorem and show that it is a direct product. Therefore if we can repeat it $\dim X$ times, then we can prove that $X=\mathbb{R}^n$. Lemma 16.3 implies that if x is an interior point of a minimal geodesic, then T_xX contains a line. Therefore if we can find n $(=\dim X)$ "independent" geodesic for which x is an interior point, then the tangent cone T_xX is isometric to $\mathbb{R}^n$. This may imply that X is a manifold in a neighborhood of x. This argument appears in Sections 17, 18 and in Sections 20, 22, 23.

We next explain an outline of the proof of splitting theorem. The main tool we use is convexity of Busemann function (it is used also in the proof of the soul theorem). Let X be a length space and $\ell:[0,\infty)\to X$ be a ray.

DEFINITION 16.6. The *Busemann function* is the limit $b_\ell(x)=\lim_{t\to\infty}(t-d(x,\ell(t)))$.

PROPOSITION 16.5. *If X satisfies either* (a) *or* (b), *then the Busemann function of its ray ℓ is convex.*

If X satisfies (c), *then the Busemann function of its ray is subharmonic.*

In the situation (d) we cannot define subharmonicity in the usual way. So the argument is more involved. See Section 23 and [26,31].

The proof of Proposition 16.5 is by a comparison theorem. Namely it follows immediately from the Laplacian and Hessian comparison theorem (Theorem 16.6) for the distance function. We remark that the Hessian Hess f of a function f on a Riemannian manifold is defined by

$$(\mathrm{Hess}_x f)(V, W) = V\big(W(f)\big) - (\nabla_V W)(f) \tag{16.1}$$

and is a symmetric bilinear map $T_xM \otimes T_xM \to \mathbb{R}$. A function f is convex if its Hessian Hess f is nonnegative everywhere.

The Laplacian Δf is its trace. Namely

$$\Delta f(x) = \sum_{i=1}^{n} (\mathrm{Hess}\, f)(e_i, e_i), \tag{16.2}$$

where e_i is an orthonormal basis of T_pM. (We remark that we are using the nonpositive Laplacian. Namely $\Delta = -(d^*d + dd^*)$.) We say a smooth function is subharmonic if its Laplacian is nonnegative.

THEOREM 16.6. *Let M be a Riemannian manifold and $p \in M$. We consider the function $d_p(x) = d(p, x)$.*

(1) *If $K_M \geqslant \kappa$ then*

$$\mathrm{Hess}_x\, d_p \leqslant \frac{s'_\kappa(d(p,x))}{s_\kappa(d(p,x))}(g_x - dd_p \otimes dd_p). \tag{16.3}$$

Here $dd_p : T_xM \to \mathbb{R}$ is the exterior derivative of d_p.

(2) *If $\mathrm{Ricci}_M \geqslant \kappa$ then*

$$\Delta f(x) \leqslant (n-1)\frac{s'_\kappa(d(p,x))}{s_\kappa(d(p,x))}. \tag{16.4}$$

Here s_κ is as in (4.1).

Remark 16.1. We remark that d_p is not differentiable outside the ball $B_p(i_M(p), M)$. So we need to be more careful to state Theorem 16.6. Precisely speaking (16.3), (16.4) hold in a barrier sense. See, for example, [26].

We omit the proof of Theorem 16.6. We remark that (16.6) implies Corollary 5.3. In fact

$$\begin{aligned}\frac{d}{dt}\mathrm{Vol}\big(B_p(t, M)\big) &= \int_{\partial B_p(t,M)} \langle \mathrm{grad}\, d_p, \mathrm{grad}\, d_p\rangle \Omega_{\partial B_p(t,M)} \\ &= \int_{B_p(t,M)} \mathrm{div}\,\mathrm{grad}\, d_p \Omega_M\end{aligned}$$

$$\leqslant (n-1)\int_{B_0(t,T_pM)} \frac{s'_\kappa(d(p,x))}{s_\kappa(d(p,x))}\Omega_{\mathbb{R}^n}$$

$$\leqslant \frac{d}{dt}\int_{B_0(t,T_pM)} s_\kappa\big(d(p,x)\big)^n \Omega_{\mathbb{R}^n}$$

$$\leqslant \frac{d}{dt}\operatorname{Vol}\big(B_{p_0}\big(t,\mathbb{S}^n_\kappa\big)\big).$$

Let us explain how we use Proposition 16.5 to prove Theorem 16.4, in cases (a), (b). We assume that X contains a line $\ell:\mathbb{R}\to X$. We then have two rays $\ell_\pm:[0,\infty)\to X$ by

$$\ell_+(t)=\ell(t), \qquad \ell_-(t)=\ell(-t).$$

We study their Busemann functions $b_{\ell_\pm}$. The triangle inequality implies

$$b_{\ell_+}(t)+b_{\ell_-}(t)\leqslant 0. \tag{16.5}$$

By Proposition 16.5 the right-hand side is convex. Since a bounded convex function is constant, it follows that $\ell_+(t)+\ell_-(t)$ is constant. (Actually it is 0.) It follows that $\ell_+(t)=\text{const}-\ell_-(t)$ is convex and is concave. Hence its level surface if totally geodesic. (Here we say $S\subset M$ is totally geodesic if any minimal geodesic of M joining two points of S are contained in S.) This implies Theorem 16.4. In case (c) we use subharmonicity in place of convexity. □

We next discuss the soul theorem.

THEOREM 16.7 (Cheeger–Gromoll [36,128]). *If a complete Riemannian manifold M has nonnegative sectional curvature then there exists a compact submanifold $S\subseteq M$ without boundary, such that M is diffeomorphic to the normal bundle of S. Moreover S is totally geodesic.*

We call S the *soul* of M. The basis of the proof of Theorem 16.7 is Proposition 16.5. It asserts that, for each ray $\ell:[0,\infty)\to M$, the Busemann function b_ℓ is convex. In particular for any c the closed set

$$H(\ell,c)=\{x\in M \mid b_\ell(x)\leqslant c\}$$

is convex. The next lemma is the key of the proof of Theorem 16.7. We fix $p\in M$ and let $\operatorname{Ray}(p)$ be the set of all rays of M such that $\ell(0)=p$.

LEMMA 16.8. *The set $C_c(p)=\bigcap_{\ell\in\operatorname{Ray}(p)}H(\ell,c)$ is compact.*

The proof is by contradiction. Namely we assume that $C_c(p)$ is not compact and let $p_i\in C_c(p)$ be a divergent sequence. We put $d(p,p_i)=t_i$, and let $\ell_i:[0,t_i]\to M$ be a minimal geodesic such that $\ell_i(0)=p$, $\ell_i(t_i)=p_i$ and that it is parameterized by arc length.

Since $\frac{d\ell_i}{dt}(0) \in T_pM$ is a unit vector, we may take subsequence so that it converges. Let $\ell : [0, \infty) \to M$ be a geodesic such that

$$\lim_{i\to\infty} \frac{d\ell_i}{dt}(0) = \frac{d\ell}{dt}(0).$$

Since $\lim_{i\to\infty} t_i = \infty$, it follows that ℓ is a ray. On the other hand, we have

$$\lim_{i\to\infty} b_\ell(p_i) = \infty.$$

This contradicts to $p_i \in C_c(p)$. □

Thus, we obtained a compact convex subset $C_c(p)$ of M. We can find a compact convex submanifold S in it. The argument to do so is rather technical and is omitted. (See [33, Chapter 8].)

Perelman [115] proved that if, in the situation of Theorem 16.7, there exists a point where $K_M > 0$, then the soul S is a one point. We refer to [63] for other topics related to the soul theorem.

We remark that we already applied Theorem 16.7 in Section 12 to construct on F-structure.

17. Alexandrov space—I

In this section and in the next sections, we discuss recent developments [22,119,113] in the theory of Alexandrov space. A good text book on the contents of this section is [138]. (See also [127].) In the following sections, we study compactly generated length spaces of finite Hausdorff dimension only. So we always assume that the length space has this property.

The Alexandrov space is a length space with curvature bounded from below. To define the notion of curvature for length space, we use a Toponogov type comparison theorem in the opposite direction. Namely we *define* the condition $K_X \geqslant 1$ by using the conclusion of a comparison theorem. However the conclusion of Theorem 4.7 does not (yet) make sense for length space, since it uses angles. So we consider the following slightly different version.

We use the notation of Theorem 4.7. Let M be a Riemannian manifold and $x, y, z, v, w \in M$. Let $x', y', z', v', w' \in \mathbb{S}^n(\kappa)$. We assume $v \in \overline{xy}$, $w \in \overline{xz}$, $v' \in \overline{x'y'}$, $w' \in \overline{x'z'}$.

THEOREM 17.1 (Alexandrov–Toponogov). *We assume $K_M \geqslant \kappa$ and $d(x, y) = d(x', y')$, $d(x, z) = d(x', z')$, $d(y, z) = d(y', z')$, $d(x, v) = d(x', v')$, $d(x, w) = d(x', w')$. $u \in \overline{xy}$, $u' \in \overline{x'y'}$, $v \in \overline{xz}$, $v' \in \overline{x'y'}$. Then we have $d(v, w) \geqslant d(v', w')$. (See Figure* 17.1.)

DEFINITION 17.1 (Alexandrov). A length space of finite dimension is said to be an *Alexandrov space* with $K \geqslant \kappa$ if the conclusion of Theorem 17.1 holds for X.

Remark 17.1. There are several other definitions equivalent to Definition 17.1. We will explain them later (Theorems 17.9 and 17.10).

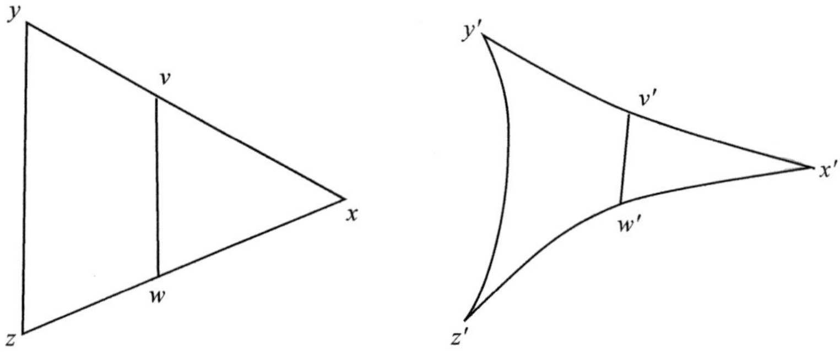

Fig. 17.1.

Remark 17.2. There is a notion of Alexandrov space with curvature bounded from *above*. We do not discuss it in this article. It is proved by Berestovskij that if a length space is an Alexandrov space with curvature bounded from above and below then it is a C^0-Riemannian manifold. This result is related to Theorem 3.4 but was proved earlier than that. See [17].

Hereafter we say Alexandrov space for Alexandrov space with $K \geqslant \kappa$ with some κ.

The notion of Alexandrov space was introduced by Alexandrov [6] more than 50 years ago. There are several related pioneering works around those old days, like Busemann [23]. In [22], Burago–Gromov–Perelman proved several fundamental theorems on Alexandrov spaces. After that the study of Alexandrov space became very active and important in metric Riemannian geometry. Their main results are

THEOREM 17.2 (Burago–Gromov–Perelman [22]). *Let X be an Alexandrov space. Then there exists a dense open subset X_0 such that, for each $p \in X_0$, there exists a neighborhood U_p and a Lipschitz homeomorphism $U_p \to V_p$ where $V_p \subset \mathbb{R}^n$ is an open set.*

THEOREM 17.3 (Burago–Gromov–Perelman). *The Hausdorff dimension of an Alexandrov space is an integer and is equal to its topological dimension.*

Remark 17.3. There are several ways to define topological dimension, that is covering dimension (big and small), inductive dimension, etc. Theorem 17.3 also implies that they coincide for Alexandrov spaces.

We do not discuss the proof of Theorem 17.3. (It will follow from Corollary 18.3 in the next section.) Before explaining some of the ideas of the proof of Theorems 17.2, we give some examples of Alexandrov spaces.

EXAMPLE 17.1. (0) A Riemannian manifold (M, g) is an Alexandrov space with $K \geqslant \kappa$, if and only if the sectional curvature of (M, g) is greater than κ everywhere.

(1) Let $\Omega \subseteq \mathbb{R}^n$ be a compact and convex domain. Let $S = \partial\Omega$. We define the length metric d on S. Namely, the distance between $x, y \in S$ is the minimum of the length of the

curves in S joining x with y. Then we can prove that (S, d) is an Alexandrov space of curvature $\geqslant 0$.

(2) Let M be a Riemannian manifold with $K_M \geqslant \kappa$ and G be a compact group acting on M by isometry. Then the quotient space M/G, equipped with the quotient metric is an Alexandrov space.

An important example of an Alexandrov space is a Gromov–Hausdorff limit of an Riemannian manifold. Actually we have

PROPOSITION 17.4. *Let X_i be a sequence of compact length spaces and $X = \lim_{i\to\infty}^{GH} X_i$. If X_i are Alexandrov spaces with $K \geqslant \kappa$, then so is X. (Here κ is independent of i.)*

The proof is elementary.

Remark 17.4. Yamaguchi [153] proved that if M a C^∞-manifold and G is a compact Lie group acting smoothly on M, then there exists a sequence of metrics g_i on M such that $K_{g_i} \geqslant \kappa$ for some κ independent and (M, g_i) converges to M/G.

Another source of examples is a cone, which we define below.

DEFINITION 17.2. Let (Y, d) be a metric space. We consider the product $T \times [0, \infty)$ and identify $(x, 0)$ and $(y, 0)$. We thus obtain a space CY. We define a cone metric on it as follows:

$$d\big((x,t),(y,s)\big) = \sqrt{t^2 + s^2 - 2st\cos d(x,y)}.$$

We denote by $\mathbf{o} \in CY$ the equivalence class of $(x, 0)$.

EXAMPLE 17.2. If $Y = S^n$ with $K_{S^n} \equiv 1$, then CS^n is isometric to $\mathbb{R}^{n+1}$.

LEMMA 17.5. *If Y is a length space and* $\mathrm{Diam}(Y) \leqslant \pi$, *then CY is a length space.*

We can prove an analogue of Myers' theorem (Theorem 5.4) for Alexandrov spaces. Namely

THEOREM 17.6 [22]. *If M is an Alexandrov space with $K \geqslant 1$, then* $\mathrm{Diam}(Y) \leqslant \pi$.

THEOREM 17.7 [22].

(1) *If CY is an Alexandrov space, then Y is an Alexandrov space with $K \geqslant 1$.*
(2) *If* $\dim Y > 1$ *and Y is an Alexandrov space with $K \geqslant 1$ then CY is an Alexandrov space with $K \geqslant 0$.*
(3) *In case* $\dim Y = 1$, *the cone CY is an Alexandrov space with $K \geqslant 0$ if and only if* $\mathrm{Diam}(Y) \leqslant \pi$.

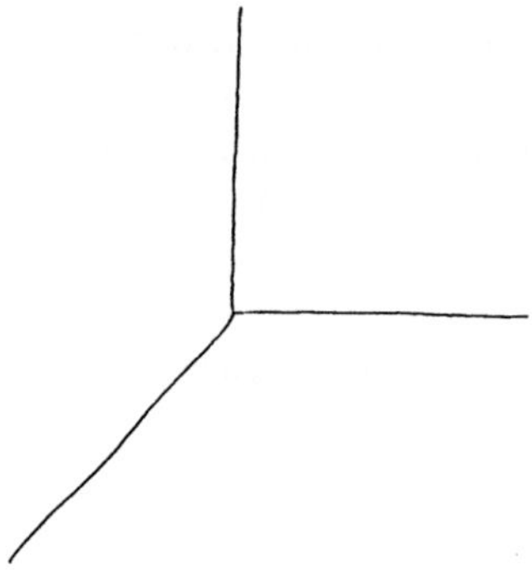

Fig. 17.2.

We do not discuss the proof.

We next discuss an example of a length space which is not an Alexandrov space.

EXAMPLE 17.3. Let us consider a simplicial complex X consisting of three arcs which are joined at one point o. (See Figure 17.2.) We can define a metric on it such that the length of each arc is 1. Let x, y, z be interior points of each of the three simplexes, respectively. We can choose $v = w$ on $\overline{xy} \cap \overline{xz} = \overline{xo}$. Then $d(v, w) = 0$. But if we choose x', y', z', v', w' as in Theorem 17.1 then $d(v', w') > 0$. (For any κ.) So the conclusion of Theorem 17.1 does not hold. Namely X is not an Alexandrov space.

The argument of Example 17.3 implies the following. We call a map $\ell : (a, b) \to X$ a *geodesic* if for each $c \in (a, b)$ there exists ϵ such that the length of the restriction of ℓ to $(c - \epsilon, c + \epsilon)$ is $d(\ell(c - \epsilon), \ell(c + \epsilon))$.

LEMMA 17.8. *If ℓ_1, ℓ_2 are geodesics on an Alexandrov space X and if they coincide on an open set, then their union is also a geodesic.*

In other words, geodesics can never branch.

We next explain some other equivalent definitions of Alexandrov spaces.

THEOREM 17.9. *Let X be a length space. We assume that for each $p \in X$ there exists a neighborhood U such that the conclusion of Theorem* 17.1 *holds for any $x, y, z, u, v \in U$. Then X is an Alexandrov space with $K \geqslant \kappa$. In other words, the same conclusion holds globally.*

In fact, usually the assumption of Theorem 17.9 is the definition of Alexandrov space.

We discuss another equivalent definition. Let X be a length space and $x, y, z \in X$. Let $\kappa \in \mathbb{R}$. In case $\kappa > 0$, we assume $d(x, y), d(x, z), d(y, z) < \pi/\sqrt{\kappa}$. We choose $x', y', z' \in \mathbb{S}^n(\kappa)$ such that $d(x, y) = d(x', y')$, $d(y, z) = d(y', z')$, $d(x, z) = d(x', z')$. We define

$$\angle_\kappa yxz = \angle y'x'z'.$$

THEOREM 17.10. *Let X be a length space.*

(1) *If, for each $p \in X$, there exists a neighborhood U of p such that*

$$\angle_\kappa bac + \angle_\kappa cac + \angle_\kappa cab \leqslant 2\pi$$

for and $a, b, c, d \in U$, then X is an Alexandrov space with $K \geqslant \kappa$.

(2) *Let X be an Alexandrov space with $K \geqslant \kappa$ and let $a, b, c, d \in X$. Then*

$$\angle_\kappa bac + \angle_\kappa cac + \angle_\kappa cab \leqslant 2\pi.$$

Remark 17.5. By Theorem 17.6 $\angle_\kappa bac$ etc. in (2) is well defined.

The idea that if the comparison theorem holds locally, then it holds globally is due to Alexandrov and Toponogov. Theorem 17.10 is proved in [22].

We next discuss the angle between geodesics. Hereafter we assume that geodesics are parameterized by arc length. Let X be an Alexandrov space with $K \geqslant \kappa$ and $\ell_1, \ell_2 : [0, c) \to X$ be geodesics such that $p = \ell_1(0) = \ell_2(0)$.

LEMMA 17.11. *If $s_1 \leqslant t_1$, $s_2 \leqslant t_2$ then*

$$\angle_\kappa \ell_1(s_1) p \ell_2(s_2) \geqslant \angle_\kappa \ell_1(t_1) p \ell_2(t_2).$$

This follows easily from the definition. Therefore we can define

DEFINITION 17.3. $\angle \ell_1 \ell_2 = \lim_{t_1, t_2 \to 0} \angle_\kappa \ell_1(t_1) p \ell_2(t_2)$.

In case ℓ_1, ℓ_2 are minimal geodesics joining p to x, y, respectively, we write $\angle xpy = \angle \ell_1 \ell_2$.

Remark 17.6. (1) The angle $\angle xpy$ is independent of κ.

(2) Two geodesics ℓ_1, ℓ_2 coincide to each other if $\angle \ell_1 \ell_2 = 0$.

THEOREM 17.12. *If X is an Alexandrov space of $K \geqslant \kappa$ and $x, y, z \in X$, then we have $d(y, z) \leqslant s(d(x, y), d(x, z), \angle yxz, \kappa)$.*

Here s is defined in (4.4). In other words, Theorem 4.7 holds for an Alexandrov space. The other version of the triangle comparison theorem also holds.

THEOREM 17.13. *If X is an Alexandrov space with $K \geqslant \kappa$ and $x, y, z \in X$, then we have $\angle yxz \geqslant \angle_\kappa yxz$.*

Remark 17.7. The Toponogov type comparison theorem holds in Alexandrov space. Hence we can generalize the argument of the last section to prove the splitting theorem (Theorem 16.4) for an Alexandrov space with $K \geqslant 0$.

As we mentioned before an Alexandrov space has a tangent cone.

THEOREM 17.14 [22]. *If* (X,d) *is an Alexandrov space with* $K \geqslant \kappa$ *and* $x \in X$, *then* $\lim_{k\to\infty}((X,kd),x)$ *converges with respect to the pointed Gromov–Hausdorff distance.*

The limit in Theorem 17.14 is the tangent cone T_xX. The tangent cone is related to the angle between geodesics as follows.

DEFINITION 17.4. Let $\tilde{\Sigma}_x^0$ be the set of all geodesics (parameterized by arc length) $\ell:[0,c)\to X$ for some c such that $\ell(0)=x$. We identify ℓ_1 and ℓ_2 if they coincide on a neighborhood of 0. We denote by Σ_x^0 the set of this equivalence relation. We can easily show that the angle $\angle$ defines a metric on it. We define the *space of directions* $\Sigma_x(X)$ as the completion of Σ_x^0.

LEMMA 17.15. *If* X *is an Alexandrov space, then* $\Sigma_x(X)$ *is an Alexandrov space with* $K \geqslant 1$ *and* T_xX *is an Alexandrov space with* $K \geqslant 0$.

THEOREM 17.16 [22]. *The tangent cone* T_xX *of an Alexandrov space* X *is isometric to the cone* $C\Sigma_x(X)$. *If* $\dim X = n$, *then* $\dim \Sigma_x(X) = n-1$ *and* $\dim T_xX = n$.

We remark that the second half of Theorem 17.16 is a consequence of Proposition 17.7.

Now we start the discussion of the proof of Theorem 17.1. As we mentioned in the last section, if $x\in X$ is an interior point of $n=\dim X$ "independent" minimal geodesics, then T_xX is isometric to $\mathbb{R}^n$, and this implies that x has neighborhood, homeomorphic to $\mathbb{R}^n$. However the condition about the existence of a geodesic is a bit too strict. So we relax it a bit. Let X be an Alexandrov space with $K\geqslant\kappa$.

DEFINITION 17.5. Let $x\in X$ and $(a_i,b_i)\in X^2$, $i=1,\dots,n$. We say that $\{(a_i,b_i)\}_{i=1,2,\dots,n}$ is an (n,δ)-*strainer* at x, if

$$\angle_\kappa a_ixb_i \geqslant \pi-\delta,$$

and

$$\angle_\kappa a_ixa_j,\ \angle_\kappa a_ixb_j,\ \angle_\kappa b_ixb_j \leqslant \delta, \quad \text{for } i\neq j.$$

A point $x\in X$ is said to be (n,δ)-*strained* if there exists an (n,δ) strainer at x. (See Figure 17.3.)

Remark 17.8. In [22] the strainer is called "explosion" and a strained point is called "burst point". The name strainer and strained point seems to be more popular now.

The main step of the proof of Theorem 17.2 is the following

PROPOSITION 17.17 ([22, Theorem 9.4] or [138, Theorem 7.4]). *If* $p\in X$ *is an* (n,δ)-*strained point, then there exists* $\rho>0$, *neighborhoods* $U\subset V$ *of* p, *and a map* $\varphi:V\to\mathbb{R}^n$ *with* $\varphi(p)=0$ *with the following properties*:

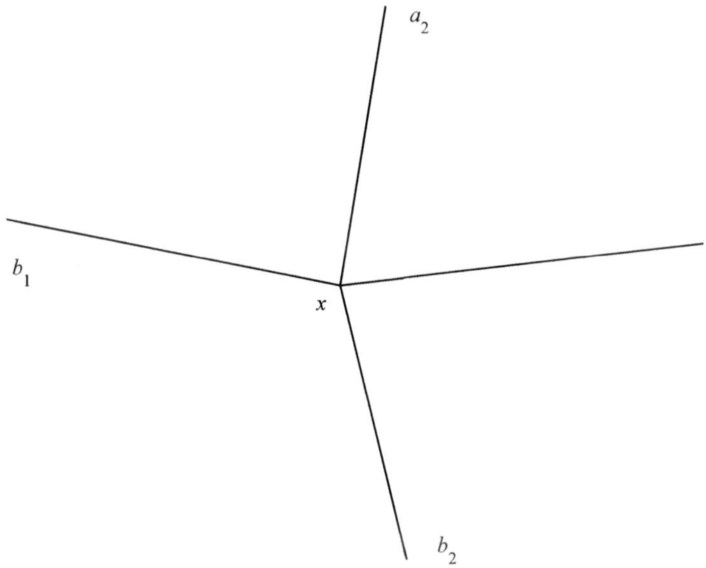

Fig. 17.3.

(1) $d(\varphi(x), \varphi(y)) < 2d(x, y)$.
(2) *Let* $x \in U$ *and* $X \in \mathbb{R}^n$, *with* $d(\varphi(x), X) < \rho$. *Then there exists* $y \in V$ *such that* $\varphi(y) = X$ *and* $d(x, y) \leqslant Cd(\varphi(x), X)$, *where* C *depends only on* n *and* δ.

We remark that (2) implies that φ is an open mapping in a neighborhood of x. Hence if φ is injective then φ gives a chart in a neighborhood of x. We can use the following to show φ is injective.

LEMMA 17.18. *We may choose* U *small enough so that if* φ *is not injective then there exists an* $(n + 1, 10\delta)$*-strained point on a small neighborhood of* U.

We remark that the set of all the (n, δ)-strained points is open. On the other hand Proposition 17.17 implies that if a (n, δ)-strained point exists, then the Hausdorff dimension is not smaller than n.

Hence Proposition 17.17 and Lemma 17.18 imply the following. For each open set U, we can find n and a nonempty open subset $U_0 \subset U$ consisting of (n, δ)-strained points such that there are no $(n + 1, 10\delta)$-strained points on U. Then U_0 is an n-dimensional manifold by Proposition 17.17 and Lemma 17.18. The proof of Theorem 17.2 then will be completed by using the next lemma.

LEMMA 17.19 [22, Corollary 6.5]. *We assume* X *is connected. If* U, V *are nonempty open subsets of* X, *then the Hausdorff dimension of* U *is equal to the Hausdorff dimension of* V.

We now sketch the proof of Proposition 17.17 and Lemmas 17.18, 17.19. We put

$$\mu = \inf\{d(p,a_1),\dots,d(p,a_n),d(p,b_1),\dots,d(p,b_n)\}.$$

We first explain the idea of the proof of Proposition 17.17. We put

$$\varphi(x) = -\big(d(x,a_1),\dots,d(x,a_n)\big) + \big(d(p,a_1),\dots,d(p,a_n)\big).$$

It is easy to see that (1) is satisfied. We show (2). For simplicity, we consider the case $x = p$, $n = 2$. For each $X = (X_1, X_2) \in B_0(\rho, \mathbb{R}^2)$, we will find w with $\varphi(w) = (X_1, X_2)$, $d(p,w) \leqslant Cd(0,X)$. We assume $X_1, X_2 > 0$. We first take the point $q_1 \in \overline{pa_1}$ such that $d(p,q_1) = X_1$. We first show

$$\frac{|\varphi(q_1) - (X_1, 0)|}{d(0,X)} \leqslant \tau(\rho,\delta|n,\kappa,\mu). \tag{17.1}$$

In fact we can prove

$$d(q_1,a_2) \geqslant d(p,a_2) - \tau(\rho,\delta|n,\kappa,\mu)X_1$$

by applying Theorem 17.1, where we put $x = a_1$, $y = p$, $z = v = a_2$, $u = q_1$.

To prove the opposite inequality we take the point $p' \in \overline{b_1q_1}$ such that $d(p',q_1) = X_1$. We have $d(p,p') \leqslant X_1\tau(\rho,\delta|n,\kappa,\mu)$. In fact, since $\angle_\kappa b_1pa_1 > \pi - \delta$, it follows that $\angle b_1q_1a_1 \geqslant \angle_\kappa b_1q_1a_1 > \pi - \delta - \tau(\rho|n,\kappa)$. Hence $\angle pq_1b_1 < \delta + \tau(\rho|n,\kappa)$. Theorem 17.12 then implies $d(p,p') \leqslant X_1\tau(\rho,\delta|n,\kappa,\mu)$.

We use $d(p,p') \leqslant X_1\tau(\rho,\delta|n,\kappa,\mu)$ to show

$$|\angle b_1q_1a_2 - \pi/2|, |\angle b_1p'a_2 - \pi/2| < \tau(\rho,\delta|n,\kappa,\mu). \tag{17.2}$$

We next apply Theorem 17.1 again by putting $x = q_1$, $y = b_1$, $z = v = a_2$, $u = p'$. Then using (17.2) have $d(p',a_2) \geqslant d(q_1,a_2) - \tau(\rho,\delta|n,\kappa,\mu)X_1$. Hence $d(p,a_2) \geqslant d(q_1,a_2) - \tau(\rho,\delta|n,\kappa,\mu)X_1$. We have proved (17.1).

We next take $w_1 \in \overline{a_2q_1}$ such that $d(w_1,q_1) = X_2$. Then we have

$$\frac{|\varphi(w_1) - (X_1, X_2)|}{d(0,X)} \leqslant \tau(\rho,\delta|n,\kappa,\mu).$$

We repeat the process replacing p by w_1 and obtain w_2 such that $d(w_1,w_2) < C|\varphi(w_1) - (X_1,X_2)|$ and

$$\frac{|\varphi(w_2) - (X_1, X_2)|}{|\varphi(w_1) - (X_1, X_2)|} \leqslant \tau(\rho,\delta|n,\kappa,\mu).$$

We can define $w_3,\dots$ in a similar way. w_i is a Cauchy sequence whose limit w has the required property. □

Let us prove Lemma 17.18. Let $\varphi(x) = \varphi(y)$. Let $z \in \overline{xy}$ with $d(x,z) = d(x,y)$. It is easy to see that (a_i, b_i), $i = 1, \dots, n$, and (x, y) is an $(n+1, 2\delta)$-strainer if $d(x, y)$ is small. □

Finally we sketch the proof of Lemma 17.19. We may assume X is compact. Take $p \in V$ and put $D = \sup\{d(p,x) \mid x \in U\}$. We take R such that $B_p(D/R, X) \subset V$. We define $\Phi : U \to V$ as follows. For $x \in V$ we take a point $\Phi(x) \in \overline{px}$ such that $Rd(p, \Phi(x)) = d(p,x)$. (Note the minimal geodesic $\overline{px}$ may not be unique. So we need some technical argument to find Φ which is measurable.) Definition 17.1 implies that there exists $\rho > 0$ such that $d(\Phi(x), \Phi(y)) \geqslant \rho d(x, y)$. It follows that the Hausdorff dimension of $\Phi(U)$ is not smaller than the Hausdorff dimension of U. Therefore the Hausdorff dimension of V is not smaller than the Hausdorff dimension of U. We can prove the opposite inequality in the same way. □

We thus finished a sketch of the proof of Theorem 17.2. □

DEFINITION 17.6. We define the *boundary* ∂X of an Alexandrov space X by induction on $\dim X$ as follows. If $\dim X = 1$, then X is either an arc or a circle. So we can define its boundary in an obvious way. Suppose ∂X is defined for X with $\dim X < k$. Let X be an Alexandrov space of $\dim X = k$. Then we say $x \in \partial X$ if $\partial \Sigma_x(X) \neq \emptyset$. (We remark that $\Sigma_x(X)$ is an Alexandrov space and $\dim \Sigma_x(X) = k - 1$.)

Theorem 17.2 is improved by Otsu–Shioya [112]. To state their results we define the notion of singular point set in an Alexandrov space more precisely.

DEFINITION 17.7. Let X be an n-dimensional Alexandrov space and $\delta > 0$. We put

$$\mathcal{S}_\delta(X) = \{x \in X \mid \mathrm{Vol}(\Sigma_x(X)) \leqslant \mathrm{Vol}(S^{n-1}) - \delta\},$$

$$\mathcal{S}(X) = \bigcup_{\delta > 0} \mathcal{S}_\delta.$$

We remark that the Alexandrov space version of the following theorem is a motivation of Definition 17.7.

THEOREM 17.20 (Otsu–Shiohama–Yamaguchi [111]). *If an n-dimensional Riemannian manifold M satisfies* $\mathrm{Vol}(M) \geqslant \mathrm{Vol}(S^n) - \epsilon_n$, $K_M \geqslant 1$ *then M is diffeomorphic to the sphere. Also M is close to S^n with respect to the Hausdorff distance.*[27]

We discuss the idea of the proof of Theorem 17.20 in Section 21.

[27] This theorem is improved later to Theorem 21.7 and to Corollary 22.4. Before [111], Shiohama [137] proved that M is homeomorphic to the sphere under a similar but different assumption $K_M \geqslant -C$, Ricci $\geqslant (n-1)$, $\mathrm{Vol}(M) \geqslant \mathrm{Vol}(S^n) - \epsilon(n, C)$.

THEOREM 17.21 (Burago–Gromov–Perelman, Otsu–Shioya). *Let X be an Alexandrov space of dimension n. Then the Hausdorff dimension of $\mathcal{S}(X)$ is not greater than $n-1$. The Hausdorff dimension of $\mathcal{S}(X) \setminus \partial X$ is not greater than $n-2$.*

THEOREM 17.22 (Otsu–Shioya [111]). *There exists a C^0-Riemannian metric on $X \setminus \mathcal{S}(X)$ which induces the metric on X. Moreover there exists $X_0 \subset X \setminus \mathcal{S}(X)$ such that the (n-dimensional Hausdorff) measure of $X \setminus X_0$ is 0 and that there exists manifold structure of $C^{1.5}$-class and a Riemannian structure is of $C^{0.5}$-class on X_0.*

Remark 17.9. Actually we need to define a $C^{1.5}$-structure etc. in the above theorem. This is because $X \setminus \mathcal{S}(X)$, X_0 are not open subsets in general. Hence they are not manifolds. See [112] for the precise statement.

Theorem 17.22 is used by Kuwae–Machigashira–Shioya [99] to develop analysis on Alexandrov spaces.

We also remark the following

THEOREM 17.23 (Fukaya–Yamaguchi [60]). *The isometry group of an Alexandrov space is a Lie group.*

18. Alexandrov space—II

In [119,113] Perelman proved the following two fundamental results on Alexandrov spaces.

THEOREM 18.1 (Perelman). *Let X be an Alexandrov space with $K \geqslant \kappa$. Then, for any $x \in X$, there exists a neighborhood of x homeomorphic to T_xX, the tangent cone.*

THEOREM 18.2 (Perelman). *Let X_i be a sequence of Alexandrov spaces with $K \geqslant \kappa$ where κ is independent of i. We assume $X = \lim_{i\to\infty}^{GH} X_i$ and $\dim X = \dim X_i$. Then X_i is homeomorphic to X for large i.*

Remark 18.1. Both of these theorems are proved in [119]. Later Perelman published another paper [113] where the proof of Theorem 18.1 is given in a simplified way. Perelman says in [113] that a similar method gives a slight simplification of the proof of Theorem 18.2, but the simplification is not so big compared with the one for Theorem 18.1. Unfortunately the paper [119] is not yet published.

In fact Theorem 18.1 follows from Theorem 18.2 (and Theorems 17.14, 17.16). However the proof of Theorem 18.2 requires Theorem 18.1.

In this section we give a review of the proof of Theorem 18.1. Before that let us mention some of the corollaries of them.

We remark that T_xX is homeomorphic to $C\Sigma_x(X)$ by Theorem 17.16. Since $\Sigma_x(X)$ is again an Alexandrov space, we can apply Theorem 18.1 again. We then find that the singularity of X is obtained locally by taking cones several times. Let us define it more precisely.

DEFINITION 18.1. We define a connected metrizable space X to be an *MCS-space* of dimension n inductively on n as follows.

(1) An MCS-space of dimension 2 is a 2-dimensional manifold with or without boundary.

(2) X is an MCS-space of dimension n if, for each $x \in X$, there exists a neighborhood U of x and an MCS-space Y_x of dimension $n-1$, such that there exists a homeomorphism F from the cone of Y_x to U such that F sends the cone point to x.

The following is immediate from Theorem 18.1.

COROLLARY 18.3. *Every Alexandrov space is an MCS-space.*

The following is also an immediate corollary.

COROLLARY 18.4. *For an Alexandrov space X, there exists X_k with $\bigcup X_k = X$ such that X_k is a k-dimensional topological manifold and $\bar{X}_k = \bigcup_{i \leqslant k} X_i$.*

COROLLARY 18.5. *An Alexandrov space X is locally contractible. If it is compact then $\pi_1(X)$ and $H_k(X)$ are finitely generated.*

Hereafter we assume our Alexandrov space X has no boundary, for simplicity.[28] An idea used in [113] to prove these result is to generalize Morse theory of the distance function to an Alexandrov space. Let us give the following definition. Hereafter X is an Alexandrov space with $K \geqslant -1$. Let $p \in X$. We put $d_p(x) = d(x, p)$.

DEFINITION 18.2. x is said to be a *regular point* of d_p if there exists $\xi \in \Sigma_x(X)$ such that for each minimal geodesic ℓ joining x to p we have $\angle \xi \ell' > \pi/2$. Here $\ell' \in \Sigma_x(X)$ is the equivalence class of ℓ in $\Sigma_x(X)$.

Definition 18.2 is a generalization of Definition 14.1. We can generalize Proposition 14.2 also and we further generalize it to Theorem 18.7. For the proof of Theorem 18.1 we need to use a bit more general function than the distance function and define the "regularity" of a map $X \to \mathbb{R}^k$ for $k > 1$ also. To state this generalization we need some notations.

DEFINITION 18.3. Let U be an open subset of X.

(1) An *admissible function* $f: U \to \mathbb{R}$ is a function of the form

$$f(x) = \sum_{i=1}^{m} a_i \phi_i\big(d(A_i, x)\big), \tag{18.1}$$

[28]The general case can be handled by taking a double $X \cup_{\partial X} X$ which is an Alexandrov space by [124].

where A_i is a compact subset of X, ϕ_i are smooth functions with $0 \leqslant \phi_i' \leqslant 1$ and $a_i \geqslant 0$, $\sum a_i \leqslant 1$.

(2) An *admissible map* $F : U \to \mathbb{R}^k$ is a composition $F = G \circ \vec{f}$, where G is a bi-Lipschitz homeomorphism and $\vec{f} = (f^1, \dots, f^k)$ with admissible functions f_i.

Remark 18.2. In [113] a more general function (map) is called an admissible function (map). But only those in Definition 18.3 are used.

For $A \subset X$ and $x \in X$ we define $\Sigma_x^0(\overline{xA}) \subset \Sigma_x(X)$ by

$$\Sigma_x^0(\overline{xA}) = \bigl(\bigl\{[\ell] \mid \ell \text{ is a minimal geodesic joining } x \text{ to a point of } A\bigr\}\bigr).$$

Let $\Sigma_x(\overline{xA}) \subset \Sigma_x(X)$ be the closure of $\Sigma_x^0(\overline{xA})$.

For $\Lambda_1, \Lambda_2 \subset \Sigma_x(X)$ we put

$$\angle \Lambda_1 \Lambda_2 = \inf\{\angle uv \mid u \in \Lambda_1,\ v \in \Lambda_2\}.$$

For an admissible function f as in (18.1) we can define its direction derivative $D_x f : \Sigma_x(X) \to \mathbb{R}$ by

$$(D_x f)(u) = \sum_i a_i \phi'\bigl(d(x, A_i)\bigr) \cos \angle\bigl(u, \Sigma_x(\overline{xA_i})\bigr).$$

In case X is a manifold $D_x f$ is the direction derivative in the usual sense.

If $f^{(1)}, f^{(2)}$ are admissible functions as in (18.1) we put

$$\begin{aligned}&\bigl\langle D_x f^{(1)}, D_x f^{(2)}\bigr\rangle \\ &\quad = \sum_{i,j} a_i^{(1)} a_j^{(2)} \phi^{(1)\prime}\bigl(d\bigl(x, A_i^{(1)}\bigr)\bigr) \phi^{(2)\prime}\bigl(d\bigl(x, A_j^{(2)}\bigr)\bigr) \cos \angle A_i^{(1)} A_j^{(2)}.\end{aligned}$$

This again coincides with the usual inner product between derivatives in case when X is a manifold and $f^{(1)}, f^{(2)}$ are differentiable.

DEFINITION 18.4. We say $F : U \to \mathbb{R}^k$ where $F = G \circ \vec{f}$ is *ϵ-regular* at $p \in U$ if the following conditions hold. Let us put $\vec{f} = (f^1, \dots, f^k)$.

(1) For each $i \neq j$, we have $\langle D_p f^i, D_p f^j\rangle < -\epsilon$.

(2) There exists $\xi \in \Sigma_p(X)$ such that $(D_p f^i)(\xi) < -\epsilon$, for each i.

We say F is regular if it is ϵ-regular for some $\epsilon > 0$. We say F is ϵ-regular on U if it is ϵ-regular at every point of U.

EXAMPLE 18.1. (1) If $f : U \to \mathbb{R}$ is defined by $f(x) = d_p(x)$. It is an admissible function and hence is an admissible map. It is ϵ-regular at x for some $\epsilon > 0$ if and only if x is a regular point of f in the sense of Definition 18.2.

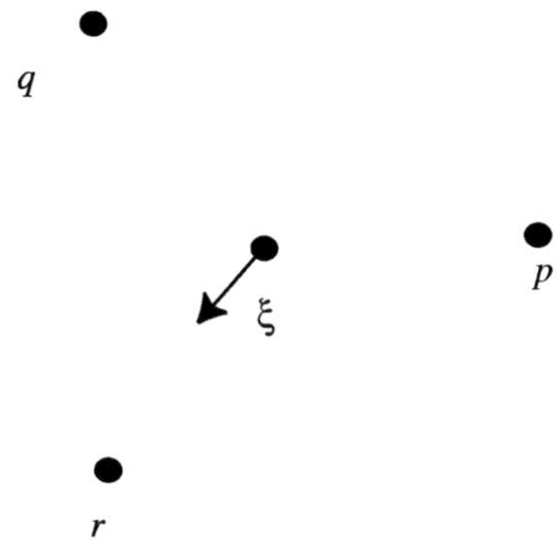

Fig. 18.1.

(2) Let X be a two-dimensional Alexandrov space and $(a_1, b_1), (a_2, b_2)$ be a $(2, \delta)$-strainer at x. Let us define $\varphi : X \to \mathbb{R}^2$ by

$$\varphi(x) = -\big(d(x, a_1), d(x, a_2)\big) + \big(d(p, a_1), (p, a_2)\big),$$

as in the proof of Proposition 17.17. Then φ is a homeomorphism in a neighborhood of x. We put

$$p = \varphi^{-1}(r, 0), \quad q = \varphi^{-1}(-r/2, r\sqrt{2}/2), \quad r = \varphi^{-1}(-r/2, -r\sqrt{2}/2)$$

and define $F = \vec{f} = (d_p, d_q)$. We also set $\xi \in \Sigma_x(\overline{xr})$. We can prove (1), (2) for sufficiently small ϵ. We can generalize this construction to the case of higher dimension and prove that if x is a (k, δ)-strained point, then there exists $F : U \to \mathbb{R}^k$ from a neighborhood of x which is ϵ-regular at x. (See Figure 18.1.)

We can prove the following in a way similar to the proof of Proposition 17.17. (See [116, Lemma 2.3 and the argument just after that].)

LEMMA 18.6. *Let $F : B_x(\rho, X) \to \mathbb{R}^n$ be an admissible map which is ϵ-regular at x. Then there exists a neighborhood $U \subseteq B_x(\rho, X)$ of x and $\delta > 0$, with the following property. If $y \in U$, $X \in \mathbb{R}^k$ with $d(F(v), X) \leqslant \delta$, then there exists $z \in B_x(\rho, X)$ such that $F(z) = X$ and $d(z, y) < Cd(F(v), X)$. Here C depends only on ρ, δ, ϵ.*

Lemma 18.6 implies that F is an open mapping. In case $\dim X = k$ and if there exists an ϵ-regular map at x, then Lemma 18.6 shows that a neighborhood of x is a manifold. In the general case, we have to study the situation where $k < \dim X$. The following Proposition 18.7 is the main result in such a case. We need a definition.

DEFINITION 18.5. A map $F : X \to Y$ between topological spaces is called a *topological submersion* at $x \in X$ if there exists a neighborhood U of x, a neighborhood V of $F(x)$, and a topological space W such that there exists a homeomorphism $\Phi : U \cong V \times W$ satisfying $F = Pr_1 \circ \Phi$ on U.

In case X, Y are smooth manifolds and F is a smooth map, F is a topological submersion if its derivative is of maximal rank.

PROPOSITION 18.7 [113, Theorem 1.4]. *An admissible map $F : X \to \mathbb{R}^k$ is a topological submersion at a regular point.*

The following result is also used in the proof of Theorem 18.1.

THEOREM 18.8 (Siebenmann [142, Corollary 6.14]). *Every proper topological submersion between MCS-spaces is a locally trivial fiber bundle.*

Remark 18.3. (1) We remark that if M, N are smooth manifolds (without boundary) and if $F : M \to N$ is a proper *smooth* submersion then F is a locally trivial fiber bundle. This fact can be proved much more easily than Theorem 18.8.

(2) The proof of Theorem 18.8 is based on the isotopy extension theory. We remark that isotopy extension theory for manifolds (see [48]) was used by Cheeger for the proof of his finiteness theorem. (See Section 6.)

We next sketch the proof of Proposition 18.7. The difficult case is when X is of dimension greater than k. We try to increase k as much as possible, we then arrive at the following situation.

DEFINITION 18.6. Let $F : U \to \mathbb{R}^k$ be a regular admissible map from an open set U of an Alexandrov space X. We say $p \in X$ is *imcomplementable* if there exists no g such that $(f^1, \dots, f^k, g)$ is regular at p.

The case $k = 0$ is included. Namely in that case $p \in X$ is imcomplementable if there exists no admissible function such that p is regular.

EXAMPLE 18.2. (1) Let us consider the domain $\{(r\cos\theta, r\sin\theta) \mid \theta \in [-\alpha, +\alpha],\ r \geqslant 0\}$. We glue $(r\cos\alpha, r\sin\alpha)$ and $(r\cos -\alpha, r\sin -\alpha)$ to obtain a space X_α. We can show that $\mathbf{o} = [0, 0]$ is imcomplementable if and only if $\alpha \leqslant \pi/2$. Actually we put $g = d_{[r,0]}$. Then g is regular if $\alpha > \pi/2$. On the other hand, if $\alpha \leqslant \pi/2$ then the diameter of $\Sigma_{\mathbf{o}} X_\alpha$ is not greater than $\pi/2$. Hence it is easy to see that (2) in Definition 18.4 can never be satisfied.[29] (See Figure 18.2.)

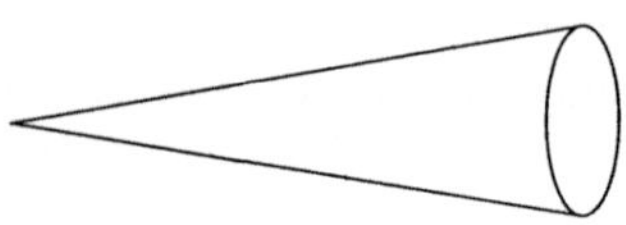

Fig. 18.2.

[29] It is easy to see from this argument that in case $k = 0$ the point $p \in X$ is imcomplementable if and only if $\operatorname{Diam} \Sigma_p(X) \leqslant \pi/2$.

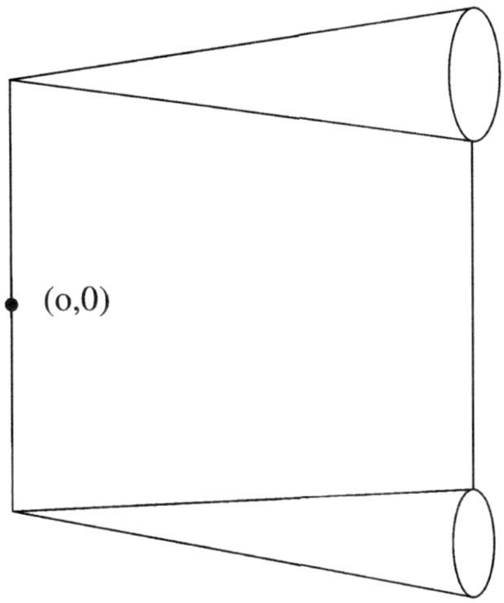

Fig. 18.3.

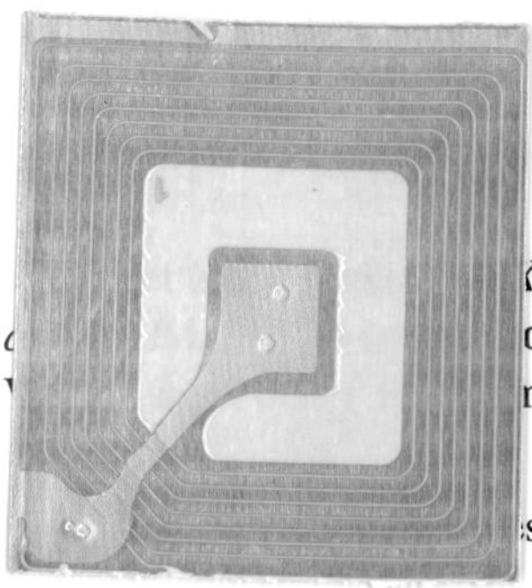

... $X_\alpha \times \mathbb{R}$, where X_α is as above. We define $f : X \to \mathbb{R}$ by $f =$
... $\mathbf{o}, 0)$ is a regular point. Actually we may take $\xi = D_{(\mathbf{o},0)} d_{(\mathbf{o},1)}$.
... nplementable at $(\mathbf{o}, 0)$ if $\alpha \leqslant \pi/2$. (See Figure 18.3.)

... sult in [113] is as follows.

LEMMA 18.9 [113, 1.3]. *If $F : U \to \mathbb{R}^k$ is admissible and regular at $p \in U$, and if p is incomplementable, then there exists an admissible function $g : V \to \mathbb{R}$ defined on an open neighborhood V of p with the following properties. We write $F = G \circ \vec{f}$, $\vec{f} = (f^1, \ldots, f^k)$.*

(1) *$g \leqslant 0$ on V and $g(p) = 0$.*
(2) *$F|_{g^{-1}(0)} : g^{-1}(0) \to \mathbb{R}^k$ defines a homeomorphism onto a neighborhood of $F(p)$.*
(3) *$(F, g) : V \to \mathbb{R}^{k+1}$ is regular on $V \setminus g^{-1}(0)$.*
(4) *There exists $\rho > 0$ such that $\{x \in V \mid d(F(x), F(p)) \leqslant \rho,\ g(x) \geqslant -\rho\}$ is compact.*

Let us show how to choose such g in the case of Example 18.2(2). Namely we have $U = X_\alpha \times \mathbb{R}$ and $F = f = d_{(\mathbf{o},-1)}$. We write a point of U as $([r\cos\theta, r\sin g\theta], t)$ and use r, θ, t as a coordinate. (We take $r \geqslant 0$, $\theta \in [-\alpha, \alpha]$.) Then $f(r, \theta, t) = \sqrt{(t+1)^2 + r^2}$. We take $q = (\delta, 0, \delta)$ and put $h = d_q$. Then $h(r, \theta, t) = \sqrt{(t-\delta)^2 + r^2 + \delta^2 - 2r\delta\cos|\theta|}$. It is easy to see that $(f, h) : U \to \mathbb{R}^2$ is regular outside on $(B_{(\mathbf{o},0)}(\rho, U) \setminus \{\mathbf{o}\}) \times \mathbb{R}$. (We remark $\{\mathbf{o}\} \times \mathbb{R}$ is the set of singular points.)

However if we put $g = h$ then (1), (2) are not satisfied. So we compose it with a homeomorphism of $\mathbb{R}^2$ so that (1), (2) will be satisfied. (See Figure 18.4.)

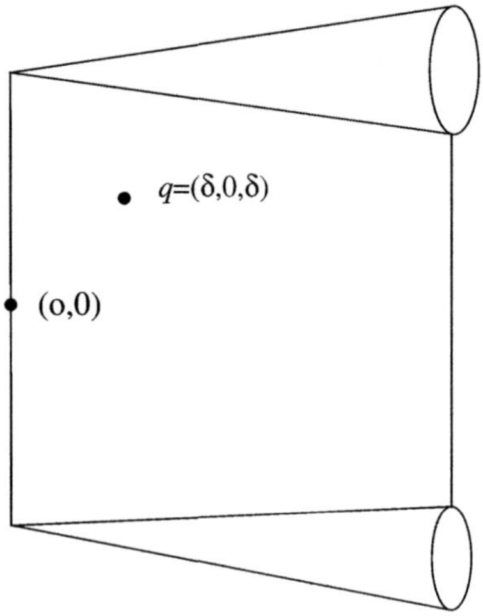

Fig. 18.4.

We consider the set $K(v) = \{x \in U \mid f(x) = v,\ r < \rho\}$ where $|\rho|$ and $|v|$ is small. We can easily check that

$$h(v, 0, 0) = \sup\{h(x) \mid x \in K(v)\}$$

if $\alpha \leqslant \pi/2$.[30] Namely

(∗) The restriction of h to $K(v)$ takes its maximum at a unique point.

We remark that $f(v, 0, 0) = 1 + v$ and $h(v, 0, 0) = \sqrt{(v - \delta)^2 + \delta^2}$. So if we put

$$g(r, v, \theta) = h(r, v, \theta) - \sqrt{\left(f(r, v, \theta) - 1 - \delta\right)^2 + \delta^2},$$

then (1), (2) are satisfied. We define $G : B_{(1,\sqrt{2}\delta)}(\rho, \mathbb{R}^2) \to \mathbb{R}^2$ by

$$G(a, b) = \left(a, b - \sqrt{(a - 1 - \delta)^2 + \delta^2}\right),$$

where $\rho \ll \delta$. Since $(f, g) = G \circ (f, h)$, it follows that (f, g) is admissible. It also satisfies (3) since (f, h) satisfies (3). Thus we constructed g in the case of Example 18.2(2).

In the general case, we need to choose h more carefully so that it is enough "concave". (Then (∗) holds.) The proof of Lemma 18.9 is in [113, Section 3]. □

We can use Lemma 18.9 to complete the proof of Proposition 18.7 as follows. We also prove the following at the same time.

[30] This condition is equivalent to the condition that $(\mathbf{o}, 0)$ is imcomplementable.

PROPOSITION 18.10. *If $F : X \to \mathbb{R}^k$ is an admissible map and if $p \in X$ is a regular point, then $F^{-1}(F(p))$ is an MCS-space near p.*

The proof is by downward induction on k. If $k = \dim X$ then both Theorem 18.7 and Proposition 18.10 follow from Lemma 18.6. Let us assume that Theorem 18.7 and Proposition 18.10 are true for $k+1$ and prove the case k. We remark that both propositions are local statements on p.[31] In case p is not imcomplementable, then we can increase k and use the induction hypothesis. So it suffices to consider the case that p is imcomplementable. We apply Lemma 18.9 to get g. Then $(F, g)|_{V \setminus g^{-1}(0)} : V \setminus g^{-1}(0) \to \mathbb{R}^{k+1}$ is regular. We can use the induction hypothesis to conclude that it is a topological submersion and the fibers are MCS-spaces. Therefore $V \setminus g^{-1}(0)$ is an MCS-space. Let $(F, g)^{-1}(B_0(\rho, \mathbb{R}^k) \times (-\rho, 0)) = W$. Since

$$F : W \to B_0(\rho, \mathbb{R}^k) \times (-\rho, 0) \tag{18.2}$$

is proper, Theorem 18.8 implies that (18.2) is a locally trivial fiber bundle. Since the base space is trivial it follows that (18.2) is a trivial bundle. Hence using Lemma 18.9(2), we can prove Propositions 18.7 and 18.10 for $F : U \to \mathbb{R}^k$. Thus the induction works.

We remark that Proposition 18.10 implies Theorem 18.1 by putting $k = 0$.

We thus sketched the proof of Theorem 18.1. The proof of Theorem 18.2 uses a similar argument but more involved. See [119].

Let us compare the results we reviewed in the last and this section so far to the one in earlier sections, where we consider the case $|K_M|$ is bounded. In Section 11, we asked two questions, Problem 11.1 for a sequence M_i converging to X. (1) was on the singularity of X and (2) was on the relation between topologies of M_i and X.

In the case $|K_{M_i}| \leqslant 1$, an answer to (1) was Theorem 11.4 and an answer to (2) was Theorem 11.5 and 12.7.

In our more general case where we assume $K_{M_i} \geqslant -1$ only, Theorem 18.1 and Corollary 18.3 give a satisfactory answer to (1).

However, results on (2) are not satisfactory. In case $X = \lim^{GH}_{i\to\infty} M_i$ satisfies $\dim X = \dim M_i$, Theorem 18.2 is a satisfactory answer. This is the noncollapsing case. On the other hand if X is a smooth Riemannian manifold, Theorem 11.3 by Yamaguchi, gives a nice answer. Namely there exists a fiber bundle $M_i \to X$ for large i.[32] However, the trick (taking frame bundles) we explained in Section 11, does not work in our more general situation to reduce the problem to the case when X is a manifold. So the result is not yet satisfactory. There are however several interesting approaches and partial results about the problem (2) in the case $M_i \geqslant -1$, which we review very briefly here.

First, Theorem 11.3 is generalized to the case when the limit X has a rather mild singularity. There are two papers about it. In [154], Yamaguchi assumed that for each $x \in X$, there exists a strainer (a_i, b_i), $i = 1, \ldots, n = \dim X$, with $d(x, a_i), d(x, b_i) > \mu > 0$ where μ is independent of x. Then he concludes that there exists a locally trivial Lipschitz fiber bundle structure $M_i \to X$.[33]

[31] So we prove them by induction without assuming completeness of X.

[32] Theorem 11.3 does not say much about the fibers. But there are various results which shows that the fibers are "of nonnegative curvature" in some sense.

[33] In the preprint version of [154] (which the author has), the locally triviality is not asserted. It is proved in [141].

Perelman in [117] assumed that X has no proper extremal set. Here

DEFINITION 18.7. $F \subset X$ is extremal if for each $p \notin F$, $x \in X$, and $u \in \Sigma_x(X)$, we have $D_x d_p(u) \leqslant 0$.

For example, $F = \{x\}$ consisting of one point is not extremal if and only if there exists an admissible function f which is regular at x.

Perelman's theorem in [117] is that, if there is no extremal set, then there exists $f_i : M_i \to X$ such that $\pi_k(M_i, f_i^{-1}(p)) \cong \pi_k(X)$ for each $p \in X$. The plan proposed by Perelman [116] then is to stratify X using an extremal set and to construct a fiber bundle structure stratawise. This plan is not yet completed.

Shioya–Yamaguchi [141] and Yamaguchi [155,156] studied the case when $\dim M_i = 3, 4$ without extra assumption on X and gave a satisfactory description in that case. In this article, we discuss the 3-dimensional case only. Let M_i be 3-dimensional Riemannian manifold with $K_{M_i} \geqslant -1$, and $X = \lim_{i\to\infty}^{GH} M_i$. We assume $\dim X \leqslant 2$. Then X is homeomorphic to a manifold with or without boundary. We assume that X is connected.

THEOREM 18.11 (Shioya–Yamaguchi [141]). *We assume* $\dim X = 2$.

(1) *If $\partial X = \emptyset$, then there exists a structure of Seifert fibered space $M_i \to X$ for large i.*
(2) *If $\partial X \neq \emptyset$, then M_i is homeomorphic to* $\mathrm{Sei}_i(X) \cup (\partial X \times D^2)$ *where* $\mathrm{Sei}_i(X)$ *is a Seifert fibered space over* $\mathrm{Int}\, X$. *We glue it with $\partial X \times D^2$ where the fibers of* $\mathrm{Sei}_i(X)$ *over the boundary point x are glued with $\{x\} \times \partial D^2$.*

In case $\dim X = 1$ there are two possibilities, $X \cong S^1$ or $[0, 1]$. In the case $X = S^1$ there exists a fiber bundle $M_i \to S^1$ by Theorem 11.3.

THEOREM 18.12 (Shioya–Yamaguchi [141]). *If $X \cong [0, 1]$ then M_i is obtained by gluing B_i and C_i along their boundaries where each of B_i, C_i is homeomorphic to one of the following* 4*-manifolds.*

(1) D^3,
(2) *A nontrivial* $[0, 1]$*-bundle over* $\mathbb{R}P^2$,
(3) $S^1 \times D^2$,
(4) *A nontrivial* $[0, 1]$*-bundle over the Klein bottle.*

The rough idea of the proofs of Theorems 18.11, 18.12 are as follows. In either cases, we can apply a generalization [154] of Theorem 11.3 except in finitely many points (plus ∂X in case (2) of Theorem 18.11). In the neighborhood of those points we scale the metric to obtain a noncompact nonpositively curved Alexandrov space. Then applying the soul theorem (the Alexandrov space analogue of Theorem 16.7). The soul S is an Alexandrov space of dimension $\leqslant 2$, so it is a manifold with or without boundary. Actually Shioya–Yamaguchi classified 3-dimensional noncompact complete Alexandrov spaces with $K \geqslant 0$. In this way, we can classify neighborhoods $\subset M_i$ of a singular point of X locally. Then the last step is to glue those local neighborhoods. □

In the case when $\dim X = 0$, we can scale the metric of M_i and obtain a limit of nonzero dimension. In this way [141] (improving [153,59]) proved the following

THEOREM 18.13 (Shioya–Yamaguchi). *There exists ϵ such that if M is a Riemannian 3-manifold with K_M Diam$(M) \geqslant -\epsilon$, then a finite cover of M is homeomorphic to $S^1 \times S^2$, T^3, a nilmanifold or a simply connected Alexandrov space with $K \geqslant 0$.*

19. First Betti number and fundamental group

So far we discussed results about sectional curvature. In this section we discuss also Ricci curvature. The recent progress mainly due to Cheeger–Colding will be discussed in later sections. In this section, we mainly concern with older results. To study Ricci curvature we need partial differential equations frequently. But we do not mention them so much.

We first review Theorem 2.3. The proof of Theorem 2.3 is based on the Bochner trick. The most famous result in metric Riemannian geometry based on the Bochner trick is the following

THEOREM 19.1 (Bochner [157]). *If an n-dimensional compact Riemannian manifold M has nonnegative Ricci curvature, then the first Betti number of M is not greater than n.*

The proof of Theorem 19.1 due to Bochner is as follows. Let u be a one-form on M. Then we have the following equality of Weitzenbeck type. (For proof see [157]. We remark that we use the nonpositive Laplacian (16.2).)

$$\langle -\Delta u, u\rangle = -\frac{1}{2}\Delta\|u\|^2 + \langle \nabla u, \nabla u\rangle + \mathrm{Ricci}(u, u). \tag{19.1}$$

Let u be a harmonic one-form. We integrate (19.1) over M. The left-hand side is zero (since u is harmonic) and the integral of the first term in the right-hand side vanish. Therefore we have

$$\int_M \langle \nabla u, \nabla u\rangle \Omega_M + \int_M \mathrm{Ricci}(u, u)\Omega_M = 0. \tag{19.2}$$

(Here Ω_M is the volume element.) The first term of (19.2) is nonnegative. If we assume that the Ricci curvature is nonnegative, then the second term also is nonnegative. Therefore the first and second term both are zero. Namely every harmonic one-form is parallel. Since a parallel one-form is determined by its value at one point (here we are assuming that M is connected), it follows that the dimension of the space of harmonic one forms on M is at most n. Theorem 19.1 follows. □

When we try to apply a similar argument to the forms of higher degree and try to estimate higher Betti numbers by Ricci curvature, we will meet a trouble. In formula (19.1), the third term involves only Ricci curvature. This is true only for one-forms. A similar formula for forms of higher degree is much more complicated. The assumption that we need to apply a similar argument to forms of higher degree is exactly the assumption in Theorem 2.3, which is much stronger than the one on Ricci curvature.[34]

[34]On the other hand, if we write a formula similar to (19.1) for the spinor and Dirac operator the second term involves only scalar curvature. (See the text book of the Atiyah–Singer index theorem.) A theorem by Lichnerow-

In Section 16, we discussed the splitting theorem of Riemannian manifolds of nonnegative Ricci curvature (Theorem 16.4). We can prove Theorem 19.1 by using this theorem also. Actually we have the following

THEOREM 19.2 (Cheeger–Gromoll). *If M is a compact manifold with nonnegative Ricci curvature, then there exists a finite cover $\tilde{M}$ of M, such that $\tilde{M}$ is isometric to the direct product $X \times T^k$, where X is simply connected and T^k is a flat torus.*

To prove Theorem 19.2, we consider the universal covering $\hat{M}$. Since we may assume that the fundamental group of M is infinite (otherwise we may take $X = \hat{M}$), we can prove that $\hat{M}$ contains a line.[35] Now by applying Theorem 16.4, we find $\hat{M} = \mathbb{R} \times Y$. We may split $\hat{M} = \mathbb{R}^k \times Y'$ so that Y' has no $\mathbb{R}$ factor. If Y' is not compact, we can show Y'contains a line by the same argument. Then, by Theorem 16.4, Y' has an $\mathbb{R}$ factor, a contradiction. Namely Y' is compact. The $\pi_1 M$ action preserves the splitting $\hat{M} = \mathbb{R}^k \times Y'$. Theorem 19.2 follows easily. □

In case $k = n$ in Theorem 19.2, or in case when the first Betti number is equal to the dimension in Theorem 19.1, we can show that M is flat. (We can show this fact either by Bochner's proof using (19.1) or by Cheeger–Gromoll's proof based on the splitting theorem.)

Theorem 19.2 is generalized by Gromov as follows.

THEOREM 19.3 [69, p. 73]. *There exists a continuous function $b(n, \rho)$ of $\rho \in \mathbb{R}$ with $b(n, 0) = n$, such that the following holds. If M is an n-dimensional Riemannian manifold with diameter* 1, *Ricci curvature $\geqslant \rho$, Then its first Betti number is not greater than $b(n, \rho)$.*

COROLLARY 19.4. *If M is an n-dimensional Riemannian manifold with diameter* 1 *and* Ricci $> -\epsilon_n$, *then its first Betti number is not greater than n. Here ϵ_n is a positive number depending only on n.*

Gromov's proof is based on the estimate of the growth function by using the Bishop–Gromov inequality (Proposition 5.2) and is closer to the study of the fundamental group we mention later in this section (Theorem 19.9, Theorem 19.10). The analytic proof, using a similar idea to Bochner's, is given by Gallot [61].

As we mentioned before, the idea of the proof of Theorem 19.1 cannot directly be applied to the study of the second or higher Betti number. In fact, a result similar to Theorem 19.3 does not hold for higher Betti numbers. Namely the statement such as:

"If M is an n-dimensional compact Riemannian manifold with diameter 1, Ricci curvature $\geqslant \rho$, then its Betti number is smaller than a number depending only on ρ and n", is

icz which asserts "The $\hat{A}$ genus of Riemannian manifold of positive scalar curvature is zero" is obtained from this fact.

[35]Let $p_i, q_i \in M$ with $d(p_i, q_i) \to \infty$. Let x_i be the midpoint of a minimal geodesic joining p_i and q_i. Moving them by an action of $\pi_1(M)$, we may assume that there exists R independent of i such that $d(x, x_i) < R$. Then a subsequence of the sequence of geodesics joining p_i, x_i, q_i has a limit. This limit is a line.

<u>false</u>. See [135,118] for counter examples. Note that if we replace Ricci $\geqslant \rho$ by $K_M \geqslant \rho$ in the statement in the parenthesis, then it is Theorem 14.6.

Let us consider the case when equality holds in Corollary 19.4, namely the case Ricci $> -\epsilon_n$ and the first Betti number is n.

THEOREM 19.5 (Yamaguchi [153]). *There exists a positive number ϵ_n such that if M is an n-dimensional Riemannian manifold with diameter* $\mathrm{Diam}(M)K_M > -\epsilon_n$, *and its first Betti number is b, then there exists a finite cover $\tilde{M}$ of M and a fiber bundle $\tilde{M} \to T^b$ over a b-dimensional torus.*

Moreover, if $b = \dim M$, then M is diffeomorphic to the torus.

Remark 19.1. (1) Yamaguchi proved the same conclusion for the fiber of Theorem 11.3.

(2) Yamaguchi [151] proved the same conclusion under a different hypothesis $K_M \leqslant 1$, $\mathrm{Diam} \leqslant D$, $\mathrm{Ricci}_M \geqslant -\epsilon(D,n)$.

To prove Theorem 19.5, Yamaguchi used Theorem 16.4 case (b). The second half of Theorem 19.5 is generalized by Colding [46] and Cheeger–Colding [29] as follows.

THEOREM 19.6 [46,29]. *If M is an n-dimensional Riemannian manifold with* $\mathrm{Diam}(M)\,\mathrm{Ricci}_M > -\epsilon_n$, *and its first Betti number is n, then M is diffeomorphic to a torus.*

Remark 19.2. The first half of the statement of Theorem 19.5 does not hold under the milder assumption $\mathrm{Diam}(M)\,\mathrm{Ricci}_M > -\epsilon_n$. Anderson [10] constructed an example of M with $\mathrm{Diam}(M)\,\mathrm{Ricci}_M > -\epsilon_n$ but that has no fibration over $T^{b_1(M)}$.

We here explain some of the ideas used by Yamaguchi in [153] to show Theorem 19.5, which is also used in [46]. (The additional ideas due to [46,29] will be explained in later sections.)

For simplicity we consider the case $b = n = \dim M$ only. The proof is by contradiction. By scaling we may assume that there exists M_i with $\mathrm{Diam}(M_i) = 1$, $K_{M_i} \geqslant -\epsilon_i$ but M_i is not diffeomorphic to T^n. We consider the covering space $\hat{M}_i \to M_i$ whose covering transformation group is $\Gamma_i = \mathbb{Z}^b$. We study the limit of the pair $(\hat{M}_i, \Gamma_i)$. Here we define

DEFINITION 19.1 [52]. A sequence of pairs $((X_i, p_i), \Gamma_i)$ of pointed metric spaces (X_i, p_i) and groups of isometries Γ_i is said to converge to $((X, p), G)$ with respect to the *equivariant pointed Hausdorff convergence* if there exists $\varphi_i : B_{p_i}(1/\epsilon_i, X_i) \to B_p(1/\epsilon_i, X)$, $\varphi_i' : B_p(1/\epsilon_i, X) \to B_{p_i}(1/\epsilon_i, X_i)$, $\psi_i : \Gamma_i \to G$, $\psi_i' : G \to \Gamma_i$ with $\epsilon_i \to 0$ such that

(1) φ_i, φ_i' are ϵ_i-Hausdorff approximations and

$$d\big(x, \varphi_i\big(\varphi_i'(x)\big)\big) < \epsilon_i, \qquad d\big(x, \varphi_i'\big(\varphi_i(x)\big)\big) < \epsilon_i.$$

(2) If $x, \gamma(x) \in B_{p_i}(1/\epsilon_i, X_i)$, $\gamma \in \Gamma_i$, then

$$d\big(\varphi_i\big(\gamma(x)\big), \psi_i(\gamma)\big(\varphi_i(x)\big)\big) < \epsilon_i.$$

(3) If $x, \gamma(x) \in B_{p_i}(1/\epsilon_i, X)$, $\gamma \in \Gamma_i$, then

$$d\big(\varphi_i'(\gamma(x)), \psi_i'(\gamma)\big(\varphi_i'(x)\big)\big) < \epsilon_i.$$

We remark that we do not assume that ψ_i, ψ_i' are homomorphisms.

We can prove a similar compactness result as Theorem 16.1. Now let us go back to the proof of Theorem 19.5. Fix $p_i \in \hat{M}_i$. We may consider the limit $((\hat{M}_i, \Gamma_i), p_i)$ with respect to the equivariant pointed Hausdorff convergence. However, then the limit may be a continuous group and is a bit hard to handle. So we use the following lemma.

LEMMA 19.7 [153]. *There exist subgroups $\Gamma_i' \subset \Gamma_i$ of finite index and η, η' (independent of i) such that*

(1) *For each $\gamma \in \Gamma_i'$ with $\gamma \neq 1$ we have $d(p_i, \gamma(p_i)) \geqslant \eta$.*

(2) *Γ_i' is generated by elements $\gamma_1, \dots, \gamma_n$ such that $d(p_i, \gamma_k(p_i)) \leqslant \eta'$. (Here $n = \dim M$.)*

Lemma 19.7 appeared in the proof by Gromov of Theorem 19.3. The fact that Γ_i is Abelian plays an important role in the proof. We omit the proof of Lemma 19.7.

Now we can consider the limit of the sequence $((\hat{M}_i, \Gamma_i'), p_i)$. We denote it by $((X, G), p)$. Using Lemma 19.7 we can easily show that $G \cong \mathbb{Z}^n$ and its action is properly discontinuous. Now we apply the splitting theorem to X and obtain $X = \mathbb{R}^k \times Y$ where Y is compact.[36] Since $\mathbb{Z}^n$ acts on it properly discontinuously, it follows that $k = n$. Since $\dim X \leqslant \dim \hat{M}_i = n$, it follows that $X = \mathbb{R}^n$. We can also prove that $\hat{M}_i/\Gamma_i'$ converges to $X/G \cong T^n$. We put $\tilde{M}_i = \hat{M}_i/\Gamma_i'$. Since $\tilde{M}_i$ is n-dimensional and converges to T^n, it follows from Theorem 11.3 that $\tilde{M}_i$ is diffeomorphic to T^n. Using $H_n(M_i, \mathbb{Q}) = n$ again, we can show that M_i is homeomorphic to T^n. Furthermore we can arrange the covering index $\tilde{M}_i \to M_i$ so that "M_i is homeomorphic to T^N and M_i' is diffeomorphic to T^n" imply that M_i is diffeomorphic to T^n, if $\geqslant 5$. (This point is a standard application of nonsimply connected surgery.) The last step in the low-dimensional case is a bit complicated and is omitted. □

Remark 19.3. The above argument can be applied to the situation of Theorem 19.6. We only need to replace the splitting theorem to the one by Cheeger–Colding and Theorem 11.3 by Theorem 22.3. Colding's argument in [46] (though using Lemma 19.7) is slightly different. This is probably because the splitting theorem we need for this purpose was not yet proved at that time.

We next remark the following corollary of Theorem 19.2.

COROLLARY 19.8. *If a compact Riemannian manifold M has nonnegative Ricci curvature, then its fundamental group $\pi_1(M)$ contains an Abelian subgroup of finite index.*

It seems that series of results related to Corollary 19.8 began with the following theorem.

[36] I think this was the first place where splitting theorem of the limit (singular) space was applied to study Riemannian manifold.

THEOREM 19.9 (Milnor [104]). *If a complete manifold M has nonnegative Ricci curvature and if G is a finitely generated subgroup of $\pi_1(M)$, then G has polynomial growth.*

The definition of a group having polynomial growth is as follows. Let G be a finitely generated group and $g_1, \dots, g_k$ generate G. Let $f_G(N)$ be the number of elements of G which can be written by a product of at most N of g_i or g_i^{-1}.

DEFINITION 19.2. We say that G has *polynomial growth*, if there exists C, K such that $f_G(N) < C(N^K + 1)$.

It is easy to see that this definition is independent of the choice of generator of G.

The proof of Theorem 19.9 is based on Proposition 5.2 and proceeds as follows. Let us assume M is compact for simplicity. Let $\tilde{M}$ be the covering space of M corresponding to G. Let $p \in \tilde{M}$. By Proposition 5.2 we have

$$\mathrm{Vol}\big(B_p(R, \tilde{M})\big) \leqslant C R^n.$$

By an elementary argument using fundamental domain, we can show the existence of C with

$$C^{-1} < \frac{\mathrm{Vol}(B_p(R, \tilde{M}))}{f_{\pi_1(M)}(R)} < C.$$

Theorem 19.9 follows. □

Roughly speaking the growth function f_G evaluates how far G is from being commutative. In fact, if G is free and non-Abelian, then there exist c, C such that

$$f_G(R) > c e^{R/C}.$$

(We say that G has exponential growth in this case.) On the other hand, $G = \mathbb{Z}^k$ has polynomial growth.

Gromov [71] proved the following

THEOREM 19.10 (Gromov). *A finitely generated group G has polynomial growth if and only if G has a nilpotent subgroup of finite index.*

Let us very briefly sketch its proof here. First we recall the following

THEOREM 19.11 (Tits [147]). *Let G be a finitely generated subgroup of $GL(n, \mathbb{R})$. Then either G contains a solvable subgroup of finite index or G contains a noncommutative free group.*

If G contains a noncommutative free group, we can show that G is not of polynomial growth. On the other hand, Milnor proved that a solvable group is of polynomial growth if and only if it contains a nilpotent group of finite index. Hence to prove Theorem 19.10 it suffices to embed G to some Lie group. Gromov's idea is to do so by using Hilbert's 5th problem. Let G a group of as in Theorem 19.10. We define a metric (the word metric) on G as follows. Let $\gamma_1, \dots, \gamma_n$ be generators. Let $\mu_1, \mu_2 \in G$. We define $d(\mu_1, \mu_2)$ to be the smallest number k such that $\mu_2 = \gamma_{i_1}^{\epsilon_1} \dots \gamma_{i_k}^{\epsilon_k} \mu_1$. Here $i_j \in \{1, \dots, n\}$, $\epsilon_j = \pm 1$.

Now we consider the limit $\lim^{GH}_{N\to\infty}(G, \frac{1}{N}d)$ as $N \to \infty$. The assumption that G is of polynomial growth is used to show that the limit exists. It is easy to see that the limit G' has the structure of a group.

We then can use the fact that G' acts as isometry on itself preserving the metric and a solution of Hilbert's 5th problem, to show that G' is a Lie group. So if we can embed G to G', we are done. But it is not so easy to embed G to G'. (Actually in case G is a discrete subgroup of a nilpotent Lie group N, the limit is N but has a strange metric called the Carnot–Carathéodory metric (see [74]).) Therefore we need to discuss it more carefully and some more technical argument is required. We omit it. □

Theorems 19.9 and 19.10 imply that the finitely generated subgroup of the fundamental group of a complete manifold of nonnegative Ricci curvature has nilpotent subgroup of finite index. This fact is generalized by Fukaya–Yamaguchi [59][37] and further by Cheeger–Colding [46] as follows.

THEOREM 19.12 (Cheeger–Colding). *There exists a positive number ϵ_n such that if an n-dimensional Riemannian manifold satisfies* $\mathrm{Diam}(M)^2 \mathrm{Ricci}_M \geqslant -\epsilon_n$, *then $\pi_1(M)$ contains a nilpotent subgroup of finite index.*

We remark that Theorem 19.12 follows Theorem 10.5. We also remark that Theorem 10.5 implies Theorem 19.9. In Theorem 19.12 we cannot replace the conclusion "nilpotent" by "Abelian". Namely we cannot replace the assumption Ricci $\geqslant 0$ of Corollary 19.8 by $\geqslant -\epsilon_n$. The counter example is an almost flat manifold (Example 10.1).

Some more results on the fundamental group is proved in [59] and [60] which we review here.

A group Γ is said to be polycyclic if there exists

$$1 = \Gamma_0 \subset \Gamma_1 \subset \cdots \subset \Gamma_k = \Gamma_k, \tag{19.3}$$

such that Γ_i is a normal subgroup of Γ_{i+1} and Γ_{i+1}/Γ_i is cyclic. The smallest such number k is called the *degree of polycyclicity* of Γ.

THEOREM 19.13 ([59, Theorem 0.6, Corollary 7.20] plus [46]). *There exists ϵ_n and w_n such that if an n-dimensional Riemannian manifold M satisfies* $\mathrm{Ricci}_M \mathrm{Diam}(M) \geqslant -\epsilon_n$ *then π_1 contains a normal subgroup Γ such that*

[37] The result of [59] is the same conclusion as Theorem 19.12, but the assumption is on sectional curvature instead of Ricci curvature.

(1) $[\pi_1(M) : \Gamma] \leqslant w_n$.
(2) *Γ is polycyclic and its degree of polycyclicity is not greater n.*

THEOREM 19.14 (Fukaya–Yamaguchi [60]). *For each D, n there exists a finite set of groups $\mathfrak{G}$ with the following properties. Let M be a manifold with $K_M \geqslant -1$,* Diam(M) $\leqslant D$. *Then there exists $G \in \mathfrak{G}$ and a surjective homomorphism $\pi_1 M \to G$ such that the kernel Γ satisfies* (1), (2) *of Theorem* 19.13.

Theorem 19.14 implies the following. For a group G let us put

$$D(G,n) = \inf\{\mathrm{Diam}(M) \mid K_M \geqslant -1,\ \dim M = n,\ \pi_1 M \supseteq G\}.$$

Then, for any sequence of noncommutative simple groups G_i with $G_i \neq G_j$ for $i \neq j$, we have $\lim_{i\to\infty} D(G_i, n) = \infty$.

Remark 19.4. Theorem 17.23 plays a key role in the proof of Theorem 19.14. So far the author does not know the proof of the conclusion of Theorem 19.14 under the milder assumption $\mathrm{Ricci}_M \geqslant -(n-1)$. The trouble is a generalization of Theorem 17.23 to the limit X of the manifolds M_i with $\mathrm{Ricci}_{M_i} \geqslant -\delta_i$, where $\delta_i \to 0$. (Namely the problem whether the isometry group of such X is a Lie group or not.) Cheeger–Colding [30] proved that the group of isometries of X is a Lie group under the additional assumption $\mathrm{Vol}(M_i) \geqslant v > 0$. Under this additional assumption there is the following result (Anderson [9]): The number of isomorphism classes of $\pi_1 M$ of n-dimensional Riemannian manifolds M with $\mathrm{Ricci}_M \geqslant -(n-1)$, $\mathrm{Vol}(M_i) \geqslant v > 0$, $\mathrm{Diam}(M) \leqslant D$, is finite.

We here sketch a part of the proof of Theorem 19.12 given in [59]. Namely we assume the splitting theorem and explain how to deduce Theorem 19.12 from it. Here we consider the case $\mathrm{Diam}(M) K_M \geqslant -\epsilon_n$[38] to simplify the argument.

We first need a lemma on the convergence of groups. If Γ acts on a metric space X by isometries and p is a base point of X we write

$$\Gamma(D) = \langle\{\gamma \mid d(\gamma(p), p) \leqslant D\}\rangle.$$

Here $\langle A\rangle$ is the subgroup generated by A.

LEMMA 19.15 [59, Theorem 3.10]. *Let (X_i, Γ_i, p_i) converge to (X, G, q) in the pointed equivariant Hausdorff distance. We assume that the connected component G_0 of G is a Lie group and G/G_0 is discrete and finitely presented. We also assume that X/G is compact. Moreover we assume that X_i is simply connected and Γ_i is properly discontinuous and free.*

[38] If we use Cheeger–Colding's splitting theorem, a similar argument works. However we need several modifications on the technical points of the arguments of [59] or given below. Unfortunately the technical details of such arguments is not written in the literature. The author and T. Yamaguchi are planning to write it and make it public in the near future. But maybe it is too technical to be included in this article.

Then there exists a sequence of normal subgroups $\Gamma_{i,0}$ converging to G_0 such that $\Gamma_i/\Gamma_{i,0} \cong G/G_0$ for large i.[39]

We omit the proof. Now we prove the following

PROPOSITION 19.16. *Let (M_i, p_i) converge to $(\mathbb{R}^k, 0)$ with respect to the pointed Hausdorff distance. Assume* $\mathrm{Ricci}_{M_i} \geqslant -(n-1)$. *Then there exists $\epsilon > 0$ such that the image of $\pi_1(B_{p_i}(\epsilon, M_i))$ in $\pi_1(B_{p_i}(1, M_i))$ has a solvable subgroup of finite index for large i.*

The solvability in Theorem 19.12 is the case $k = 0$ of Proposition 19.16. (The proof of a more precise statement as in Theorem 19.13 and nilpotency is omitted.)

The proof of Proposition 19.16 is by downward induction on k. The case $k = \dim M_i$ follows from Theorem 11.3. We assume Proposition 19.16 is correct for $k + m$ $(m > 0)$ and show it for k by contradiction.

Let (M_i, p_i) as in Proposition 19.16. We use Theorem 11.3 to find $V_i \subseteq M_i$ and a fiber bundle $f_i : V_i \to B_0(C_i, \mathbb{R}^n)$ with $C_i \to \infty$. Here $V_i \supseteq B_{p_i}(C_i/2, M_i)$. Let $\delta_i = \mathrm{Diam}(f_i^{-1}(0))$. We take the metric $g_{i,1} = g_i/\sqrt{\delta_i}$. The limit of $(V_i, g_{i,1})$ with respect to the pointed Hausdorff distance is $\mathbb{R}^k \times Z$ where Z is an Alexandrov space with $K \geqslant 0$. Let $\Gamma_i = \pi_1(F_i) = \pi_1(V_i)$. We take $((\tilde{V}_i, \tilde{g}_{i,1}), \Gamma_i, \tilde{p}_i)$ where $(\tilde{V}_i, \tilde{g}_{i,1})$ is the covering space of V_i equipped with metric induced from $g_{i,1}$. Let us take a subsequence and let (V_∞, G, q) be the limit. We apply the splitting Theorem 16.4 to V_∞ and find $V_\infty = \mathbb{R}^\ell \times Y$ where Y is compact. Since $(\mathbb{R}^\ell \times Y)/G \cong \mathbb{R}^k \times Z$ we find that $V_\infty = \mathbb{R}^k \times \mathbb{R}^{\ell-k} \times Y$ such that G acts only on $\mathbb{R}^{\ell-k} \times Y$ and $(\mathbb{R}^{\ell-k} \times Y)/G = Z$.

Since G is a Lie group by Theorem 17.23, it follows that we can take its connected component G_0. Since G/G_0 is discrete and $(\mathbb{R}^{\ell-k} \times Y)/G$ is compact we can prove that G/G_0 has Abelian subgroup of finite index. (This is easy to see if G acts effectively on $\mathbb{R}^{\ell-k}$. The compact factor Y only contributes a finite group.) To apply Lemma 19.15 we replace $V_\infty = \mathbb{R}^k \times \mathbb{R}^{\ell-k} \times Y$ by $X = B_0(D, \mathbb{R}^k) \times \mathbb{R}^{\ell-k} \times Y$ for large but fixed D and find a sequence $((X_i, d_{X_i}), \Gamma_i, p_i)$ converging to $((X, d_X), G, q)$. (We can find such $X_i \subset V_i$ easily by using the fiber bundle f_i.)

We now apply Lemma 19.15 to obtain $\Gamma_{i,0}$.

Since (X, d_X) is an Alexandrov space, it follows from Theorem 17.2 that we can find q' near q and $r_i \to \infty$ such that $((X, r_i d_X), q')$ converges to $(\mathbb{R}^{k+m}, 0)$ with $m > 0$. (Note that since $\mathrm{Diam}(Z) = 1$, it follows that $\mathbb{R}^{\ell-k} \times Y$ is not a point.)

We may replace $((X_i, d_{X_i}), \Gamma_i, p_i)$ by a subsequence which converges to $((X_i, d_{X_i}), \Gamma_i, p_i)$ very quickly compared to $1/r_i$. Then we find q_i such that $((X_i, r_i d_{X_i}), \Gamma_i, q_i)$ converges to $(\mathbb{R}^{k+m}, G', 0)$ for some G'. Since we can use the fact that $\Gamma_{i,0}$ converges to G_0, the connected Lie group and the convergence is quick compared to r_i to show that $\Gamma_{i,0}$ is generated by $\Gamma_{i,0}(\delta_i) = \{\gamma \in \Gamma_{i,0} \mid d(\gamma(q_i), q_i) < \delta_i\}$ where $\delta_i \to 0$. Now we apply induction hypothesis. Then if $\epsilon > \delta_i$ (where ϵ is as in Proposition 19.16), we find that $\Gamma_{i,0}(\delta_i)$ has a solvable subgroup with finite index. This is a contradiction since $\Gamma_i/\Gamma_{i,0} \cong G/G_0$ has an Abelian subgroup with finite index. □

---

[39]This lemma is actually weaker than [59, Theorem 3.10]. But it is enough for the present purpose, since we now have Theorem 17.23.

20. Hausdorff convergence of Einstein manifolds

In the last four sections, we discuss Gromov–Hausdorff convergence under the assumption $\mathrm{Ricci} \geqslant -(n-1)$.

We first remark that, when we work under the assumption $\mathrm{Ricci} \geqslant -(n-1)$, the topology can change when we go to the limit, even in the noncollapsing situation, namely in the situation where we assume $\mathrm{Vol} \geqslant v > 0$. Such a phenomenon was first observed in the study of 4-dimensional Einstein (or complex 2-dimensional Kähler–Einstein) manifold (at least around 20 years ago as far as I know).

Let us start by a review of the case of Einstein manifolds. Let $\Gamma \subset SU(2)$ be a finite subgroup. We consider the quotient $\mathbb{C}^2/\Gamma$. It is a Kähler orbifold with isolated singularity at the origin. (This singularity is called the Kleinian singularity.) There is a resolution called minimal resolution of the Kleinian singularity which we denote by $\widetilde{\mathbb{C}^2/\Gamma} \to \mathbb{C}^2/\Gamma$. Eguchi–Hanson [49] and others constructed a Ricci flat Kähler metric $g_{\widetilde{\mathbb{C}^2/\Gamma}}$ on $\widetilde{\mathbb{C}^2/\Gamma}$ which is asymptotically locally Euclidean (in the sense we define later in Definition 20.1). (Such a metric is called a gravitational instanton.) Asymptotically locally Euclidean metrics on $\widetilde{\mathbb{C}^2/\Gamma}$ are classified by Kronheimer [98].

Suppose (X, g_X) is a 4-dimensional Riemannian orbifold locally of Kleinian type. (Namely X is locally a quotient of $\mathbb{C}^2$ by a finite group $\Gamma \subset SU(2)$. We assume also that the metric on X is a quotient metric with respect to a certain Γ-invariant metric locally.) We assume that X is Ricci flat and Kähler. (Namely its Ricci curvature at regular points is 0 and the metric is Kähler at regular point.) We can locally glue the metric g_X on X and the Ricci flat Kähler metric $\epsilon g_{\widetilde{\mathbb{C}^2/\Gamma}}$ on X to obtain a metric g'_ϵ on the resolution $\tilde{X}$ of X. g'_ϵ is almost Ricci flat. We can use the technique of Yau's proof of the Calabi conjecture [158] to show that there exists a Ricci flat Kähler metric on $\tilde{X}$ near g_ϵ. (See [96,13].) We remark that $(\tilde{X}, g'_\epsilon)$ and $(\tilde{X}, g_\epsilon)$ converges to (X, g) with respect to the Gromov–Hausdorff distance. A typical example is a Kummer surface where we take $X = T^4/\mathbb{Z}_2$ (and $\Gamma = \mathbb{Z}_2$).

Thus, we have

OBSERVATION 20.1. *There exists a family of Riemannian manifolds* (X, g_ϵ), *such that* $\mathrm{Ricci}_{g_\epsilon} \equiv 0$, $\mathrm{Vol}(X, g_\epsilon) \geqslant v > 0$ *and the limit of* (X, g_ϵ) *as* $\epsilon \to 0$ *converges to a compact metric space* X *which is not a manifold.*

The construction here is an analogue of Taubes' construction [145] of anti-self dual connections on 4 manifolds.

Later Joyce (see [89]) generalized this construction and used it to construct (higher-dimensional) Riemannian manifolds with exceptional holonomy. (They are in particular Ricci flat.) Namely Joyce started, for example, with a 7-dimensional flat orbifold $X = T^7/\Gamma$, which is obtained by T^7, divided by a finite group of isometries Γ. In his example, the singular locus of X is a codimension 4 totally geodesic smooth submanifold (actually it is a disjoint union of T^3). Then Joyce glued $T^3 \times \widetilde{\mathbb{C}^2/\mathbb{Z}_2}$ (equipped with direct product metric) along a singularity to obtain a Riemannian manifold and used the implicit function theorem to obtain a manifold with exceptional holonomy. In his construction, we also have a family of metrics g_ϵ which is of exceptional holonomy (and in particular is Ricci flat) and which converges to X.

A converse to Observation 20.1 is proved by Nakajima and others as follows.

THEOREM 20.2 (Nakajima [106]). *Let g_i be a sequence of Einstein metrics with* Ricci $=$ ± 1 *or* 0, *on a 4 manifolds M, such that* $\mathrm{Vol}(M, g_i) \geqslant v > 0$. (*Here v is independent of i.*) *Let $X = \lim^{GH}_{i\to\infty}(M, g_i)$. Then there exists a finite subset $S \subset X$ such that $X \setminus S$ is an Einstein 4-manifold.*

Moreover, for every $\delta > 0$ there exists a diffeomorphism $\Phi_i : X \setminus N_\delta S \to M$ such that the pullback Riemannian metric $\Phi_i^* g_i$ converges to the Riemannian metric on X in C^∞-topology.

Remark 20.1. In case of 4-dimensional Einstein manifolds, the L^2-norm of the curvature

$$\int_{M_i} |R_{M_i}|^{n/2} \Omega_{M_i} = \int_M |R_M|^2 \Omega_M \tag{20.1}$$

is a topological invariant and is estimated by the Euler number. This fact is essential in the proof of Theorem 20.2.[40]

In case $\dim M_i > 4$, the same conclusion as Theorem 20.2 holds under the additional hypothesis

$$\int_{M_i} |R_{M_i}|^{n/2} \Omega_{M_i} \leqslant C. \tag{20.2}$$

(In case we assume (20.2) we do not need to fix a topological type of M.) Namely under assumption (20.2) and $\mathrm{Vol}(M_i) \geqslant v > 0$, the limit space X of a sequence of Einstein manifolds M_i has only finitely many singular points.

We remark however the assumption (20.2) is too restrictive to handle the limit of Einstein manifolds. In the example of Joyce mentioned above the limit of a sequence of 7-dimensional Einstein manifolds is $T^7/\mathbb{Z}_2$ whose singularity is 3-dimensional. In this example, the L^2-norm of the Ricci curvature is bounded but the $L^{3.5}$ norm is not bounded.

To study the structure of M_i or X near a singular point $\in S$, we use the scaling argument as follows. For completeness we include the case when $\dim M$ is general. Namely we assume we have a sequence of Einstein manifolds M_i converging to X. We assume (20.2) and $\mathrm{Vol}(M_i) \geqslant v > 0$. (Then the singular point set S of X is of finite order.) Let $p_i \in M_i$ which converges to $p_\infty \in S$. We scale the metric g_{M_i} to $R_i g_{M_i}$ so that $|K_{R_i g_{M_i}}|$ becomes 1 at p_i. We then consider the limit $((M_i, R_i g_{M_i}), p_i)$ with respect to the pointed Gromov–Hausdorff distance. Theorem 16.1 implies that it has a limit, which we denote by (X, g_X).

[40] (20.1) is scale invariant if and only if $\dim M = 4$. (We do not need the Einstein condition for this.) In this sense also the situation is very much similar to the study of the Yang–Mills equation in dimension 4. (Compare also the footnote 15 at the beginning of Section 7.)

Using the injectivity radius bound we can show that (X, g_X) is a Ricci flat Riemannian manifold. (It is noncompact but complete.) It also satisfies the following condition:

$$\begin{cases} \int_X |R_X|^{n/2} \Omega_X \leqslant C_1, \\ \mathrm{Vol}(B_p(R, X)) \geqslant C_2 R^n. \end{cases} \tag{20.3}$$

(See [14].) We then can apply the following Theorem 20.3. We define

DEFINITION 20.1. A complete pointed Riemannian manifold $((X, g), p)$ is said to be *locally almost Euclidean* (abbreviated by ALE hereafter) of order $\tau > 0$, if there exists a finite group $\Gamma \subset O(n)$ and a diffeomorphism $\Phi : X \setminus B_p(R, X) \to (\mathbb{R}^n \setminus B_0(R, \mathbb{R}^n))/\Gamma$ such that

$$\left|\left(\Phi^{-1}\right)^* g_X - g_{\mathrm{can}})(x)\right| \leqslant C|x|^{-\tau}, \tag{20.4a}$$

$$\frac{|(\nabla^k \Phi^{-1})^* g_X(x) - (\nabla^k \Phi^{-1})^* g_X(y)|}{|x - y|^{\alpha}} \leqslant C \min\left(|x|, |y|\right)^{-1-\tau-\alpha}, \tag{20.4b}$$

holds for some α and R. Here g_{can} is the metric on $B_p(R, X) \geqslant C_2 R^n$ induced by the Euclidean metric on $\mathbb{R}^n$.

THEOREM 20.3 (Bando–Kasue–Nakajima [14]). *If (X, g_X) is an n-dimensional Einstein manifold satisfying* (20.3) *then it is ALE of order $n - 1$. If (X, g_X) is Einstein–Kähler and $n = 4$ then it is ALE of order n.*

Combining them we have

THEOREM 20.4 ([14], Anderson [11]). *The limit space X in Theorem* 20.2 *is an Einstein orbifold.*[41]

In higher dimensions Theorems 23.16, 23.17 give a natural generalization of the results we explained here. If we remove the assumption $\mathrm{Vol}(M_i) \geqslant v > 0$ (namely if we study the collapsing situation), then even in the case of Einstein manifold, not so many things are known. This problem is related to mirror symmetry in string theory and is calling attention of several differential geometers working on it. There is a result by Gross–Wilson [76] which discusses the case of K3-surfaces in the collapsing situation and obtains a singular torus fibration.

We now consider more general Riemannian manifolds under the condition of Ricci curvature below. To obtain a result similar to Theorem 3.4 we need to avoid the phenomenon we described in Observation 20.1. There are several results assuming a lower bound of injectivity radius, for this purpose. We denote by $\mathfrak{S}_n(D, i > \rho)$ the set of all isometry classes of n-dimensional compact Riemannian manifolds (without boundary) such that $\mathrm{Ricci}_M \geqslant -(n-1)$, $\mathrm{Diam}(M) \leqslant D$ and $i_M \geqslant \rho$ everywhere. Let $\alpha \in (0, 1)$.

[41] A similar result holds under an additional assumption (20.2).

THEOREM 20.5 (Anderson–Cheeger [12]). *Let $M_i \in \mathfrak{S}_n(D, i > \rho)$ and $X = \lim^{GH}_{i\to\infty} M_i$. Then X is a Riemannian manifold of C^α-class and there exist diffeomorphisms $\varphi_i : M_i \to X$ such that $(\varphi_i^{-1})^* g_{M_i}$ converges to g_X with respect to the C^α-norm.*

Remark 20.2. Under the stronger assumption $|\mathrm{Ricci}_M| \leqslant (n-1)$, $\mathrm{Diam}(M) \leqslant D$ and $i_M \geqslant \rho$, Anderson [7] proved a stronger result. Namely the limit space X is a $C^{1,\alpha}$-Riemannian manifold and $(\varphi_i^{-1})^* g_{M_i}$ converges to g_X with respect to $C^{1,\alpha}$-norm. It was applied (in [7]) to prove a sphere theorem and a pinching theorem for an almost Einstein metric.

COROLLARY 20.6. *The number of diffeomorphism classes represented by elements of $\mathfrak{S}_n(D, i > \rho)$ is finite.*

The proof of Theorem 20.5 is quite similar to the arguments in Sections 6–8. Namely we construct harmonic coordinate and obtain an appropriate estimate. Then the proof is completed by using the diffeotopy extension theorem (or the center of mass technique which we can apply to a smooth metric near the limit C^α metric). So the new result in [12] is the following

THEOREM 20.7 [12, Theorem 0.1]. *There exists $C(n,\rho), \epsilon(n,\rho) > 0$ with the following property. Let $M \in \mathfrak{S}_n(D, i > \rho)$. We can then cover M by harmonic coordinates U_i such that the $C^{1,\alpha}$-norm of the coordinate transformation is smaller than $C(n,\rho)$ and the C^α-norm of the metric tensor written in this coordinate is smaller than $C(n,\rho)$. Moreover for any $p \in M$, the metric ball $B_p(\epsilon(n,\rho), M)$ is contained in some U_i.*

21. Sphere theorem and L^2 comparison theorem

In the last three sections, we concern with the class of Riemannian manifolds with Ricci curvature bounded from below. Especially we discuss results obtained by Colding and Cheeger–Colding recently. The surveys [45,47,62] and the book [26] are recommended for their results. The basic tool to study such Riemannian manifolds is Theorem 5.2. So we first draw some of its consequences. We put

$$\begin{aligned} A_p(a,b;M) &= \{x \in M \mid a \leqslant d(p,x) \leqslant b\}, \\ S_p(a;M) &= \{x \in M \mid d(p,x) = a\}. \end{aligned} \tag{21.1}$$

LEMMA 21.1. *If $\mathrm{Ricci}_M \geqslant \kappa$, $a < b < c$, then*

$$\frac{\mathrm{Vol}(A_p(a,b;M))}{\mathrm{Vol}(A_{p_0}(a,b;\mathbb{S}^n(\kappa)))} \geqslant \frac{\mathrm{Vol}(A_p(b,c;M))}{\mathrm{Vol}(A_{p_0}(b,c;\mathbb{S}^n(\kappa)))} \tag{21.2}$$

and

$$\frac{\mathrm{Vol}(S_p(a;M))}{\mathrm{Vol}(S_{p_0}(a;\mathbb{S}^n(\kappa)))} \geqslant \frac{\mathrm{Vol}(A_p(a,b;M))}{\mathrm{Vol}(A_{p_0}(a,b;\mathbb{S}^n(\kappa)))} \geqslant \frac{\mathrm{Vol}(S_p(b;M))}{\mathrm{Vol}(S_{p_0}(b;\mathbb{S}^n(\kappa)))}. \tag{21.3}$$

(21.2) follows from

$$\frac{\mathrm{Vol}(A_p(0,a;M)) + \mathrm{Vol}(A_p(a,b;M))}{\mathrm{Vol}(A_{p_0}(0,a;\mathbb{S}^n(\kappa))) + \mathrm{Vol}(A_{p_0}(a,b;\mathbb{S}^n(\kappa)))}$$
$$\geqslant \frac{\mathrm{Vol}(A_p(0,a;M)) + \mathrm{Vol}(A_p(a,b;M)) + \mathrm{Vol}(A_p(b,c;M))}{\mathrm{Vol}(A_{p_0}(0,a;\mathbb{S}^n(\kappa))) + \mathrm{Vol}(A_{p_0}(a,b;\mathbb{S}^n(\kappa))) + \mathrm{Vol}(A_{p_0}(b,c;\mathbb{S}^n(\kappa)))}.$$

By taking the limit $b \to a$ and $b \to c$ in (21.2) we obtain (21.3). □

LEMMA 21.2. *If* $\mathrm{Ricci}_M \geqslant (n-1) = \dim M - 1$ *and if* $p, q \in M$ *with* $d(p,q) > \pi - \epsilon$, *then for each* $x \in M$ *we have*

$$d(p,x) + d(q,x) - d(p,q) \leqslant \tau(\epsilon|n).$$

To show the lemma, let $\delta = d(p,x) + d(q,x) - d(p,q)$, $r = d(p,x) - \delta/2$, $s = d(p,q) - r = d(q,x) - \delta/2$. Then $(B_p(r,M) \cup B_q(s,M)) \cap B_x(\delta/2,M) = \emptyset$ and $B_p(r,M) \cap B_q(s,M) = \emptyset$. Therefore, by Theorem 5.2, we have

$$\frac{\mathrm{Vol}(B_p(r,M) \cup B_q(s,M))}{\mathrm{Vol}(M)} \geqslant 1 - \tau(\epsilon|n), \qquad \frac{B_x(\delta/2,M)}{\mathrm{Vol}(M)} \geqslant C\delta^n.$$

Hence $\delta < \tau(\epsilon|n)$ as required. □

COROLLARY 21.3. *If* $\mathrm{Ricci}_M \geqslant (n-1) = \dim M - 1$ *and if* $p, q \in M$ *with* $d(p,q) > \pi - \epsilon$, *then*

$$\mathrm{Diam}\big(M \setminus B_p(\pi - \epsilon, M)\big) < \tau(\epsilon|n).$$

Corollary 21.3 is an immediate consequence of Lemma 21.2 and Myers' Theorem 5.4. We remark that Corollary 21.3 is a version of Proposition 4.4. Namely the conclusion of Corollary 21.3 is weaker than that of Proposition 4.4, but it holds under milder assumption.

LEMMA 21.4. *If* $\mathrm{Ricci}_M \geqslant (n-1) = \dim M - 1$ *and if* $p, q \in M$ *with* $d(p,q) > \pi - \epsilon$, *then*

$$\frac{\mathrm{Vol}(S_p(\delta;M))}{\mathrm{Vol}(S_{p_0}(\delta;\mathbb{S}^n(1)))} \leqslant \frac{\mathrm{Vol}(S_p(\pi-\delta;M))}{\mathrm{Vol}(S_{p_0}(\pi-\delta;\mathbb{S}^n(1)))} + \tau(\epsilon|\delta,n). \tag{21.4}$$

We remark

$$\frac{\mathrm{Vol}(S_p(\delta;M))}{\mathrm{Vol}(S_{p_0}(\delta;\mathbb{S}^n(1)))} \geqslant \frac{\mathrm{Vol}(S_p(\pi-\delta;M))}{\mathrm{Vol}(S_{p_0}(\pi-\delta;\mathbb{S}^n(1)))}$$

is a consequence of (21.1). Hence (21.4) implies that the ratio of volume $\mathrm{Vol}(S_p(t;M))/\mathrm{Vol}(S_{p_0}(t;\mathbb{S}^n(1)))$ is almost constant for $t \in [\delta, \pi - \delta]$.

Let us prove Lemma 21.4. Let $\epsilon \ll \rho \ll \delta$. By Corollary 21.3, we have

$$\begin{aligned} &A_q(\delta - 2\rho, \delta - \rho; M) \\ &\quad \subseteq A_p\big(\pi - \delta + \rho - \tau(\epsilon|n), \pi - \delta + 2\rho + \tau(\epsilon|n); M\big). \end{aligned} \tag{21.5}$$

We may assume $\rho - \tau(\epsilon|n) \geqslant 0$. We remark

$$\mathrm{Vol}\big(A_{p_0}\big(\delta - 2\rho, \delta - \rho; \mathbb{S}^n(1)\big)\big) = \mathrm{Vol}\big(A_{p_0}\big(\pi - \delta + \rho, \pi - \delta + 2\rho; \mathbb{S}^n(1)\big)\big).$$

Therefore (21.5) (together with Lemma 21.1) implies the first inequality of

$$\begin{aligned} &\frac{\mathrm{Vol}(A_q(\delta - 2\rho, \delta - \rho; M))}{\mathrm{Vol}(A_{p_0}(\delta - 2\rho, \delta - \rho; \mathbb{S}^n(1)))} \\ &\quad \leqslant \frac{\mathrm{Vol}(A_p(\pi - \delta, \pi - \delta + \rho; M))}{\mathrm{Vol}(A_{p_0}(\pi - \delta, \pi - \delta + \rho; \mathbb{S}^n(1)))} + \tau(\epsilon|\delta, \rho, n) \\ &\quad \leqslant \frac{\mathrm{Vol}(A_p(\delta - 2\rho, \delta - \rho; M))}{\mathrm{Vol}(A_{p_0}(\delta - 2\rho, \delta - \rho; \mathbb{S}^n(1)))} + \tau(\epsilon|\delta, \rho, n). \end{aligned} \tag{21.6}$$

Here the second inequality is a consequence of Lemma 21.1. Changing the role of p and q we have

$$\frac{\mathrm{Vol}(A_p(\delta - \rho, \delta; M))}{\mathrm{Vol}(A_{p_0}(\delta - \rho, \delta; \mathbb{S}^n(1)))} \leqslant \frac{\mathrm{Vol}(A_q(\delta - \rho, \delta; M))}{\mathrm{Vol}(A_{p_0}(\delta - \rho, \delta; \mathbb{S}^n(1)))} + \tau(\epsilon|\delta, \rho, n). \tag{21.7}$$

Therefore by (21.6), (21.7) and Lemma 21.1 we have

$$\frac{\mathrm{Vol}(S_p(\pi - \delta; M))}{\mathrm{Vol}(S_{p_0}(\pi - \delta; \mathbb{S}^n(1)))} + \tau(\epsilon|\delta, \rho, n) \geqslant \frac{\mathrm{Vol}(S_p(\delta; M))}{\mathrm{Vol}(S_{p_0}(\delta; \mathbb{S}^n(1)))}$$

as required. □

LEMMA 21.5. *If* $\mathrm{Ricci}_M \geqslant \kappa$ *and* $p \in M$ *then*

$$\mathrm{Vol}\big(B_p(R, M)\big) \leqslant \mathrm{Vol}\big(B_{p_0}\big(R, \mathbb{S}^n(\kappa)\big)\big).$$

This is an immediate consequence of Theorem 5.2. □

We next discuss sphere theorems. The sphere theorem appearing here can be regarded as a generalization of Theorem 17.20. So we first sketch its proof. We first remark:

LEMMA 21.6 [79,111]. *If* $K_M \geqslant 1$, $\mathrm{Vol}(M) > \mathrm{Vol}(S^n) - \epsilon$, *then the Gromov–Hausdorff distance between* M *and* S^n *is smaller than* $\tau(\epsilon|n)$.

Let $p \in M$. We identify T_pM with $T_{p_*}S^n$ for a point $p_* \in S^n$. We then define $\Phi : M \to S^n$ by $\Phi = \mathrm{Exp}_{p_*} \circ \mathrm{Exp}_p^{-1}$. Note Exp_p^{-1} is discontinuous. Using Corollary 21.3, Lemma 21.5 and the Toponogov comparison Theorem 4.7, we can show that Φ is a $\tau(\epsilon|n)$-Hausdorff approximation. We omit the details since we discuss a sharper result (Theorem 21.8) later. $\square$

Now under the assumption of Lemma 21.6 we can find points $p_0, \ldots, p_n, q_0, \ldots, q_n \in M$ such that

$$\begin{aligned} &\left|d(p_i, p_j) - \pi/2\right| < \tau(\epsilon|n), \qquad \left|d(p_i, q_j) - \pi/2\right| < \tau(\epsilon|n), \\ &\left|d(p_i, q_i) - \pi\right| < \tau(\epsilon|n). \end{aligned} \tag{21.8}$$

In fact, if $S^n \subseteq \mathbb{R}^{n+1}$, the points $p_i' = (0, \ldots, 0, \overset{i}{1}, 0, \ldots, 0)$, $q_i' = -p_i'$, satisfy (21.8). Hence we can choose $p_i = \Phi(p_i')$ where $\Phi : S^n \to M$ is an ϵ-Hausdorff approximation.

Moreover, in case of S^n, the canonical embedding $I_{S^n} : S^n \to \mathbb{R}^{n+1}$ is obtained by

$$I_{S^n}(x) = \big(\cos d(p_0, x), \ldots, \cos d(p_n, x)\big). \tag{21.9}$$

Now the idea is to embed M in a neighborhood of S^n by using a formula similar to (21.9). Namely we first take a smooth function φ_i which is close to $d(x, p_i)$ up to the first derivative, if $x \notin B_{p_i}(o(\epsilon), M) \cup B_{q_i}(o(\epsilon), M)$. We then define $I_M : M \to \mathbb{R}^{n+1}$ by

$$I_M(x) = \big(\varphi_0(x), \ldots, \varphi_n(x)\big). \tag{21.10}$$

We can then prove that $d(I_M\Phi(x), I(x)) < o(\epsilon)$ and

$$\mathrm{dist}\big(T_{I(x)}S^n, T_{I_M\Phi(x)}\big(I_M(M)\big)\big) < \tau(\epsilon|n).$$

Here dist in the above formula is a distance as a codimension one linear subspace in $\mathbb{R}^{n+1}$. We can use these two formulas to prove that M is diffeomorphic to S^n (in a similar way to Section 9). $\square$

Theorem 17.20 is generalized by Perelman as follows.

THEOREM 21.7 [114]. *There exists $\epsilon_n > 0$ such that if M satisfies* $\mathrm{Ricci}_M \geqslant (n-1)$, $\mathrm{Vol}(M) \geqslant \mathrm{Vol}(S^n) - \epsilon_n$, *then M is homeomorphic to a sphere.*

Actually Perelman proved that $\pi_k(M) = 1$ for $k < n$ under the assumption of Theorem 21.7 and applied the generalized Poincaré conjecture. The idea of the proof is hard to explain for the author in this kind of article. So we refer [114] or [159]. We will discuss the proof of a sharper version, Corollary 22.4, in Section 22.

Remark 21.1. We remark that a similar sphere theorem replacing volume by diameter does not holds. Actually Anderson [8] and Otsu [110] found examples of manifolds (M, g_i) such that $\mathrm{Ricci}_{g_i} \geqslant (n-1)$, $\mathrm{Vol}(M, g_i) \geqslant v > 0$ and $\mathrm{Diam}(M, g_i) \to \pi$ but $M \neq S^n$. (Otsu's example is $S^m \times S^{n-m}$ and Anderson's example is $\mathbb{C}P^n$ or $\mathbb{C}P^2 \sharp \mathbb{C}P^2$.)

We remark that $\mathrm{Vol}(M_i) \to \mathrm{Vol}(S^n)$ implies $\mathrm{Diam}\, M_i \to \pi$ (under the assumption $\mathrm{Ricci}_{g_i} \geqslant (n-1)$) by the Bishop–Gromov comparison Theorem 5.2.

Now we start the review of the works of Colding, who began with the following theorem closely related to Theorems 17.20 and 21.7.

THEOREM 21.8 (Colding [43,44]). *Let M be an n-dimensional Riemannian manifold with* $\mathrm{Ricci}_M \geqslant (n-1)$.

(1) *If* $\mathrm{Vol}(M) \geqslant \mathrm{Vol}(S^n) - \epsilon$ *then* $d_{GH}(M, S^n) < \tau(\epsilon|n)$.

(2) *If* $d_{GH}(M, S^n) < \epsilon$ *then* $\mathrm{Vol}(M) \geqslant \mathrm{Vol}(S^n) - \tau(\epsilon|n)$.

The proof is somewhat similar to the proof of Theorem 17.20. However we need several new ideas. Especially we need to develop some method to compare I_M (21.10) with I_{S^n} (21.9). In the situation of the proof of Theorem 17.20, this was done by Toponogov's comparison theorem. In our situation, Toponogov's comparison theorem does not apply since there is no sectional curvature bound. Colding developed the L^2-comparison theorem for this purpose. We describe it below.

We consider $p_* \in S^n$ and $\ell_* : [0, \alpha] \to S^n$ be a geodesic parameterized by arc length. We put $\ell_*(0) = q_*$, $(d\ell_*/dt)(0) = v_* \in T_{q_*}S^n$. We then put $h_{p_*,\alpha}(v_*, t) = \cos d(\ell_*(t), p_*)$. We can calculate it easily as

$$h_{p_*,\alpha}(v_*, t) = \frac{1}{\sin\alpha}\bigl(d\bigl(p_*, \ell(\alpha)\bigr)\sin(\alpha - t) + d\bigl(p_*, \ell(0)\bigr)\sin t\bigr). \tag{21.11}$$

Now we use (21.11) to define a function on M with which we compare the distance function. Let $p \in M$ and $\ell : [0, \alpha] \to M$ be a geodesic parameterized by arc length. We put $\ell(0) = q$, $(d\ell/dt)(0) = v \in T_qM$. ($\ell$ is determined by v so we write $\ell = \ell_v$.) Let $f : M \to \mathbb{R}$ be a function. We then define

$$h_{f,\alpha}(v, t) = \frac{1}{\sin\alpha}\bigl(f\bigl(\ell_v(\alpha)\bigr)\sin(\alpha - t) + f\bigl(\ell_v(0)\bigr)\sin t\bigr). \tag{21.12}$$

We remark that $h_{f,\alpha}$ may be regarded as a function of $(v, t) \in SM \times [0, \alpha]$, where SM is the unit tangent bundle $SM = \{v \in TM \mid |v| = 1\}$. In case $f(x) = d(p, x)$, we put $h_{f,\alpha} = h_{p,\alpha}$.

Now the L^2-Toponogov theorem in [44] is as follows.

THEOREM 21.9 [44, Proposition 1.15]. *Let $a_0 \in [\pi/2, \pi)$. We assume* $\mathrm{Ricci}_M \geqslant (n-1)$ *and $p, q \in M$ with $d(p, q) \geqslant \pi - \epsilon$. Then, for $\alpha \leqslant \alpha_0$, we have*

$$\begin{aligned}&\frac{1}{\alpha\,\mathrm{Vol}(SM)} \int_{v\in SM} \int_0^{\alpha} \bigl|\cos d\bigl(p, \ell_v(t)\bigr) - h_{p,\alpha}(v, t)\bigr|^2 \Omega_{SM}\, dt \\ &\quad < \tau(\epsilon|n, \alpha_0),\end{aligned} \tag{21.13}$$

$$\frac{1}{\alpha \operatorname{Vol}(SM)} \int_{v \in SM} \int_0^{\alpha} \left| \frac{d}{dt} \cos d\big(p, \ell_v(t)\big) - \frac{dh_{p,\alpha}}{dt}(v,t) \right|^2 \Omega_{SM}\, dt$$
$$< \tau(\epsilon|n, \alpha_0). \tag{21.14}$$

Here Ω_{SM} is the Liouville measure. (Hereafter we omit the symbol of volume form in case it is clear which volume form we use.)

Remark 21.2. We remark that (21.13) means that the length $d(p, \ell_v(t))$ is close to the length of the corresponding triangle in S^n in L^2-sense.

(21.14) means that the angle $\angle p\ell_v(t)\ell_v(0)$ is close to the angle in corresponding triangle in S^n in L^2-sense.

Let us explain a part of the ideas of the proof of Theorem 21.9.

We first recall the following. Let $\lambda_1(M)$ denotes the first nonzero eigenvalue of the Laplacian on (the functions of) M.

THEOREM 21.10 (Lichnerowicz [101], Obata [109]). *If an n-dimensional Riemannian manifold satisfies* $\mathrm{Ricci}_M \geqslant (n-1)$ *then $\lambda_1(M) \leqslant -n$. The equality holds if and only if M is isometric to the sphere.*

The proof can be done by the Bochner formula, in the same way as the argument of Step 1 below.

We also remark the following theorem by Cheng which is closely related to Theorem 21.10.

THEOREM 21.11 [41]. *Let M be a compact Riemannian manifolds with* $\mathrm{Ricci}_M \geqslant (n-1)$. *If* $\mathrm{Diam}(M) \geqslant \pi$, *then M is isometric to S^n.*

What is important for us is that the first eigenfunction of S^n is $\cos d(p,\cdot)$ and is exactly the function we want study in Theorem 21.10. So the idea of the proof of Theorem 21.10 goes as follows.

Step 1. Let f satisfy $\|\Delta f + nf\| < \delta$, $\|f\| = 1$. (Here $\|\ \|$ is the L^2-norm.) We prove

$$\frac{1}{\alpha \operatorname{Vol}(SM)} \int_{v \in SM} \int_0^{\alpha} \left| f\big(\ell_v(t)\big) - h_{f,\alpha}(v,t) \right|^2 < \tau(\delta|n), \tag{21.15}$$

and a similar estimate for the t-derivative of $\cos f(\ell_v(t)) - h_{f,\alpha}(v,t)$ [44, Lemma 1.4].

This step uses the Bochner–Weitzenbeck formula

$$\frac{1}{2}\Delta|\nabla f|^2 = \left|\mathrm{Hess}(f)\right|^2 + \langle \nabla\Delta f, \nabla f\rangle + \mathrm{Ricci}(\nabla f, \nabla f). \tag{21.16}$$

Here $\mathrm{Hess}(f)(X,Y) = X(Y(f)) - (\nabla_X Y)(f)$. (Note we are using the positive Laplacian.) The proof is a kind of an "almost version" of the proofs of Theorems 21.10 and 21.11. To clarify the geometric ideas, avoiding analytic details, we consider the case $\Delta f = \lambda f$, $n \geqslant -\lambda > 0$ and prove $f(\ell_v(t)) = h_{f,\alpha}(v,t)$. We integrate (21.16) and using $\int_M \langle \nabla f_1, \nabla f_2 \rangle = -\int_M \langle \Delta f_1, f_2 \rangle$, we find

$$\int_M \left(\left|\mathrm{Hess}(f)\right|^2 - |\Delta f|^2 + (n-1)|\nabla f|^2 \right) \leqslant 0.$$

Since $\lambda \int_M |\nabla f|^2 = \int_M \langle \nabla \Delta f, \nabla f \rangle = -\int_M |\Delta f|^2$, it follows that

$$\int_M \left(\left|\mathrm{Hess}(f)\right|^2 - \frac{\lambda + n - 1}{\lambda} |\Delta f|^2 \right) \leqslant 0.$$

By Trace $\mathrm{Hess}(f) = \Delta f$ and elementary linear algebra, we find $\lambda = -n$ and

$$\mathrm{Hess}(f) = -f g_M. \tag{21.17}$$

Using the fact $d^2 f(\ell_v(t))/dt^2 = \mathrm{Hess}\, f(\dot{\ell}_v(t), \dot{\ell}_v(t))$ we have

$$\frac{d^2}{dt^2} f\big(\ell_v(t)\big) = -f\big(\ell_v(t)\big). \tag{21.18}$$

$f(\ell_v(t)) = h_{f,\alpha}(v,t)$ follows.

Step 2. Let $p, q \in M$ with $d(p,q) > \pi - \delta$. We consider $g(x) = \cos d(p,x)$. We then find f with $\|\Delta f + nf\| < \delta$ and $\|f - g\|_{L^2_1} < \delta$. ($\| \ \|_{L^2_1}$ is the Sobolev norm, that is an L^2-norm up to the first derivative [44, Lemma 1.10].)

The essential part of this step (which is explained below) is to show

$$\left| n \int_M g^2 - \int_M |\nabla g|^2 \right| \leqslant \tau(\delta)\, \mathrm{Vol}(M), \tag{21.19}$$

$$\left| \int g \right| < \tau(\delta)\, \mathrm{Vol}(M). \tag{21.20}$$

In fact, (21.20) implies that g is almost perpendicular to the 0th eigenfunction of the Laplacian (the constant). Then we can use (21.19) and $\lambda_1 \geqslant n$ to get the conclusion.

Let $a(v,t)$ be as in the proof of Theorem 5.2. We extend it as 0 outside V. (So precisely speaking $a(v,t)$ is the function which we wrote $a'(v,t)$ in the proof of Theorem 5.2.) By Lemma 21.4 we have

$$\int_{v \in S^{n-1}} a(v,\delta) \leqslant \int_{v \in S^{n-1}} a(v, \pi - \delta) + \tau(\epsilon | \delta, n). \tag{21.21}$$

On the other hand, the map $t \mapsto a(v,t)$ is nondecreasing by the proof of Theorem 5.2. It follows that

$$\left| \int_{v \in S^{n-1}} a(v,s) - \int_{v \in S^{n-1}} a(v,s') \right| \leqslant \tau(\epsilon \mid \delta, n) \tag{21.22}$$

for $s, s' \in [\delta, \pi - \delta]$. Therefore

$$\begin{aligned}\left|\int_M g\right| &= \left|\int_{v\in S^{n-1}}\int_{t=0}^{\pi} a(v,t)\cos t \sin^{n-1} t\right| \\ &= \left|\int_{v\in S^{n-1}}\int_{t=0}^{\pi/2} \big(a(v,t) - a(v,\pi - t)\big)\cos t \sin^{n-1} t\right| \\ &\leqslant \tau(\epsilon,\delta|n)\,\mathrm{Vol}(M). \end{aligned} \tag{21.23}$$

Moreover using $|\nabla g|^2(x) = \sin^2 d(p,x)$ we have

$$\int_M |\nabla g|^2 = \int_{v\in S^{n-1}}\int_{t=0}^{\pi} a(v,t)\sin^{n+1} t.$$

On the other hand

$$\int_M |g|^2 = \int_{v\in S^{n-1}}\int_{t=0}^{\pi} a(v,t)\cos^2 t \sin^{n-1} t.$$

We remark $\int_0^\pi \sin^{n+1} t\, dt = n\int_0^\pi \cos^2 t \sin^{n-1} t\, dt$. Hence using (21.22) we can easily show

$$\left|\int_M |\nabla g|^2 - n\int_M |g|^2\right| < \tau(\epsilon,\delta \mid n)\,\mathrm{Vol}(M). \tag{21.24}$$

(21.23) and (21.24) complete this step as we mentioned before. □

These two steps and some more arguments imply Theorem 21.9. (The integral in Theorem 21.9 is taken with respect to the Liouville measure on the unit sphere bundle. In the argument so far the measure is taken with respect to the measure on M itself (or its products). They are equivalent by Theorem 5.2.) □

We remark that in Theorem 21.9 we use only a weaker assumption $\mathrm{Diam}(M) \sim \pi$ and not yet $\mathrm{Vol}(M) \sim \mathrm{Vol}(S^n)$. (Compare Remark 21.1, which shows that $\mathrm{Diam}(M) \sim \pi$ does not imply that $d_{GH}(M, S^n)$ is small.)

Now using Theorem 21.9, the proof of Theorem 21.8 goes roughly as follows.

We first explain (1). Let us assume $\mathrm{Vol}(M) \geqslant \mathrm{Vol}(S^n) - \delta$. It then implies that for each $p \in M$ there exists $q \in M$ such that $d(p,q) > \pi - \tau(\delta)$.[42] (This follows from the Bishop–Gromov Theorem 5.2.) Now we claim

LEMMA 21.12 [44, Lemma 2.25]. *Under the assumption of Theorem* 21.8(1) *there exist* p_i, q_i $(i = 0,\dots,n)$ *such that* (21.8) *holds.*

[42] Shiohama–Yamaguchi [140] introduced the notion of radius of M that is $\inf_p \sup_q d(p,q)$. This assertion means that the radius of M is close to π.

Once we have Lemma 21.12, we can construct a Hausdorff approximation $\Phi : M \to S^n$ by perturbing $x \to (\cos d(x, p_0), \dots, \cos d(x, p_n))$. In fact, by Theorem 21.8, we can prove that the function $x \mapsto \cos d(x, p_0)$ behaves in a similar way (modulo $\tau(\delta)$) outside the set of measure $\tau(\delta)$. This is enough to show that it is a Hausdorff approximation. □

Remark 21.3. As we mentioned before we can use the L^2-comparison theorem directly to show that a map is a Hausdorff approximation. However we cannot use it directly to find a homeomorphism. This is because the L^2-comparison theorem does not tell what happens on a set of small measure. This point is very different from the Toponogov comparison theorem, which however works only under the assumption of sectional curvature. We can use several 'indirect' arguments to obtain various topological information using the L^2-comparison theorem. (See the next two sections.)

The proof of Lemma 21.12 uses Theorem 21.7 and goes as follows. We construct p_i, q_i $(i = 0, \dots, k)$ satisfying (21.8) by induction on k. Suppose we have p_i, q_i $(i = 0, \dots, k)$. We then construct a map $\Phi_k : M \to \mathbb{R}^k$ by $x \to (\cos d(x, p_0), \dots, \cos d(x, p_k))$. We construct a set A_k from of $p_0, \dots, p_k, q_0, \dots, q_k$. In case $M = S^n$ and $p_i = (0, \dots, 0, \overset{i}{1}, 0, \dots, 0)$ $A_k = \mathbb{R}^{k+1} \cap S^n = S^k$ is obtained by joining p_i, q_i several times along minimal geodesics. We imitate the construction of A_k from p_i, q_j in M to obtain $A_k \subset M$. (Actually we need to join only by good geodesics ℓ_v that is a geodesic such that $\cos d(p, \ell_v(t)) - h_{p,a}(v, t)$ is small. Theorem 21.9 implies that there are enough such geodesics.)

Now the restriction of Φ_k to A_k is similar to the one for S^n. Hence $\Phi_k(A_k)$ lies in a neighborhood of S^k and we may regard $A_k \cong S^k$. Since $k < n$, Theorem 21.7 implies that A_k is homotopic to zero in M. This implies that there exists $p_{k+1} \in M$ such that $\Phi_k(p_{k+1}) = 0$. We take q_{k+1} with $d(p_{k+1}, q_{k+1}) > \pi - \delta$. Thus induction works. □

To prove (2) of Theorem 21.8 we proceed as follows. We take p_i, q_i $(i = 0, \dots, n)$ such that (21.8) holds. (Since $d_H(M, S^n)$ is small we can take such p_i, q_i.) We use it to construct $\Phi : M \to S^n$ by $\tilde{\Phi}(x) = (\cos d(p_0, x), \dots, \cos d(p_n, x))$, $\Phi(x) = \tilde{\Phi}(x)/|\tilde{\Phi}(x)|$. Using Theorem 21.9, we find that the determinant of the Jacobi matrix of Φ is almost everywhere close to 1. It follows that $|\mathrm{Vol}(M) - \mathrm{Vol}(\Phi(M))| < \tau(\delta|n)$. We need another idea to show that $\mathrm{Vol}(S^n \setminus \Phi(M)) < \tau(\delta|n)$. Actually for this purpose we need a "local version" of Theorem 21.13 [43, Proposition 4.5]. We omit it. □

The argument of the proof of Theorem 21.8 is a prototype of the argument which is used by Colding and Cheeger–Colding at several other places. We explain them more in the last two sections where the argument is combined with other arguments which are of more analytic nature.

22. Hausdorff convergence and Ricci curvature—I

In Section 21, we compared the distance function of a manifold of positive Ricci curvature to the one of the round sphere, in the sense of L_1^2-norm. In this section, we compare the distance function of a manifold of almost nonnegative Ricci curvature to the one of Euclidean space.

THEOREM 22.1 (Colding [46, Theorem 0.1]). *Let M_i be a sequence of n-dimensional Riemannian manifolds with* $\mathrm{Ricci}_{M_i} \geqslant -(n-1)$ *and let M_∞ be another n-dimensional Riemannian manifold. We assume* $\lim^{GH}_{i\to\infty} M_i = M_\infty$. *Then we have*

$$\lim_{i\to\infty} \mathrm{Vol}(M_i) = \mathrm{Vol}(M_\infty).$$

Remark 22.1. Actually Colding proved the following stronger (local) result in [46]. Let M_i and M_∞ be complete Riemannian manifolds. We assume $\mathrm{Ricci}_{M_i} \geqslant -(n-1)$. Let $p_i \in M_i$, $p_\infty \in M_\infty$, and $r > 0$. We assume that $\lim^{GH}_{i\to\infty} B_{p_i}(r, M_i) = B_{p_\infty}(r, M_\infty)$. Then $\lim_{i\to\infty} \mathrm{Vol}(B_{p_i}(r, M_i)) = \mathrm{Vol}(B_{p_\infty}(r, M_\infty))$.

Together with a result by Perelman and using results of controlled surgery, Theorem 22.1 implies the following

THEOREM 22.2 [46]. *In the situation of Theorem* 22.1, *M_i is homotopy equivalent to M_∞ for large i. Moreover M_i is homeomorphic to M_∞ for large i if $n \neq 3$.*

Remark 22.2. In case the limit space is singular we cannot prove a result similar to Theorem 22.2 because of Example 21.1 by Anderson and Otsu.

The Gromov–Hausdorff limit of the metrics Otsu constructed on $S^3 \times S^2$ is a suspension of $S^2 \times S^2$ and hence is not a topological manifold.

Theorem 22.1 follows from Theorem 22.2 roughly in the following way. Choose $p_j^\infty \in M_\infty$, $j = 1, \dots, N$, and small $r > 0$ such that

$$\bigcup_{i=1}^{N} B_{p_j^\infty}(r, M_\infty) = M_\infty,$$

and

$$1 - \delta \leqslant \frac{\mathrm{Vol}(B_{p_j^\infty}(r, M_\infty))}{\mathrm{Vol}_0(B_0(r, \mathbb{R}^n))} \leqslant 1 + \delta. \tag{22.1}$$

Let $\Phi_i : M_\infty \to M_i$ be an ϵ_i Hausdorff approximation with $\epsilon_i \to 0$. We take $p^i_j = \Phi_i(p_j^\infty) \in M_i$. Since $d_H(B_{p_i^j}(r, M_i), B_{p_\infty^j}(r, M_\infty))$ is small it follows from Theorem 22.1 (more precisely its local version stated in Remark 22.1) together with (22.1) that

$$1 - 2\delta \leqslant \frac{\mathrm{Vol}(B_{p_j^i}(r, M_\infty))}{\mathrm{Vol}_0(B_0(r, \mathbb{R}^n))} \leqslant 1 + 2\delta. \tag{22.2}$$

We can then apply the method of Perelman that appeared in the proof of Theorem 21.6. It may[43] imply that $B_{p_j^i}((1-\epsilon)r, M_\infty)$ is contractible in $B_{p_j^i}(r, M_\infty)$. This will imply that M_i is homotopy equivalent to M in a way similar to the proof of Theorem 3.5 in Section 15. Using control surgery in a way similar to [122] we can prove that M_i is homeomorphic to M. □

The proof of Theorem 22.2 is not worked out in so much detail in [46]. However we do not need to worry about it at all now, since Cheeger–Colding [29] improved Theorem 22.2 as follows.

THEOREM 22.3 [29, Theorem A.1.12]. *In the situation of Theorem* 22.1, M_i *is diffeomorphic to* M_∞ *for large* i.

We discuss its proof later in this section. Theorem 22.3 together with Theorem 21.8 immediately imply the following sharpening of Theorem 21.7. (We stated it as Theorem 2.6 in Section 2.)

COROLLARY 22.4 (Cheeger–Colding [29, Theorem A.1.10]). *There exists* $\epsilon_n > 0$ *such that if* M *satisfies* $\mathrm{Ricci}_M \geqslant (n-1)$, $\mathrm{Vol}(M) \geqslant \mathrm{Vol}(S^n) - \epsilon_n$, *then* M *is diffeomorphic to a sphere.*

Remark 22.3. We remark Theorem 21.8 is used in the proof of Corollary 22.4. The proof of Theorem 21.8 we sketched in the last section uses Theorem 21.7. However we can avoid it as follows. Let $\mathrm{Ricci}_{M_i} \geqslant (n-1)$, $\mathrm{Vol}(M_i) \geqslant \mathrm{Vol}(S^n) - \epsilon_i$, where $\epsilon_i \to 0$. We may assume that M_i converges to a metric space X. Then, by Theorem 23.11, X is isometric to a metric suspension SY, where SY is defined in Example 23.1(3). Using the assumption on M_i and Theorem 22.5, we can show that the tangent cone T_xX of X at any point $x \in X$ is $\mathbb{R}^n$. Therefore, since $X = SY$, it follows that $Y = S^{n-1}$. Hence $X = S^n$ (isometric) as required.

Remark 22.4. The assumption of Theorem 22.2 plus an additional assumption $\mathrm{Ricci}_{M_i} \leqslant \lambda$ implies that the Riemannian metric of M_i converges to the one of M in $C^{1,\alpha}$-topology (after identifying the manifolds by an appropriate diffeomorphism). ([46, Theorem 0.6].)

We now explain some of the ideas of the proof of Theorem 22.1. The main part of the proof is the proof of (2) of the following theorem.

THEOREM 22.5 [46, Theorem 0.8 and Corollary 2.19]. *Let* M *be an* n*-dimensional Riemannian manifold with* $\mathrm{Ricci}_M \geqslant -\lambda$ *and* $p \in M$.

(1) *If* $\mathrm{Vol}(B_p(1, M)) \geqslant \mathrm{Vol}(B_0(1, \mathbb{R}^n)) - \epsilon$, *then we have*

$$d_{GH}\big(B_p(1, M), B_0(1, \mathbb{R}^n)\big) < \tau(\epsilon, \lambda|n).$$

[43] I wrote "may" here since Perelman did not state this result explicitly and only says that "The Main Lemma can obviously be modified..." at [114, p. 300]. Indeed it is very likely so. But I did not check it in detail. By the way, Colding quote [119] in place of [114] at [46, p. 478] just before Theorem 0.4. I believe it is a misprint.

(2) *If* $d_{GH}(B_p(1,M), B_0(1,\mathbb{R}^n)) < \epsilon$, *then we have*

$$\mathrm{Vol}\big(B_p(1,M)\big) \geqslant \mathrm{Vol}\big(B_0(1,\mathbb{R}^n)\big) - \tau(\epsilon,\lambda|n).$$

An argument to show Theorem 22.1 by using Theorem 22.5(2) is omitted.

Let us sketch how to prove Theorem 22.5(2). We will discuss the proof of Theorem 22.5(1) in the next section. We only show the following version.

LEMMA 22.6. *If* M *satisfies* $d_{GH}(B_p(2R,B), B_0(2R,\mathbb{R}^n)) < \epsilon$ *and* $\mathrm{Ricci}_M \geqslant -\lambda$ *then we have*

$$\mathrm{Vol}\big(B_p(1,M)\big) \geqslant \big(1-\tau(\epsilon,\lambda,1/R|n)\big)\,\mathrm{Vol}\big(B_1(1,\mathbb{R}^n)\big).$$

The argument to show Theorem 22.5(2) using Lemma 22.6 is tricky but technical. (See [46, p. 494].) (Note the inequality of opposite direction

$$(1-\tau)\,\mathrm{Vol}\big(B_p(1,M)\big) \leqslant \mathrm{Vol}\big(B_1(1,\mathbb{R}^n)\big)$$

is a consequence of Theorem 5.2.)

Theorem 22.5 looks similar to Theorem 21.8. The proof of Lemma 22.6 also is similar. We first need a result corresponding to Theorem 21.13. In the proof of Theorem 21.13 we consider the function $x \mapsto \cos d(p,x)$ in case there exists q with $d(p,q) \geqslant \pi - \delta$. Here we consider the following function b_+^i, $i = 1,\dots,n$, instead.

Let $\Phi : B_0(2R,\mathbb{R}^n) \to B_p(2R,M)$ be an ϵ-Hausdorff approximation. Let $q_i = \Phi(0,\dots,0,\overset{i}{1},0,\dots,0) \in M$. We put

$$b_i(x) = d(x,q_i) - d(p,q_i), \tag{22.3}$$

and study it in the ball $B_p(1,M)$. We remark that b_i may be regarded as an approximation of the Busemann function (Definition 16.6). In the proof of the Cheeger–Gromoll splitting Theorem 16.4, subharmonicity of the Busemann function is the main point.

We choose ρ with $1 \ll \rho \ll R$. We consider $\mathbf{b}_i : B_p(\rho,M) \to \mathbb{R}$ such that

$$\Delta \mathbf{b}_i = 0, \tag{22.4a}$$

$$\mathbf{b}_i = b_i \quad \text{on } \partial B_p(\rho,M). \tag{22.4b}$$

In the case of Euclidean space, the Busemann function is nothing but a linear function. So we compare b_i with a linear function. We put $g_i(v,t) = \mathbf{b}_i(\ell_v(t))$.

PROPOSITION 22.7. *For* $r \leqslant \alpha < 1$, *we have*

$$\|\mathbf{b}_i - b_i\|_{L_1^2(B_p(1,M))} \leqslant \tau, \tag{22.5a}$$

$$\int_{v\in SB_p(1,M)} \left| \frac{dg_i(v,\cdot)}{dt}(r) - \frac{g_i(v,\alpha) - g_i(v,0)}{\alpha} \right| < \tau, \tag{22.5b}$$

$$\int_{B_p(1,M)} \left|\langle \nabla \mathbf{b}_i, \nabla \mathbf{b}_j \rangle - \delta_{ij}\right| < \tau, \tag{22.5c}$$

$$\int_{B_p(1,M)} \left|\mathrm{Hess}(\mathbf{b}_i)\right| < \tau. \tag{22.5d}$$

Here $\tau = \tau(\lambda, \rho/R, 1/\rho|n)$ *and* $\|\ \|_{L_1^2}$ *is the* L^2*-norm up to the first derivative.*

The proof of (22.5a) is based on Li–Shoen's Poincaré inequality [100] (estimate of the first eigenvalue of $B_p(\rho, M)$), and the proof of (22.5d) is based on the Bochner–Weitzenbeck formula (19.1) and Cheng–Yau's gradient estimate [42]. Then (22.5b) follows in a way similar as the proof of Theorem 21.9. We can use it to prove (22.5c). □

We put

$$\tilde{\Phi} = (\mathbf{b}_1, \ldots, \mathbf{b}_n) : B_p(1, M) \to \mathbb{R}^n.$$

(22.5a), (22.5b) imply that it induces an τ-Hausdorff approximation to $B_0(1, \mathbb{R}^n)$. (22.5c) implies that $\tilde{\Phi}$ almost preserves volume.

To complete the proof of Lemma 22.6 we need to show that $\mathrm{Vol}(B_0(1, \mathbb{R}^n) \setminus \tilde{\Phi}(B_p(1, M)))$ is small. We can prove it as follows.[44] Using (22.5) we can find a point $p_0 \in B_p(1/2, M)$ such that $\tilde{\Phi}^{-1}(\tilde{\Phi}(p_0)) = \{p_0\}$. (See [26, pp. 53–54] for the proof of this fact.) On the other hand, since $\tilde{\Phi}$ is a τ-Hausdorff approximation, it follows that $\tilde{\Phi}(\partial B_p(1, M)) \subset B_{1+\tau}(0, \mathbb{R}^n) \setminus B_{1-\tau}(0, \mathbb{R}^n)$. Hence

$$\tilde{\Phi}_* : H_n\big(B_p(1, M), \partial B_p(1, M); \mathbb{Z}_2\big) \to H_n\big(B_{1+\tau}(0, \mathbb{R}^n), B_{1-\tau}(0, \mathbb{R}^n); \mathbb{Z}_2\big)$$

is well defined. Note

$$H_n\big(B_p(1, M), \partial B_p(1, M); \mathbb{Z}_2\big) \cong H_n\big(B_{1+\tau}(0, \mathbb{R}^n), B_{1-\tau}(0, \mathbb{R}^n); \mathbb{Z}_2\big) \cong \mathbb{Z}_2.$$

Using $\tilde{\Phi}^{-1}(\tilde{\Phi}(p_0)) = \{p_0\}$ we can show

$$\tilde{\Phi}_* : H_n\big(B_p(1, M), \partial B_p(1, M); \mathbb{Z}_2\big) \to H_n\big(B_{1+\tau}(0, \mathbb{R}^n), B_{1-\tau}(0, \mathbb{R}^n); \mathbb{Z}_2\big)$$

is nonzero. This implies $\tilde{\Phi}(B_p(1, M)) \supset B_{1-\tau}(0, \mathbb{R}^n)$. This completes the proof of Lemma 22.6. □

We next sketch the proof of Theorem 22.3 given in [29, Appendix A]. As is mentioned there this proof is similar to the proof by Cheeger [25] of his finiteness theorem using the diffeotopy extension theorem (which we explained briefly in Section 6).

[44] Here we follow [26, pp. 53–54]. Colding's argument in [46] is a bit different.

Let us begin with a definition. Let Z be a complete metric space and $\epsilon, r > 0$. (n is a positive integer.)

DEFINITION 22.1. We say that Z satisfies the $\mathcal{R}_{\epsilon,r,n}$ *condition* if for each $x \in Z$ there exists $s < r$ such that

$$d_{GH}\big(B_x(s, Z), B_0(s, \mathbb{R}^n)\big) < \epsilon s. \tag{22.6}$$

THEOREM 22.8 (Cheeger–Colding [29, Theorems A.1.2, A.1.3]). *For each n there exists ϵ_n, independent of r, such that the following holds. If Z satisfies the $\mathcal{R}_{\epsilon,r,n}$ condition with $\epsilon < \epsilon_n$, then, for each $s < r$, we can associate a smooth Riemannian manifold $Z(s)$ with the following properties:*

(1) *There exists a homeomorphism $\Phi_{Z,s} : Z \to Z(s)$ which is $C^{1-\tau(\epsilon|n)}$-Hölder continuous. Namely*

$$C^{-1}d(x, y)^{1+\tau(\epsilon|n)} \leqslant \big(d\big(\Phi_{Z,s}(x), \Phi_{Z,s}(y)\big)\big) \leqslant Cd(x, y)^{1-\tau(\epsilon|n)} \tag{22.7}$$

for each $x, y \in Z$. Moreover $\Phi_{Z,s}$ is an $s\tau(\epsilon|n)$ Hausdorff approximation.

(2) *$Z(s)$ is 'well-defined' and 'independent' of s in the following sense. If $u \leqslant s$ then there exists a diffeomorphism which is $C^{1-\tau(\epsilon|n)}$-Hölder continuous in a way independent of t, u. Namely we have*

$$C^{-1}d(x, y)^{1+\tau(\epsilon|n)} \leqslant d\big(\Phi_{Z,u,s}(x), \Phi_{Z,u,s}(y)\big) \leqslant Cd(x, y)^{1-\tau(\epsilon|n)}, \tag{22.8}$$

where C is independent of u, s, x, y. Moreover $\Phi_{Z,u,s}$ is an $s\tau(\epsilon|n)$ approximation and satisfies

$$d\big(\Phi_{Z,u,s} \circ \Phi_{Z,s}(x), \Phi_{Z,u}(x)\big) < s\tau(\epsilon|n). \tag{22.9}$$

(3) *If Z is a Riemannian manifold then we may choose $\Phi_{Z,s}$ to be a diffeomorphism for sufficiently small s.*

(4) *There exists $\delta(n, r) > 0$ depending n and r such that if Z, Z' both satisfy $\mathcal{R}_{\epsilon,r,n}$ condition with $\epsilon < \epsilon_n$, and if $d_{GH}(Z, Z') < \delta(n, r)$ then there exists a diffeomorphism $\Psi : Z(r/2) \to Z'(r/2)$ such that*

$$e^{-\tau(\epsilon,\delta|r,n)} \leqslant \frac{d(\Psi(x), \Psi(y))}{d(x, y)} \leqslant e^{\tau(\epsilon,\delta|r,n)}, \tag{22.10}$$

$$d\big(\Psi \circ \Phi_{Z,r/2}(x), \Phi_{Z',r/2}(x)\big) < \tau(\epsilon, \delta|r, n). \tag{22.11}$$

To apply Theorem 22.8 for the proof of Theorem 22.3 we need the following

PROPOSITION 22.9. *Let M_i be a sequence of n-dimensional Riemannian manifolds and let M_∞ be another Riemannian manifold of the same dimension. We assume*

$\lim^{GH}_{i\to\infty} M_i = M_\infty$. Then for each ϵ there exists r such that M_i for large i and M_∞ satisfy the $\mathcal{R}_{\epsilon,r,n}$ condition.

Proposition 22.9 and Theorem 22.8 immediately imply Theorem 22.3.

Let us prove Proposition 22.9. Under the assumption we have $r = r(\mu)$ for each μ such that

$$1-\mu \leqslant \frac{\mathrm{Vol}(B_p(r, M_i))}{\mathrm{Vol}(B_0(r, \mathbb{R}^n))} \leqslant 1+\mu$$

for large i and $i = \infty$ and any $p \in M_i$. (See (22.2).) Then we apply Theorem 5.2 to obtain

$$1-\mu \leqslant \frac{\mathrm{Vol}(B_p(s, M_i))}{\mathrm{Vol}(B_0(s, \mathbb{R}^n))} \leqslant 1+\mu$$

for *any* $s \leqslant r$. We now apply Theorem 22.5(1) after scaling to obtain

$$d_{GH}\big(B_p(s, M_i), B_0(s, \mathbb{R}^n)\big) < s\tau(\mu|r, n)$$

as required. (Note that since we scale the metric by a factor $1/s > 1/r$, the curvature will be Ricci $\geqslant -(1-n)r^2$. So the curvature assumption in Theorem 22.5 is satisfied if r is small enough.) □

We remark that the independence of ϵ_n of r in Theorem 22.8 played a key role here.

We now prove of Theorem 22.8. Let $100s < r$. We will construct $Z(s)$ first. We remark that by using assumption (22.6) we can find subsets $\{x_i \mid i \in I\} \in Z$ such that

$$\bigcup_{i\in I} B_{x_i}(s, Z) = Z, \tag{22.12a}$$

$$\sharp\big\{i \in I \mid B_{x_i}(30s, Z) \cap B_{x_j}(30s, Z) \neq \emptyset\big\} \leqslant N(n) \tag{22.12b}$$

for each $j \in I$. Here $N(n)$ is independent of i, s. Let $\varphi_i : B_{x_i}(100s, Z) \to B_0(100s, \mathbb{R}^n)$ be a $\tau(\epsilon|n)s$-Hausdorff approximation. We have a $\tau(\epsilon|n)s$ Hausdorff approximation $\varphi_i' : B_0(100s, \mathbb{R}^n) \to B_{x_i}(100s, Z)$ such that $\mathrm{dist}(\varphi_i' \circ \varphi_i, id) < \tau(\epsilon|n)s$ and $\mathrm{dist}(\varphi_i \circ \varphi_i', id) < \tau(\epsilon|n)s$. We consider

$$\varphi_{ji} = \varphi_j \circ \varphi_i'|_{B_0(10s, \mathbb{R}^n)} : B_0(10s, \mathbb{R}^n) \to B_0(35s, \mathbb{R}^n) \tag{22.13}$$

for $i \cap j$ with $B_{x_i}(30s, Z) \cap B_{x_j}(30s, Z) \neq \emptyset$. It satisfies

$$\big|d\big(\varphi_{ji}(x), \varphi_{ji}(x)\big) - d(x, y)\big| < \tau(\epsilon|n)s. \tag{22.14}$$

We here remark the following simple lemma.

LEMMA 22.10. *If φ_{ji} satisfies* (22.14) *then there exists $\psi'_{ji} : B_0(10s, \mathbb{R}^n) \to B_0(35s, \mathbb{R}^n)$ satisfying* (22.14) *and*

$$e^{-\tau(\epsilon|n)} \leqslant \frac{d(\psi'_{ji}(x), \psi'_{ji}(x))}{d(x, y)} \leqslant e^{\tau(\epsilon|n)}, \tag{22.15a}$$

$$d\big(\varphi_{ji}(x), \psi'_{ji}(x)\big) < s\tau(\epsilon|n), \tag{22.15b}$$

$$|\psi'_{ji}|_{C^k} < s^{-k} C_{k,n}. \tag{22.15c}$$

Here $C_{k,n}$ depends only on k and n.

The proof is an elementary smoothing argument.

We want to construct a smooth manifold by using ψ_{ji} as a coordinate transformation. It does *not* satisfy $\psi'_{kj} \circ \psi'_{ji} = \psi'_{ki}$ but the following holds if $B_{x_i}(20s, Z) \cap B_{x_j}(20s, Z) \cap B_{x_k}(20s, Z) \neq \emptyset$:

$$d\big(\psi'_{kj} \circ \psi'_{ji}(x), \psi'_{ki}(x)\big) \leqslant s\tau(\epsilon|n) \tag{22.16}$$

for $x \in B_0(20s, \mathbb{R}^n)$. We can now use the argument of [25] to approximate ψ'_{ji} by ψ_{ji} which satisfies (22.15) and

$$\psi_{kj} \circ \psi_{ji} = \psi_{ki}. \tag{22.17}$$

(Note that the number of steps we need to take to achieve (22.17) is controlled by (22.12b).)

We thus constructed a manifold $Z(t)$ whose coordinate transformation is ψ_{ji}. We can use a partition of unity to modify the standard metric on $\mathbb{R}^n$ so that it is compatible with ψ_{ji}. Hence $Z(t)$ is a Riemannian manifold. We will construct $\Phi_{Z,s} : Z \to Z(s)$ later. At this stage we have $\Psi_{Z,s} : Z \to Z(s)$ which is an $s\tau(\epsilon|n)$-Hausdorff approximation.

We next show the 'well-definedness' property (2). We first consider the case $u \in [s/2, s]$. Let us suppose we have $Z(u)$ for $u \geqslant s$. We use the symbol $\tilde{}$ for points, maps, etc. used to construct $Z(u)$. (Namely we write $\tilde{\varphi}_{\tilde{i}}$, $\tilde{x}_{\tilde{i}}$, etc.)

Let $B_{\tilde{x}_{\tilde{i}}}(30u, Z) \cap B_{x_j}(30s, Z) \neq \emptyset$. We define $\Psi_{j\tilde{i}} : B_0(20u, \mathbb{R}^n) \to B_0(30s, \mathbb{R}^n)$ by $\Psi_{j\tilde{i}} = \varphi_j \circ \tilde{\varphi}^{\prime}_{\tilde{i}}$. It satisfies

$$\big|d\big(\Psi_{j\tilde{i}}(x), \Psi_{j\tilde{i}}(y)\big) - d(x, y)\big| \leqslant \tau(\epsilon)s \leqslant 2\tau(\epsilon)u.$$

Hence we can approximate it by a smooth map $\Phi'_{j\tilde{i}}$ satisfying (22.15c). It is almost compatible with the coordinate transformations $\psi_{ji}, \tilde{\psi}_{\tilde{j}\tilde{i}}$. Hence again by an argument similar to [25] (or by using the center of mass technique) we can approximate it by a diffeomorphism $\Phi_{j\tilde{i}}(x)$ which is exactly compatible with the coordinate transformation. We thus obtain $\Phi_{Z,s,u}$, if $u \in [s/2, s]$. It is also an $s\tau(\epsilon|n)$-Hausdorff approximation. Let $\Phi_{Z,u,s}$ be the inverse of it. We remark that we have an inequality

$$e^{-\tau(\epsilon|n)} \leqslant \frac{d(\Phi_{Z,u,s}(x), \Phi_{Z,u,s}(y))}{d(x,y)} \leqslant e^{\tau(\epsilon|n)} \tag{22.18}$$

which is sharper than (22.10) in case $u \in [s/2, s]$.

We remark here that the proof of Theorem 22.8(4) is almost the same as this argument. (So we do not discuss it.)

Now we continue the proof of (2) for the general u, s. We may assume $u = 2^{-k}s$ and put

$$\Phi_{Z,u,s} = \Phi_{Z,u,2u} \circ \cdots \circ \Phi_{Z,s/2,s}. \tag{22.19}$$

It is a diffeomorphism. We will check (22.10). Let $\rho > 0$. We first remark that $\Phi_{Z,a,b}$ is a $b\tau(\epsilon|n)$-Hausdorff approximation for $a \leqslant b$. (This is because if $b = 2^k a$ then $\Phi_{Z,a,b}$ is a $\sum_{j=0}^{k} \tau(\epsilon|n)2^{-j}b$-Hausdorff approximation.)

We first take ℓ such that

$$e^{-\rho} < \frac{1+2^{-\ell-1}}{1-2^{-\ell-1}} < e^{\rho}. \tag{22.20}$$

Now we take $x, y \in Z(s)$. We take k_1 such that $2^{-k_1-1}s < d(x,y) \leqslant 2^{-k_1}s$. (In case $d(x,y) \geqslant s$, we put $k_1 = 0$.) Note, if $d(x,y) \leqslant s$, we have

$$k_1 \leqslant -C \log \frac{d(x,y)}{s}. \tag{22.21}$$

We use (22.18) and (22.21) and obtain

$$\frac{d(\Phi_{Z,2^{-k_1}s,s}(x), \Phi_{Z,2^{-k_1}s,s}(y))}{d(x,y)} \leqslant e^{k_1\tau(\epsilon|n)} \leqslant Cd(x,y)^{-\tau(\epsilon|n)}. \tag{22.22}$$

Combining the inequality of the opposite direction, which can be proved in a similar way, we have

$$\begin{aligned} C^{-1}d(x,y)^{1+\tau(\epsilon|n)} &\leqslant d\big(\Phi_{Z,2^{-k_1}s,s}(x), \Phi_{Z,2^{-k_1}s,s}(y)\big) \\ &\leqslant Cd(x,y)^{1-\tau(\epsilon|n)}. \end{aligned} \tag{22.23}$$

We next put $k_2 = k_1 + \ell$. We may take ϵ so small that $\ell\tau(\epsilon|n) < \rho$. Then we have

$$e^{-\rho} \leqslant \frac{d(\Phi_{Z,2^{-k_2}s,s}(x), \Phi_{Z,2^{-k_2}s,s}(y))}{d(\Phi_{Z,2^{-k_1}s,s}(x), \Phi_{Z,2^{-k_1}s,s}(y))} \leqslant e^{\rho}. \tag{22.24}$$

Finally (22.20) implies

$$e^{-\rho} \leqslant \frac{d(\Phi_{Z,u,s}(x), \Phi_{Z,u,s})}{d(\Phi_{Z,2^{-k_2}s,s}(x), \Phi_{Z,2^{-k_2}s,s}(y))} \leqslant e^{\rho} \tag{22.25}$$

since $\Phi_{Z,u,2^{-k_2}s}$ is an $2^{-\ell-k_1}s\tau(\epsilon|n)$-Hausdorff approximation and $d(x,y) \geqslant 2^{-k_1-1}s$. The proof of (2) is complete.

We remark here that once the well-definedness property is established (3) is actually obvious. We only need to take s much smaller than the injectivity radius of Z.

We finally construct $\Phi_{Z,s} : Z \to Z(s)$. We put $\Phi_{ji} = \Phi_{Z;2^{-j}s,2^{-i}s} : Z(2^{-is}) \to Z(2^{-js})$. By construction we have $\Phi_{kj} \circ \Phi_{ji} = \Phi_{ki}$. Thus we have an inductive system. Using (2), it is easy to see that the inductive limit $\lim Z(2^{-i}s)$ is isometric to Z and there exists a map $Z(s) \to \lim_{i\to\infty} Z(2^{-i}s)$ satisfying the condition of (1).

We thus finished the proof of Theorem 22.8. □

23. Hausdorff convergence and Ricci curvature—II

In this section we continue the discussion about the Gromov–Hausdorff limit of a sequence of manifolds M_i with $\mathrm{Ricci}_{M_i} \geqslant -(n-1)$. It is known that the limit space X can be very wild. For example, it may not be locally contractible [102,118]. Nevertheless various results are known for such limit spaces, some of which we describe in this section.

We first state Theorem 16.4 again.

THEOREM 23.1 (Cheeger–Colding [28]). *Let M_i be a sequence of n-dimensional Riemannian manifolds with* $\mathrm{Ricci}_{M_i} > -\lambda_i$ *with $\lambda_i \to 0$ and let $(X,p) = \lim^{pGH}_{i\to\infty}(M_i,p_i)$. Suppose X contains a line. Then X is isometric to the direct product $\mathbb{R} \times X'$.*

Remark 23.1. The following slightly more general statement is proved.

Let $\mathrm{Ricci}_{M_i} \geqslant -\lambda_i$ with $\lambda_i \to 0$, and $(X,p) = \lim^{pGH}_{i\to\infty}(M_i,p_i)$. We assume $X \cong \mathbb{R}^k \times Y$ and that Y contains a line. Then $Y \cong \mathbb{R} \times Y'$.

Before explaining the outline of the proof, we mention several of its applications. One important application is Theorems 19.12 and 10.5, which we explained already.

To state other applications, we need some definitions.

DEFINITION 23.1. A *measured metric space* (X,μ) is a pair of a metric space X and a Borel measure μ on it. In this article we always assume that $\mu(X) = 1$. For a pointed measured metric space (X,p,μ) we assume $\mu(B_p(1,X)) = 1$.

For a Riemannian manifold M we use the renormalized volume form $\mu_M = \Omega_M/\mathrm{Vol}(M)$ and regard it as a measured metric space (unless another measure is specified explicitly). For a pointed Riemannian manifold (M,p), we use the renormalized volume form $\mu_M = \Omega_M/\mathrm{Vol}(B_p(1,M))$.

DEFINITION 23.2 [54]. A sequence of measured metric spaces (X_i,μ_i) is said to converge to (X,μ) with respect to the *measured Gromov–Hausdorff topology* which we write as $\lim^{mGH}_{i\to\infty}(X_i,\mu_i) = (X,\mu)$, if there exists a sequence of ϵ_i-Hausdorff approximations

$\varphi_i : X_i \to X$ with $\epsilon_i \to 0$, which are Borel measurable and such that, for any continuous function f on X, we have

$$\lim_{i\to\infty} \int_{X_i} (f \circ \varphi_i)\, d\mu_i = \int_X f\, d\mu.$$

The pointed measured Gromov–Hausdorff convergence is defined in the same way. (To be precise we need a net in place of a sequence to define a topology. It is an obvious modification and is omitted.)

Remark 23.2. In [75, Chapter $3\frac{1}{2}$D], Gromov defined a notion of $\square_\lambda$ convergence for measured metric spaces. It is similar to, but slightly different from, measured Gromov–Hausdorff topology defined above. Namely there is a situation where the support $\operatorname{supp}\mu$ of the limit measure is different from X. In that case for $\square_\lambda$ convergence the limit of (X_i, μ_i) is $(\operatorname{supp}\mu, \mu)$, and is different from the limit (X, μ) of the measured Gromov–Hausdorff topology. However if $(X_i, \mu_i) = (M_i, \mu_{M_i})$ is a Riemannian manifold and if $\mathrm{Ricci}_{M_i} \geqslant -(n-1)$, then the support of the limit measure is always X itself. (We can prove it using the Bishop–Gromov inequality.) So the two definitions coincide to each other.

Measured Gromov–Hausdorff convergence was introduced to study spectra of the Laplace operator. We mention it later.

LEMMA 23.2 [54]. *If $\lim^{GH}_{i\to\infty} X_i = X$, and if μ_i is a probability Borel measure on X_i, then there exists a subsequence k_i such that $\lim^{mGH}_{i\to\infty}(X_{k_i}, \mu_{k_i}) = (X, \mu)$.*

The proof is elementary.

We remark that the limit measure μ depends on the choice of the subsequence in general. In fact, let us consider $T^2 = S^1 \times S^1$ with metric $g^f_\epsilon = dt^2 + \epsilon^2 f(t)^2\, ds^2$, where $f : S^1 \to \mathbb{R}_+$ is a smooth function. Then (T^2, g_ϵ) converges to S^1 with standard metric and measure $f\,dt$ with respect to the measured Gromov–Hausdorff topology. On the other hand, the limit in the Gromov–Hausdorff distance is independent of f.

We denote by $\mathfrak{S}_n(D)$ the set of n-dimensional Riemannian manifolds M with $\mathrm{Ricci}_M \geqslant -(n-1)$, Diam(M) $\leqslant D$. We denote by $\mathfrak{S}_n(\infty)$ the set of n-dimensional pointed Riemannian manifold (M, p) such that $\mathrm{Ricci}_M \geqslant -(n-1)$. Let $\bar{\mathfrak{S}}_n(D)$, $\bar{\mathfrak{S}}_n(\infty)$ be the closure of $\mathfrak{S}_n(D)$, $\mathfrak{S}_n(\infty)$ with respect to the Gromov–Hausdorff distance, the pointed Gromov–Hausdorff distance, respectively.

We next define the singularity set and the regular set of a length space $X \in \bar{\mathfrak{S}}_n(D)$. We recall that the sequence $(X, R_i d_X, x)$ with $R_i \to \infty$ always has a subsequence such that $(X, R_i d_X, x)$ converges with respect to the pointed Gromov–Hausdorff distance (Proposition 16.2). However the limit is not unique. (Such an example is constructed in [29, Section 8].)

DEFINITION 23.3. We say that T_xX is a *tangent cone* of X at x if there exists a sequence $R_i \to \infty$ such that $(X, R_i d_X, x)$ converges to $(T_xX, \mathbf{o})$ with respect to the pointed Gromov–Hausdorff distance.[45]

DEFINITION 23.4. Let $X \in \bar{\mathfrak{S}}_n(\infty)$. We say that a point $x \in X$ is in $\mathcal{R}_k$ if $\mathbb{R}^k$ is a tangent cone T_xX of x.

We say x is *regular* if it is in $\mathcal{R} = \bigcup_k \mathcal{R}_k$. Otherwise it is said to be *singular* and we denote by $\mathcal{S}$ the set of all singular points.

Remark 23.3. This definition coincides with $\mathcal{S}(X)$ in Definition 17.7 by Otsu–Shioya in case when X is an Alexandrov space, because of Theorem 22.5.

One of the main results by Cheeger–Colding on the limit space X (in the collapsing situation) is the following

THEOREM 23.3 (Cheeger–Colding [29]). $\mu(\mathcal{S}) = 0$ *for any limit measure* μ.

Remark 23.4. (1) We remark that Theorem 23.3 implies $\mu(X \setminus \bigcup_k \mathcal{R}_k) = 0$, but does not imply the existence of k such that $\mu(X \setminus \mathcal{R}_k) = 0$.

(2) In Theorem 23.3 the limit measure μ is used. We do not know how to use the Hausdorff measure since it is not known whether the Hausdorff dimension of $X \in \bar{\mathfrak{S}}_n(D)$ is an integer or not.

Here is some very brief idea how a statement like Theorem 23.3 follows from Theorem 23.1. We want to find many points x on $X \in \bar{\mathfrak{S}}_n(D)$ such that T_xX is a Euclidean space. A naive idea to find such a point may be as follows. First we consider a minimal geodesic $\overline{xy_1}$ and take an interior point on it and put it x_1. Then any tangent cone $T_{x_1}X$ contains a line and hence splits as $T_{x_1}X \cong \mathbb{R} \times X_1$. We may next take a point x_2 near $\mathbf{o} \in T_{x_1}X$ which is a midpoint of the minimal geodesic. Then $T_{x_2}X_1$ contains a line and hence $T_{x_2}X_1$ splits. This process stops after finitely many stages since we can estimate the dimension of the tangent cone by the Bishop–Gromov inequality. Thus we find near x some kind of 'point' for which a tangent cone is $\mathbb{R}^k$. This argument however is too much naive to prove Theorem 23.3. So we need to work more seriously. See [29, Section 2]. □

We remark that Theorem 23.3 can be applied also to the collapsing situation. Namely it can be applied to the limit X of M_i such that $\mathrm{Vol}(M_i) \to 0$. Several other results are proved by Cheeger–Colding in [29,30]. Nevertheless there are yet many things unclear in the collapsing situation. (In other words, the result in the collapsing situation does not seem to be in the final form.) So we do not discuss it here. (We will discuss one of the main results of [31] later.)

[45] I am sorry that this terminology is inconsistent with one in Definition 16.4, where T_xX is called tangent cone when *any* such sequence $(X, R_i d_X, x)$ converges to $(T_x, 0)$. In this section we follow Cheeger–Colding and use this terminology. In Definition 16.4 we followed Burago–Gromov–Perelman.

In the noncollapsing case, Cheeger–Colding obtained more precise results. We discuss some of them here. The following theorem is a generalization of Theorem 22.1. We denote by $\mathfrak{S}_n(D, v)$ the set of n-dimensional Riemannian manifolds M with $\mathrm{Ricci}_M \geqslant -(n-1)$, $\mathrm{Diam(M)} \leqslant D$, $\mathrm{Vol}(M) \geqslant v$. We also denote by $\mathfrak{S}_n(\infty, v)$ the set of n-dimensional pointed Riemannian manifold (M, p) such that $\mathrm{Ricci}_M \geqslant -(n-1)$, $\mathrm{Vol}(B_p(1, M)) \geqslant v$. Let $\bar{\mathfrak{S}}_n(D, v)$, $\bar{\mathfrak{S}}_n(\infty, v)$ be the closure of $\mathfrak{S}_n(D, v)$, $\mathfrak{S}_n(\infty, v)$ with respect to the Gromov–Hausdorff distance, the pointed Gromov–Hausdorff distance, respectively.

THEOREM 23.4 [29, Theorem 5.9]. *Let $M_i \in \mathfrak{S}_n(\infty, v)$. We assume that* $\lim^{pGH}_{i\to\infty}(M_i, p_i) = (X, p)$. *Then for any R we have*

$$\lim_{i\to\infty} \mathrm{Vol}\big(B_{p_i}(R, M_i)\big) = \mathcal{H}^n\big(B_p(R, X)\big).$$

Here $\mathcal{H}^n$ denotes the n-dimensional Hausdorff measure.

COROLLARY 23.5 [29]. *If $X \in \bar{\mathfrak{S}}_n(\infty, v)$ then the Hausdorff dimension of X is n. Moreover any limit measure μ is equal to a multiple of the n-dimensional Hausdorff measure.*

Corollary 23.5 follows from Theorem 23.4 easily. We explain an idea of the proof of Theorem 23.4 later in this section.

THEOREM 23.6 [28, Theorem 5.2]. *Let $X \in \bar{\mathfrak{S}}_n(\infty, v)$ and $x \in X$. Then any tangent cone $T_x X$ is isometric to a cone CY of some length space Y of diameter $\leqslant \pi$.*

Remark 23.5. This result is a kind of generalization of the corresponding result Theorem 17.16 on Alexandrov spaces. However it is not asserted that CY is unique. Actually there is a counter example [29, 8.41]. The conclusion of Theorem 23.6 does not hold in the collapsing situation [29, 8.95].

To prove Theorem 23.6 we need another kind of comparison theorem, which we will explain later.

To state the next result we need a definition.

DEFINITION 23.5. Let $X \in \bar{\mathfrak{S}}_n(\infty, v)$. We say that $x \in \mathcal{R}_\epsilon$ if every tangent cone $T_x X$ satisfies

$$d_{GH}\big(B_0(1, T_x X), B_0(1, \mathbb{R}^n)\big) < \epsilon.$$

We put $\mathcal{S}_\epsilon = X \setminus \mathcal{R}_\epsilon$.

Remark 23.6. (1) Using Theorems 22.5, 23.4, we can prove that there exists δ such that $\mathrm{Vol}(B_x(r, X)) \geqslant (1-\delta)\,\mathrm{Vol}(B_0(r, \mathbb{R}^n))$ implies $x \in \mathcal{R}_\epsilon$. Thus Definition 23.5 is equivalent to $\mathcal{S}_\delta(X)$ in Definition 17.7.

(2) We can easily see that if $\epsilon' < \epsilon$, then $\mathcal{R}_{\epsilon'}$ is contained in the interior $\mathrm{Int}\,\mathcal{R}_\epsilon$ of $\mathcal{R}_\epsilon$.

(3) In the case of $X \in \bar{\mathfrak{S}}_n(\infty, v)$, we can easily show $\mathcal{R} = \mathcal{R}_n$. Using it, we can easily prove $\mathcal{S} = \bigcup_{\epsilon>0} \mathcal{S}_\epsilon$, $\mathcal{R} = \bigcap_{\epsilon>0} \mathcal{R}_\epsilon$.

The following is an analogue of Theorems 17.2 and 17.22.

THEOREM 23.7 [29, Theorem 5.14]. *There exists $\epsilon_0(n)$ such that if $X \in \bar{\mathfrak{S}}_n(\infty, v)$ and if $\epsilon < \epsilon_0(n)$, then there exists a smooth Riemannian manifold $Z(\epsilon)$ and a homeomorphism $\Phi_\epsilon : Z(\epsilon) \to \operatorname{Int} \mathcal{R}_\epsilon$ such that*

$$C^{-1} d(x, y)^{1+\tau(\epsilon|n)} < d\big(\Phi_\epsilon(x), \Phi_\epsilon(y)\big) < C d(x, y)^{1-\tau(\epsilon|n)}.$$

Theorem 23.7 actually follows easily from Theorem 22.8. Namely we find, for each $\epsilon' > \epsilon$ and $x \in \mathcal{R}_\epsilon$, a positive number r such that $d_{GH}(B_x(r, X), B_0(r, \mathbb{R}^n)) < \epsilon'$. Therefore any compact subset of $\operatorname{Int} \mathcal{R}_\epsilon$ satisfies condition $\mathcal{R}_{\epsilon', r, n}$ for some r. Theorem 23.7 then follows from Theorem 22.8. □

We next prove Theorem 23.4. For simplicity of notation we assume that X is compact. We first prove $\mathcal{H}^n(\mathcal{S}) = 0$. Let μ be a limit measure. We remark that there exists C_1, C_2 such that for $0 < r \leqslant 1$ we have

$$C_1 r^n \leqslant \operatorname{Vol}\big(B_p(r, M_i)\big) \leqslant C_2 r^n, \tag{23.1a}$$

$$C_1 r^n \leqslant \mu\big(B_p(r, X)\big) \leqslant C_2 r^n. \tag{23.1b}$$

In fact (23.1a) is a consequence of the Bishop–Gromov inequality and $\operatorname{Vol}(M_i) \geqslant v > 0$. Then (23.1b) follows, since μ is a limit measure. By Theorem 23.3 we have $\mu(\mathcal{S}) = 0$. Therefore by (23.1b) and the definition of Hausdorff measure we have $\mathcal{H}^n(\mathcal{S}) = 0$.

It follows that

$$\lim_{\epsilon \to 0} \mathcal{H}^n(\operatorname{Int} \mathcal{R}_\epsilon) = \mathcal{H}^n(X). \tag{23.2}$$

We can take the disjoint union of finitely many balls $U_\epsilon = \bigcup_j B_{y_j}(r_j, X) \subset \mathcal{R}_\epsilon$ such that

$$\mathcal{H}^n(X \setminus U_\epsilon) < \tau(\epsilon), \tag{23.3a}$$

$$\left| \omega_n \sum_j r_j^n - \mathcal{H}^n(X) \right| < \tau(\epsilon), \tag{23.3b}$$

$$d_{GH}\big(B_{y_j}(r_j, X), B_0(r_j, \mathbb{R}^n)\big) < 2\epsilon r_j, \tag{23.3c}$$

where $\omega_n = \operatorname{Vol}(B_0(1, \mathbb{R}^n))$. Here (23.3c) is a consequence of $y_j \subset \mathcal{R}_\epsilon$. Then, for large i, we have a disjoint union of balls $U_{\epsilon,i} = \bigcup_j B_{y_{j,i}}(r_j, M_i) \subset M_i$ with

$$d_{GH}\big(B_{y_{j,i}}(r_j, M_i), B_0(r_j, X)\big) < 3\epsilon r_j. \tag{23.4}$$

Therefore, by (23.3b), (23.3c), (23.4), and Theorem 22.5 we have

$$\left|\mathcal{H}^n(U_\epsilon) - \mathrm{Vol}(U_{\epsilon,i})\right| < \tau(\epsilon, 1/i|n). \tag{23.5}$$

We thus proved

$$\mathcal{H}^n(X) \leqslant \mathrm{Vol}(M_i) + \tau(\epsilon, 1/i|n).$$

To prove the opposite inequality, we take finitely many balls $B_{z_a}(t_a, X)$ such that

$$X \subseteq \bigcup_j B_{y_j}(r_j, X) \cup \bigcup_j B_{z_a}(t_a, X), \tag{23.6a}$$

$$\sum_j t_a^n \leqslant \tau(\epsilon). \tag{23.6b}$$

Then for large i we find $z_{a,i}$ such that

$$M_i = \bigcup_j B_{y_{j,i}}(r_j, M_i) \cup \bigcup_j B_{z_{a,i}}(t_{a,i}, M_i). \tag{23.7}$$

Since

$$\mathrm{Vol}\left(\bigcup_j B_{z_{a,i}}(t_{a,i}, M_i)\right) < C_n t_a^n,$$

it follows that

$$\mathrm{Vol}(M_i \setminus U_{\epsilon,i}) \leqslant \tau(\epsilon|n). \tag{23.8}$$

Therefore, $\mathcal{H}^n(X) \geqslant \mathrm{Vol}(M_i) - \tau(\epsilon, 1/i|n)$, as required. □

We now sketch the proof of Theorem 23.1. We start with the following situation.

(A) M is a Riemannian manifold with $\mathrm{Ricci}_M \geqslant -\lambda$ with small λ.

(B) We assume $d_{GH}(B_L(z, M), B_L(z', X)) < \rho/10$ and there is a line containing z'. (Here L is large.)

(C) Let $p, q \in M$ with $d(p,q) = 2L$ with large L.

(D) $d(z, \overline{pq}) \leqslant \rho/3$, $|d(z,p) - L| \leqslant \rho/3$, $|d(z,q) - L| \leqslant \rho/3$.

Here $M = M_i$, where M_i is as in Theorem 23.1 for large i. Such pair of points p, q exists because of (B).

We want to find a length space X' such that $B_z(R, M)$ is close to a D ball $B_{(x',0)} \times (R, \mathbb{R} \times X')$ in $\mathbb{R} \times X'$ with respect to the Hausdorff distance. (Here $1 \ll R \ll L$.)

We use the following function which is an approximation of the Busemann function:

$$b_+(x) = d(p,x) - d(p,z), \qquad b_-(x) = d(q,x) - d(q,z). \tag{23.9}$$

The argument to control them is similar to the proof of Proposition 22.7. However our problem is a bit different from the situation of Proposition 22.7 where Hausdorff approximation is given by assumption. Our situation is similar to Theorem 22.5(1), where we use another assumption (which was the almost maximality of the volume in case of Theorem 22.5(1)) to find Hausdorff approximation. In our case, we use the following theorem by Abresch–Gromoll to obtain some information on $b_\pm$ and improve it by using a similar argument as in the proof of Proposition 22.7. To state the result by Abresch–Gromoll we need a notation.

DEFINITION 23.6. For $x, p, q \in M$, an *excess* $E(x; p, q)$ is by definition

$$E(x; p, q) = d(x, p) + d(x, q) - d(p, q).$$

THEOREM 23.8 (Abresch–Gromoll [2]). *If* $\mathrm{Ricci}_M \geqslant -(n-1)\lambda, d(z, p) \geqslant L, d(z, q) \geqslant L$ *and if* $E(z; p, q) < \rho$, *then*

$$E(x; p, q) < \tau(\rho, \lambda, 1/L|n, R)$$

for any $x \in B_z(R, M)$.

Remark 23.7. Abresch–Gromoll stated Theorem 23.8 in the case $E(z; p, q) = \rho$ namely the case $z \in \overline{pq}$. The above form is a modification by Cheeger–Colding [28, Proposition 6.2]. ([26, Theorem 9.1].)

Remark 23.8. Abresch–Gromoll used Theorem 23.8 to show the following Theorem 23.9. It seems that Theorem 23.8 is the first comparison theorem established assuming conditions on Ricci curvature only.

For a length space M and $B \subset A \subseteq M$ we denote by $\mathrm{Diam}(B \subset A)$ the following number

$$\sup_{p,q \in B} \{\text{the length of the shortest curve joining } p \text{ and } q \text{ in A}\}.$$

THEOREM 23.9 (Abresch–Gromoll [2]). *If M is a complete manifold with* $\mathrm{Ricci}_M \geqslant 0$, $\inf K_M > -\infty$, *and*

$$\mathrm{Diam}\big(S_p(3R, M) \subseteq A_p(2R, 4R; M)\big) \leqslant C/R,$$

then M is homotopy equivalent to an interior of a compact manifold with boundary.

Theorem 23.9 is proved by Theorem 23.8 and Morse theory of the distance function in a way similar to Theorem 14.6.

Let us go back to the discussion of the proof of Theorem 23.1. Theorem 23.8 and $b_+ + b_- = E(x; p, q) - E(z; p, q)$ imply

$$-\rho \leqslant b_+ + b_- \leqslant \tau\big(L^{-1}, \rho, \lambda|R, n\big). \tag{23.10}$$

Using the fact that $b_{\pm}$ is "almost subharmonic" we have the following formula (23.12). We define $\mathbf{b}_+ : B_z(R, M) \to \mathbb{R}$ by

$$\Delta \mathbf{b}_+ = 0, \tag{23.11a}$$

$$\mathbf{b}_+ = b_+ \quad \text{on } \partial B_z(R, M). \tag{23.11b}$$

Then, we can prove

$$\|\mathbf{b}_+ - b_+\|_{L_1^2(B_z(R,M))} \leqslant \tau\big(L^{-1}, \rho, \lambda | R, n\big). \tag{23.12}$$

(Here the right-hand side will become small by taking L large, λ, ρ small.)

We now consider the Bochner formula

$$\frac{1}{2}\Delta\big(|\nabla \mathbf{b}_+|\big) = |\operatorname{Hess} \mathbf{b}_+|^2 + \operatorname{Ricci}(\nabla \mathbf{b}_+, \nabla \mathbf{b}_+). \tag{23.13}$$

We remark $|\nabla b_+| = 1$. Hence using (23.12) the integral of the left-hand side of (23.13) is small. Since $\operatorname{Ricci}_M \geqslant -(n-1)\lambda$ it follows from (23.13)

$$\int_{B_z(R,M)} |\operatorname{Hess} \mathbf{b}_+| \leqslant \tau\big(L^{-1}, \rho, \lambda | R, n\big), \tag{23.14}$$

$$\int_{B_z(R,M)} \big(|\nabla \mathbf{b}_+| - 1\big) \leqslant \tau\big(L^{-1}, \rho, \lambda | R, n\big). \tag{23.15}$$

We put $X' = \mathbf{b}_+^{-1}(0)$. Now we will use (23.14), (23.15) to show that $B_x(R, M)$ is close to a R ball in $X' \times \mathbb{R}$ with respect to the Gromov–Hausdorff distance as follows.

Let us take $y, z \in B_x(R, M)$. Let $y_0, z_0 \in X'$ such that

$$d(y, y_0) = d(y, X'), \qquad d(z, z_0) = d(z, X').$$

We will prove

$$\big|d(y,z)^2 - d(y_0, z_0)^2 - \big(\mathbf{b}_+(y) - \mathbf{b}_+(z)\big)^2\big| \leqslant \tau\big(L^{-1}, \rho, \lambda | R, n\big). \tag{23.16}$$

(23.16) obviously implies that $y \mapsto (y_0, \mathbf{b}_+(y)) : B_x(R, M) \to B_{(x,0)}(R, X' \times \mathbb{R})$ is a Hausdorff approximation and hence

$$d_{GH}\big(B_x(R, M), B_{(x,0)}(R, X' \times \mathbb{R})\big) \leqslant \tau\big(L^{-1}, \rho, \lambda | R, n\big),$$

which is enough to complete the proof of Theorem 23.1.

Let us sketch the proof of (23.16). For simplicity we take $\mathbf{b}_+(z) = 0$ and $z = z_0$.

Let $\ell : [0, l] \to M$ be a minimal geodesic joining y_0 to y. We put $Q(t) = d(\ell(t), z)$. Let $\gamma_t : [0, Q(t)] \to M$ be a minimal geodesic joining z to $\ell(t)$. (See Figure 23.1.)

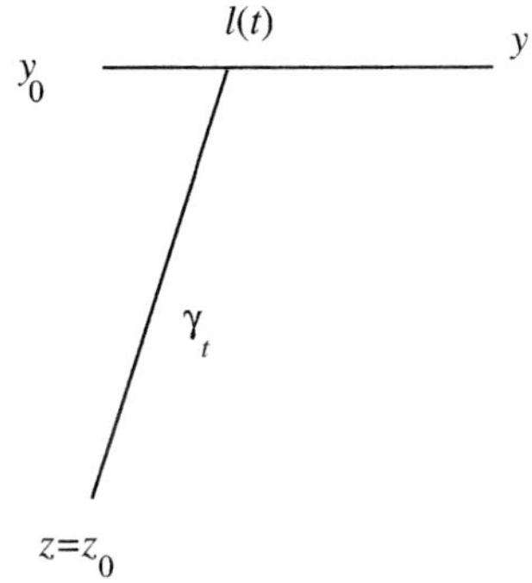

Fig. 23.1.

We put $h_t(s) = \mathbf{b}_+(\gamma_t(s))$. We remark that

$$\frac{d^2 h_t}{ds^2}(s) = (\mathrm{Hess}\,\mathbf{b}_+)\big(\dot\gamma_t(s), \dot\gamma_t(s)\big) \ll 1. \tag{23.17}$$

On the other hand $h_t(Q(t)) = \mathbf{b}_+(\gamma(t))$ is almost equal to t.[46] Hence

$$\left\| \frac{dh_t}{ds}(s) - \frac{t}{Q(t)} \right\| \ll 1. \tag{23.18}$$

(Here we remark that (23.17), (23.18) do not hold pointwise but only after integrating over some domain. We omit the technical difficulty which arises from this point.) By the first variational formula, we have

$$\frac{dh_t}{ds}(s) = \big\langle \dot\gamma_t(s), \nabla \mathbf{b}_+ \big\rangle \doteq \cos \angle y_0 \gamma(t) z = \frac{dQ}{dt}(t). \tag{23.19}$$

(Here and hereafter $\doteq$ means "almost" equal.)

Hence $Q(t)$ "almost" satisfies the following differential equation:

$$\frac{dQ}{dt} \doteq \frac{t}{Q(t)}. \tag{23.20}$$

The solution of (23.20) with initial value $Q(0) = d(y_0, z)$ is $Q(t) = \sqrt{d(y_0, z)^2 + t^2}$. Hence at $t = b_+(y)$ we have (23.16) with $z = z_0$. (See [26, Chapter 9] or [28, Section 6] for the details of the proof.) □

Here we say a few words about the proof of Remark 23.1. In this situation we can take not only p, q but also p_i, q_i, $i = 1, \dots, k$. Namely p, q are points close to the line on Y and p_i, q_i are taken as points close to the point on the coordinate axis of $\mathbb{R}^k$. Using them we

[46] We can find $(d^2/dt^2)(\mathbf{b}_+ \circ \gamma)$ is small in the same way as (23.17). Moreover $(d^2/dt^2)(\mathbf{b}_+ \circ \gamma)(0)$ is close to 1 by definition and (23.15).

obtain $\mathbf{b}_+$ together with $\mathbf{b}_+^i$, $i = 1, \ldots, k$. They all satisfy (23.14), (23.15). Moreover we have $\langle \nabla \mathbf{b}_+^i, \nabla \mathbf{b}_+^j \rangle \doteq \delta_{ij}$ where $\mathbf{b}_+^0 = \mathbf{b}_+$. We define $\Phi : X \to \mathbb{R}^{k+1}$ by $\Phi = (\mathbf{b}_+^0, \ldots, \mathbf{b}_+^k)$. Using it we can construct a pointed Hausdorff approximation $X \to \Phi^{-1}(0) \times \mathbb{R}^k$ in a way similar to the proof of Theorem 23.1. See [29, pp. 425–426], where similar arguments appears. □

In Section 20 we reviewed several results obtained by the L^2 comparison theorem where we compared a manifold with round sphere. In Section 22 we used L^2 comparison theorem where the model space was flat Euclidean space. In the proof Theorem 23.1, we compared a manifold with direct product $\mathbb{R} \times X'$. In [28], Cheeger–Colding developed a comparison theorem where the model space is a warped product (hereafter we call it the warped product comparison theorem) and gave various applications. We first review some of its applications.

One of its applications is Theorem 22.5(1). The following is closely related to it. (Theorem 22.5(1) corresponds to the case when $Y = S^{n-1}$.) Cheeger–Colding called this theorem ‘volume cone implies metric cone theorem’.

THEOREM 23.10 (Cheeger–Colding [28]). *For each ϵ there exists $\delta = \delta(\epsilon, n)$ with the following property. Let M be an n-dimensional Riemannian manifold with* $\mathrm{Ricci}_M \geqslant -\delta(n-1)$. *We assume*

$$\frac{\mathrm{Vol}(B_p(1, M))}{\mathrm{Vol}(S_p(1, M))} \leqslant (1+\delta) \frac{\mathrm{Vol}(B_0(1, \mathbb{R}^n))}{\mathrm{Vol}(S_0(1, \mathbb{R}^n))}.$$

Then there exists a length space Y with $\mathrm{Diam}(Y) \leqslant \pi$ *such that*

$$d_{GH}\big(B_p(1, M), B_0(1, CY)\big) \leqslant \epsilon.$$

We remark that $CY = ([0, \infty) \times Y)/\sim$ where $(0, x) \sim (0, y)$ with metric defined in Definition 17.2.

Another application of the warped product comparison theorem is Theorem 23.11. To state it we define a warped product.

DEFINITION 23.7. Let (X, g_X) be a Riemannian manifold and $f : (a, b) \to \mathbb{R}_+$ be a smooth function. Then the *warped product* $(a, b) \times_f X$ is by definition a product $(a, b) \times X$ equipped with the metric $dr^2 \oplus f(r)^2 g_X$, where r is the coordinate of the interval (a, b).

We need to define the warped product for general length space also. Let X be a length space and $f : (a, b) \to \mathbb{R}_+$ be a smooth function. Let $\ell : [\alpha, \beta] \to (a, b) \times X$ be a path which is, say, Lipschitz continuous. We put $\ell(t) = (r(t), \ell_X(t))$. We may change the parameter so that $\ell_X(t) : [\alpha, \beta] \to X$ is parameterized by arc length. We then define the length $L(\ell)$ of $\ell : [\alpha, \beta] \times_f X$ by

$$L(\ell) = \int_\alpha^\beta \sqrt{\big((dr/dt)(t)\big)^2 + f\big(r(t)\big)^2}\, dt.$$

We thus defined the length space $(a, b) \times_f X$.

EXAMPLE 23.1. (1) The simplest case is $f \equiv 1$. Then the warped product is the direct product.

(2) If $(a,b) = (0,\infty)$ and $f(r) = r$, then the warped product $(0,\infty) \times_r X$ is the cone CX minus 0. If moreover $X = S^{n-1}$ then it is $\mathbb{R}^n \setminus \{0\}$.

(3) We take $(a,b) = (0,\pi)$ and $f(r) = \sin r$. In this case the warped product $(0,\pi) \times_{\sin r} X$ is called the *metric suspension* SX. In particular the metric suspension SS^{n-1} of a round sphere S^{n-1} is the round sphere S^n.

THEOREM 23.11 [28, Theorem 5.14]. *If* $\mathrm{Ricci}_M \geqslant (n-1)$, $\dim M = n$ *and if* $\mathrm{Diam(M)} \geqslant \pi - \epsilon$ *then there exists a length space* X *such that* $d_{GH}(M, SX) < \tau(\epsilon|n)$.

Remark 23.9. It is not true in general that M is homeomorphic (or homotopy equivalent to) SX. The counter examples are the ones by Anderson and Otsu we mentioned already.

There are several other applications, for example, to the study of the cone at infinity. We omit it.

We now explain the idea of the proofs of these theorems. The main idea is to use the warped product comparison theorem. To state it we need some preliminary discussion. We begin with a characterization of a warped product. Let $f : (a,b) \to \mathbb{R}_+$ be a smooth function, we put

$$\mathcal{F}(r) = \int_a^r f(t)\,dt, \qquad k(r) = \frac{df}{dr}(r). \tag{23.21}$$

LEMMA 23.12. *Let* X *be a Riemannian manifold and* $M = (a,b) \times_f X$. *Then we have*

$$\mathrm{Hess}(\mathcal{F}) = k(r) g_M. \tag{23.22}$$

EXAMPLE 23.2. (1) In case $M = \mathbb{R} \times_1 X$ the direct product. $\mathcal{F}$ is linear and $k = 0$.

(2) In case $M = \mathbb{R}^n = CX \setminus 0 = (0,\infty) \times_r X$, we have $\mathcal{F} = r^2/2$ and $k(r) = 1$. If $X = S^{n-1}$, $M = \mathbb{R}^n$ then $\mathcal{F}(x_1,\dots,x_n) = \frac{1}{2}(x_1^2 + \cdots + x_n^2)$ and (23.22) is obvious.

(3) In case $M = (0,\pi) \times_{\sin r} S^{n-1}$ we have $\mathcal{F}(r) = -k(r) = \cos r$. Formula (23.22) is (21.17).

Let us prove Lemma 23.12. We put $\partial_r = \partial/\partial r$. $\mathrm{Hess}(\mathcal{F})(\partial_r, \partial_r) = k$ is obvious since $t \mapsto (t,p)$ is a geodesic. Let V be a vector filed of X, which we regard a vector field on M. We have $[V, \partial_r] = 0$. Since $g_M(V,V) = f^2 g_X(V,V)$ it follows that

$$-g_M(\partial_r, \nabla_V V) = g_M(\nabla_V \partial_r, V) = g_M(\nabla_{\partial_r} V, V) = fk g_X(V,V).$$

On the other hand, $V(\mathcal{F}) = 0$, $\partial_r(\mathcal{F}) = f$. Therefore

$$\mathrm{Hess}(\mathcal{F})(V,V) = -(\nabla_V V)(\mathcal{F}) = f^2 k g_X(V,V) = k g_M(V,V),$$

as required. □

The warped product comparison theorem is an 'almost version' of the following converse to Lemma 23.12.

PROPOSITION 23.13. *If M is a Riemannian manifold $\mathcal{F}: M \to (\alpha, \beta)$ is a fiber bundle. Suppose that there exists a function which $k: M \to \mathbb{R}$ such that*

$$\mathrm{Hess}_x(\mathcal{F}) = k(x) g_M. \tag{23.23}$$

We put $X = \{x \in M \mid \mathcal{F}(p) = \mathcal{F}(x)\}$. Then there exists a function $f:(a,b) \to \mathbb{R}_+$ such that

$$M \cong (a,b) \times_f X \quad (\textit{isometry}), \tag{23.24a}$$

$$\mathcal{F}(x) = \int_{r(p)}^{r(x)} f(t)\,dt, \tag{23.24b}$$

$$k(x) = \frac{df}{dr}\big(r(x)\big). \tag{23.24c}$$

Here $r: M \cong (a,b) \times_f X \to (a,b)$ is the projection to the first factor.

We now state the warped product comparison theorem. Let M be a complete Riemannian manifold and K be a compact subset. We put

$$r(x) = d(x, K) = \inf\big\{y \in K \mid d(x,y)\big\}, \tag{23.25a}$$

$$A_K(a,b,M) = \big\{x \in M \mid a < r(x) < b\big\}. \tag{23.25b}$$

Let $f:(a,b) \to \mathbb{R}_+$ be a smooth function and we define $\mathcal{F}(r)$ and $k(r)$ as in (23.21). We regard r as a function on $A_K(a,b,M)$ then $\mathcal{F}$ and k are functions on $A_K(a,b,M)$ as well. The following assumptions are a generalization of similar formulae we met several times already. For example, (23.14), (23.14) where $k(r) = 0$, and (21.17) where $k(r) = \cos r$.

ASSUMPTION 23.1. There exists $\tilde{\mathcal{F}}: A_K(a,b,M) \to (a,b)$ such that

$$\sup|\tilde{\mathcal{F}} - \mathcal{F}| \leqslant \epsilon, \tag{23.26a}$$

$$\frac{1}{\mathrm{Vol}(A_K(a,b,M))} \int_{A_K(a,b,M)} |\nabla\tilde{\mathcal{F}} - \nabla\mathcal{F}| \leqslant \epsilon, \tag{23.26b}$$

$$\frac{1}{\mathrm{Vol}(A_K(a,b,M))} \int_{A_K(a,b,M)} |\mathrm{Hess}\,\tilde{\mathcal{F}} - k g_M| \leqslant \epsilon. \tag{23.26c}$$

Theorem 23.14 asserts under Assumption 23.1 plus some more (which will follow), that $A_K(a,b,M)$ is Gromov–Hausdorff close to some warped product $(a,b) \times_f X$.

ASSUMPTION 23.2. M is an n-dimensional complete Riemannian manifold with $K_M \geqslant -\Lambda$. $\mathrm{Diam}(A_K(a,b,M)) \leqslant D$. $0 < \alpha' < \alpha$, $0 < \xi < \alpha - \alpha'$.

For each $x \in r^{-1}(a+\alpha')$ there exists $y \in r^{-1}(b-\alpha')$ such that

$$d'(x,y) \leqslant b-a-2\alpha'+\epsilon. \tag{23.27}$$

THEOREM 23.14 (Cheeger–Colding [28, Theorem 3.6]). *Under Assumptions* 23.1 *and* 23.2, *there exists a length space X such that*

$$d_{GH}\big((A_K(a+\alpha,b-\alpha,M),d'),(a+\alpha,b-\alpha)\times_f X\big) \leqslant \tau(\epsilon|\alpha',\xi,n,f,D).$$

Remark 23.10. In Assumption 23.2 and Theorem 23.14 we use the symbol d' for the metric of subsets of M. Note that the space $A_K(a,b,M)$ is not complete. So when we define the metric function $d\colon A_K(a,b,M)\times A_K(a,b,M)\to\mathbb{R}$ using the Riemannian metric, we need to be a bit careful. Namely for $p,q\in A_K(a,b,M)$ we need to take the infimum of the length of the curves joining them in a slightly larger domain. The metric d' stands for such a metric. We do not define it since it is too technical. See [28, pp. 205–206].

Let us explain the idea of the proof of Theorem 23.14. Actually the idea is quite similar to one of the proof of (23.16) we discussed already.

We take $X=r^{-1}(a+\alpha)$. To define a metric on it we consider a broken geodesic on a small neighborhood and take the infimum of the length of them. Now we construct the Hausdorff approximation $\Phi\colon A_K(a+\alpha,b-\alpha,M)\to(a+\alpha,b-\alpha)\times_f X$. Let $y,z\in A_K(a+\alpha,b-\alpha,M)$. We take $y_0,z_0\in X$ so that $d(y,y_0)=d(y,X)$, $d(z,z_0)=d(z,X)$. We remark $r(x)=d(x,X)-a-\alpha'$. We put

$$\Phi(y)=\big(r(y),y_0\big),$$

and will prove that Φ is a Hausdorff approximation.

We assume $z_0=z$ for simplicity. Let $\ell\colon[0,l]\to M$ be a minimal geodesic joining y_0 to y. We put $Q(t)=d(\ell(t),z)$. Let $\gamma_t\colon[0,Q(t)]\to M$ be a minimal geodesic joining z to $\ell(t)$.

Actually there is a technical trouble here. Namely since $A_K(a+\alpha,b-\alpha,M)$ is not complete, we may not be able to take γ_t. By this reason, we need to take a broken geodesic. (See Figure 23.2.) However since this is a technical point, we forget it and assume that we can take γ_t.

We put $h_t(s)=\tilde{\mathcal{F}}(\gamma_t(s))$. By (23.26), h_t 'almost' satisfies the differential equation

$$\frac{d^2h_t}{ds^2}(s)\doteq H\big(h_t(s)\big), \tag{23.28}$$

where $H(c)=k(\tilde{F}^{-1}(c))$. We remark that (23.28) is an ordinary differential equation of second order and hence has unique solution under an appropriate boundary condition. Note that $h_t(0)\doteq\mathcal{F}(a+\alpha)$, $h_t(Q(t))=\mathcal{F}(t+a+\alpha)$. Thus h_t is determined by $Q(t)$. (Precisely we have to say that h_t is 'almost' determined by $Q(t)$ since (23.28) is only 'almost' satisfied.) Moreover we have

$$\frac{dh_t}{ds}\big(Q(t)\big)\doteq\frac{dQ}{dt}(t) \tag{23.29}$$

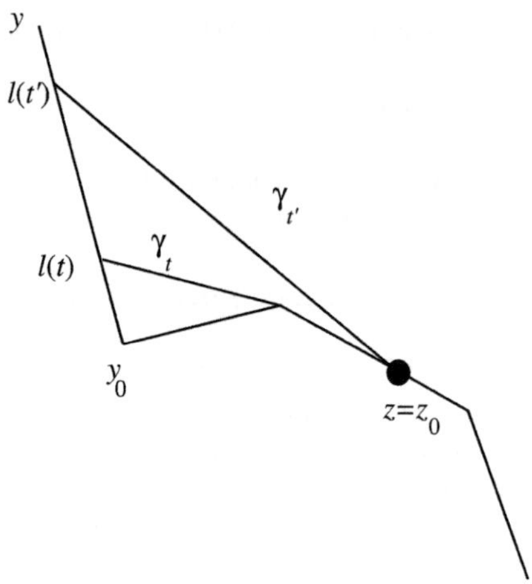

Fig. 23.2.

by the same reason as (23.19). Thus, (23.29) becomes a differential equation of first order on Q and is determined by f. We remark that Q satisfies an initial value condition $Q(0) = d(z, y_0)$. Therefore, the value of Q at $t = r(y) - a - \alpha$ is determined by this equation and the initial value $d(z, y_0)$. (Precisely speaking, we can only say the value of Q is almost determined.) By definition $Q(r(y) - a - \alpha) = d(y, z)$. Since it is almost determined by $d(z, y_0)$ and $r(y)$ and $r(z)$ (which we assumed to be zero for simplicity), it follows that Φ 'almost' preserves the length.

The fact that a small neighborhood of the image of Φ contains $(a + \alpha, b - \alpha) \times_f X$ follows from (23.27). This is a sketch of the proof of Theorem 23.14. □

We now discuss applications of Theorem 23.14.

We first show how we can use Theorem 23.14 to prove Theorem 22.5(1). Let us assume $\mathrm{Vol}(B_p(1, M)) \geqslant \mathrm{Vol}(B_0(1, \mathbb{R}^n)) - \epsilon$ and $\mathrm{Ricci}_M \geqslant -\lambda$. We put $f(t) = t$. Then $k(t) \equiv 1$, $\mathcal{F}(t) = t^2/2$. We need to check Assumptions 23.1, 23.2. Put $r(x) = d(x, p)$. We calculate

$$(2 - n) \int_{S_p(R,M)} r^{1-n} = \int_{S_p(R,M)} \operatorname{grad} r^{2-n} \cdot dn = \int_{B_p(R,M)} \Delta r^{2-n}. \tag{23.30}$$

Since $\mathrm{Vol}(B_p(1, M)) \geqslant \mathrm{Vol}(B_0(1, \mathbb{R}^n)) - \epsilon$ it follows from Lemma 21.1 that $\int_{S_p(R,M)} r^{1-n} = c_n \mathrm{Vol}(S_p(R, M))/\mathrm{Vol}(S_0(R, \mathbb{R}^n))$ is almost independent of R. Hence (23.30) implies $\int_{B_p(R,M)\setminus B_p(\delta,M)} \Delta r^{2-n}$ is small. Namely r^{2-n} is almost a harmonic function on $B_p(R, M) \setminus B_p(\delta, M)$. (We remark that r^{2-n} is harmonic on $\mathbb{R}^n$.) Then we have

$$\Delta r^{2-n} = (2 - n) \operatorname{div}\bigl(r^{1-n} \operatorname{grad} r\bigr) = (2 - n) r^{1-n} \Delta r + (2 - n)(1 - n) r^{-n}.$$

Hence

$$\Delta r \doteq (n - 1) r^{-1}. \tag{23.31}$$

And hence

$$\Delta r^2 = 2\,\mathrm{div}(r\,\mathrm{grad}\,r) = 2 + 2r\Delta r \doteq 2n. \tag{23.32}$$

We now apply (21.16) to $r^2/2 = \mathcal{F}$ and obtain

$$n \doteq \frac{1}{2}\Delta r^2 \doteq |\,\mathrm{Hess}\,\mathcal{F}|^2 + r^2\,\mathrm{Ricci}(\nabla r, \nabla r). \tag{23.33}$$

Since $\mathrm{Ricci}_M \geqslant -\lambda$ it follows from (23.33) and (23.32) that

$$\mathrm{Hess}\,\mathcal{F} \doteq g_M.$$

Hence if $\tilde{\mathcal{F}}$ is an harmonic function which approximates $\mathcal{F}$ we can check Assumptions 23.1, 23.2. Therefore Theorem 23.14 implies that there exists X such that

$$d_{GH}\big((A_p(2\delta, 1-2\delta, M), d'), (2\delta, 1-2\delta) \times_r X\big) \leqslant \tau(\epsilon|\delta, \lambda, n).$$

To complete the proof it suffices to show that X is close to S^{n-1} with respect to the Gromov–Hausdorff distance. We can do it by looking at the proof of Theorem 23.14 in our case a bit more carefully. Alternatively we can proceed as follows. Take $\rho \ll 1$, with $\delta \ll \rho^n$. By the assumption and the Bishop–Gromov inequality we can find p_1, q_1 such that $2d(p, p_1) = 2d(p, q_1) = d(p_1, q_1)$. We use it in the same way as in the proof of Theorem 23.1 to find $V_1 \supset B_p(\rho, M)$ such that $d_{GH}(V_1, [-\rho, \rho] \times X_1) \leqslant \rho\tau(\epsilon, \lambda|n)$. We then take points p_2, q_2 in a neighborhood of X_1 such that $2d(p_2, p) \doteq 2d(q_2, p) \doteq d(p_2, q_2)$. Then we use it in the same way as in the proof of Remark 23.1 to find $V_2 \supset B_p(\rho^2, M)$ such that $d_{GH}(V_2, [-\rho, \rho]^2 \times X_2) \leqslant \rho^2\tau(\epsilon, \lambda|n)$. Repeating this n times, we obtain $V_n \supset B_p(\rho^n, M)$ such that $d_{GH}(V_n, [-\rho, \rho]^n) \leqslant \rho^n\tau(\epsilon, \lambda|n)$. Since $\delta \ll \rho^n$ it then follows that $d_{GH}(X, S^{n-1}) \leqslant \tau(\epsilon, \lambda|n)$. This implies Theorem 22.5(1). □

The proof of Theorem 23.10 is similar to the first half of the proof of Theorem 22.5(1) and is omitted. □

We next explain the proof of Theorem 23.11. Let $\mathrm{Ricci}_M \geqslant (n-1)$ and $p, q \in M$ with $d(p,q) \geqslant \pi - \epsilon$. Put $f(r) = \sin r$, $r(x) = d(x, p)$, $\mathcal{F}(r) = -k(r) = -\cos r$. By (the proof of) Theorem 21.9 (see (21.17); we remark that f there is our $\mathcal{F}$),

$$\mathrm{Hess}(\mathcal{F}) \doteq k(r) g_M.$$

In this way we can check Assumption 23.1. Assumption 23.2 follows from the Bishop–Gromov inequality in this case. We thus can apply Theorem 23.14 and prove Theorem 23.11. □

We next explain the idea of the proof of Theorem 23.6. Let $((X, d_X), x) = \lim^{pGH}_{i\to\infty}(M_i, x_i)$ with $(M_i, x_i) \in \mathfrak{S}_n(\infty, v)$. We suppose that a tangent cone $T_xX = \lim^{pGH}_{i\to\infty}((X, r_i d_X), x)$ is not a cone.

Then there exist δ, R, ρ (and a subsequence of r_i which we denote by the same symbols) such that

$$d_{GH}\big(A_x(\delta, R; (X, r_i d_X)), A_{\mathbf{o}}(\delta, R; CY)\big) > \rho$$

for a cone CY. We can take $j_i \to \infty$ such that

$$d_{GH}\big(A_{x_{j_i}}(\delta, R; (M_{j_i}, r_i g_{M_{j_i}})), A_{\mathbf{o}}(\delta, R; CY)\big) > \rho/2 \tag{23.34}$$

for a cone CY. We now claim that

$$\begin{aligned}&\frac{\mathrm{Vol}(S_{x_{j_{i+1}}}(\delta/r_{i+1}, (M_{j_{i+1}}, g_{M_{j_{i+1}}})))}{\mathrm{Vol}(S_0(\delta/r_{i+1}, \mathbb{R}^n))}\\ &\quad > (1+\epsilon)\frac{\mathrm{Vol}(S_{x_{j_i}}(R/r_i, (M_{j_i}, g_{M_{j_i}})))}{\mathrm{Vol}(S_0(R/r_i, \mathbb{R}^n))}\end{aligned} \tag{23.35}$$

for ϵ independent of i. In fact, if (23.35) does not hold, then we can apply the argument of the first half of the proof of Theorem 22.5(1) to $A_{x_{j_i}}(\delta, R; (M_{j_i}, r_i g_{M_{j_i}}))$ and using Theorem 23.14, we can show that (23.34) does not hold.

Now it is easy to deduce a contradiction from (23.35). By taking a subsequence we may assume that $\delta/r_i > R/r_{i+1}$. Then (23.35) and the Bishop–Gromov inequality implies that

$$\begin{aligned}&\frac{\mathrm{Vol}(S_{x_{j_i}}(R/r_i, (M_{j_i}, g_{M_{j_i}})))}{\mathrm{Vol}(S_0(R/r_i, \mathbb{R}^n))}\\ &\quad > (1+\epsilon)^{i-1}\frac{\mathrm{Vol}(S_{x_{j_1}}(R/r_1, (M_{j_1}, g_{M_{j_1}})))}{\mathrm{Vol}(S_0(R/r_1, \mathbb{R}^n))}.\end{aligned} \tag{23.36}$$

This is a contradiction since the left-hand side is bounded as $i \to \infty$. □

In [31], Cheeger–Colding studied a convergence of the eigenvalue of the Laplace operator using the results explained so far. We state their result (without outline of the proof) here.

We start with a simple example to illustrate that the measured Hausdorff convergence is related to the eigenvalue of the Laplace operator. Let us consider $T^2 = S^1 \times S^2$ with Riemannian metric $g_\epsilon^f = dt^2 + \epsilon^2 f(t)^2 ds^2$. Here $f : S^1 \to \mathbb{R}_+$. We assume $\int f\,dt = 1$. As we mentioned before the limit of (T^2, g_ϵ^f) with respect to the measured Hausdorff topology is S^1 with standard metric and measure $f\,dt$. The Dirichlet integral on (T^2, g_ϵ^f) is

$$D(h,h) = \epsilon \int f(t)\left(\left(\frac{dh}{dt}\right)^2 + \frac{1}{\epsilon f(t)}\left(\frac{dh}{ds}\right)^2\right) dt\,ds.$$

In case we consider the eigenvalues of the Laplacian which stay bounded as $\epsilon \to 0$, it suffices to consider h which is constant along the s direction. Hence we are to consider the bilinear form on $L^2(S^1)$ defined by

$$D(h,h) = \int f(t)\left(\frac{dh}{dt}\right)^2 dt.$$

In [54] the author proved that a similar phenomenon occurs in the situation we discussed in Section 11. Cheeger–Colding generalized it much and proved the following Theorem 23.15.

THEOREM 23.15 [31, Theorem 7.9]. *Let $M_i \in \mathfrak{S}_n(D)$. We assume that it converges to (X, μ) with respect to the measured Hausdorff topology. Then there exists a (unbounded) symmetric bilinear form D on $L^2(X, \mu)$ with discrete spectrum $\lambda_0(D) = 0 < \lambda_1(D) \leqslant \lambda_2(D) \leqslant \cdots$ such that kth eigenvalue $\lambda_k(-\Delta_{M_i})$ of the Laplace operator (on functions) on M_i converges to $\lambda_k(D)$.*

Remark 23.11. (1) In case the multiplicity of eigenvalue $\lambda_k(D)$ is m then we put $\lambda_k(D) = \cdots = \lambda_{k+m-1}(D)$.

(2) The eigenfunction of $-\Delta_{M_i}$ converges to the eigenfunctions of D in an appropriate sense.

We finally remark that the study of limits of Einstein manifolds (or manifolds with integral bounds of the curvature tensor) we discussed in Section 20 is improved by [32,27], etc. Here we restrict ourselves to quote the following Theorems 23.16, 23.17. Let M_i be a sequence of n-dimensional Riemannian manifolds. We consider the following integral bounds of the curvature for $p_i \in M_i$:

$$\int_{B_{p_i}(1,M_i)} |R_{M_i}|^p \Omega_{M_i} < C, \tag{23.37}$$

where C is independent of i. Let $\mathcal{S}$, $\mathcal{S}_k$ be as in Definitions 23.4. $\mathcal{H}^m$ is the m-dimensional Hausdorff measure.

We say $x \in \mathcal{S}$ is $(n-4k)$-*nonexceptional* if there exists a tangent cone T_xX which is *not* isometric to $\mathbb{R}^{n-4k} \times C(S^{4k-1}/\Gamma)$ where $\Gamma \subset O(4k)$ is a finite group acting freely on S^{4k-1}. Otherwise x is said to be $(n-4k)$-exceptional. Let $\mathcal{N}_{n-4k} \subset \mathcal{S}_{n-4k}$ be the set of all $(n-4k)$-nonexceptional points.

THEOREM 23.16 ([32, Theorems 1.15, 1.20], [27, Theorem 6.10]). *Let $M_i \in \mathfrak{S}_n(\infty, v)$ and $\lim^{pGH}_{i\to\infty}(M_i, p_i) = (X, p)$. We assume (23.37) for $1 \geqslant p \geqslant n/2 = \dim M_i/2$:*

(1) *If p is not an integer then $\mathcal{H}^{n-2p}(\mathcal{S}) = 0$.*

(2) *The Hausdorff dimension of $\mathcal{S}$ is not greater than $n-2p$.*

(3) *If $p = 2$, then $\mathcal{H}^{n-4}(\mathcal{N}_{n-4}) = 0$.*

(4) *If M_i are Kähler and p is an integer, then $\mathcal{H}^{n-2p}(\mathcal{S} \cap B_p(R, X)) < \infty$ for any R.*

We remark that in case $n = 2$ and the manifolds M_i are Einstein, Theorem 23.16(3) is Theorem 20.4.

THEOREM 23.17 [27, Theorem 11.1]. *In the situation of Theorem* 23.16 *we have*

(1) *If* $p = 1$, *then compact subsets of* $\mathcal{S}$ *are* $(n-2)$ *rectifiable.*

(2) *If either* $p = 2k$ *is an even integer, then* $\mathcal{N}_{n-4k} \cap B_p(R, X)$ *are* $(n-4k)$*-rectifiable.*

(3) M_i *are Kähler and* p *is integer, then* $\mathcal{S} \cap B_p(R, X)$ *are* $(n-2p)$*-rectifiable.*

We remark that as in 4-dimensional case, if $p = 2$ and the M_i are Einstein, the condition (23.37) can be written in terms of characteristic classes and hence is a topological one.

These results are parallel to the corresponding results in (higher-dimensional) gauge theory [105,146].

References

[1] U. Abresch, *Über das glatten Riemannisher Metriken*, Habilitationsschrift, Reinishcehn Friedrich-Willhelms-Universität, Bonn (1988).

[2] U. Abresch and D. Gromoll, *On complete manifolds with nonnegative Ricci curvature*, J. Amer. Math. Soc. **3** (1990), 355–374.

[3] U. Abresch and W. Meyer, *Injectivity radius estimates and sphere theorems*, Comparison Geometry, Math. Sci. Res. Inst. Publ., vol. 30, Cambridge Univ. Press, Cambridge (1997), 1–47.

[4] U. Abresch and V. Schroeder, *Graph manifolds, ends of negatively curved spaces and the hyperbolic* 120-*cell space*, J. Differential Geom. **35** (1992), 299–336.

[5] A.D. Alexandrov, *A theorem on triangles in a metric space and some of its applications*, Trudy Mat. Inst. Steklov., vol. 38, Izdat. Akad. Nauk SSSR, Moscow (1951), 5–23 (in Russian).

[6] A.D. Alexandrov, *Über eine Verallgemeinerung der Riemannschen Geometrie*, Schriftenreiche der Institut für Mathematik **1** (1957), 33–84.

[7] M.T. Anderson, *Convergence and rigidity of manifolds under Ricci curvature bounds*, Invent. Math. **102** (1990), 429–445.

[8] M.T. Anderson, *Metrics of positive Ricci curvature with large diameter*, Manuscripta Math. **68** (1990), 405–415.

[9] M.T. Anderson, *Short geodesics and gravitational instantons*, J. Differential Geom. **31** (1990), 265–275.

[10] M.T. Anderson, *Hausdorff perturbation of Ricci flat manifolds and the splitting theorem*, Duke Math. J. **68** (1992), 67–82.

[11] M.T. Anderson, *The L^2 structure of moduli spaces of Einstein metrics on* 4-*manifolds*, Geom. Funct. Anal. **2** (1) (1992), 29–89.

[12] M.T. Anderson and J. Cheeger, *C^α-compactness for manifolds with Ricci curvature and injectivity radius bounded below*, J. Differential Geom. **35** (1992), 265–281.

[13] S. Bando, *Bubbling out of Einstein manifolds*, Tohoku Math. J. (2) **42** (1990), 205–216; Correction: Tohoku Math. J. (2) **42** (1990), 587–588.

[14] S. Bando, A. Kasue and H. Nakajima, *On a construction of coordinates at infinity on manifolds with fast curvature decay and maximal volume growth*, Invent. Math. **97** (1989), 313–349.

[15] J. Bemelmans, Min-Oo and E. Ruh, *Smoothing Riemannian metrics*, Math. Z. **188** (1984), 69–74.

[16] P. Bérard, G. Besson and S. Gallot, *Embedding Riemannian manifolds by their heat kernel*, Geom. Funct. Anal. **4** (1994), 373–398.

[17] V.N. Berestovskij and I.G. Nikolaev, *Multidimensional generalized Riemannian spaces*, Geometry, IV, Encyclopaedia Math. Sci., vol. 70, Springer, Berlin (1993), 165–243, 245–250.

[18] M. Berger, *Les variétés Riemanniennes* (1/4)-*pincées*, Ann. Scuola Norm. Sup. Pisa (3) **14** (1960), 161–170.

[19] M. Berger, *Riemannian Geometry During the Second Half of the Twentieth Century*, University Lecture Series, vol. 17, American Mathematical Society, Providence, RI (2000).

[20] M. Berger, *A Panoramic View of Riemannian Geometry*, Springer (2003).

[21] G. Besson, G. Courtrois and S. Gallot, *Entropies et rigidités des espaces localement symétrique de courbure strictment négative*, Geom. Funct. Anal. **5** (1995), 731–799.

[22] Yu. Burago, M. Gromov and G. Perelman, *A.D. Alexandrov's spaces with curvature bounded from below*, Uspekhi Mat. Nauk **47** (1992), 3–51, 222 (in Russian); English transl.: Russian Math. Surveys **47** (1992), 1–58.

[23] H. Busemann, *The Geometry of Geodesics*, Academic Press, New York (1955).

[24] P. Buser and H. Karcher, *Gromov's almost flat manifolds*, Astérisque, vol. 81, Société Mathématique de France, Paris (1981), 148 pp.

[25] J. Cheeger, *Finiteness theorems for Riemannian manifolds*, Amer. J. Math. **92** (1970), 61–74.

[26] J. Cheeger, *Degeneration of Riemannian Metrics under Ricci Curvature Bounds*, Scuola Normale Superiore, Pisa (2001).

[27] J. Cheeger, *Integral bounds on curvature elliptic estimates and rectifiability of singular sets*, Geom. Funct. Anal. **13** (2003), 20–72.

[28] J. Cheeger and T. Colding, *Lower bounds on Ricci curvature and the almost rigidity of warped products*, Ann. of Math. **144** (1996), 189–237.

[29] J. Cheeger and T. Colding, *On the structure of spaces with Ricci curvature bounded below. I*, J. Differential Geom. **46** (1997), 406–480.

[30] J. Cheeger and T. Colding, *On the structure of spaces with Ricci curvature bounded below. II*, J. Differential Geom. **54** (2000), 13–35.

[31] J. Cheeger and T. Colding, *On the structure of spaces with Ricci curvature bounded below. III*, J. Differential Geom. **54** (2000), 37–74.

[32] J. Cheeger, T.H. Colding and G. Tian, *On the singularities of spaces with bounded Ricci curvature*, Geom. Funct. Anal. **12** (2002), 873–914.

[33] J. Cheeger and G. Ebin, *Comparison Theorems in Riemannian Geometry*, North-Holland Mathematical Library, vol. 9, North-Holland/American Elsevier, Amsterdam/New York (1975).

[34] J. Cheeger, K. Fukaya and M. Gromov, *Nilpotent structures and invariant metrics on collapsed manifolds*, J. Amer. Math. Soc. **5** (1992), 327–372.

[35] J. Cheeger and D. Gromoll, *The splitting theorem for manifolds of nonnegative Ricci curvature*, J. Differential Geom. **6** (1971–1972), 119–128.

[36] J. Cheeger and D. Gromoll, *On the structure of complete manifolds of nonnegative curvature*, Ann. of Math. **96** (1972), 413–443.

[37] J. Cheeger and D. Gromoll, *On the lower bound for the injectivity radius of* $1/4$*-pinched Riemannian manifolds*, J. Differential Geom. **15** (1980), 437–442.

[38] J. Cheeger and M. Gromov, *Collapsing Riemannian manifolds while keeping their curvature bounded. I*, J. Differential Geom. **23** (1986), 309–346.

[39] J. Cheeger and M. Gromov, *Collapsing Riemannian manifolds while keeping their curvature bounded. II*, J. Differential Geom. **32** (1990), 269–298.

[40] J. Cheeger and X. Rong, *Existence of polarized* F*-structures on collapsed manifolds with bounded curvature and diameter*, Geom. Funct. Anal. **6** (1996), 411–429.

[41] S.Y. Cheng, *Eigenvalue comparison theorems and its geometric applications*, Math. Z. **143** (1975), 289–297.

[42] S.Y. Cheng and S.T. Yau, *Differential equations on Riemannian manifolds and their geometric applications*, Comm. Pure Appl. Math. **28** (1975), 333–354.

[43] T. Colding, *Large manifolds with positive Ricci curvature*, Invent. Math. **124** (1996), 193–214.

[44] T. Colding, *Shape of manifolds with positive Ricci curvature*, Invent. Math. **124** (1996), 175–191.

[45] T. Colding, *Aspects of Ricci curvature*, Comparison Geometry, Math. Sci. Res. Inst. Publ., vol. 30, Cambridge Univ. Press, Cambridge (1997), 83–98.

[46] T. Colding, *Ricci curvature and volume convergence*, Ann. of Math. (2) **145** (1997), 477–501.

[47] T. Colding, *Space with Ricci curvature bound*, Documenta Mathematica, Extra Volume, ICM 1998 II, Doc. Math. J. DMV (1999), 299–308.

[48] R.D. Edwards and R.C. Kirby, *Deformations of spaces of imbeddings*, Ann. of Math. **93** (1971), 63–88.

[49] T. Eguchi and A. Hanson, *Asymptotically flat self-dual solutions to Euclidean gravity*, Phys. Lett. B **74** (1978), 249–251.

[50] J.H. Eschenburg, *New examples of manifolds with strictly positive curvature*, Invent. Math. **66** (1982), 469–480.

[51] F. Fang and X. Rong, *Positive pinching, volume and second Betti number*, Geom. Funct. Anal. **9** (1999), 641–674.

[52] K. Fukaya, *Theory of convergence of Riemannian orbifolds*, Japan. J. Math. **12** (1984), 121–160.

[53] K. Fukaya, *Collapsing Riemannian manifolds to ones of lower dimensions*, J. Differential Geom. **25** (1987), 139–156.

[54] K. Fukaya, *Collapsing of Riemannian manifolds and eigenvalues of Laplace operator*, Invent. Math. **87** (1987), 517–547.

[55] K. Fukaya, *A boundary of the set of Riemannian manifolds with bounded curvature and diameter*, J. Differential Geom. **28** (1988), 1–21.

[56] K. Fukaya, *Collapsing Riemannian manifolds to ones with lower dimension. II*, J. Math. Soc. Japan **41** (1989), 333–356.

[57] K. Fukaya, *Hausdorff convergence of Riemannian manifolds and its applications*, Recent Topics in Differential and Analytic Geometry, Adv. Stud. Pure Math., vol. 18-I, Academic Press, Boston, MA (1990), 143–238.

[58] K. Fukaya and T. Yamaguchi, *Almost nonpositively curved manifolds*, J. Differential Geom. **33** (1991), 67–90.

[59] K. Fukaya and T. Yamaguchi, *The fundamental groups of almost non-negatively curved manifolds*, Ann. of Math. **136** (1992), 253–333.

[60] K. Fukaya and T. Yamaguchi, *Isometry groups of singular spaces*, Math. Z. **216** (1994), 31–44.

[61] S. Gallot, *A Sobolev inequality and some geometric applications*, Spectra of Riemannian Manifolds, Berger, Murakami and Ochiai, eds, Kaigai, Kyoto (1981).

[62] S. Gallot, *Volumes, courbure de Ricci et convergence des variétés d'après T.H. Colding et Cheeger–Colding*, Séminaire Bourbaki, 50èm année, vol. 835 (1997–1998), 1–33.

[63] R.E. Greene, *Noncompact manifolds of nonnegative curvature*, Comparison Geometry, Math. Sci. Res. Inst. Publ., vol. 30, Cambridge Univ. Press, Cambridge (1997), 99–131.

[64] R.E. Greene and H. Wu, *Lipschitz convergence of Riemannian manifolds*, Pacific J. Math. **131** (1988), 119–141.

[65] D. Gromoll, *Differenzierbare Strukturen und Metriken positiver Krümmung auf Sphären*, Math. Ann. **164** (1966), 353–371.

[66] D. Gromoll and W. Myers, *On complete open manifolds of positive curvature*, Ann. of Math. **90** (1969), 75–90.

[67] M. Gromov, *Manifolds of negative curvature*, J. Differential Geom. **13** (1978), 223–230.

[68] M. Gromov, *Almost flat manifolds*, J. Differential Geom. **13** (1978), 231–241.

[69] M. Gromov, *Structures métriques pour les variétés riemannienes*, J. Lafontaine and P. Pansu, eds, Textes Mathématiques, vol. 1, CEDIC, Paris (1981).

[70] M. Gromov, *Curvature, diameter and Betti numbers*, Comment. Math. Helv. **56** (1981), 179–195.

[71] M. Gromov, *Groups of polynomial growth and expanding maps*, Comment. Math. Helv. **56** (1981), 179–195.

[72] M. Gromov, *Volume and bounded cohomology*, Inst. Hautes Etudes Sci. Publ. Math. **56** (1982), 5–99.

[73] M. Gromov, *Filling Riemannian manifolds*, J. Differential Geom. **18** (1983), 1–147.

[74] M. Gromov, *Carnot–Carathéodory spaces seen from within*, Sub-Riemannian Geometry, Progress in Mathematics, vol. 144, Birkhäuser, Basel (1996), 79–323.

[75] M. Gromov, *Metric structures for Riemannian and non-Riemannian spaces*, With appendices by M. Katz, P. Pansu and S. Semmes, Translated from the French by Sean Michael Bates, Progress in Mathematics, vol. 152, Birkhäuser Boston, Boston, MA (1999).

[76] M. Gross and P. Wilson, *Large complex structure limit of $K3$ surfaces*, J. Differential Geom. **55** (2000), 475–546.

[77] K. Grove and H. Karcher, *How to conjugate C^1-close group actions*, Math. Z. **132** (1973), 11–20.

[78] K. Grove and P. Petersen, *Bounding homotopy types by geometry*, Ann. of Math. (2) **128** (1988), 195–206.

[79] K. Grove and P. Petersen, *Homotopy types of positively curved manifolds with large volume*, Amer. J. Math. **110** (1988), 1183–1188.

[80] K. Grove and P. Petersen, *Manifolds near the boundary of existence*, J. Differential Geom. **33** (1991), 379–394.
[81] K. Grove, P. Petersen and J.Y. Wu, *Geometric finiteness theorems via controlled topology*, Invent. Math. **99** (1990), 205–213.
[82] K. Grove and K. Shiohama, *A generalized sphere theorem*, Ann. of Math. (2) **106** (1977), 201–211.
[83] K. Grove and W. Ziller, *Cohomogeneity one manifolds with positive Ricci curvature*, Invent. Math. **149** (2002), 619–646.
[84] T. Hattori, *Asymptotic geometry of arithmetic quotients of symmetric spaces*, Math. Z. **222** (1996), 247–277.
[85] N. Hitchin, *Harmonic spinor*, Adv. Math. **14** (1974), 1–55.
[86] H. Im Hof and E. Ruh, *An equivariant pinching theorem*, Comment. Math. Helv. **50** (1975), 389–401.
[87] J. Jost and H. Karcher, *Geometrische Methoden zur Gewinnung von a-priori-Schranken für harmonische Abbildungen*, Manuscripta Math. **40** (1982), 27–77.
[88] J. Jost and H. Karcher, *Almost linear functions and a priori estimates for harmonic maps*, Global Riemannian Geometry (Durham, 1983), Ellis Horwood Ser. Math. Appl., Horwood, Chichester (1984), 148–155.
[89] D. Joyce, *Compact Manifolds with Special Holonomy*, Oxford Univ. Press (2000).
[90] A. Kasue, *A convergence theorem for Riemannian manifolds and some applications*, Nagoya Math. J. **114** (1989), 21–51.
[91] A. Kasue, *Spectral convergence of Riemannian manifolds*, Tohoku Math. J. (2) **46** (1994), 147–179.
[92] A. Kasue, Y. Otsu, T. Sakai, T. Shioya and T. Yamaguchi, *Riemannian manifolds and its limit*, Mem. Math. Soc. Japan (in Japanese), in press.
[93] A. Katsuda, *Gromov's convergence theorem and its application*, Nagoya Math. J. **100** (1985), 11–48.
[94] W. Klingenberg, *Contributions to Riemannian geometry in the large*, Ann. of Math. **69** (1959), 654–666.
[95] W. Klingenberg and T. Sakai, *Injectivity radius estimate for* $1/4$*-pinched manifolds*, Arch. Math. (Basel) **34** (1980), 371–376.
[96] R. Kobayashi and A. Todorov, *Polarized period map for generalized* $K3$ *surface and moduli of Einstein metric*, Tohoku Math. J. **39** (1987), 341–363.
[97] S. Kobayashi and K. Nomizu, *Foundation of Differential Geometry, I, II*, Interscience (1963–1969).
[98] P. Kronheimer, *The construction of ALE spaces as hyper Kähler quotients*, J. Differential Geom. **29** (1989), 665–683.
[99] K. Kuwae, Y. Machigashira and T. Shioya, *Sobolev spaces, Laplacian, and heat kernel on Alexandrov spaces*, Math. Z. **238** (2001), 269–316.
[100] P. Li and R. Schoen, L^p *and mean value properties of subharmonic functions on Riemannian manifolds*, Acta Math. **153** (1984), 279–301.
[101] A. Lichnerowicz, *Geometric des Groupes des transformationes*, Dunod, Paris (1958).
[102] X. Menguy, *Noncollapsing examples with positive Ricci curvature and infinite topological type*, Geom. Funct. Anal. **10** (2000), 600–627.
[103] J. Milnor, *Morse Theory*, Annals of Mathematics Studies, vol. 51, Princeton Univ. Press, Princeton, NJ (1963).
[104] J. Milnor, *A note on curvature and fundamental group*, J. Differential Geom. **2** (1968), 1–7.
[105] H. Nakajima, *Compactness of the moduli space of Yang–Mills connections in higher dimensions*, J. Math. Soc. Japan **40** (1988), 383–392.
[106] H. Nakajima, *Hausdorff convergence of Einstein 4-manifolds*, J. Fac. Sci. Univ. Tokyo Sect. IA Math. **35** (1988), 411–424.
[107] I.G. Nikolaev, *Parallel translation and smoothness of the metric of spaces with bounded curvature*, Dokl. Akad. Nauk SSSR **250** (1980), 1056–1058.
[108] I.G. Nikolaev, *A metric characterization of Riemannian spaces*, Siberian Adv. Math. **9** (1999), 1–58.
[109] M. Obata, *Certain condition for a Riemannian manifold to be isometric to a sphere*, J. Math. Soc. Japan **14** (1962), 333–340.
[110] Y. Otsu, *On manifolds of positive Ricci curvature with large diameter*, Math. Z. **206** (1991), 255–264.
[111] Y. Otsu, K. Shiohama and T. Yamaguchi, *A new version of differentiable sphere theorem*, Invent. Math. **98** (1989), 219–228.

[112] Y. Otsu and T. Shioya, *The Riemannian structure of Alexandrov spaces*, J. Differential Geom. **39** (1994), 629–658.

[113] G. Perelman, *Elements of Morse theory on Aleksandrov spaces*, Algebra i Analiz **5** (1993), 232–241 (in Russian); English transl.: St. Petersburg Math. J. **5** (1994), 205–213.

[114] G. Perelman, *Manifolds of positive Ricci curvature with almost maximal volume*, J. Amer. Math. Soc. **7** (1994), 299–305.

[115] G. Perelman, *Proof of the soul conjecture of Cheeger and Gromoll*, J. Differential Geom. **40** (1) (1994), 209–212.

[116] G. Perelman, *Spaces with curvature bounded below*, Proceedings of the International Congress of Mathematicians, vols. 1, 2 (Zürich, 1994), Birkhäuser, Basel (1995), 517–525.

[117] G. Perelman, *Collapsing with no proper extremal subsets*, Comparison Geometry (Berkeley, CA, 1993–1994), Math. Sci. Res. Inst. Publ., vol. 30, Cambridge Univ. Press, Cambridge (1997), 149–155.

[118] G. Perelman, *Construction of manifolds of positive Ricci curvature with big volume and large Betti numbers*, Comparison Geometry, Math. Sci. Res. Inst. Publ., vol. 30, Cambridge Univ. Press, Cambridge (1997), 157–163.

[119] G. Perelman, *A.D. Alexandrov's spaces with curvatures bounded from below II*, preprint.

[120] S. Peters, *Cheeger's finiteness theorem for diffeomorphism classes of Riemannian manifolds*, J. Reine Angew. Math. **349** (1984), 77–82.

[121] S. Peters, *Convergence of Riemannian manifolds*, Compositio Math. **62** (1987), 3–16.

[122] P. Petersen, *A finiteness theorem for metric spaces*, J. Differential Geom. **31** (1990), 387–395.

[123] P. Petersen, *Convergence theorems in Riemannian geometry*, Comparison Geometry, Math. Sci. Res. Inst. Publ., vol. 30, Cambridge Univ. Press, Cambridge (1997), 167–202.

[124] A. Petrunin, *Applications of quasigeodesics and gradient curves*, Comparison Geometry (Berkeley, CA, 1993–1994), Math. Sci. Res. Inst. Publ., vol. 30, Cambridge Univ. Press, Cambridge (1997), 203–219.

[125] A. Petrunin, X. Rong and W. Tuschmann, *Collapsing vs. positive pinching*, Geom. Funct. Anal. **9** (1999), 699–735.

[126] A. Petrunin and W. Tuschmann, *Diffeomorphism finiteness, positive pinching, and second homotopy*, Geom. Funct. Anal. **9** (1999), 736–774.

[127] C. Plaut, *Metric spaces of curvature* $\geqslant k$, Handbook of Geometric Topology, North-Holland, Amsterdam (2002), 819–898.

[128] W.A. Poor, *Some results on nonnegatively curved manifolds*, J. Differential Geom. **9** (1974), 583–600.

[129] H.E. Rauch, *A contribution to differential geometry in the large*, Ann. of Math. (2) **54** (1951), 38–55.

[130] X. Rong, *The existence of polarized* F*-structures on volume collapsed* 4*-manifolds*, Geom. Funct. Anal. **3** (1993), 474–501.

[131] X. Rong, *The limiting eta invariants of collapsed three-manifolds*, J. Differential Geom. **37** (1993), 535–568.

[132] X. Rong, *On the fundamental groups of manifolds of positive sectional curvature*, Ann. of Math. (2) **143** (1996), 397–411.

[133] E. Ruh, *Almost flat manifolds*, J. Differential Geom. **17** (1982), 1–14.

[134] V. Schroeder, *Rigidity of nonpositively curved graphmanifolds*, Math. Ann. **274** (1986), 19–26.

[135] J. Sha and D. Yang, *Examples of manifolds of positive Ricci curvature*, J. Differential Geom. **29** (1989), 95–103.

[136] Y. Shikata, *On the differentiable pinching problem*, Osaka J. Math. **4** (1967), 279–287.

[137] K. Shiohama, *A sphere theorem for manifolds of positive Ricci curvature*, Trans. Amer. Math. Soc. **275** (1983), 811–819.

[138] K. Shiohama, *An Introduction to the Geometry of Alexandrov Spaces*, Lecture Notes Series, vol. 8, Seoul National University, Seoul, Korea (1993).

[139] K. Shiohama, *Sphere theorems*, Handbook of Differential Geometry, vol. I, North-Holland, Amsterdam (2000), 865–903.

[140] K. Shiohama and T. Yamaguchi, *Positively curved manifolds with restricted diameters*, Geometry of Manifolds (Matsumoto, 1988), Perspect. Math., vol. 8, Academic Press, Boston, MA (1989), 345–350.

[141] T. Shioya and T. Yamaguchi, *Collapsing three-manifolds under a lower curvature bound*, J. Differential Geom. **56** (2000), 1–66.

[142] L.C. Siebenmann, *Deformation of homeomorphisms on stratified sets. I, II*, Comment. Math. Helv. **47** (1972), 123–136; ibid. **47** (1972), 137–163.

[143] M. Sugimoto and K. Shiohama, *On the differentiable pinching problem*, Math. Ann. **195** (1971), 1–16.

[144] Y. Suyama, *Differentiable sphere theorem by curvature pinching*, J. Math. Soc. Japan **43** (3) (1991), 527–553.

[145] C. Taubes, *Self-dual Yang–Mills connections on non-self-dual 4-manifolds*, J.Differential Geom. **17** (1982), 139–170.

[146] G. Tian, *Gauge theory and calibrated geometry. I*, Ann. of Math. (2) **151** (2000), 193–268.

[147] J. Tits, *Free subgroups in linear groups*, J. Algebra **20** (1972), 250–270.

[148] V.A. Toponogov, *Riemann spaces with curvature bounded below*, Uspekhi Mat. Nauk **14** (1) (1959), 87–130.

[149] V.A. Toponogov, *Riemannian spaces containing straight lines*, Dokl. Akad. Nauk SSSR **127** (1959), 977–979.

[150] A. Weinstein, *On the homotopy type of positively-pinched manifolds*, Arch. Math. (Basel) **18** (1967), 523–524.

[151] T. Yamaguchi, *Manifolds of almost nonnegative Ricci curvarture*, J. Differential Geom. **28** (1988), 157–167.

[152] T. Yamaguchi, *Lipschitz convergence of manifolds of positive Ricci curvature with large volume*, Math. Ann. **284** (1989), 423–436.

[153] T. Yamaguchi, *Collapsing and pinching under a lower curvature bound*, Ann. of Math. (2) **133** (1991), 317–357.

[154] T. Yamaguchi, *A convergence theorem in the geometry of Alexandrov spaces*, Actes de la Table Ronde de Géométrie Différentielle (Luminy, 1992), Sémin. Congr., vol. 1, Soc. Math. France, Paris (1996), 601–642.

[155] T. Yamaguchi, *Collapsing Riemannian 4-manifolds*, Suugaku **52** (2000), 172–186 (in Japanese).

[156] T. Yamaguchi, *Collapsing 4-manifolds under a Lower curvature bound*, Preprint (2002).

[157] K. Yano and S. Bochner, *Curvature and Betti Numbers*, Annals of Mathematics Studies, vol. 32, Princeton Univ. Press, Princeton, NJ (1953).

[158] S.T. Yau, *On the Ricci curvature of a compact Kähler manifold and the complex Monge–Ampère equation. I*, Comm. Pure Appl. Math. **31** (1978), 339–411.

[159] S. Zhu, *The comparison geometry of Ricci curvature*, Comparison Geometry, Math. Sci. Res. Inst. Publ., vol. 30, Cambridge Univ. Press, Cambridge (1997), 221–262.

CHAPTER 5

Contact Geometry

Hansjörg Geiges*

Mathematisches Institut, Universität zu Köln, Weyertal 86-90, 50931 Köln, Germany

E-mail: *geiges@math.uni-koeln.de*

Contents

*I am grateful to Fan Ding, Jesús Gonzalo and Federica Pasquotto for their attentive reading of the original manuscript. I also thank John Etnyre and Stephan Schönenberger for allowing me to use a couple of their figures (viz., Figures 2 and 1 of the present text, respectively).

HANDBOOK OF DIFFERENTIAL GEOMETRY, VOL. II
Edited by F.J.E. Dillen and L.C.A. Verstraelen

1. Introduction

Over the past two decades, contact geometry has undergone a veritable metamorphosis: once the ugly duckling known as 'the odd-dimensional analogue of symplectic geometry', it has now evolved into a proud field of study in its own right. As is typical for a period of rapid development in an area of mathematics, there are a fair number of folklore results that every mathematician working in the area knows, but no references that make these results accessible to the novice. I therefore take the present article as an opportunity to take stock of some of that folklore.

There are many excellent surveys covering specific aspects of contact geometry (e.g., classification questions in dimension 3, dynamics of the Reeb vector field, various notions of symplectic fillability, transverse and Legendrian knots and links). All these topics deserve to be included in a comprehensive survey, but an attempt to do so here would have left this article in the 'to appear' limbo for much too long.

Thus, instead of adding yet another survey, my plan here is to cover in detail some of the more fundamental differential topological aspects of contact geometry. In doing so, I have not tried to hide my own idiosyncrasies and preoccupations. Owing to a relatively leisurely pace and constraints of the present format, I have not been able to cover quite as much material as I should have wished. Nonetheless, I hope that the reader of the present handbook chapter will be better prepared to study some of the surveys I alluded to—a guide to these surveys will be provided—and from there to move on to the original literature.

A book chapter with comparable aims is Chapter 8 in [1]. It seemed opportune to be brief on topics that are covered extensively there, even if it is done at the cost of leaving out some essential issues. I hope to return to the material of the present chapter in a yet to be written more comprehensive monograph.

2. Contact manifolds

Let M be a differential manifold and $\xi \subset TM$ a field of hyperplanes on M. Locally such a hyperplane field can always be written as the kernel of a non-vanishing 1-form α. One way to see this is to choose an auxiliary Riemannian metric g on M and then to define $\alpha = g(X, .)$, where X is a local non-zero section of the line bundle $\xi^{\perp}$ (the orthogonal complement of ξ in TM). We see that the existence of a globally defined 1-form α with $\xi = \ker \alpha$ is equivalent to the orientability (hence triviality) of $\xi^{\perp}$, i.e. the coorientability of ξ. Except for an example below, I shall always assume this condition.

If α satisfies the Frobenius integrability condition

$$\alpha \wedge d\alpha = 0,$$

then ξ is an integrable hyperplane field (and vice versa), and its integral submanifolds form a codimension 1 foliation of M. Equivalently, this integrability condition can be written as

$$X, Y \in \xi \quad \Longrightarrow \quad [X, Y] \in \xi.$$

An integrable hyperplane field is locally of the form $dz = 0$, where z is a coordinate function on M. Much is known, too, about the global topology of foliations, cf. [100].

Contact structures are in a certain sense the exact opposite of integrable hyperplane fields.

DEFINITION 2.1. Let M be a manifold of odd dimension $2n+1$. A *contact structure* is a maximally non-integrable hyperplane field $\xi = \ker\alpha \subset TM$, that is, the defining 1-form α is required to satisfy

$$\alpha \wedge (d\alpha)^n \neq 0$$

(meaning that it vanishes nowhere). Such a 1-form α is called a *contact form*. The pair (M, ξ) is called a *contact manifold*.

REMARK 2.2. Observe that in this case $\alpha \wedge (d\alpha)^n$ is a volume form on M; in particular, M needs to be orientable. The condition $\alpha \wedge (d\alpha)^n \neq 0$ is independent of the specific choice of α and thus is indeed a property of $\xi = \ker\alpha$: Any other 1-form defining the same hyperplane field must be of the form $\lambda\alpha$ for some smooth function $\lambda : M \to \mathbb{R} \setminus \{0\}$, and we have

$$(\lambda\alpha) \wedge \big(d(\lambda\alpha)\big)^n = \lambda\alpha \wedge (\lambda\, d\alpha + d\lambda \wedge \alpha)^n = \lambda^{n+1}\alpha \wedge (d\alpha)^n \neq 0.$$

We see that if n is odd, the sign of this volume form depends only on ξ, not the choice of α. This makes it possible, given an orientation of M, to speak of *positive* and *negative* contact structures.

REMARK 2.3. An equivalent formulation of the contact condition is that we have $(d\alpha)^n|_\xi \neq 0$. In particular, for every point $p \in M$, the $2n$-dimensional subspace $\xi_p \subset T_pM$ is a vector space on which $d\alpha$ defines a skew-symmetric form of maximal rank, that is, $(\xi_p, d\alpha|_{\xi_p})$ is a *symplectic* vector space. A consequence of this fact is that there exists a complex bundle structure $J : \xi \to \xi$ compatible with $d\alpha$ (see [92, Proposition 2.63]), i.e. a bundle endomorphism satisfying

- $J^2 = -\mathrm{id}_\xi$,
- $d\alpha(JX, JY) = d\alpha(X, Y)$ for all $X, Y \in \xi$,
- $d\alpha(X, JX) > 0$ for $0 \neq X \in \xi$.

REMARK 2.4. The name 'contact structure' has its origins in the fact that one of the first historical sources of contact manifolds are the so-called spaces of contact elements (which in fact have to do with 'contact' in the differential geometric sense), see [7] and [45].

In the 3-dimensional case the contact condition can also be formulated as

$$X, Y \in \xi \text{ linearly independent} \quad \Longrightarrow \quad [X, Y] \notin \xi;$$

this follows immediately from the equation

$$d\alpha(X,Y)=X\big(\alpha(Y)\big)-Y\big(\alpha(X)\big)-\alpha\big([X,Y]\big)$$

and the fact that the contact condition (in dimension 3) may be written as $d\alpha|_\xi\neq 0$.

In the present article I shall take it for granted that contact structures are worthwhile objects of study. As I hope to illustrate, this is fully justified by the beautiful mathematics to which they have given rise. For an apology of contact structures in terms of their origin (with hindsight) in physics and the multifarious connections with other areas of mathematics I refer the reader to the historical surveys [87] and [45]. Contact structures may also be justified on the grounds that they are generic objects: A generic 1-form α on an odd-dimensional manifold satisfies the contact condition outside a smooth hypersurface, see [89]. Similarly, a generic 1-form α on a $2n$-dimensional manifold satisfies the condition $\alpha\wedge(d\alpha)^{n-1}\neq 0$ outside a submanifold of codimension 3; such 'even-contact manifolds' have been studied in [51], for instance, but on the whole their theory is not as rich or well motivated as that of contact structures.

DEFINITION 2.5. Associated with a contact form α one has the so-called *Reeb vector field* R_α, defined by the equations
(i) $d\alpha(R_\alpha,.)\equiv 0$,
(ii) $\alpha(R_\alpha)\equiv 1$.

As a skew-symmetric form of maximal rank $2n$, the form $d\alpha|_{T_pM}$ has a 1-dimensional kernel for each $p\in M^{2n+1}$. Hence equation (i) defines a unique line field $\langle R_\alpha\rangle$ on M. The contact condition $\alpha\wedge(d\alpha)^n\neq 0$ implies that α is non-trivial on that line field, so a global vector field is defined by the additional normalisation condition (ii).

2.1. *Contact manifolds and their submanifolds*

We begin with some examples of contact manifolds; the simple verification that the listed 1-forms are contact forms is left to the reader.

EXAMPLE 2.6. On $\mathbb{R}^{2n+1}$ with Cartesian coordinates $(x_1,y_1,\dots,x_n,y_n,z)$, the 1-form

$$\alpha_1=dz+\sum_{j=1}^n x_j\,dy_j$$

is a contact form.

EXAMPLE 2.7. On $\mathbb{R}^{2n+1}$ with polar coordinates (r_j,φ_j) for the (x_j,y_j)-plane, $j=1,\dots,n$, the 1-form

$$\alpha_2=dz+\sum_{j=1}^n r_j^2\,d\varphi_j=dz+\sum_{j=1}^n(x_j\,dy_j-y_j\,dx_j)$$

is a contact form.

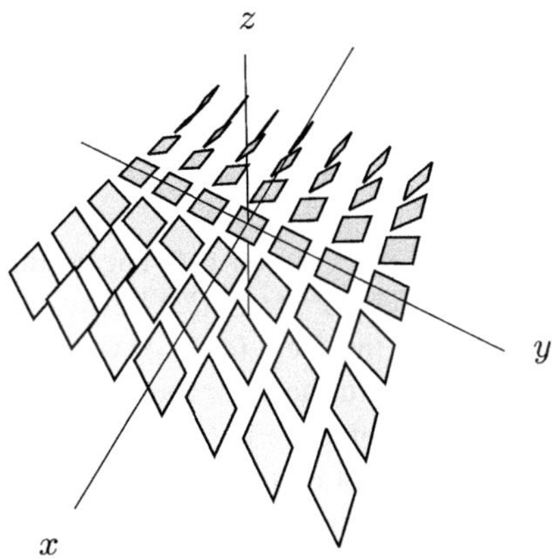

Fig. 1. The contact structure $\ker(dz + x\,dy)$.

DEFINITION 2.8. Two contact manifolds (M_1, ξ_1) and (M_2, ξ_2) are called *contactomorphic* if there is a diffeomorphism $f : M_1 \to M_2$ with $Tf(\xi_1) = \xi_2$, where $Tf : TM_1 \to TM_2$ denotes the differential of f. If $\xi_i = \ker \alpha_i$, $i = 1, 2$, this is equivalent to the existence of a nowhere zero function $\lambda : M_1 \to \mathbb{R}$ such that $f^*\alpha_2 = \lambda\alpha_1$.

EXAMPLE 2.9. The contact manifolds $(\mathbb{R}^{2n+1}, \xi_i = \ker \alpha_i)$, $i = 1, 2$, from the preceding examples are contactomorphic. An explicit contactomorphism f with $f^*\alpha_2 = \alpha_1$ is given by

$$f(x, y, z) = \big((x + y)/2, (y - x)/2, z + xy/2\big),$$

where x and y stand for $(x_1, \dots, x_n)$ and $(y_1, \dots, y_n)$, respectively, and xy stands for $\sum_j x_j y_j$. Similarly, both these contact structures are contactomorphic to $\ker(dz - \sum_j y_j\,dx_j)$. Any of these contact structures is called the *standard contact structure on* $\mathbb{R}^{2n+1}$.

EXAMPLE 2.10. The *standard contact structure on the unit sphere* S^{2n+1} in $\mathbb{R}^{2n+2}$ (with Cartesian coordinates $(x_1, y_1, \dots, x_{n+1}, y_{n+1})$) is defined by the contact form

$$\alpha_0 = \sum_{j=1}^{n+1} (x_j\,dy_j - y_j\,dx_j).$$

With r denoting the radial coordinate on $\mathbb{R}^{2n+2}$ (that is, $r^2 = \sum_j (x_j^2 + y_j^2)$) one checks easily that $\alpha_0 \wedge (d\alpha_0)^n \wedge r\,dr \neq 0$ for $r \neq 0$. Since S^{2n+1} is a level surface of r (or r^2), this verifies the contact condition.

Alternatively, one may regard S^{2n+1} as the unit sphere in $\mathbb{C}^{n+1}$ with complex structure J (corresponding to complex coordinates $z_j = x_j + iy_j$, $j = 1, \dots, n + 1$). Then $\xi_0 = \ker \alpha_0$

defines at each point $p \in S^{2n+1}$ the complex (i.e. J-invariant) subspace of $T_p S^{2n+1}$, that is,

$$\xi_0 = TS^{2n+1} \cap J\big(TS^{2n+1}\big).$$

This follows from the observation that $\alpha = -r\,dr \circ J$. The hermitian form $d\alpha(.,J.)$ on ξ_0 is called the *Levi form* of the hypersurface $S^{2n+1} \subset \mathbb{C}^{n+1}$. The contact condition for ξ corresponds to the positive definiteness of that Levi form, or what in complex analysis is called the *strict pseudoconvexity* of the hypersurface. For more on the question of pseudoconvexity from the contact geometric viewpoint see [1, Section 8.2]. Beware that the 'complex structure' in their Proposition 8.14 is not required to be integrable, i.e. constitutes what is more commonly referred to as an 'almost complex structure'.

DEFINITION 2.11. Let (V, ω) be a *symplectic manifold* of dimension $2n+2$, that is, ω is a closed ($d\omega = 0$) and non-degenerate ($\omega^{n+1} \neq 0$) 2-form on V. A vector field X is called a *Liouville vector field* if $\mathcal{L}_X \omega = \omega$, where $\mathcal{L}$ denotes the Lie derivative.

With the help of Cartan's formula $\mathcal{L}_X = d \circ i_X + i_X \circ d$ this may be rewritten as $d(i_X\omega) = \omega$. Then the 1-form $\alpha = i_X\omega$ defines a contact form on any hypersurface M in V transverse to X. Indeed,

$$\alpha \wedge (d\alpha)^n = i_X\omega \wedge \big(d(i_X\omega)\big)^n = i_X\omega \wedge \omega^n = \frac{1}{n+1} i_X\big(\omega^{n+1}\big),$$

which is a volume form on $M \subset V$ provided M is transverse to X.

EXAMPLE 2.12. With $V = \mathbb{R}^{2n+2}$, symplectic form $\omega = \sum_j dx_j \wedge dy_j$, and Liouville vector field $X = \sum_j (x_j\,\partial_{x_j} + y_j\,\partial_{y_j})/2 = r\,\partial_r/2$, we recover the standard contact structure on S^{2n+1}.

For finer issues relating to hypersurfaces in symplectic manifolds transverse to a Liouville vector field I refer the reader to [1, Section 8.2].

Here is a further useful example of contactomorphic manifolds.

PROPOSITION 2.13. *For any point $p \in S^{2n+1}$, the manifold $(S^{2n+1} \setminus \{p\}, \xi_0)$ is contactomorphic to $(\mathbb{R}^{2n+1}, \xi_2)$.*

PROOF. The contact manifold (S^{2n+1}, ξ_0) is a homogeneous space under the natural $\mathrm{U}(n+1)$-action, so we are free to choose $p = (0, \dots, 0, -1)$. Stereographic projection from p does almost, but not quite yield the desired contactomorphism. Instead, we use a map that is well known in the theory of Siegel domains (cf. [3, Chapter 8]) and that looks a bit like a complex analogue of stereographic projection; this was suggested in [92, Exercise 3.64].

Regard S^{2n+1} as the unit sphere in $\mathbb{C}^{n+1} = \mathbb{C}^n \times \mathbb{C}$ with Cartesian coordinates $(z_1, \dots, z_n, w) = (z, w)$. We identify $\mathbb{R}^{2n+1}$ with $\mathbb{C}^n \times \mathbb{R} \subset \mathbb{C}^n \times \mathbb{C}$ with coordinates

$(\zeta_1,\dots,\zeta_n,s)=(\zeta,s)=(\zeta,\operatorname{Re}\sigma)$, where $\zeta_j=x_j+iy_j$. Then

$$\alpha_2=ds+\sum_{j=1}^{n}(x_j\,dy_j-y_j\,dx_j)=ds+\frac{i}{2}(\zeta\,d\bar\zeta-\bar\zeta\,d\zeta)$$

and

$$\alpha_0=\frac{i}{2}(z\,d\bar z-\bar z\,dz+w\,d\bar w-\bar w\,dw).$$

Now define a smooth map $f:S^{2n+1}\setminus\{(0,-1)\}\to\mathbb{R}^{2n+1}$ by

$$(\zeta,s)=f(z,w)=\left(\frac{z}{1+w},-\frac{i(w-\bar w)}{2|1+w|^2}\right).$$

Then

$$\begin{aligned}f^*\,ds&=-\frac{i\,dw}{2|1+w|^2}+\frac{i\,d\bar w}{2|1+w|^2}+\frac{i(w-\bar w)}{2(1+w)}\frac{dw}{|1+w|^2}+\frac{i(w-\bar w)}{2(1+\bar w)}\frac{d\bar w}{|1+w|^2}\\&=\frac{i}{2|1+w|^2}\left(-dw+d\bar w+\frac{w-\bar w}{1+w}\,dw+\frac{w-\bar w}{1+\bar w}\,d\bar w\right)\end{aligned}$$

and

$$\begin{aligned}f^*(\zeta\,d\bar\zeta-\bar\zeta\,d\zeta)&=\frac{z}{1+w}\left(\frac{d\bar z}{1+\bar w}-\frac{\bar z}{(1+\bar w)^2}\,d\bar w\right)\\&\quad-\frac{\bar z}{1+\bar w}\left(\frac{dz}{1+w}-\frac{z}{(1+w)^2}\,dw\right)\\&=\frac{1}{|1+w|^2}\left(z\,d\bar z-\bar z\,dz+|z|^2\left(\frac{dw}{1+w}-\frac{d\bar w}{1+\bar w}\right)\right).\end{aligned}$$

Along S^{2n+1} we have

$$|z|^2=1-|w|^2=(1-w)(1+\bar w)+(w-\bar w)=(1-\bar w)(1+w)-(w-\bar w),$$

whence

$$|z|^2\left(\frac{dw}{1+w}-\frac{d\bar w}{1+\bar w}\right)=(1-\bar w)\,dw-\frac{w-\bar w}{1+w}\,dw-(1-w)\,d\bar w-\frac{w-\bar w}{1+\bar w}\,d\bar w.$$

From these calculations we conclude $f^*\alpha_2=\alpha_0/|1+w|^2$. So it only remains to show that f is actually a diffeomorphism of $S^{2n+1}\setminus\{(0,-1)\}$ onto $\mathbb{R}^{2n+1}$. To that end, consider the map

$$\tilde f:(\mathbb{C}^n\times\mathbb{C})\setminus\big(\mathbb{C}^n\times\{-1\}\big)\to(\mathbb{C}^n\times\mathbb{C})\setminus\big(\mathbb{C}^n\times\{-i/2\}\big)$$

defined by

$$(\zeta, \sigma) = \tilde{f}(z, w) = \left(\frac{z}{1+w}, -\frac{i}{2}\frac{w-1}{w+1} \right).$$

This is a biholomorphic map with inverse map

$$(\zeta, \sigma) \mapsto \left(\frac{2\zeta}{1-2i\sigma}, \frac{1+2i\sigma}{1-2i\sigma} \right).$$

We compute

$$\begin{aligned}
\operatorname{Im}\sigma &= -\frac{w-1}{4(w+1)} - \frac{\bar{w}-1}{4(\bar{w}+1)} = -\frac{(w-1)(\bar{w}+1)+(\bar{w}-1)(w+1)}{4|1+w|^2} \\
&= \frac{1-|w|^2}{2|1+w|^2}.
\end{aligned}$$

Hence for $(z, w) \in S^{2n+1} \setminus \{(0, -1)\}$ we have

$$\operatorname{Im}\sigma = \frac{|z|^2}{2|1+w|^2} = \frac{1}{2}|\zeta|^2;$$

conversely, any point (ζ, σ) with $\operatorname{Im}\sigma = |\zeta|^2/2$ lies in the image of $\tilde{f}|_{S^{2n+1}\setminus\{(0,-1)\}}$, that is, $\tilde{f}$ restricted to $S^{2n+1} \setminus \{(0, -1)\}$ is a diffeomorphism onto $\{\operatorname{Im}\sigma = |\zeta|^2/2\}$. Finally, we compute

$$\begin{aligned}
\operatorname{Re}\sigma &= -\frac{i(w-1)}{4(w+1)} + \frac{i(\bar{w}-1)}{4(\bar{w}+1)} = -i\frac{(w-1)(\bar{w}+1)-(\bar{w}-1)(w+1)}{4|1+w|^2} \\
&= -\frac{i(w-\bar{w})}{2|1+w|^2},
\end{aligned}$$

from which we see that for $(z, w) \in S^{2n+1} \setminus \{(0, -1)\}$ and with $(\zeta, \sigma) = \tilde{f}(z, w)$ we have $f(z, w) = (\zeta, \operatorname{Re}\sigma)$. This concludes the proof. □

At the beginning of this section I mentioned that one may allow contact structures that are not coorientable, and hence not defined by a global contact form.

EXAMPLE 2.14. Let $M = \mathbb{R}^{n+1} \times \mathbb{R}P^n$ with Cartesian coordinates $(x_0, \ldots, x_n)$ on the $\mathbb{R}^{n+1}$-factor and homogeneous coordinates $[y_0 : \ldots : y_n]$ on the $\mathbb{R}P^n$-factor. Then

$$\xi = \ker\left(\sum_{j=0}^{n} y_j \, dx_j \right)$$

is a well-defined hyperplane field on M, because the 1-form on the right-hand side is well defined up to scaling by a non-zero real constant. On the open submanifold $U_k = \{y_k \neq 0\} \cong \mathbb{R}^{n+1} \times \mathbb{R}^n$ of M we have $\xi = \ker \alpha_k$ with

$$\alpha_k = dx_k + \sum_{j \neq k} \left(\frac{y_j}{y_k}\right) dx_j$$

an honest 1-form on U_k. This is the standard contact form of Example 2.6, which proves that ξ is a contact structure on M.

If n is even, then M is not orientable, so there can be no global contact form defining ξ (cf. Remark 2.2), i.e. ξ is *not coorientable*. Notice, however, that a contact structure on a manifold of dimension $2n+1$ with n even is always *orientable*: the sign of $(d\alpha)^n|_\xi$ does not depend on the choice of local 1-form defining ξ.

If n is odd, then M is orientable, so it would be possible that ξ is the kernel of a globally defined 1-form. However, since the sign of $\alpha \wedge (d\alpha)^n$, for n odd, is independent of the choice of local 1-form defining ξ, it is also conceivable that no global contact form exists. (In fact, this consideration shows that any manifold of dimension $2n+1$, with n odd, admitting a contact structure (coorientable or not) needs to be orientable.) This is indeed what happens, as we shall prove now.

PROPOSITION 2.15. *Let (M, ξ) be the contact manifold of the preceding example. Then TM/ξ can be identified with the canonical line bundle on $\mathbb{R}P^n$ (pulled back to M). In particular, TM/ξ is a non-trivial line bundle, so ξ is not coorientable.*

PROOF. For given $y = [y_0 : \ldots : y_n] \in \mathbb{R}P^n$, the vector $y_0\, \partial_{x_0} + \cdots + y_n\, \partial_{x_n} \in T_x\mathbb{R}^{n+1}$ is well defined up to a non-zero real factor (and independent of $x \in \mathbb{R}^{n+1}$), and hence defines a line ℓ_y in $T_x\mathbb{R}^{n+1} \cong \mathbb{R}^{n+1}$. The set

$$\begin{aligned} E &= \{(t, x, y)\colon\, x \in \mathbb{R}^{n+1},\ y \in \mathbb{R}P^n,\ t \in \ell_y\} \\ &\subset T\mathbb{R}^{n+1} \times \mathbb{R}P^n \subset T(\mathbb{R}^{n+1} \times \mathbb{R}P^n) = TM \end{aligned}$$

with projection $(t, x, y) \mapsto (x, y)$ defines a line sub-bundle of TM that restricts to the canonical line bundle over $\{x\} \times \mathbb{R}P^n \equiv \mathbb{R}P^n$ for each $x \in \mathbb{R}^{n+1}$. The canonical line bundle over $\mathbb{R}P^n$ is well known to be non-trivial [95, p. 16], so the same holds for E.

Moreover, E is clearly complementary to ξ, i.e. $TM/\xi \cong E$, since

$$\sum_{j=0}^{n} y_j\, dx_j \left(\sum_{k=0}^{n} y_k\, \partial_{x_k}\right) = \sum_{j=0}^{n} y_j^2 \neq 0.$$

This proves that ξ is not coorientable. □

To sum up, in the example above we have one of the following two situations:

- If n is odd, then M is orientable; ξ is neither orientable nor coorientable.
- If n is even, then M is not orientable; ξ is not coorientable, but it is orientable.

We close this section with the definition of the most important types of submanifolds.

DEFINITION 2.16. Let (M, ξ) be a contact manifold.

(i) A submanifold L of (M, ξ) is called an *isotropic* submanifold if $T_xL \subset \xi_x$ for all $x \in L$.

(ii) A submanifold M' of M with contact structure ξ' is called a *contact submanifold* if $TM' \cap \xi|_{M'} = \xi'$.

Observe that if $\xi = \ker\alpha$ and $i : M' \to M$ denotes the inclusion map, then the condition for (M', ξ') to be a contact submanifold of (M, ξ) is that $\xi' = \ker(i^*\alpha)$. In particular, $\xi' \subset \xi|_{M'}$ is a symplectic sub-bundle with respect to the symplectic bundle structure on ξ given by $d\alpha$.

The following is a manifestation of the maximal non-integrability of contact structures.

PROPOSITION 2.17. *Let (M, ξ) be a contact manifold of dimension $2n+1$ and L an isotropic submanifold. Then* $\dim L \leqslant n$.

PROOF. Write i for the inclusion of L in M and let α be an (at least locally defined) contact form defining ξ. Then the condition for L to be isotropic becomes $i^*\alpha \equiv 0$. It follows that $i^*\,d\alpha \equiv 0$. In particular, $T_pL \subset \xi_p$ is an isotropic subspace of the symplectic vector space $(\xi_p, d\alpha|_{\xi_p})$, i.e. a subspace on which the symplectic form restricts to zero. From Linear Algebra we know that this implies $\dim T_pL \leqslant (\dim \xi_p)/2 = n$. □

DEFINITION 2.18. An isotropic submanifold $L \subset (M^{2n+1}, \xi)$ of maximal possible dimension n is called a *Legendrian submanifold*.

In particular, in a 3-dimensional contact manifold there are two distinguished types of knots: *Legendrian knots* on the one hand, *transverse*[1] *knots* on the other, i.e. knots that are everywhere transverse to the contact structure. If ξ is cooriented by a contact form α and $\gamma : S^1 \to (M, \xi = \ker\alpha)$ is oriented, one can speak of a *positively* or *negatively* transverse knot, depending on whether $\alpha(\dot\gamma) > 0$ or $\alpha(\dot\gamma) < 0$.

2.2. *Gray stability and the Moser trick*

The Gray stability theorem that we are going to prove in this section says that there are no non-trivial deformations of contact structures on closed manifolds. In fancy language, this means that contact structures on closed manifolds have discrete moduli. First a preparatory lemma.

LEMMA 2.19. *Let ω_t, $t \in [0, 1]$, be a smooth family of differential k-forms on a manifold M and $(\psi_t)_{t\in[0,1]}$ an isotopy of M. Define a time-dependent vector field X_t on M by*

[1] Some people like to call them 'transversal knots', but I adhere to J.H.C. Whitehead's dictum, as quoted in [64]: "*Transversal* is a noun; the adjective is *transverse*."

$X_t \circ \psi_t = \dot{\psi}_t$, where the dot denotes derivative with respect to t (so that ψ_t is the flow of X_t). Then

$$\frac{d}{dt}(\psi_t^*\omega_t) = \psi_t^*(\dot{\omega}_t + \mathcal{L}_{X_t}\omega_t).$$

PROOF. For a time-independent k-form ω we have

$$\frac{d}{dt}(\psi_t^*\omega) = \psi_t^*(\mathcal{L}_{X_t}\omega).$$

This follows by observing that

(i) the formula holds for functions,
(ii) if it holds for differential forms ω and ω', then also for $\omega \wedge \omega'$,
(iii) if it holds for ω, then also for $d\omega$,
(iv) locally functions and differentials of functions generate the algebra of differential forms.

We then compute

$$\begin{aligned}
\frac{d}{dt}(\psi_t^*\omega_t) &= \lim_{h\to 0} \frac{\psi_{t+h}^*\omega_{t+h} - \psi_t^*\omega_t}{h} \\
&= \lim_{h\to 0} \frac{\psi_{t+h}^*\omega_{t+h} - \psi_{t+h}^*\omega_t + \psi_{t+h}^*\omega_t - \psi_t^*\omega_t}{h} \\
&= \lim_{h\to 0} \psi_{t+h}^*\left(\frac{\omega_{t+h} - \omega_t}{h}\right) + \lim_{h\to 0} \frac{\psi_{t+h}^*\omega_t - \psi_t^*\omega_t}{h} \\
&= \psi_t^*(\dot{\omega}_t + \mathcal{L}_{X_t}\omega_t).
\end{aligned}$$

For that last equality observe (regarding the second summand) that $\psi_{t+h} = \psi_h^t \circ \psi_t$, where ψ_h^t denotes, for fixed t and time-variable h, the flow of the time-dependent vector field $X_h^t := X_{t+h}$; then apply the result for time-independent k-forms. □

THEOREM 2.20 (Gray stability). *Let ξ_t, $t \in [0, 1]$, be a smooth family of contact structures on a closed manifold M. Then there is an isotopy $(\psi_t)_{t\in[0,1]}$ of M such that*

$$T\psi_t(\xi_0) = \xi_t \quad \textit{for each } t \in [0, 1].$$

PROOF. The simplest proof of this result rests on what is known as the *Moser trick*, introduced by J. Moser [96] in the context of stability results for (equicohomologous) volume and symplectic forms. J. Gray's original proof [61] was based on deformation theory à la Kodaira–Spencer. The idea of the Moser trick is to assume that ψ_t is the flow of a time-dependent vector field X_t. The desired equation for ψ_t then translates into an equation for X_t. If that equation can be solved, the isotopy ψ_t is found by integrating X_t; on a closed manifold the flow of X_t will be globally defined.

Let α_t be a smooth family of 1-forms with $\ker \alpha_t = \xi_t$. The equation in the theorem then translates into

$$\psi_t^* \alpha_t = \lambda_t \alpha_0,$$

where $\lambda_t : M \to \mathbb{R}^+$ is a suitable smooth family of smooth functions. Differentiation of this equation with respect to t yields, with the help of the preceding lemma,

$$\psi_t^* (\dot{\alpha}_t + \mathcal{L}_{X_t} \alpha_t) = \dot{\lambda}_t \alpha_0 = \frac{\dot{\lambda}_t}{\lambda_t} \psi_t^* \alpha_t,$$

or, with the help of Cartan's formula $\mathcal{L}_X = d \circ i_X + i_X \circ d$ and with $\mu_t = \frac{d}{dt}(\log \lambda_t) \circ \psi_t^{-1}$,

$$\psi_t^* \big(\dot{\alpha}_t + d\big(\alpha_t(X_t)\big) + i_{X_t} d\alpha_t\big) = \psi_t^* (\mu_t \alpha_t).$$

If we choose $X_t \in \xi_t$, this equation will be satisfied if

$$\dot{\alpha}_t + i_{X_t} d\alpha_t = \mu_t \alpha_t. \tag{2.1}$$

Plugging in the Reeb vector field R_{α_t} gives

$$\dot{\alpha}_t (R_{\alpha_t}) = \mu_t. \tag{2.2}$$

So we can use (2.2) to define μ_t, and then the non-degeneracy of $d\alpha_t|_{\xi_t}$ and the fact that $R_{\alpha_t} \in \ker(\mu_t \alpha_t - \dot{\alpha}_t)$ allow us to find a unique solution $X_t \in \xi_t$ of (2.1). □

REMARK 2.21. (1) Contact *forms* do *not* satisfy stability, that is, in general one cannot find an isotopy ψ_t such that $\psi_t^* \alpha_t = \alpha_0$. For instance, consider the following family of contact forms on $S^3 \subset \mathbb{R}^4$:

$$\alpha_t = (x_1 \, dy_1 - y_1 \, dx_1) + (1+t)(x_2 \, dy_2 - y_2 \, dx_2),$$

where $t \geqslant 0$ is a real parameter. The Reeb vector field of α_t is

$$R_{\alpha_t} = (x_1 \, \partial_{y_1} - y_1 \, \partial_{x_1}) + \frac{1}{1+t} (x_2 \, \partial_{y_2} - y_2 \, \partial_{x_2}).$$

The flow of R_{α_0} defines the Hopf fibration, in particular all orbits of R_{α_0} are closed. For $t \in \mathbb{R}^+ \setminus \mathbb{Q}$, on the other hand, R_{α_t} has only two periodic orbits. So there can be no isotopy with $\psi_t^* \alpha_t = \alpha_0$, because such a ψ_t would also map R_{α_0} to R_{α_t}.

(2) Y. Eliashberg [25] has shown that on the open manifold $\mathbb{R}^3$ there are likewise no non-trivial deformations of contact structures, but on $S^1 \times \mathbb{R}^2$ there does exist a continuum of non-equivalent contact structures.

(3) For further applications of this theorem it is useful to observe that at points $p \in M$ with $\dot{\alpha}_{t,p}$ identically zero in t we have $X_t(p) \equiv 0$, so such points remain stationary under the isotopy ψ_t.

2.3. *Contact Hamiltonians*

A vector field X on the contact manifold $(M, \xi = \ker\alpha)$ is called an *infinitesimal automorphism* of the contact structure if the local flow of X preserves ξ (the study of such automorphisms was initiated by P. Libermann, cf. [80]). By slight abuse of notation, we denote this flow by ψ_t; if M is not closed, ψ_t (for a fixed $t \neq 0$) will not in general be defined on all of M. The condition for X to be an infinitesimal automorphism can be written as $T\psi_t(\xi) = \xi$, which is equivalent to $\mathcal{L}_X\alpha = \lambda\alpha$ for some function $\lambda : M \to \mathbb{R}$ (notice that this condition is independent of the choice of 1-form α defining ξ). The local flow of X preserves α if and only if $\mathcal{L}_X\alpha = 0$.

THEOREM 2.22. *With a fixed choice of contact form α there is a one-to-one correspondence between infinitesimal automorphisms X of $\xi = \ker\alpha$ and smooth functions $H : M \to \mathbb{R}$. The correspondence is given by*

- $X \mapsto H_X = \alpha(X)$;
- $H \mapsto X_H$, *defined uniquely by* $\alpha(X_H) = H$ *and* $i_{X_H}\, d\alpha = dH(R_\alpha)\alpha - dH$.

The fact that X_H is uniquely defined by the equations in the theorem follows as in the preceding section from the fact that $d\alpha$ is non-degenerate on ξ and $R_\alpha \in \ker(dH(R_\alpha)\alpha - dH)$.

PROOF. Let X be an infinitesimal automorphism of ξ. Set $H_X = \alpha(X)$ and write $dH_X + i_X\, d\alpha = \mathcal{L}_X\alpha = \lambda\alpha$ with $\lambda : M \to \mathbb{R}$. Applying this last equation to R_α yields $dH_X(R_\alpha) = \lambda$. So X satisfies the equations $\alpha(X) = H_X$ and $i_X\, d\alpha = dH_X(R_\alpha)\alpha - dH_X$. This means that $X_{H_X} = X$.

Conversely, given $H : M \to \mathbb{R}$ and with X_H as defined in the theorem, we have

$$\mathcal{L}_{X_H}\alpha = i_{X_H}\, d\alpha + d\big(\alpha(X_H)\big) = dH(R_\alpha)\alpha,$$

so X_H is an infinitesimal automorphism of ξ. Moreover, it is immediate from the definitions that $H_{X_H} = \alpha(X_H) = H$. □

COROLLARY 2.23. *Let $(M, \xi = \ker\alpha)$ be a closed contact manifold and $H_t : M \to \mathbb{R}$, $t \in [0, 1]$, a smooth family of functions. Let $X_t = X_{H_t}$ be the corresponding family of infinitesimal automorphisms of ξ (defined via the correspondence described in the preceding theorem). Then the globally defined flow ψ_t of the time-dependent vector field X_t is a contact isotopy of (M, ξ), that is, $\psi_t^*\alpha = \lambda_t\alpha$ for some smooth family of functions $\lambda_t : M \to \mathbb{R}^+$.*

PROOF. With Lemma 2.19 and the preceding proof we have

$$\frac{d}{dt}(\psi_t^*\alpha) = \psi_t^*(\mathcal{L}_{X_t}\alpha) = \psi_t^*\big(dH_t(R_\alpha)\alpha\big) = \mu_t\psi_t^*\alpha$$

with $\mu_t = dH_t(R_\alpha) \circ \psi_t$. Since $\psi_0 = \mathrm{id}_M$ (whence $\psi_0^*\alpha = \alpha$) this implies that, with

$$\lambda_t = \exp\left(\int_0^t \mu_s \, ds\right),$$

we have $\psi_t^*\alpha = \lambda_t \alpha$. □

This corollary will be used in Section 2.5 to prove various isotopy extension theorems from isotopies of special submanifolds to isotopies of the ambient contact manifold. In a similar vein, contact Hamiltonians can be used to show that standard general position arguments from differential topology continue to hold in the contact geometric setting. Another application of contact Hamiltonians is a proof of the fact that the contactomorphism group of a connected contact manifold acts transitively on that manifold [12]. (See [8] for more on the general structure of contactomorphism groups.)

2.4. *Darboux's theorem and neighbourhood theorems*

The flexibility of contact structures inherent in the Gray stability theorem and the possibility to construct contact isotopies via contact Hamiltonians results in a variety of theorems that can be summed up as saying that there are no local invariants in contact geometry. Such theorems form the theme of the present section.

In contrast with Riemannian geometry, for instance, where the local structure coming from the curvature gives rise to a rich theory, the interesting questions in contact geometry thus appear only at the global level. However, it is actually that local flexibility that allows us to prove strong global theorems, such as the existence of contact structures on certain closed manifolds.

2.4.1. *Darboux's theorem*

THEOREM 2.24 (Darboux's theorem). *Let α be a contact form on the $(2n+1)$-dimensional manifold M and p a point on M. Then there are coordinates $x_1, \dots, x_n, y_1, \dots, y_n, z$ on a neighbourhood $U \subset M$ of p such that*

$$\alpha|_U = dz + \sum_{j=1}^n x_j \, dy_j.$$

PROOF. We may assume without loss of generality that $M = \mathbb{R}^{2n+1}$ and $p = 0$ is the origin of $\mathbb{R}^{2n+1}$. Choose linear coordinates $x_1, \dots, x_n, y_1, \dots, y_n, z$ on $\mathbb{R}^{2n+1}$ such that

$$\text{on } T_0\mathbb{R}^{2n+1}: \quad \begin{cases} \alpha(\partial_z) = 1, \quad i_{\partial_z} d\alpha = 0, \\ \partial_{x_j}, \partial_{y_j} \in \ker\alpha \quad (j = 1, \dots, n), \qquad d\alpha = \sum_{j=1}^n dx_j \wedge dy_j. \end{cases}$$

This is simply a matter of linear algebra (the normal form theorem for skew-symmetric forms on a vector space).

Now set $\alpha_0 = dz + \sum_j x_j\,dy_j$ and consider the family of 1-forms

$$\alpha_t = (1-t)\alpha_0 + t\alpha, \quad t \in [0,1],$$

on $\mathbb{R}^{2n+1}$. Our choice of coordinates ensures that

$$\alpha_t = \alpha, \quad d\alpha_t = d\alpha \quad \text{at the origin.}$$

Hence, on a sufficiently small neighbourhood of the origin, α_t is a contact form for all $t \in [0,1]$.

We now want to use the Moser trick to find an isotopy ψ_t of a neighbourhood of the origin such that $\psi_t^*\alpha_t = \alpha_0$. This aim seems to be in conflict with our earlier remark that contact forms are not stable, but as we shall see presently, locally this equation can always be solved.

Indeed, differentiating $\psi_t^*\alpha_t = \alpha_0$ (and assuming that ψ_t is the flow of some time-dependent vector field X_t) we find

$$\psi_t^*(\dot\alpha_t + \mathcal{L}_{X_t}\alpha_t) = 0,$$

so X_t needs to satisfy

$$\dot\alpha_t + d\big(\alpha_t(X_t)\big) + i_{X_t}\,d\alpha_t = 0. \tag{2.3}$$

Write $X_t = H_t R_{\alpha_t} + Y_t$ with $Y_t \in \ker\alpha_t$. Inserting R_{α_t} in (2.3) gives

$$\dot\alpha_t(R_{\alpha_t}) + dH_t(R_{\alpha_t}) = 0. \tag{2.4}$$

On a neighbourhood of the origin, a smooth family of functions H_t satisfying (2.4) can always be found by integration, provided only that this neighbourhood has been chosen so small that none of the R_{α_t} has any closed orbits there. Since $\dot\alpha_t$ is zero at the origin, we may require that $H_t(0) = 0$ and $dH_t|_0 = 0$ for all $t \in [0,1]$. Once H_t has been chosen, Y_t is defined uniquely by (2.3), i.e. by

$$\dot\alpha_t + dH_t + i_{Y_t}\,d\alpha_t = 0.$$

Notice that with our assumptions on H_t we have $X_t(0) = 0$ for all t.

Now define ψ_t to be the local flow of X_t. This local flow fixes the origin, so there it is defined for all $t \in [0,1]$. Since the domain of definition in $\mathbb{R} \times M$ of a local flow on a manifold M is always open (cf. [15, 8.11]), we can infer[2] that ψ_t is actually defined for all $t \in [0,1]$ on a sufficiently small neighbourhood of the origin in $\mathbb{R}^{2n+1}$. This concludes the proof of the theorem (strictly speaking, the local coordinates in the statement of the theorem are the coordinates $x_j \circ \psi_1^{-1}$, etc.). □

[2] To be absolutely precise, one ought to work with a family α_t, $t \in \mathbb{R}$, where $\alpha_t \equiv \alpha_0$ for $t \leqslant \varepsilon$ and $\alpha_t \equiv \alpha_1$ for $t \geqslant 1-\varepsilon$, i.e. a *technical homotopy* in the sense of [15]. Then X_t will be defined for all $t \in \mathbb{R}$, and the reasoning of [15] can be applied.

REMARK 2.25. The proof of this result given in [1] is incomplete: It is not possible, as is suggested there, to prove the Darboux theorem for contact *forms* if one requires $X_t \in \ker \alpha_t$.

2.4.2. *Isotropic submanifolds* Let $L \subset (M, \xi = \ker \alpha)$ be an isotropic submanifold in a contact manifold with cooriented contact structure. Write $(TL)^\perp \subset \xi|_L$ for the sub-bundle of $\xi|_L$ that is symplectically orthogonal to TL with respect to the symplectic bundle structure $d\alpha|_\xi$. The conformal class of this symplectic bundle structure depends only on the contact structure ξ, not on the choice of contact form α defining ξ: If α is replaced by $\lambda\alpha$ for some smooth function $\lambda : M \to \mathbb{R}^+$, then $d(\lambda\alpha)|_\xi = \lambda\, d\alpha|_\xi$. So the bundle $(TL)^\perp$ is determined by ξ.

The fact that L is isotropic implies $TL \subset (TL)^\perp$. Following Weinstein [105], we call the quotient bundle $(TL)^\perp/TL$ with the conformal symplectic structure induced by $d\alpha$ the *conformal symplectic normal bundle* of L in M and write

$$\mathrm{CSN}(M, L) = (TL)^\perp/TL.$$

So the normal bundle $NL = (TM|_L)/TL$ of L in M can be split as

$$NL \cong (TM|_L)/(\xi|_L) \oplus (\xi|_L)/(TL)^\perp \oplus \mathrm{CSN}(M, L).$$

Observe that if $\dim M = 2n + 1$ and $\dim L = k \leqslant n$, then the ranks of the three summands in this splitting are 1, k and $2(n - k)$, respectively. Our aim in this section is to show that a neighbourhood of L in M is determined, up to contactomorphism, by the isomorphism type (as a conformal symplectic bundle) of $\mathrm{CSN}(M, L)$.

The bundle $(TM|_L)/(\xi|_L)$ is a trivial line bundle because ξ is cooriented. The bundle $(\xi|_L)/(TL)^\perp$ can be identified with the cotangent bundle T^*L via the well-defined bundle isomorphism

$$\begin{aligned} \Psi : (\xi|_L)/(TL)^\perp &\to T^*L, \\ Y &\mapsto i_Y\, d\alpha|_{TL}. \end{aligned}$$

(Ψ is obviously injective and well defined by the definition of $(TL)^\perp$, and the ranks of the two bundles are equal.)

Although Ψ is well defined on the quotient $(\xi|_L)/(TL)^\perp$, to proceed further we need to choose an isotropic complement of $(TL)^\perp$ in $\xi|_L$. Restricted to each fibre ξ_p, $p \in L$, such an isotropic complement of $(T_pL)^\perp$ exists. There are two ways to obtain a smooth bundle of such isotropic complements. The first would be to carry over Arnold's corresponding discussion of Lagrangian subbundles of symplectic bundles [6] to the isotropic case in order to show that the space of isotropic complements of $U^\perp \subset V$, where U is an isotropic subspace in a symplectic vector space V, is convex. (This argument uses generating functions for isotropic subspaces.) Then by a partition of unity argument the desired complement can be constructed on the bundle level.

A slightly more pedestrian approach is to define this isotropic complement with the help of a complex bundle structure J on ξ compatible with $d\alpha$ (cf. Remark 2.3). The condition

$d\alpha(X, JX) > 0$ for $0 \neq X \in \xi$ implies that $(T_pL)^\perp \cap J(T_pL) = \{0\}$ for all $p \in L$, and so a dimension count shows that $J(TL)$ is indeed a complement of $(TL)^\perp$ in $\xi|_L$. (In a similar vein, $\mathrm{CSN}(M, L)$ can be identified as a sub-bundle of ξ, viz., the orthogonal complement of $TL \oplus J(TL) \subset \xi$ with respect to the bundle metric $d\alpha(., J.)$ on ξ.)

On the Whitney sum $TL \oplus T^*L$ (for any manifold L) there is a canonical symplectic bundle structure Ω_L defined by

$$\Omega_{L,p}(X + \eta, X' + \eta') = \eta(X') - \eta'(X) \quad \text{for } X, X' \in T_pL,\ \eta, \eta' \in T_p^*L.$$

LEMMA 2.26. *The bundle map*

$$\mathrm{id}_{TL} \oplus \Psi : \big(TL \oplus J(TL), d\alpha\big) \to (TL \oplus T^*L, \Omega_L)$$

is an isomorphism of symplectic vector bundles.

PROOF. We only need to check that $\mathrm{id}_{TL} \oplus \Psi$ is a symplectic bundle map. Let $X, X' \in T_pL$ and $Y, Y' \in J_p(T_pL)$. Write $Y = J_pZ$, $Y' = J_pZ'$ with $Z, Z' \in T_pL$. It follows that

$$d\alpha(Y, Y') = d\alpha(JZ, JZ') = d\alpha(Z, Z') = 0,$$

since L is an isotropic submanifold. For the same reason $d\alpha(X, X') = 0$. Hence

$$\begin{aligned} d\alpha(X + Y, X' + Y') &= d\alpha(Y, X') - d\alpha(Y', X) = \Psi(Y)(X') - \Psi(Y')(X) \\ &= \Omega_L\big(X + \Psi(Y), X' + \Psi(Y')\big). \end{aligned}$$ □

THEOREM 2.27. *Let (M_i, ξ_i), $i = 0, 1$, be contact manifolds with closed isotropic submanifolds L_i. Suppose there is an isomorphism of conformal symplectic normal bundles $\Phi : \mathrm{CSN}(M_0, L_0) \to \mathrm{CSN}(M_1, L_1)$ that covers a diffeomorphism $\phi : L_0 \to L_1$. Then ϕ extends to a contactomorphism $\psi : \mathcal{N}(L_0) \to \mathcal{N}(L_1)$ of suitable neighbourhoods $\mathcal{N}(L_i)$ of L_i such that $T\psi|_{\mathrm{CSN}(M_0, L_0)}$ and Φ are bundle homotopic (as conformal symplectic bundle isomorphisms).*

COROLLARY 2.28. *Diffeomorphic (closed) Legendrian submanifolds have contactomorphic neighbourhoods.*

PROOF. If $L_i \subset M_i$ is Legendrian, then $\mathrm{CSN}(M_i, L_i)$ has rank 0, so the conditions in the theorem, apart from the existence of a diffeomorphism $\phi : L_0 \to L_1$, are void. □

EXAMPLE 2.29. Let $S^1 \subset (M^3, \xi)$ be a Legendrian knot in a contact 3-manifold. Then with a coordinate $\theta \in [0, 2\pi]$ along S^1 and coordinates x, y in slices transverse to S^1, the contact structure

$$\cos\theta\, dx - \sin\theta\, dy = 0$$

provides a model for a neighbourhood of S^1.

PROOF OF THEOREM 2.27. Choose contact forms α_i for ξ_i, $i = 0, 1$, scaled in such a way that Φ is actually an isomorphism of symplectic vector bundles with respect to the symplectic bundle structures on $\mathrm{CSN}(M_i, L_i)$ given by $d\alpha_i$. Here we think of $\mathrm{CSN}(M_i, L_i)$ as a sub-bundle of $TM_i|_{L_i}$ (rather than as a quotient bundle).

We identify $(TM_i|_{L_i})/(\xi_i|_{L_i})$ with the trivial line bundle spanned by the Reeb vector field R_{α_i}. In total, this identifies

$$NL_i = \langle R_{\alpha_i}\rangle \oplus J_i(TL_i) \oplus \mathrm{CSN}(M_i, L_i)$$

as a sub-bundle of $TM_i|_{L_i}$.

Let $\Phi_R : \langle R_{\alpha_0}\rangle \to \langle R_{\alpha_1}\rangle$ be the obvious bundle isomorphism defined by requiring that $R_{\alpha_0}(p)$ map to $R_{\alpha_1}(\phi(p))$.

Let $\Psi_i : J_i(TL_i) \to T^*L_i$ be the isomorphism defined by taking the interior product with $d\alpha_i$. Notice that

$$T\phi \oplus (\phi^*)^{-1} : (TL_0 \oplus T^*L_0, \Omega_{L_0}) \to (TL_1 \oplus T^*L_1, \Omega_{L_1})$$

is an isomorphism of symplectic vector bundles. With Lemma 2.26 it follows that

$$T\phi \oplus \Psi_1^{-1} \circ (\phi^*)^{-1} \circ \Psi_0 : \big(TL_0 \oplus J_0(TL_0), d\alpha_0\big) \to \big(TL_1 \oplus J_1(TL_1), d\alpha_1\big)$$

is an isomorphism of symplectic vector bundles.

Now let

$$\tilde{\Phi} : NL_0 \to NL_1$$

be the bundle isomorphism (covering ϕ) defined by

$$\tilde{\Phi} = \Phi_R \oplus \Psi_1^{-1} \circ (\phi^*)^{-1} \circ \Psi_0 \oplus \Phi.$$

Let $\tau_i : NL_i \to M_i$ be tubular maps, that is, the τ (I suppress the index i for better readability) are embeddings such that $\tau|_L$—where L is identified with the zero section of NL—is the inclusion $L \subset M$, and $T\tau$ induces the identity on NL along L (with respect to the splittings $T(NL)|_L = TL \oplus NL = TM|_L$).

Then $\tau_1 \circ \tilde{\Phi} \circ \tau_0^{-1} : \mathcal{N}(L_0) \to \mathcal{N}(L_1)$ is a diffeomorphism of suitable neighbourhoods $\mathcal{N}(L_i)$ of L_i that induces the bundle map

$$T\phi \oplus \tilde{\Phi} : TM_0|_{L_0} \to TM_1|_{L_1}.$$

By construction, this bundle map pulls α_1 back to α_0 and $d\alpha_1$ to $d\alpha_0$. Hence, α_0 and $(\tau_1 \circ \tilde{\Phi} \circ \tau_0^{-1})^*\alpha_1$ are contact forms on $\mathcal{N}(L_0)$ that coincide on $TM_0|_{L_0}$, and so do their differentials.

Now consider the family of 1-forms

$$\beta_t = (1-t)\alpha_0 + t\big(\tau_1 \circ \tilde{\Phi} \circ \tau_0^{-1}\big)^*\alpha_1, \quad t \in [0,1].$$

On $TM_0|_{L_0}$ we have $\beta_t \equiv \alpha_0$ and $d\beta_t \equiv d\alpha_0$. Since the contact condition $\alpha \wedge (d\alpha)^n \neq 0$ is an open condition, we may assume—shrinking $\mathcal{N}(L_0)$ if necessary—that β_t is a contact form on $\mathcal{N}(L_0)$ for all $t \in [0, 1]$. By the Gray stability theorem (Theorem 2.20) and Remark 2.21(3) following its proof, we find an isotopy ψ_t of $\mathcal{N}(L_0)$, fixing L_0, such that $\psi_t^* \beta_t = \lambda_t \alpha_0$ for some smooth family of smooth functions $\lambda_t : \mathcal{N}(L_0) \to \mathbb{R}^+$.

(Since $\mathcal{N}(L_0)$ is not a closed manifold, ψ_t is a priori only a local flow. But on L_0 it is stationary and hence defined for all t. As in the proof of the Darboux theorem (Theorem 2.24) we conclude that ψ_t is defined for all $t \in [0, 1]$ in a sufficiently small neighbourhood of L_0, so shrinking $\mathcal{N}(L_0)$ once again, if necessary, will ensure that ψ_t is a global flow on $\mathcal{N}(L_0)$.)

We conclude that $\psi = \tau_1 \circ \tilde{\Phi} \circ \tau_0^{-1} \circ \psi_1$ is the desired contactomorphism. □

REMARK 2.30. With a little more care one can actually achieve $T\psi_1 = \mathrm{id}$ on $TM_0|_{L_0}$, which implies in particular that $T\psi|_{\mathrm{CSN}(M_0, L_0)} = \Phi$, cf. [105]. (Remember that there is a certain freedom in constructing an isotopy via the Moser trick if the condition $X_t \in \xi_t$ is dropped.) The key point is the generalised Poincaré lemma, cf. [80, p. 361], which allows us to write a closed differential form γ given in a neighbourhood of the zero section of a bundle and vanishing along that zero section as an exact form $\gamma = d\eta$ with η and its partial derivatives with respect to all coordinates (in any chart) vanishing along the zero section. This lemma is applied first to $\gamma = d(\beta_1 - \beta_0)$, in order to find (with the symplectic Moser trick) a diffeomorphism σ of a neighbourhood of $L_0 \subset M_0$ with $T\sigma = \mathrm{id}$ on $TM_0|_{L_0}$ and such that $d\beta_0 = d(\sigma^* \beta_1)$. It is then applied once again to $\gamma = \beta_0 - \sigma^* \beta_1$.

(The proof of the symplectic neighbourhood theorem in [92] appears to be incomplete in this respect.)

EXAMPLE 2.31. Let $M_0 = M_1 = \mathbb{R}^3$ with contact forms $\alpha_0 = dz + x\,dy$ and $\alpha_1 = dz + (x + y)\,dy$ and $L_0 = L_1 = 0$ the origin in $\mathbb{R}^3$. Thus

$$\mathrm{CSN}(M_0, L_0) = \mathrm{CSN}(M_1, L_1) = \mathrm{span}\{\partial_x, \partial_y\} \subset T_0\mathbb{R}^3.$$

We take $\Phi = \mathrm{id}_{\mathrm{CSN}}$.

Set $\alpha_t = dz + (x + ty)\,dy$. The Moser trick with $X_t \in \ker \alpha_t$ yields $X_t = -y\,\partial_x$, and hence $\psi_t(x, y, z) = (x - ty, y, z)$. Then

$$T\psi_1 = \begin{pmatrix} 1 & -1 & 0 \\ 0 & 1 & 0 \\ 0 & 0 & 1 \end{pmatrix},$$

which does not restrict to Φ on CSN.

However, a different solution for $\psi_t^* \alpha_t = \alpha_0$ is $\psi_t(x, y, z) = (x, y, z - ty^2/2)$, found by integrating $X_t = -y^2\,\partial_z/2$ (a multiple of the Reeb vector field of α_t). Here we get

$$T\psi_1 = \begin{pmatrix} 1 & 0 & 0 \\ 0 & 1 & 0 \\ 0 & -y & 1 \end{pmatrix},$$

hence $T\psi_1|_{T_0\mathbb{R}^3} = \mathrm{id}$, so in particular $T\psi_1|_{\mathrm{CSN}} = \Phi$.

2.4.3. *Contact submanifolds* Let $(M', \xi' = \ker \alpha') \subset (M, \xi = \ker \alpha)$ be a contact submanifold, that is, $TM' \cap \xi|_{M'} = \xi'$. As before we write $(\xi')^{\perp} \subset \xi|_{M'}$ for the symplectically orthogonal complement of ξ' in $\xi|_{M'}$. Since M' is a contact submanifold (so ξ' is a symplectic sub-bundle of $(\xi|_{M'}, d\alpha)$), we have

$$TM' \oplus (\xi')^{\perp} = TM|_{M'},$$

i.e. we can identify $(\xi')^{\perp}$ with the normal bundle NM'. Moreover, $d\alpha$ induces a conformal symplectic structure on $(\xi')^{\perp}$, so we call $(\xi')^{\perp}$ the *conformal symplectic normal bundle* of M' in M and write

$$\mathrm{CSN}(M, M') = (\xi')^{\perp}.$$

THEOREM 2.32. *Let (M_i, ξ_i), $i = 0, 1$, be contact manifolds with compact contact submanifolds (M_i', ξ_i'). Suppose there is an isomorphism of conformal symplectic normal bundles $\Phi : \mathrm{CSN}(M_0, M_0') \to \mathrm{CSN}(M_1, M_1')$ that covers a contactomorphism $\phi : (M_0', \xi_0') \to (M_1', \xi_1')$. Then ϕ extends to a contactomorphism ψ of suitable neighbourhoods $\mathcal{N}(M_i')$ of M_i' such that $T\psi|_{\mathrm{CSN}(M_0, M_0')}$ and Φ are bundle homotopic (as conformal symplectic bundle isomorphisms).*

EXAMPLE 2.33. A particular instance of this theorem is the case of a transverse knot in a contact manifold (M, ξ), i.e. an embedding $S^1 \hookrightarrow (M, \xi)$ transverse to ξ. Since the symplectic group $\mathrm{Sp}(2n)$ of linear transformations of $\mathbb{R}^{2n}$ preserving the standard symplectic structure $\omega_0 = \sum_{i=1}^n dx_i \wedge dy_i$ is connected, there is only one conformal symplectic $\mathbb{R}^{2n}$-bundle over S^1 up to conformal equivalence. A model for the neighbourhood of a transverse knot is given by

$$\left(S^1 \times \mathbb{R}^{2n}, \xi = \ker\left(d\theta + \sum_{i=1}^{n} (x_i\, dy_i - y_i\, dx_i)\right)\right),$$

where θ denotes the S^1-coordinate; the theorem says that in suitable local coordinates the neighbourhood of any transverse knot looks like this model.

PROOF OF THEOREM 2.32. As in the proof of Theorem 2.27 it is sufficient to find contact forms α_i on M_i and a bundle map $TM_0|_{M_0'} \to TM_1|_{M_1'}$, covering ϕ and inducing Φ, that pulls back α_1 to α_0 and $d\alpha_1$ to $d\alpha_0$; the proof then concludes as there with a stability argument.

For this we need to make a judicious choice of α_i. The essential choice is made separately on each M_i, so I suppress the subscript i for the time being. Choose a contact form α' for ξ' on M'. Write R' for the Reeb vector field of α'. Given any contact form α for ξ on M we may first scale it such that $\alpha(R') \equiv 1$ along M'. Then $\alpha|_{TM'} = \alpha'$, and hence $d\alpha|_{TM'} = d\alpha'$. We now want to scale α further such that its Reeb vector field R coincides with R' along M'. To this end it is sufficient to find a smooth function $f : M \to \mathbb{R}^+$ with

$f|_{M'} \equiv 1$ and $i_{R'}d(f\alpha) \equiv 0$ on $TM|_{M'}$. This last equation becomes

$$0 = i_{R'}d(f\alpha) = i_{R'}(df \wedge \alpha + f\,d\alpha) = -df + i_{R'}d\alpha \quad \text{on } TM|_{M'}.$$

Since $i_{R'}d\alpha|_{TM'} = i_{R'}d\alpha' \equiv 0$, such an f can be found.

The choices of α_0' and α_1' cannot be made independently of each other; we may first choose α_1', say, and then define $\alpha_0' = \phi^*\alpha_1'$. Then define α_0, α_1 as described and scale Φ such that it is a symplectic bundle isomorphism of

$$\big((\xi_0')^\perp, d\alpha_0\big) \to \big((\xi_1')^\perp, d\alpha_1\big).$$

Then

$$T\phi \oplus \Phi : TM_0|_{M_0'} \to TM_1|_{M_1'}$$

is the desired bundle map that pulls back α_1 to α_0 and $d\alpha_1$ to $d\alpha_0$. □

REMARK 2.34. The condition that $R_i \equiv R_i'$ along M' is necessary for ensuring that $(T\phi \oplus \Phi)(R_0) = R_1$, which guarantees (with the other stated conditions) that $(T\phi \oplus \Phi)^*(d\alpha_1) = d\alpha_0$. The condition $d\alpha_i|_{TM_i'} = d\alpha_i'$ and the described choice of Φ alone would only give $(T\phi \oplus \Phi)^*(d\alpha_1|_{\xi_1}) = d\alpha_0|_{\xi_0}$.

2.4.4. *Hypersurfaces* Let S be an oriented hypersurface in a contact manifold $(M, \xi = \ker\alpha)$ of dimension $2n+1$. In a neighbourhood of S in M, which we can identify with $S \times \mathbb{R}$ (and S with $S \times \{0\}$), the contact form α can be written as

$$\alpha = \beta_r + u_r\,dr,$$

where β_r, $r \in \mathbb{R}$, is a smooth family of 1-forms on S and $u_r : S \to \mathbb{R}$ a smooth family of functions. The contact condition $\alpha \wedge (d\alpha)^n \neq 0$ then becomes

$$\begin{aligned} 0 \neq \alpha \wedge (d\alpha)^n &= (\beta_r + u_r\,dr) \wedge (d\beta_r - \dot{\beta}_r \wedge dr + du_r \wedge dr)^n \\ &= (-n\beta_r \wedge \dot{\beta}_r + n\beta_r \wedge du_r + u_r\,d\beta_r) \wedge (d\beta_r)^{n-1} \wedge dr, \end{aligned} \tag{2.5}$$

where the dot denotes derivative with respect to r. The intersection $TS \cap (\xi|_S)$ determines a distribution (of non-constant rank) of subspaces of TS. If α is written as above, this distribution is given by the kernel of β_0, and hence, at a given $p \in S$, defines either the full tangent space T_pS (if $\beta_{0,p} = 0$) or a 1-codimensional subspace both of T_pS and ξ_p (if $\beta_{0,p} \neq 0$). In the former case, the symplectically orthogonal complement $(T_pS \cap \xi_p)^\perp$ (with respect to the conformal symplectic structure $d\alpha$ on ξ_p) is $\{0\}$; in the latter case, $(T_pS \cap \xi_p)^\perp$ is a 1-dimensional subspace of ξ_p contained in $T_pS \cap \xi_p$.

From that it is intuitively clear what one should mean by a 'singular 1-dimensional foliation', and we make the following somewhat provisional definition.

Fig. 2. The characteristic foliation on $S^2 \subset (\mathbb{R}^3, \xi_2)$.

DEFINITION 2.35. The *characteristic foliation* S_ξ of a hypersurface S in (M, ξ) is the singular 1-dimensional foliation of S defined by $(TS \cap \xi|_S)^\perp$.

EXAMPLE 2.36. If $\dim M = 3$ and $\dim S = 2$, then $(T_pS \cap \xi_p)^\perp = T_pS \cap \xi_p$ at the points $p \in S$ where $T_pS \cap \xi_p$ is 1-dimensional. Figure 2 shows the characteristic foliation of the unit 2-sphere in $(\mathbb{R}^3, \xi_2)$, where ξ_2 denotes the standard contact structure of Example 2.7: The only singular points are $(0, 0, \pm 1)$; away from these points the characteristic foliation is spanned by

$$(y - xz)\,\partial_x - (x + yz)\,\partial_y + \left(x^2 + y^2\right)\partial_z.$$

The following lemma helps to clarify the notion of singular 1-dimensional foliation.

LEMMA 2.37. *Let β_0 be the 1-form induced on S by a contact form α defining ξ, and let Ω be a volume form on S. Then S_ξ is defined by the vector field X satisfying*

$$i_X\Omega = \beta_0 \wedge (d\beta_0)^{n-1}.$$

PROOF. First of all, we observe that $\beta_0 \wedge (d\beta_0)^{n-1} \neq 0$ outside the zeros of β_0: Arguing by contradiction, assume $\beta_{0,p} \neq 0$ and $\beta_0 \wedge (d\beta_0)^{n-1}|_p = 0$ at some $p \in S$. Then $(d\beta_0)^n|_p \neq 0$ by (2.5). On the codimension 1 subspace $\ker \beta_{0,p}$ of T_pS the symplectic form $d\beta_{0,p}$ has maximal rank $n - 1$. It follows that $\beta_0 \wedge (d\beta_0)^{n-1}|_p \neq 0$ after all, a contradiction.

Next we want to show that $X \in \ker \beta_0$. We observe

$$0 = i_X(i_X\Omega) = \beta_0(X)(d\beta_0)^{n-1} - (n-1)\beta_0 \wedge i_X\, d\beta_0 \wedge (d\beta_0)^{n-2}. \tag{2.6}$$

Taking the exterior product of this equation with β_0 we get

$$\beta_0(X)\beta_0 \wedge (d\beta_0)^{n-1} = 0.$$

By our previous consideration this implies $\beta_0(X) = 0$.

It remains to show that for $\beta_{0,p} \neq 0$ we have

$$d\beta_0\big(X(p), v\big) = 0 \quad \text{for all } v \in T_pS \cap \xi_p.$$

For $n = 1$ this is trivially satisfied, because in that case v is a multiple of $X(p)$. I suppress the point p in the following calculation, where we assume $n \geqslant 2$. From (2.6) and with $\beta_0(X) = 0$ we have

$$\beta_0 \wedge i_X \, d\beta_0 \wedge (d\beta_0)^{n-2} = 0. \tag{2.7}$$

Taking the interior product with $v \in TS \cap \xi$ yields

$$-d\beta_0(X, v)\beta_0 \wedge (d\beta_0)^{n-2} + (n-2)\beta_0 \wedge i_X \, d\beta_0 \wedge i_v \, d\beta_0 \wedge (d\beta_0)^{n-3} = 0.$$

(Thanks to the coefficient $n-2$ the term $(d\beta_0)^{n-3}$ is not a problem for $n = 2$.) Taking the exterior product of that last equation with $d\beta_0$, and using (2.7), we find

$$d\beta_0(X, v)\beta_0 \wedge (d\beta_0)^{n-1} = 0,$$

and thus $d\beta_0(X, v) = 0$. □

REMARK 2.38. (1) We can now give a more formal definition of 'singular 1-dimensional foliation' as an equivalence class of vector fields $[X]$, where X is allowed to have zeros and $[X] = [X']$ if there is a nowhere zero function on all of S such that $X' = fX$. Notice that the non-integrability of contact structures and the reasoning at the beginning of the proof of the lemma imply that the zero set of X does not contain any open subsets of S.

(2) If the contact structure ξ is cooriented rather than just coorientable, so that α is well defined up to multiplication with a *positive* function, then this lemma allows to give an orientation to the characteristic foliation: Changing α to $\lambda\alpha$ with $\lambda : M \to \mathbb{R}^+$ will change $\beta_0 \wedge (d\beta_0)^{n-1}$ by a factor λ^n.

We now restrict attention to surfaces in contact 3-manifolds, where the notion of characteristic foliation has proved to be particularly useful.

The following theorem is due to E. Giroux [52].

THEOREM 2.39 (Giroux). *Let S_i be closed surfaces in contact 3-manifolds (M_i, ξ_i), $i = 0, 1$ (with ξ_i coorientable), and $\phi : S_0 \to S_1$ a diffeomorphism with $\phi(S_{0,\xi_0}) = S_{1,\xi_1}$ as oriented characteristic foliations. Then there is a contactomorphism $\psi : \mathcal{N}(S_0) \to \mathcal{N}(S_1)$ of suitable neighbourhoods $\mathcal{N}(S_i)$ of S_i with $\psi(S_0) = S_1$ and such that $\psi|_{S_0}$ is isotopic to ϕ via an isotopy preserving the characteristic foliation.*

PROOF. By passing to a double cover, if necessary, we may assume that the S_i are orientable hypersurfaces. Let α_i be contact forms defining ξ_i. Extend ϕ to a diffeomorphism (still denoted ϕ) of neighbourhoods of S_i and consider the contact forms α_0 and $\phi^*\alpha_1$ on a neighbourhood of S_0, which we may identify with $S_0 \times \mathbb{R}$.

By rescaling α_1 we may assume that α_0 and $\phi^*\alpha_1$ induce the same form β_0 on $S_0 \equiv S_0 \times \{0\}$, and hence also the same form $d\beta_0$.

Observe that the expression on the right-hand side of Eq. (2.5) is linear in $\dot{\beta}_r$ and u_r. This implies that convex linear combinations of solutions of (2.5) (for $n=1$) with the same β_0 (and $d\beta_0$) will again be solutions of (2.5) for sufficiently small r. This reasoning applies to

$$\alpha_t := (1-t)\alpha_0 + t\phi^*\alpha_1, \quad t \in [0,1].$$

(I hope the reader will forgive the slight abuse of notation, with α_1 denoting both a form on M_1 and its pull-back $\phi^*\alpha_1$ to M_0.) As is to be expected, we now use the Moser trick to find an isotopy ψ_t with $\psi_t^*\alpha_t = \lambda_t\alpha_0$, just as in the proof of Gray stability (Theorem 2.20). In particular, we require as there that the vector field X_t that we want to integrate to the flow ψ_t lie in the kernel of α_t.

On TS_0 we have $\dot{\alpha}_t \equiv 0$ (thanks to the assumption that α_0 and $\phi^*\alpha_1$ induce the same form β_0 on S_0). In particular, if v is a vector in S_{0,ξ_0}, then by Eq. (2.1) we have $d\alpha_t(X_t, v) = 0$, which implies that X_t is a multiple of v, hence tangent to S_{0,ξ_0}. This shows that the flow of X_t preserves S_0 and its characteristic foliation. More formally, we have

$$\mathcal{L}_{X_t}\alpha_t = d\big(\alpha_t(X_t)\big) + i_{X_t}\,d\alpha_t = i_{X_t}\,d\alpha_t,$$

so with v as above we have $\mathcal{L}_{X_t}\alpha_t(v) = 0$, which shows that $\mathcal{L}_{X_t}\alpha_t|_{TS_0}$ is a multiple of $\alpha_0|_{TS_0} = \beta_0$. This implies that the (local) flow of X_t changes β_0 by a conformal factor.

Since S_0 is closed, the local flow of X_t restricted to S_0 integrates up to $t=1$, and so the same is true[3] in a neighbourhood of S_0. Then $\psi = \phi \circ \psi_1$ will be the desired diffeomorphism $\mathcal{N}(S_0) \to \mathcal{N}(S_1)$. □

As observed previously in the proof of Darboux's theorem for contact *forms*, the Moser trick allows more flexibility if we drop the condition $\alpha_t(X_t) = 0$. We are now going to exploit this extra freedom to strengthen Giroux's theorem slightly. This will be important later on when we want to extend isotopies of hypersurfaces.

THEOREM 2.40. *Under the assumptions of the preceding theorem we can find ψ: $\mathcal{N}(S_0) \to \mathcal{N}(S_1)$ satisfying the stronger condition that $\psi|_{S_0} = \phi$.*

PROOF. We want to show that the isotopy ψ_t of the preceding proof may be assumed to fix S_0 pointwise. As there, we may assume $\dot{\alpha}_t|_{TS_0} \equiv 0$.

If the condition that X_t be tangent to $\ker\alpha_t$ is dropped, the condition X_t needs to satisfy so that its flow will pull back α_t to $\lambda_t\alpha_0$ is

$$\dot{\alpha}_t + d\big(\alpha_t(X_t)\big) + i_{X_t}\,d\alpha_t = \mu_t\alpha_t, \tag{2.8}$$

where μ_t and λ_t are related by $\mu_t = \frac{d}{dt}(\log\lambda_t) \circ \psi_t^{-1}$, cf. the proof of the Gray stability theorem (Theorem 2.20).

[3] Cf. the proof (and the footnote therein) of Darboux's theorem (Theorem 2.24).

Write $X_t = H_t R_t + Y_t$ with R_t the Reeb vector field of α_t and $Y_t \in \xi_t = \ker \alpha_t$. Then condition (2.8) translates into

$$\dot{\alpha}_t + dH_t + i_{Y_t} d\alpha_t = \mu_t \alpha_t. \tag{2.9}$$

For given H_t one determines μ_t from this equation by inserting the Reeb vector field R_t; the equation then admits a unique solution $Y_t \in \ker \alpha_t$ because of the non-degeneracy of $d\alpha_t|_{\xi_t}$.

Our aim now is to ensure that $H_t \equiv 0$ on S_0 and $Y_t \equiv 0$ along S_0. The latter we achieve by imposing the condition

$$\dot{\alpha}_t + dH_t = 0 \quad \text{along } S_0 \tag{2.10}$$

(which entails with (2.9) that $\mu_t|_{S_0} \equiv 0$). The conditions $H_t \equiv 0$ on S_0 and (2.10) can be simultaneously satisfied thanks to $\dot{\alpha}_t|_{TS_0} \equiv 0$.

We can therefore find a smooth family of smooth functions H_t satisfying these conditions, and then define Y_t by (2.9). The flow of the vector field $X_t = H_t R_t + Y_t$ then defines an isotopy ψ_t that fixes S_0 pointwise (and thus is defined for all $t \in [0, 1]$ in a neighbourhood of S_0). Then $\psi = \phi \circ \psi_1$ will be the diffeomorphism we wanted to construct. □

2.4.5. *Applications* Perhaps the most important consequence of the neighbourhood theorems proved above is that they allow us to perform differential topological constructions such as surgery or similar cutting and pasting operations in the presence of a contact structure, that is, these constructions can be carried out on a contact manifold in such a way that the resulting manifold again carries a contact structure.

One such construction that I shall explain in detail in Section 3 is the surgery of contact 3-manifolds along transverse knots, which enables us to construct a contact structure on every closed, orientable 3-manifold.

The concept of *characteristic foliation* on surfaces in contact 3-manifolds has proved seminal for the classification of contact structures on 3-manifolds, although it has recently been superseded by the notion of *dividing curves*. It is clear that Theorem 2.39 can be used to cut and paste contact manifolds along hypersurfaces with the same characteristic foliation. What actually makes this useful in dimension 3 is that there are ways to manipulate the characteristic foliation of a surface by isotoping that surface inside the contact 3-manifold.

The most important result in that direction is the *Elimination Lemma* proved by Giroux [52]; an improved version is due to D. Fuchs, see [26]. This lemma says that under suitable assumptions it is possible to cancel singular points of the characteristic foliation in pairs by a C^0-small isotopy of the surface (specifically: an elliptic and a hyperbolic point of the same sign—the sign being determined by the matching or non-matching of the orientation of the surface S and the contact structure ξ at the singular point of S_ξ). For instance, Eliashberg [24] has shown that if a contact 3-manifold (M, ξ) contains an embedded disc D' such that D'_ξ has a limit cycle, then one can actually find a so-called *overtwisted disc*: an embedded disc D with boundary ∂D tangent to ξ (but D transverse to ξ along ∂D, i.e. no singular points of D_ξ on ∂D) and with D_ξ having exactly one singular point (of elliptic type); see Section 3.6.

Moreover, in the generic situation it is possible, given surfaces $S \subset (M, \xi)$ and $S' \subset (M', \xi')$ with S_ξ *homeomorphic* to $S'_{\xi'}$, to perturb one of the surfaces so as to get *diffeomorphic* characteristic foliations.

Chapter 8 of [1] contains a section on surfaces in contact 3-manifolds, and in particular a proof of the Elimination Lemma. Further introductory reading on the matter can be found in the lectures of J. Etnyre [35]; of the sources cited above I recommend [26] as a starting point.

In [52] Giroux initiated the study of *convex surfaces* in contact 3-manifolds. These are surfaces S with an infinitesimal automorphism X of the contact structure ξ with X transverse to S. For such surfaces, it turns out, much less information than the characteristic foliation S_ξ is needed to determine ξ in a neighbourhood of S, viz., only the so-called *dividing set* of S_ξ. This notion lies at the centre of most of the recent advances in the classification of contact structures on 3-manifolds [55,71,72]. A brief introduction to convex surface theory can be found in [35].

2.5. *Isotopy extension theorems*

In this section we show that the isotopy extension theorem of differential topology—an isotopy of a closed submanifold extends to an isotopy of the ambient manifold—remains valid for the various distinguished submanifolds of contact manifolds. The neighbourhood theorems proved above provide the key to the corresponding isotopy extension theorems. For simplicity, I assume throughout that the ambient contact manifold M is closed; all isotopy extension theorems remain valid if M has non-empty boundary ∂M, provided the isotopy stays away from the boundary. In that case, the isotopy of M found by extension keeps a neighbourhood of ∂M fixed. A further convention throughout is that our ambient isotopies ψ_t are understood to start at $\psi_0 = \mathrm{id}_M$.

2.5.1. *Isotropic submanifolds*

An embedding $j : L \to (M, \xi = \ker\alpha)$ is called *isotropic* if $j(L)$ is an isotropic submanifold of (M, ξ), i.e. everywhere tangent to the contact structure ξ. Equivalently, one needs to require $j^*\alpha \equiv 0$.

THEOREM 2.41. *Let $j_t : L \to (M, \xi = \ker\alpha)$, $t \in [0, 1]$, be an isotopy of isotropic embeddings of a closed manifold L in a contact manifold (M, ξ). Then there is a compactly supported contact isotopy $\psi_t : M \to M$ with $\psi_t(j_0(L)) = j_t(L)$.*

PROOF. Define a time-dependent vector field X_t along $j_t(L)$ by

$$X_t \circ j_t = \frac{d}{dt} j_t.$$

To simplify notation later on, we assume that L is a submanifold of M and j_0 the inclusion $L \subset M$. Our aim is to find a (smooth) family of compactly supported, smooth functions

$\tilde{H}_t : M \to \mathbb{R}$ whose Hamiltonian vector field $\tilde{X}_t$ equals X_t along $j_t(L)$. Recall that $\tilde{X}_t$ is defined in terms of $\tilde{H}_t$ by

$$\alpha(\tilde{X}_t) = \tilde{H}_t, \quad i_{\tilde{X}_t} d\alpha = d\tilde{H}_t(R_\alpha)\alpha - d\tilde{H}_t,$$

where, as usual, R_α denotes the Reeb vector field of α.

Hence, we need

$$\alpha(X_t) = \tilde{H}_t, \quad i_{X_t} d\alpha = d\tilde{H}_t(R_\alpha)\alpha - d\tilde{H}_t \quad \text{along } j_t(L). \tag{2.11}$$

Write $X_t = H_t R_\alpha + Y_t$ with $H_t : j_t(L) \to \mathbb{R}$ and $Y_t \in \ker\alpha$. To satisfy (2.11) we need

$$\tilde{H}_t = H_t \quad \text{along } j_t(L). \tag{2.12}$$

This implies

$$d\tilde{H}_t(v) = dH_t(v) \quad \text{for } v \in T\big(j_t(L)\big).$$

Since j_t is an isotopy of isotropic embeddings, we have $T(j_t(L)) \subset \ker\alpha$. So a prerequisite for (2.11) is that

$$d\alpha(X_t, v) = -dH_t(v) \quad \text{for } v \in T\big(j_t(L)\big). \tag{2.13}$$

We have

$$\begin{aligned} d\alpha(X_t, v) + dH_t(v) &= d\alpha(X_t, v) + d\big(\alpha(X_t)\big)(v) = i_v\big(i_{X_t} d\alpha + d(i_{X_t}\alpha)\big) \\ &= i_v(\mathcal{L}_{X_t}\alpha), \end{aligned}$$

so Eq. (2.13) is equivalent to

$$i_v(\mathcal{L}_{X_t}\alpha) = 0 \quad \text{for } v \in T\big(j_t(L)\big).$$

But this is indeed tautologically satisfied: The fact that j_t is an isotopy of isotropic embeddings can be written as $j_t^*\alpha \equiv 0$; this in turn implies $0 = \frac{d}{dt}(j_t^*\alpha) = j_t^*(\mathcal{L}_{X_t}\alpha)$, as desired.

This means that we can define $\tilde{H}_t$ by prescribing the value of $\tilde{H}_t$ along $j_t(L)$ (with (2.12)) and the differential of $\tilde{H}_t$ along $j_t(L)$ (with (2.11)), where we are free to impose $d\tilde{H}_t(R_\alpha) = 0$, for instance. The calculation we just performed shows that these two requirements are consistent with each other. Any function satisfying these requirements along $j_t(L)$ can be smoothed out to zero outside a tubular neighbourhood of $j_t(L)$, and the Hamiltonian flow of this $\tilde{H}_t$ will be the desired contact isotopy extending j_t.

One small technical point is to ensure that the resulting family of functions $\tilde{H}_t$ will be smooth in t. To achieve this, we proceed as follows. Set $\hat{M} = M \times [0, 1]$ and

$$\hat{L} = \bigcup_{q \in L,\ t \in [0,1]} \big(j_t(q), t\big),$$

so that $\hat{L}$ is a submanifold of $\hat{M}$. Let g be an auxiliary Riemannian metric on M with respect to which R_α is orthogonal to $\ker\alpha$. Identify the normal bundle $N\hat{L}$ of $\hat{L}$ in $\hat{M}$ with a sub-bundle of $T\hat{M}$ by requiring its fibre at a point $(p,t)\in\hat{L}$ to be the g-orthogonal subspace of $T_p(j_t(L))$ in T_pM. Let $\tau: N\hat{L}\to\hat{M}$ be a tubular map.

Now define a smooth function $\hat{H}: N\hat{L}\to\mathbb{R}$ as follows, where (p,t) always denotes a point of $\hat{L}\subset N\hat{L}$.

- $\hat{H}(p,t)=\alpha(X_t)$,
- $d\hat{H}_{(p,t)}(R_\alpha)=0$,
- $d\hat{H}_{(p,t)}(v)=-d\alpha(X_t,v)$ for $v\in\ker\alpha_p\subset T_pM\subset T_{(p,t)}\hat{M}$,
- $\hat{H}$ is linear on the fibres of $N\hat{L}\to\hat{L}$.

Let $\chi:\hat{M}\to[0,1]$ be a smooth function with $\chi\equiv 0$ outside a small neighbourhood $\mathcal{N}_0\subset\tau(N\hat{L})$ of $\hat{L}$ and $\chi\equiv 1$ in a smaller neighbourhood $\mathcal{N}_1\subset\mathcal{N}_0$ of $\hat{L}$. For $(p,t)\in\hat{M}$, set

$$\tilde{H}_t(p)=\begin{cases}\chi(p,t)\hat{H}(\tau^{-1}(p,t)) & \text{for } (p,t)\in\tau(N\hat{L}),\\ 0 & \text{for } (p,t)\notin\tau(N\hat{L}).\end{cases}$$

This is smooth in p and t, and the Hamiltonian flow ψ_t of $\tilde{H}_t$ (defined globally since $\tilde{H}_t$ is compactly supported) is the desired contact isotopy. □

2.5.2. *Contact submanifolds* An embedding $j:(M',\xi')\to(M,\xi)$ is called a *contact embedding* if

$$\big(j(M'),Tj(\xi')\big)$$

is a contact submanifold of (M,ξ), i.e.

$$T\big(j(M)\big)\cap\xi|_{j(M)}=Tj(\xi').$$

If $\xi=\ker\alpha$, this can be reformulated as $\ker j^*\alpha=\xi'$.

THEOREM 2.42. *Let $j_t:(M',\xi')\to(M,\xi)$, $t\in[0,1]$, be an isotopy of contact embeddings of the closed contact manifold (M',ξ') in the contact manifold (M,ξ). Then there is a compactly supported contact isotopy $\psi_t: M\to M$ with $\psi_t(j_0(M'))=j_t(M')$.*

PROOF. In the proof of this theorem we follow a slightly different strategy from the one in the isotropic case. Instead of directly finding an extension of the Hamiltonian $H_t: j_t(M')\to\mathbb{R}$, we first use the neighbourhood theorem for contact submanifolds to extend j_t to an isotopy of contact embeddings of tubular neighbourhoods.

Again we assume that M' is a submanifold of M and j_0 the inclusion $M'\subset M$. As earlier, NM' denotes the normal bundle of M' in M. We also identify M' with the zero section of NM', and we use the canonical identification

$$T(NM')|_{M'}=TM'\oplus NM'.$$

By the usual isotopy extension theorem from differential topology we find an isotopy

$$\phi_t : NM' \to M$$

with $\phi_t|_{M'} = j_t$.

Choose contact forms α, α' defining ξ and ξ', respectively. Define $\alpha_t = \phi_t^*\alpha$. Then $TM' \cap \ker \alpha_t = \xi'$. Let R' denote the Reeb vector field of α'. Analogous to the proof of Theorem 2.32, we first find a smooth family of smooth functions $g_t : M' \to \mathbb{R}^+$ such that $g_t\alpha_t|_{TM'} = \alpha'$, and then a family $f_t : NM' \to \mathbb{R}^+$ with $f_t|_{M'} \equiv 1$ and

$$df_t = i_{R'} d(g_t\alpha_t) \quad \text{on } T(NM')|_{M'}.$$

Then $\beta_t = f_t g_t \alpha_t$ is a family of contact forms on NM' representing the contact structure $\ker(\phi_t^*\alpha)$ and with the properties

$$\beta_t|_{TM'} = \alpha',$$
$$d\beta_t|_{TM'} = d\alpha',$$
$$\ker(d\beta_t) = \langle R' \rangle \quad \text{along } M'.$$

The family $(NM', d\beta_t)$ of symplectic vector bundles may be thought of as a symplectic vector bundle over $M' \times [0, 1]$, which is necessarily isomorphic to a bundle pulled back from $M' \times \{0\}$ (cf. [74, Corollary 3.4.4]). In other words, there is a smooth family of symplectic bundle isomorphisms

$$\Phi_t : (NM', d\beta_0) \to (NM', d\beta_t).$$

Then

$$\mathrm{id}_{TM'} \oplus \Phi_t : T(NM')|_{M'} \to T(NM')|_{M'}$$

is a bundle map that pulls back β_t to β_0 and $d\beta_t$ to $d\beta_0$.

By the now familiar stability argument we find a smooth family of embeddings

$$\varphi_t : \mathcal{N}(M') \to NM'$$

for some neighbourhood $\mathcal{N}(M')$ of the zero section M' in NM' with $\varphi_0 =$ inclusion, $\varphi_t|_{M'} = \mathrm{id}_{M'}$ and $\varphi_t^*\beta_t = \lambda_t\beta_0$, where $\lambda_t : \mathcal{N}(M') \to \mathbb{R}^+$. This means that

$$\phi_t \circ \varphi_t : \mathcal{N}(M') \to M$$

is a smooth family of contact embeddings of $(\mathcal{N}(M'), \ker\beta_0)$ in (M, ξ).

Define a time-dependent vector field X_t along $\phi_t \circ \varphi_t(\mathcal{N}(M'))$ by

$$X_t \circ \phi_t \circ \varphi_t = \frac{d}{dt}(\phi_t \circ \varphi_t).$$

This X_t is clearly an infinitesimal automorphism of ξ: By differentiating the equation $\varphi_t^*\phi_t^*\alpha = \mu_t\phi_0^*\alpha$ (where $\mu_t : \mathcal{N}(M') \to \mathbb{R}^+$) with respect to t we get

$$\varphi_t^*\phi_t^*(\mathcal{L}_{X_t}\alpha) = \dot{\mu}_t\phi_0^*\alpha = \frac{\dot{\mu}_t}{\mu_t}\varphi_t^*\phi_t^*\alpha,$$

so $\mathcal{L}_{X_t}\alpha$ is a multiple of α (since $\phi_t \circ \varphi_t$ is a diffeomorphism onto its image).

By the theory of contact Hamiltonians, X_t is the Hamiltonian vector field of a Hamiltonian function $\hat{H}_t$ defined on $\phi_t \circ \varphi_t(\mathcal{N}(M'))$. Cut off this function with a bump function so as to obtain $H_t : M \to \mathbb{R}$ with $H_t \equiv \hat{H}_t$ near $\phi_t \circ \varphi_t(M')$ and $H_t \equiv 0$ outside a slightly larger neighbourhood of $\phi_t \circ \varphi_t(M')$. Then the Hamiltonian flow ψ_t of H_t satisfies our requirements. □

2.5.3. *Surfaces in 3-manifolds*

THEOREM 2.43. *Let $j_t : S \to (M, \xi = \ker\alpha)$, $t \in [0,1]$, be an isotopy of embeddings of a closed surface S in a 3-dimensional contact manifold (M, ξ). If all j_t induce the same characteristic foliation on S, then there is a compactly supported isotopy $\psi_t : M \to M$ with $\psi_t(j_0(S)) = j_t(S)$.*

PROOF. Extend j_t to a smooth family of embeddings $\phi_t : S \times \mathbb{R} \to M$, and identify S with $S \times \{0\}$. The assumptions say that all $\phi_t^*\alpha$ induce the same characteristic foliation on S. By the proof of Theorem 2.40 and in analogy with the proof of Theorem 2.42 we find a smooth family of embeddings

$$\varphi_t : S \times (-\varepsilon, \varepsilon) \to S \times \mathbb{R}$$

for some $\varepsilon > 0$ with $\varphi_0 =$ inclusion, $\varphi_t|_{S\times\{0\}} = \mathrm{id}_S$ and $\varphi_t^*\phi_t^*\alpha = \lambda_t\phi_0^*\alpha$, where $\lambda_t : S \times (-\varepsilon, \varepsilon) \to \mathbb{R}^+$. In other words, $\phi_t \circ \varphi_t$ is a smooth family of contact embeddings of $(S \times (-\varepsilon, \varepsilon), \ker\phi_0^*\alpha)$ in (M, ξ).

The proof now concludes exactly as the proof of Theorem 2.42. □

2.6. *Approximation theorems*

A further manifestation of the (local) flexibility of contact structures is the fact that a given submanifold can, under fairly weak (and usually obvious) topological conditions, be approximated (typically C^0-closely) by a contact submanifold or an isotropic submanifold, respectively. The most general results in this direction are best phrased in M. Gromov's language of *h-principles*. For a recent text on h-principles that puts particular emphasis on symplectic and contact geometry see [30]; a brief and perhaps more gentle introduction to h-principles can be found in [47].

In the present section I restrict attention to the 3-dimensional situation, where the relevant approximation theorems can be proved by elementary geometric ad hoc techniques.

THEOREM 2.44. *Let $\gamma : S^1 \to (M, \xi)$ be a knot, i.e. an embedding of S^1, in a contact 3-manifold. Then γ can be C^0-approximated by a Legendrian knot isotopic to γ. Alternatively, it can be C^0-approximated by a positively as well as a negatively transverse knot.*

In order to prove this theorem, we first consider embeddings $\gamma : (a, b) \to (\mathbb{R}^3, \xi)$ of an open interval in $\mathbb{R}^3$ with its standard contact structure $\xi = \ker \alpha$, where $\alpha = dz + x\,dy$. Write $\gamma(t) = (x(t), y(t), z(t))$. Then

$$\alpha(\dot{\gamma}) = \dot{z} + x\dot{y},$$

so the condition for a Legendrian curve reads $\dot{z} + x\dot{y} \equiv 0$; for a positively (respectively negatively) transverse curve, $\dot{z} + x\dot{y} > 0$ (respectively < 0).

There are two ways to visualise such curves. The first is via its *front projection*

$$\gamma_F(t) = \big(y(t), z(t)\big),$$

the second via its *Lagrangian projection*

$$\gamma_L(t) = \big(x(t), y(t)\big).$$

2.6.1. *Legendrian knots* If $\gamma(t) = (x(t), y(t), z(t))$ is a Legendrian curve in $\mathbb{R}^3$, then $\dot{y} = 0$ implies $\dot{z} = 0$, so there the front projection has a singular point. Indeed, the curve $t \mapsto (t, 0, 0)$ is an example of a Legendrian curve whose front projection is a single point. We call a Legendrian curve *generic* if $\dot{y} = 0$ only holds at isolated points (which we call *cusp points*), and there $\ddot{y} \neq 0$.

LEMMA 2.45. *Let $\gamma : (a, b) \to (\mathbb{R}^3, \xi)$ be a Legendrian immersion. Then its front projection $\gamma_F(t) = (y(t), z(t))$ does not have any vertical tangencies. Away from the cusp points, γ is recovered from its front projection via*

$$x(t) = -\frac{\dot{z}(t)}{\dot{y}(t)} = -\frac{dz}{dy},$$

i.e. $x(t)$ is the negative slope of the front projection. The curve γ is embedded if and only if γ_F has only transverse self-intersections.

By a C^∞-small perturbation of γ we can obtain a generic Legendrian curve $\tilde{\gamma}$ isotopic to γ; by a C^2-small perturbation we may achieve that the front projection has only semi-cubical cusp singularities, i.e. around a cusp point at $t = 0$ the curve $\tilde{\gamma}$ looks like

$$\tilde{\gamma}(t) = \big(t + a, \lambda t^2 + b, -\lambda\big(2t^3/3 + at^2\big) + c\big)$$

with $\lambda \neq 0$, see Figure 3.

Any regular curve in the (y, z)-plane with semi-cubical cusps and no vertical tangencies can be lifted to a unique Legendrian curve in $\mathbb{R}^3$.

Fig. 3. The cusp of a front projection.

PROOF. The Legendrian condition is $\dot{z} + x\dot{y} = 0$. Hence $\dot{y} = 0$ forces $\dot{z} = 0$, so γ_F cannot have any vertical tangencies.

Away from the cusp points, the Legendrian condition tells us how to recover x as the negative slope of the front projection. (By continuity, the equation $x = \frac{dz}{dy}$ also makes sense at generic cusp points.) In particular, a self-intersecting front projection lifts to a non-intersecting curve if and only if the slopes at the intersection point are different, i.e. if and only if the intersection is transverse.

That γ can be approximated in the C^∞-topology by a generic immersion $\tilde{\gamma}$ follows from the usual transversality theorem (in its most simple form, viz., applied to the function $y(t)$; the function $x(t)$ may be left unchanged, and the new $z(t)$ is then found by integrating the new $-x\dot{y}$).

At a cusp point of $\tilde{\gamma}$ we have $\dot{y} = \dot{z} = 0$. Since $\tilde{\gamma}$ is an immersion, this forces $\dot{x} \neq 0$, so $\tilde{\gamma}$ can be parametrised around a cusp point by the x-coordinate, i.e. we may choose the curve parameter t such that the cusp lies at $t = 0$ and $x(t) = t + a$. Since $\ddot{y}(0) \neq 0$ by the genericity condition, we can write $y(t) = t^2 g(t) + y(0)$ with a smooth function $g(t)$ satisfying $g(0) \neq 0$ (this is proved like the 'Morse lemma' in Appendix 2 of [77]). A C^0-approximation of $g(t)$ by a function $h(t)$ with $h(t) \equiv g(0)$ for t near zero and $h(t) \equiv g(t)$ for $|t|$ greater than some small $\varepsilon > 0$ yields a C^2-approximation of $y(t)$ with the desired form around the cusp point. □

EXAMPLE 2.46. Figure 4 shows the front projection of a Legendrian trefoil knot.

PROOF OF THEOREM 2.44 (*Legendrian case*). First of all, we consider a curve γ in standard $\mathbb{R}^3$. In order to find a C^0-close approximation of γ, we simply need to choose a C^0-close approximation of its front projection γ_F by a regular curve without vertical tangencies and with isolated cusps (we call such a curve a *front*) in such a way, that the slope of the front at time t is close to $-x(t)$ (see Figure 5). Then the Legendrian lift of this front is the desired C^0-approximation of γ.

If γ is defined on an interval (a, b) and is already Legendrian near its endpoints, then the approximation of γ_F may be assumed to coincide with γ_F near the endpoints, so that the Legendrian lift coincides with γ near the endpoints.

Hence, given a knot in an arbitrary contact 3-manifold, we can cut it (by the Lebesgue lemma) into little pieces that lie in Darboux charts. There we can use the preceding recipe to find a Legendrian approximation. Since, as just observed, one can find such approximations on intervals with given boundary condition, this procedure yields a Legendrian approximation of the full knot.

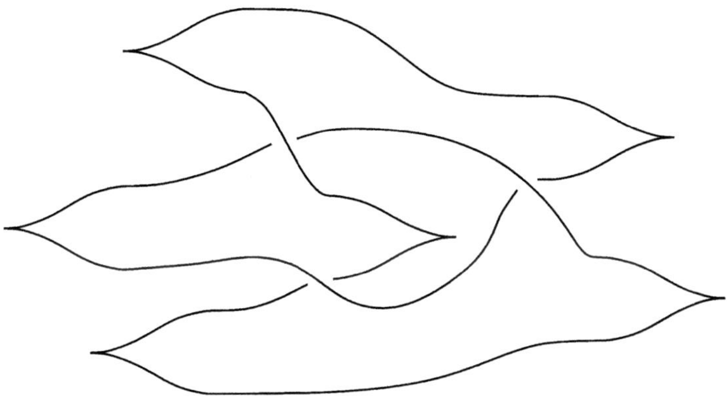

Fig. 4. Front projection of a Legendrian trefoil knot.

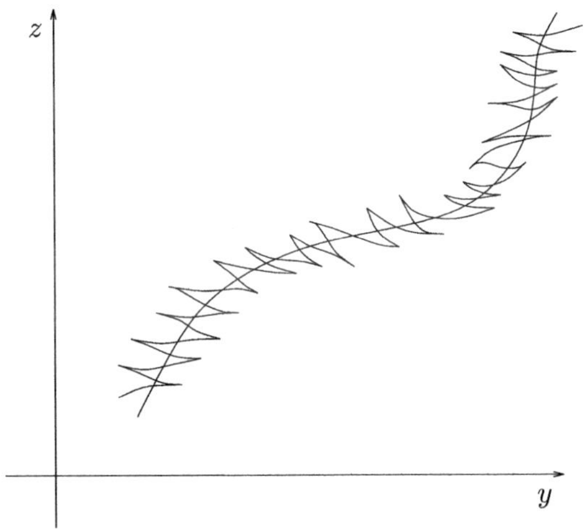

Fig. 5. Legendrian C^0-approximation via front projection.

Locally (i.e. in $\mathbb{R}^3$) the described procedure does not introduce any self-intersections in the approximating curve, provided we approximate γ_F by a front with only transverse self-intersections. Since the original knot was embedded, the same will then be true for its Legendrian C^0-approximation. □

The same result may be derived using the Lagrangian projection:

LEMMA 2.47. *Let $\gamma : (a, b) \to (\mathbb{R}^3, \xi)$ be a Legendrian immersion. Then its Lagrangian projection $\gamma_L(t) = (x(t), y(t))$ is also an immersed curve. The curve γ is recovered from γ_L via*

$$z(t_1) = z(t_0) - \int_{t_0}^{t_1} x\, dy.$$

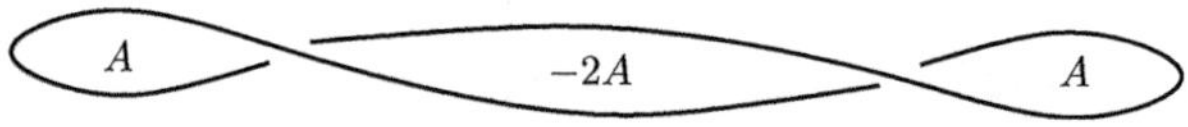

Fig. 6. Lagrangian projection of a Legendrian unknot.

A Legendrian immersion $\gamma : S^1 \to (\mathbb{R}^3, \xi)$ has a Lagrangian projection that encloses zero area. Moreover, γ is embedded if and only if every loop in γ_L (except, in the closed case, the full loop γ_L) encloses a non-zero oriented area.

Any immersed curve in the (x, y)-plane is the Lagrangian projection of a Legendrian curve in $\mathbb{R}^3$, unique up to translation in the z-direction.

PROOF. The Legendrian condition $\dot{z} + x\dot{y}$ implies that if $\dot{y} = 0$ then $\dot{z} = 0$, and hence, since γ is an immersion, $\dot{x} \neq 0$. So γ_L is an immersion.

The formula for z follows by integrating the Legendrian condition. For a closed curve γ_L in the (x, y)-plane, the integral $\oint_{\gamma_L} x\,dy$ computes the oriented area enclosed by γ_L. From that all the other statements follow. □

EXAMPLE 2.48. Figure 6 shows the Lagrangian projection of a Legendrian unknot.

ALTERNATIVE PROOF OF THEOREM 2.44 (*Legendrian case*). Again we consider a curve γ in standard $\mathbb{R}^3$ defined on an interval. The generalisation to arbitrary contact manifolds and closed curves is achieved as in the proof using front projections.

In order to find a C^0-approximation of γ by a Legendrian curve, one only has to approximate its Lagrangian projection γ_L by an immersed curve whose 'area integral'

$$z(t_0) - \int_{t_0}^{t} x\,dy$$

lies as close to the original $z(t)$ as one wishes. This can be achieved by using small loops oriented positively or negatively (see Figure 7). If γ_L has self-intersections, this approximating curve can be chosen in such a way that along loops properly contained in that curve the area integral is non-zero, so that again we do not introduce any self-intersections in the Legendrian approximation of γ. □

2.6.2. *Transverse knots* The quickest proof of the transverse case of Theorem 2.44 is via the Legendrian case. However, it is perfectly feasible to give a direct proof along the lines of the preceding discussion, i.e. using the front or the Lagrangian projection. Since this picture is useful elsewhere, I include a brief discussion, restricting attention to the front projection.

Thus, let $\gamma(t) = (x(t), y(t), z(t))$ be a curve in $\mathbb{R}^3$. The condition for γ to be positively transverse to the standard contact structure $\xi = \ker(dz + x\,dy)$ is that $\dot{z} + x\dot{y} > 0$. Hence,

$$\begin{cases} \text{if } \dot{y} = 0, & \text{then } \dot{z} > 0, \\ \text{if } \dot{y} > 0, & \text{then } x > -\dot{z}/\dot{y}, \\ \text{if } \dot{y} < 0, & \text{then } x < -\dot{z}/\dot{y}. \end{cases}$$

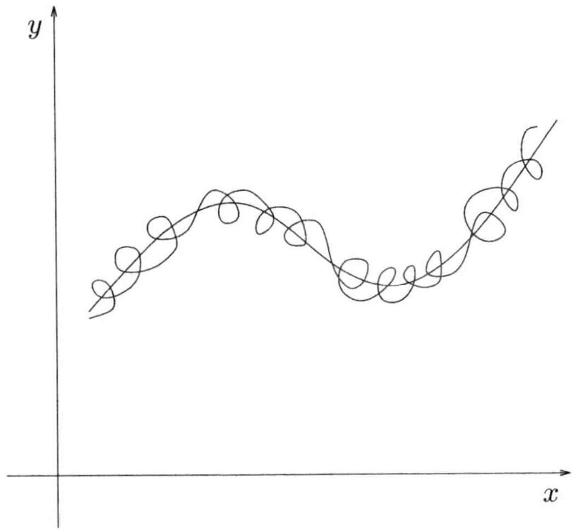

Fig. 7. Legendrian C^0-approximation via Lagrangian projection.

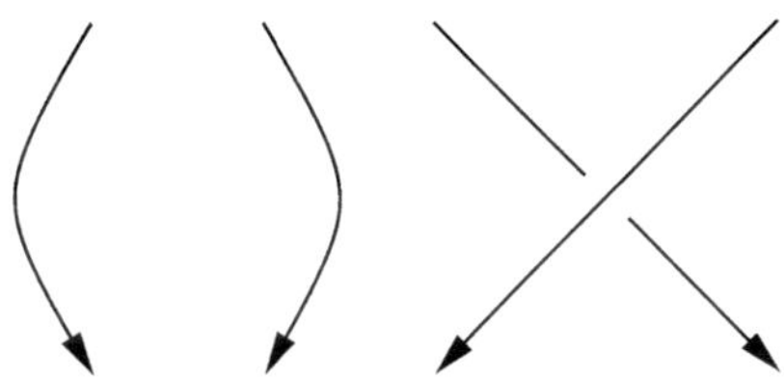

Fig. 8. Impossible front projections of positively transverse curve.

The first statement says that there are no vertical tangencies oriented downwards in the front projection. The second statement says in particular that for $\dot{y} > 0$ and $\dot{z} < 0$ we have $x > 0$; the third, that for $\dot{y} < 0$ and $\dot{z} < 0$ we have $x < 0$. This implies that the situations shown in Figure 8 are not possible in the front projection of a positively transverse curve. I leave it to the reader to check that all other oriented crossings are possible in the front projection of a positively transverse curve, and that any curve in the (y, z)-plane without the forbidden crossing or downward vertical tangencies admits a lift to a positively transverse curve.

EXAMPLE 2.49. Figure 9 shows the front projection of a positively transverse trefoil knot.

PROOF OF THEOREM 2.44 (*Transverse case*). By the Legendrian case of this theorem, the given knot γ can be C^0-approximated by a Legendrian knot γ_1. By Example 2.29, a neighbourhood of γ_1 in (M, ξ) looks like a solid torus $S^1 \times D^2$ with contact structure $\cos\theta\, dx - \sin\theta\, dy = 0$, where $\gamma_1 \equiv S^1 \times \{0\}$. Then the curve

$$\gamma_2(t) = (\theta = t, x = \delta \sin t, y = \delta \cos t), \quad t \in [0, 2\pi],$$

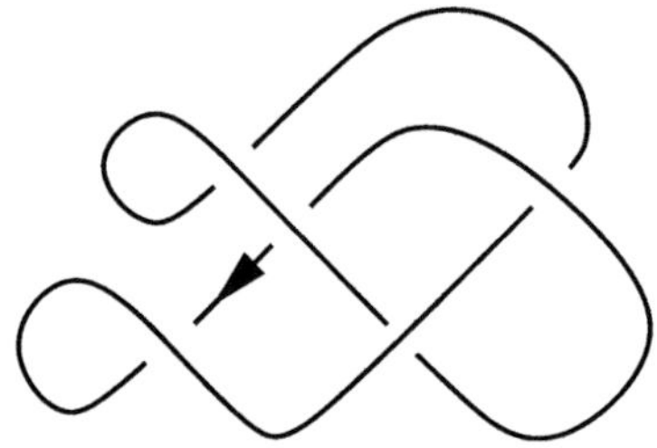

Fig. 9. Front projection of a positively transverse trefoil knot.

is a positively (respectively negatively) transverse knot if $\delta > 0$ (respectively < 0). By choosing $|\delta|$ small we obtain as good a C^0-approximation of γ_1 and hence of γ as we wish. □

3. Contact structures on 3-manifolds

Here is the main theorem proved in this section:

THEOREM 3.1 (Lutz–Martinet). *Every closed, orientable 3-manifold admits a contact structure in each homotopy class of tangent 2-plane fields.*

In Section 3.2 I present what is essentially J. Martinet's [90] proof of the existence of a contact structure on every 3-manifold. This construction is based on a surgery description of 3-manifolds due to R. Lickorish and A. Wallace. For the key step, showing how to extend over a solid torus certain contact forms defined near the boundary of that torus (which then makes it possible to perform Dehn surgeries), we use an approach due to W. Thurston and H. Winkelnkemper; this allows us to simplify Martinet's proof slightly.

In Section 3.3 we show that every orientable 3-manifold is parallelisable and then build on this to classify (co-)oriented tangent 2-plane fields up to homotopy.

In Section 3.4 we study the so-called Lutz twist, a topologically trivial Dehn surgery on a contact manifold (M, ξ) which yields a contact structure ξ' on M that is not homotopic (as 2-plane field) to ξ. We then complete the proof of the main theorem stated above. These results are contained in R. Lutz's thesis [84]. Of Lutz's published work, [83] only deals with the 3-sphere (and is only an announcement); [85] also deals with a more restricted problem. I learned the key steps of the construction from an exposition given in V. Ginzburg's thesis [50]. I have added proofs of many topological details that do not seem to have appeared in a readily accessible source before.

In Section 3.5 I indicate two further proofs for the existence of contact structures on every 3-manifold (and provide references to others). The one by Thurston and Winkelnkemper uses a description of 3-manifolds as open books due to J. Alexander; the crucial idea in their proof is the one we also use to simplify Martinet's argument. Indeed, my discussion of the Lutz twist in the present section owes more to the paper by Thurston–Winkelnkemper than to any other reference. The second proof, by J. Gonzalo, is based on a branched cover description of 3-manifolds found by H. Hilden, J. Montesinos and T. Thickstun. This branched cover description also yields a very simple geometric proof that every orientable 3-manifold is parallelisable.

In Section 3.6 we discuss the fundamental dichotomy between tight and overtwisted contact structures, introduced by Eliashberg, as well as the relation of these types of contact structures with the concept of symplectic fillability. The chapter concludes in Section 3.7 with a survey of classification results for contact structures on 3-manifolds.

But first we discuss, in Section 3.1, an invariant of transverse knots in $\mathbb{R}^3$ with its standard contact structure. This invariant will be an ingredient in the proof of the Lutz–Martinet theorem, but is also of independent interest.

I do not feel embarrassed to use quite a bit of machinery from algebraic and differential topology in this chapter. However, I believe that nothing that cannot be found in such standard texts as [14,77,95] is used without proof or an explicit reference.

Throughout this third section, M denotes a closed, orientable 3-manifold.

3.1. *An invariant of transverse knots*

Although the invariant in question can be defined for transverse knots in arbitrary contact manifolds (provided the knot is homologically trivial), for the sake of clarity I restrict attention to transverse knots in $\mathbb{R}^3$ with its standard contact structure $\xi_0 = \ker(dz + x\,dy)$. This will be sufficient for the proof of the Lutz–Martinet theorem. In Section 3.6 I say a few words about the general situation and related invariants for Legendrian knots.

Thus, let γ be a transverse knot in $(\mathbb{R}^3, \xi_0)$. Push γ a little in the direction of ∂_x—notice that this is a nowhere zero vector field contained in ξ_0, and in particular transverse to γ—to obtain a knot γ'. An orientation of γ induces an orientation of γ'. The orientation of $\mathbb{R}^3$ is given by $dx \wedge dy \wedge dz$.

DEFINITION 3.2. The *self-linking number* $l(\gamma)$ of the transverse knot γ is the linking number of γ and γ'.

Notice that this definition is independent of the choice of orientation of γ (but it changes sign if the orientation of $\mathbb{R}^3$ is reversed). Furthermore, in place of ∂_x we could have chosen any nowhere zero vector field X in ξ_0 to define $l(\gamma)$: The difference between the self-linking number defined via ∂_x and that defined via X is the degree of the map $\gamma \to S^1$ given by associating to a point on γ the angle between ∂_x and X. This degree is computed with the induced map $\mathbb{Z} \cong H_1(\gamma) \to H_1(S^1) \cong Z$. But the map $\gamma \to S^1$ factors through $\mathbb{R}^3$, so the induced homomorphism on homology is the zero homomorphism.

Observe that $l(\gamma)$ is an invariant under isotopies of γ within the class of transverse knots.

We now want to compute $l(\gamma)$ from the front projection of γ. Recall that the *writhe* of an oriented knot diagram is the signed number of self-crossings of the diagram, where the sign of the crossing is given in Figure 10.

LEMMA 3.3. *The self-linking number $l(\gamma)$ of a transverse knot is equal to the writhe $w(\gamma)$ of its front projection.*

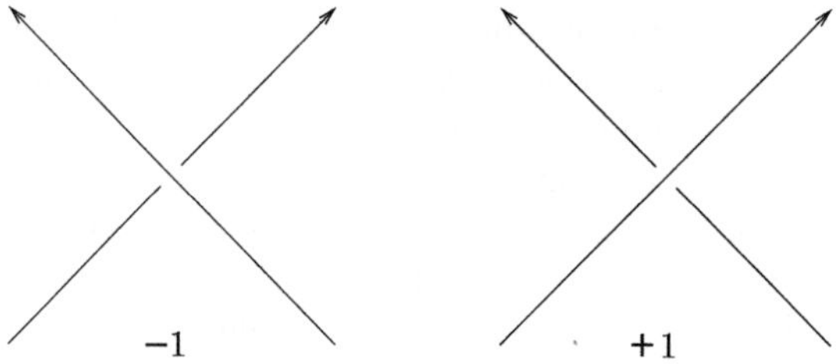

Fig. 10. Signs of crossings in a knot diagram.

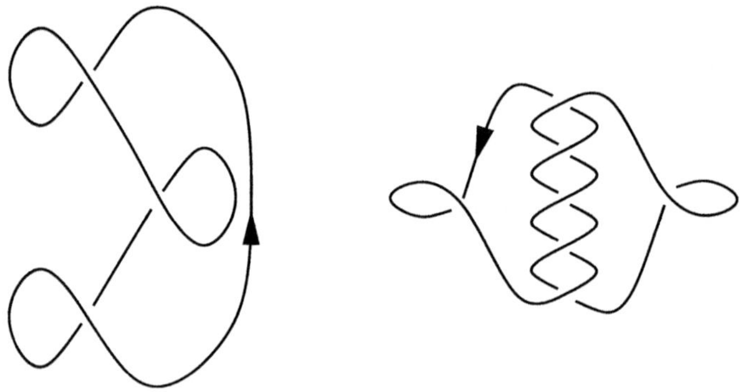

Fig. 11. Transverse knots with self-linking number ± 3.

PROOF. Let γ' be the push-off of γ as described. Observe that each crossing of the front projection of γ contributes a crossing of γ' underneath γ of the corresponding sign. Since the linking number of γ and γ' is equal to the signed number of times that γ' crosses underneath γ (cf. [98, p. 37]), we find that this linking number is equal to the signed number of self-crossings of γ, that is, $l(\gamma) = w(\gamma)$. □

PROPOSITION 3.4. *Every self-linking number is realised by a transverse link in standard* $\mathbb{R}^3$.

PROOF. Figure 11 shows front projections of positively transverse knots (cf. Section 2.6.2) with self-linking number ± 3. From that the construction principle for realising any odd integer should be clear. With a two component link any even integer can be realised. □

REMARK 3.5. It is no accident that I do not give an example of a transverse knot with *even* self-linking number. By a theorem of Eliashberg [26, Proposition 2.3.1] that relates $l(\gamma)$ to the Euler characteristic of a Seifert surface S for γ and the signed number of singular points of the characteristic foliation S_ξ, the self-linking number $l(\gamma)$ of a knot can only take *odd* values.

3.2. *Martinet's construction*

According to Lickorish [81] and Wallace [103] M can be obtained from S^3 by Dehn surgery along a link of 1-spheres. This means that we have to remove a disjoint set of embedded solid tori $S^1 \times D^2$ from S^3 and glue back solid tori with suitable identification by a diffeomorphism along the boundaries $S^1 \times S^1$. The effect of such a surgery (up to diffeomorphism of the resulting manifold) is completely determined by the induced map in homology

$$H_1(S^1 \times \partial D^2) \to H_1(S^1 \times \partial D^2),$$
$$\mathbb{Z} \oplus \mathbb{Z} \to \mathbb{Z} \oplus \mathbb{Z},$$

which is given by a unimodular matrix $\left(\begin{smallmatrix} n & q \\ m & p \end{smallmatrix}\right) \in \mathrm{GL}(2, \mathbb{Z})$. Hence, denoting coordinates in $S^1 \times S^1$ by (θ, φ), we may always assume the identification maps to be of the form

$$\begin{pmatrix} \theta \\ \varphi \end{pmatrix} \mapsto \begin{pmatrix} n & q \\ m & p \end{pmatrix} \begin{pmatrix} \theta \\ \varphi \end{pmatrix}.$$

The curves μ and λ on $\partial(S^1 \times D^2)$ given respectively by $\theta = 0$ and $\varphi = 0$ are called *meridian* and *longitude*. We keep the same notation μ, λ for the homology classes these curves represent. It turns out that the diffeomorphism type of the surgered manifold is completely determined by the class $p\mu + q\lambda$, which is the class of the curve that becomes homotopically trivial in the surgered manifold (cf. [98, p. 28]). In fact, the Dehn surgery is completely determined by the surgery coefficient p/q, since the diffeomorphism of $\partial(S^1 \times D^2)$ given by $(\lambda, \mu) \mapsto (\lambda, -\mu)$ extends to a diffeomorphism of the solid torus that we glue back.

Similarly, the diffeomorphism given by $(\lambda, \mu) \mapsto (\lambda + k\mu, \mu)$ extends. By such a change of longitude in $S^1 \times D^2 \subset M$, which amounts to choosing a different trivialisation of the normal bundle (i.e. *framing*) of $S^1 \times \{0\} \subset M$, the gluing map is changed to $\left(\begin{smallmatrix} n & q \\ m-kn & p-kq \end{smallmatrix}\right)$. By a change of longitude in the solid torus that we glue back, the gluing map is changed to $\left(\begin{smallmatrix} n+kq & q \\ m+kp & p \end{smallmatrix}\right)$. Thus, a Dehn surgery is a so-called handle surgery (or 'honest surgery' or simply 'surgery') if and only if the surgery coefficient is an integer, that is, $q = \pm 1$. For in exactly this case we may assume $\left(\begin{smallmatrix} n & q \\ m & p \end{smallmatrix}\right) = \left(\begin{smallmatrix} 0 & 1 \\ 1 & 0 \end{smallmatrix}\right)$, and the surgery is given by cutting out $S^1 \times D^2$ and gluing back $S^1 \times D^2$ with the identity map

$$\partial(D^2 \times S^1) \to \partial(S^1 \times D^2).$$

The theorem of Lickorish and Wallace remains true if we only allow handle surgeries. However, this assumption does not entail any great simplification of the existence proof for contact structures, so we shall work with general Dehn surgeries.

Our aim in this section is to use this topological description of 3-manifolds for a proof of the following theorem, first proved by Martinet [90]. The proof presented here is in spirit the one given by Martinet, but, as indicated in the introduction to this third section, amalgamated with ideas of Thurston and Winkelnkemper [101], whose proof of the same theorem we shall discuss later.

THEOREM 3.6 (Martinet). *Every closed, orientable 3-manifold M admits a contact structure.*

In view of the theorem of Lickorish and Wallace and the fact that S^3 admits a contact structure, Martinet's theorem is a direct consequence of the following result.

THEOREM 3.7. *Let ξ_0 be a contact structure on a 3-manifold M_0. Let M be the manifold obtained from M_0 by a Dehn surgery along a knot K. Then M admits a contact structure ξ which coincides with ξ_0 outside the neighbourhood of K where we perform surgery.*

PROOF. By Theorem 2.44 we may assume that K is positively transverse to ξ_0. Then, by the contact neighbourhood theorem (Example 2.33), we can find a tubular neighbourhood of K diffeomorphic to $S^1 \times D^2(\delta_0)$, where K is identified with $S^1 \times \{0\}$ and $D^2(\delta_0)$ denotes a disc of radius δ_0, such that the contact structure ξ_0 is given as the kernel of $d\bar{\theta} + \bar{r}^2\,d\bar{\varphi}$, with $\bar{\theta}$ denoting the S^1-coordinate and $(\bar{r}, \bar{\varphi})$ polar coordinates on $D^2(\delta_0)$.

Now perform a Dehn surgery along K defined by the unimodular matrix $\left(\begin{smallmatrix} n & q \\ m & p \end{smallmatrix}\right)$. This corresponds to cutting out $S^1 \times D^2(\delta_1) \subset S^1 \times D^2(\delta_0)$ for some $\delta_1 < \delta_0$ and gluing it back in the way described above.

Write $(\theta; r, \varphi)$ for the coordinates on the copy of $S^1 \times D^2(\delta_1)$ that we want to glue back. Then the contact form $d\bar{\theta} + \bar{r}^2\,d\bar{\varphi}$ given on $S^1 \times D^2(\delta_0)$ pulls back (along $S^1 \times \partial D^2(\delta_1)$) to

$$d(n\theta + q\varphi) + r^2\,d(m\theta + p\varphi).$$

This form is defined on all of $S^1 \times (D^2(\delta_1) - \{0\})$, and to complete the proof it only remains to find a contact form on $S^1 \times D^2(\delta_1)$ that coincides with this form near $S^1 \times \partial D^2(\delta_1)$. It is at this point that we use an argument inspired by the Thurston–Winkelnkemper proof (but which goes back to Lutz).

LEMMA 3.8. *Given a unimodular matrix $\left(\begin{smallmatrix} n & q \\ m & p \end{smallmatrix}\right)$, there is a contact form on $S^1 \times D^2(\delta)$ that coincides with $(n + mr^2)\,d\theta + (q + pr^2)\,d\varphi$ near $r = \delta$ and with $\pm d\theta + r^2\,d\varphi$ near $r = 0$.*

PROOF. We make the ansatz

$$\alpha = h_1(r)\,d\theta + h_2(r)\,d\varphi$$

with smooth functions $h_1(r), h_2(r)$. Then

$$d\alpha = h_1'\,dr \wedge d\theta + h_2'\,dr \wedge d\varphi$$

and

$$\alpha \wedge d\alpha = \begin{vmatrix} h_1 & h_2 \\ h_1' & h_2' \end{vmatrix}\, d\theta \wedge dr \wedge d\varphi.$$

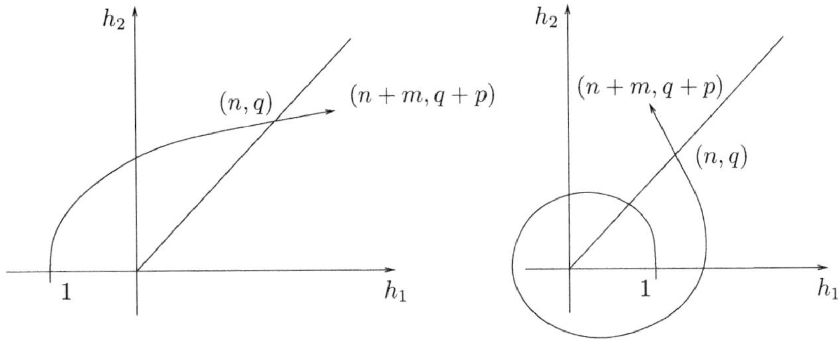

Fig. 12. Dehn surgery.

So to satisfy the contact condition $\alpha \wedge d\alpha \neq 0$, all we have to do is to find a parametrised curve

$$r \mapsto \big(h_1(r), h_2(r)\big), \quad 0 \leqslant r \leqslant \delta,$$

in the plane satisfying the following conditions:

1. $h_1(r) = \pm 1$ and $h_2(r) = r^2$ near $r = 0$,
2. $h_1(r) = n + mr^2$ and $h_2(r) = q + pr^2$ near $r = \delta$,
3. $(h_1(r), h_2(r))$ is never parallel to $(h_1'(r), h_2'(r))$.

Since $np - mq = \pm 1$, the vector (m, p) is not a multiple of (n, q). Figure 12 shows possible solution curves for the two cases $np - mq = \pm 1$.

This completes the proof of the lemma and hence that of Theorem 3.7. □

REMARK 3.9. On S^3 we have the standard contact forms $\alpha_\pm = x\,dy - y\,dx \pm (z\,dt - t\,dz)$ defining opposite orientations (cf. Remark 2.2). Performing the above surgery construction either on $(S^3, \ker\alpha_+)$ or on $(S^3, \ker\alpha_-)$ we obtain both positive and negative contact structures on any given M. The same is true for the Lutz construction that we study in the next two sections. Hence: *A closed oriented 3-manifold admits both a positive and a negative contact structure in each homotopy class of tangent 2-plane fields.*

3.3. *2-plane fields on 3-manifolds*

First we need the following well-known fact.

THEOREM 3.10. *Every closed, orientable 3-manifold M is parallelisable.*

REMARK. The most geometric proof of this theorem can be given based on a structure theorem of Hilden, Montesinos and Thickstun. This will be discussed in Section 3.5.2. Another proof can be found in [76]. Here we present the classical algebraic proof.

PROOF. The main point is to show the vanishing of the second Stiefel–Whitney class $w_2(M) = w_2(TM) \in H^2(M; \mathbb{Z}_2)$. Recall the following facts, which can be found in [14]; for the interpretation of Stiefel–Whitney classes as obstruction classes see also [95].

There are Wu classes $v_i \in H^i(M; \mathbb{Z}_2)$ defined by

$$\langle \mathrm{Sq}^i(u), [M] \rangle = \langle v_i \cup u, [M] \rangle$$

for all $u \in H^{3-i}(M; \mathbb{Z}_2)$, where Sq denotes the Steenrod squaring operations. Since $\mathrm{Sq}^i(u) = 0$ for $i > 3 - i$, the only (potentially) non-zero Wu classes are $v_0 = 1$ and v_1. The Wu classes and the Stiefel–Whitney classes are related by $w_q = \sum_j \mathrm{Sq}^{q-j}(v_j)$. Hence $v_1 = \mathrm{Sq}^0(v_1) = w_1$, which equals zero since M is orientable. We conclude $w_2 = 0$.

Let $V_2(\mathbb{R}^3) = \mathrm{SO}(3)/\mathrm{SO}(1) = \mathrm{SO}(3)$ be the Stiefel manifold of oriented, orthonormal 2-frames in $\mathbb{R}^3$. This is connected, so there exists a section over the 1-skeleton of M of the 2-frame bundle $V_2(TM)$ associated with TM (with a choice of Riemannian metric on M understood[4]). The obstruction to extending this section over the 2-skeleton is equal to w_2, which vanishes as we have just seen. The obstruction to extending the section over all of M lies in $H^3(M; \pi_2(V_2(\mathbb{R}^3)))$, which is the zero group because of $\pi_2(\mathrm{SO}(3)) = 0$.

We conclude that TM has a trivial 2-dimensional sub-bundle ε^2. The complementary 1-dimensional bundle $\lambda = TM/\varepsilon^2$ is orientable and hence trivial since $0 = w_1(TM) = w_1(\varepsilon^2) + w_1(\lambda) = w_1(\lambda)$. Thus $TM = \varepsilon^2 \oplus \lambda$ is a trivial bundle. □

Fix an arbitrary Riemannian metric on M and a trivialisation of the unit tangent bundle $STM \cong M \times S^2$. This sets up a one-to-one correspondence between the following sets, where all maps, homotopies, etc. are understood to be smooth.

- Homotopy classes of unit vector fields X on M,
- Homotopy classes of (co-)oriented 2-plane distributions ξ in TM,
- Homotopy classes of maps $f: M \to S^2$.

(I use the term '2-plane distribution' synonymously with '2-dimensional sub-bundle of the tangent bundle'.) Let ξ_1, ξ_2 be two arbitrary 2-plane distributions (always understood to be cooriented). By elementary obstruction theory there is an obstruction

$$d^2(\xi_1, \xi_2) \in H^2\bigl(M; \pi_2(S^2)\bigr) \cong H^2(M; \mathbb{Z})$$

for ξ_1 to be homotopic to ξ_2 over the 2-skeleton of M and, if $d^2(\xi_1, \xi_2) = 0$ and after homotoping ξ_1 to ξ_2 over the 2-skeleton, an obstruction (which will depend, in general, on that first homotopy)

$$d^3(\xi_1, \xi_2) \in H^3\bigl(M; \pi_3(S^2)\bigr) \cong H^3(M; \mathbb{Z}) \cong \mathbb{Z}$$

for ξ_1 to be homotopic to ξ_2 over all of M. (The identification of $H^3(M; \mathbb{Z})$ with $\mathbb{Z}$ is determined by the orientation of M.) However, rather than relying on general obstruction theory, we shall interpret d^2 and d^3 geometrically, which will later allow us to give a

[4] This is not necessary, of course. One may also work with arbitrary 2-frames without reference to a metric. This does not affect the homotopical data.

geometric proof that every homotopy class of 2-plane fields ξ on M contains a contact structure.

The only fact that I want to quote here is that, by the Pontrjagin–Thom construction, homotopy classes of maps $f: M \to S^2$ are in one-to-one correspondence with framed cobordism classes of framed (and oriented) links of 1-spheres in M. The general theory can be found in [14] and [77]; a beautiful and elementary account is given in [94].

For given f, the correspondence is defined by choosing a regular value $p \in S^2$ for f and a positively oriented basis $\mathfrak{b}$ of T_pS^2, and associating with it the oriented framed link $(f^{-1}(p), f^*\mathfrak{b})$, where $f^*\mathfrak{b}$ is the pull-back of $\mathfrak{b}$ under the fibrewise bijective map $Tf: T(f^{-1}(p))^{\perp} \to T_pS^2$. The orientation of $f^{-1}(p)$ is the one which together with the frame $f^*\mathfrak{b}$ gives the orientation of M.

For a given framed link L the corresponding f is defined by projecting a (trivial) disc bundle neighbourhood $L \times D^2$ of L in M onto the fibre $D^2 \cong S^2 - p^*$, where 0 is identified with p and p^* denotes the antipode of p, and sending $M - (L \times D^2)$ to p^*. Notice that the orientations of M and the components of L determine that of the fibre D^2, and hence determine the map f.

Before proceeding to define the obstruction classes d^2 and d^3 we make a short digression and discuss some topological background material which is fairly standard but not contained in our basic textbook references [14] and [77].

3.3.1. *Hopf's Umkehrhomomorphismus* If $f: M^m \to N^n$ is a continuous map between smooth manifolds, one can define a homomorphism $\varphi: H_{n-p}(N) \to H_{m-p}(M)$ on homology classes represented by submanifolds as follows. Given a homology class $[L]_N \in H_{n-p}(N)$ represented by a codimension p submanifold L, replace f by a smooth approximation transverse to L and define $\varphi([L]_N) = [f^{-1}(L)]_M$. This is essentially Hopf's *Umkehrhomomorphismus* [73], except that he worked with combinatorial manifolds of equal dimension and made no assumptions on the homology class. The following theorem, which in spirit is contained in [41], shows that φ is independent of choices (of submanifold L representing a class and smooth transverse approximation to f) and actually a homomorphism of intersection rings. This statement is not as well known as it should be, and I only know of a proof in the literature for the special case where L is a point [60]. In [14] this map is called *transfer map* (more general transfer maps are discussed in [60]), but is only defined indirectly via Poincaré duality (though implicitly the statement of the following theorem is contained in [14], see, for instance, p. 377).

THEOREM 3.11. *Let $f: M^m \to N^n$ be a smooth map between closed, oriented manifolds and $L^{n-p} \subset N^n$ a closed, oriented submanifold of codimension p such that f is transverse to L. Write $u \in H^p(N)$ for the Poincaré dual of $[L]_N$, that is, $u \cap [N] = [L]_N$. Then $[f^{-1}(L)]_M = f^*u \cap [M]$. In other words: If u is Poincaré dual to $[L]_N$, then $f^*u \in H^p(M)$ is Poincaré dual to $[f^{-1}(L)]_M$.*

PROOF. Since f is transverse to L, the differential Tf induces a fibrewise isomorphism between the normal bundles of $f^{-1}(L)$ and L, and we find (closed) tubular neighbourhoods $W \to L$ and $V = f^{-1}(W) \to f^{-1}(L)$ (considered as disc bundles) such that $f: V \to W$ is a fibrewise isomorphism. Write $[V]_0$ and $[W]_0$ for the orientation classes in $H_m(V, V - f^{-1}(L))$ and $H_n(W, W - L)$, respectively. We can identify these homology groups with $H_m(V, \partial V)$ and $H_n(W, \partial W)$, respectively. Let $\tau_W \in H^p(W, \partial W)$ and $\tau_V \in H^p(V, \partial V)$ be the Thom classes of these disc bundles defined by

$$\tau_W \cap [W]_0 = [L]_N,$$
$$\tau_V \cap [V]_0 = \left[f^{-1}(L)\right]_M.$$

Notice that $f^*\tau_W = \tau_V$ since $f : W \to V$ is fibrewise isomorphic and the Thom class of an oriented disc bundle is the unique class whose restriction to each fibre is a positive generator of $H^p(D^p, \partial D^p)$. Writing $i : M \to (M, M - f^{-1}(L))$ and $j : N \to (N, N - L)$ for the inclusion maps we have

$$\left[f^{-1}(L)\right]_M = \tau_V \cap [V]_0 = f^*\tau_W \cap [V]_0 = f^*\tau_W \cap i_*[M],$$

where we identify $H_m(M, M - f^{-1}(L))$ with $H_m(V, V - f^{-1}(L))$ under the excision isomorphism. Then we have further

$$\left[f^{-1}(L)\right]_M = i^* f^*\tau_W \cap [M] = f^* j^*\tau_W \cap [M].$$

So it remains to identify $j^*\tau_W$ as the Poincaré dual u of $[L]_N$. Indeed,

$$j^*\tau_W \cap [N] = \tau_W \cap j_*[N] = \tau_W \cap [W]_0 = [L]_N,$$

where we have used the excision isomorphism between the groups $H_n(W, W - L)$ and $H_n(N, N - L)$. □

3.3.2. *Representing homology classes by submanifolds* We now want to relate elements in $H_1(M; \mathbb{Z})$ to cobordism classes of links in M.

THEOREM 3.12. *Let M be a closed, oriented 3-manifold. Any $c \in H_1(M; \mathbb{Z})$ is represented by an embedded, oriented link (of 1-spheres) L_c in M. Two links L_0, L_1 represent the same class $[L_0] = [L_1]$ if and only if they are cobordant in M, that is, there is an embedded, oriented surface S in $M \times [0, 1]$ with*

$$\partial S = L_1 \sqcup (-L_0) \subset M \times \{1\} \sqcup M \times \{0\},$$

where $\sqcup$ denotes disjoint union.

PROOF. Given $c \in H_1(M; \mathbb{Z})$, set $u = PD(c) \in H^2(M; \mathbb{Z})$, where PD denotes the Poincaré duality map from homology to cohomology. There is a well-known isomorphism

$$H^2(M; \mathbb{Z}) \cong \left[M, K(\mathbb{Z}, 2)\right] = [M, \mathbb{C}P^\infty],$$

where brackets denote homotopy classes of maps (cf. [14, VII.12]). So u corresponds to a homotopy class of maps $[f] : M \to \mathbb{C}P^\infty$ such that $f^*u_0 = u$, where u_0 is the positive generator of $H^2(\mathbb{C}P^\infty)$ (that is, the one that pulls back to the Poincaré dual of $[\mathbb{C}P^{k-1}]_{\mathbb{C}P^k}$ under the natural inclusion $\mathbb{C}P^k \subset \mathbb{C}P^\infty$). Since $\dim M = 3$, any map $f: M \to \mathbb{C}P^\infty$ is homotopic to a smooth map $f_1: M \to \mathbb{C}P^1$. Let p be a regular value of f_1. Then

$$PD(c) = u = f_1^* u_0 = f_1^* PD[p] = PD\big[f_1^{-1}(p)\big]$$

by our discussion above, and hence $c = [f_1^{-1}(p)]$. So $L_c = f_1^{-1}(p)$ is the desired link.

It is important to note that in spite of what we have just said it is not true that $[M, \mathbb{C}P^\infty] = [M, \mathbb{C}P^1]$, since a map $F: M \times [0,1] \to \mathbb{C}P^\infty$ with $F(M \times \{0,1\}) \subset \mathbb{C}P^1$ is not, in general, homotopic rel$(M \times \{0,1\})$ to a map into $\mathbb{C}P^1$. However, we do have $[M, \mathbb{C}P^\infty] = [M, \mathbb{C}P^2]$.

If two links L_0, L_1 are cobordant in M, then clearly

$$[L_0] = [L_1] \in H_1\big(M \times [0,1]; \mathbb{Z}\big) \cong H_1(M; \mathbb{Z}).$$

For the converse, suppose we are given two links $L_0, L_1 \subset M$ with $[L_0] = [L_1]$. Choose arbitrary framings for these links and use this, as described at the beginning of this section, to define smooth maps $f_0, f_1: M \to S^2$ with common regular value $p \in S^2$ such that $f_i^{-1}(p) = L_i$, $i = 0, 1$. Now identify S^2 with the standardly embedded $\mathbb{C}P^1 \subset \mathbb{C}P^2$. Let $P \subset \mathbb{C}P^2$ be a second copy of $\mathbb{C}P^1$, embedded in such a way that $[P]_{\mathbb{C}P^2} = [\mathbb{C}P^1]_{\mathbb{C}P^2}$ and P intersects $\mathbb{C}P^1$ transversely in p only. This is possible since $\mathbb{C}P^1 \subset \mathbb{C}P^2$ has self-intersection one. Then the maps f_0, f_1, regarded as maps into $\mathbb{C}P^2$, are transverse to P and we have $f_i^{-1}(P) = L_i$, $i = 0, 1$. Hence

$$f_i^* u_0 = f_i^*\big(PD[P]_{\mathbb{C}P^2}\big) = PD\big[f_i^{-1}(P)\big]_M = PD[L_i]_M$$

is the same for $i = 0$ or 1, and from the identification

$$\begin{aligned}\big[M, \mathbb{C}P^2\big] &\stackrel{\cong}{\to} H^2(M, \mathbb{Z}),\\ [f] &\mapsto f^* u_0\end{aligned}$$

we conclude that f_0 and f_1 are homotopic as maps into $\mathbb{C}P^2$.

Let $F: M \times [0,1] \to \mathbb{C}P^2$ be a homotopy between f_0 and f_1, which we may assume to be constant near 0 and 1. This F can be smoothly approximated by a map $F': M \times [0,1] \to \mathbb{C}P^2$ which is transverse to P and coincides with F near $M \times 0$ and $M \times 1$ (since there the transversality condition was already satisfied). In particular, F' is still a homotopy between f_0 and f_1, and $S = (F')^{-1}(P)$ is a surface with the desired property $\partial S = L_1 \sqcup (-L_0)$. □

Notice that in the course of this proof we have observed that cobordism classes of links in M (equivalently, classes in $H_1(M; \mathbb{Z})$) correspond to homotopy classes of maps $M \to \mathbb{C}P^2$, whereas framed cobordism classes of framed links correspond to homotopy classes of maps $M \to \mathbb{C}P^1$.

By forming the connected sum of the components of a link representing a certain class in $H_1(M; \mathbb{Z})$, one may actually always represent such a class by a link with only one component, that is, a knot.

3.3.3. *Framed cobordisms* We have seen that if $L_1, L_2 \subset M$ are links with $[L_1] = [L_2] \in H_1(M; \mathbb{Z})$, then L_1 and L_2 are cobordant in M. In general, however, a given framing on L_1 and L_2 does not extend over the cobordism. The following observation will be useful later on.

Write (S^1, n) for a contractible loop in M with framing $n \in \mathbb{Z}$ (by which we mean that S^1 and a second copy of S^1 obtained by pushing it away in the direction of one of the vectors in the frame have linking number n). When writing $L = L' \sqcup (S^1, n)$ it is understood that (S^1, n) is not linked with any component of L'.

Suppose we have two framed links $L_0, L_1 \subset M$ with $[L_0] = [L_1]$. Let $S \subset M \times [0, 1]$ be an embedded surface with

$$\partial S = L_1 \sqcup (-L_0) \subset M \times \{1\} \sqcup M \times \{0\}.$$

With D^2 a small disc embedded in S, the framing of L_1 and L_2 in M extends to a framing of $S - D^2$ in $M \times [0, 1]$ (since $S - D^2$ deformation retracts to a 1-dimensional complex containing L_0 and L_1, and over such a complex an orientable 2-plane bundle is trivial). Now we embed a cylinder $S^1 \times [0, 1]$ in $M \times [0, 1]$ such that

$$S^1 \times [0, 1] \cap M \times \{0\} = \emptyset,$$
$$S^1 \times [0, 1] \cap M \times \{1\} = S^1 \times \{1\},$$

and

$$S^1 \times [0, 1] \cap \left(S - D^2\right) = S^1 \times \{0\} = \partial D^2,$$

see Figure 13. This shows that L_0 is framed cobordant in M to $L_1 \sqcup (S^1, n)$ for suitable $n \in \mathbb{Z}$.

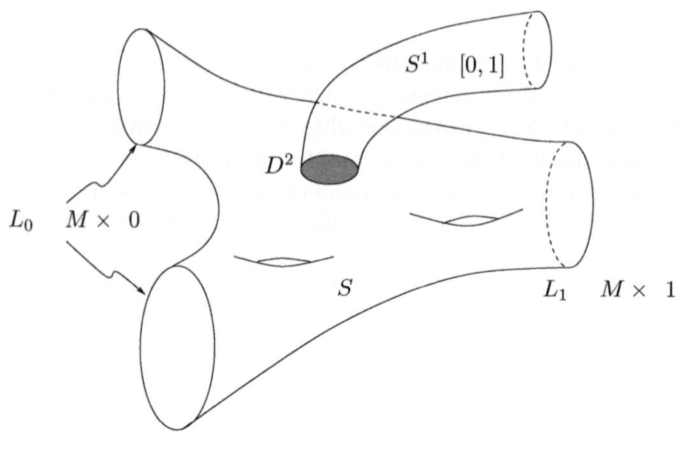

Fig. 13. The framed cobordism between L_0 and $L_1 \sqcup (S^1, n)$.

3.3.4. *Definition of the obstruction classes* We are now in a position to define the obstruction classes d^2 and d^3. With a choice of Riemannian metric on M and a trivialisation of STM understood, a 2-plane distribution ξ on M defines a map $f_\xi : M \to S^2$ and hence an oriented framed link L_ξ as described above. Let $[L_\xi] \in H_1(M; \mathbb{Z})$ be the homology class represented by L_ξ. This only depends on the homotopy class of ξ, since under homotopies of ξ or choice of different regular values of f_ξ the cobordism class of L_ξ remains invariant. We define

$$d^2(\xi_1, \xi_2) = PD[L_{\xi_1}] - PD[L_{\xi_2}].$$

With this definition d^2 is clearly additive, that is,

$$d^2(\xi_1, \xi_2) + d^2(\xi_2, \xi_3) = d^2(\xi_1, \xi_3).$$

The following lemma shows that d^2 is indeed the desired obstruction class.

LEMMA 3.13. *The* 2*-plane distributions* ξ_1 *and* ξ_2 *are homotopic over the* 2*-skeleton* $M^{(2)}$ *of* M *if and only if* $d^2(\xi_1, \xi_2) = 0$.

PROOF. Suppose $d^2(\xi_1, \xi_2) = 0$, that is, $[L_{\xi_1}] = [L_{\xi_2}]$. By Theorem 3.12 we find a surface S in $M \times [0, 1]$ with

$$\partial S = L_{\xi_2} \sqcup (-L_{\xi_1}) \subset M \times \{1\} \sqcup M \times \{0\}.$$

From the discussion on framed cobordism above we know that for suitable $n \in \mathbb{Z}$ we find a *framed* surface S' in $M \times [0, 1]$ such that

$$\partial S' = \left(L_{\xi_2} \sqcup \left(S^1, n\right)\right) \sqcup (-L_{\xi_1}) \subset M \times \{1\} \sqcup M \times \{0\}$$

as framed manifolds.

Hence ξ_1 is homotopic to a 2-plane distribution ξ_1' such that $L_{\xi_1'}$ and L_{ξ_2} differ only by one contractible framed loop (not linked with any other component). Then the corresponding maps f_1', f_2 differ only in a neighbourhood of this loop, which is contained in a 3-ball, so f_1' and f_2 (and hence ξ_1' and ξ_2) agree over the 2-skeleton.

Conversely, if ξ_1 and ξ_2 are homotopic over $M^{(2)}$, we may assume $\xi_1 = \xi_2$ on $M - D^3$ for some embedded 3-disc $D^3 \subset M$ without changing $[L_{\xi_1}]$ and $[L_{\xi_2}]$. Now $[L_{\xi_1}] = [L_{\xi_2}]$ follows from $H_1(D^3, S^2) = 0$. □

REMARK 3.14. By [99, §37] the obstruction to homotopy between ξ and ξ_0 (corresponding to the constant map $f_{\xi_0} : M \to S^2$) over the 2-skeleton of M is given by $f_\xi^* u_0$, where u_0 is the positive generator of $H^2(S^2; \mathbb{Z})$. So $u_0 = PD[p]$ for any $p \in S^2$, and taking p to be a regular value of f_ξ we have

$$f_\xi^* u_0 = f_\xi^* PD[p] = PD\bigl[f_\xi^{-1}(p)\bigr] = PD[L_\xi] = d^2(\xi, \xi_0).$$

This gives an alternative way to see that our geometric definition of d^2 does indeed coincide with the class defined by classical obstruction theory.

Now suppose $d^2(\xi_1, \xi_2) = 0$. We may then assume that $\xi_1 = \xi_2$ on $M - \operatorname{int}(D^3)$, and we define $d^3(\xi_1, \xi_2)$ to be the Hopf invariant $H(f)$ of the map $f : S^3 \to S^2$ defined as $f_1 \circ \pi_+$ on the upper hemisphere and $f_2 \circ \pi_-$ on the lower hemisphere, where π_+, π_- are the orthogonal projections of the upper respectively lower hemisphere onto the equatorial disc, which we identify with $D^3 \subset M$. Here, given an orientation of M, we orient S^3 in such a way that π_+ is orientation-preserving and π_- orientation-reversing; the orientation of S^2 is inessential for the computation of $H(f)$. Recall that $H(f)$ is defined as the linking number of the preimages of two distinct regular values of a smooth map homotopic to f. Since the Hopf invariant classifies homotopy classes of maps $S^3 \to S^2$ (it is in fact an isomorphism $\pi_3(S^2) \to \mathbb{Z}$), this is a suitable definition for the obstruction class d^3. Moreover, the homomorphism property of $H(f)$ and the way addition in $\pi_3(S^2)$ is defined entail the additivity of d^3 analogous to that of d^2.

For $M = S^3$ there is another way to interpret d^3. Oriented 2-plane distributions on M correspond to sections of the bundle associated to TM with fibre $\mathrm{SO}(3)/\mathrm{U}(1)$, hence to maps $M \to \mathrm{SO}(3)/\mathrm{U}(1) \cong S^2$ since TM is trivial. Similarly, almost complex structures on D^4 correspond to maps $D^4 \to \mathrm{SO}(4)/\mathrm{U}(2) \cong \mathrm{SO}(3)/\mathrm{U}(1)$ (cf. [61] for this isomorphism). A cooriented 2-plane distribution on M can be interpreted as a triple (X, ξ, J), where X is a vector field transverse to ξ defining the coorientation, and J a complex structure on ξ defining the orientation. Such a triple is called an *almost contact structure*. (This notion generalises to higher (odd) dimensions, and by Remark 2.3 every *cooriented* contact structure induces an almost contact structure, and in fact a unique one up to homotopy as follows from the result cited in that remark.) Given an almost contact structure (X, ξ, J) on S^3, one defines an almost complex structure $\tilde{J}$ on $TD^4|S^3$ by $\tilde{J}|\xi = J$ and $\tilde{J}N = X$, where N denotes the outward normal vector field. So there is a canonical way to identify homotopy classes of almost contact structures on S^3 with elements of $\pi_3(\mathrm{SO}(3)/\mathrm{U}(1)) \cong \mathbb{Z}$ such that the value zero corresponds to the almost contact structure that extends as almost complex structure over D^4.

3.4. *Let's twist again*

Consider a 3-manifold M with cooriented contact structure ξ and an oriented 1-sphere $K \subset M$ embedded transversely to ξ such that the positive orientation of K coincides with the positive coorientation of ξ. Then in suitable local coordinates we can identify K with $S^1 \times \{0\} \subset S^1 \times D^2$ such that $\xi = \ker(d\theta + r^2\, d\varphi)$ and ∂_θ corresponds to the positive orientation of K (see Example 2.33). Strictly speaking, if, as we shall always assume, S^1 is parametrised by $0 \leqslant \theta \leqslant 2\pi$, then this formula for ξ holds on $S^1 \times D^2(\delta)$ for some,

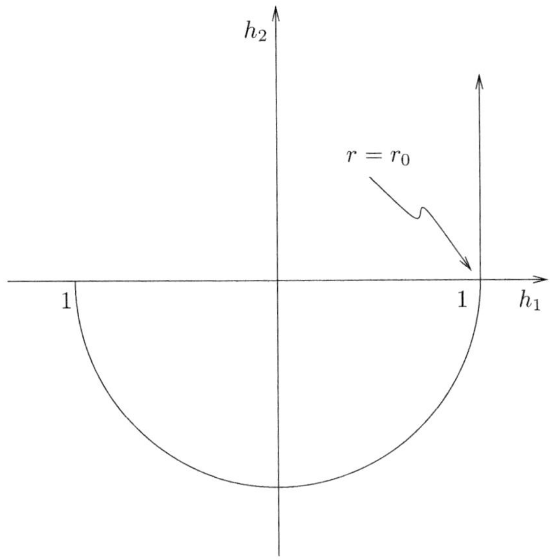

Fig. 14. Lutz twist.

possibly small, $\delta > 0$. However, to simplify notation we usually work with $S^1 \times D^2$ as local model.

We say that ξ' is obtained from ξ by a *Lutz twist* along K and write $\xi' = \xi^K$ if on $S^1 \times D^2$ the new contact structure ξ' is defined by

$$\xi' = \ker\bigl(h_1(r)\, d\theta + h_2(r)\, d\varphi\bigr)$$

with $(h_1(r), h_2(r))$ as in Figure 14, and ξ' coincides with ξ outside $S^1 \times D^2$.

More precisely, $(h_1(r), h_2(r))$ is required to satisfy the conditions

1. $h_1(r) = -1$ and $h_2(r) = -r^2$ near $r = 0$,
2. $h_1(r) = 1$ and $h_2(r) = r^2$ near $r = 1$,
3. $(h_1(r), h_2(r))$ is never parallel to $(h_1'(r), h_2'(r))$.

This is the same as applying the construction of Section 3.2 to the topologically trivial Dehn surgery given by the matrix $\left(\begin{smallmatrix}-1 & 0\\ 0 & -1\end{smallmatrix}\right)$.

It will be useful later on to understand more precisely the behaviour of the map $f_{\xi'}: M \to S^2$. For the definition of this map we assume—this assumption will be justified below—that on $S^1 \times D^2$ the map f_ξ was defined in terms of the standard metric $d\theta^2 + du^2 + dv^2$ (with u, v Cartesian coordinates on D^2 corresponding to the polar coordinates r, φ) and the trivialisation $\partial_\theta, \partial_u, \partial_v$ of $T(S^1 \times D^2)$. Since ξ' is spanned by ∂_r and $h_2(r)\, \partial_\theta - h_1(r)\, \partial_\varphi$ (respectively ∂_u, ∂_v for $r = 0$), a vector positively orthogonal to ξ' is given by

$$h_1(r)\, \partial_\theta + h_2(r)\, \partial_\varphi,$$

which makes sense even for $r = 0$. Observe that the ratio $h_1(r)/h_2(r)$ (for $h_2(r) \neq 0$) is a strictly monotone decreasing function since by the third condition above we have

$$(h_1/h_2)' = (h_1'h_2 - h_1h_2')/h_2^2 < 0.$$

This implies that any value on S^2 other than $(1,0,0)$ (corresponding to ∂_θ) is regular for the map $f_{\xi'}$; the value $(1,0,0)$ is attained along the torus $\{r = r_0\}$, with $r_0 > 0$ determined by $h_2(r_0) = 0$, and hence not regular.

If $S^1 \times D^2$ is endowed with the orientation defined by the volume form $d\theta \wedge r\,dr \wedge d\varphi = d\theta \wedge du \wedge dv$ (so that ξ and ξ' are positive contact structures) and $S^2 \subset \mathbb{R}^3$ is given its 'usual' orientation defined by the volume form $x\,dy \wedge dz + y\,dz \wedge dx + z\,dx \wedge dy$, then

$$f_{\xi'}^{-1}(-1,0,0) = S^1 \times \{0\}$$

with orientation given by $-\partial_\theta$, since $f_{\xi'}$ maps the slices $\{\theta\} \times D^2(r_0)$ orientation-reversingly onto S^2.

More generally, for any $p \in S^2 - \{(1,0,0)\}$ the preimage $f_{\xi'}^{-1}(p)$ (inside the domain $\{(\theta, r, \varphi)\colon h_2(r) < 0\} = \{r < r_0\}$) is a circle $S^1 \times \{\mathbf{u}\}$, $\mathbf{u} \in D^2$, with orientation given by $-\partial_\theta$.

We are now ready to show how to construct a contact structure on M in any given homotopy class of 2-plane distributions by starting with an arbitrary contact structure and performing suitable Lutz twists. First we deal with homotopy over the 2-skeleton. One way to proceed would be to prove directly, with notation as above, that $d^2(\xi^K, \xi) = -PD[K]$. However, it is somewhat easier to compute $d^2(\xi^K, \xi)$ in the case where ξ is a trivial 2-plane bundle and the trivialisation of STM is adapted to ξ. Since I would anyway like to present an alternative argument for computing the effect of a Lutz twist on the Euler class of the contact structure, and thus relate $d^2(\xi_1, \xi_2)$ with the Euler classes of ξ_1 and ξ_2, it seems opportune to do this first and use it to show the existence of a contact structure with Euler class zero. In the next section we shall actually discuss a direct geometric proof, due to Gonzalo, of the existence of a contact structure with Euler class zero.

Recall that the Euler class $e(\xi) \in H^2(B;\mathbb{Z})$ of a 2-plane bundle over a complex B (of arbitrary dimension) is the obstruction to finding a nowhere zero section of ξ over the 2-skeleton of B. Since $\pi_i(S^1) = 0$ for $i \geqslant 2$, all higher obstruction groups $H^{i+1}(B; \pi_i(S^1))$ are trivial, so a 2-dimensional orientable bundle ξ is trivial if and only if $e(\xi) = 0$, no matter what the dimension of B.

Now let ξ be an arbitrary cooriented 2-plane distribution on an oriented 3-manifold M. Then $TM \cong \xi \oplus \varepsilon^1$, where ε^1 denotes a trivial line bundle. Hence $w_2(\xi) = w_2(\xi \oplus \varepsilon^1) = w_2(TM) = 0$, and since $w_2(\xi)$ is the mod 2 reduction of $e(\xi)$ we infer that $e(\xi)$ has to be even.

PROPOSITION 3.15. *For any even element $e \in H^2(M;\mathbb{Z})$ there is a contact structure ξ on M with $e(\xi) = e$.*

PROOF. Start with an arbitrary contact structure ξ_0 on M with $e(\xi_0) = e_0$ (which we know to be even). Given any even $e_1 \in H^2(M;\mathbb{Z})$, represent the Poincaré dual of $(e_0 - e_1)/2$ by

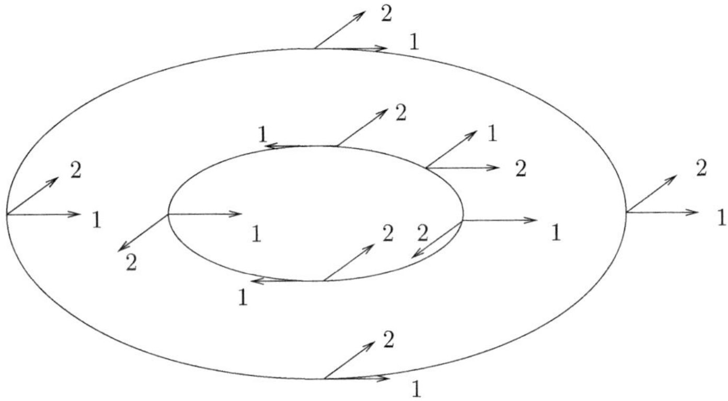

Fig. 15. Effect of Lutz twist on Euler class.

a collection of embedded oriented circles positively transverse to ξ_0. (Here by $(e_0 - e_1)/2$ I mean some class whose double equals $e_0 - e_1$; in the presence of 2-torsion there is of course a choice of such classes.) Choose a section of ξ_0 transverse to the zero section of ξ_0, that is, a vector field in ξ_0 with generic zeros. We may assume that there are no zeros on the curves representing $PD^{-1}(e_0 - e_1)/2$. Now perform a Lutz twist as described above along these curves and call ξ_1 the resulting contact structure. It is easy to see that in the local model for the Lutz twist a constant vector field tangent to ξ_0 along $\partial(S^1 \times D^2(r_0))$ extends to a vector field tangent to ξ_1 over $S^1 \times D^2(r_0)$ with zeros of index $+2$ along $S^1 \times \{0\}$ (Figure 15). So the vector field in ξ_0 extends to a vector field in ξ_1 with new zeros of index $+2$ along the curves representing $PD^{-1}(e_1 - e_0)/2$ (notice that a Lutz twist along a positively transverse knot K turns K into a negatively transverse knot). Since the self-intersection class of M in the total space of a vector bundle is Poincaré dual to the Euler class of that bundle, this proves $e(\xi_1) = e(\xi_0) + e_1 - e_0 = e_1$. □

We now fix a contact structure ξ_0 on M with $e(\xi_0) = 0$ and give M the orientation induced by ξ_0 (i.e. the one for which ξ_0 is a positive contact structure). Moreover, we fix a Riemannian metric on M and define X_0 as the vector field positively orthonormal to ξ_0. Since ξ_0 is a trivial plane bundle, X_0 extends to an orthonormal frame X_0, X_1, X_2, hence a trivialisation of STM, with X_1, X_2 tangent to ξ_0 and defining the orientation of ξ_0. With these choices, ξ_0 corresponds to the constant map $f_{\xi_0} : M \to (1, 0, 0) \in S^2$.

PROPOSITION 3.16. *Let $K \subset M$ be an embedded, oriented circle positively transverse to ξ_0. Then $d^2(\xi_0^K, \xi_0) = -PD[K]$.*

PROOF. Identify a tubular neighbourhood of $K \subset M$ with $S^1 \times D^2$ with framing defined by X_1, and ξ_0 given in this neighbourhood as the kernel of $d\theta + r^2\, d\varphi = d\theta + u\, dv - v\, du$. We may then change the trivialisation X_0, X_1, X_2 by a homotopy, fixed outside $S^1 \times D^2$, such that $X_0 = \partial_\theta$, $X_1 = \partial_u$ and $X_2 = \partial_v$ near K; this does not change the homotopical data of 2-plane distributions computed via the Pontrjagin–Thom construction. Then f_{ξ_0} is no longer constant, but its image still does not contain the point $(-1, 0, 0)$.

Now perform a Lutz twist along $K \times \{0\}$. Our discussion at the beginning of this section shows that $(-1, 0, 0)$ is a regular value of the map $f_\xi : M \to S^2$ associated with $\xi = \xi_0^K$ and $f_\xi^{-1}(-1, 0, 0) = -K$. Hence, by definition of the obstruction class d^2 we have $d^2(\xi_0^K, \xi_0) = -PD[K]$. □

PROOF OF THEOREM 3.1. Let η be a 2-plane distribution on M and ξ_0 the contact structure on M with $e(\xi_0) = 0$ that we fixed earlier on. According to our discussion in Section 3.3.2 and Theorem 2.44, we can find an oriented knot K positively transverse to ξ_0 with $-PD[K] = d^2(\eta, \xi_0)$. Then $d^2(\eta, \xi_0) = d^2(\xi_0^K, \xi_0)$ by the preceding proposition, and therefore $d^2(\xi_0^K, \eta) = 0$.

We may then assume that $\eta = \xi_0^K$ on $M - D^3$, where we choose D^3 so small that ξ_0^K is in Darboux normal form there (and identical with ξ_0). By Proposition 3.4 we can find a link K' in D^3 transverse to ξ_0^K with self-linking number $l(K')$ equal to $d^3(\eta, \xi_0^K)$.

Now perform a Lutz twist of ξ_0^K along each component of K' and let ξ be the resulting contact structure. Since this does not change ξ_0^K over the 2-skeleton of M, we still have $d^2(\xi, \eta) = 0$.

Observe that $f_{\xi_0^K}|_{D^3}$ does not contain the point $(-1, 0, 0) \in S^2$, and—since $f_{\xi_0^K}(D^3)$ is compact—there is a whole neighbourhood $U \subset S^2$ of $(-1, 0, 0)$ not contained in $f_{\xi_0^K}(D^3)$. Let $f : S^3 \to S^2$ be the map used to compute $d^3(\xi, \xi_0^K)$, that is, f coincides on the upper hemisphere with $f_\xi|_{D^3}$ and on the lower hemisphere with $f_{\xi_0^K}|_{D^3}$. By the discussion in Section 3.3, the preimage $f^{-1}(u)$ of any $u \in U - \{(-1, 0, 0)\}$ will be a push-off of $-K'$ determined by the trivialisation of $\xi_0^K|_{D^3} = \xi_0|_{D^3}$. So the linking number of $f^{-1}(u)$ with $f^{-1}(-1, 0, 0)$, which is by definition the Hopf invariant $H(f) = d^3(\xi, \xi_0^K)$, will be equal to $l(K')$. By our choice of K' and the additivity of d^3 this implies $d^3(\xi, \eta) = 0$. So ξ is a contact structure that is homotopic to η as a 2-plane distribution. □

3.5. *Other existence proofs*

Here I briefly summarise the other known existence proofs for contact structures on 3-manifolds, mostly by pointing to the relevant literature. In spirit, most of these proofs are similar to the one by Lutz–Martinet: start with a structure theorem for 3-manifolds and show that the topological construction can be performed compatibly with a contact structure.

3.5.1. *Open books* According to a theorem of Alexander [4], cf. [97], every closed, orientable 3-manifold M admits an *open book decomposition*. This means that there is a link $L \subset M$, called the *binding*, and a fibration $f : M - L \to S^1$, whose fibres are called the *pages*, see Figure 16. It may be assumed that L has a tubular neighbourhood $L \times D^2$ such that f restricted to $L \times (D^2 - \{0\})$ is given by $f(\theta, r, \varphi) = \varphi$, where θ is the coordinate along L and (r, φ) are polar coordinates on D^2.

At the cost of raising the genus of the pages, one may decrease the number of components of L, and in particular one may always assume L to be a knot. Another way

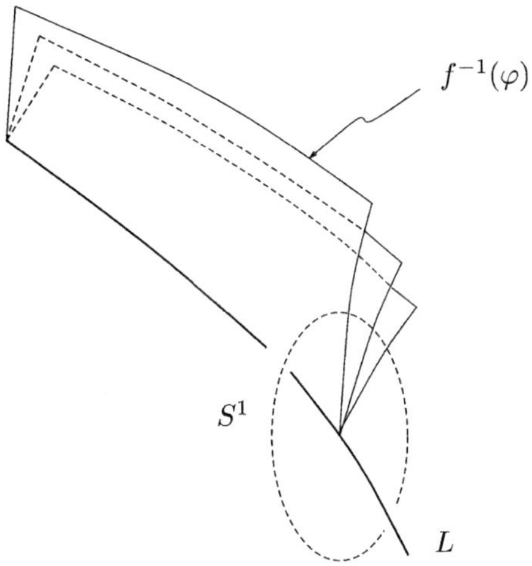

Fig. 16. An open book near the binding.

to think of such an open book is as follows. Start with a surface Σ with boundary $\partial\Sigma = K \cong S^1$ and a self-diffeomorphism h of Σ with $h = \mathrm{id}$ near K. Form the mapping torus $T_h = \Sigma \times [0, 2\pi]/\sim$, where '$\sim$' denotes the identification $(p, 2\pi) \sim (h(p), 0)$. Define a 3-manifold M by

$$M = T_h \cup_{K \times S^1} \left(K \times D^2\right).$$

This M carries by construction the structure of an open book with binding K and pages diffeomorphic to Σ.

Here is a slight variation on a simple argument of Thurston and Winkelnkemper [101] for producing a contact structure on such an open book (and hence on any closed, orientable 3-manifold):

Start with a 1-form β_0 on Σ with $\beta_0 = e^t\, d\theta$ near $\partial\Sigma = K$, where θ denotes the coordinate along K and t is a collar parameter with $K = \{t = 0\}$ and $t < 0$ in the interior of Σ. Then $d\beta_0$ integrates to 2π over Σ by Stokes's theorem. Given any area form ω on Σ (with total area equal to 2π) satisfying $\omega = e^t\, dt \wedge d\theta$ near K, the 2-form $\omega - d\beta_0$ is, by de Rham's theorem, an exact 1-form, say $d\beta_1$, where we may assume $\beta_1 \equiv 0$ near K.

Set $\beta = \beta_0 + \beta_1$. Then $d\beta = \omega$ is an area form (of total area 2π) on Σ and $\beta = e^t\, d\theta$ near K. The set of 1-forms satisfying these two properties is a convex set, so we can find a 1-form (still denoted β) on T_h which has these properties when restricted to the fibre over any $\varphi \in S^1 = [0, 2\pi]/_{0 \sim 2\pi}$. We may (and shall) require that $\beta = e^t\, d\theta$ near ∂T_h.

Now a contact form α on T_h is found by setting $\alpha = \beta + C\, d\varphi$ for a sufficiently large constant $C \in \mathbb{R}^+$, so that in

$$\alpha \wedge d\alpha = (\beta + C\, d\varphi) \wedge d\beta$$

the non-zero term $d\varphi \wedge d\beta = d\varphi \wedge \omega$ dominates. This contact form can be extended to all of M by making the ansatz $\alpha = h_1(r)\,d\theta + h_2(r)\,d\varphi$ on $K \times D^2$, as described in our discussion of the Lutz twist. The boundary conditions in the present situation are, say,

1. $h_1(r) = 2$ and $h_2(r) = r^2$ near $r = 0$,
2. $h_1(r) = e^{1-r}$ and $h_2(r) = C$ near $r = 1$.

Observe that subject to these boundary conditions a curve $(h_1(r), h_2(r))$ can be found that does not pass the h_2-axis (i.e. with $h_1(r)$ never being equal to zero). In the 3-dimensional setting this is not essential (and the Thurston–Winkelnkemper ansatz lacked that feature), but it is crucial when one tries to generalise this construction to higher dimensions. This has recently been worked out by Giroux and J.-P. Mohsen [57]. This, however, is only the easy part of their work. Rather strikingly, they have shown that a converse of this result holds: Given a compact contact manifold of arbitrary dimension, it admits an open book decomposition that is adapted to the contact structure in the way described above. Full details have not been published at the time of writing, but see Giroux's talk [56] at the ICM 2002.

3.5.2. *Branched covers* A theorem of Hilden, Montesinos and Thickstun [63] states that every closed, orientable 3-manifold M admits a branched covering $\pi : M \to S^3$ such that the upstairs branch set is a simple closed curve that bounds an embedded disc. (Moreover, the cover can be chosen 3-fold and simple, i.e. the monodromy representation of $\pi_1(S^3 - K)$, where K is the branching set downstairs (a knot in S^3), represents the meridian of K by a transposition in the symmetric group S_3. This, however, is not relevant for our discussion.)

It follows immediately, as announced in Section 3.3, that every closed, orientable 3-manifold is parallelisable: First of all, S^3 is parallelisable since it carries a Lie group structure (as the unit quaternions, for instance). Given M and a branched covering $\pi : M \to S^3$ as above, there is a 3-ball $D^3 \subset M$ containing the upstairs branch set. Outside of D^3, the covering π is unbranched, so the 3-frame on S^3 can be lifted to a frame on $M - D^3$. The bundle $TM|_{D^3}$ is trivial, so the frame defined along ∂D^3 defines an element of SO(3) (cf. the footnote in the proof of Theorem 3.10). Since $\pi_2(\mathrm{SO}(3)) = 0$, this frame extends over D^3.

In [59], Gonzalo uses this theorem to construct a contact structure on every closed, orientable 3-manifold M, in fact one with zero Euler class: Away from the branching set one can lift the standard contact structure from S^3 (which has Euler class zero: a trivialisation is given by two of the three (quaternionic) Hopf vector fields). A careful analysis of the branched covering map near the branching set then shows how to extend this contact structure over M (while keeping it trivial as 2-plane bundle).

A branched covering construction for higher-dimensional contact manifolds is discussed in [43].

3.5.3. ... *and more* The work of Giroux [52], in which he initiated the study of convex surfaces in contact 3-manifolds, also contains a proof of Martinet's theorem.

An entirely different proof, due to S. Altschuler [5], finds contact structures from solutions to a certain parabolic differential equation for 1-forms on 3-manifolds. Some of these ideas have entered into the more far-reaching work of Eliashberg and Thurston on so-called 'confoliations' [32], that is, 1-forms satisfying $\alpha \wedge d\alpha \geqslant 0$.

3.6. *Tight and overtwisted*

The title of this section describes the fundamental dichotomy of contact structures in dimension 3 that has proved seminal for the development of the field.

In order to motivate the notion of an overtwisted contact structure, as introduced by Eliashberg [21], we consider a 'full' Lutz twist as follows. Let (M, ξ) be a contact 3-manifold and $K \subset M$ a knot transverse to ξ. As before, identify K with $S^1 \times \{0\} \subset S^1 \times D^2 \subset M$ such that $\xi = \ker(d\theta + r^2\, d\varphi)$ on $S^1 \times D^2$. Now define a new contact structure ξ' as in Section 3.4, with $(h_1(r), h_2(r))$ now as in Figure 17, that is, the boundary conditions are now

$$h_1(r) = 1 \quad \text{and} \quad h_2(r) = r^2 \quad \text{for } r \in [0, \varepsilon] \cup [1 - \varepsilon, 1]$$

for some small $\varepsilon > 0$.

LEMMA 3.17. *A full Lutz twist does not change the homotopy class of ξ as a 2-plane field.*

PROOF. Let $(h_1^t(r), h_2^t(r))$, $r, t \in [0, 1]$, be a homotopy of paths such that

1. $h_1^0 \equiv 1$, $h_2^0(r) = r^2$,
2. $h_i^1 \equiv h_i$, $i = 1, 2$,
3. $h_i^t(r) = h_i(r)$ for $r \in [0, \varepsilon] \cup [1 - \varepsilon, 1]$.

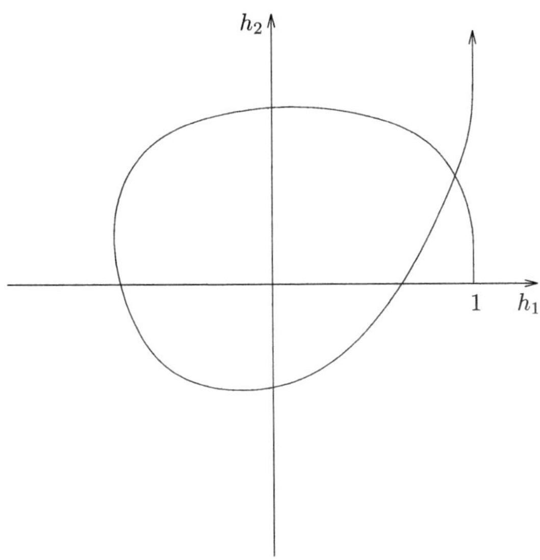

Fig. 17. A full Lutz twist.

Let $\chi : [0, 1] \to \mathbb{R}$ be a smooth function which is identically zero near $r = 0$ and $r = 1$ and $\chi(r) > 0$ for $r \in [\varepsilon, 1 - \varepsilon]$. Then

$$\alpha_t = t(1-t)\chi(r)\,dr + h_1^t(r)\,d\theta + h_2^t(r)\,d\varphi$$

is a homotopy from $\alpha_0 = d\theta + r^2\,d\varphi$ to $\alpha_1 = h_1(r)\,d\theta + h_2(r)\,d\varphi$ through non-zero 1-forms. This homotopy stays fixed near $r = 1$, and so it defines a homotopy between ξ and ξ' as 2-plane fields. □

Let r_0 be the smaller of the two positive radii with $h_2(r_0) = 0$ and consider the embedding

$$\begin{aligned} \phi : D^2(r_0) &\to S^1 \times D^2, \\ (r, \varphi) &\mapsto \big(\theta(r), r, \varphi\big), \end{aligned}$$

where $\theta(r)$ is a smooth function with $\theta(r_0) = 0$, $\theta(r) > 0$ for $0 \leqslant r < r_0$, and $\theta'(r) = 0$ only for $r = 0$. We may require in addition that $\theta(r) = \theta(0) - r^2$ near $r = 0$. Then

$$\phi^*\big(h_1(r)\,d\theta + h_2(r)\,d\varphi\big) = h_1(r)\theta'(r)\,dr + h_2(r)\,d\varphi$$

is a differential 1-form on $D^2(r_0)$ which vanishes only for $r = 0$, and along $\partial D^2(r_0)$ the vector field ∂_φ tangent to the boundary lies in the kernel of this 1-form, see Figure 18. In other words, the contact planes $\ker(h_1(r)\,d\theta + h_2(r)\,d\varphi)$ intersected with the tangent planes to the embedded disc $\phi(D^2(r_0))$ induce a singular 1-dimensional foliation on this disc with the boundary of this disc as closed leaf and precisely one singular point in the

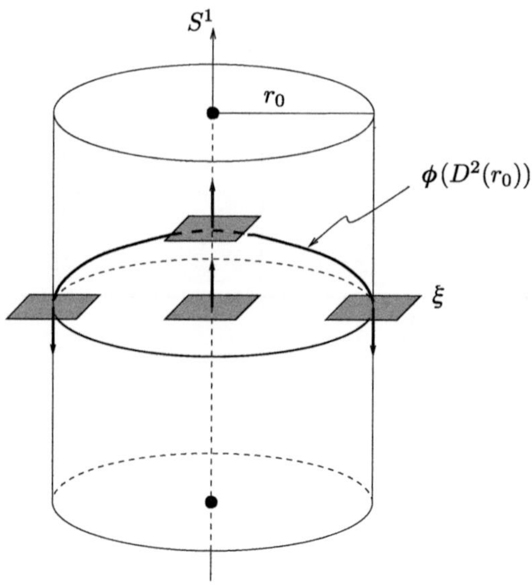

Fig. 18. An overtwisted disc.

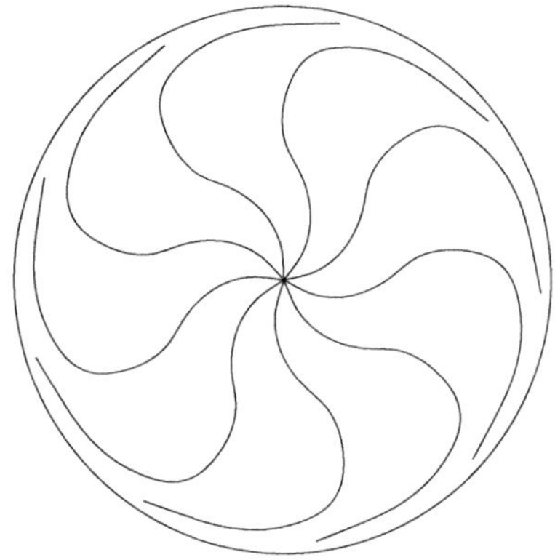

Fig. 19. Characteristic foliation on an overtwisted disc.

interior of the disc (Figure 19; notice that the leaves of this foliation are the integral curves of the vector field $h_1(r)\theta'(r)\,\partial_\varphi - h_2(r)\,\partial_r$). Such a disc is called an *overtwisted disc*.

A contact structure ξ on a 3-manifold M is called *overtwisted* if (M,ξ) contains an embedded overtwisted disc. In order to justify this terminology, observe that in the radially symmetric standard contact structure of Example 2.7, the angle by which the contact planes turn approaches $\pi/2$ asymptotically as r goes to infinity. By contrast, any contact manifold which has been constructed using at least one (simple) Lutz twist contains a similar cylindrical region where the contact planes twist by more than π in radial direction (at the smallest positive radius r_0 with $h_2(r_0)=0$ the twisting angle has reached π).

We have shown the following:

PROPOSITION 3.18. *Let ξ be a contact structure on M. By a full Lutz twist along any transversely embedded circle one obtains an overtwisted contact structure ξ' that is homotopic to ξ as a 2-plane distribution.*

Together with the theorem of Lutz and Martinet we find that M contains an *overtwisted* contact structure in every homotopy class of 2-plane distributions. In fact, Eliashberg [21] has proved the following much stronger theorem.

THEOREM 3.19 (Eliashberg). *On a closed, orientable 3-manifold there is a one-to-one correspondence between homotopy classes of overtwisted contact structures and homotopy classes of 2-plane distributions.*

This means that two overtwisted contact structures which are homotopic as 2-plane fields are actually homotopic as contact structures and hence isotopic by Gray's stability theorem.

Thus, it 'only' remains to classify contact structures that are not overtwisted. In [24] Eliashberg defined *tight* contact structures on a 3-manifold M as contact structures ξ for which there is no embedded disc $D \subset M$ such that D_ξ contains a limit cycle. So, by definition, overtwisted contact structures are not tight. In that same paper, as mentioned above in Section 2.4.5, Eliashberg goes on to show the converse with the help of the Elimination Lemma: non-overtwisted contact structures are tight.

There are various ways to detect whether a contact structure is tight. Historically the first proof that a certain contact structure is tight is due to D. Bennequin [9, Corollary 2, p. 150]:

THEOREM 3.20 (Bennequin). *The standard contact structure ξ_0 on S^3 is tight.*

The steps of the proof are as follows: (i) First, Bennequin shows that if γ_0 is a transverse knot in (S^3, ξ_0) with Seifert surface Σ, then the self-linking number of γ satisfies the inequality

$$l(\gamma_0) \leqslant -\chi(\Sigma).$$

(ii) Second, he introduces an invariant for Legendrian knots; nowadays this is called the *Thurston–Bennequin invariant*: Let γ be a Legendrian knot in (S^3, ξ_0). Take a vector field X along γ transverse to ξ_0, and let γ' be the push-off of γ in the direction of X. Then the Thurston–Bennequin invariant $tb(\gamma)$ is defined to be the linking number of γ and γ'. (This invariant has an extension to homologically trivial Legendrian knots in arbitrary contact 3-manifolds.)

(iii) By pushing γ in the direction of $\pm X$, one obtains transverse curves $\gamma^{\pm}$ (either of which is a candidate for γ' in (ii)). One of these curves is positively transverse, the other negatively transverse to ξ_0. The self-linking number of $\gamma^{\pm}$ is related to the Thurston–Bennequin invariant and a further invariant (the rotation number) of γ. The equation relating these three invariants implies $tb(\gamma) \leqslant -\chi(\Sigma)$, where Σ again denotes a Seifert surface for γ. In particular, a Legendrian unknot γ satisfies $tb(\gamma) < 0$. This inequality would be violated by the vanishing cycle of an overtwisted disc (which has $tb = 0$), which proves that (S^3, ξ_0) is tight.

REMARK 3.21. (1) Eliashberg [25] generalised the Bennequin inequality $l(\gamma_0) \leqslant -\chi(\Sigma)$ for transverse knots (and the corresponding inequality for the Thurston–Bennequin invariant of Legendrian knots) to arbitrary tight contact 3-manifolds. Thus, whereas Bennequin established the tightness (without that name) of the standard contact structure on S^3 by proving the inequality that bears his name, that inequality is now seen, conversely, as a consequence of tightness.

(2) In [9] Bennequin denotes the positively (respectively negatively) transverse push-off of the Legendrian knot γ by γ^- (respectively γ^+). This has led to some sign errors in the literature. Notably, the $\pm$ in Proposition 2.2.1 of [25], relating the described invariants of γ and $\gamma^{\pm}$, needs to be reversed.

COROLLARY 3.22. *The standard contact structure on $\mathbb{R}^3$ is tight.*

PROOF. This is immediate from Proposition 2.13. □

Here are further tests for tightness:

1. A closed contact 3-manifold (M, ξ) is called *symplectically fillable* if there exists a compact symplectic manifold (W, ω) bounded by M such that

- the restriction of ω to ξ does not vanish anywhere,
- the orientation of M defined by ξ (i.e. the one for which ξ is positive) coincides with the orientation of M as boundary of the symplectic manifold (W, ω) (which is oriented by ω^2).

We then have the following result of Eliashberg [20, Theorem 3.2.1], [22] and Gromov [62, 2.4.D_2'(b)], cf. [10]:

THEOREM 3.23 (Eliashberg–Gromov). *A symplectically fillable contact structure is tight.*

EXAMPLE 3.24. The 4-ball $D^4 \subset \mathbb{R}^4$ with symplectic form $\omega = dx_1 \wedge dy_1 + dx_2 \wedge dy_2$ is a symplectic filling of S^3 with its standard contact structure ξ_0. This gives an alternative proof of Bennequin's theorem.

2. Let $(\tilde{M}, \tilde{\xi}) \to (M, \xi)$ be a covering map and contactomorphism. If $(\tilde{M}, \tilde{\xi})$ is tight, then so is (M, ξ), for any overtwisted disc in (M, ξ) would lift to an overtwisted disc in $(\tilde{M}, \tilde{\xi})$.

EXAMPLE 3.25. The contact structures ξ_n, $n \in \mathbb{N}$, on the 3-torus T^3 defined by

$$\alpha_n = \cos(n\theta_1)\, d\theta_2 + \sin(n\theta_1)\, d\theta_3 = 0$$

are tight: Lift the contact structure ξ_n to the universal cover $\mathbb{R}^3$ of T^3; there the contact structure is defined by the same equation $\alpha_n = 0$, but now $\theta_i \in \mathbb{R}$ instead of $\theta_i \in \mathbb{R}/2\pi\mathbb{Z} \cong S^1$. Define a diffeomorphism f of $\mathbb{R}^3$ by

$$f(x, y, z) = (y/n, z \cos y + x \sin y, z \sin y - x \cos y) =: (\theta_1, \theta_2, \theta_3).$$

Then $f^*\alpha_n = dz + x\, dy$, so the lift of ξ_n to $\mathbb{R}^3$ is contactomorphic to the tight standard contact structure on $\mathbb{R}^3$.

Notice that it is possible for a tight contact structure to be finitely covered by an overtwisted contact structure. The first such examples were due to S. Makar-Limanov [88]. Other examples of this kind have been found by V. Colin [18] and R. Gompf [58].

3. The following theorem of H. Hofer [65] relates the dynamics of the Reeb vector field to overtwistedness.

THEOREM 3.26 (Hofer). *Let α be a contact form on a closed 3-manifold such that the contact structure* $\ker\alpha$ *is overtwisted. Then the Reeb vector field of α has at least one contractible periodic orbit.*

EXAMPLE 3.27. The Reeb vector field R_n of the contact form α_n of the preceding example is

$$R_n = \cos(n\theta_1)\, \partial_{\theta_2} + \sin(n\theta_1)\, \partial_{\theta_3}.$$

Thus, the orbits of R_n define constant slope foliations of the 2-tori $\{\theta_1 = \text{const}\}$; in particular, the periodic orbits of R_n are even homologically non-trivial. It follows, again, that the ξ_n are tight contact structures on T^3. (This, admittedly, amounts to attacking starlings with rice puddings fired from catapults.[5])

3.7. *Classification results*

In this section I summarise some of the known classification results for contact structures on 3-manifolds. By Eliashberg's Theorem 3.19 it suffices to list the tight contact structures, up to isotopy or diffeomorphism, on a given closed 3-manifold.

THEOREM 3.28 (Eliashberg [24]). *Any tight contact structure on S^3 is isotopic to the standard contact structure ξ_0.*

This theorem has a remarkable application in differential topology, viz., it leads to a new proof of Cerf's theorem [16] that any diffeomorphism of S^3 extends to a diffeomorphism of the 4-ball D^4. The idea is that the above theorem implies that any diffeomorphism of S^3 is isotopic to a contactomorphism of ξ_0. Eliashberg's technique [22] of filling by holomorphic discs can then be used to show that such a contactomorphism extends to a diffeomorphism of D^4.

As remarked earlier (Remark 2.21), Eliashberg has also classified contact structures on $\mathbb{R}^3$. Recall that homotopy classes of 2-plane distributions on S^3 are classified by $\pi_3(S^2) \cong \mathbb{Z}$. By Theorem 3.19, each of these classes contains a unique (up to isotopy) overtwisted contact structure. When a point of S^3 is removed, each of these contact structures induces one on $\mathbb{R}^3$, and Eliashberg [25] shows that they remain non-diffeomorphic there. Eliashberg shows further that, apart from this integer family of overtwisted contact structures, there is a unique tight contact structure on $\mathbb{R}^3$ (the standard one), and a single overtwisted one that is 'overtwisted at infinity' and cannot be compactified to a contact structure on S^3.

In general, the classification of contact structures up to diffeomorphism will differ from the classification up to isotopy. For instance, on the 3-torus T^3 we have the following diffeomorphism classification due to Y. Kanda [75]:

THEOREM 3.29 (Kanda). *Every (positive) tight contact structure on T^3 is contactomorphic to one of the ξ_n, $n \in \mathbb{N}$, described above. For $n \neq m$, the contact structures ξ_n and ξ_m are not contactomorphic.*

Giroux [54] had proved earlier that ξ_n for $n \geqslant 2$ is not contactomorphic to ξ_1.

On the other hand, all the ξ_n are homotopic as 2-plane fields to $\{d\theta_1 = 0\}$. This shows one way how Eliashberg's classification Theorem 3.19 for overtwisted contact structures can fail for tight contact structures:

- There are tight contact structures on T^3 that are homotopic as plane fields but not contactomorphic.

[5] This turn of phrase originates from [93].

P. Lisca and G. Matić [82] have found examples of the same kind on homology spheres by applying Seiberg–Witten theory to Stein fillings of contact manifolds, cf. also [78].

Eliashberg and L. Polterovich [31] have determined the isotopy classes of diffeomorphisms of T^3 that contain a contactomorphism of ξ_1: they correspond to exactly those elements of $\mathrm{SL}(3,\mathbb{Z}) = \pi_0(\mathrm{Diff}(T^3))$ that stabilise the subspace $0 \oplus \mathbb{Z}^2$ corresponding to the coordinates (θ_2, θ_3). In combination with Kanda's result, this allows one to give an isotopy classification of tight contact structures on T^3. One particular consequence of the result of Eliashberg and Polterovich is the following:

- There are tight contact structures on T^3 that are contactomorphic and homotopic as plane fields, but not isotopic (i.e. not homotopic through contact structures).

Again, such examples also exist on homology spheres, as S. Akbulut and R. Matveyev [2] have shown.

Another aspect of Eliashberg's classification of overtwisted contact structures that fails to hold for tight structures is of course the existence of such a structure in every homotopy class of 2-plane fields, as is already demonstrated by the classification of contact structures on S^3. Etnyre and K. Honda [37] have recently even found an example of a manifold—the connected some of two copies of the Poincaré sphere with opposite orientations—that does not admit any tight contact structure at all.

For the classification of tight contact structures on lens spaces and T^2-bundles over S^1 see [55,71,72]. A partial classification of tight contact structures on lens spaces had been obtained earlier in [34].

As further reading on 3-dimensional contact geometry I can recommend the lucid Bourbaki talk by Giroux [53]. This covers the ground up to Eliashberg's classification of overtwisted contact structures and the uniqueness of the tight contact structure on S^3.

4. A guide to the literature

In this concluding section I give some recommendations for further reading, concentrating on those aspects of contact geometry that have not (or only briefly) been touched upon in earlier sections.

Two general surveys that emphasise historical matters and describe the development of contact geometry from some of its earliest origins are the one by Lutz [87] and one by the present author [45].

One aspect of contact geometry that I have neglected in these notes is the Riemannian geometry of contact manifolds (leading, for instance, to Sasakian geometry). The survey by Lutz has some material on that; D. Blair [11] has recently published a monograph on the topic.

There have also been various ideas for defining interesting families of contact structures. Again, the survey by Lutz has something to say on that; one such concept that has exhibited very intriguing ramifications—if this commercial break be permitted—was introduced in [48].

4.1. *Dimension* 3

As mentioned earlier, Chapter 8 in [1] is in parts complementary to the present notes and has some material on surfaces in contact 3-manifolds. Other surveys and introductory texts on 3-dimensional contact geometry are the introductory lectures by Etnyre [35] and the Bourbaki talk by Giroux [52]. Good places to start further reading are two papers by Eliashberg: [24] for the classification of tight contact structures and [26] for knots in contact 3-manifolds. Concerning the latter, there is also a chapter by Etnyre [36] in a companion *Handbook* and an article by Etnyre and Honda [38] with an extensive introduction to that subject.

The surveys [20] and [27] by Eliashberg are more general in scope, but also contain material about contact 3-manifolds.

3-dimensional contact topology has now become a fairly wide arena; apart from the work of Eliashberg, Giroux, Etnyre–Honda and others described earlier, I should also mention the results of Colin, who has, for instance, shown that surgery of index one (in particular: taking the connected sum) on a tight contact 3-manifold leads again to a tight contact structure [17].

Finally, Etnyre and Ng [40] have compiled a useful list of problems in 3-dimensional contact topology.

4.2. *Higher dimensions*

The article [46] by the present author contains a survey of what was known at the time of writing about the existence of contact structures on higher-dimensional manifolds. One of the most important techniques for constructing contact manifolds in higher dimensions is the so-called contact surgery along isotropic spheres developed by Eliashberg [23] and Weinstein [105]. The latter is a very readable paper. For a recent application of this technique see [49]. Other constructions of contact manifolds (branched covers, gluing along codimension 2 contact submanifolds) are described in my paper [43].

Odd-dimensional tori are of course amongst the manifolds with the simplest global description, but they do not easily lend themselves to the construction of contact structures. In [86] Lutz found a contact structure on T^5; since then it has been one of the prize questions in contact geometry to find a contact structure on higher-dimensional tori. That prize, as it were, recently went to F. Bourgeois [13], who showed that indeed all odd-dimensional tori do admit a contact structure. His construction uses the result of Giroux and Mohsen [56,57] about open book decompositions adapted to contact structures in conjunction with the original proof of Lutz. With the help of the branched cover theorem described in [43] one can conclude further that every manifold of the form $M \times \Sigma$ with M a contact manifold and Σ a surface of genus at least 1 admits a contact structure.

Concerning the classification of contact structures in higher dimensions, the first steps have been taken by Eliashberg [28] with the development of contact homology, which has been taken further in [29]. This has been used by Ustilovsky [102] to show that on S^{4n+1} there exist infinitely many non-isomorphic contact structures that are homotopically equivalent (in the sense that they induce the same almost contact structure, i.e. reduction of the structure group of TS^{4n+1} to $1 \times U(2n)$). Earlier results in this direction can be found in [44] in the context of various applications of contact surgery.

4.3. *Symplectic fillings*

A survey on the various types of symplectic fillings of contact manifolds is given by Etnyre [33], cf. also the survey by Bennequin [10]. Etnyre and Honda [39] have recently shown that certain Seifert fibred 3-manifolds M admit tight contact structures ξ that are not symplectically semi-fillable, i.e. there is no symplectic filling W of (M, ξ) even if W is allowed to have other contact boundary components. That paper also contains a good update on the general question of symplectic fillability.

A related question is whether symplectic manifolds can have disconnected boundary of contact type (this corresponds to a stronger notion of symplectic filling defined via a Liouville vector field transverse to the boundary and pointing outwards). For (boundary) dimension 3 this is discussed by McDuff [91]; higher-dimensional symplectic manifolds with disconnected boundary of contact type have been constructed in [42].

Note added in proof: Eliashberg (Geom. Topol. **8** (2004), 277–293) has shown recently that every contact 3-manifold has a concave filling. This implies, in particular, that semi-fillable contact manifolds are always fillable.

4.4. *Dynamics of the Reeb vector field*

In a seminal paper, Hofer [65] applied the method of pseudo-holomorphic curves, which had been introduced to symplectic geometry by Gromov [62], to solve (for large classes of contact 3-manifolds) the so-called Weinstein conjecture [104] concerning the existence of periodic orbits of the Reeb vector field of a given contact form. (In fact, one of the remarkable aspects of Hofer's work is that in many instances it shows the existence of a periodic orbit of the Reeb vector field of any contact form defining a given contact structure.) A Bourbaki talk on the state of the art around the time when Weinstein formulated the conjecture that bears his name was given by Desolneux-Moulis [19]; another Bourbaki talk by Laudenbach describes Hofer's contribution to the problem.

The textbook by Hofer and Zehnder [70] addresses these issues, although its main emphasis, as is clear from the title, lies more in the direction of symplectic geometry and Hamiltonian dynamics. Two surveys by Hofer [66,67], and one by Hofer and Kriener [68], are more directly concerned with contact geometry. Of the three, [66] may be the best place to start, since it derives from a course of five lectures. In collaboration with Wysocki and Zehnder, Hofer has expanded his initial ideas into a far-reaching project on the characterisation of contact manifolds via the dynamics of the Reeb vector field, see, e.g., [69].

References

[1] B. Aebischer, M. Borer, M. Kälin, Ch. Leuenberger and H.M. Reimann, *Symplectic Geometry*, Progr. Math., vol. 124, Birkhäuser, Basel (1994).

[2] S. Akbulut and R. Matveyev, *A note on contact structures*, Pacific J. Math. **182** (1998), 201–204.

[3] D.N. Akhiezer, *Homogeneous complex manifolds*, Several Complex Variables IV, S.G. Gindikin and G.M. Khenkin, eds, Encyclopaedia Math. Sci., vol. 10, Springer, Berlin (1990), 195–244.

[4] J.W. Alexander, *A lemma on systems of knotted curves*, Proc. Nat. Acad. Sci. U.S.A. **9** (1923), 93–95.

[5] S. Altschuler, *A geometric heat flow for one-forms on three-dimensional manifolds*, Illinois J. Math. **39** (1995), 98–118.

[6] V.I. Arnold, *Characteristic class entering in quantization conditions*, Funct. Anal. Appl. **1** (1967), 1–13.

[7] V.I. Arnold, *Mathematical Methods of Classical Mechanics*, Grad. Texts in Math., vol. 60, Springer, Berlin (1978).

[8] A. Banyaga, *The Structure of Classical Diffeomorphism Groups*, Math. Appl., vol. 400, Kluwer, Dordrecht (1997).

[9] D. Bennequin, *Entrelacements et équations de Pfaff*, IIIe Rencontre de Géométrie du Schnepfenried, vol. 1, Astérisque **107–108** (1983), 87–161.

[10] D. Bennequin, *Topologie symplectique, convexité holomorphe et structures de contact* (d'après Y. Eliashberg, D. McDuff et al.), Séminaire Bourbaki, vol. 1989/1990, Astérisque **189–190** (1990), 285–323.

[11] D.E. Blair, *Riemannian Geometry of Contact and Symplectic Manifolds*, Progr. Math., vol. 203, Birkhäuser, Basel (2002).

[12] W.M. Boothby, *Transitivity of the automorphisms of certain geometric structures*, Trans. Amer. Math. Soc. **137** (1969), 93–100.

[13] F. Bourgeois, *Odd dimensional tori are contact manifolds*, Internat. Math. Res. Notices **2002** (2002), 1571–1574.

[14] G.E. Bredon, *Topology and Geometry*, Grad. Texts in Math., vol. 139, Springer, Berlin (1993).

[15] Th. Bröcker and K. Jänich, *Einführung in die Differentialtopologie*, Springer, Berlin (1973).

[16] J. Cerf, *Sur les difféomorphismes de la sphère de dimension trois* ($\Gamma_4 = 0$), Lecture Notes in Math., vol. 53, Springer, Berlin (1968).

[17] V. Colin, *Chirurgies d'indice un et isotopies de sphères dans les variétés de contact tendues*, C. R. Acad. Sci. Paris Sér. I Math. **324** (1997), 659–663.

[18] V. Colin, *Recollement de variétés de contact tendues*, Bull. Soc. Math. France **127** (1999), 43–69.

[19] N. Desolneux-Moulis, *Orbites périodiques des systèmes hamiltoniens autonomes* (d'après Clarke, Ekeland-Lasry, Moser, Rabinowitz and Weinstein), Séminaire Bourbaki, vol. 1979/1980, Lecture Notes in Math., vol. 842, Springer, Berlin (1981), 156–173.

[20] Y. Eliashberg, *Three lectures on symplectic topology in Cala Gonone: Basic notions, problems and some methods*, Conference on Differential Geometry and Topology (Sardinia, 1988), Rend. Sem. Fac. Sci. Univ. Cagliari, vol. 58 (1988), 27–49, suppl.

[21] Y. Eliashberg, *Classification of overtwisted contact structures on 3-manifolds*, Invent. Math. **98** (1989), 623–637.

[22] Y. Eliashberg, *Filling by holomorphic discs and its applications*, Geometry of Low-Dimensional Manifolds (Durham, 1989), vol. 2, London Math. Soc. Lecture Note Ser., vol. 151, Cambridge Univ. Press (1990), 45–67.

[23] Y. Eliashberg, *Topological characterization of Stein manifolds of dimension* > 2, Internat. J. Math. **1** (1990), 29–46.

[24] Y. Eliashberg, *Contact 3-manifolds twenty years since J. Martinet's work*, Ann. Inst. Fourier (Grenoble) **42** (1–2) (1992), 165–192.

[25] Y. Eliashberg, *Classification of contact structures on* $\mathbb{R}^3$, Internat. Math. Res. Notices **1993** (1993), 87–91.

[26] Y. Eliashberg, *Legendrian and transversal knots in tight contact 3-manifolds*, Topological Methods in Modern Mathematics (Stony Brook, 1991), Publish or Perish, Houston (1993), 171–193.

[27] Y. Eliashberg, *Symplectic topology in the nineties*, Differential Geom. Appl. **9** (1998), 59–88.

[28] Y. Eliashberg, *Invariants in contact topology*, Proceedings of the International Congress of Mathematicians (Berlin, 1998), vol. II, Doc. Math., Extra vol. II (1998), 327–338.

[29] Y. Eliashberg, A. Givental and H. Hofer, *Introduction to symplectic field theory*, Geom. Funct. Anal., vol. 2000, Special vol., Part II (2000), 560–673.

[30] Y. Eliashberg and N. Mishachev, *Introduction to the h-principle*, Grad. Stud. Math., vol. 48, American Mathematical Society, Providence (2002).

[31] Y. Eliashberg and L. Polterovich, *New applications of Luttinger's surgery*, Comment. Math. Helv. **69** (1994), 512–522.

[32] Y. Eliashberg and W.P. Thurston, *Confoliations*, Univ. Lecture Ser., vol. 13, American Mathematical Society, Providence (1998).

[33] J.B. Etnyre, *Symplectic convexity in low-dimensional topology*, Topology Appl. **88** (1998), 3–25.

[34] J.B. Etnyre, *Tight contact structures on lens spaces*, Commun. Contemp. Math. **2** (2000), 559–577; erratum: ibid. **3** (2001), 649–652.

[35] J.B. Etnyre, *Introductory lectures on contact geometry*, Topology and Geometry of Manifolds (Athens, GA, 2001), Proc. Sympos. Pure Math., vol. 71, American Mathematical Society, Providence (2003), 81–107.

[36] J.B. Etnyre, *Legendrian and transverse knots*, Handbook of Knot Theory, W. Menasco and M. Thistlethwaite, eds, in press.

[37] J.B. Etnyre and K. Honda, *On the nonexistence of tight contact structures*, Ann. of Math. (2) **153** (2001), 749–766.

[38] J.B. Etnyre and K. Honda, *Knots and contact geometry I: Torus knots and the figure eight knot*, J. Symplectic Geom. **1** (2001), 63–120.

[39] J.B. Etnyre and K. Honda, *Tight contact structures with no symplectic fillings*, Invent. Math. **148** (2002), 609–626.

[40] J.B. Etnyre and L.L. Ng, *Problems in low dimensional contact topology*, Topology and Geometry of Manifolds (Athens, GA, 2001), Proc. Sympos. Pure Math., vol. 71, American Mathematical Society, Providence (2003), 337–357.

[41] H. Freudenthal, *Zum Hopfschen Umkehrhomomorphismus*, Ann. of Math. (2) **38** (1937), 847–853.

[42] H. Geiges, *Symplectic manifolds with disconnected boundary of contact type*, Internat. Math. Res. Notices **1994** (1994), 23–30.

[43] H. Geiges, *Constructions of contact manifolds*, Math. Proc. Cambridge Philos. Soc. **121** (1997), 455–464.

[44] H. Geiges, *Applications of contact surgery*, Topology **36** (1997), 1193–1220.

[45] H. Geiges, *A brief history of contact geometry and topology*, Exposition. Math. **19** (2001), 25–53.

[46] H. Geiges, *Contact topology in dimension greater than three*, European Congress of Mathematics (Barcelona, 2000), vol. 2, Progress in Math., vol. 202, Birkhäuser, Basel (2001), 535–545.

[47] H. Geiges, *h-principles and flexibility in geometry*, Mem. Amer. Math. Soc. **164** (2003), 779.

[48] H. Geiges and J. Gonzalo, *Contact geometry and complex surfaces*, Invent. Math. **121** (1995), 147–209.

[49] H. Geiges and C.B. Thomas, *Contact structures, equivariant spin bordism, and periodic fundamental groups*, Math. Ann. **320** (2001), 685–708.

[50] V.L. Ginzburg, *On closed characteristics of 2-forms*, Ph.D. thesis, Berkeley (1990).

[51] V.L. Ginzburg, *Calculation of contact and symplectic cobordism groups*, Topology **31** (1992), 767–773.

[52] E. Giroux, *Convexité en topologie de contact*, Comment. Math. Helv. **66** (1991), 637–677.

[53] E. Giroux, *Topologie de contact en dimension 3 (autour des travaux de Yakov Eliashberg)*, Séminaire Bourbaki, vol. 1992/1993, Astérisque **216** (1993), 7–33.

[54] E. Giroux, *Une structure de contact, même tendue, est plus ou moins tordue*, Ann. Sci. École Norm. Sup. (4) **27** (1994), 697–705.

[55] E. Giroux, *Structures de contact en dimension trois et bifurcations des feuilletages de surfaces*, Invent. Math. **141** (2000), 615–689.

[56] E. Giroux, *Géométrie de contact: de la dimension trois vers les dimensions supérieures*, Proceedings of the International Congress of Mathematicians (Beijing, 2002), vol. II, Higher Education Press, Beijing (2002), 405–414.

[57] E. Giroux and J.-P. Mohsen, *Structures de contact et fibrations symplectiques au-dessus du cercle*, in preparation.

[58] R.E. Gompf, *Handlebody construction of Stein surfaces*, Ann. of Math. (2) **148** (1998), 619–693.

[59] J. Gonzalo, *Branched covers and contact structures*, Proc. Amer. Math. Soc. **101** (1987), 347–352.

[60] D.H. Gottlieb, *Partial transfers*, Geometric Applications of Homotopy Theory I, Lecture Notes in Math., vol. 657, Springer, Berlin (1978), 255–266.

[61] J.W. Gray, *Some global properties of contact structures*, Ann. of Math. (2) **69** (1959), 421–450.

[62] M. Gromov, *Pseudoholomorphic curves in symplectic manifolds*, Invent. Math. **82** (1985), 307–347.

[63] H.M. Hilden, J.M. Montesinos and T. Thickstun, *Closed oriented 3-manifolds as 3-fold branched coverings of S^3 of special type*, Pacific J. Math. **65** (1976), 65–76.

[64] M.W. Hirsch, *Differential Topology*, Grad. Texts in Math., vol. 33, Springer, Berlin (1976).

[65] H. Hofer, *Pseudoholomorphic curves in symplectizations with applications to the Weinstein conjecture in dimension* 3, Invent. Math. **114** (1993), 515–563.

[66] H. Hofer, *Holomorphic curves and dynamics in dimension three*, Symplectic Geometry and Topology (Park City, 1997), IAS/Park City Math. Ser., vol. 7, American Mathematical Society, Providence (1999), 35–101.

[67] H. Hofer, *Holomorphic curves and real three-dimensional dynamics*, Geom. Funct. Anal., vol. 2000, Special vol., Part II (2000), 674–704.

[68] H. Hofer and M. Kriener, *Holomorphic curves in contact dynamics*, Differential Equations (La Pietra, 1996), Proc. Sympos. Pure Math., vol. 65, American Mathematical Society, Providence (1999), 77–131.

[69] H. Hofer, K. Wysocki and E. Zehnder, *A characterisation of the tight three-sphere*, Duke Math. J. **81** (1995), 159–226; erratum: ibid. **89** (1997), 603–617.

[70] H. Hofer and E. Zehnder, *Symplectic Invariants and Hamiltonian Dynamics*, Birkhäuser, Basel (1994).

[71] K. Honda, *On the classification of tight contact structures I*, Geom. Topol. **4** (2000), 309–368.

[72] K. Honda, *On the classification of tight contact structures II*, J. Differential Geom. **55** (2000), 83–143.

[73] H. Hopf, *Zur Algebra der Abbildungen von Mannigfaltigkeiten*, J. Reine Angew. Math. **105** (1930), 71–88.

[74] D. Husemoller, *Fibre Bundles*, 3rd ed., Grad. Texts in Math., vol. 20, Springer, Berlin (1994).

[75] Y. Kanda, *The classification of tight contact structures on the* 3*-torus*, Comm. Anal. Geom. **5** (1997), 413–438.

[76] R.C. Kirby, *The Topology of* 4*-Manifolds*, Lecture Notes in Math., vol. 1374, Springer, Berlin (1989).

[77] A.A. Kosinski, *Differential Manifolds*, Academic Press, Boston (1993).

[78] P.B. Kronheimer and T.S. Mrowka, *Monopoles and contact structures*, Invent. Math. **130** (1997), 209–255.

[79] F. Laudenbach, *Orbites périodiques et courbes pseudo-holomorphes, application à la conjecture de Weinstein en dimension* 3 (d'après H. Hofer et al.), Séminaire Bourbaki, vol. 1993/1994, Astérisque **227** (1995), 309–333.

[80] P. Libermann and C.-M. Marle, *Symplectic Geometry and Analytical Mechanics*, Math. Appl., vol. 35, Reidel, Dordrecht (1987).

[81] W.B.R. Lickorish, *A representation of orientable combinatorial* 3*-manifolds*, Ann. of Math. (2) **76** (1962), 531–540.

[82] P. Lisca and G. Matić, *Tight contact structures and Seiberg–Witten invariants*, Invent. Math. **129** (1997), 509–525.

[83] R. Lutz, *Sur l'existence de certaines formes différentielles remarquables sur la sphère* S^3, C. R. Acad. Sci. Paris Sér. A–B **270** (1970), A1597–A1599.

[84] R. Lutz, *Sur quelques propriétés des formes differentielles en dimension trois*, Thèse, Strasbourg (1971).

[85] R. Lutz, *Structures de contact sur les fibrés principaux en cercles de dimension trois*, Ann. Inst. Fourier (Grenoble) **27** (3) (1977), 1–15.

[86] R. Lutz, *Sur la géométrie des structures de contact invariantes*, Ann. Inst. Fourier (Grenoble) **29** (1) (1979), 283–306.

[87] R. Lutz, *Quelques remarques historiques et prospectives sur la géométrie de contact*, Conference on Differential Geometry and Topology (Sardinia, 1988), Rend. Sem. Fac. Sci. Univ. Cagliari, vol. 58 (1988), 361–393, suppl.

[88] S. Makar-Limanov, *Tight contact structures on solid tori*, Trans. Amer. Math. Soc. **350** (1998), 1013–1044.

[89] J. Martinet, *Sur les singularités des formes différentielles*, Ann Inst. Fourier (Grenoble) **20** (1) (1970), 95–178.

[90] J. Martinet, *Formes de contact sur les variétés de dimension* 3, Proc. Liverpool Singularities Sympos. II, Lecture Notes in Math., vol. 209, Springer, Berlin (1971), 142–163.

[91] D. McDuff, *Symplectic manifolds with contact type boundaries*, Invent. Math. **103** (1991), 651–671.

[92] D. McDuff and D. Salamon, *Introduction to Symplectic Topology*, 2nd ed., Oxford Univ. Press (1998).

[93] S. Milligan, *The Starlings*, The Goon Show, Series 4, Special programme, broadcast 31st August 1954.

[94] J.W. Milnor, *Topology from the Differentiable Viewpoint*, The University Press of Virginia, Charlottesville (1965).

[95] J.W. Milnor and J.D. Stasheff, *Characteristic Classes*, Ann. of Math. Studies, vol. 76, Princeton Univ. Press, Princeton (1974).

[96] J. Moser, *On the volume elements on a manifold*, Trans. Amer. Math. Soc. **120** (1965), 286–294.

[97] D. Rolfsen, *Knots and Links*, Publish or Perish, Houston (1976).

[98] N. Saveliev, *Lectures on the Topology of* 3-*Manifolds*, de Gruyter, Berlin (1999).
[99] N. Steenrod, *The Topology of Fibre Bundles*, Princeton Univ. Press, Princeton (1951).
[100] I. Tamura, *Topology of Foliations: An Introduction*, Transl. Math. Monogr., vol. 97, American Mathematical Society, Providence (1992).
[101] W.P. Thurston and H.E. Winkelnkemper, *On the existence of contact forms*, Proc. Amer. Math. Soc. **52** (1975), 345–347.
[102] I. Ustilovsky, *Infinitely many contact structures on* S^{4m+1}, Internat. Math. Res. Notices **1999** (1999), 781–791.
[103] A.H. Wallace, *Modifications and cobounding manifolds*, Canad. J. Math. **12** (1960), 503–528.
[104] A. Weinstein, *On the hypotheses of Rabinowitz' periodic orbit theorem*, J. Differential Equations **33** (1979), 353–358.
[105] A. Weinstein, *Contact surgery and symplectic handlebodies*, Hokkaido Math. J. **20** (1991), 241–251.

CHAPTER 6

Complex Differential Geometry

Ion Mihai*

Faculty of Mathematics, University of Bucharest, Str. Academiei 14, 010014 Bucharest, Romania
E-mail: imihai@fmi.unibuc.ro

Contents

*This chapter was written while the author has visited Michigan State University, Department of Mathematics, supported by a Fulbright grant. He expresses his hearty thanks to Professor B.Y. Chen for the hospitality he received during this visit. The author is very obliged to Professors D. Blair, B.Y. Chen, F. Defever, F. Dillen, G.D. Ludden, K. Matsumoto, R. Rosca, B. Suceavă, Y. Tazawa and L. Verstraelen for valuable discussions on the topics presented therein.

HANDBOOK OF DIFFERENTIAL GEOMETRY, VOL. II
Edited by F.J.E. Dillen and L.C.A. Verstraelen

Introduction

The geometry of complex manifolds, in particular Kaehler manifolds, is an important research topic in Differential Geometry.

In the present chapter, we present the basic notions and certain important results in complex differential geometry.

First, we define complex and almost complex manifolds and give standard examples.

The most interesting class of complex manifolds are the Kaehler manifolds. Locally, a Kaehlerian metric differs from the Euclidean metric on the complex space $\mathbf{C}^n$ starting with the second power of the Taylor series. There are topological obstructions to the existence of Kaehlerian metrics on a compact complex manifold. We give examples and counterexamples. We introduce complex space forms and we state a Schur-like theorem for complex space forms.

Next, we study the differential forms on a Kaehler manifold. Then, we prove the complex version of Hodge theorem. As an application, it follows that the Betti numbers of odd order on a compact Kaehler manifold are even. An interesting example of an almost Kaehler manifold which do not admit any Kaehlerian metric is the Thurston–Abbena manifold. The Iwasawa manifold is a complex manifold which does not carry any Kaehlerian metric.

The Chern classes are introduced axiomatically and their construction is given. Some applications are derived.

Last section deals with the deformation of complex structures in the sense of Kodaira. We state the theorems of existence and completeness. The number of moduli of a compact complex manifold is introduced.

1. Complex manifolds

An n-dimensional *complex manifold* is a pairing $(M, \mathcal{A})$, where M is a non-empty set and $\mathcal{A} = \{(U_\alpha, h_\alpha) \mid \alpha \in A\}$ is a family of mappings satisfying the following properties:

(i) For each $\alpha \in A$, U_α is a subset of M and $h_\alpha : U_\alpha \to \mathbf{C}^n$ is one-to-one.

(ii) The family $\{U_\alpha\}_{\alpha \in A}$ is a covering of M, i.e.,

$$M = \bigcup_{\alpha \in A} U_\alpha.$$

(iii) For each $\alpha, \beta \in A$, the set $h_\alpha(U_\alpha \cap U_\beta)$ is an open subset in $\mathbf{C}^n$ and the mapping

$$h_\beta \circ h_\alpha^{-1} : h_\alpha(U_\alpha \cap U_\beta) \to \mathbf{C}^n$$

is holomorphic.

(iv) If $(M, \mathcal{A}')$ satisfies the properties (i)–(iii) and $\mathcal{A} \subset \mathcal{A}'$, then $\mathcal{A}' = \mathcal{A}$ (i.e., $(M, \mathcal{A})$ is maximal).

REMARK. If $(M, \mathcal{B}) = \{(U_\beta, h_\beta) \mid \beta \in B\}$ is a pairing satisfying the properties (i)–(iii) and we denote by $\mathcal{A}$ the set of all the pairs (U, h), with $U \subset M$, $h : U \to \mathbf{C}^n$ one-to-one such that for each $\beta \in B$, $h(U \cap U_\beta), h_\beta(U \cap U_\beta) \subset \mathbf{C}^n$ are open sets and the mappings $h_\beta \circ h^{-1} : h(U \cap U_\beta) \to \mathbf{C}^n$ and $h \circ h_\beta^{-1} : h_\beta(U \cap U_\beta) \to \mathbf{C}^n$ respectively are holomorphic, then $(M, \mathcal{A})$ is an n-dimensional complex manifold.

Thus, in order to define a complex structure on a non-empty set M, it is sufficient to construct a family $\mathcal{B}$ satisfying the properties (i)–(iii).

EXAMPLES. 1. Each open subset U of $\mathbf{C}^n$ admits a canonical complex structure, taking $\mathcal{B} = \{(U, i_U)\}$, where $i_U : U \to \mathbf{C}^n$ is the inclusion map of U into $\mathbf{C}^n$. In particular, $GL(n, \mathbf{C})$ is a complex Lie group.

2. The complex projective space $P^n(\mathbf{C})$ is an n-dimensional complex manifold. On $\mathbf{C}^{n+1} - \{0\}$, we define the equivalence relation

$$\begin{aligned}&(z^1, \dots, z^{n+1}) \sim (w^1, \dots, w^{n+1})\\ &\quad \Leftrightarrow \quad \exists \lambda \in \mathbf{C} - \{0\} \quad \text{such that } w^i = \lambda z^i,\ i \in \{1, \dots, n+1\}.\end{aligned}$$

We denote by $P^n(\mathbf{C})$ the quotient space

$$P^n(\mathbf{C}) = \left\{ \left[z^1, z^2, \dots, z^{n+1}\right] \mid \left(z^1, z^2, \dots, z^{n+1}\right) \in \mathbf{C}^{n+1} - \{0\} \right\}.$$

For any $\alpha \in \{1, \dots, n+1\}$, we put $U_\alpha = \{[z^1, z^2, \dots, z^{n+1}] \mid z^\alpha \neq 0\}$ and $h_\alpha : U_\alpha \to \mathbf{C}^n$,

$$h_\alpha\left[z^1, \dots, z^{n+1}\right] = \left(\frac{z^1}{z^\alpha}, \dots, \frac{z^{\alpha-1}}{z^\alpha}, \frac{z^{\alpha+1}}{z^\alpha}, \dots, \frac{z^{n+1}}{z^\alpha} \right).$$

Then the family $\mathcal{B} = \{(U_\alpha, h_\alpha),\ \alpha \in \{1, \dots, n+1\}\}$ satisfies the properties (i)–(iii). For $\alpha \neq \beta$, the mapping $h_\beta \circ h_\alpha^{-1}$ is given by

$$\bar{z}^\gamma = \frac{z^\gamma}{z^\beta}, \quad \gamma \notin \{\alpha, \beta\}; \qquad \bar{z}^\alpha = \frac{1}{z^\beta}.$$

EXERCISE. Determine the holomorphic vector fields on the complex projective spaces $P^1(\mathbf{C})$ and $P^2(\mathbf{C})$.

Solution. Locally, a holomorphic vector field v on $P^1(\mathbf{C})$ has the form

$$v = \left(az^2 + bz + c\right) \frac{d}{dz}, \quad a, b, c \in \mathbf{C},$$

and a holomorphic vector field v on $P^2(\mathbf{C})$ is given by

$$\begin{aligned} v &= \left(c_1 + c_2 z^1 + c_3 z^2 + c_7 \left(z^1\right)^2 + c_8 z^1 z^2\right) \frac{\partial}{\partial z^1} \\ &\quad + \left(c_4 + c_5 z^1 + c_6 z^2 + c_7 z^1 z^2 + c_8 \left(z^2\right)^2\right) \frac{\partial}{\partial z^2}, \end{aligned}$$

where $c_1, \dots, c_8 \in \mathbf{R}$.

3. The complex Grassmann manifold.

Let $G_p(\mathbf{C}^n)$ be the set of all p-dimensional linear subspaces of $\mathbf{C}^n$. We define on $G_p(\mathbf{C}^n)$ a complex structure of dimension $p(n-p)$.

Let $\{e_1, \dots, e_n\}$ be the standard basis of $\mathbf{C}^n$. For any sequence $1 \leqslant i_1 < i_2 < \dots < i_p \leqslant n$, let $\pi_{i_1 \dots i_p} : \mathbf{C}^n \to E_{i_1 \dots i_p} = \mathrm{Span}\{e_{i_1}, \dots, e_{i_p}\}$ be the orthogonal projection. We denote by

$$U_{i_1 \dots i_p} = \{V \in G_p(\mathbf{C}^n);\ \pi_{i_1 \dots i_p}|_V \text{ is a linear isomorphism}\}.$$

Let $V \in U_{i_1 \dots i_p}$. Then, for each $k \in \{1, \dots, p\}$, there exists a unique $f_k \in V$ such that $\pi_{i_1 \dots i_p}(f_k) = e_{i_k}$. Obviously,

$$f_k = e_{i_k} + \sum_{l \notin \{i_1, \dots, i_p\}} c_k^l e_l.$$

Consider the mapping $h_{i_1 \dots i_p} : U_{i_1 \dots i_p} \to \mathbf{C}^{p(n-p)}$,

$$h_{i_1 \dots i_p}(V) = \left(c_k^l\right)_{k \in \{i_1, \dots, i_p\}, l \notin \{i_1, \dots, i_p\}}.$$

If we put $\mathcal{B} = \{h_{i_1 \dots i_p} \mid 1 \leqslant i_1 < \dots < i_p \leqslant n\}$, then $G_p(\mathbf{C}^n)$ becomes a complex manifold of dimension $p(n-p)$.

REMARK. $G_1(\mathbf{C}^{n+1}) = P^n(\mathbf{C})$.

4. The Calabi manifolds $S^{2m+1} \times S^{2n+1}$ [9].

The differentiable manifold $M = S^{2m+1} \times S^{2n+1}$ has real dimension $2m + 2n + 2$. We may consider

$$S^{2m+1} = \left\{\xi \in \mathbf{C}^{m+1} \mid \sum_{\alpha=1}^{m+1} |\xi^\alpha|^2 = 1\right\},$$

$$S^{2n+1} = \left\{\eta \in \mathbf{C}^{n+1} \mid \sum_{\beta=1}^{n+1} |\eta^\beta|^2 = 1\right\},$$

then $M \subset \mathbf{C}^{m+1} \times \mathbf{C}^{n+1}$.

Let $j \in \{1, \dots, m+1\}$ and $k \in \{1, \dots, n+1\}$ and denote by

$$U_{jk} = \{(\xi, \eta) \in M \mid \xi^j \eta^k \neq 0\}.$$

We will construct a family of mappings $h_{jk} : U_{jk} \to \mathbf{C}^{m+n+1}$, which defines a complex structure on M. Put

$$h_{jk}(\xi,\eta) = \left(\frac{\xi^1}{\xi^j},\ldots,\frac{\xi^{j-1}}{\xi^j},\frac{\xi^{j+1}}{\xi^j},\ldots,\frac{\xi^{m+1}}{\xi^j},\frac{\eta^1}{\eta^k},\ldots,\frac{\eta^{k-1}}{\eta^k},\frac{\eta^{k+1}}{\eta^k},\ldots,\frac{\eta^{n+1}}{\eta^k},t\right),$$

where

$$t = \frac{1}{2\pi i}(\log \xi^j + i \log \eta^k) \quad \mod (1, i).$$

5. Each orientable surface M is a 1-dimensional complex manifold.

Let g be a Riemannian metric on M. By a theorem of Lichtenstein, locally, the metric can be written as $g = \lambda^2(dx^2+dy^2)$, with $\lambda > 0$. Putting $z = x+iy$, M becomes a complex manifold of dimension 1.

In particular, the 2-sphere S^2 admits a complex structure.

We will give a direct construction of a complex structure on S^2.

It is known that S^2 is a 2-dimensional differentiable manifold. Let

$$S^2 = \{(u^1,u^2,u^3) \in \mathbf{E}^3 \mid (u^1)^2 + (u^2)^2 + (u^3)^2 = 1\}.$$

Denote by $U_N = S^2 - \{N\}$ and $U_S = S^2 - \{S\}$, where $N = (0,0,1)$ and $S = (0,0,-1)$ are the north and south poles, respectively. Consider the stereographic projections

$$h_N : U_N \to \mathbf{R}^2, \qquad h_S : U_S \to \mathbf{R}^2$$

from the north and south poles, respectively.

Their equations are

$$h_N(u^1,u^2,u^3) = \left(\frac{u^1}{1-u^3}, \frac{u^2}{1-u^3}\right),$$

$$h_S(u^1,u^2,u^3) = \left(\frac{u^1}{1+u^3}, \frac{u^2}{1+u^3}\right),$$

respectively.

We define $\tilde{h}_N : U_N \to \mathbf{C}$ and $\tilde{h}_S : U_S \to \mathbf{C}$, by

$$\tilde{h}_N(u^1,u^2,u^3) = \frac{u^1}{1-u^3} + i\frac{u^2}{1-u^3}$$

and

$$\tilde{h}_S(u^1,u^2,u^3) = \frac{u^1}{1+u^3} - i\frac{u^2}{1+u^3}.$$

The mapping

$$h_S \circ h_N^{-1} : \mathbf{C} - \{0\} \to \mathbf{C}, \quad \left(h_S \circ h_N^{-1}\right)(z) = \frac{1}{z},$$

is holomorphic. Thus the family $\mathcal{B} = \{(U_N, \tilde{U}_N), (U_S, \tilde{h}_S)\}$ satisfies the conditions (i)–(iii), and therefore the sphere S^2 is a 1-dimensional complex manifold.

Next, we will indicate one procedure for obtaining new complex manifolds.

Let M be a complex manifold. The set of all automorphisms of M is a group $\mathcal{G}$ endowed with the composition of mappings. Any subgroup G of $\mathcal{G}$ is called a *group of automorphisms* of M. For any $p \in M$, the set $G_p = \{g(p) \mid g \in G\}$ is the *orbit* of G at p. Obviously $G_p \cap G_q \neq \emptyset$ if and only if $q \in G_p$. The set of all orbits, denoted by $M/_G$, is the quotient space of M via G.

EXAMPLE. Let $M = \mathbf{C}^{n+1} - \{0\}$. Each $g \in \mathbf{C}^*$ defines an automorphism of M by

$$g : \left(z^1, \dots, z^{n+1}\right) \mapsto \left(gz^1, \dots, gz^{n+1}\right).$$

Thus $\mathbf{C}^*$ becomes a group of automorphisms of M and $M/_{\mathbf{C}^*} = P^n(\mathbf{C})$.

DEFINITION. A group G of automorphisms of the complex manifold M is called *properly discontinuous* if for any compact subsets K_1 and K_2 in M, the set $\{g \in G \mid g(K_1) \cap K_2 \neq \emptyset\}$ is finite.

In this case, each orbit G_p is a discrete set.

We say that G is fixed points free if each $g \in G - \{1_M\}$ has no fixed points.

REMARK. In the above example, $\mathbf{C}^*$ is fixed points free, but it is not properly discontinuous.

THEOREM 1.1. *Let M be an n-dimensional complex manifold and G a fixed points free and properly discontinuous group of automorphisms. Then $M/_G$ carries a natural n-dimensional complex structure induced by the complex structure of M.*

Using this theorem, we can obtain other examples of complex manifolds.

6. The complex torus is defined as follows. Let $\{\omega_1, \dots, \omega_{2n}\}$ be $2n$ vectors in $\mathbf{C}^n$ linearly independent over $\mathbf{R}$. We define a fixed points free and properly discontinuous group of automorphisms of $\mathbf{C}^n$, by

$$G = \left\{ z \mapsto z + \sum_{j=1}^{2n} m_j \omega_j \,\middle|\, m_j \in \mathbf{Z},\ j \in \{1, \dots, 2n\} \right\}.$$

The quotient space $T^n = \mathbf{C}^n/_G$, which we call the n-dimensional complex torus, becomes a complex manifold.

REMARK. For $n = 1$, T^1 is an algebraic curve. If $n \geqslant 2$, then T^n is not necessarily an algebraic manifold. T^n is an algebraic manifold if and only if its period matrix is a Riemann matrix (see [54]).

An n-dimensional complex manifold M is said to be complex parallelizable if there exist n holomorphic vector fields which are linearly independent at every point of M. Every complex torus is complex parallelizable.

H.G. Wang [61] proved that every compact complex parallelizable manifold can be written as a quotient space $G/_D$ of a complex Lie group G by a discrete subgroup D.

7. The Hopf manifolds.

Let $M = \mathbf{C}^n - \{0\}$ and $\alpha_1, \ldots, \alpha_n \in \mathbf{C}$ such that $|\alpha_j| > 1$ $(j \in \{1, \ldots, n\})$ and let G be the cyclic group spanned by the automorphism g of M, given by

$$g : z = (z^1, \ldots, z^n) \mapsto g(z) = (\alpha_1 z^1, \ldots, \alpha_n z^n).$$

Applying the above theorem, $M/_G$ has a complex structure; it is called Hopf manifold. It is known (see [37]) that $M/_G$ is diffeomorphic with $S^1 \times S^{2n-1}$.

Thus, Hopf manifolds are a particular case of Calabi manifolds. We want to point out that they are not algebraic manifolds.

EXERCISE. Let $L = \{([z], \xi) \in P^n(\mathbf{C}) \times \mathbf{C}^{n+1} \mid \xi \in [z]\}$. Prove that $(L, pr_1, P^n(\mathbf{C}))$ is a holomorphic vector bundle of rank 1 (called the *tautological vector bundle* over the complex projective space).

8. The blowing-up at a point.

Consider a small ball $B = B_r(0) \subset \mathbf{C}^n$, with $n \geqslant 2$. Let $z = (z^1, \ldots, z^n)$ be the standard coordinates of $\mathbf{C}^n$. The *blowing-up* of B at 0, denoted by $\bar{B}$, is the complex n-dimensional manifold defined by

$$\bar{B} = \{(z, w) \in B \times P^{n-1}(\mathbf{C}) \mid z^i w^j = z^j w^i, \ \forall 1 \leqslant i < j \leqslant n\},$$

where $w = [w^1, \ldots, w^n] \in P^{n-1}(\mathbf{C})$. If we cover $P^{n-1}(\mathbf{C})$ by the open subsets $U_i = \{w \in P^{n-1}(\mathbf{C}) \mid w^i \neq 0\}$, $i = 1, \ldots, n$, then $B \times P^{n-1}(\mathbf{C})$ is covered by the open subsets $B \times U_i$, $1 \leqslant i \leqslant n$. In each $B \times U_i$, holomorphic coordinates are given by

$$\left(z^1, \ldots, z^n, \frac{w^1}{w^i}, \ldots, \frac{w^{i-1}}{w^i}, \frac{w^{i+1}}{w^i}, \ldots, \frac{w^n}{w^i}\right)$$

and $\bar{B} \cap (B \times U_i)$ is defined by the $n - 1$ equations

$$z^j = z^i \frac{w^j}{w^i}, \quad 1 \leqslant j \leqslant n, \ j \neq i.$$

Hence $\bar{B}$ is a complex submanifold of dimension n in $B \times P^{n-1}(\mathbf{C})$.

9. A *Stein manifold* is a closed complex submanifold in some $\mathbf{C}^m$. It is clear that any closed submanifold of a Stein manifold is also Stein and the product of two Stein manifolds is Stein.

Stein manifold can also be defined intrinsically. For this purpose, let $M \subset \mathbf{C}^m$ be such a manifold of dimension n and denote by $\mathcal{O}(M)$ the ring of global holomorphic functions on M. Then M satisfies the following:

(i) M is *holomorphically convex*, that is, for any compact $K \subset M$, the set

$$\hat{K} := \left\{x \in M \,\middle|\, |f(x)| \leqslant \sup_K |f|,\ \forall f \in \mathcal{O}(M)\right\}$$

is also compact.

(ii) Given any two distinct points $x, y \in M$, there exists $f \in \mathcal{O}(M)$ such that $f(x) \neq f(y)$.

(iii) Given any $x \in M$, there exist $f_1, \dots, f_n$ in $\mathcal{O}(M)$ such that $(f_1, \dots, f_n)$ gives holomorphic coordinates in a neighborhood of x.

Conversely, any complex manifold M satisfying conditions (i)–(iii) is Stein (see [26] for a proof).

A Stein manifold is always non-compact (by the maximum principle, a compact complex manifold does not admit any non-constant holomorphic function). Any non-compact Riemann surface is Stein [26]. So, in a way, the Stein manifolds are generalizations in high dimensions of the non-compact Riemann surfaces (while the projective manifolds generalize the compact Riemann surfaces).

2. Almost complex structures

In this section, we define the notion of an almost complex structure and we investigate the existence of such structures on spheres.

An *almost complex* structure on a (real) differentiable manifold M is an anti-involutive endomorphism J of the tangent bundle TM (i.e., the map $J : TM \to TM$ is differentiable and at each $p \in M$, the linear endomorphism $J_p = J|_{T_pM} : T_pM \to T_pM$ satisfies $J_p^2 = -1_{T_pM}$). The pairing (M, J) is called an almost complex manifold.

PROPOSITION 2.1. *Any almost complex manifold has even dimension and is orientable.*

The existence of an almost complex structure on an orientable $2n$-dimensional manifold M means that the tangent bundle TM admits a reduction from the $GL(2n, \mathbf{R})$-structure to the $GL(n, \mathbf{C})$-structure. So the existence problem for an almost complex structure is a purely algebraic topological one.

An infinitesimal automorphism of an almost complex structure J on M is a vector field X such that $\mathcal{L}_X J = 0$, where $\mathcal{L}_X$ denotes the Lie differentiation with respect to X. It is known that a vector field X is an infinitesimal automorphism of J if and only if it generates a local 1-parameter group of local almost complex transformations.

PROPOSITION 2.2. *A vector field X is an infinitesimal automorphism of an almost complex structure J on a manifold M if and only if*

$$[X, JY] = J[X, Y], \quad \forall Y \in \Gamma(TM).$$

PROPOSITION 2.3. *Each complex manifold admits a canonical almost complex structure.*

PROOF. Let M be an n-dimensional complex manifold and $(z^1, \ldots, z^n)$ a system of local holomorphic coordinates. We put $z^k = x^k + iy^k$, $k \in \{1, \ldots, n\}$.

The canonical almost complex structure J on U is given by

$$J\left(\frac{\partial}{\partial x^k}\right) = \frac{\partial}{\partial y^k}, \quad J\left(\frac{\partial}{\partial y^k}\right) = -\frac{\partial}{\partial x^k} \quad \bigl(k \in \{1, \ldots, n\}\bigr).$$

It is easy to see that the definition of J does not depend on the local holomorphic coordinates and J is an almost complex structure. □

PROPOSITION 2.4. *On a complex manifold M, the Lie algebra of infinitesimal automorphisms of the complex structure J is isomorphic with the Lie algebra of holomorphic vector fields, the isomorphism being given by $X \mapsto \frac{1}{2}(X - iJX)$.*

EXAMPLES. 1. Let S^2 be the 2-sphere and $\mathbf{H}$ the quaternion algebra. We denote by $\{e_0, e_1, e_2, e_3\}$ the standard basis of $\mathbf{H} \cong \mathbf{R}^4$.

The quaternion multiplication has the following table:

$\cdot$	e_0	e_1	e_2	e_3
e_0	e_0	e_1	e_2	e_3
e_1	e_1	$-e_0$	e_3	$-e_2$
e_2	e_2	$-e_3$	$-e_0$	e_1
e_3	e_3	e_2	$-e_1$	$-e_0$

Consider $\mathbf{R}^3 = \{q \in \mathbf{H} \mid q = x^1e_1 + x^2e_2 + x^3e_3,\ x^1, x^2, x^3 \in \mathbf{R}\}$. For $q, q' \in \mathbf{R}^3$, one has

$$qq' = -\langle q, q'\rangle e_0 + q \times q',$$

where $\langle q, q'\rangle$ is the Euclidean inner product in $\mathbf{R}^3$ and $q \times q' \in \mathbf{R}^3$ is the vector product.

For each $p \in S^2$, the tangent space T_pS^2 can be identified with

$$p^\perp = \bigl\{y \in \mathbf{R}^3 \mid \langle p, y\rangle = 0\bigr\}.$$

An almost complex structure $J_p : T_pS^2 \to T_pS^2$ is given by

$$J_p(y) = yp, \quad y \in T_pS^2.$$

REMARK. It is easy to see that J is the canonical almost complex structure associated with the standard complex structure of S^2.

2. Let S^6 be the 6-sphere and $\mathbf{Ca} = \mathbf{H} \times \mathbf{H} \cong \mathbf{R}^8$ the Cayley algebra of octanions. An almost complex structure on S^6 is defined by

$$J_x : T_x S^6 \to T_x S^6, \quad J_x(y) = yx,$$

where $x \in S^6$, $y \in T_x S^6$ and yx is the Cayley multiplication.

Clearly this almost complex structure is not associated with a complex structure on S^6.

REMARK. The existence of a complex structure on S^6 is still an open problem.

3. Let M be a 6-dimensional orientable manifold and $g : M \to S^6$ the spherical map of Gauss. The tangent spaces $T_p M$ and $T_{g(p)} S^6$ are parallel in $\mathbf{R}^7$ and can be naturally identified. Hence every almost complex structure on S^6 induces an almost complex structure on M. When J is the standard complex structure on S^6, the induced almost complex structure on M coincides with the one constructed by Calabi [8].

THEOREM 2.5 [34]. *If the sphere S^n admits an almost complex structure, then the tangent bundle $T S^{n+1}$ is trivial.*

On the other hand, it is known that the spheres having trivial tangent bundle are S^1, S^3 and S^7 (see [3]). Thus, only the spheres S^2 and S^6 carry almost complex structures.

On an almost complex manifold (M, J), the *Nijenhuis* tensor field N_j is defined by

$$N_J(X, Y) = [JX, JY] - J[JX, Y] - J[X, JY] - [X, Y],$$

for any vector fields X, Y tangent to M, where $[\cdot, \cdot]$ denotes the Lie bracket.

The Nijenhuis tensor field of every almost complex structure on a 2-dimensional orientable manifold vanishes identically.

The converse of Proposition 2.3 is false. We have the following

THEOREM 2.6 (Newlander–Nirenberg [51]). *Let J be an almost complex structure on a $2n$-dimensional differentiable manifold M. The necessary and sufficient condition for M to be a complex manifold with associated almost complex structure J is the vanishing of the Nijenhuis tensor field.*

REMARK. If (M, J) is analytic, an elegant proof is given in [36].

3. Dolbeault Lemma

Let M be an n-dimensional complex manifold and $(z^1, \dots, z^n)$ local holomorphic coordinates on M. If $z^k = x^k + i y^k$ $(k \in \{1, \dots, n\})$, then:

$$dz^k = dx^k + i\, dy^k, \qquad d\bar{z}^k = dx^k - i\, dy^k,$$

$$\frac{\partial}{\partial z^k} = \frac{1}{2}\left(\frac{\partial}{\partial x^k} - i\frac{\partial}{\partial y^k}\right), \qquad \frac{\partial}{\partial \bar{z}^k} = \frac{1}{2}\left(\frac{\partial}{\partial x^k} + i\frac{\partial}{\partial y^k}\right),$$

for any $k \in \{1, \ldots, n\}$.

Locally, a (p, q)-differential form ω on M is written as

$$\omega = \sum_{\substack{j_1<\cdots<j_p \\ k_1<\cdots<k_q}} \omega_{j_1\ldots j_p\bar{k}_1\ldots\bar{k}_q} dz^{j_1} \wedge \cdots \wedge dz^{j_p} \wedge d\bar{z}^{k_1} \wedge \cdots \wedge d\bar{z}^{k_q}.$$

The exterior derivative $d\omega$ of a (p, q)-differential form ω is decomposed as

$$d\omega = \partial\omega + \bar{\partial}\omega,$$

where

$$\partial\omega = \frac{\partial\omega_{j_1\ldots j_p\bar{k}_1\ldots\bar{k}_q}}{\partial z^j} dz^j \wedge dz^{j_1} \wedge \cdots \wedge dz^{j_p} \wedge d\bar{z}^{k_1} \wedge \cdots \wedge d\bar{z}^{k_q},$$

$$\bar{\partial}\omega = \frac{\partial\omega_{j_1\ldots j_p\bar{k}_1\ldots\bar{k}_q}}{\partial \bar{z}^k} d\bar{z}^k \wedge dz^{j_1} \wedge \cdots \wedge dz^{j_p} \wedge d\bar{z}^{k_1} \wedge \cdots \wedge d\bar{z}^{k_q}.$$

By analogy with the de Rham cohomology on a differentiable manifold, one defines the *Dolbeault* cohomology on a complex manifold M with respect to the differential operator $\bar{\partial}$.

Denote by $\mathcal{E}^{p,q}(M)$ the set of all (p, q)-differential forms on M and

$$Z^{p,q}(M, \bar{\partial}) = \{\omega \in \mathcal{E}^{p,q}(M) \mid \bar{\partial}\omega = 0\},$$

$$B^{p,q}(M, \bar{\partial}) = \{\omega \in \mathcal{E}^{p,q}(M) \mid \exists\theta \in \mathcal{E}^{p,q-1}(M) \text{ such that } \omega = \bar{\partial}\theta\}.$$

Obviously $Z^{p,q}(M, \bar{\partial})$ and $B^{p,q}(M, \bar{\partial})$ are submodules of $\mathcal{E}^{p,q}(M)$ over the ring of holomorphic functions $\mathcal{O}(M)$ and $B^{p,q}(M, \bar{\partial}) \subset Z^{p,q}(M, \bar{\partial})$.

The *Dolbeault* cohomology groups are defined by

$$H^{p,q}(M, \bar{\partial}) = Z^{p,q}(M, \bar{\partial})/_{B^{p,q}(M,\bar{\partial})}.$$

The operator $\bar{\partial}$ satisfies a Poincaré-like Lemma, known as Dolbeault Lemma.

THEOREM 3.1 (Dolbeault Lemma). *Let $D' \subset D \subset \mathbf{C}^n$ be complex polydiscs of radii $r' < r$ and $\omega \in \mathcal{E}^{p,q}(D)$, $q \geqslant 1$, such that $\bar{\partial}\omega = 0$. Then there exists $\theta \in \mathcal{E}^{p,q-1}(D)$ satisfying $\omega = \bar{\partial}\theta$ on D'.*

Using the abstract theorem of de Rham, one may prove, by the help of Dolbeault Lemma, that

$$H^{p,q}(M, \bar{\partial}) \cong H^q(M, \Omega^p),$$

where Ω^p is the sheaf of the holomorphic p-forms on the complex manifold M (see [62]).

4. Kaehler manifolds

Let (M, J) be an almost complex manifold. A Hermitian metric on M is a Riemannian metric g invariant by J, i.e.,

$$g(JX, JY) = g(X, Y), \quad \forall X, Y \in \Gamma(TM).$$

Every almost complex manifold admits a Hermitian metric.

Any Hermitian metric g on the almost complex manifold M determines a non-degenerate 2-form $\omega(X, Y) = g(JX, Y)$, $X, Y \in \Gamma(TM)$, called the *fundamental* (*or Kaehler*) 2-form. Clearly $\omega(JX, JY) = \omega(X, Y)$.

An (almost) complex manifold M endowed with a Hermitian metric g is called an (almost) Hermitian manifold.

For any vector fields X, Y, Z on an almost Hermitian manifold M one has the following formula:

$$4g\big((\nabla_X J)Y, Z\big) = 6d\omega(X, Y, Z) - 6d\omega(X, JY, JZ) + g\big(N_J(Y, Z), JX\big),$$

where ∇ is the Levi-Civita connection with respect to g.

Let (M, g) be an n-dimensional Hermitian manifold and ω its fundamental 2-form. Locally, we may write:

$$g = \sum_{j,k=1}^{n} g_{j\bar{k}}\, dz^j\, d\bar{z}^k = \sum_{j=1}^{n} \varphi_j \otimes \bar{\varphi}_j,$$

$$\omega = i \sum_{j,k=1}^{n} g_{j\bar{k}}\, dz^j \wedge d\bar{z}^k,$$

where $\{\varphi_1, \ldots, \varphi_n\} \subset \mathcal{E}^{1,0}(M)$ is a local orthonormal frame.

LEMMA 4.1. *There exists a unique matrix ψ of 1-forms such that*
(i) $\psi^i_j + \bar{\psi}^j_i = 0$,
(ii) $d\varphi_i = \psi^j_i \wedge \varphi_j + \tau_i$,
where τ_j are $(2, 0)$-forms, called torsion 2-forms.

J.A. Schouten, D. van Dantzig [53] and E. Kähler [32] discovered an important class of Hermitian manifolds, known as Kaehler manifolds.

DEFINITION. An (almost) Hermitian manifold is said to be (*almost*) *Kaehlerian* if the fundamental 2-form ω is closed.

The following theorem provides equivalent conditions to the definition of a Kaehler manifold.

THEOREM 4.2. *Let (M,g) be an n-dimensional Hermitian manifold and ∇ the Levi-Civita connection with respect to g. Then the following assertions are equivalent to each other*:

(i) *M is a Kaehler manifold.*

(ii) *The canonical almost complex structure J on M is parallel with respect to ∇, i.e., $\nabla J = 0$.*

(iii) *The torsion 2-forms τ_j vanish identically, for all $j \in \{1,\dots,n\}$.*

(iv) *For any $z_0 \in M$, there exist local holomorphic coordinates $(z_1,\dots,z_n)$ in a neighborhood of z_0 such that*

$$g_{j\bar{k}} = \delta_{jk} + h_{jk},$$

with $h_{jk}(z_0) = \frac{\partial h_{jk}}{\partial z^l}(z_0) = 0$.

(v) *Locally, there exists a real differentiable function F such that the fundamental 2-form ω is given by*

$$\omega = i\partial\bar{\partial}F.$$

PROOF. (i) $\Leftrightarrow$ (ii) Straightforward calculations lead to

$$2g\big((\nabla_X J)Y, Z\big) = 3\,d\omega(X,Y,Z) - 3\,d\omega(X,JY,JZ)$$

and

$$3\,d\omega(X,Y,Z) = -g\big(X,(\nabla_Z J)Y\big) + g\big(Y,(\nabla_X J)Z\big) - g\big(Z,(\nabla_Y J)X\big),$$

respectively, for any vector fields X, Y, Z on M.

(i) $\Leftrightarrow$ (iii) Computing, we find

$$\begin{aligned} -i\,d\omega &= d\varphi_j \wedge \bar{\varphi}_j - \varphi_j \wedge d\bar{\varphi}_j \\ &= \psi_j^k \wedge \varphi_k \wedge \bar{\varphi}_j - \varphi_j \wedge \bar{\psi}_j^k \wedge \bar{\varphi}_k + \tau_j \wedge \bar{\varphi}_j - \varphi_j \wedge \bar{\tau}_j. \end{aligned}$$

Since

$$\psi_j^k \wedge \varphi_k \wedge \bar{\varphi}_j - \varphi_j \wedge \bar{\psi}_j^k \wedge \bar{\varphi}_k = \psi_j^k \wedge \varphi_k \wedge \bar{\varphi}_j + \varphi_j \wedge \psi_k^j \wedge \bar{\varphi}_k = 0$$

and $\varphi_j \in \mathcal{E}^{1,0}(M)$ and $\bar{\varphi}_j \in \mathcal{E}^{0,1}$ are linearly independent at every point, one gets $d\omega = 0$ if and only if $\tau_j = 0$, $\forall j \in \{1,\dots,n\}$.

(iv) $\Rightarrow$ (i) is trivial.

(i) $\Rightarrow$ (iv) We assume that (M,g) is a Kaehler manifold. Locally, we may write:

$$\omega = i\big(\delta_{jk} + a_{jkl}z^l + a_{jk\bar{l}}\bar{z}^l + h'_{jk}\big)\,dz^j \wedge d\bar{z}^k,$$

with $h'_{jk}(z_0) = \frac{\partial h'_{jk}}{\partial z^l}(z_0) = 0$.

Then, one has

$$g_{k\bar{j}} = \overline{g_{j\bar{k}}} \quad \Rightarrow \quad a_{jk\bar{l}} = \overline{a_{kj\bar{l}}},$$
$$d\omega = 0 \quad \Rightarrow \quad a_{jkl} = a_{lkj}.$$

If we change the local holomorphic coordinates by

$$z^j = w^j - \frac{1}{2} a_{mjl} w^l w^m,$$

we get

$$dz^j = dw^j - a_{mjl} w^l \, dw^m$$

and

$$\begin{aligned}
-i\omega &= \left(dw^j - a_{mjl} w^l \, dw^m\right) \wedge \left(d\bar{w}^j - a_{jq\bar{p}} \bar{w}^p \, d\bar{w}^q\right) \\
&\quad + \left(a_{jkl} w^l + a_{jk\bar{l}} \bar{w}^l\right) dw^j \wedge d\bar{w}^k + h''_{jk} \, dw^j \wedge d\bar{w}^k \\
&= \left(\delta_{jk} + a_{jkl} w^l + a_{jk\bar{l}} \bar{w}^l - a_{jkl} w^l - a_{jk\bar{l}} \bar{w}^l + h_{jk}\right) dw^j \wedge d\bar{w}^k \\
&= (\delta_{jk} + h_{jk}) \, dw^j \wedge d\bar{w}^k,
\end{aligned}$$

with $h''_{jk}(z_0) = \frac{\partial h''_{jk}}{\partial w^l}(z_0) = 0$ and $h_{jk}(z_0) = \frac{\partial h_{jk}}{\partial w^l}(z_0) = 0$.

(v) $\Rightarrow$ (i) is trivial.

(i) $\Rightarrow$ (v) Let (M, g) be a Kaehler manifold. Since its fundamental 2-form ω is closed, locally, there exists a real differentiable 1-form φ such that $d\varphi = \omega$. Then we have $\varphi = \theta + \theta'$, with θ a $(1,0)$-form and θ' a $(0,1)$-form. But φ is real, then $\theta' = \bar{\theta}$ and so $\varphi = \theta + \bar{\theta}$. It follows that

$$d\varphi = \partial\theta + \bar{\partial}\theta + \partial\bar{\theta} + \bar{\partial}\bar{\theta}.$$

Since ω is a $(1,1)$-form, we obtain $\partial\theta = 0$ and $\bar{\partial}\bar{\theta} = 0$. By Dolbeault Lemma, there exists a complex function f such that $\bar{\theta} = \bar{\partial} f$. Therefore, we have

$$\omega = d\varphi = \partial\bar{\partial} f + \bar{\partial}\partial \bar{f} = \partial\bar{\partial}(f - \bar{f});$$

but $f - \bar{f}$ is purely imaginary, so $F = -i(f - \bar{f})$ is real and $\omega = i\partial\bar{\partial}F$.

COROLLARY 4.3. *Let M be a Kaehler manifold, R its curvature tensor field and S its Ricci tensor field, respectively. Then*:

(i) $R(X,Y)J = JR(X,Y)$, $R(JX,JY) = R(X,Y)$;

(ii) $S(JX,JY) = S(X,Y)$, $S(X,Y) = \frac{1}{2}\,\mathrm{Trace}\, JR(X,JY)$,

for any $X, Y \in \Gamma(TM)$.

REMARK 1. The condition (iv) of Theorem 4.2 is equivalent to the existence of an orthonormal frame $\{\varphi_1, \dots, \varphi_n\}$ of $(1,0)$-forms in the neighborhood of any point $z_0 \in M$ such that $d\varphi_j(z_0) = 0$.

We obtain the following *general principle*:

Any identity in terms of the Hermitian metric and its first order derivatives which holds true on $\mathbf{C}^n$ equipped with the canonic Euclidean metric, holds also on any Kaehler manifold endowed with the corresponding metric.

REMARK 2. If the local expression of the fundamental 2-form of a Kaehler manifold is

$$\omega = i\partial\bar{\partial}F = i\frac{\partial^2 F}{\partial z^j \partial \bar{z}^k} dz^j \wedge d\bar{z}^k,$$

then the coefficients of the Hermitian metric g are

$$g_{j\bar{k}} = \frac{\partial^2 F}{\partial z^j \partial \bar{z}^k}, \quad \forall j, k \in \{1, \dots, n\}.$$

EXAMPLES OF KAEHLER MANIFOLDS. 1. The complex space $\mathbf{C}^n$ endowed with the Euclidean metric $g = \sum_{k=1}^n dz^k \, d\bar{z}^k$ is a complete, flat Kaehler manifold.

2. Let $T^n = \mathbf{C}^n/G$ be a complex torus (see Example 6 in Section 1) endowed with the Hermitian metric induced by the Euclidean metric on $\mathbf{C}^n$. The complex tori are the only complex parallelizable manifolds which admit Kaehler metrics [61].

3. The complex projective space $P^n(\mathbf{C})$.

Let $(z_j^1, \dots, z_j^{j-1}, z_j^{j+1}, \dots, z_j^{n+1})$ be local holomorphic coordinates on $U_j \subset P^n(\mathbf{C})$. We put

$$f_j(z) = \sum_{k=1}^{n+1} |z_j^k|^2,$$

where $z_j^j = 1$. Then the 2-form ω, defined on U_j by

$$\omega = i\partial\bar{\partial} \log f_j,$$

is globally defined on $P^n(\mathbf{C})$ and closed.

For $j = n+1$, $f_{n+1} = 1 + \sum_{k=1}^n |z^k|^2$ and

$$\omega = i\frac{(1 + z^s \bar{z}^s) \, dz^k \wedge d\bar{z}^k - \bar{z}^k \, dz^k \wedge z^j \, d\bar{z}^j}{(1 + z^s \bar{z}^s)^2}.$$

Thus the Kaehlerian metric g associated to ω has the coefficients

$$g_{j\bar{k}} = \frac{(1 + z^s \bar{z}^s)\delta_{jk} - z^k \bar{z}^j}{(1 + z^s \bar{z}^s)^2}.$$

This metric is known as the *Fubini–Study* metric.

4. The complex Grassmann manifold $G_p(\mathbf{C}^{p+q})$ endowed with a generalized Fubini–Study metric (see [36]).

5. Let $D^n = \operatorname{Int} S^{2n-1}$ be the complex unit disk in $\mathbf{C}^n$, i.e.,

$$D^n = \left\{ z \in \mathbf{C}^n;\ \sum_{j=1}^{n} |z^j|^2 < 1 \right\}.$$

Put $\omega = -i\partial\bar{\partial}\log(1 - \sum_{j=1}^{n} |z^j|^2)$. The associated Kaehlerian metric is

$$g_{j\bar{k}} = \frac{(1 - z^s \bar{z}^s)\delta_{jk} + \bar{z}^j z^k}{(1 - z^s \bar{z}^s)^2}.$$

This metric is called the *Bergman* metric.

6. Any complex submanifold of a Kaehler manifold is a Kaehler manifold. Moreover it is a minimal submanifold. In particular, any algebraic manifold is Kaehlerian.

7. Any orientable surface is a Kaehler manifold.

A topological obstruction to the existence of Kaehlerian metrics on a compact complex manifold is given by the following

THEOREM 4.4. *On a compact Kaehler manifold, the de Rham cohomology groups of even order are non-trivial.*

PROOF. Let M be an n-dimensional compact Kaehler manifold. Denote by $\omega^n = \omega \wedge \cdots \wedge \omega$ (n times).

Since ω^n nowhere vanishes, M is orientable. We consider the orientation on M such that ω^n be positive. Then

$$\int_M \omega^n > 0.$$

On the other hand, since M is compact, Stokes theorem implies that ω^n cannot be exact. Thus $[\omega^n] \neq 0$ in $H^{2n}(M, d)$.

We will prove that $[\omega^k] \neq 0$ in $H^{2k}(M, d)$, for any $k \in \{1, \ldots, n-1\}$.

We assume there exists $\theta \in \mathcal{E}^{2k-1}(M)$ such that $\omega^k = d\theta$. Then

$$\omega^n = \omega^k \wedge \omega^{n-k} = d\theta \wedge \omega^{n-k} = d\left(\theta \wedge \omega^{n-k}\right),$$

which is a contradiction. Therefore $H^{2k}(M, d) \neq 0$, $\forall k \in \{1, \ldots, n\}$. Obviously $H^0(M, d) \neq 0$. □

COROLLARY 4.5. *The Calabi manifolds $S^{2m+1} \times S^{2n+1}$ do not admit any Kaehlerian structure if $(m, n) \neq (0, 0)$. In particular, Hopf manifolds are not Kaehler manifolds.*

In Section 7, we will state an important result concerning the Betti numbers of odd order on a compact Kaehler manifold. As an application, we will construct an example of an almost Kaehler manifold which does not admit any Kaehlerian metric.

By using the decomposition theorem of de Rham and a result of J.I. Hano and Y. Matsushima [27], we state the following theorem of decomposition for Kaehler manifolds.

THEOREM 4.6. *Let M be a simply connected complete Kaehler manifold. Then it is holomorphically isometric to a direct product $M_0 \times M_1 \times \cdots \times M_k$, where M_0 is a complex Euclidean space and $M_1, \ldots, M_k$ are simply connected, complete, irreducible Kaehler manifolds. Moreover, such a decomposition is unique.*

A sufficient condition for a compact Kaehler manifold to be isometric to a complex projective space was proved by M. Berger [5].

THEOREM 4.7. *Let M be a compact Kaehler manifold with positive sectional curvature and constant scalar curvature. Then M is isometric to a complex projective space.*

THEOREM 4.8 (Kodaira–Yau). *Any compact Kaehler manifold homeomorphic to $P^n(\mathbf{C})$ is analytic diffeomorphic to $P^n(\mathbf{C})$. When $n = 2$, one can drop the Kaehlerness assumption.*

T. Frankel [21] proved that a compact Kaehler surface with positive bisectional curvature must be analytic isometric to the complex projective plane $P^2(\mathbf{C})$. He conjectured that the same be true in higher dimension. This is known as the Frankel's conjecture. It was proved by Y.T. Siu and S.T. Yau [55] using a differential geometric method. An algebraic proof of Frankel's conjecture is due to S. Mori [49].

Recall a nice result of H. Wu, which says that the universal covering space of any complete Kaehler manifold with non-positive sectional curvature is always a Stein manifold (see [65]).

THEOREM 4.9 (Wu). *Any simply-connected, complete Kaehler manifold with non-positive sectional curvature is a Stein manifold.*

5. Complex space forms

We recall the geometric interpretation of the *sectional curvature* of the plane section spanned by the linearly independent vectors $u, v \in T_pM$, $p \in M$. It is the Gauss curvature of the surface

$$(\lambda, \mu) \mapsto \exp_p(\lambda u + \mu v).$$

PROPOSITION 5.1. *Let M be a Kaehler manifold of complex dimension $n > 1$. If M has constant sectional curvature, then it is flat.*

REMARK. It is known that in this case M is locally isometric to the complex Euclidean space $\mathbf{C}^n$.

Thus, the notion of constant sectional curvature for a Kaehler manifold is not essential. For this reason, one introduces the notion of holomorphic sectional curvature.

Let M be a Kaehler manifold and J its canonical almost complex structure. The sectional curvature of a holomorphic plane section π (i.e., $J\pi = \pi$) is called a *holomorphic sectional curvature* of M.

Since the plane section π is invariant by J, we may choose an orthonormal basis $\{X, JX\}$ of π, with unit X. Then, the holomorphic sectional curvature $K(\pi)$ is given by $K(\pi) = R(X, JX, X, JX)$.

A version of the well-known Schur theorem holds for Kaehler manifolds.

The curvature tensor field of a Kaehler manifold satisfies the following identities:

(i) $R(X, Y, Z, W) = -R(Y, X, Z, W) = -R(X, Y, W, Z)$;

(ii) $R(X, Y, Z, W) = R(Z, W, X, Y)$;

(iii) $R(X, Y, Z, W) + R(X, Z, W, Y) + R(X, W, Y, Z) = 0$;

(iv) $R(JX, JY, Z, W) = R(X, Y, JZ, JW) = R(X, Y, Z, W)$,

for any vector fields X, Y, Z, W on M.

DEFINITION. Let M be a Kaehler manifold. If the holomorphic sectional curvature function is constant for all holomorphic plane sections π in T_pM and all points $p \in M$, then M is said to be a *complex space form* or a space of constant holomorphic sectional curvature.

A complex space form having constant holomorphic sectional curvature c is denoted by $M(c)$.

Using similar ideas as in the proof of Schur theorem, we state the following

THEOREM 5.2. *Let M be an n-dimensional ($n \geqslant 2$) connected Kaehler manifold. If the holomorphic sectional curvature depends only on the point $p \in M$ (and does not depend on the holomorphic plane sections π in T_pM), then M is a complex space form.*

The proof can be read in [36] or [63].

The curvature tensor R of a complex space form $M(c)$ is given by

$$
\begin{aligned}
R(X,Y,Z,W) = \frac{c}{4}\{&g(X,Z)g(Y,W) - g(X,W)g(Y,Z) \\
&- g(JX,W)g(JY,Z) + g(JX,Z)g(JY,W) \\
&+ 2g(X,JY)g(Z,JW)\},
\end{aligned}
$$

for any vector fields X, Y, Z, W on $M(c)$.

Then its Ricci tensor is given by $S = \frac{1}{2}(n+1)cg$. Thus each complex space form is an Einstein space.

EXAMPLES. 1. The complex Euclidean space $\mathbf{C}^n$ endowed with the Euclidean metric is a flat complex space form ($c = 0$).

2. The complex projective space $P^n(\mathbf{C})$ endowed with the Fubini–Study metric has positive constant holomorphic sectional curvature ($c = 4$).

3. The complex unit disk D^n endowed with the Bergman metric has negative constant holomorphic sectional curvature ($c = -4$).

Conversely, one has the following result of [28] and [31] (see also [36]).

THEOREM 5.3. *Let M be an n-dimensional simply connected complete complex space form. Then M is isometric to either the complex Euclidean space $\mathbf{C}^n$, the complex projective space $P^n(\mathbf{C})$, or the complex unit disk D^n, according as $c = 0$, $c > 0$ or $c < 0$, respectively.*

We mention that a complete Kaehler manifold of positive constant holomorphic sectional curvature is necessarily simply connected (see [57,35]).

Necessary and sufficient conditions for a Kaehler manifold to be a complex space form are given by the following theorem of K. Nomizu [52].

THEOREM 5.4. *Let M be a $2n$-dimensional Kaehler manifold. The following assertions are equivalent*:

(i) *M is a complex space form*;

(ii) *M satisfies the axiom of holomorphic $2k$-planes, for some k, $1 \leqslant k \leqslant n-1$, i.e., for any $2k$-dimensional holomorphic subspace S of T_pM, there exists a $2k$-dimensional totally geodesic submanifold V of M containing p, such that $T_pV = S$, for all $p \in M$*;

(iii) *M satisfies the axiom of totally real k-planes, for some k, $2 \leqslant k \leqslant n$, i.e., for any k-dimensional totally real subspace S of T_pM, there exists a k-dimensional totally geodesic submanifold V of M containing p, such that $T_pV = S$, for all $p \in M$.*

The Bochner curvature tensor B on an n-dimensional Kaehler manifold M is defined by

$$
B(X,Y)Z = R(X,Y)Z - \frac{1}{2n+4}\big[g(Y,Z)QX - g(QX,Z)Y + g(JY,Z)QJX
$$

$$
\begin{aligned}
&- g(QJX, Z)JY + g(QY, Z)X - g(X, Z)QY + g(QJY, Z)JX \\
&- g(JX, Z)QJY - 2g(JX, QY)JZ - 2g(JX, Y)QJZ\big] \\
&+ \frac{\tau}{(2n+2)(2n+4)}\big[g(Y, Z)X - g(X, Z)Y + g(JY, Z)JX \\
&- g(JX, Z)JY - 2g(JX, Y)JZ\big],
\end{aligned}
$$

where Q and τ denote the Ricci operator and scalar curvature, respectively.

M. Matsumoto and S. Tanno [41] proved the following result.

THEOREM 5.5. *If a Kaehler manifold with vanishing Bochner curvature tensor has constant scalar curvature, then either*

(i) *M is a complex space form, or*

(ii) *M is locally the Riemannian product of two complex space forms $M_1(c)$ and $M_2(-c)$.*

Recently, T. Adachi and S. Maeda [2] gave characterizations of complex space forms in terms of geodesics and circles on their geodesic spheres.

THEOREM 5.6. *A Kaehler manifold M of complex dimension $n \geqslant 2$ is a complex space form if and only if, at an arbitrary point $p \in M$, every sufficiently small geodesic sphere has constant structure torsion.*

THEOREM 5.7. *A Kaehler manifold M of complex dimension $n \geqslant 2$ is flat if and only if, at an arbitrary point $p \in M$ and for a sufficiently small $r > 0$, there exists $k_{p,r} > 0$ such that every circle of curvature $k_{p,r}$ on a geodesic sphere $G_{p,r}$ with radius r has constant first curvature as a curve in M.*

For complex submanifolds in the complex Euclidean space, we state the following results.

THEOREM 5.8 (Smyth–Chern). *Let M be an n-dimensional complex submanifold of $\mathbf{C}^{n+1}$. If M has constant Ricci curvature, then it must be flat and totally geodesic.*

THEOREM 5.9 [59]. *Let M be an n-dimensional complex submanifold of $\mathbf{C}^{n+2}$. If M has constant Ricci curvature, then it must be flat and totally geodesic.*

More generally, any Kaehler–Einstein submanifold M in a complex Euclidean space $\mathbf{C}^m$ is totally geodesic [60].

6. Laplace–Beltrami operator on a Hermitian manifold

In this section, we define the Laplace–Beltrami operator on a Hermitian manifold and state its general properties.

Let (M, J, g) be an n-dimensional Hermitian manifold and ω its fundamental 2-form. Then, locally, one has

$$\omega^n = i^n n!(\det g)\, dz^1 \wedge d\bar{z}^1 \wedge \cdots \wedge dz^n \wedge d\bar{z}^n.$$

If we put, $z^\alpha = x^{2\alpha-1} + ix^{2\alpha}$, then

$$\frac{\omega^n}{n!} = 2^n (\det g)\, dx^1 \wedge \cdots \wedge dx^{2n}$$

is the standard volume form on M.

In particular, the volume of M is given by $\frac{1}{n!}\int_M \omega^n$. In general, if N is a k-dimensional complex submanifold of M, the volume of N is given by

$$\operatorname{vol}(N) = \frac{1}{k!}\int_N \omega^k.$$

This result is known as Wirtinger's theorem (see [65]).

We can define the integral of a continuous function f on M by

$$\int_M f(z)\frac{\omega^n}{n!} = \int_M f(z) 2^n (\det g)\, dx^1 \ldots dx^{2n}.$$

Let $\varphi, \psi \in \mathcal{E}^{p,q}(M)$. Locally, we may write

$$\varphi(z) = \varphi_{\alpha_1\ldots\alpha_p\bar{\beta}_1\ldots\bar{\beta}_q}(z)\, dz^{\alpha_1} \wedge \cdots \wedge dz^{\alpha_p} \wedge d\bar{z}^{\beta_1} \wedge \cdots \wedge d\bar{z}^{\beta_q},$$

$$\psi(z) = \psi_{\alpha_1\ldots\alpha_p\bar{\beta}_1\ldots\bar{\beta}_q}(z)\, dz^{\alpha_1} \wedge \cdots \wedge dz^{\alpha_p} \wedge d\bar{z}^{\beta_1} \wedge \cdots \wedge d\bar{z}^{\beta_q},$$

with $1 \leqslant \alpha_1 < \cdots < \alpha_p \leqslant n$, $1 \leqslant \beta_1 < \cdots < \beta_q \leqslant n$.

Sometimes, we will use the abbreviated notation

$$\varphi(z) = \varphi_{A_p\bar{B}_q}(z)\, dz^{A_p} \wedge d\bar{z}^{B_q}.$$

The scalar product at a point $z \in M$ is given by

$$(\varphi, \psi)(z) = \varphi_{\alpha_1\ldots\alpha_p\bar{\beta}_1\ldots\bar{\beta}_q}(z)\bar{\psi}^{\alpha_1\ldots\alpha_p\bar{\beta}_1\ldots\bar{\beta}_q}(z),$$

where

$$\bar{\psi}^{\alpha_1\ldots\alpha_p\bar{\beta}_1\ldots\bar{\beta}_q}(z) = g^{\bar{\lambda}_1\alpha_1}\ldots g^{\bar{\lambda}_p\alpha_p} g^{\bar{\beta}_1\mu_1}\ldots g^{\bar{\beta}_q\mu_q}\overline{\psi_{\lambda_1\ldots\lambda_p\bar{\mu}_1\ldots\bar{\mu}_q}(z)}.$$

The (global) scalar product is defined by

$$\langle\varphi, \psi\rangle = \int_M (\varphi, \psi)(z)\frac{\omega^n}{n!}.$$

For any $\varphi, \psi \in \mathcal{E}^{p,q}(M)$, we have:

(i) $\langle \varphi, \psi \rangle = \overline{\langle \psi, \varphi \rangle}$;
(ii) $\langle \varphi, \varphi \rangle \geqslant 0$,
$\langle \varphi, \varphi \rangle = 0 \Leftrightarrow \varphi = 0$.
The norm of φ is given by $\|\varphi\|^2 = \langle \varphi, \varphi \rangle$.

PROPOSITION 6.1. *For each $\psi \in \mathcal{E}^{p,q}(M)$, there exists a unique $*\bar{\psi} \in \mathcal{E}^{n-p,n-q}(M)$ such that*

$$(\varphi, \psi)(z)\frac{\omega^n}{n!} = \varphi(z) \wedge *\bar{\psi}(z), \quad \forall \varphi \in \mathcal{E}^{p,q}(M).$$

If $\psi(z) = \psi_{A_p \bar{B}_q}(z)\, dz^{A_p} \wedge d\bar{z}^{B_q}$, we denote $A_{n-p} = (\alpha_{p+1}, \dots, \alpha_n)$, with $\alpha_{p+1} < \cdots < \alpha_n$ and $\{\alpha_1, \dots, \alpha_p, \alpha_{p+1}, \dots, \alpha_n\} = \{1, \dots, n\}$. Then

$$*\bar{\psi}(z) = i^n(-1)^k \sum_{A_p, B_q} \varepsilon \begin{pmatrix} A_p & A_{n-p} \\ B_q & B_{n-q} \end{pmatrix} (\det g)\bar{\psi}^{A_p \bar{B}_q}(z)\, dz^{A_{n-p}} \wedge d\bar{z}^{B_{n-q}},$$

where $k = \frac{1}{2}n(n-1) + (n-p)q$.

COROLLARY 6.2. *For each $\psi \in \mathcal{E}^{p,q}(M)$, one has*

$$*\psi(z) = i^n(-1)^{\frac{n(n-1)}{2}+np} \sum_{A_p, B_q} \varepsilon \begin{pmatrix} A_p & A_{n-p} \\ B_q & B_{n-q} \end{pmatrix} (\det g)\psi^{\bar{A}_p B_q}\, dz^{B_{n-q}} \wedge d\bar{z}^{A_{n-p}}.$$

The linear operator $*: \mathcal{E}^{p,q}(M) \to \mathcal{E}^{n-q,n-p}(M)$ has the following properties:
(i) $\overline{*\psi} = *\bar{\psi}$;
(ii) $**\psi = (-1)^{p+q}\psi$,
for each $\psi \in \mathcal{E}^{p,q}(M)$.

We may write

$$\langle \varphi, \psi \rangle = \int_M \varphi \wedge *\bar{\psi}, \quad \forall \varphi, \psi \in \mathcal{E}^{p,q}(M).$$

DEFINITION. Let (M, g) be a compact Hermitian manifold. The operator

$$\delta: \mathcal{E}^{p,q}(M) \to \mathcal{E}^{p,q-1}(M), \quad q \geqslant 1,$$

defined by

$$\delta = - * \partial *,$$

is a differential operator of order 1.

THEOREM 6.3. *Let M be a compact Hermitian manifold. Then the differential operator δ is the formal adjoint of the operator $\bar{\partial}$, i.e.,*

$$\langle \bar{\partial}\varphi, \psi \rangle = \langle \varphi, \delta\psi \rangle, \quad \forall \varphi \in \mathcal{E}^{p,q-1}(M),\ \psi \in \mathcal{E}^{p,q}(M).$$

PROOF. Stokes Theorem implies:

$$0 = \int_M d(\varphi \wedge *\bar{\psi}) = \int_M \bar{\partial}(\varphi \wedge *\bar{\psi}) = \int_M \bar{\partial}\varphi \wedge *\bar{\psi} - (-1)^{p+q} \int_M \varphi \wedge \bar{\partial} * \bar{\psi}$$
$$= \int_M \bar{\partial}\varphi \wedge *\bar{\psi} - \int_M \varphi \wedge *\overline{\delta\psi} = \langle \bar{\partial}\varphi, \psi \rangle - \langle \varphi, \delta\psi \rangle.$$

If $\psi \in \mathcal{E}^{p,q}(M)$, then

$$(\delta\psi)^{\bar{A}_p \beta_2 \ldots \beta_q}(z) = -\frac{1}{\det g} \frac{\partial}{\partial z^\beta} \big[(\det g)(z) \psi^{\beta \bar{A}_p \beta_2 \ldots \beta_q}(z)\big].$$ □

DEFINITION. The Laplace–Beltrami operator

$$\Box : \mathcal{E}^{p,q}(M) \to \mathcal{E}^{p,q}(M)$$

is defined by

$$\Box = \bar{\partial}\delta + \delta\bar{\partial}.$$

COROLLARY 6.4. *For each $\varphi \in \mathcal{E}^{p,q}(M)$,*

$$(\Box\varphi)(z) = -g^{\bar{\beta}\alpha}(z) \frac{\partial^2}{\partial z^\alpha \partial \bar{z}^\beta} \varphi_{A_p \bar{B}_q}(z)\, dz^{A_p} \wedge d\bar{z}^{B_q} + L\varphi(z),$$

where L is a differential operator of order 1 whose coefficients are polynomials of $g_{\alpha\bar{\beta}}$, $g^{\bar{\beta}\alpha}$ and their first order partial derivatives.

THEOREM 6.5. *The Laplace–Beltrami operator on a compact Hermitian manifold is self-adjoint and positive definite, i.e.,*

(i) $\langle \Box\varphi, \psi \rangle = \langle \varphi, \Box\psi \rangle$,

(ii) $\langle \Box\varphi, \varphi \rangle = \|\bar{\partial}\varphi\|^2 + \|\delta\varphi\|^2$,

for any $\varphi, \psi \in \mathcal{E}^{p,q}(M)$.

A differential form $\varphi \in \mathcal{E}^{p,q}(M)$ is called *harmonic* if $\Box\varphi = 0$.

Obviously, a differential form $\varphi \in \mathcal{E}^{p,q}(M)$ is harmonic if and only if it is $\bar{\partial}$-closed and δ-closed.

We denote by $\mathcal{H}^{p,q} = \{\varphi \in \mathcal{E}^{p,q}(M) \mid \varphi \text{ harmonic}\}$.

Any self-adjoint and positive definite differential operator satisfies the following kernel-image theorem.

THEOREM 6.6 [37]. *$\mathcal{E}^{p,q}(M) = \mathcal{H}^{p,q} \oplus \Box\mathcal{E}^{p,q}(M)$, with mutually orthogonal factors.*

COROLLARY 6.7. *$\mathcal{E}^{p,q}(M) = \mathcal{H}^{p,q} \oplus \bar{\partial}\mathcal{E}^{p,q-1}(M) \oplus \delta\mathcal{E}^{p,q+1}(M)$, all factors being mutually orthogonal.*

Using the above results, we may prove the following important theorem.

THEOREM 6.8 (Hodge). *Let M be a compact Hermitian manifold. Then the Dolbeault cohomology group $H^{p,q}(M, \bar{\partial})$ is isomorphic to $\mathcal{H}^{p,q}$.*

Dolbeault theorem (see, for instance, [20]) implies the following

THEOREM 6.9 (Hodge–Dolbeault). *Let M be a compact Hermitian manifold. Then*

$$H^q(M, \Omega^p) \cong \mathcal{H}^{p,q}, \quad \forall q \in \mathbf{N}.$$

COROLLARY 6.10. *Let M be a compact Hermitian manifold. Then $H^q(M, \Omega^p)$ is a finite-dimensional linear space.*

7. Harmonic differential forms on Kaehler manifolds

Let M be a Kaehler manifold and $\Box$ and Δ the Laplacians with respect to $\bar{\partial}$ and d, respectively.

On the complex Euclidean space endowed with the Euclidean metric, by a straightforward computation, we get

$$\Box = \frac{1}{2}\Delta.$$

We will show that this relation holds on any Kaehler manifold.

Define an operator $L : \mathcal{E}^{p,q}(M) \to \mathcal{E}^{p+1,q+1}(M)$, by

$$L(\eta) = \omega \wedge \eta.$$

Let $\Lambda = L^* : \mathcal{E}^{p,q}(M) \to \mathcal{E}^{p-1,q-1}(M)$ be its adjoint. Also, define the operator

$$d^c = \frac{i}{4\pi}(\bar{\partial} - \partial).$$

Obviously

$$dd^c = -d^c d = \frac{i}{2\pi}\partial\bar{\partial}.$$

PROPOSITION 7.1. *On any Kaehler manifold, we have the following relations:*

(i) $[\Lambda, d] = -4\pi d^{c*}$;
(ii) $[L, d^*] = 4\pi d^c$.

PROOF. It is easy to see that (i) and (ii) are equivalent each other. By decomposition, the relation (i) is equivalent to

$$\begin{cases}[\Lambda, \bar{\partial}] = -i\partial^*,\\ [\Lambda, \partial] = i\delta.\end{cases}$$

Since Λ, d and d^c are real operators, each of these relations implies the others. We will prove that

$$[\Lambda, \partial] = i\delta.$$

First, we prove it on $\mathbf{C}^n$ equipped with the Euclidean metric.

We introduce new operators acting on the differential forms with compact support on $\mathbf{C}^n$. For any $k \in \{1, \dots, n\}$, let $e_k : \mathcal{E}_0^{p,q}(\mathbf{C}^n) \to \mathcal{E}_0^{p+1,q}(\mathbf{C}^n)$, given by

$$e_k(\varphi) = dz^k \wedge \varphi,$$

and $\bar{e}_k : \mathcal{E}_0^{p,q}(\mathbf{C}^n) \to \mathcal{E}_0^{p,q+1}(\mathbf{C}^n)$, given by

$$\bar{e}_k(\varphi) = d\bar{z}^k \wedge \varphi.$$

Let i_k and $\bar{i}_k$ be the adjoint operators of e_k and $\bar{e}_k$, respectively. Clearly $e_k, \bar{e}_k, i_k$ and $\bar{i}_k$ are linear over $C^\infty(\mathbf{C}^n)$.

One has $i_k(dz^J \wedge d\bar{z}^K) = 0$, if $k \notin J$.

Since $(dz^k, dz^k) = 2$, it follows that

$$i_k\big(dz^k \wedge dz^J \wedge d\bar{z}^K\big) = 2\, dz^J \wedge d\bar{z}^K.$$

Indeed,

$$\big(i_k\big(dz^J \wedge d\bar{z}^K\big), dz^L \wedge d\bar{z}^M\big) = \big(dz^J \wedge d\bar{z}^K, dz^k \wedge dz^L \wedge d\bar{z}^M\big) = 0$$

and

$$\begin{aligned}\big(i_k\big(dz^k \wedge dz^J \wedge d\bar{z}^K\big), dz^L \wedge d\bar{z}^M\big) &= \big(dz^k \wedge dz^J \wedge d\bar{z}^K, dz^k \wedge dz^L \wedge d\bar{z}^M\big)\\ &= 2\big(dz^J \wedge d\bar{z}^K, dz^L \wedge d\bar{z}^M\big),\end{aligned}$$

respectively.

Similarly

$$\bar{i}_k\big(dz^J \wedge d\bar{z}^K\big) = 0, \quad k \notin K,$$

and

$$\bar{i}_k\left(d\bar{z}^k \wedge dz^J \wedge d\bar{z}^K\right) = 2\,dz^J \wedge d\bar{z}^K.$$

Also

$$i_k e_k\left(dz^J \wedge d\bar{z}^K\right) = \begin{cases} 0, & k \in J, \\ 2\,dz^J \wedge d\bar{z}^K, & k \notin J, \end{cases}$$

$$e_k i_k\left(dz^J \wedge d\bar{z}^K\right) = \begin{cases} 2\,dz^J \wedge d\bar{z}^K, & k \in J, \\ 0, & k \notin J. \end{cases}$$

Then $i_k e_k + e_k i_k = 2\,\mathrm{id}$ and similarly $\bar{i}_k \bar{e}_k + \bar{e}_k \bar{i}_k = 2\,\mathrm{id}$.

For $k \neq l$, we have

$$\begin{aligned} i_k e_l\left(dz^k \wedge dz^J \wedge d\bar{z}^K\right) &= i_k\left(dz^l \wedge dz^k \wedge dz^J \wedge d\bar{z}^K\right) \\ &= i_k\left(-dz^k \wedge dz^l \wedge dz^J \wedge d\bar{z}^K\right) = -2\,dz^l \wedge dz^J \wedge d\bar{z}^K \\ &= -2e_l\left(dz^J \wedge d\bar{z}^K\right) = -e_l i_k\left(dz^k \wedge dz^J \wedge d\bar{z}^K\right), \\ i_k e_l\left(dz^J \wedge d\bar{z}^K\right) &= e_l i_k\left(dz^J \wedge d\bar{z}^K\right) = 0, \quad k \notin J. \end{aligned}$$

Thus $e_k i_l + i_l e_k = 0$.

We consider the differential operators ∂_k and $\bar{\partial}_k$ on $\mathcal{E}_0^{p,q}(\mathbf{C}^n)$, given by

$$\partial_k\left(\varphi_{A\bar{B}}\,dz^A \wedge d\bar{z}^B\right) = \frac{\partial \varphi_{A\bar{B}}}{\partial z^k}\,dz^A \wedge d\bar{z}^B,$$

$$\bar{\partial}_k\left(\varphi_{A\bar{B}}\,dz^A \wedge d\bar{z}^B\right) = \frac{\partial \varphi_{A\bar{B}}}{\partial \bar{z}^k}\,dz^A \wedge d\bar{z}^B.$$

We remark that ∂_k and $\bar{\partial}_k$ commute with e_l, $\bar{e}_l$, i_l and $\bar{i}_l$.

Let $\psi \in C^\infty(\mathbf{C}^n)$. Then, integrating by parts, we obtain

$$\begin{aligned} \left\langle -\bar{\partial}_k \varphi, \psi\,dz^L \wedge d\bar{z}^M \right\rangle &= \left\langle -\frac{\partial \varphi_{L\bar{M}}}{d\bar{z}^k}\,dz^L \wedge d\bar{z}^M, \psi\,dz^L \wedge d\bar{z}^M \right\rangle \\ &= 2^{|L|+|M|} \int_{\mathbf{C}^n} -\frac{\partial \varphi_{L\bar{M}}}{\partial \bar{z}^k}\,\bar{\psi} = 2^{|L|+|M|} \int_{\mathbf{C}^n} \varphi_{L\bar{M}} \frac{\partial \bar{\psi}}{\partial \bar{z}^k} \\ &= 2^{|L|+|M|} \int_{\mathbf{C}^n} \varphi_{L\bar{M}} \overline{\frac{\partial \psi}{\partial z^k}} = \left\langle \varphi_{L\bar{M}}\,dz^L \wedge d\bar{z}^M, \partial_k\left(\psi\,dz^L \wedge d\bar{z}^M\right)\right\rangle \\ &= \left\langle \varphi, \partial_k\left(\psi\,dz^L \wedge d\bar{z}^M\right)\right\rangle. \end{aligned}$$

Thus the adjoint of ∂_k is $-\bar{\partial}_k$ and the adjoint of $\bar{\partial}_k$ is $-\partial_k$, respectively.

One has

$$\partial = \partial_k e_k = e_k \partial_k,$$
$$\bar{\partial} = \bar{\partial}_k \bar{e}_k = \bar{e}_k \bar{\partial}_k$$

and

$$\delta = -\partial_k \bar{i}_k,$$
$$\partial^* = -\bar{\partial}_k i_k,$$

respectively. Also

$$L = \frac{i}{2} e_k \bar{e}_k,$$
$$\Lambda = -\frac{i}{2} \bar{i}_k i_k.$$

Therefore

$$\Lambda\partial = -\frac{i}{2} \bar{i}_k i_k \partial_l e_l = -\frac{i}{2} \partial_l \bar{i}_k i_k e_l = -\frac{i}{2}\left(\sum_k \partial_k \bar{i}_k i_k e_k + \sum_{k \neq l} \partial_l \bar{i}_k i_k e_l\right)$$
$$= \frac{i}{2} \sum_k \partial_k \bar{i}_k e_k i_k - i \sum_k \partial_k \bar{i}_k + \frac{i}{2} \sum_{k \neq l} \partial_l \bar{i}_k e_l i_k$$
$$= -\frac{i}{2} \sum_k \partial_k e_k \bar{i}_k i_k - i \partial_k \bar{i}_k - \frac{i}{2} \sum_{k \neq l} \partial_l e_l \bar{i}_k i_k = -\frac{i}{2} \partial_l e_l \bar{i}_k i_k - i \partial_k \bar{i}_k.$$

Consequently,

$$\Lambda\partial = \partial\Lambda + i\delta,$$

i.e., the desired equation is proved on $\mathbf{C}^n$.

If M is a Kaehler manifold, in a neighborhood of any $z_0 \in M$ we can choose an orthonormal frame $\{\varphi_1, \ldots, \varphi_n\}$ of (1,0)-forms such that $d\varphi_j(z_0) = 0$. The expression of Λ is the same, substituting dz^J by φ^J. Also the computation of $[\Lambda, \bar{\partial}]$ is the same as above, by using $\bar{\partial}\varphi_j$.

Since $[\Lambda, \bar{\partial}]$ contains only first order partial derivatives, all the other terms will contain $\bar{\partial}\varphi_j$, which vanishes at z_0. Therefore the identity to prove holds at z_0, and thus everywhere on M. □

COROLLARY 7.2. *Let M be a Kaehler manifold. Then*

(i) $[L, \Delta] = 0$;

(ii) $[\Lambda, \Delta] = 0$.

PROOF. Since ω is closed, one has

$$d(\omega \wedge \eta) = \omega \wedge d\eta,$$

i.e., $[L, d] = 0$.

Taking the adjoint, we find $[\Lambda, d^*] = 0$. Also

$$\begin{aligned}\Lambda\Delta &= \Lambda(dd^* + d^*d) = (d\Lambda d^* - 4\pi d^{c*}d^*) + d^*\Lambda d \\ &= d\Lambda d^* + (4\pi d^* d^{c*} + d^*\Lambda d) = (dd^* + d^*d)\Lambda = \Delta\Lambda.\end{aligned}$$ □

COROLLARY 7.3. *On any Kaehler manifold, the following identity holds*:

$$\Box = \frac{1}{2}\Delta.$$

PROOF. First, we prove that

$$\partial\delta + \delta\partial = 0.$$

We know that $\Lambda\bar{\partial} - \bar{\partial}\Lambda = i\delta$. Then

$$i(\partial\delta + \delta\partial) = \partial(\Lambda\partial - \partial\Lambda) + (\Lambda\partial - \partial\Lambda)\partial = \partial\Lambda\partial - \partial\Lambda\partial = 0.$$

Next

$$\begin{aligned}\Delta &= (\partial + \bar{\partial})(\partial^* + \delta) + (\partial^* + \delta)(\partial + \bar{\partial}) \\ &= (\partial\partial^* + \partial^*\partial) + (\bar{\partial}\delta + \delta\bar{\partial}) + (\partial\delta + \bar{\partial}\partial^* + \partial^*\bar{\partial} + \delta\partial) \\ &= (\partial\partial^* + \partial^*\partial) + (\bar{\partial}\delta + \delta\bar{\partial}) = \Box_\partial + \Box.\end{aligned}$$

We have to show that $\Box_\partial = \Box$. One has

$$\begin{aligned}-i\Box_\partial &= \partial(\Lambda\bar{\partial} - \bar{\partial}\Lambda) + (\Lambda\bar{\partial} - \bar{\partial}\Lambda)\partial = \partial\Lambda\bar{\partial} - \partial\bar{\partial}\Lambda + \Lambda\bar{\partial}\partial - \bar{\partial}\Lambda\partial, \\ i\Box &= \bar{\partial}(\Lambda\partial - \partial\Lambda) + (\Lambda\partial - \partial\Lambda)\bar{\partial} = \bar{\partial}\Lambda\partial - \bar{\partial}\partial\Lambda + \Lambda\partial\bar{\partial} - \partial\Lambda\bar{\partial}.\end{aligned}$$

The last two equations imply $\Box_\partial = \Box$. □

COROLLARY 7.4. *On any compact Kaehler manifold, Δ preserves the degree of complex differential forms, i.e.*,

$$[\Delta, \pi^{p,q}] = 0,$$

where $\pi^{p,q} : \mathcal{E}(M) \to \mathcal{E}^{p,q}(M)$ is the standard projection.

NOTATIONS.

$$\mathcal{H}_d^{p,q}(M) = \{\eta \in \mathcal{E}^{p,q}(M) \mid \Delta\eta = 0\},$$

$$\mathcal{H}_d^{r}(M) = \{\eta \in \mathcal{E}^{r}(M) \mid \Delta\eta = 0\},$$

$$H_d^{p,q}(M) = Z_d^{p,q}(M)/_{Z_d^{p,q}(M)\cap \operatorname{Im} d}.$$

COROLLARY 7.5. *On any compact Kaehler manifold, one has*:
(i) $\mathcal{H}_d^r(M) = \bigoplus_{p+q=r} \mathcal{H}_d^{p,q}(M)$;
(ii) $\mathcal{H}_d^{p,q}(M) = \overline{\mathcal{H}_d^{q,p}(M)}$.

Moreover, there exists an isomorphism $H_d^{p,q}(M) \cong \mathcal{H}_d^{p,q}(M)$.
Applying the above results, we find the following

THEOREM 7.6 (Hodge). *If M is a compact Kaehler manifold, then*:
(i) $H_d^r(M, \mathbf{C}) \cong \bigoplus_{p+q=r} H_d^{p,q}(M)$;
(ii) $H_d^{p,q}(M) \cong H_d^{q,p}(M)$.

Since $\Delta = 2\Box$, one has $\mathcal{H}_d^{p,q}(M) = \mathcal{H}^{p,q}$, and so,

$$H_d^{p,q}(M) \cong H^{p,q}(M, \bar\partial) \cong H^q(M, \Omega^p).$$

In particular, for $q = 0$,

$$H_d^{p,0}(M) = H^0(M, \Omega^p).$$

Therefore, the holomorphic differential forms are harmonic on any compact Kaehler manifold.

THEOREM 7.7. *The Betti numbers of odd order of a compact Kaehler manifold are even.*

PROOF. Define the Hodge numbers by

$$h^{p,q}(M) = \dim H_d^{p,q}(M).$$

Hodge theorem implies

$$b_r(M) = \sum_{p+q=r} h^{p,q}(M),$$

$$h^{p,q}(M) = h^{q,p}(M).$$

Therefore

$$b_{2q+1}(M) = 2\sum_{p=0}^{q} h^{p,2q+1-p}(M)$$

is an even number. □

REMARK. The converse of Theorem 7.7 is not true. However, for $\dim M = 2$, using a result of Miyaoka which says that any elliptic surface with first Betti number even is Kaehler, the following result holds (see [65]).

THEOREM 7.8. *A compact complex surface is Kaehler if and only if its first Betti number $b_1(M)$ is even.*

COROLLARY 7.9. *On a compact Kaehler manifold, each holomorphic differential form is d-harmonic, in particular d-closed.*

REMARK. Corollary 7.8 is not true if M is not compact.

Indeed, a non-constant holomorphic function on $\mathbf{C}^n$ is a holomorphic form, but it is not d-closed.

COROLLARY 7.10.

$$H^q\big(P^n(\mathbf{C}), \Omega^p\big) = H^{p,q}\big(P^n(\mathbf{C}), \bar{\partial}\big) = \begin{cases} 0, & p \neq q, \\ \mathbf{C}, & p = q. \end{cases}$$

COROLLARY 7.11. *There do not exist non-zero global holomorphic forms on the complex projective space $P^n(\mathbf{C})$.*

Finally, we state the $\partial\bar{\partial}$-lemma (see [65]).

THEOREM 7.12 ($\partial\bar{\partial}$-lemma). *Let M be an n-dimensional compact Kaehler manifold and $\psi \in \mathcal{E}^{p,q}(M)$ a d-closed differential form, with $p, q > 0$. If ψ is exact with respect to either d, ∂ or $\bar{\partial}$, then ψ is $\partial\bar{\partial}$-exact, i.e., there exists $\varphi \in \mathcal{E}^{p-1,q-1}(M)$ such that $\psi = \partial\bar{\partial}\varphi$. If, in addition, $p = q$ and ψ is real, then we can choose φ such that $i\varphi$ be real.*

REMARKS. This lemma is global, unlike the Poincaré Lemma. Both the compactness and Kaehlerness are crucial. However, if a compact complex manifold M is birational to a Kaehler manifold, then the $\partial\bar{\partial}$-lemma still holds true even if M is not Kaehler.

8. Applications

An interesting application of Theorem 7.7 is the construction of an almost Kaehler manifold which do not admit any Kaehlerian metric (see [1]).

Let $H \subset GL(3, \mathbf{R})$ be the *Heisenberg group*, i.e.,

$$H = \left\{ \begin{pmatrix} 1 & x & z \\ 0 & 1 & y \\ 0 & 0 & 1 \end{pmatrix};\ x, y, z \in \mathbf{R} \right\}.$$

We denote by Γ the maximal discrete subgroup of H defined as the set of all matrices of H with integer entries. Since Γ is closed, $H/_\Gamma$ is a homogeneous space. Let $S^1 = \{e^{2\pi i t} \mid t \in \mathbf{R}\}$ be the unit circle.

We investigate the compact homogeneous space $M = H/_\Gamma \times S^1$.

A basis in the Lie algebra of the Heisenberg group H is

$$X_1 = \frac{\partial}{\partial x}, \quad X_2 = \frac{\partial}{\partial y} + x\frac{\partial}{\partial z}, \quad X_3 = \frac{\partial}{\partial z}.$$

These vector fields are invariant under the action of Γ, thus they induce linearly independent vector fields e_1, e_2, e_3 on $H/_\Gamma$. Let $e_4 = \frac{d}{dt}$ be the standard vector field on S^1.

The dual 1-forms to the vector fields X_1, X_2, X_3 are

$$\theta_1 = dx, \quad \theta_2 = dy, \quad \theta_3 = dz - x\,dy.$$

Together with dt, they induce the linearly independent 1-forms $\alpha_1, \alpha_2, \alpha_3, \alpha_4$ on M.

We define a 2-form ω of maximum rank on M, by

$$\omega = \alpha_4 \wedge \alpha_1 + \alpha_2 \wedge \alpha_3.$$

Moreover, ω is closed because

$$\pi^*(d\omega) = d(\pi^*\omega) = d\big(dt \wedge dx + dy \wedge (dz - x\,dy)\big) = 0,$$

where $\pi : H \times S^1 \to M$ is the standard projection.

We consider the invariant Riemannian metric $\tilde{g}$ on the manifold $H \times S^1$, defined by

$$\tilde{g} = dx^2 + dy^2 + (dz - x\,dy)^2 + dt^2.$$

With respect to $\tilde{g}$, the vector fields $\{X_1, X_2, X_3, \frac{d}{dt}\}$ form an orthonormal frame.

The Riemannian metric g induced by $\tilde{g}$ on M is given by

$$g = \alpha_1^2 + \alpha_2^2 + \alpha_3^2 + \alpha_4^2.$$

The frame $\{e_1, e_2, e_3, e_4\}$ is orthonormal with respect to g.

An almost complex structure J on M can be defined by

$$Je_j = (-1)^j e_{5-j}, \quad j \in \{1, 2, 3, 4\}.$$

A straightforward computation shows that

$$\omega(e_i, e_j) = g(Je_i, e_j), \quad \forall i, j \in \{1, 2, 3, 4\},$$

i.e., ω is the fundamental 2-form on the almost Hermitian manifold (M, J, g). Since ω is closed, M becomes an almost Kaehler manifold.

Because the dimension of $H^1(M, \mathbf{R})$ is 3 (see [1]), it follows that M does not admit any Kaehlerian metric.

One can prove that M is the total space of a locally trivial fibering over the torus T^2 with fibers T^2. Some authors call M a Thurston torus (see [1,58]).

Finally, we will give an example of a compact complex manifold which does not admit any Kaehlerian metric.

This example shows that Theorem 7.8 is not true if $\dim M > 2$.

EXAMPLE (*The Iwasawa manifold, see* [65]). We consider the complex Lie group

$$G = \left\{ \begin{pmatrix} 1 & x & z \\ 0 & 1 & y \\ 0 & 0 & 1 \end{pmatrix} ; \ x, y, z \in \mathbf{C} \right\}$$

in $GL(3, \mathbf{C})$. Obviously G is biholomorphic to $\mathbf{C}^3$.

Let Γ be the subgroup of G of all matrices with x, y, z belonging to the ring of Gauss integers $\mathbf{Z}[i]$. Γ acts on G by left multiplication. This action has no fixed points and is properly discontinuous, then the quotient space $M = G/_{\Gamma}$ is a complex manifold, called the *Iwasawa manifold*.

We consider the global 1-forms

$$\psi_1 = dx, \quad \psi_2 = dy, \quad \psi_3 = dz - x\,dy$$

on G. Obviously ψ_1, ψ_2, ψ_3 are invariant by Γ, then they induce global holomorphic 1-forms on M. We see that

$$d\psi_3 = -dx \wedge dy \neq 0.$$

Therefore Corollary 7.9 implies that the Iwasawa manifold does not admit any Kaehlerian metric.

We remark that the above defined holomorphic 1-forms are linearly independent on M, so they give a trivialization of the cotangent bundle of M. Such a manifold is called *complex parallelizable*.

One can prove (see [65]) that

$$h^{1,0} = 3, \quad h^{3,0} = h^{0,3} = 1, \quad h^{2,0} = 3,$$
$$h^{0,1} = h^{0,2} = 2, \quad h^{1,1} = h^{1,2} = h^{2,1} = 6.$$

Also, we can compute $b_1(M) = 4$.

It is clear that the map $(x, y, z) \mapsto (x, y)$ makes M a holomorphic fiber bundle over a complex 2-torus B, with fibers F being a complex 1-torus.

REMARK [65]. By the Fröhlicher spectral sequence, the Hodge numbers and Betti numbers of any compact complex manifold M always satisfy

$$\sum_{p+q=r} h^{p,q}(M) \geqslant b_r(M), \qquad \sum_{p,q}(-1)^{p+q} h^{p,q}(M) = \sum_r (-1)^r b_r(M).$$

Moreover, one has the following monotonicity of the Betti numbers on an n-dimensional Kaehler manifold:

$$1 \leqslant b_2(M) \leqslant b_4(M) \leqslant \cdots \leqslant b_{2s}(M),$$
$$b_1(M) \leqslant b_3(M) \leqslant \cdots \leqslant b_{2s+1}(M),$$

where $s = [\frac{n}{2}]$ is the largest integer less than or equal to $\frac{n}{2}$.

For compact complex surfaces, we mention the following results [65].

LEMMA 8.1. *On a compact complex surface M, $H^{r,0}(M) \cap \bar{H}^{r,0}(M) = 0$ in $H^r(M)$, for $r = 1, 2$. Also*

$$2h^{1,0}(M) \leqslant b_1(M) \leqslant h^{1,0}(M) + h^{0,1}(M).$$

COROLLARY 8.2. *For any compact complex surface M, $b_1(M) = h^{1,0}(M) + h^{0,1}(M)$. Furthermore, either $h^{1,0}(M) = h^{0,1}(M)$ or $h^{0,1}(M) = h^{1,0}(M) + 1$. In the first case $b_1(M)$ is even, while in the second case $b_1(M)$ is odd.*

9. Chern classes

Let (E, π, M) be a holomorphic line bundle (i.e., a holomorphic vector bundle of rank 1). We consider a locally finite covering $\mathcal{U} = (U_j)_{j \in I}$ of M such that $\pi^{-1}(U_j)$ and $U_j \times \mathbf{C}$ are analytic isomorphic.

Then $(z, \xi_j) \in U_j \times \mathbf{C}$ and $(z, \xi_k) \in U_k \times \mathbf{C}$ define the same point in E if $\xi_j = f_{jk}(z)\xi_k$, where the transition function f_{jk} is holomorphic and does not vanish on $U_j \cap U_k$. It is easily seen that $f_{ik}(z) = f_{ij}(z) f_{jk}(z)$.

Let $\mathcal{O}_p^*$ be the set of the germs of holomorphic functions which do not vanish at $p \in M$. Then $\mathcal{O}_p^*$ is a group with respect to the multiplication of functions. If we regard it as a $\mathbf{Z}$-module, we define the sheaf $\mathcal{O}^* = \bigcup_{p \in M} \mathcal{O}_p^*$ of the germs of holomorphic functions nowhere zero on M. Obviously $\mathcal{O}^*$ is an open subset of $\mathcal{O}$.

Let g be a holomorphic function defined on the open subset W of M. We denote $e(g) = f$, where $f(z) = e^{2\pi i g(z)}$. Thus e defines a sheaf morphism $e : \mathcal{O} \to \mathcal{O}^*$. Since $\operatorname{Ker} e = \mathbf{Z}$, one has the following exact sequence:

$$0 \to \mathbf{Z} \to^{i} \mathcal{O} \to^{e} \mathcal{O}^* \to 1.$$

It determines the long exact cohomology sequence

$$\cdots \to H^1(M,\mathcal{O}) \to H^1(M,\mathcal{O}^*) \to^{\delta^*} H^2(M,\mathbf{Z}) \to \cdots.$$

Considering the transition functions f_{jk} of a holomorphic line bundle (E,π,M) as sections $f_{jk} \in \Gamma(U_j \cap U_k, \mathcal{O}^*)$ of $\mathcal{O}^*$ over $U_j \cap U_k \neq \emptyset$, $\{f_{jk}\}$ determines a 1-cocycle; then $\{f_{jk}\} \in Z^1(\mathcal{U},\mathcal{O}^*)$.

We may assume that every U_j is a coordinate polydisc. Then the sequence

$$\cdots \to H^1(U_j,\mathcal{O}) \to H^1(U_j,\mathcal{O}^*) \to^{\delta^*} H^2(U_j,\mathbf{Z}) \to \cdots$$

is exact.

By applying Dolbeault theorem, one has

$$H^1(U_j,\mathcal{O}) \cong H^{0,1}(U_j,\bar\partial).$$

Using Dolbeault Lemma, for any $\bar\partial$-closed form $\varphi \in \mathcal{E}^{0,1}(U_j)$, there exists $\psi \in C^\infty(U_j)$ such that $\varphi = \bar\partial\psi$. Therefore $H^1(U_j,\mathcal{O}) = 0$.

But $H^2(U_j,\mathbf{Z}) = 0$, thus $H^1(U_j,\mathcal{O}^*) = 0$. It follows that

$$H^1(\mathcal{U},\mathcal{O}^*) = H^1(M,\mathcal{O}^*).$$

Thus, the 1-cocycles $\{f_{jk}\}$ and $\{g_{jk}\}$ in $Z^1(\mathcal{U},\mathcal{O}^*)$ belong to the same cohomology class if and only if there exists $\{h_j\} \in C^0(\mathcal{U},\mathcal{O}^*)$ such that $\{f_{jk}g_{jk}^{-1}\} = \delta\{h_j\}$, i.e., $f_{jk}g_{jk}^{-1} = h_k h_j^{-1}$, or equivalently,

$$g_{jk} = h_j f_{jk} h_k^{-1}.$$

Consequently, $\{f_{jk}\}$ and $\{g_{jk}\}$ belong to the same cohomology class if and only if are transition functions for the same holomorphic line bundle E.

Thus, we may identify one holomorphic line bundle with the cohomology class of the 1-cocycle $\{f_{jk}\}$, i.e.,

$$E \mapsto \big[\{f_{jk}\}\big] \in H^1(M,\mathcal{O}^*).$$

Then $H^1(M,\mathcal{O}^*)$ becomes the group of holomorphic line bundles over M, having the tensor product as group law.

DEFINITION. For each holomorphic line bundle $E \in H^1(M,\mathcal{O}^*)$, $\delta^*E \in H^2(M,\mathbf{Z})$ is called the *Chern class* of E. We denote $c(E) = \delta^*E$.

Let $\mathcal{U} = (U_j)_{j\in I}$ be a locally finite covering of M such that $U_j \cap U_k$ be simply connected and $U_j \cap U_k \cap U_l$ connected.

Let $E = [\{f_{jk}\}] \in H^1(M,\mathcal{O}^*)$, with $\{f_{jk}\} \in Z^1(\mathcal{U},\mathcal{O}^*)$.

Since f_{jk} is holomorphic and does not vanish on a simply connected open set, each leaf of $\log f_{jk}$ is a well-defined holomorphic function on $U_j \cap U_k$. Put

$$g_{jk}(z) = \frac{1}{2\pi i} \log f_{jk}(z).$$

One has $g_{jk}(z) = -g_{kj}(z)$ and $e(g_{jk}(z) - g_{ik}(z) + g_{ij}(z)) = 1$. Thus $c_{ijk} = g_{jk}(z) - g_{ik}(z) + g_{ij}(z)$ is a constant integer on $U_i \cap U_j \cap U_k \neq \emptyset$. Because $e\{g_{jk}\} = \{f_{jk}\}$ and $\delta\{g_{jk}\} = \{c_{ijk}\} \in Z^2(\mathcal{U}, \mathbf{Z})$, then $\delta^* E$ is represented by the 2-cocycle $\{c_{ijk}\}$.

Therefore, the Chern class of E is given by $c(E) = [\{c_{ijk}\}]$, where

$$c_{ijk} = \frac{1}{2\pi i}(\log f_{jk} - \log f_{ik} + \log f_{ij}).$$

Next, we will define axiomatically the Chern classes of a complex vector bundle of rank l over a differentiable manifold (see [22]).

AXIOM 1. For each complex vector bundle E over M and any $i \in \mathbf{N}$, the Chern class $c_i(E) \in H^{2i}(M, \mathbf{R})$ and $c_0(E) = 1$.

The total Chern class of E is

$$c(E) = \sum_{i=0}^{\infty} c_i(E).$$

AXIOM 2. Let E be a complex vector bundle over M and $f: M' \to M$ a differentiable map. Then

$$c\big(f^{-1}E\big) = f^*\big(c(E)\big) \in H^*(M', \mathbf{R}),$$

where $f^{-1}E$ is the complex vector bundle over M' induced by f from E.

AXIOM 3 (Whitney sum formula). Let $E_1, \ldots, E_q$ be complex line bundles (i.e., complex vector bundles of rank 1) over the differentiable manifold M (their fibre is $\mathbf{C}$) and $E_1 \oplus \cdots \oplus E_q$ their Whitney sum, i.e., $E_1 \oplus \cdots \oplus E_q = d^{-1}(E_1 \times \cdots \times E_q)$, with $d: M \to M \times \cdots \times M$ (q times), $d(x) = (x, \ldots, x)$. Then

$$c(E_1 \oplus \cdots \oplus E_q) = c(E_1) \ldots c(E_q).$$

Consider the complex projective space $P^n(\mathbf{C})$ as the set of the complex lines in $\mathbf{C}^{n+1}$. The group $\mathbf{C}^* = \mathbf{C} - \{0\}$ acts on $\mathbf{C}^{n+1} - \{0\}$ by complex multiplication. Thus $P^n(\mathbf{C}) = (\mathbf{C}^{n+1} - \{0\})/_{\mathbf{C}^*}$. We equip $P^n(\mathbf{C})$ with the quotient topology and denote by L_n the tautological vector bundle over $P^n(\mathbf{C})$.

$$L_n = \big\{\big([z], \zeta\big) \in P^n(\mathbf{C}) \times \mathbf{C}^{n+1} \mid \zeta \in [z]\big\}.$$

Its dual vector bundle L_n^* is called the hyperplane bundle.

Let $(z^0, \dots, z^n)$ be the coordinates on $\mathbf{C}^{n+1}$ and

$$U_j = \{[z] \in P^n(\mathbf{C}) \mid z^j \neq 0\}.$$

The transition functions of the tautological vector bundle are:

$$f_{jk}([z]) = \frac{z^j}{z^k} \quad \text{on } U_j \cap U_k.$$

AXIOM 4 (*Normalization*). The Chern class $-c_1(L_1)$ is the generator of $H^2(P^1(\mathbf{C}), \mathbf{Z})$, i.e., the integral of $c_1(L_1)$ on $P^1(\mathbf{C})$ is equal to -1.

We will construct the Chern classes of a complex vector bundle E of rank l over a differentiable manifold M.

A connection ∇ on E is a differential operator of first order $\nabla : \Gamma(E) \to \Gamma(T^*M \otimes E)$, such that

$$\nabla(f\sigma) = df \otimes \sigma + f\nabla\sigma,$$

for every $f \in C^\infty(M)$ and $\sigma \in \Gamma(E)$.

Let $\{\sigma_1, \dots, \sigma_l\}$ be a local frame of sections on E. Then, each section $\sigma \in \Gamma(E)$ can be written, locally, as

$$\sigma(x) = f^i(x)\sigma_i(x), \quad f^i \in C^\infty(M).$$

Computing, we get

$$\nabla\sigma = df^i \otimes \sigma_i + f^i\nabla\sigma_i = df^i \otimes \sigma_i + f^i\omega_i^j \otimes \sigma_j,$$

where $\nabla\sigma_i = \omega_i^j \otimes \sigma_j$.

The connection 1-form ω, defined by $\omega = (\omega_i^j)$, is a matrix of 1-forms. The connection ∇ is uniquely determined by the connection 1-form ω.

With respect to another local frame, $\sigma_i' = h_i^j\sigma_j$, one has

$$\omega_i'^j = dh_i^k (h^{-1})_k^j + h_i^k\omega_k^l (h^{-1})_l^j,$$

or matriceally,

$$\omega' = dh \cdot h^{-1} + h\omega h^{-1}.$$

The connection ∇ can be extended to

$$\Gamma(\wedge^p T^*M \otimes E) \to \Gamma(\wedge^{p+1} T^*M \otimes E),$$

such that

$$\nabla(\theta_p \otimes \sigma) = d\theta_p \otimes \sigma + (-1)^p \theta_p \wedge \nabla\sigma.$$

By a straightforward calculation, we find

$$\nabla^2(f\sigma) = \nabla(df \otimes \sigma + f\nabla\sigma) = f\nabla^2\sigma,$$

i.e., ∇^2 is C^∞-linear.

We can write $(\nabla^2\sigma)(x_0) = \Omega(x_0)\sigma(x_0)$, where the curvature form Ω is a section in the bundle $\wedge^2 T^*M \otimes \mathrm{End}(E)$.

In local coordinates, one has

$$\Omega_i^j = d\omega_i^j - \omega_i^k \wedge \omega_k^j \quad \text{or} \quad \Omega = d\omega - \omega \wedge \omega.$$

Thus Ω transforms as a tensor, i.e., $\Omega'^j_i = h_i^k \Omega_k^l (h^{-1})_l^j$ or $\Omega' = h\Omega h^{-1}$.

LEMMA 9.1. *Let ∇ be a connection on a complex vector bundle E. Then there exists a local frame $\{\sigma_1, \dots, \sigma_l\}$ such that at a fixed point x_0 one has*

$$\omega(x_0) = 0, \qquad d\Omega(x_0) = 0.$$

REMARK. Generally we cannot find a parallel frame such that $\omega = 0$ in a neighborhood of x_0 (this would imply $\nabla^2 = 0$).

Let $(A_{ij})_{i,j=1,\dots,l} \in \mathrm{End}(\mathbf{C}^l)$ and $P : \mathrm{End}(\mathbf{C}^l) \to \mathbf{C}$ a polynomial map. We assume that P is invariant (that is, $P(hAh^{-1}) = P(A)$, $\forall h \in GL(l, \mathbf{C})$). We define $P(\Omega)$ as an even differential form on M. Since P is invariant, we can put $P(\nabla) = P(\Omega)$, independently of the frame.

DEFINITION. $c(A) = \det(I - \frac{1}{2\pi i} A) = 1 + c_1(A) + \cdots + c_l(A)$ is called *total Chern form*.

$\mathrm{ch}(A) = \mathrm{Tr}\, e^{-\frac{A}{2\pi i}} = \mathrm{Tr} \sum_{j=0}^{\infty} (-\frac{A}{2\pi i})^j / j!$ is the *Chern character*.

Thus $c_j(A)$ represents the homogeneous part of order j in $c(A)$, $c_j(\nabla) \in \wedge^{2j}(M)$. Analogously

$$\mathrm{ch}(A) = \sum_{j=0}^{\infty} \mathrm{ch}_j(A),$$

with

$$\mathrm{ch}_j(A) = \frac{1}{j!} \mathrm{Tr}\left(-\frac{A}{2\pi i}\right)^j.$$

Generally, $\mathrm{ch}(A)$ is not polynomial, but substituting A by Ω, the sum becomes finite, because $\operatorname{Tr}\Omega^j = 0$, for $2j > \dim M$.

The differential form $P(\nabla)$ depends on the connection ∇. But it is easily proved that it is closed and as an element of the de Rham cohomology $P(\nabla)$ does not depend on ∇ and defines a cohomology class denoted by $P(E)$.

Let $\nabla_t = t\nabla_1 + (1-t)\nabla_0$. Its connection 1-form is $\omega_t = \omega_0 + t\theta$, with $\theta = \omega_1 - \omega_0$. Then

$$\theta' = \omega_1' - \omega_0' = h(\omega_1 - \omega_0)h^{-1} = h\theta h^{-1}.$$

Let Ω_t be the curvature form of the connection ∇_t. Then

$$P(\theta, \Omega_t, \ldots, \Omega_t) \in \wedge^{2k-1}(T^*M)$$

is invariant and

$$P(\nabla_1) - P(\nabla_0) = \int_0^1 \frac{d}{dt} P(\Omega_t, \ldots, \Omega_t)\, dt = k\int_0^1 P(\Omega_t', \Omega_t, \ldots, \Omega_t)\, dt.$$

Putting

$$TP(\nabla_0, \nabla_1) = k\int_0^1 P(\theta, \Omega_t, \ldots, \Omega_t)\, dt,$$

one finds

$$dP(\theta, \Omega_t, \ldots, \Omega_t) = P(\Omega_t', \Omega_t, \ldots, \Omega_t).$$

We assume that the matrix $A = \operatorname{diag}(\lambda_1, \ldots, \lambda_l)$ be diagonal. Then $c_j(A)$ is an elementary symmetric function of order j with respect to $\lambda_1, \ldots, \lambda_l$, because

$$\det(I + A) = \prod_{j=1}^{k}(1 + \lambda_j) = 1 + s_1(\lambda) + \cdots + s_l(\lambda).$$

By applying the fundamental theorem of symmetric polynomials, there exists a unique polynomial Q such that $P(A) = Q(c_1, \ldots, c_l)(A)$.

Since the diagonalizable matrices are dense, we obtain the following

THEOREM 9.2. *For any complex vector bundles E, E_1 and E_2, we have*:

(i) $c(E_1 \oplus E_2) = c(E_1)c(E_2)$;

(ii) $c(E^*) = 1 - c_1(E) + c_2(E) - \cdots + (-1)^l c_l(E)$;

(iii) $\mathrm{ch}(E_1 \oplus E_2) = \mathrm{ch}(E_1) + \mathrm{ch}(E_2)$;

(iv) $\mathrm{ch}(E_1 \otimes E_2) = \mathrm{ch}(E_1)\,\mathrm{ch}(E_2)$.

Summarizing, one has:

Let E be a complex vector bundle over M, with fiber $\mathbf{C}^l$.

We define the polynomial functions $f_0, \ldots, f_l$ on the Lie algebra $gl(l, \mathbf{C})$, by

$$\det\left(\lambda I_l - \frac{1}{2\pi i}X\right) = \sum_{k=0}^{l} f_k(X)\lambda^{l-k}, \quad X \in gl(l, \mathbf{C}).$$

Then there exists a unique $\gamma_k \in Z^{2k}(M, \mathbf{R})$ such that

$$\pi^*\gamma_k = f_k(\Omega).$$

Therefore we may write

$$\det\left(I_l - \frac{1}{2\pi i}\Omega\right) = \pi^*(1 + \gamma_1 + \cdots + \gamma_l).$$

THEOREM 9.3. *The Chern class $c_k(E)$ of a complex vector bundle E over M is represented by the closed $2k$-form γ_k defined above.*

If we express the curvature form Ω by a matrix-valued 2-form (Ω^i_j), then the $2k$-form γ_k representing the kth Chern class $c_k(E)$ can be written as follows:

$$\pi^*\gamma_k = \frac{(-1)^k}{(2\pi i)^k k!}\sum \delta^{j_1 \ldots j_k}_{i_1 \ldots i_k} \Omega^{i_1}_{j_1} \wedge \cdots \wedge \Omega^{i_k}_{j_k},$$

where the summation is taken over all ordered subsets $(i_1, \ldots, i_k)$ of k elements from $(1, \ldots, l)$ and all permutations $(j_1, \ldots, j_k)$ of $(i_1, \ldots, i_k)$ and the symbol $\delta^{j_1 \ldots j_k}_{i_1 \ldots i_k}$ denotes the sign of the permutation $(i_1, \ldots, i_k) \to (j_1, \ldots, j_k)$.

REMARK. Let E be a Hermitian vector bundle over a complex manifold M with fibre $\mathbf{C}^l$ and fibre metric h. Since the first Chern class $c_1(E)$ can be represented by a closed 2-form γ_1 on M such that $\pi^*\gamma_1 = -\frac{1}{2\pi i}\operatorname{trace}\Omega$, we may prove that $c_1(E)$ can be represented by the closed 2-form $\frac{1}{2\pi i}\bar{\partial}\partial \log H$, where $H = \det(h_{\alpha\bar{\beta}})$. Also

$$\gamma_2 = -\frac{1}{8\pi^2}\sum_{\alpha,\beta} \Omega^\alpha_\alpha \wedge \Omega^\beta_\beta - \Omega^\alpha_\beta \wedge \Omega^\beta_\alpha.$$

For real vector bundles one defines an analogue of Chern classes, called *Pontrjagin classes*.

Let E be a real vector bundle over M and $E^c = E \otimes_{\mathbf{R}} \mathbf{C}$ its complexified vector bundle. The Pontrjagin classes are given by

$$p_j(E) = (-1)^j c_{2j}(E^c) \in H^{4j}(M, \mathbf{R}),$$

where $c_{2k}(E^c)$ denotes the $2k$th Chern class of the complex vector bundle E^c. The total Pontrjagin class $p(E)$ is defined to be

$$1 + p_1(E) + p_2(E) + \cdots \in H^*(M).$$

In particular, the Chern classes over the complex projective space $P^n(\mathbf{C})$ are determined. Let $s^j = (z_j^0, \ldots, z_j^n)$ be a section of the tautological bundle L_n over U_j. Then $s^j = (z_k^j)^{-1} s^k$.

The coordinate functions z_j on $\mathbf{C}^{n+1}$ are global holomorphic sections on L_n^*. On the trivial vector bundle $P^n(\mathbf{C}) \times \mathbf{C}^{n+1}$ there exists a natural Hermitian product, which gives rise to a metric on the fibres of L_n. Define

$$x = -c_1(L_n) = -\frac{1}{2\pi i} \partial \bar{\partial} \log\bigl(1 + |z_j|^2\bigr).$$

Obviously x is a closed 2-form on $P^n(\mathbf{C})$ invariant under the action of $U(n+1)$.

THEOREM 9.4. *If* $x = -c_1(L_n)$, *then*:
(a) $\int_{P^n(\mathbf{C})} x^n = 1$;
(b) $H^*(P^n(\mathbf{C}), \mathbf{C})$ *is a polynomial ring with generators* $\{1, x, \ldots, x^n\}$;
(c) *If* $i : P^{n-1}(\mathbf{C}) \to P^n(\mathbf{C})$ *is the standard inclusion, then* $i^* x = x$.

LEMMA 9.5.
(a) *There exists a short exact sequence of holomorphic vector bundles*

$$0 \to \wedge^{1,0}\bigl(P^n(\mathbf{C})\bigr) \to L_n \oplus \cdots \oplus L_n \to \theta_1 \to 0,$$

where θ_1 *is the trivial vector bundle of rank* 1.
(b) *There exists an isomorphism of complex vector bundles*

$$T_{\mathbf{C}} P^n(\mathbf{C}) \oplus \theta_1 \cong L_n^* \oplus \cdots \oplus L_n^* \quad (n+1 \textit{ times}).$$

COROLLARY 9.6 [22].
(a) $c(T_{\mathbf{C}} P^n(\mathbf{C})) = (1 + x)^{n+1}$;
(b) $p(T P^n(\mathbf{C})) = (1 + x^2)^{n+1}$.

B.Y. Chen and K. Ogiue [16] proved the following results on complex space forms, in terms of their Chern numbers (the Chern numbers of the tangent bundle).

THEOREM 9.7. *If* M *is an* n*-dimensional complex space form* $(n \geqslant 2)$, *then* $c_2 = \frac{1}{2}(n/(n+1))c_1^2$.

THEOREM 9.8. *Let* M *be an* n*-dimensional compact Kaehler–Einstein manifold* $(n \geqslant 2)$. *If* $c_2 = \frac{1}{2}(n/(n+1))c_1^2$, *then* M *is a complex space form.*

THEOREM 9.9. *Let M be an n-dimensional compact Kaehler manifold with vanishing Bochner curvature tensor* ($n \geqslant 2$). *If* $c_2 = \frac{1}{2}(n/(n+1))c_1^2$, *then M is a complex space form.*

As an application, we study the nucleus of a nearly-Kaehler manifold. Recall that an almost Hermitian manifold (M, J, g) is *nearly-Kaehler* if $(\nabla_X J)X = 0$, $\forall X \in \Gamma(TM)$, where ∇ is the Levi-Civita connection with respect to g.

We recall the following results (see [23]).

THEOREM 9.10. *Let M be a real n-dimensional nearly-Kaehler manifold. Then we have*:

(i) *If $n = 4$, then M is Kaehlerian.*
(ii) *If $n = 6$, then M is Einstein.*
(iii) *If $n = 8$ and M is complete and simply connected, then M is $M_1 \times M_2$, where M_1 is a 6-dimensional Einstein nearly-Kaehler manifold and M_2 is a 2-dimensional Kaehler manifold.*

As an example of a non-Kaehler nearly-Kaehler manifold we mention the 6-dimensional sphere S^6.

On a nearly-Kaehler manifold M, A. Gray [23] considered the distribution defined by

$$D_x = \{X \in T_x M \mid (\nabla_X J)Y = 0,\ \forall Y \in T_x M\},$$

for each $x \in M$. He proved that this distribution is integrable and its integral submanifolds are Kaehler manifolds. Such an integral submanifold is called a nucleus of M. The nucleus M' of a nearly-Kaehler manifold is a minimal submanifold (see [19]).

The first Chern class $c_1(TM)$ is represented by

$$\gamma_1(X, Y) = \frac{1}{2\pi}\sum_{i=1}^{n} R(X, Y, e_i, Je_i) = \frac{1}{2}g\big((\nabla_X J)e_i, J\big((\nabla_Y J)e_i\big)\big),$$

where $\{e_1, \ldots, e_n\}$ is a local orthonormal frame on M.

On the other hand, since M' is a Kaehler manifold, $c_1(TM')$ is represented by

$$\gamma_1'(X, Y) = \frac{1}{4\pi}\operatorname{Ric}(JX, Y).$$

By a straightforward computation we find $c_1(TM|_{M'}) = 2c_1(TM')$.

But $c_1(TM|_{M'}) = c_1(TM') + c(T^{\perp}M')$, thus

$$c_1(TM') = c_1(T^{\perp}M').$$

A holomorphic line bundle L over a compact complex manifold M is said to be *positive*, denoted by $L > 0$, if it admits a Hermitian metric whose curvature is positive everywhere.

THEOREM 9.11 (Kodaira–Nakano vanishing). *If M is an n-dimensional compact Kaehler manifold and L is a positive line bundle over M, then $H^{p,q}(M, L) = 0$, whenever $p+q > n$.*

THEOREM 9.12 (Kodaira vanishing). *Let M be a compact complex manifold and L a positive line bundle over M. Then for any holomorphic vector bundle E over M, there exists a positive integer m_0 such that $H^q(M, L^{\otimes m} \otimes E) = 0$, for all $q > 0$ and all $m \geqslant m_0$.*

THEOREM 9.13 (Kodaira embedding). *On a compact complex manifold M, any positive line bundle L is ample, i.e., there exists a positive integer m such that $L^{\otimes m}$ gives an embedding of M into some $P^N(\mathbf{C})$.*

E. Calabi [7] raised the famous conjecture about prescribing the Ricci curvature on a compact Kaehler manifold. That is, suppose M is a compact Kaehler manifold and ψ a real $(1, 1)$-form representing $c_1(M)$. Then does there exist a Kaehler metric g on M whose Ricci form is ψ?

S.T. Yau [64] solved this conjecture. More precisely, he proved the following

THEOREM 9.14. *Let (M, h) be a compact Kaehler manifold and ψ a real $(1, 1)$-form representing $c_1(M)$. Then there exists a unique Kaehler metric g on M such that its Kaehler 2-form ω_g satisfies $[\omega_g] = [\omega_h]$ in $H^{1,1}(M)$ and its Ricci form is ψ.*

COROLLARY 9.15. *Let M be a compact Kaehler manifold. If $c_1 > 0$ (or $c_1 < 0$), i.e., the $(1, 1)$-cohomology class c_1 (or $-c_1$) is represented by some positive $(1, 1)$-form, then in any given Kaehler class, there exists a Kaehler metric with positive (or negative) Ricci curvature. If $c_1 = 0$ in $H^2(M)$, then in any given Kaehler class, there exists a Ricci flat Kaehler metric.*

THEOREM 9.16 [4]. *Let M be a compact Kaehler manifold with $c_1 < 0$. Then there exists a unique Kaehler–Einstein metric with Ricci curvature -1.*

The Chern numbers of a compact Kaehler manifold M satisfy the following inequality, due to S.T. Yau (see [65]).

THEOREM 9.17. *Let M be an n-dimensional ($n \geqslant 2$) compact Kaehler manifold. If $c_1 < 0$, then*

$$c_1^n \leqslant (-1)^n \frac{2(n+1)}{n} c_1^{n-2} c_2,$$

and the equality holds if and only if the universal cover of M is the unit ball in $\mathbf{C}^n$. If $c_1 = 0$, then the cup product $c_2[\omega]^{n-2} \geqslant 0$, for any Kaehler class $[\omega]$, and the equality occurs when and only when a finite cover of M is a complex torus.

When $n = 2$, this means, if $c_1 = 0$, then $c_2 \geqslant 0$; and $c_2 = 0$ only when M is covered by a complex 2-torus. If $c_1 < 0$, then $c_1^2 \leqslant 3c_2$, with equality holding if and only if M is covered by the unit ball B^2 in $\mathbf{C}^2$.

It turns out that for surfaces, the inequality $c_1^2 \leqslant 3c_2$ holds even without the assumption $c_1 < 0$. It is called the Miyaoka–Yau inequality.

10. Deformation of complex structures

This section is based on the monograph of K. Kodaira [37]. We state some basic notions and the theorems of existence and completeness, respectively. We give several examples.

DEFINITION. Let $B \subset \mathbf{C}^m$ a domain (connected and open) and a family $\{M_t \mid t \in B\}$ of compact complex manifolds (M_t depends on $t = (t^1, \dots, t^m) \in B$). We call $\{M_t \mid t \in B\}$ a *complex analytic family of compact complex manifolds* if there exists a complex manifold $\mathcal{M}$ and a holomorphic map $\tilde{\omega}: \mathcal{M} \to B$ satisfying the following conditions:

(i) For all $t \in B$, $\tilde{\omega}^{-1}(t)$ is a compact complex submanifold of $\mathcal{M}$;

(ii) $M_t = \tilde{\omega}^{-1}(t)$;

(iii) $\tilde{\omega}$ is a submersion, i.e., the rank of the Jacobian of $\tilde{\omega}$ is m at every point of $\mathcal{M}$.

Locally, condition (iii) may be written as follows. Let $(z_q^1, \dots, z_q^n, z_q^{n+1}, \dots, z_q^{n+m})$ be local coordinates on $\mathcal{M}$ and $(t^1, \dots, t^m) = \tilde{\omega}(z_q^1, \dots, z_q^n, z_q^{n+1}, \dots, z_q^{n+m})$.

Then, condition (iii) becomes:

$$\operatorname{rank} \frac{\partial(t^1, \dots, t^m)}{\partial(z_q^1, \dots, z_q^n, z_q^{n+1}, \dots, z_q^{n+m})} = m.$$

Therefore one may choose a locally finite covering $\mathcal{U} = (U_j)_{j \geqslant 1}$ with coordinate polydiscs such that

(a) $z_j(p) = (z_j^1(p), \dots, z_j^n(p), t^1, \dots, t^m)$, where $(t^1, \dots, t^m) = \tilde{\omega}(p)$.

(b) A system of local holomorphic coordinates on M_t is given by

$$p \mapsto \left(z_j^1(p), \dots, z_j^n(p)\right), \qquad U_j \cap M_t \neq \emptyset.$$

For $j, k \geqslant 1$ with $U_j \cap U_k \neq \emptyset$, the coordinate transformation is denoted by

$$f_{jk}: \left(z_k^1, \dots, z_k^n, t\right) \mapsto \left(z_j^1, \dots, z_j^n, t\right),$$

or, equivalently,

$$z_j^\alpha = f_{jk}^\alpha\left(z_k^1, \dots, z_k^n, t^1, \dots, t^m\right), \quad \alpha \in \{1, \dots, n\}.$$

The above definition can be extended to the case when B is a complex manifold.

In the following, a complex analytic family of compact complex manifolds is denoted by $(\mathcal{M}, B, \tilde{\omega})$.

EXAMPLES. 1. Let $B = \{\omega \in \mathbf{C} \mid \operatorname{Im}\omega > 0\}$ and $G_\omega = \{m\omega + n \mid m, n \in \mathbf{Z}\}$. The family $\{C_\omega \mid \operatorname{Im}\omega > 0\}$, where $C_\omega = \mathbf{C}/_{G_\omega}$ is a complex analytic family.

We want to remark that C_ω and $C_{\omega'}$ are analytic isomorphic if and only if

$$\omega' = \frac{a\omega + b}{c\omega + d}, \quad a, b, c, d \in \mathbf{Z},\ ad - bc = 1.$$

DEFINITION. Let M and N be two compact complex manifolds. Then N is said to be a *deformation* of M if there exists a complex analytic family $(\mathcal{M}, B, \tilde{\omega})$ such that $M = \tilde{\omega}^{-1}(t_0)$ and $N = \tilde{\omega}^{-1}(t_1)$, with $t_0, t_1 \in B$.

Two complex analytic families $(\mathcal{M}, B, \tilde{\omega})$ and $(\mathcal{N}, B, \pi)$ are called *holomorphically equivalent* if there exists an analytic isomorphism $\Phi : \mathcal{M} \to \mathcal{N}$ such that $\pi \circ \Phi = \tilde{\omega}$. In this case, M_t and N_t are analytic isomorphic via Φ, for all $t \in B$.

Let M be a compact complex manifold and B an arbitrary complex manifold. Then $(M \times B, B, pr_2)$ is a complex analytic family, where $pr_2 : M \times B \to B$ is the projection on the second factor.

A complex analytic family holomorphically equivalent to $(M \times B, B, pr_2)$, with $M = \tilde{\omega}^{-1}(t_0)$, $t_0 \in B$, is called *trivial*. If $(\mathcal{M}, B, \tilde{\omega})$ is trivial, then $\forall t \in B$, $M_t = \tilde{\omega}^{-1}(t)$ is analytic isomorphic to M.

Let $(\mathcal{M}, B, \tilde{\omega})$ be a complex analytic family and $U \subset B$ a subdomain. Denoting by $\mathcal{M}_U = \tilde{\omega}^{-1}(U)$ and $\tilde{\omega}_U = \tilde{\omega}|_U$, the complex analytic family $(\mathcal{M}_U, U, \tilde{\omega}_U)$ is the *restriction* of $(\mathcal{M}, B, \tilde{\omega})$ to U.

The complex analytic family $(\mathcal{M}, B, \tilde{\omega})$ is called *locally trivial* if for each $t \in B$, there exists $t \in U \subset B$ such that $(\mathcal{M}_U, U, \tilde{\omega}_U)$ is trivial.

THEOREM 10.1 [37]. *Let $(\mathcal{M}, B, \tilde{\omega})$ be a complex analytic family of compact complex manifolds and $t_0 \in B$. Then $M_t = \tilde{\omega}^{-1}(t)$ is C^∞-diffeomorphic to M_{t_0}, for any $t \in B$.*

2. Let $W = \mathbf{C}^2 - \{0\}$ and g_t an automorphism of W defined by

$$g_t(z^1, z^2) = (\alpha z^1 + t z^2, \alpha z^2),$$

where $0 < |\alpha| < 1$ and $t \in \mathbf{C}$. Let $G_t = \{g_t^m \mid m \in \mathbf{Z}\}$ be the infinite cyclic group spanned by g_t. We consider the complex surface $M_t = W/_{G_t}$. We see that M_0 is a Hopf surface.

It is easy to prove that $\{M_t \mid t \in \mathbf{C}\}$ is a complex analytic family.

Let $U = \mathbf{C} - \{0\}$. Then the restriction of $(\mathcal{M}, \mathbf{C}, \tilde{\omega})$ to U is trivial. Thus M_t is analytic isomorphic to M_1, $\forall t \neq 0$. But M_0 and M_1 are not analytic isomorphic (see [37]). On the other hand, by Theorem 10.1 they are C^∞-diffeomorphic.

3. We write $P^1(\mathbf{C}) = \mathbf{C} \cup \{\infty\}$ as $P^1(\mathbf{C}) = U_1 \cup U_2$, with $U_1 = \mathbf{C}$ and $U_2 = P^1(\mathbf{C}) - \{0\}$. We denote by z_1, z_2 the non-homogeneous coordinates on U_1 and U_2, respectively. On $U_1 \cap U_2$, one has $z_1 z_2 = 1$.

For any $m \in \mathbf{N}$, we define $\tilde{M}_m = (U_1 \times P^1(\mathbf{C})) \cup (U_2 \times P^1(\mathbf{C}))$, where $(z_1, \zeta_1) \in U_1 \times P^1(\mathbf{C})$ and $(z_2, \zeta_2) \in U_2 \times P^1(\mathbf{C})$ are identified if and only if

$$z_1 z_2 = 1, \qquad \zeta_1 = z_2^m \zeta_2.$$

We can define a complex analytic family $\{M_t \mid t \in \mathbf{C}\}$ with $M_0 = \tilde{M}_m$. Fix a natural number $k \leqslant \frac{m}{2}$ and put

$$M_t = \big[\big(U_1 \times P^1(\mathbf{C})\big) \cup \big(U_2 \times P^1(\mathbf{C})\big)\big]/_{\sim},$$

where

$$(z_1, \zeta_1) \sim (z_2, \zeta_2) \quad \Leftrightarrow \quad \begin{cases} z_1 z_2 = 1, \\ \zeta_1 = z_2^m \zeta_2 + t z_2^k. \end{cases}$$

For $t = 0$, $M_0 = \tilde{M}_m$ and for $t \neq 0$, $M_t = \tilde{M}_{m-2k}$. Thus, for any natural number $k \leqslant \frac{m}{2}$, $\tilde{M}_m$ is a deformation of $\tilde{M}_{m-2k}$. Therefore, putting $k = \frac{m}{2}$, if m is even, and $k = \frac{m}{2} - \frac{1}{2}$, if m is odd, it follows that $\tilde{M}_m$ is a deformation of $\tilde{M}_0 = P^1(\mathbf{C}) \times P^1(\mathbf{C})$, if m is even, and a deformation of $\tilde{M}_1$, if m is odd, respectively. One can prove that $\tilde{M}_1$ and $P^1(\mathbf{C}) \times P^1(\mathbf{C})$ are not diffeomorphic.

Consequently, $\tilde{M}_m$ and $\tilde{M}_n$ are diffeomorphic if $m \equiv n \pmod 2$, but they are not analytic isomorphic, if $m \neq n$.

We will state a necessary and sufficient condition for a complex analytic family to be locally trivial.

Let $(\mathcal{M}, B, \tilde{\omega})$ be a complex analytic family of compact complex manifolds and $\{(z_j, t)\}$ its system of local coordinates. The transition maps

$$z_j^\alpha = f_{jk}^\alpha\big(z_k^1, \dots, z_k^n, t^1, \dots, t^m\big), \quad \alpha \in \{1, \dots, n\},$$

are holomorphic in $z_k^1, \dots, z_k^n, t^1, \dots, t^m$.

On $U_i \cap U_j \cap U_k \neq \emptyset$, one has $f_{ik}(z_k, t) = f_{ij}(f_{jk}(z_k, t), t)$.

Assume $m = 1$.

The above relation can be written as

$$f_{ik}^\alpha(z_k, t) = f_{ij}^\alpha\big(f_{jk}^1(z_k, t), \dots, f_{jk}^n(z_k, t), t\big), \quad \alpha \in \{1, \dots, n\},$$

or equivalently, using holomorphic vector fields,

$$\sum_{\alpha=1}^n \frac{\partial f_{ik}^\alpha(z_k, t)}{\partial t} \cdot \frac{\partial}{\partial z_i^\alpha} = \sum_{\alpha=1}^n \frac{\partial f_{ij}^\alpha(z_j, t)}{\partial t} \cdot \frac{\partial}{\partial z_i^\alpha} + \sum_{\beta=1}^n \frac{\partial f_{jk}^\beta(z_k, t)}{\partial t} \cdot \frac{\partial}{\partial z_j^\beta}.$$

We introduce the holomorphic vector fields

$$\theta_{jk}(t) = \sum_{\alpha=1}^n \frac{\partial f_{jk}^\alpha(z_k, t)}{\partial t} \cdot \frac{\partial}{\partial z_j^\alpha},$$

where we put $z_k = f_{kj}(z_j, t)$. Thus, one gets

$$\theta_{ik}(t) = \theta_{ij}(t) + \theta_{jk}(t).$$

Let Θ_t be the sheaf of the germs of holomorphic vector fields on M_t. Then

$$\theta_{jk}(t) \in \Gamma(U_j \cap U_k, \Theta_t)$$

defines a 1-cocycle with respect to the covering $\mathcal{U}_t = (U_j)_{j \geqslant 1}$ on M_t, i.e., $\{\theta_{jk}(t)\} \in Z^1(\mathcal{U}_t, \Theta_t)$. Denote by $\theta(t) \in H^1(M_t, \Theta_t)$ the cohomology class of the 1-cocycle $\{\theta_{jk}(t)\}$.

Intuitively, $\theta(t)$ represents the "derivative" of the complex structure of M_t with respect to t.

DEFINITION. $\theta(t)$ is called the *infinitesimal deformation* of the complex manifold M_t. We will use the notation

$$\frac{dM_t}{dt} = \theta(t).$$

It is easy to see that the infinitesimal deformation does not depend on the local coordinate system.

Let now $m \geqslant 1$. For the tangent vector

$$\frac{\partial}{\partial t} = \sum_{\lambda=1}^{m} c_\lambda \frac{\partial}{\partial t^\lambda} \in T_t B, \quad c_\lambda \in \mathbf{C},$$

we put

$$\theta_{jk}(t) = \sum_{\alpha=1}^{n} \frac{\partial f_{jk}^\alpha(z_k, t^1, \ldots, t^m)}{\partial t} \cdot \frac{\partial}{\partial z_j^\alpha}, \qquad z_k = f_{kj}(z_j, t).$$

The cohomology class $\theta(t) \in H^1(M_t, \Theta_t)$ of the 1-cocycle $\{\theta_{jk}(t)\}$ is the *infinitesimal deformation* along $\frac{\partial}{\partial t}$ and is denoted by $\frac{\partial M_t}{\partial t}$. The map

$$\rho_t : T_t B \to H^1(M_t, \Theta_t),$$
$$\rho_t : \frac{\partial}{\partial t} \mapsto \rho_t\left(\frac{\partial}{\partial t}\right) = \frac{\partial M_t}{\partial t}$$

is **C**-linear.

Obviously, if a complex analytic family $(\mathcal{M}, B, \tilde{\omega})$ is locally trivial, then $\rho_t = 0$, $\forall t \in B$. Conversely, we have the following

THEOREM 10.2 [37]. *Let $(\mathcal{M}, B, \tilde{\omega})$ be a complex analytic family of compact complex manifolds. If $\dim H^1(M_t, \Theta_t)$ does not depend on $t \in B$ and $\rho_t = 0$, $\forall t \in B$, then $(\mathcal{M}, B, \tilde{\omega})$ is locally trivial.*

Let $(\mathcal{M}, B, \tilde{\omega}) = \{M_t \mid t \in B\}$ be a complex analytic family of compact complex manifolds and $h : D \to B$ a holomorphic map defined on the domain $D \subset \mathbf{C}^r$. Define a holomorphic map $\Pi : \mathcal{M} \times D \to B \times D$, by $\Pi(p, s) = (\tilde{\omega}(p), s) = (t, s)$. Then $(\mathcal{M} \times D,$

$B \times D, \Pi)$ is a complex analytic family over $B \times D$. Obviously $\Pi^{-1}(t,s) = M_t \times \{s\}$. The graph of h

$$G = \{(h(s), s) \in B \times D \mid s \in D\}$$

is a complex submanifold in $B \times D$. Since $pr_2 : B \times D \to D$ induces an analytic isomorphism of G on D, we identify G and D, via pr_2. $\mathcal{N} = \Pi^{-1}(G)$ is a complex submanifold of $\mathcal{M} \times D$. Thus, we obtain the complex analytic family $(\mathcal{N}, D, \pi)$, where $\pi = pr_2 \circ \Pi$, which is called the complex analytic family *induced* by h from $(\mathcal{M}, B, \tilde{\omega})$.

The relationship between the infinitesimal deformations of the complex analytic families $(\mathcal{M}, B, \tilde{\omega})$ and $(\mathcal{N}, D, \pi)$, respectively, is given by

$$\frac{\partial M_{h(s)}}{\partial s} = \sum_{\lambda=1}^{m} \frac{\partial t^\lambda}{\partial s} \cdot \frac{\partial M_t}{\partial t^\lambda}, \quad t = h(s),$$

for any tangent vector $\frac{\partial}{\partial s} \in T_s D$.

REMARK. The assumption in Theorem 10.2 is essential.

Indeed, consider the complex analytic family $(\mathcal{M}, \mathbf{C}, \tilde{\omega})$ given in Example 2. We saw that it is trivial over $U = \mathbf{C} - \{0\}$, but M_0 and M_t, with $t \neq 0$, are not analytic isomorphic.

Let $(\mathcal{N}, \mathbf{C}, \pi)$ be the complex analytic family induced by the holomorphic map $h(s) = s^2$ from $(\mathcal{M}, \mathbf{C}, \tilde{\omega})$. Then, one has

$$\rho_s\left(\frac{d}{ds}\right) = \frac{dM_{s^2}}{ds} = \frac{dt}{ds} \cdot \frac{dM_t}{dt} = 2s\frac{dM_t}{dt}.$$

Therefore, $\rho_s = 0$, $\forall s \in \mathbf{C}$, but the family $(\mathcal{N}, \mathbf{C}, \pi)$ is not locally trivial.

Computing, one gets

$$\dim H^1(M_t, \Theta_t) = \begin{cases} 4, & t = 0, \\ 2, & t \neq 0. \end{cases}$$

If $(\mathcal{M}, B, \tilde{\omega})$ is a complex analytic family of compact complex manifolds, with $0 \in B \subset \mathbf{C}$ a domain, the infinitesimal deformation $\theta = (\frac{dM_t}{dt})_{t=0} \in H^1(M, \Theta)$, where $M = \tilde{\omega}^{-1}(0)$.

Conversely, we may state the following problem. Let M be a compact complex manifold and $\theta \in H^1(M, \Theta)$. Does there exist a complex analytic family $(\mathcal{M}, B, \tilde{\omega})$, with $0 \in B \subset \mathbf{C}$, such that

$$\tilde{\omega}^{-1}(0) = M, \qquad \left(\frac{dM_t}{dt}\right)_{t=0} = \theta?$$

We denote by $[\cdot, \cdot]$ the Lie bracket of vector fields and by $[\theta, \theta]$ the cohomology class of $[\theta_{ij}, \theta_{jk}]$.

THEOREM 10.3. *Let M be a compact complex manifolds and $\theta \in H^1(M, \Theta)$. A necessary condition for the existence of a complex analytic family of compact complex manifolds $(\mathcal{M}, B, \tilde{\omega})$ such that $\tilde{\omega}^{-1}(0) = M$ and $(\frac{dM_t}{dt})_{t=0} = \theta$, is $[\theta, \theta] = 0$.*

We call $[\theta, \theta]$ the primary obstruction.

THEOREM OF EXISTENCE [37]. *Let M be a compact complex manifold. If $H^2(M, \Theta) = 0$, then, for each $\theta \in H^1(M, \Theta)$, there exists a complex analytic family of compact complex manifolds $(\mathcal{M}, B, \tilde{\omega})$, with $0 \in B \subset \mathbf{C}$, such that*

$$\tilde{\omega}^{-1}(0) = M, \qquad \left(\frac{dM_t}{dt}\right)_{t=0} = \theta.$$

A complex analytic family $(\mathcal{M}, B, \tilde{\omega})$ is said to be *effectively parametrized* if for any $t \in B$, $\rho_t : T_t B \to H^1(M_t, \Theta_t)$ is one-to-one.

Let $(\mathcal{M}, B, \tilde{\omega})$ be a complex analytic family of compact complex manifolds and $t_0 \in B$. Then $(\mathcal{M}, B, \tilde{\omega})$ is called *complete* at t_0 if for any complex analytic family $(\mathcal{N}, D, \pi)$ such that $0 \in D \subset \mathbf{C}^l$ and $\pi^{-1}(0) = \tilde{\omega}^{-1}(t_0)$, there exists a domain E with $0 \in E \subset D$ and a holomorphic map $h : E \to B$, $s \mapsto t = h(s)$, with $h(0) = t_0$, such that $(\mathcal{N}_E, E, \pi)$ is the complex analytic family induced by h from $(\mathcal{M}, B, \tilde{\omega})$, where $\mathcal{N}_E = \pi^{-1}(E)$.

Since $N_s = \pi^{-1}(s)$ is a deformation of $N_0 = M_{t_0}$, if $(\mathcal{M}, B, \tilde{\omega})$ is complete at $t_0 \in B$, then it contains all deformations N_s of M_{t_0}, for s sufficiently small.

A complex analytic family $(\mathcal{M}, B, \tilde{\omega})$ of compact complex manifolds is called *complete* if it is complete at any point $t \in B$.

DEFINITION. Let M be a compact complex manifold. If there exists a complete and effectively parametrized complex analytic family $(\mathcal{M}, B, \tilde{\omega})$, with $0 \in B \subset \mathbf{C}^m$ and $\tilde{\omega}^{-1}(0) = M$, then $m = \dim B$ is called the *moduli number* of M, which is denoted by $m(M)$.

If such a complex analytic family does not exist, the moduli number of M cannot be defined.

THEOREM OF COMPLETENESS [37]. *Let $(\mathcal{M}, B, \tilde{\omega})$ be a complex analytic family of compact complex manifolds, with $t_0 \in B \subset \mathbf{C}^m$, $\rho_t : T_t B \to H^1(M_t, \Theta_t)$,*

$$\rho_t : \frac{\partial}{\partial t} \mapsto \rho_t\left(\frac{\partial}{\partial t}\right) = \frac{\partial M_t}{\partial t}.$$

If ρ_{t_0} is surjective, i.e., $\rho_{t_0}(T_{t_0} B) = H^1(M_{t_0}, \Theta_{t_0})$, then $(\mathcal{M}, B, \tilde{\omega})$ is complete at t_0.

The following result follows.

If $(\mathcal{M}, B, \tilde{\omega})$ is an effectively parametrized complex analytic family and $m = \dim B = \dim H^1(M_t, \Theta_t)$, then $m = m(M)$.

4. Let $P^n(\mathbf{C})$ be the complex projective space. Since $H^1(P^n(\mathbf{C})) = 0$, then $m(P^n(\mathbf{C})) = 0$.

5. For the complex torus $T^n = \mathbf{C}^n/_G$ one has $\dim H^1(T^n, \Theta) = n^2$. Thus the moduli number of T^n is n^2 (see [37]).

We derive some applications of the theorems of existence and completeness.

THEOREM 10.4. *Let M be a compact complex manifold such that $H^2(M, \Theta) = 0$ and the moduli number $m(M)$ is defined. Then $m(M) = \dim H^1(M, \Theta)$.*

THEOREM 10.5. *Let M be a compact complex manifold with $H^2(M, \Theta) = 0$. Then its moduli number $m(M)$ is defined if and only $\dim H^1(M_t, \Theta_t)$ does not depend on $t \in \Delta$, where $0 \in \Delta \subset B$ is sufficiently small. In this case, the complex analytic family $(\mathcal{M}_\Delta, \Delta, \tilde{\omega})$ is complete and effectively parametrized.*

THEOREM 10.6. *Let M be a compact complex manifold, with $H^0(M, \Theta) = 0$ and $H^2(M, \Theta) = 0$. Then its moduli number $m(M)$ is defined and $m(M) = \dim H^1(M, \Theta)$.*

EXAMPLE. Consider the complex analytic family constructed in Example 3 above. One has

$$\dim H^1(\tilde{M}_m, \Theta) = \begin{cases} 0, & m = 0, 1, \\ m - 1, & m \geqslant 2. \end{cases}$$

Put $m = 2$ and $k = 1$, i.e.,

$$M_t = \left[\left(U_1 \times P^1(\mathbf{C})\right) \cup \left(U_2 \times P^1(\mathbf{C})\right)\right]/_\sim, \quad U_1 = U_2 = \mathbf{C},$$

where $(z_1, \zeta_1) \sim (z_2, \zeta_2)$ if and only if

$$z_1 z_2 = 1, \qquad \zeta_1 = z_2^2 \zeta_2 + t z_2.$$

Since $M_0 = \tilde{M}_2$ and $M_t = \tilde{M}_0$, it follows that

$$\dim H^1(M_t, \Theta_t) = \begin{cases} 1, & t = 0, \\ 0, & t \neq 0. \end{cases}$$

The infinitesimal deformation $\frac{dM_t}{dt} \in H^1(M_t, \Theta_t)$ is the cohomology class of the 1-cocycle

$$\theta_{12}(t) = \frac{\partial f_{12}^2(z_2, \zeta_2, t)}{\partial t} \cdot \frac{\partial}{\partial \zeta_1} = z_2 \frac{\partial}{\partial \zeta_1}.$$

For $t = 0$, $\{z_2 \frac{\partial}{\partial \zeta_1}\}$ is a basis in $H^1(M_0, \Theta_0) = H^1(\tilde{M}_2, \Theta)$. Thus $\rho_0 : T_0\mathbf{C} \to H^1(M_0, \Theta_0)$ is surjective. By the Theorem of completeness, the family $\{M_t \mid t \in \mathbf{C}\}$ is complete at $t = 0$.

One has $H^2(\tilde{M}_2, \Theta) = 0$. Applying the Theorem of existence, there exists a complex analytic family $(\mathcal{M}, B, \tilde{\omega})$ such that $\tilde{\omega}^{-1}(0) = \tilde{M}_2$ and $\rho_0 : T_0 B \to H^1(\tilde{M}_2, \Theta)$ is an isomorphism, where $0 \in B \subset \mathbf{C}$. This family is nothing but $\{M_t \mid t \in \mathbf{C}\}$.

For $t \neq 0$, $H^1(M_t, \Theta_t) = 0$. It follows that $\frac{dM_t}{dt} = 0$, thus the family $\{M_t \mid t \in \mathbf{C}\}$ is complete at each t.

Summarizing, the complex analytic family $\{M_t \mid t \in \mathbf{C}\}$, satisfies $H^1(M_0, \Theta_0) = 1$ and $H^1(M_t, \Theta_t) = 0$, for $t \neq 0$. Thus, by Theorem 10.5, the moduli number $m(\tilde{M}_2)$ cannot be defined (although $H^2(\tilde{M}_2, \Theta) = 0$). Consequently, the assumption $H^0(M, \Theta) = 0$ in Theorem 10.6 cannot be omitted.

Later, M. Kuranishi [38] proved the Theorem of existence for general case.

THEOREM 10.7. *For any compact complex manifold M, there exists a complete complex analytic family $\{M_t \mid t \in B\}$, $0 \in B$, with $M_0 = M$.*

In this case, one has

$$\dim B \geqslant \dim H^1(M, \Theta) - \dim H^2(M, \Theta).$$

From this inequality, it follows that

$$m(M) \geqslant \dim H^1(M, \Theta) - \dim H^2(M, \Theta),$$

provided $m(M)$ is defined.

The Kodaira's conjecture $m(M) = \dim H^1(M, \Theta)$ [37] turned out to be false.

First, D. Mumford [50] constructed a counterexample to this conjecture. He constructed a 3-dimensional complex manifold $M = \mu_C(P^3(\mathbf{C}))$ obtained from $P^3(\mathbf{C})$ by a monoidal transformation μ_C, whose center is a certain curve $C \subset P^3(\mathbf{C})$ of genus 24 and degree 14.

Moreover, A. Kas found a 2-dimensional counterexample (see [33]).

References

[1] E. Abbena, *An example of an almost Kähler manifold which is not Kählerian*, Boll. Un. Mat. Ital. A (6) **3** (1984), 383–392.

[2] T. Adachi and S. Maeda, *Characterization of complex space forms in terms of geodesics and circles on their geodesic spheres*, Nihonkai Math. J. **14** (2003), 95–98.

[3] J.F. Adams, *On the non-existence of elements of Hopf invariant one*, Ann. of Math. **72** (1960), 20–104.

[4] T. Aubin, *Equations du type de Monge–Ampère sur les variétés Kähleriennes compactes*, C. R. Acad. Sci. Paris **283** (1976), 119–121.

[5] M. Berger, *Sur les variétés d'Einstein compactes*, C. R. III-e Reunion Math. Expression Latine, Namur (1965), 35–55.

[6] D. Blair, *Riemannian Geometry of Contact and Symplectic Manifolds*, Birkhäuser, Boston (2002).

[7] E. Calabi, *The space of Kähler metrics*, Proc. Int. Congress Math. Amsterdam **2** (1954), 206–207.

[8] E. Calabi, *Constructions and properties of some 6-dimensional almost complex manifolds*, Trans. Amer. Math. Soc. **87** (1958), 407–438.

[9] E. Calabi and B. Eckmann, *A class of compact complex manifolds which are not algebraic*, Ann. of Math. **58** (1953), 494–500.

[10] B.Y. Chen, *Notes on Characteristic Classes*, Michigan State Univ. (1971).

[11] B.Y. Chen, *Geometry of Submanifolds*, Dekker, New York (1973).

[12] B.Y. Chen, *Geometry of Submanifolds and Its Applications*, Sci. Univ. Tokyo (1981).

[13] B.Y. Chen, *Total Mean Curvature and Submanifolds of Finite Type*, World Scientific (1984).
[14] B.Y. Chen, *Geometry of Slant Submanifolds*, Katholieke Univ. Leuven (1990).
[15] B.Y. Chen, *Riemannian submanifolds*, Handbook of Differential Geometry, F. Dillen and L. Verstraelen, eds, Elsevier Science, Amsterdam (2000), 187–418.
[16] B.Y. Chen and K. Ogiue, *Some characterisations of complex space forms in terms of Chern classes*, Quart. J. Math. **26** (1975), 459–464.
[17] B.Y. Chen and L. Verstraelen, *Laplace Transformation on Submanifolds*, Katholieke Univ. Leuven, Katholieke Univ. Brussel (1995).
[18] F. Defever, I. Mihai and L. Verstraelen, *B.Y. Chen's inequality for C-totally real submanifolds of Sasakian space forms*, Boll. Un. Mat. Ital. **11** (1997), 365–374.
[19] S. Deshmukh and M.H. Shahid, *Nucleus of nearly-Kähler manifolds*, J. Univ. Kuwait **16** (1989), 7–11.
[20] O. Forster, *Introduction to Complex Analysis of Several Variables*, Lecture Notes, Regensburg (1973).
[21] T. Frankel, *Manifolds with positive curvatures*, Pacific J. Math. **11** (1961), 165–174.
[22] P. Gilkey, *Invariance Theory, Heat Equation and the Atiyah–Singer Index Theorem*, Publish or Perish (1984).
[23] A. Gray, *Nearly Kaehler manifolds*, J. Differential Geometry **4** (1970), 283–309.
[24] S. Greenfield, *Cauchy–Riemann equations in several variables*, Ann. Scuola Norm. Sup. Pisa **22** (1968), 275–314.
[25] Ph. Griffiths and J. Harris, *Principles of Algebraic Geometry*, Wiley (1978).
[26] R. Gunning and Rossi, *Analytic Functions of Several Complex Variables*, Prentice Hall, NJ (1965).
[27] J.I. Hano and Y. Matsushima, *Some studies of Kaehlerian homogeneous spaces*, Nagoya Math. J. **9** (1957), 77–92.
[28] N.S. Hawley, *Constant holomorphic curvature*, Canad. J. Math. **5** (1953), 53–56.
[29] W.V.D. Hodge, *Theory and Applications of Harmonic Integrals*, Cambridge Univ. Press, London (1941).
[30] H. Hopf, *Zur Topologie der Komplexen Mannigfeltigkeiten*, Essays presented to R. Courant, Interscience, New York (1948), 167–185.
[31] J. Igusa, *On the structure of a certain class of Kähler manifolds*, Amer. J. Math. **76** (1954), 669–678.
[32] E. Kähler, *Über eine bemerkenswerte Hermitiesche metrik*, Abh. Math. Sem. Univ. Hamburg **9** (1933), 173–186.
[33] A. Kas, *On obstructions to deformations of complex analytic surfaces*, Proc. Nat. Acad. Sci. USA **58** (1967), 402–404.
[34] A. Kirchhoff, *Sur l'existence de certains champs tensoriels sur les sphères à n dimensions*, C. R. Acad. Sci. Paris **225** (1947), 1258–1260.
[35] S. Kobayashi, *On compact Kähler manifolds with positive Ricci tensor*, Ann. of Math. **74** (1961), 570–574.
[36] S. Kobayashi and K. Nomizu, *Foundations of Differential Geometry*, vols. I, II, Interscience, New York (1963, 1969).
[37] K. Kodaira, *Complex Manifolds and Deformation of Complex Structures*, Springer, Berlin (1985).
[38] M. Kuranishi, *On the locally complete families of complex analytic structures*, Ann. of Math. **75** (1962), 536–577.
[39] K. Matsumoto, I. Mihai and R. Miron, *Almost Kählerian structures determined by Riemannian structures*, Publ. Math. Debrecen **62** (2003), 497–509.
[40] K. Matsumoto, I. Mihai and Y. Tazawa, *Ricci tensor of slant submanifolds in complex space forms*, Kodai Math. J. **26** (2003), 85–94.
[41] M. Matsumoto and S. Tanno, *Kaehlerian spaces with parallel or vanishing Bochner curvature tensor*, Tensor (N.S.) **27** (1973), 291–294.
[42] I. Mihai, *Differential geometry of (pseudo-) Riemannian manifolds admitting certain endomorphisms of its tangent bundles and their submanifolds*, Doctoral thesis, Katholieke Univ. Leuven (1993).
[43] I. Mihai, *Special Chapters on Geometry of Complex Manifolds*, Ed. Univ. Bucureşti (1995) (in Romanian).
[44] I. Mihai, *Geometry of Submanifolds in Complex Manifolds*, Ed. Univ. Bucureşti (2001) (in Romanian).
[45] I. Mihai, *Ideal Kaehlerian slant submanifolds in complex space forms*, Rocky Mountain J. Math. **35** (2005), 941–951.
[46] I. Mihai, R. Rosca and L. Verstraelen, *Some Aspects of the Differential Geometry of Vector Fields*, PADGE, vol. 2, Katholieke Univ. Leuven, Katholieke Univ. Brussel (1996).

[47] R. Miron, *Geometry of Higher Order Lagrange Spaces. Applications to Mechanics and Physics*, Kluwer Academic (1997).

[48] R. Miron, *Geometry of Higher Order Finsler Spaces*, Hadronic Press (1998).

[49] S. Mori, *Projective manifolds with ample tangent bundles*, Ann. of Math. **110** (1979), 593–606.

[50] D. Mumford, *Further pathologies in algebraic geometry*, Amer. J. Math. **84** (1962), 642–648.

[51] A. Newlander and L. Nirenberg, *Complex analytic coordinates in almost complex manifolds*, Ann. of Math. **65** (1957), 391–404.

[52] K. Nomizu, *Conditions for constancy of the holomorphic sectional curvature*, J. Differential Geometry **8** (1973), 335–339.

[53] J.A. Schouten and D. van Dantzig, *Über unitäre Geometrie*, Math. Ann. **103** (1930), 319–346.

[54] C.L. Siegel, *Discontinuous groups*, Ann. of Math. **44** (1946), 674–689.

[55] Y.T. Siu and S.T. Yau, *Compact Kähler manifolds of positive bisectional curvature*, Invent. Math. **59** (1980), 189–204.

[56] B. Suceavă, *Fundamental inequalities and strongly minimal submanifolds*, Contemp. Math. **337** (2003), 155–170.

[57] J.L. Synge, *On the connectivity of spaces of positive curvature*, Quart. J. Math. **7** (1936), 316–320.

[58] W. Thurston, *Some simple examples of symplectic manifolds*, Proc. Amer. Math. Soc. **55** (1976), 467–468.

[59] K. Tsukada, *Einstein Kaehler submanifolds with codimension* 2 *in a complex space form*, Math. Ann. **274** (1986), 503–516.

[60] M. Umehara, *Einstein Kaehler submanifolds of a complex linear or hyperbolic space*, Tohoku Math. J. **39** (1987), 385–389.

[61] H.C. Wang, *Complex parallelisable manifolds*, Proc. Amer. Math. Soc. **5** (1954), 771–776.

[62] R.O. Wells, *Differential Analysis on Complex Spaces*, Springer, Berlin (1980).

[63] K. Yano and M. Kon, *Structures on Manifolds*, World Scientific, Singapore (1984).

[64] S.T. Yau, *Calabi's conjecture and some new results in algebraic geometry*, Proc. Nat. Acad. Sci. USA **74** (1977), 1798–1799.

[65] F. Zheng, *Complex Differential Geometry*, AMS/IP Studies Adv. Math. (2000).

CHAPTER 7

Compendium on the Geometry of Lagrange Spaces

Radu Miron*

Faculty of Mathematics, "Al.I. Cuza" University Iasi, Bul. Copou 11, 6600 Iasi, Romania
E-mail: *radu.miron@uaic.ro*

Contents

*The author expresses his gratitude to professors F.J.E. Dillen, I. Mihai and L.C.A. Verstraelen for their scientific and moral help.

HANDBOOK OF DIFFERENTIAL GEOMETRY, VOL. II
Edited by F.J.E. Dillen and L.C.A. Verstraelen

Introduction

The purpose of the present compendium is a short presentation of the geometrical theory of Lagrange spaces. In the last 20 years, geometers, mechanicians and physicists from all over the world worked in this field. I mention only some of them: P.L. Antonelli, M. Anastasiei, G.S. Asanov, M. Crampin, R.S. Ingarden, M. Matsumoto, R. Miron, H. Rund, S. Ikeda, H. Shimada, L. Tamassy and P. Stavrinos.

After the explicit formulation, in 1980, of the notion of Lagrange space, due to the present author [15,16], some excellent books on the geometrical study of these spaces, as well as their applications to Mechanics, Physics and Mathematical Biology, have been published. The bibliography therein is a selection of books and papers on these topics.

Therefore, in the present compendium I sketch the general framework of the Lagrange geometry, based on the books of R. Miron [16–18], R. Miron and M. Anastasiei [19,20], R. Miron, D. Hrimiuc, H. Shimada and S. Sabău [22] and R. Miron, M. Anastasiei and I. Bucătaru [21]. A short presentation of the geometry of higher-order Lagrange spaces is presented at the end of this compendium.

Since the geometry of the tangent bundle (TM, π, M) of a manifold M is a basic tool in the study of Lagrange geometry, I devote the first section to the geometry of TM, pointing-out the main geometrical objects, as: Liouville vector field $\mathbb{C}$, almost tangent structure J, semispray and nonlinear connection.

In the second section, I introduce the notion of a Lagrange space $L^n = (M, L(x, y))$, with Lagrangian $L: TM \to \mathbb{R}$ and fundamental tensor g_{ij}, assumed nonsingular and of constant signature. The known Lagrangian from Electrodynamics assures the existence of Lagrange spaces.

The variational problem associated to the integral of action

$$I(c) = \int L(x, \dot{x})\, dt$$

allows to determine the Euler–Lagrange equations, conservation law of the energy E_L, as well as the canonical semispray S of L^n. The canonical semispray S determines the canonical nonlinear connection N and the metrical N-linear connection D, given by the generalized Christoffel symbols. The structure equations of D are derived. This theory is applied to the study of the electromagnetic and gravitational fields of the space L^n. An almost Kählerian model is constructed and the notion of a generalized Lagrange space GL^n is defined.

The third section is entitled *Finsler spaces* $F^n = (M, F(x, y))$. The class of these spaces is a subclass of that of Lagrange spaces. It follows that the geometry of Finsler spaces F^n can be constructed only by means of Analytical Mechanics principles.

But, since a Riemann space $\mathcal{R}^n = (M, g)$ is a particular Finsler space $F^n = (M, F(x, y))$, we get the following remarkable sequence of inclusions:

(I) $\{\mathcal{R}^n\} \subset \{F^n\} \subset \{L^n\} \subset \{GL^n\}$.

Thus, the Lagrangian geometry is the geometrical study of this sequence.

Sections 4 and 5 are devoted to the natural extension of the Lagrangian geometries to higher order, extremely important in the Lagrangian Mechanics of the accelerations of order $k \geqslant 1$ ($k \in \mathbb{N}^*$). We introduce the bundle (T^kM, π^k, M), the Lagrange spaces $L^{(k)n} = (M, L)$ and study the solution of the classical problem of the prolongation of Riemannian structures given on the base manifold M to T^kM. It gives some interesting examples of spaces $L^{(k)n}$. The variational calculus applied to the integral of action

$$I(c) = \int_0^1 L\left(x(t), \frac{dx}{dt}, \ldots, \frac{1}{k!}\frac{d^kx}{dt^k}\right) dt$$

leads to the Euler–Lagrange equations and to the k-semispray S, which determines a nonlinear connection N and an N-linear connection D. These two geometrical object fields govern the geometry of Lagrange spaces of order k, $L^{(k)n}$.

Also, the generalized Lagrange spaces of order k, $G^{(k)n} = (M, g_{ij})$ are defined and studied. Following the classical theory, from the case $k = 1$, one defines the notion of Finsler space of order k, $F^{(k)n} = (M, F)$, and show that the sequence (I) holds for $k \geqslant 1$.

I want to remark that this compendium is not only an introduction to the Lagrangian geometries, it is also an useful geometrical instrument for their applications to Variational Calculus, Analytical Mechanics, Physics, Biology, Optimal Control, etc.

1. Tangent bundle

The geometry of a Lagrange space over a finite-dimensional manifold M has been introduced and studied as a subgeometry of the geometry of the tangent manifold TM by R. Miron and his coworkers [16–18]. We start with the study of the geometry of the tangent bundle (TM, π, M). The tangent manifold TM carries some natural object fields, as: the Liouville vector field $\mathbb{C}$, the tangent structure J, the vertical distribution V, the notion of a semispray S. We can develop a very consistent geometry of TM based on this concept of semispray.

In the following, we assume all the geometrical object fields and mappings to be C^∞-differentiable and we express this by the words "differentiable" or "smooth".

1.1. *The manifold TM*

The differentiable structure on TM is induced by that of the base manifold M such that the natural projection $\pi : TM \to M$ is a differentiable submersion and the triple (TM, π, M) is a differentiable vector bundle. Let M be a real, n-dimensional differentiable manifold and $(U, \varphi = (x^i))$ a local chart in a neighborhood of a point $x \in M$. Then any curve $\sigma : I \to M$, $\operatorname{Im}\sigma \subset U$, which passes through x at $t = 0$ is analytically represented by $x^i = x^i(t)$, $t \in I$, $\varphi(x) = (x^i(0))$ $(i, j, \ldots = 1, \ldots, n)$. The tangent vector $[\sigma]_x$ is determined by the coefficients

$$x^i = x^i(0), \qquad y^i = \frac{dx^i}{dt}(0).$$

Then the pair $(\pi^{-1}(U), \Phi)$, with $\Phi([\sigma]_x) = (x^i, y^i) \in \mathbb{R}^{2n}$, $\forall[\sigma]_x \in \pi^{-1}(U)$, is a local chart. It will be denoted by $(\pi^{-1}(U), \phi = (x^i, y^i))$. The set of these "induced" local charts determines a differentiable structure on TM such that (TM, π, M) is a differentiable vector bundle.

A change of coordinates on M, $(U, \varphi = x^i) \to (V, \psi = \tilde{x}^i)$, with $\operatorname{rank}(\frac{\partial \tilde{x}^i}{\partial x^j}) = n$, determines the corresponding change of coordinates on $TM\colon (\pi^{-1}(U), \Phi = (x^i, y^i)) \to (\pi^{-1}(V), \Psi = (\tilde{x}^i, \tilde{y}^i))$, given by

$$\begin{cases} \tilde{x}^i = \tilde{x}^i(x^j), \quad \operatorname{rank}\left(\frac{\partial \tilde{x}^i}{\partial x^j}\right) = n, \\ \tilde{y}^i = \frac{\partial \tilde{x}^i}{\partial x^j} y^j. \end{cases} \tag{1.1.1}$$

The determinant of the Jacobian of $\Psi_\circ \Phi^{-1}$ is $\det(\frac{\partial \tilde{x}^i}{\partial x^j})^2 > 0$. Thus the manifold TM is orientable and of dimension $2n$.

The tangent space $T_u TM$ at a point $u \in TM$ to TM is a $2n$-dimensional vector space, having the natural basis $\{\frac{\partial}{\partial x^i}, \frac{\partial}{\partial y^i}\}$ at u. A change of coordinates (1.1.1) on TM implies the change of natural basis, at u as follows:

$$\begin{cases} \frac{\partial}{\partial x^i} = \frac{\partial \tilde{x}^j}{\partial x^i} \frac{\partial}{\partial \tilde{x}^j} + \frac{\partial \tilde{y}^j}{\partial x^i} \frac{\partial}{\partial \tilde{y}^j}, \\ \frac{\partial}{\partial y^i} = \frac{\partial \tilde{x}^j}{\partial x^i} \frac{\partial}{\partial \tilde{y}^j}. \end{cases} \tag{1.1.2}$$

A vector $X_u \in T_u TM$ is given by $X = X^i(u)\frac{\partial}{\partial x^i} + Y^i(u)\frac{\partial}{\partial y^i}$. Then a vector field X on TM is a section $X\colon TM \to TTM$ of the bundle $\tau\colon TTM \to TM$, given by $\tau(x, y, X, Y) = (x, y)$. We denote by π_* the projection $\pi_*\colon TTM \to TM$ with $\pi_*(x, y, X, Y) = (x, X)$.

From the formula (1.1.2) we can see that $(\frac{\partial}{\partial y^i})$ at a point $u \in TM$ span an n-dimensional vector subspace $V(u)$ of $T_u TM$. It is called the vertical subspace. The map $V\colon u \in TM \to V(u) \subset T_u TM$ is an integrable distribution called the vertical distribution. Then $VTM = \bigcup_{u \in TM} V(u)$ is a subbundle of the tangent bundle (TTM, π, TM) to TM. Since $\pi\colon TM \to M$ is a submersion, it follows that $\pi_{*,u}\colon T_u TM \to T_{\pi(u)} M$ is an epimorphism of linear spaces. The kernel of $\pi_{*,u}$ is exactly the vertical subspace $V(u)$.

We denote by $\mathcal{X}^v(TM)$ the set of all vertical vector field on TM. It is a real subalgebra of Lie algebra of vector fields on TM, $\mathcal{X}(TM)$.

Consider the cotangent space $T_u^* TM$, $u \in TM$. It is the dual of the space $T_u TM$ and $(dx^i, dy^i)_u$ is the natural cobasis with respect to (1.1.1) we have

$$d\tilde{x}^i = \frac{\partial \tilde{x}^i}{\partial x^j} dx^j, \qquad d\tilde{y}^i = \frac{\partial \tilde{x}^i}{\partial x^j} dy^j + \frac{\partial \tilde{y}^i}{\partial x^j} dx^j. \tag{1.1.3}$$

The almost tangent structure J of the tangent bundle is defined by

$$J = \frac{\partial}{\partial y^i} \otimes dx^i. \tag{1.1.4}$$

By means of (1.1.2) and (1.1.3) we can prove that J is globally defined on TM and that we have

$$J\left(\frac{\partial}{\partial x^i}\right)=\frac{\partial}{\partial y^i},\qquad J\left(\frac{\partial}{\partial y^i}\right)=0. \tag{1.1.4'}$$

It follows that the following formulae hold:

$$J^2=J\circ J=0,\qquad \operatorname{Ker} J=\operatorname{Im} J=VTM.$$

The almost cotangent structure J^* is defined by

$$J^*=dx^i\otimes\frac{\partial}{\partial y^i}.$$

Therefore, we obtain

$$J^*(dx^i)=0,\qquad J^*(dy^i)=dx^i.$$

The Liouville vector field on $\widetilde{TM}=TM\setminus\{0\}$ is given by

$$\mathbb{C}=y^i\frac{\partial}{\partial y^i}. \tag{1.1.5}$$

It is globally defined on $\widetilde{TM}$ and $\mathbb{C}\neq 0$.

A smooth function $f:TM\to\mathbb{R}$ is called r-homogeneous ($r\in\mathbb{Z}$) with respect to the variables y^i if $f(x,ay)=a^r f(x,y)$, $\forall a\in\mathbb{R}^+$. The Euler theorem holds: A function $f\in\mathcal{F}(TM)$ differentiable on $\widetilde{TM}$ is r-homogeneous with respect to y^i if and only if

$$\mathcal{L}_{\mathbb{C}}f=\mathbb{C}f=y^i\frac{\partial f}{\partial y^i}=rf, \tag{1.1.6}$$

$\mathcal{L}_{\mathbb{C}}$ being the Lie derivation with respect to $\mathbb{C}$.

A vector field $X\in\mathcal{X}(TM)$ is r-homogeneous with respect to y^i if $\mathcal{L}_{\mathbb{C}}X=(r-1)X$, where $\mathcal{L}_{\mathbb{C}}X=[\mathbb{C},X]$.

Finally a 1-form $\omega\in\mathcal{X}^*(TM)$ is r-homogeneous if $\mathcal{L}_{\mathbb{C}}\omega=r\omega$.

Obviously, the notion of homogeneity can be extended to a tensor field T of type (r,s) on the manifold TM.

1.2. *Semisprays on the manifold TM*

The notion of semispray on the total space TM of the tangent bundle is strongly related to the second-order differential equations on the base manifold M:

$$\frac{d^2x^i}{dt^2}+2G^i\left(x,\frac{dx}{dt}\right)=0. \tag{1.2.1}$$

Writing Eq. (1.2.1), on TM, in the equivalent form

$$\frac{dy^i}{dt^2} + 2G^i(x, y) = 0, \qquad y^i = \frac{dx^i}{dt}, \tag{1.2.2}$$

we remark that with respect to the change of coordinates (1.1.1) on TM, the functions $G^i(x, y)$ transform according to:

$$2\tilde{G}^i = \frac{\partial \tilde{x}^i}{\partial x^j} 2G^j - \frac{\partial \tilde{y}^i}{\partial x^j} y^j. \tag{1.2.3}$$

But (1.2.2) are the integral curve of the vector field:

$$S = y^i \frac{\partial}{\partial x^i} - 2G^i(x, y)\frac{\partial}{\partial y^i}. \tag{1.2.4}$$

By means of (1.2.3) one proves that S is a vector field globally defined on TM. It is called a semispray on TM and G^i are called the coefficients of S.

S is homogeneous of degree 2 if and only if its coefficients G^i are homogeneous functions of degree 2. If S is 2-homogeneous then we say that S is a spray.

If the base manifold M is paracompact, then always there exist semisprays on TM.

1.3. *Nonlinear connections*

As we have seen in Section 1.1, the vertical distribution VTM is a regular, n-dimensional, integrable distribution on TM. Then it is naturally to search for a complementary distribution of VTM. It will be called a horizontal distribution. Such a distribution is equivalent to a nonlinear connection.

Consider the tangent bundle (TM, π, M) of the base manifold M and the tangent bundle (TTM, π_*, TM) of the manifold TM. As we know the kernel of π_* is the vertical subbundle (VTM, π_V, TM). Its fibres are the vertical spaces $V(u)$, $u \in TM$.

A vector field $X \in \mathcal{X}(TM)$, is given locally by

$$X = X^i(x, y)\frac{\partial}{\partial x^i} + Y^i(x, y)\frac{\partial}{\partial y^i},$$

or shorter $X = (x^i, y^i, X^i, Y^i)$.

The mapping $\pi_* : TTM \to TM$ has the local form $\pi_*(x, y, X, Y) = (x, X)$. The points of the submanifold VTM are of the form (x, y, O, Y).

Let us consider the pull-back bundle

$$\pi^*(TM) = TM \times_\pi TM = \{(u, v) \in TM \times TM \mid \pi(u) = \pi(v)\}.$$

The fibres $\pi_u^*(TM)$ of $\pi^*(TM)$ are isomorphic to $T_{\pi(u)}M$. Then, we define the following morphism: $\pi! : TTM \longrightarrow \pi^*(TM)$ by $\pi!(X_u) = (u, \pi_{*,u}(X_u))$. Therefore we have $\operatorname{Ker} \pi! = \operatorname{Ker} \pi_* = VTM$.

The following sequence is exact:

$$0 \longrightarrow VTM \stackrel{i}{\longrightarrow} TTM \stackrel{\pi!}{\longrightarrow} \pi^*(TM) \longrightarrow 0. \tag{1.3.1}$$

DEFINITION 1.3.1. A nonlinear connection on the tangent manifold TM is a left splitting of the exact sequence (1.3.1).

Consequently, a nonlinear connection on TM is a vector bundle morphism $C : TTM \to VTM$ such that $C \circ i = 1_{VTM}$.

The kernel of the morphism C is a vector subbundle of the tangent bundle (TTM, π_*, TM), denoted by (NTM, π_N, TM) and called the *horizontal* subbundle. Its fibres $N(u)$ determine a regular n-dimensional distribution $N : u \in TM \to N(u) \subset T_u TM$, complementary to the vertical distribution $V : u \in TM \to V(u) \subset T_u TM$, i.e.,

$$T_u TM = N(u) \oplus V(u), \quad \forall u \in TM. \tag{1.3.2}$$

Therefore, a nonlinear connection on TM induces the Whitney sum

$$TTM = NTM \oplus VTM. \tag{1.3.2'}$$

The converse statement is also true.

An adapted local basis to the direct decomposition (1.3.2) has the form $(\frac{\delta}{\delta x^i}, \frac{\partial}{\partial y^j})_u$, where

$$\frac{\delta}{\delta x^i} = \frac{\partial}{\partial x^i} - N_i^j(x, y)\frac{\partial}{\partial y^j} \tag{1.3.3}$$

and $\frac{\delta}{\delta x^i}|_u$ $(i = 1, \ldots, n)$ are vector fields belonging to $N(u)$.

They are n-linearly independent vector fields and are independent from the vector fields $(\frac{\partial}{\partial y^i})_u$, $i = 1, \ldots, n$, which belong to $V(u)$.

The functions $N_i^j(x, y)$ are called the coefficients of the nonlinear connection, denoted in the following by N.

Remarking that $\pi_{*,u} : T_u TM \to T_{\pi(u)} M$ is an epimorphism and the restriction of $\pi_{*,u}$ to $N(u)$ is an isomorphism, we can take the inverse map $l_{h,u}$, the horizontal lift determined by the nonlinear connection N.

Consequently, the vector fields $\frac{\delta}{\delta x^i}|_u$ are given by

$$\left(\frac{\delta}{\delta x^i}\right)_u = l_{h,u}\left(\frac{\partial}{\partial x^i}\right)_{\pi(u)}.$$

With respect to a change of local coordinates on the base manifold M we have $\frac{\partial}{\partial x^i} = \frac{\partial \tilde{x}^j}{\partial x^i} \frac{\theta}{\partial \tilde{x}^j}$.

Then, with respect to (1.1.1), $(\frac{\delta}{\delta x^i})_u$ are changed by

$$\frac{\delta}{\delta x^i} = \frac{\partial \tilde{x}^j}{\partial x^i}\frac{\delta}{\delta \tilde{x}^j}. \tag{1.3.4}$$

It follows from (1.3.3) that the coefficients $N^i_j(x, y)$ of the nonlinear connection N, with respect to a change of local coordinates on the manifold TM, are transformed by the rule:

$$\frac{\partial \tilde{x}^j}{\partial x^k}N^k_i = \tilde{N}^j_k\frac{\partial \tilde{x}^k}{\partial x^i} + \frac{\partial \tilde{y}^j}{\partial x^i} \tag{1.3.5}$$

and conversely.

It is known [22] that there exists a nonlinear connection on TM if M is a paracompact manifold.

THEOREM 1.3.1. *If S is a semispray with the coefficients $G^i(x, y)$, then the functions*

$$N^i_j(x, y) = \frac{\partial G^i}{\partial y^j} \tag{1.3.6}$$

are the coefficients of a nonlinear connection N.

Indeed, the formula (1.2.3) and $\frac{\partial}{\partial y^i} = \frac{\partial \tilde{x}^j}{\partial x^i}\frac{\partial}{\partial \tilde{y}^j}$ give the rule of transformation (1.3.5) for N^i_j defined by (1.3.6).

The adapted dual basis $\{dx^i, \delta y^i\}$ of the basis $(\frac{\delta}{\delta x^i}, \frac{\partial}{\partial y^i})$ has the 1-forms δy^i as follows:

$$\delta y^i = dy^i + N^i_j dx^j. \tag{1.3.7}$$

With respect to a change of coordinates, (1.1.1), we have

$$d\tilde{x}^i = \frac{\partial \tilde{x}^i}{\partial x^j}dx^j, \qquad \delta\tilde{y}^i = \frac{\partial \tilde{x}^i}{\partial x^j}\delta y^j. \tag{1.3.7$'$}$$

Now, we can consider the horizontal and vertical projectors h and v with respect to the direct decomposition (1.3.2):

$$h = \frac{\delta}{\delta x^i} \otimes dx^i, \qquad v = \frac{\partial}{\partial y^i} \otimes \delta y^i. \tag{1.3.8}$$

Some remarkable geometric structures, as the almost product structure $\mathbb{P}$ and almost complex structure $\mathbb{F}$, are determined by the nonlinear connection N:

$$\mathbb{P} = \frac{\delta}{\delta x^i} \otimes dx^i - \frac{\partial}{\partial y^i} \otimes \delta y^i = h - v, \tag{1.3.9}$$

$$\mathbb{F} = \frac{\delta}{\delta x^i} \otimes \delta y^i - \frac{\partial}{\partial y^i} \otimes dx^i. \tag{1.3.9$'$}$$

It easily seen that $\mathbb{P}$ and $\mathbb{F}$ are globally defined on $\widetilde{TM}$ and

$$\mathbb{P} \circ \mathbb{P} = \mathrm{Id}, \qquad \mathbb{F} \circ \mathbb{F} = -\mathrm{Id}. \tag{1.3.10}$$

With respect to (1.3.2) a vector field $X \in \mathcal{X}(TM)$ can be uniquely written as

$$X = hX + vX = X^H + X^V, \tag{1.3.11}$$

with $X^H = hX$ and $X^V = vX$.

A 1-form $\omega \in \mathcal{X}^*(TM)$ has the similar form

$$\omega = h\omega + v\omega,$$

where $h\omega(X) = \omega(X^H)$, $v\omega(X) = \omega(X^V)$.

A tensor field T on TM of type (r, s) is called a distinguished tensor field (shortly a d-tensor) if

$$T(\omega_1, \ldots, \omega_r, X_1, \ldots, X_s) = T(\varepsilon_1\omega_1, \ldots, \varepsilon_r\omega_r, \varepsilon_1 X_1, \ldots, \varepsilon_s X_s),$$

where $\varepsilon_1, \ldots, \varepsilon_r, \ldots$ are h or v.

Therefore $hX = X^H$, $vX = X^V$, $h\omega = \omega^H$, $v\omega = \omega^V$ are d-vectors and d-covectors, respectively. With respect to the adapted basis $(\frac{\delta}{\delta x^i}, \frac{\partial}{\partial y^i})$, we have

$$X^H = X^i(x, y)\frac{\delta}{\delta x^i}, \qquad X^V = \dot{X}^i \frac{\partial}{\partial y^i}$$

and

$$\omega^H = \omega_j(x, y)\, dx^j, \qquad \omega^V = \dot{\omega}_j\, \delta y^j.$$

A change of local coordinates on TM, $(x, y) \longrightarrow (\tilde{x}, \tilde{y})$, leads to the change of coordinates of $X^H, X^V, \omega^H, \omega^V$, by using the classical rules of transformation:

$$\tilde{X}^i = \frac{\partial \tilde{x}^i}{\partial x^j} X^j, \quad \omega_j = \frac{\partial \tilde{x}^i}{\partial x^j} \tilde{\omega}_i, \quad \text{etc.}$$

So, a d-tensor T of type (r, s) can be written as

$$T = T^{i_1 \ldots i_r}_{j_1 \ldots j_s}(x, y) \frac{\delta}{\delta x^{i_1}} \otimes \cdots \otimes \frac{\partial}{\partial y^{i_r}} \otimes dx^{j_1} \otimes \cdots \otimes \delta y^{j_s}. \tag{1.3.12}$$

A change of coordinates (1.1.1) implies

$$\tilde{T}^{i_1 \ldots i_r}_{j_1 \ldots j_s}(\tilde{x}, \tilde{y}) = \frac{\partial \tilde{x}^{i_1}}{\partial x^{h_1}} \cdots \frac{\partial \tilde{x}^{i_r}}{\partial x^{h_r}} \frac{\partial x^{k_1}}{\partial \tilde{x}^{j_1}} \cdots \frac{\partial x^{k_s}}{\partial \tilde{y}^{j_s}} = T^{h_1 \ldots h_2}_{k_1 \ldots k_s}. \tag{1.3.12$'$}$$

Next, we study the integrability of the nonlinear connection N and of the structures $\mathbb{P}$ and $\mathbb{F}$.

Since $(\frac{\delta}{\delta x^i})$, $i = 1, \ldots, n$, is an adapted basis to N, according to the Frobenius theorem it follows that N is integrable if and only if the Lie brackets $[\frac{\delta}{\delta x^i}, \frac{\delta}{\delta x^j}]$, $i, j = 1, \ldots, n$, are vector fields on the distribution N.

But we have

$$\left[\frac{\delta}{\delta x^i}, \frac{\delta}{\delta x^j}\right] = R^h_{ij}\frac{\partial}{\partial y^h}, \tag{1.3.13}$$

where

$$R^h_{ij} = \frac{\delta N^h_i}{\delta x^j} - \frac{\delta N^h_j}{\delta x^i}. \tag{1.3.13$'$}$$

It is not difficult to prove that R^h_{ij} are the coordinates of a d-tensor field of type $(1, 2)$, called the *curvature* tensor of the nonlinear connection N.

We state the following

THEOREM 1.3.1. *The nonlinear connection N is integrable if and only if its curvature tensor R^h_{ij} vanishes.*

The weak torsion t^h_{ij} of N is defined by

$$t^h_{ij} = \frac{\partial N^h_i}{\partial y^i} - \frac{\partial N^h_j}{\partial y^i}. \tag{1.3.14}$$

It is a d-tensor field of type $(1, 2)$ too. We say that N is a symmetric nonlinear connection if its weak torsion t^h_{ij} vanishes.

THEOREM 1.3.2.

(1) *The almost product structure $\mathbb{P}$ is integrable if and only if the nonlinear connection N is integrable.*
(2) *The almost complex structure $\mathbb{F}$ is integrable if and only if the symmetric nonlinear connection N is integrable.*

The proof is simple using the Nijenhuis tensors $N_{\mathbb{P}}$ and $N_{\mathbb{F}}$. The expression of $N_{\mathbb{P}}$ is

$$N_{\mathbb{P}}(X, Y) = \mathbb{P}^2[X, Y] + [\mathbb{P}X, \mathbb{P}Y] - \mathbb{P}[\mathbb{P}X, Y] - \mathbb{P}[X, \mathbb{P}Y],$$
$$\forall X, Y \in \chi(TM).$$

Also we can see that any structure $\mathbb{P}$ or $\mathbb{F}$ characterizes the nonlinear connection N [21].

Autoparallel curves of a nonlinear connection can be obtained considering the horizontal curves as follows.

A curve $c : t \in I \subset R \to (x^i(t), y^i(t)) \in TM$ has the tangent vector $\dot{c}$ given by

$$\dot{c} = \dot{c}^H + \dot{c}^V = \frac{dx^i}{dt}\frac{\delta}{\delta x^i} + \frac{\delta y^i}{dt}\frac{\partial}{\partial y^i}, \tag{1.3.15}$$

where

$$\frac{\delta y^i}{dt} = \frac{dy^i}{dt} + N^i_j(x, y)\frac{dx^j}{dt}. \tag{1.3.15'}$$

The curve c is a *horizontal curve* if $\dot{c}^V = 0$ or $\frac{\delta y^i}{dt} = 0$.

Obviously, if the functions $x^i(t)$, $t \in I$, are given and $y^i(t)$ are the solutions of the system of differential equations, then we have an horizontal curve $x^i = x^i(t)$, $y^i = y^i(t)$ in TM with respect to N.

In the case $y^i = \frac{dx^i}{dt}$, the horizontal curves are called the autoparallel curves of the nonlinear connection N. They are characterized by the system of differential equations

$$\frac{dy^i}{dt} + N^i_j(x, y)\frac{dx^j}{dt} = 0, \qquad y^i = \frac{dx^i}{dt}. \tag{1.3.16}$$

A theorem of existence and uniqueness of the autoparallel curves of a nonlinear connection N, given by its coefficients $N^i_j(x, y)$, holds.

1.4. *N-linear connections*

An N-linear connection on the manifold TM is a special linear connection D on TM, which preserves by parallelism the horizontal distribution N and the vertical distribution V. We study such linear connections determining the curvature, torsion and structure equations.

Throughout this subsection N is an a priori given nonlinear connection with the coefficients N^i_j.

DEFINITION 1.4.1. A linear connection D on the manifold TM is called a *distinguished connection* (shortly d-connection) if it preserves by parallelism the horizontal distribution N.

Thus, we have $Dh = 0$. It follows that we also have $Dv = 0$ and $D\mathbb{P} = 0$.
If $Y = Y^H + Y^V$ we get

$$D_X Y = \left(D_X Y^H\right)^H + \left(D_X Y^V\right)^V, \quad \forall X, Y \in \chi(TM).$$

PROPOSITION 1.4.1. *For a d-connection the following conditions are equivalent*:
(1) $DJ = 0$;
(2) $D\mathbb{F} = 0$.

DEFINITION 1.4.2. A d-connection D is called an *N-linear connection* if the structure J (or $\mathbb{F}$) is absolute parallel with respect to D, i.e., $DJ = 0$.

With respect to an adapted basis, an N-linear connection has the form:

$$\begin{cases} D_{\frac{\delta}{\delta x^j}} \frac{\delta}{\delta x^i} = L^h_{ij} \frac{\delta}{\delta x^h}, & D_{\frac{\delta}{\delta x^j}} \frac{\partial}{\partial y^i} = L^h_{ij} \frac{\partial}{\partial y^h}, \\ D_{\frac{\partial}{\partial y^j}} \frac{\delta}{\delta x^i} = C^h_{ij} \frac{\delta}{\delta x^h}, & D_{\frac{\partial}{\partial y^j}} \frac{\partial}{\partial y^i} = C^h_{ij} \frac{\partial}{\partial y^h}. \end{cases} \tag{1.4.1}$$

The set of functions $D\Gamma = (N^i_j(x, y), L^h_{ij}(x, y), C^h_{ij}(x, y))$ are called the local coefficients of the N-linear connection D. Since N is fixed, we denote sometimes by $D\Gamma(N) = (L^h_{ij}(x, y), C^h_{ij}(x, y))$ the coefficients of D.

For instance, the $B\Gamma(N) = (\frac{\partial N^h_i}{\partial y^j}, 0)$ are the coefficients of a special N-linear connection, derived only by the nonlinear connection N. It is called the *Berwald connection* of the nonlinear connection N. We can prove this affirmation showing that the system of functions $(\frac{\partial N^h_i}{\partial y^j})$ has the same rule of transformation, with respect to (1.1.1), as the coefficients L^h_{ij}. Indeed, under a change of coordinates (1.1.1) on TM, the coefficients (L^h_{ij}, C^h_{ij}) are transformed by the rules:

$$\begin{cases} \tilde{L}^h_{ij} = \frac{\partial \tilde{x}^h}{\partial x^s} L^s_{pq} \frac{\partial x^p}{\partial \tilde{x}^i} \frac{\partial x^q}{\partial \tilde{x}^j} - \frac{\partial^2 \tilde{x}^h}{\partial x^p \partial x^q} \frac{\partial x^p}{\partial \tilde{x}^i} \frac{\partial x^q}{\partial \tilde{x}^j}, \\ \tilde{C}^h_{ij} = \frac{\partial \tilde{x}^h}{\partial x^s} C^s_{pq} \frac{\partial x^p}{\partial \tilde{x}^i} \frac{\partial x^q}{\partial \tilde{x}^j}. \end{cases} \tag{1.4.2}$$

So, the horizontal coefficients L^h_{ij} of D have the same rule of transformation as the local coefficients of a linear connection on the base manifold M. The vertical coefficients C^h_{ij} are the components of a d-tensor field of type $(1, 2)$.

But, conversely, if the set of functions $(L^i_{jk}(x, y), C^i_{jk}(x, y))$ are given, having the property (1.4.2), then the equalities (1.4.1) determine an N-linear connection D on TM.

To an N-linear connection D on TM we shall associate two operators of h- and v-covariant derivation on the algebra of d-tensor fields. For each $X \in \chi(TM)$ we set:

$$D^H_X Y = D_{X^H} Y, \quad D^H_X f = X^H f, \quad \forall Y \in \chi(TM), \ \forall f \in \mathcal{F}(TM). \tag{1.4.3}$$

If $\omega \in \chi^*(TM)$, we obtain

$$\left(D^H_X \omega\right)(Y) = X^H\left(\omega(Y)\right) - \omega\left(D^H_X Y\right). \tag{1.4.3'}$$

Thus we may extend the action of the operator D^H_X to any d-tensor field by asking that D^H_X preserves the type of d-tensor fields, is $\mathbb{R}$-linear, satisfies the Leibniz rule with respect to tensor product and commutes with the operation of contraction. D^H_X will be called the *h-covariant derivation* operator.

In a similar way, we set

$$D^V_X Y = D_{X^V} Y, \quad D^V_X f = X^V(f), \quad \forall Y \in \chi(TM), \ \forall f \in \mathcal{F}(TM), \tag{1.4.4}$$

and

$$D^V_X\omega = X^V\big(\omega(Y)\big) - \omega\big(D^V_X Y\big), \quad \forall\omega \in \chi(TM).$$

Also, we extend the action of the operator D^V_X to any d-tensor field. D^V_X is called the *v-covariant derivation* operator.

Consider now a d-tensor T given by (1.3.12). According to (1.4.1), the h-covariant derivative of T is given by

$$D^H_X T = X^k T^{i_1\dots i_r}_{j_1\dots j_s|k} \frac{\delta}{\delta x^{i_1}} \otimes \cdots \otimes \frac{\partial}{\partial y^{i_r}} \otimes dx^{j_1} \otimes \cdots \otimes dx^{j_s}, \tag{1.4.5}$$

where

$$\begin{aligned} T^{i_1\dots i_r}_{j_1\dots j_s|k} &= \frac{\delta T^{i_1\dots i_r}_{j_1\dots j_s}}{\delta x^k} + L^{i_1}_{pk} T^{pi_2\dots i_r}_{j_1\dots j_s} + \cdots + L^{i_r}_{pk} T^{i_1\dots i_{r-1}p}_{j_1\dots j_s} \\ &\quad - L^p_{j_1k} T^{i_1\dots i_r}_{pj_2\dots j_s} - \cdots - L^p_{j_sk} T^{i_1\dots i_r}_{j_1\dots j_{s-1}p}. \end{aligned} \tag{1.4.5'}$$

The v-covariant derivative $D^V_X T$ is

$$D^V_X T = X^k T^{i_1\dots i_r}_{j_1\dots j_s}\big|_k \frac{\delta}{\delta x^{i_1}} \otimes \cdots \otimes \frac{\partial}{\partial y^{i_r}} \otimes dx^{j_1} \otimes \cdots \otimes dx^{j_s}, \tag{1.4.6}$$

where

$$\begin{aligned} T^{i_1\dots i_r}_{j_1\dots j_s}\big|_k &= \frac{\partial T^{i_1\dots i_r}_{j_1\dots j_s}}{\partial y^k} + C^{i_1}_{pk} T^{pi_2\dots i_r}_{j_1\dots j_s} + \cdots + C^{i_r}_{pk} T^{i_1\dots i_{r-1}p}_{j_1\dots j_s} \\ &\quad - C^p_{j_1k} T^{i_1\dots i_r}_{pj_2\dots j_s} - \cdots - C^p_{j_rk} T^{i_1\dots i_r}_{j_1\dots j_{s-1}p}. \end{aligned} \tag{1.4.6'}$$

For instance, if g is a d-tensor of type (0, 2) having the components $g_{ij}(x, y)$, we have

$$\begin{aligned} g_{ij|k} &= \frac{\delta g_{ij}}{\delta x^k} - L^p_{ik} g_{pj} - L^p_{jk} g_{ip}, \\ g_{ij}|k &= \frac{\partial g_{ij}}{\partial y^k} - C^p_{ik} g_{pj} - C^p_{jk} g_{ip}. \end{aligned} \tag{1.4.7}$$

For the operators "$_|$" and "$|$" we use the same denominations of h- and v-covariant derivations.

The torsion T of an N-linear connection is given by

$$T(X, Y) = D_X Y - D_Y X - [X, Y], \quad \forall X, Y \in \chi(TM). \tag{1.4.8}$$

The horizontal part $hT(X,Y)$ and the vertical part $vT(X,Y)$, for $X \in \{X^H, X^V\}$ and $Y \in \{Y^H, Y^V\}$ determine five d-tensor fields associated to the torsion T:

$$\begin{cases} hT(X^H,Y^H) = D^H_X Y^H - D^H_Y X^H - [X^H,Y^H]^H, \\ vT(X^H,Y^H) = -v[X^H,Y^H]^V, \\ hT(X^H,Y^V) = -D^V_Y X^H - [X^H,Y^V]^V, \\ vT(X^H,Y^V) = D^H_X Y^V - [X^H,Y^V]^V, \\ vT(X^V,Y^V) = D^V_X Y^V - D^V_Y X^V - [X^V,Y^V]^V. \end{cases} \tag{1.4.9}$$

With respect to an adapted basis, the components of the torsion are

$$\begin{aligned} &hT\left(\frac{\delta}{\delta x^i},\frac{\delta}{\delta x^j}\right) = T^k_{ji}\frac{\delta}{\delta x^k}, \qquad vT\left(\frac{\delta}{\delta x^i},\frac{\delta}{\delta x^j}\right) = R^k_{ji}\frac{\partial}{\partial y^k}, \\ &hT\left(\frac{\partial}{\partial y^i},\frac{\delta}{\delta x^j}\right) = C^k_{ji}\frac{\delta}{\delta x^k}, \qquad vT\left(\frac{\partial}{\partial y^i},\frac{\delta}{\delta x^j}\right) = P^k_{ji}\frac{\partial}{\partial y^k}, \\ &vT\left(\frac{\partial}{\partial y^i},\frac{\partial}{\partial y^j}\right) = S^k_{ji}\frac{\partial}{\partial y^k}, \end{aligned} \tag{1.4.9'}$$

where C^i_{jk} are the v-coefficients of D, R^i_{jk} is the curvature tensor of the nonlinear connection N and

$$T^i_{jk} = L^i_{jk} - L^i_{kj}, \quad S^i_{jk} = C^i_{jk} - C^i_{kj}, \quad P^i_{jk} = \frac{\partial N^i_j}{\partial y^k} - L^i_{kj}. \tag{1.4.10}$$

The N-linear connection D is said to be symmetric if $T^i_{jk} = S^i_{jk} = 0$.

Next, we study the curvature of an N-linear connection D:

$$R(X,Y)Z = D_X D_Y Z - D_Y D_X Z - D_{[X,Y]}Z, \quad \forall X,Y,Z \in \chi(TM). \tag{1.4.11}$$

Since D preserves by parallelism the distributions N and V, it follows that the operator $R(X,Y) = D_X D_Y - D_Y D_X - D_{[X,Y]}$ carries the horizontal vector fields Z^H into horizontal vector fields and vertical vector fields into vertical vector fields. Consequently we have the formula

$$R(X,Y)Z = hR(X,Y)Z^H + vR(X,Y)Z^V, \quad \forall X,Y,Z \in \chi(TM).$$

Since $R(X,Y) = -R(Y,X)$, we obtain

THEOREM 1.4.3. *The curvature R of an N-linear connection D on the tangent manifold TM is completely determined by the following six d-tensor fields*:

$$\begin{cases} R(X^H,Y^H)Z^H = D_X^H D_Y^H Z^H - D_Y^H D_X^H Z^H - D_{[X^H,Y^H]}Z^H, \\ R(X^H,Y^H)Z^V = D_X^H D_Y^H Z^V - D_Y^H D_X^H Z^V - D_{[X^H,Y^H]}Z^V, \\ R(X^V,Y^H)Z^H = D_X^V D_Y^H Z^H - D_Y^H D_X^V Z^H - D_{[X^V,Y^H]}Z^H, \\ R(X^V,Y^H)Z^V = D_X^V D_Y^H Z^V - D_Y^H D_X^V Z^V - D_{[X^V,Y^H]}Z^V, \\ R(X^V,Y^V)Z^H = D_X^V D_Y^V Z^H - D_Y^V D_X^V Z^H - D_{[X^V,Y^V]}Z^H, \\ R(X^V,Y^V)Z^V = D_X^V D_Y^V Z^V - D_Y^V D_X^V Z^V - D_{[X^V,Y^V]}Z^V. \end{cases} \tag{1.4.12}$$

The tangent structure J is absolutely parallel with respect to D. Then we have

$$JR(X,Y)Z = R(X,Y)JZ.$$

Thus $R(X,Y)Z$ has only three essential components

$$R(X^H,Y^H)Z^H, \quad R(X^V,Y^H)Z^H, \quad R(X^V,Y^V)Z^H.$$

With respect to an adapted basis, these are:

$$\begin{aligned} &R\left(\frac{\delta}{\delta x^k},\frac{\delta}{\delta x^j}\right)\frac{\delta}{\delta x^h} = R^i_{h\,kj}\frac{\delta}{\delta x^i}, \\ &R\left(\frac{\partial}{\partial y^k},\frac{\delta}{\delta x^j}\right)\frac{\delta}{\delta x^h} = P^i_{h\,kj}\frac{\delta}{\delta x^i}, \\ &R\left(\frac{\partial}{\partial y^k},\frac{\partial}{\partial y^j}\right)\frac{\delta}{\delta x^h} = S^i_{h\,kj}\frac{\delta}{\delta x^i}. \end{aligned} \tag{1.4.13}$$

The other three components are obtained by applying the operator J to the previous ones. So, we have $R(\frac{\delta}{\delta x^h},\frac{\delta}{\delta x^j})\frac{\partial}{\partial y^h} = R^i_{h\,jk}\frac{\partial}{\partial y^h}$, etc. Therefore, an N-linear connection $D\Gamma = (N^i_j, L^i_{jk}, C^i_{jk})$ has only three local components $R^i_{h\,jk}$, $P^i_{h\,jk}$ and $S^i_{h\,jk}$. They are given by

$$\begin{aligned} R^i_{h\,jk} &= \frac{\delta L^i_{hj}}{\delta x^k} - \frac{\delta L^i_{hk}}{\delta x^j} + L^s_{hj}L^i_{sk} - L^s_{hk}L^i_{sj} + C^i_{hs}R^s_{jk}, \\ P^i_{h\,jk} &= \frac{\partial L^i_{hj}}{\partial y^k} - C^i_{hk|j} + C^i_{hs}P^s_{jk}, \\ S^i_{h\,jk} &= \frac{\partial C^i_{hj}}{\partial y^k} - \frac{\partial C^i_{hk}}{\partial y^j} + C^s_{hj}C^i_{sk} - C^s_{hk}C^i_{sj}. \end{aligned} \tag{1.4.14}$$

If $X^i(x, y)$ are the components of a d-vector field on TM then from (1.4.12) we may derive the Ricci identities with respect to an N-linear connection D,

$$\begin{aligned}
&X^i_{|j|k} - X^i_{|k|j} = X^s R^i_{s\ jk} - X^i_{|s} T^s_{jk} - X^i|_s R^s_{jk},\\
&X^i_{|j}|_k - X^i|_{k|j} = X^s P^i_{s\ jk} - X^i_{|s} C^s_{jk} - X^i|_s P^s_{jk},\\
&X^i|_j|_k - X^i|_k|_j = X^s S^i_{s\ jk} - X^i|_s S^s_{jk}.
\end{aligned} \tag{1.4.15}$$

We deduce some fundamental identities for the N-linear connection D, by applying the Ricci identities to the Liouville vector field $\mathbb{C} = y^i \frac{\partial}{\partial y^i}$. Considering the d-tensors

$$D^i_j = y^i_{|j}, \qquad d^i_j = y^i|_j, \tag{1.4.16}$$

called h- and v-*deflection* tensors of D, we obtain

THEOREM 1.4.2. *For any N-linear connection D the following identities hold*:

$$\begin{aligned}
&D^i_{k|h} - D^i_{h|k} = y^s R^i_{s\ kh} - D^i_s T^s_{kh} - d^i_s R^s_{kh},\\
&D^i_k|_h - d^i_{h|k} = y^s P^i_{s\ kh} - D^i_s C^s_{kh} - d^i_s P^s_{kh},\\
&d^i_k|_h - d^i_h|_k = y^s S^i_{s\ kh} - d^i_s S^s_{kh}.
\end{aligned} \tag{1.4.17}$$

Other fundamental identities are the Bianchi identities obtained writing in the adapted basis the following Bianchi identities:

$$\begin{aligned}
&\Sigma\big[D_X T(Y, Z) - R(X, Y)Z + T\big(T(X, Y), Z\big)\big] = 0,\\
&\Sigma\big[(D_X R)(U, Y, Z) + R\big(T(X, Y), Z\big)U\big] = 0,
\end{aligned} \tag{1.4.18}$$

where Σ means cyclic summation over X, Y, Z.

1.5. *Parallelism. Structure equations*

Let $D\Gamma(N) = (L^i_{jk}, C^i_{jk})$ be an N-linear connection and an adapted basis $(\frac{\delta}{\delta x^i}, \frac{\partial}{\partial y^i})$, $i = \overline{1, n}$.

As we know, a curve $c : t \in I \to (x^i(t), y^i(t)) \in TM$, has the tangent vector $\dot{c} = \dot{c}^H + \dot{c}^V$ given by (1.3.15), i.e., $\dot{c} = \frac{dx^i}{dt}\frac{\delta}{\delta x^i} + \frac{\delta y^i}{dt}\frac{\partial}{\partial y^i}$. The curve c is horizontal if $\frac{\delta y^i}{dt} = 0$, and it is an autoparallel curve with respect to the nonlinear connection N if $\frac{\delta y^i}{dt} = 0$, $y^i = \frac{dx^i}{dt}$.

We denote

$$\frac{DX}{dt} = D_{\dot{c}} X, \quad DX = \frac{DX}{dt}\, dt, \quad \forall X \in \chi(TM). \tag{1.5.1}$$

Here $\frac{DX}{dt}$ is the covariant differential of X along the curve c with respect to the N-linear connection D.

Setting $X = X^H + X^V$, $X^H = X^i \frac{\delta}{\delta x^i}$, $X^V = \dot{X}^i \frac{\partial}{\partial y^i}$, we have

$$\frac{DX}{dt} = \frac{DX^H}{dt} + \frac{DX^V}{dt}$$
$$= \left\{ X^i_{|k} \frac{dx^k}{dt} + X^i|_k \frac{\delta y^k}{dt} \right\} \frac{\delta}{\delta x^i} + \left\{ \dot{X}^i_{|k} \frac{dx^k}{dt} + \dot{X}^i|_k \frac{\delta y^k}{dt} \right\}. \tag{1.5.2}$$

Consider the *connection* 1*-forms* of D,

$$\omega^i_j = L^i_{jk}\, dx^k + C^i_{jk}\, \delta y^k. \tag{1.5.3}$$

Then $\frac{DX}{dt}$ takes the form

$$\frac{DX}{dt} = \left\{ \frac{dX^i}{dt} + X^s \frac{\omega^i_s}{dt} \right\} \frac{\delta}{\delta x^i} + \left\{ \frac{d\dot{X}^i}{dt} + \dot{X}^s \frac{\omega^i_s}{dt} \right\} \frac{\partial}{\partial y^i}. \tag{1.5.4}$$

The vector field X on TM is said to be parallel along the curve c with respect to the N-linear connection if $\frac{DX}{dt} = 0$. Using (1.5.2), the equation $\frac{DX}{dt} = 0$ is equivalent to $\frac{DX^H}{dt} = 0$, $\frac{DX^V}{dt} = 0$. According to (1.5.4) we obtain the following

PROPOSITION 1.5.1. *The vector field $X = X^i \frac{\delta}{\delta x^i} + \dot{X}^i \frac{\partial}{\partial y^i} \in \chi(TM)$ is parallel along the parametrized curve c in TM with respect to the N-linear connection $D\Gamma(N) = (L^i_{jk}, C^i_{jk})$ if and only if its coefficients $X^i(x(t), y(t))$ and $\dot{X}^i(x(t), y(t))$ are solutions of the linear system of differential equations*

$$\frac{dZ^i}{dt} + Z^s\bigl(x(t), y(t)\bigr) \frac{\omega^i_s(x(t), y(t))}{dt} = 0.$$

A theorem of existence and uniqueness for parallel vector fields along a curve c on the manifold TM can be formulated in the classical way.

The *horizontal geodesic* of D are the horizontal curves $c : I \to TM$ with the property $D_{\dot{c}}\dot{c} = 0$. Taking $X^i = \frac{dx^i}{dt}$, $\dot{X}^i = \frac{\delta y^i}{dt} = 0$, we get

THEOREM 1.5.1. *The horizontal geodesics of an N-linear connection are characterized by the system of differential equations*:

$$\frac{d^2x^i}{dt^2} + L^i_{jk}(x, y) \frac{dx^j}{dt} \frac{dx^k}{dt} = 0, \qquad \frac{dy^i}{dt} + N^i_j(x, y) \frac{dx^j}{dt} = 0. \tag{1.5.5}$$

Now we can consider a curve $c^V_{x_0}$ on the fibre $T_{x_0}M = \pi^{-1}(x_0)$. It is represented by

$$x^i = x^i_0, \qquad y^i = y^i(t), \quad t \in I,$$

$c^v_{x_0}$ is called a *vertical* curve of TM at the point $x_0 \in M$.

The *vertical geodesic* of D are the vertical curves $c^v_{x_0}$ with the property $D_{\dot{c}^v_{x_0}} \dot{c}^v_{x_0} = 0$.

THEOREM 1.5.2. *The vertical geodesics at a point $x_0 \in M$ of an N-linear connection $D\Gamma(N) = (L^i_{jk}, C^i_{jk})$ are characterized by the following system of differential equations*:

$$x^i = x^i_0, \qquad \frac{d^2y^i}{dt^2} + C^i_{jk}(x_0, y)\frac{dy^j}{dt}\frac{dy^k}{dt} = 0. \tag{1.5.6}$$

Obviously, the local existence and uniqueness of horizontal or vertical geodesics are assured by giving initial conditions.

Now, we determine the structure equations of an N-linear connection D, considering the connection 1-forms ω^i_j, defined by (1.5.3).

LEMMA 1.5.1. *The exterior differential of the* 1*-forms $\delta y^i = dy^i + N^i_j dx^j$ are given by*

$$d(\delta y^i) = \frac{1}{2}R^i_{js}\,dx^s \wedge dx^j + B^i_{js}\,\delta y^s \wedge dx^j, \tag{1.5.7}$$

where

$$B^i_{jk} = \frac{\partial N^i_j}{\partial y^k}. \tag{1.5.7'}$$

REMARK. B^i_{jk} are the coefficients of the Berwald connection.

LEMMA 1.5.2. *With respect to a change of local coordinates on the manifold TM, the following* 2*-forms*:

$$d(dx^i) - dx^s \wedge \omega^i_s, \qquad d(\delta y^i) - \delta y^s \wedge \omega^i_s$$

transform as the components of a d-vector field.

The 2*-forms*

$$d\omega^i_j - \omega^s_j \wedge \omega^i_s$$

transform as the components of a d-tensor field of type $(1, 1)$.

THEOREM 1.5.3. *The structure equations of an N-linear connection $(L^i_{jk}, C^i_{jk}) = D\Gamma(N)$ on the manifold TM are given by*

$$d(dx^i) - dx^s \wedge \omega^i_s = -\overset{(0)}{\Omega}{}^i,$$

$$d(\delta y^i) - \delta y^s \wedge \omega^i_s = -\overset{(1)}{\Omega}{}^i,$$

$$d\omega^i_j - \omega^s_j \wedge \omega^i_s = -\Omega^i_j, \tag{1.5.8}$$

where $\overset{(0)}{\Omega}{}^i$ *and* $\overset{(1)}{\Omega}{}^i$ *are the torsion* 2*-forms*

$$\overset{(0)}{\Omega}{}^i = \frac{1}{2} T^i_{jk}\, dx^j \wedge dx^k + C^i_{jk}\, dx^j \wedge \delta y^k,$$
$$\overset{(1)}{\Omega}{}^i = \frac{1}{2} R^i_{jk}\, dx^j \wedge dx^k + P^i_{jk}\, dx^j \wedge \delta y^k + \frac{1}{2} S^i_{jk}\, \delta y^j \wedge \delta y^k, \tag{1.5.9}$$

and the curvature 2*-forms* Ω^i_j *are given by*

$$\Omega^i_j = \frac{1}{2} R^i_{j\,kh}\, dx^k \wedge dx^h + P^i_{j\,kh}\, dx^k \wedge \delta y^h + \frac{1}{2} S^i_{j\,kh}\, \delta y^j \wedge \delta y^h. \tag{1.5.10}$$

PROOF. By means of Lemma 1.5.2, for an N-linear connection D, the general structure equations of a linear connection on TM take the form (1.5.8). Using the connection 1-forms ω^i_j and formula (1.5.7), we can calculate the forms $\overset{(0)}{\Omega}{}^i$, $\overset{(1)}{\Omega}{}^i$ and Ω^i_j.

Then it is very easy to determine the structure equations (1.5.9). □

REMARK. The Bianchi identities of an N-linear connection D can be obtained from (1.5.8) by calculating the exterior differential of (1.5.8), modulo the same system (1.5.8) and using the exterior differential of $\overset{(0)}{\Omega}{}^i$, $\overset{(1)}{\Omega}{}^i$ and Ω^i_j.

2. Lagrange spaces

The notion of Lagrange spaces was introduced and studied by the present author [15,19,20]. The term "Lagrange geometry" belongs to J. Kern [11]. We study the geometry of Lagrange spaces as a subgeometry of the geometry of the tangent bundle (TM, π, M) of a manifold M, using the principles of Analytical Mechanics given by variational problem on the integral of an action of a regular Lagrangian, the law of conservation, Nöther theorem, etc. Remarking that the Euler–Lagrange equations determine a canonical semispray S on the manifold TM, we can study the geometry of a Lagrange space using this canonical semi-spray S and the methods given in the first section.

Starting 1987, the author, alone or in collaboration, published some books [19] on the Lagrange spaces and the Hamilton spaces [18–20,22], as well as the higher-order Lagrange and Hamilton spaces [22].

The present section is based on the above mentioned books.

2.1. *The notion of Lagrange space*

First we define the notion of a differentiable Lagrangian over the tangent manifolds TM and $\widetilde{TM} = TM \setminus \{0\}$, M being a real n-dimensional manifold.

DEFINITION 2.1.1. A differentiable Lagrangian is a C^∞-mapping $L:(x,y)\in TM\to L(x,y)\in\mathbb{R}$ on $\widetilde{TM}$ and continuous on the null section $0:M\to TM$ of the bundle $\pi:TM\to M$.

The Hessian of a differentiable Lagrangian L, with respect to y^i, has the components

$$g_{ij}=\frac{1}{2}\frac{\partial^2 L}{\partial y^i\partial y^j}. \tag{2.1.1}$$

The set of functions $g_{ij}(x,y)$ are the components of a symmetric d-tensor field covariant of order 2.

DEFINITION 2.1.2. A differentiable Lagrangian L is called *regular* if:

$$\operatorname{rank}\big(g_{ij}(x,y)\big)=n, \quad \text{on } \widetilde{TM}. \tag{2.1.2}$$

Now we can give the definition of a Lagrange space.

DEFINITION 2.1.3. A Lagrange space is a pair $L^n=(M,L(x,y))$, where M is a smooth, real n-dimensional manifold M and $L(x,y)$ a regular Lagrangian $L(x,y)$, for which the d-tensor g_{ij} has a constant signature over the manifold $\widetilde{TM}$.

We say that $L(x,y)$ is the *fundamental function* and $g_{ij}(x,y)$ is the *fundamental* (or metric) tensor of the Lagrange space.

EXAMPLES.

(1) Every Riemannian manifold $(M,g_{ij}(x))$ determines a Lagrange space $L^n=(M,L(x,y))$, where

$$L(x,y)=g_{ij}(x)y^iy^j. \tag{2.1.3}$$

Therefore, if a manifold M is paracompact, then there exists a Lagrangian $L(x,y)$ such that $L^n=(M,L(x,y))$ is a Lagrange space.

(2) Let L be the following Lagrangian from Electrodynamics:

$$L(x,y)=mc\gamma_{ij}(x)y^iy^j+\frac{2e}{m}A_i(x)y^i+\mathcal{U}(x), \tag{2.1.4}$$

where $\gamma_{ij}(x)$ is a pseudo-Riemannian metric, $A_i(x)$ a covector field and $\mathcal{U}(x)$ a smooth function, m,c,e are constants. The corresponding Lagrange space is called the Lagrange space of Electrodynamics.

We already have seen that $g_{ij}(x,y)$ is a d-tensor field, i.e.,

$$\tilde{g}_{ij}(\tilde{x},\tilde{y})=\frac{\partial x^h}{\partial\tilde{x}^i}\frac{\partial x^k}{\partial\tilde{x}^j}, g_{hk}(x,y).$$

THEOREM 2.1.1. *For a Lagrange space L^n the following properties hold*:

(1) *The system of functions*

$$p_i = \frac{1}{2}\frac{\partial L}{\partial y^i} \tag{2.1.5}$$

determine a d-covector field.

(2) *The functions*

$$C_{ijk} = \frac{1}{4}\frac{\partial^3 L}{\partial y^i \partial y^j \partial y^k} = \frac{1}{2}\frac{\partial g_{ij}}{\partial y^k} \tag{2.1.6}$$

are the components of a symmetric d-tensor field of type $(0, 3)$.

(3) *The* 1*-forms*

$$\omega = p_i\, dx^i = \frac{1}{2}\frac{\partial L}{\partial y^i}\, dx^i \tag{2.1.7}$$

depend only on the Lagrangian L and is globally defined on the manifold $\widetilde{TM}$.

(4) *The* 2*-form*

$$\theta = d\omega = dp_i \wedge dx^i \tag{2.1.8}$$

is globally defined on $\widetilde{TM}$ *and is a symplectic structure on* TM.

2.2. *Variational problem. Euler–Lagrange equations*

The variational problem can be formulated for differentiable Lagrangians and can be solved in the case when the integral of action is defined on parametrized curves.

Let $L : TM \to R$ be a differentiable Lagrangian and $c : t \in [0, 1] \to (x^i(t)) \in U \subset M$ be a smooth curve, with a fixed parametrization, having $\operatorname{Im} c \subset U$, where U is a domain of a local chart on the manifold M. The curve c can be extended to $\pi^{-1}(U) \subset \widetilde{TM}$ by

$$\tilde{c} : t \in [0, 1] \to \left(x^i(t), \frac{dx^i}{dt}(t)\right) \in \pi^{-1}(U).$$

So, $\operatorname{Im} \tilde{c} \subset \pi^{-1}(U)$.

The integral of action of the Lagrangian L on the curve c is given by the functional

$$I(c) = \int_0^1 L\left(x, \frac{dx}{dt}\right) dt. \tag{2.2.1}$$

Consider the curves

$$c_\varepsilon : t \in [0, 1] \to \left(x^i(t) + \varepsilon V^i(t)\right) \in M \tag{2.2.2}$$

which have the same end points $x^i(0)$ and $x^i(1)$ as the curve c, $V^i(t) = V^i(x^i(t))$ being a regular vector field on the curve c, with the property $V^i(0) = V^i(1) = 0$ and ε is a real number, sufficiently small in absolute value, so that $\operatorname{Im} c_\varepsilon \subset U$.

An extension of the curve c_ε to $\widetilde{TM}$ is given by

$$\tilde{c}_\varepsilon : t \in [0,1] \mapsto \left(x^i(t) + \varepsilon V^i(t), \frac{dx^i}{dt} + \varepsilon \frac{dV^i}{dt} \right) \in \pi^{-1}(U).$$

The integral of action of the Lagrangian L on the curve c_ε is given by

$$I(c_\varepsilon) = \int_0^1 L\left(x + \varepsilon V, \frac{dx}{dt} + \varepsilon \frac{dV}{dt} \right) dt. \tag{2.2.1$'$}$$

A necessary condition for $I(c)$ to be an extremal value of $I(c_\varepsilon)$ is

$$\left. \frac{dI(c_\varepsilon)}{d\varepsilon} \right|_{\varepsilon=0} = 0. \tag{2.2.3}$$

Under our conditions of differentiability, the operator $\frac{d}{d\varepsilon}$ and the operator of integration commute.

From (2.2.1) we obtain

$$\frac{dI(c_\varepsilon)}{d\varepsilon} = \int_0^1 \frac{d}{d\varepsilon} L\left(x + \varepsilon V, \frac{dx}{dt} + \varepsilon \frac{dV}{dt} \right) dt. \tag{2.2.4}$$

But we have

$$\frac{d}{d\varepsilon} L\left(x + \varepsilon V, \frac{dx}{dt} + \varepsilon \frac{dV}{dt} \right)\Bigg|_{\varepsilon=0} = \frac{\partial L}{\partial x^i} V^i + \frac{\partial L}{\partial y^i} \frac{dV^i}{dt}$$
$$= \left\{ \frac{\partial L}{\partial x^i} - \frac{d}{dt} \frac{\partial L}{\partial y^i} \right\} V^i + \frac{d}{dt} \left\{ \frac{\partial L}{\partial y^i} V^i \right\}, \qquad y^i = \frac{dx^i}{dt}.$$

Substituting in (2.2.4) and taking into account the fact that $V^i(x(t))$ is arbitrary, we obtain the following theorem.

THEOREM 2.2.1. *A necessary condition for the functional $I(c)$ to be an extremal value of $I(c_\varepsilon)$ is that the curve $c(t) = (x^i(t))$ satisfy the Euler–Lagrange equations*:

$$E_i(L) := \frac{\partial L}{\partial x^i} - \frac{d}{dt} \frac{\partial L}{\partial y^i} = 0, \qquad y^i = \frac{dx^i}{dt}. \tag{2.2.5}$$

For the Euler–Lagrange operator $E_i = \frac{\partial}{\partial x^i} - \frac{d}{dt} \frac{\partial}{\partial y^i}$, we have

THEOREM 2.2.2. *The following properties hold true*:

(1) $E_i(L)$ *is a* d*-covector field*;
(2) $E_i(L+L') = E_i(L) + E_i(L')$;
(3) $E_i(aL) = aE_i(L)$, $a \in \mathbb{R}$;
(4) $E_i(\frac{dF}{dt}) = 0$, $\forall F \in \mathcal{F}(TM)$ *with* $\frac{\partial F}{\partial y^i} = 0$.

The notion of *energy of a Lagrangian* L can be introduced as in Theoretical Mechanics [26,27], by

$$E_L = y^i \frac{\partial L}{\partial y^i} - L. \tag{2.2.6}$$

THEOREM 2.2.3. *For every smooth curve* c *on the base manifold* M, *one has*

$$\frac{dE_L}{dt} = -\frac{dx^i}{dt} E_i(L), \qquad y^i = \frac{dx^i}{dt}. \tag{2.2.7}$$

Consequently:

THEOREM 2.2.4. *For any differentiable Lagrangian* $L(x,y)$, *the energy* E_L *is conserved along every solution* c *of the Euler–Lagrange equations*

$$E_i(L) = 0, \qquad \frac{dx^i}{dt} = y^i.$$

A Nöther theorem can be proved [20]:

THEOREM 2.2.5. *For any infinitesimal symmetry on* $M \times \mathbb{R}$ *of the Lagrangian* $L(x,y)$ *and for any smooth function* $\phi(x)$, *the function*

$$\mathcal{F}(L,\phi) = V^i \frac{\partial L}{\partial y^i} - \tau E_L - \phi(x)$$

is conserved on every curve c, *solution of the Euler–Lagrange equations* $E_i(L) = 0$, $y^i = \frac{dx^i}{dt}$.

REMARK. An infinitesimal symmetry on $M \times \mathbb{R}$ is given by $x'^i = x^i + \varepsilon V^i(x,t)$, $t' = t + \varepsilon\tau(x,t)$.

2.3. *Canonical semispray. Nonlinear connection*

Now we can apply the previous theory to the study of the Lagrange space $L^n = (M, L(x,y))$. As we shall see, L^n determines a canonical semispray S and S gives a canonical nonlinear connection on the manifold $\widetilde{TM}$.

As we know, the fundamental tensor g_{ij} of the space L^n is nondegenerate and $E_i(L)$ is a d-covector field, so the equations $g^{ij}E_j(L) = 0$ have a geometrical meaning.

THEOREM 2.3.1. *If $L^n = (M, L)$ is a Lagrange space, then the system of differential equations*

$$g^{ij} E_j(L) = 0, \qquad y^i = \frac{dx^i}{dt} \tag{2.3.1}$$

can be written in the form:

$$\frac{d^2x^i}{dt^2} + 2G^i\left(x, \frac{dx}{dt}\right) = 0, \qquad y^i = \frac{dx^i}{dt}, \tag{2.3.1$'$}$$

where

$$2G^i(x, y) = \frac{1}{2} g^{ij} \left\{ \frac{\partial^2 L}{\partial y^j \partial x^k} y^k - \frac{\partial L}{\partial x^j} \right\}. \tag{2.3.2}$$

PROOF. We have

$$E_i(L) = \frac{\partial L}{\partial x^i} - \left\{ \frac{\partial^2 L}{\partial y^i \partial x^k} + 2g_{ij} \frac{dy^j}{dt} \right\}, \qquad y^i = \frac{dx^i}{dt}.$$

So, (2.3.1) implies (2.3.1$'$), (2.3.2). □

The previous theorem shows that the Euler–Lagrange equations for a Lagrange space are given by a system of n second-order ordinary differential equations. According to Section 1.2, it follows that Eqs. (2.3.1) determine a semispray with the coefficients $G^i(x, y)$:

$$S = y^i \frac{\partial}{\partial x^i} - 2G^i(x, y) \frac{\partial}{\partial y^i}. \tag{2.3.3}$$

S is called the canonical semispray of the Lagrange space L^n.

By means of Theorem 1.3.1, it follows that:

THEOREM 2.3.2. *Every Lagrange space $L^n = (M, L)$ has a canonical nonlinear connection N which depends only on the fundamental function L. The local coefficients of N are given by*

$$N^i_j = \frac{\partial G^i}{\partial y^j} = \frac{1}{4} \frac{\partial}{\partial y^j} \left\{ g^{ik} \left(\frac{\partial^2 L}{\partial y^k \partial x^h} y^h - \frac{\partial L}{\partial x^k} \right) \right\}. \tag{2.3.4}$$

PROPOSITION 2.3.1. *The canonical nonlinear connection N is symmetric, i.e., $t^i_{jk} = \frac{\partial N^i_j}{\partial y^k} - \frac{\partial N^i_k}{\partial y^j} = 0$, and is invariant with respect to the Carathéodory transformation*

$$L'(x, y) = L(x, y) + \frac{\partial \varphi(x)}{\partial x^i} y^i.$$

Indeed, we have

$$E_i(L') = E_i\left(L(x,y) + \frac{d\varphi}{dt}\right) = E_i(L).$$

Thus, $E_i(L'(x,y)) = 0$ determines the same canonical semispray as the one determined by $E_i(L(x,y)) = 0$. Therefore Carathéodory transformation [19] preserves the nonlinear connection N.

EXAMPLE. On the Lagrange space of Electrodynamics, $L^n = (M, L(x,y))$, where $L(x,y)$ is given by (2.1.4) with $U(x) = 0$, the canonical semispray has the coefficients:

$$G^i(x,y) = \frac{1}{2}\gamma^i_{jk}(x)y^j y^k - g^{ij}(x)F_{jk}(x)y^k, \tag{2.3.5}$$

where $\gamma^i_{jk}(x)$ are the Christoffel symbols of the metric tensor $g_{ij}(x) = mc\gamma_{ij}(x)$ of the space L^n and F_{jk} is the electromagnetic tensor

$$F_{jk}(x,y) = \frac{e}{2m}\left(\frac{\partial A_k}{\partial x^j} - \frac{\partial A_j}{\partial x^k}\right). \tag{2.3.6}$$

Therefore, the integral curves of the Euler–Lagrange equation are given by the solutions of the *Lorentz equations*:

$$\frac{d^2x^i}{dt^2} + \gamma^i_{jk}(x)\frac{dx^j}{dt}\frac{dx^k}{dt} = g^{ij}(x)F_{jk}(x)\frac{dx^k}{dt}. \tag{2.3.7}$$

According to (2.3.4), the canonical nonlinear connection of the Lagrange space of Electrodynamics L^n has the local coefficients

$$N^i_j(x,y) = \gamma^i_{jk}(x)y^k - g^{ik}(x)F_{kj}(x). \tag{2.3.8}$$

We remark that the coefficients N^i_j are linear with respect to y^i.

PROPOSITION 2.3.2. *The Berwald connection of the canonical nonlinear connection N has the coefficients $B\Gamma(N) = (\gamma^i_{jk}(x), 0)$.*

PROPOSITION 2.3.3. *The solutions of the Euler–Lagrange equations and the autoparallel curves of the canonical nonlinear connection N are given by the Lorentz equations* (2.3.7).

THEOREM 2.3.3. *The autoparallel curves of the canonical nonlinear connection N are given by the following system*:

$$\frac{d^2x^i}{dt^2} + N^i_j\left(x, \frac{dx}{dt}\right)\frac{dx^j}{dt} = 0,$$

where N^i_j are given by (2.3.4).

2.4. *Hamilton–Jacobi equations*

Consider a Lagrange space $L^n = (M, L(x, y))$ and $N(N^i_j)$ its canonical nonlinear connection. The adapted basis $(\frac{\delta}{\delta x^i}, \frac{\partial}{\partial y^i})$ to the horizontal distribution N and the vertical distribution V has the horizontal vector fields:

$$\frac{\delta}{\delta x^i} = \frac{\partial}{\partial x^i} - N^j_i \frac{\partial}{\partial y^j}. \tag{2.4.1}$$

Its dual basis is $(dx^i, \delta y^i)$, with

$$\delta y^i = dy^i + N^i_j \, dx^j. \tag{2.4.2}$$

Theorem 2.1.1 gives the momenta

$$p_i = \frac{1}{2} \frac{\partial L}{\partial y^i}, \tag{2.4.3}$$

the 1-form

$$\omega = p_i \, dx^i \tag{2.4.4}$$

and the 2-form

$$\theta = d\omega = dp_i \wedge dx^i. \tag{2.4.5}$$

These geometrical object fields are globally defined on $\widetilde{TM}$. Clearly θ is a *symplectic structure* on the manifold $\widetilde{TM}$.

PROPOSITION 2.4.1. *With respect to an adapted basis, the 2-form θ is given by*

$$\theta = g_{ij} \, \delta y^i \wedge dx^j. \tag{2.4.6}$$

Indeed,

$$\begin{aligned}\theta = dp_i \wedge dx^i &= \frac{1}{2}\left(\frac{\delta}{\delta x^s}\frac{\partial L}{\partial y^i} dx^s + \frac{\partial}{\partial y^s}\frac{\partial L}{\partial y^i} \delta y^s\right) \wedge dx^i \\ &= \frac{1}{4}\left(\frac{\delta}{\delta x^s}\frac{\partial L}{\partial y^i} - \frac{\delta}{\delta x^i}\frac{\partial L}{\partial y^s}\right) dx^s \wedge dx^i + g_{is} \, \delta y^s \wedge dx^i.\end{aligned}$$

But is easily seen that the coefficient of $dx^s \wedge dx^i$ vanishes.

The triple $(\widetilde{TM}, \theta, L)$ is called a Lagrangian system.

The energy E_L of the space L^n is given by (2.2.6). Denoting $\mathcal{H} = \frac{1}{2}E_L$, $\mathcal{L} = \frac{1}{2}L$, then (2.2.6) can be written as

$$\mathcal{H} = p_i y^i - \mathcal{L}(x, y). \tag{2.4.7}$$

Along the integral curves of the Euler–Lagrange equations (2.2.5), we have

$$\frac{\partial \mathcal{H}}{\partial x^i} = -\frac{\partial \mathcal{L}}{\partial x^i} = -\frac{dp_i}{dt},$$

and from (2.4.7), we get

$$\frac{\partial \mathcal{H}}{\partial p_i} = y^i = \frac{dx^i}{dt}.$$

So, we obtain

THEOREM 2.4.1. *Any integral curve of the Euler–Lagrange equations satisfies the Hamilton–Jacobi equations*:

$$\frac{dx^i}{dt} = \frac{\partial \mathcal{H}}{\partial p_i}, \qquad \frac{dp_i}{dt} = -\frac{\partial \mathcal{H}}{\partial x^i}, \tag{2.4.8}$$

where $\mathcal{H}$ *is given by* (2.4.7) *and* $p_i = \frac{1}{2}\frac{\partial L}{\partial y^i}$.

EXAMPLE. For the Lagrange space from Electrodynamics with the fundamental function $L(x,y)$, from (2.1.4) and $U(x)=0$, we obtain

$$\mathcal{H} = \frac{1}{2mc}\gamma^{ij}(x)p_i p_j - \frac{e}{mc^2}A^i(x)p_i + \frac{e^2}{2mc^3}A^i(x)A_i(x)$$

$(A^i = \gamma^{ij}A_j)$.

Then, we may write the Hamilton–Jacobi equations.

Since θ is a symplectic structure on $\widetilde{TM}$, then its exterior differential $d\theta$ vanishes. With respect to an adapted basis

$$d\theta = dg_{ij}\wedge \delta y^i \wedge dx^j + g_{ij} d\,\delta y^i \wedge dx^j = 0$$

gives

$$\frac{1}{2}\left(\frac{\delta g_{ij}}{\delta x^k} - \frac{\delta g_{ik}}{\delta x^j}\right)\delta y^i \wedge dx^j \wedge dx^k + \frac{1}{2}\left(\frac{\partial g_{ij}}{\partial x^k} - \frac{\partial g_{kj}}{\partial y^i}\right)\delta y^k \wedge \delta y^i \wedge dx^j$$
$$+ g_{ij}\left(\frac{1}{2}R^i_{rs}\, dx^s \wedge dx^r + B^i_{rs}\,\delta y^s \wedge dx^r\right)\wedge dx^j = 0.$$

Thus, we obtain

THEOREM 2.4.2. *For any Lagrange space* L^n *the following identities hold*:

$$g_{ij\|k} - g_{ik\|j} = 0, \qquad g_{ij}\|_k - g_{ik}\|_j = 0. \tag{2.4.9}$$

Indeed, taking into account the h- and v-covariant derivations of the metric g_{ij} with respect to Berwald connection $B\Gamma(N) = (\frac{\partial N^i_j}{\partial y^k}, 0)$, i.e.,

$$g_{ij\|k} = \frac{\delta g_{ij}}{\delta x^k} - B^r_{ik} g_{kj} - B^r_{jk} g_{ir}$$

and $g_{ij}\|_k = \frac{\partial g_{ij}}{\partial y^k}$, and using the equations $\frac{\partial g_{ij}}{\partial y^k} = 2C_{ijk}$, $B^i_{jk} = B^i_{kj}$, we obtain (2.4.9).

2.5. *Metrical N-linear connections*

Let $N(N^i_j)$ be the canonical nonlinear connection of the Lagrange space $L^n = (M, L)$ and D an N-linear connection with the coefficients $D\Gamma(N) = (L^i_{jk}, C^i_{jk})$. Then, the h- and v-covariant derivations of the fundamental tensor g_{ij}, $g_{ij|k}$ and $g_{ij}|_k$ are given by (1.4.7).

Applying the theory of N-linear connections, one proves the following

THEOREM 2.5.1.

(1) *On the manifold $\widetilde{TM}$ there exist only one N-linear connections D which verifies the following axioms*:

(A_1) *N is canonical nonlinear connection of the space L^n*;

(A_2) $g_{ij|k} = 0$ (*D is h-metrical*);

(A_3) $g_{ij}|_k = 0$ (*D is v-metrical*);

(A_4) $T^i_{jk} = 0$ (*D is h-torsion free*);

(A_5) $S^i_{jk} = 0$ (*D is v-torsion free*).

(2) *The coefficients $D\Gamma(N) = (L^i_{jk}, C^i_{jk})$ of D are expressed by the following generalized Christoffel symbols*:

$$L^i_{jk} = \frac{1}{2} g^{ir} \left(\frac{\delta g_{rk}}{\delta x^j} + \frac{\delta g_{rj}}{\delta x^k} - \frac{\delta g_{jk}}{\delta x^r} \right),$$

$$C^i_{jk} = \frac{1}{2} g^{ir} \left(\frac{\partial g_{rk}}{\partial y^j} + \frac{\partial g_{rj}}{\partial y^k} - \frac{\partial g_{jk}}{\partial y^r} \right). \tag{2.5.1}$$

(3) *This connection depends only on the fundamental function $L(x, y)$ of the Lagrange space L^n.*

The N-linear connection D given by the previous theorem is called the *canonical metrical connection* and is denoted by $C\Gamma(N) = (L^i_{jk}, C^i_{jk})$.

The connection 1-forms $\omega^i{}_j$ of $C\Gamma(N)$ are

$$\omega^i_j = L^i_{jk}\, dx^k + C^i_{jk}\, \delta y^k. \tag{2.5.2}$$

THEOREM 2.5.2. *The canonical metrical connection $C\Gamma(N)$ satisfies the following structure equations*:

$$
\begin{aligned}
&d\left(dx^i\right) - dx^k \wedge \omega^i_k = -\overset{(0)}{\Omega}{}^i,\\
&d\left(\delta y^i\right) - \delta y^k \wedge \omega^i_k = -\overset{(1)}{\Omega}{}^i,\\
&d\omega^i_j - \omega^k_j \wedge \omega^i_k = -\Omega^i_j,
\end{aligned}
\tag{2.5.3}
$$

where the torsion 2-*forms* $\overset{(0)}{\Omega}{}^i$ *and* $\overset{(1)}{\Omega}{}^i$ *are*

$$
\begin{aligned}
&\overset{(0)}{\Omega}{}^i = C^i_{jk}\, dx^j \wedge \delta y^k,\\
&\overset{(1)}{\Omega}{}^i = \frac{1}{2} R^i_{jk}\, dx^j \wedge dx^k + P^i_{jk}\, dx^j \wedge \delta y^k
\end{aligned}
\tag{2.5.4}
$$

and the curvature 2-*forms* Ω^i_j *are*

$$
\Omega^i_j = \frac{1}{2} R^i_{j\,kh}\, dx^k \wedge dx^h + P^i_{j\,kh}\, dx^k \wedge \delta y^h + \frac{1}{2} S^i_{j\,kh}\, \delta y^k \wedge \delta y^h. \tag{2.5.4$'$}
$$

The torsion d-tensors R^i_{jk}, P^i_{jk} are given by (1.3.13′) and (1.4.10), and the curvature d-tensors $R^i_{j\,kh}$, $P^i_{j\,kh}$, $S^i_{j\,kh}$ have the expressions (1.4.14).

Starting from the canonical metrical connection $C\Gamma(N) = (L^i_{jk}, C^i_{jk})$, we can derive other N-linear connections depending only on the space L^n: Berwald connection $B\Gamma(N) = (\frac{\partial N^i_j}{\partial y^k}, 0)$, Chern–Rund connection $R\Gamma(N) = (L^i_{jk}, 0)$ and Hashiguchi connection $H\Gamma(N) = (\frac{\partial N^i_j}{\partial y^k}, C^i_{jk})$. The following commutative diagram holds [19,20]:

$$
\begin{array}{ccccc}
 & & R\Gamma(N) & & \\
 & \nearrow & & \searrow & \\
C\Gamma(N) & & \longrightarrow & & B\Gamma(N)\\
 & \searrow & & \nearrow & \\
 & & H\Gamma(N) & &
\end{array}
$$

Some properties of the canonical metrical connection $C\Gamma(N)$ are given by

PROPOSITION 2.5.1. *We have*:

(1) $\sum_{(ijk)} R_{ijk} = 0$ $(R_{ijk} = g_{ih} R^h_{jk})$.

(2) $P_{ijk} = g_{ih} P^h_{jk}$ *is totally symmetric*.

(3) *The covariant curvature d-tensors* $R_{ijkh} = g_{jr} R^r_{i\,kh}$, $P_{ijkh} = g_{jr} P^r_{i\,kh}$ *and* $S_{ijkh} = g_{ir} S^r_{i\,kh}$ *are skew-symmetric with respect to the first two indices*.

(4) $S_{ijkh} = C_{iks}C^s_{jh} - C_{ihs}C^s_{jk}$.
(5) $C_{ikh} = g_{is}C^s_{jh}$.

These properties can be proved using $d\theta = 0$, Ricci identities applied for the fundamental tensor g_{ij} and the equations $g_{ij|k} = 0$, $g_{ij}|_k = 0$.

By the same method we can study the metrical connections with a priori given h- and v-torsions.

THEOREM 2.5.3.

(1) *There exists a unique N-linear connection $\bar{D}\Gamma(N) = (\bar{L}^i_{jk}, \bar{C}^i_{jk})$ which satisfies the following axioms*:
 (A'_1) *N is canonical nonlinear connection of the space L^n*;
 (A'_2) $g_{ij|k} = 0$ (*$\bar{D}$ is h-metrical*);
 (A'_3) $g_{ij}|_k = 0$ (*$\bar{D}$ is v-metrical*);
 (A'_4) *The torsion h-tensor $\bar{T}^i_{jk}$ is a priori given*;
 (A'_5) *The torsion v-tensor $\bar{S}^i_{jk}$ is a priori given.*

(2) *The coefficients $(\bar{L}^i_{jk}, \bar{C}^i_{jk})$ of $\bar{D}$ are given by*

$$\bar{L}^i_{jk} = L^i_{jk} + \frac{1}{2}g^{ih}\left(g_{jr}\bar{T}^r_{kh} + g_{kr}\bar{T}^r_{jh} - g_{hr}\bar{T}^r_{kj}\right),$$
$$\bar{C}^i_{jk} = C^i_{jk} + \frac{1}{2}g^{ih}\left(g_{jr}\bar{S}^r_{kh} + g_{kr}\bar{S}^r_{jh} - g_{hr}\bar{S}^r_{kj}\right), \tag{2.5.5}$$

where (L^i_{jk}, C^i_{jk}) are the coefficients of the canonical metrical connection.

From now on, $\bar{T}^i_{jk}, \bar{S}^i_{jk}$ will be denoted by T^i_{jk}, S^i_{jk} and the N-linear connection given by the previous theorem will be called *metrical N-connection* of the Lagrange space L^n.

Some particular cases can be studied using the expressions of the coefficients $\bar{L}^i_{jk}$ and $\bar{C}^i_{jk}$. For instance, the semi-symmetric case will be obtained taking $T^i_{jk} = \delta^i_j\sigma_k - \delta^i_k\sigma_j$, $S^i_{jk} = \delta^i_j\tau_k - \delta^i_k\tau_j$.

PROPOSITION 2.5.2. *The Ricci identities of the metrical N-linear connection $D\Gamma(N)$ are given by*:

$$X^i_{|j|k} - X^i_{|k|j} = X^r R^{\ i}_{r\ jk} - X^i_{|r}T^r_{jk} - X^i|_r R^r_{jk},$$
$$X^i_{|j}|_k - X^i|_{k|j} = X^r P^{\ i}_{r\ jk} - X^i_{|r}C^r_{jk} - X^i|_r P^r_{jk},$$
$$X^i|_j|_k - X^i|_k|_j = X^r S^{\ i}_{r\ jk} - X^i|_r S^r_{jk}. \tag{2.5.6}$$

Of course these identities can be extended to a d-tensor field of type (r, s).
Denoting

$$D^i_j = y^i_{|j}, \qquad d^i_j = y^i|_j, \tag{2.5.7}$$

we have the h- and v-deflection tensors. They have the known expressions:

$$D^i_j = y^s L^i_{sj} - N^i_j, \qquad d^i_j = \delta^i_j + y^s C^i_{sj}. \tag{2.5.7'}$$

According to Ricci identities (2.5.6), we obtain

THEOREM 2.5.4. *For any metrical N-linear connection the following identities hold*:

$$\begin{aligned} &D^i_{j|k} - D^i_{k|j} = y^s R^i_{s\,jk} - D^i_s T^s_{jk} - d^i_s R^s_{jk}, \\ &D^i_j|_k - d^i_{k|j} = y^s P^i_{s\,jk} - D^i_s C^s_{jk} - d^i_s P^s_{jk}, \\ &d^i_j|_k - d^i_k|_j = y^s S^i_{s\,jk} - d^i_s S^s_{jk}. \end{aligned} \tag{2.5.8}$$

2.6. *The electromagnetic fields and the gravitational fields*

Let consider a Lagrange spaces $L^n = (M, L)$ endowed with the canonical nonlinear connection N and with the canonical metrical N-connection $C\Gamma(N) = (L^i_{jk}, C^i_{jk})$.

The covariant deflection tensors D_{ji} and d_{ji} are given by $D_{ij} = g_{is} D^s_j$, $d_{ij} = g_{is} d^s_j$. We have

$$D_{ij|k} = g_{is} D^s_{j|k}, \qquad d_{ij|k} = g_{is} d^s_{j|k},$$

etc. Then one has

PROPOSITION 2.6.1. *The covariant deflection tensors D_{ij} and d_{ij} of the canonical metrical N-connection $C\Gamma(N)$ satisfy the identities*

$$\begin{aligned} &D_{ij|k} - D_{ik|j} = y^s R_{sijk} - d_{is} R^s_{jk}, \\ &D_{ij}|_k - d_{ik|j} = y^s P_{sijk} - D_{is} C^s_{jk} - d_{is} P^s_{jk}, \\ &d_{ij}|_k - d_{ik}|_j = y^s S_{sijk}. \end{aligned} \tag{2.6.1}$$

The Lagrangian theory of Electrodynamics leads to introduce the electromagnetic tensor fields [19,20].

DEFINITION 2.6.1. The d-tensor fields

$$F_{ij} = \frac{1}{2}(D_{ij} - D_{ji}), \qquad f_{ij} = \frac{1}{2}(d_{ij} - d_{ji}) \tag{2.6.2}$$

are the h- and v-*electromagnetic tensor* of the Lagrange space $L^n = (M, L)$.

Therefore, the Bianchi identities for $C\Gamma(N)$ and the identities (2.6.1) imply the following important result.

THEOREM 2.6.1. *The following generalized Maxwell equations hold*:

$$F_{ij|k} + F_{jk|i} + F_{ki|j} = -\sum_{(ijk)} C_{ios} R^s_{jk},$$

$$F_{ij}|_k + F_{jk}|_i + F_{ki}|_j = 0, \tag{2.6.3}$$

where $C_{ios} = C_{ijs} y^j$ *and* $\sum_{(ijk)}$ *is cyclic sum.*

COROLLARY 2.6.1. *If the canonical nonlinear connection* N *of the space* L^n *is integrable, then Eqs.* (6.3) *reduce to*:

$$\sum_{(ijk)} F_{ij|k} = 0, \qquad \sum_{(ijk)} F_{ij}|_k = 0. \tag{2.6.3'}$$

If we put

$$F^{ij} = g^{is} g^{jr} F_{sr} \tag{2.6.4}$$

and

$$hJ^i = F^{ij}_{\ |j}, \qquad vJ^i = F^{ij}|_j, \tag{2.6.5}$$

then one can prove

THEOREM 2.6.2. *The following laws of conservation hold*:

$$hJ^i_{|i} = \frac{1}{2}\{F^{ij}(R_{ij} - R_{ji}) + F^{ij}|_r R^r_{ij}\},$$

$$vJ^i|_i = 0, \tag{2.6.6}$$

where R_{ij} *is the Ricci tensor* $R^h_{i\,jh}$.

REMARK. On the Lagrange space of Electrodynamics the tensor F_{jk} is given by (2.3.6). Then $F_{ij}(x)$ satisfy the Maxwell equations $\sum_{(ijk)} F_{ij|k} = 0$ and $F_{ij}|_k = 0$, $hj^i_{|i} = 0$, $vj^i = 0$.

Now, considering the lift to $\widetilde{TM}$ of the fundamental tensor $g_{ij}(x, y)$ of the space L^n, given by

$$\mathbb{G} = g_{ij}\, dx^i \otimes dx^j + g_{ij}\, \delta y^i \otimes \delta y^j$$

we obtain the Einstein equations of the canonical metrical connection $C\Gamma(N)$ [19,20]. The Ricci curvature and scalar curvature satisfy

$$R_{ij} = R^h_{i\,jh}, \quad S_{ij} = S^h_{i\,jh}, \quad P'_{ij} = P^h_{i\,jh}, \quad P''_{ir} = P^h_{i\,hj},$$

$$R = g^{ij} R_{ij}, \quad S = g^{ij} S_{ij}. \tag{2.6.7}$$

Let denote by $\overset{H}{T}_{ij}$, $\overset{V}{T}_{ij}$, $\overset{1}{T}_{ij}$ and $\overset{2}{T}_{ij}$ the components w.r.t. an adapted basis $(\frac{\delta}{\delta x^i}, \frac{\partial}{\partial y^i})$ of the energy momentum tensor on the manifold $\widetilde{TM}$. Thus we obtain [19,20]

THEOREM 2.6.3.

(1) *The Einstein equations of the Lagrange space* $L^n = (M, L(x, y))$ *with respect to the canonical metrical connection* $C\Gamma(N) = (L^i_{jk}, C^i_{jk})$ *are as follows*:

$$R_{ij} - \frac{1}{2} R g_{ij} = \kappa \overset{H}{T}_{ij}, \qquad P'_{ij} = \kappa \overset{1}{T}_{ij},$$

$$S_{ij} - \frac{1}{2} S g_{ij} = \kappa \overset{V}{T}_{(i)(j)}, \qquad P''_{ij} = \kappa \overset{2}{T}_{ij}, \tag{2.6.8}$$

where κ *is a real constant.*

(2) *The energy momentum tensors* $\overset{H}{T}_{ij}$ *and* $\overset{V}{T}_{ij}$ *satisfy the following laws of conservation*:

$$\kappa \overset{H}{T}_{ij} = -\frac{1}{2}\left(P^{ih}_{js} R^s_{hi} + 2 R^s_{ij} P^i_s\right), \qquad \kappa \overset{V}{T}{}^i_j|_i = 0. \tag{2.6.9}$$

The physical background of the previous theory is discussed by S. Ikeda in the last chapter of [20].

The previous theory is very simple for the particular Lagrange spaces L^n having vanishing tensor $P^h_{i\,jk}$.

COROLLARY 2.6.2.

(1) *If the canonical metrical connection* $C\Gamma(N)$ *has the property* $P^i_{j\,kh} = 0$, *then the Einstein equations are*

$$R_{ij} - \frac{1}{2} R g_{ij} = \kappa \overset{H}{T}_{ij}, \qquad S_{ij} - \frac{1}{2} S g_{ij} = \kappa \overset{V}{T}_{(i)(j)}. \tag{2.6.10}$$

(2) *The following laws of conservation hold*:

$$\overset{H}{T}{}^i_{j|i} = 0, \qquad \overset{V}{T}{}^i_j|_i = 0.$$

REMARK. The Lagrange space of Electrodynamics L^n has $C\Gamma(N) = (\gamma^i_{jk}(x), 0)$, $P^i_{j\,kh} = 0$, $S^i_{j\,kh} = 0$. The Einstein equations (2.6.10) reduce to the classical Einstein equations of the space L^n.

2.7. *The almost Kählerian model of a Lagrange space* L^n

A Lagrange space $L^n = (M, L)$ can be thought as an almost Kähler space on the manifold $\widetilde{TM} = TM \setminus \{0\}$, called the geometrical model of the space L^n.

The canonical nonlinear connection N determines an almost complex structure $\mathbb{F}(\widetilde{TM})$, expressed in (1.3.9′), i.e.,

$$\mathbb{F} = \frac{\delta}{\delta x^i} \otimes \delta y^i - \frac{\partial}{\partial y^i} \otimes dx^i. \tag{2.7.1}$$

We know that $\mathbb{F}$ is integrable if and only if $R^i_{\ jk} = 0$.

$\mathbb{F}$ is globally defined on $\widetilde{TM}$ and it can be considered as an $\mathcal{F}(\widetilde{TM})$-linear mapping from $\chi(\widetilde{TM})$ to $\chi(\widetilde{TM})$:

$$\mathbb{F}\left(\frac{\delta}{\delta x^i}\right) = -\frac{\partial}{\partial y^i}, \quad \mathbb{F}\left(\frac{\partial}{\partial y^i}\right) = \frac{\delta}{\delta x^i} \quad (i = 1, \dots, n). \tag{2.7.1′}$$

The lift of the fundamental tensor g_{ij} of the space L^n with respect to N is defined by

$$\mathbb{G} = g_{ij}\, dx^i \otimes dx^j + g_{ij}\, \delta y^i \otimes \delta y^j. \tag{2.7.2}$$

Obviously $\mathbb{G}$ is a (pseudo-)Riemannian metric on the manifold $\widetilde{TM}$.

THEOREM 2.7.1.

(1) *The pair* $(\mathbb{G}, \mathbb{F})$ *is an almost Hermitian structure on* $\widetilde{TM}$, *determined only by the fundamental function* $L(x, y)$ *of* L^n.

(2) *The almost symplectic structure associated to the structure* $(\mathbb{G}, \mathbb{F})$ *is given by*

$$\theta = g_{ij}\, \delta y^i \wedge dx^j. \tag{2.7.3}$$

(3) *The space* $(\widetilde{TM}, \mathbb{G}, \mathbb{F})$ *is almost Kählerian.*

Indeed:

(1) $N, \mathbb{G}, \mathbb{F}$ are determined only by $L(x, y)$.
We have $\mathbb{G}(\mathbb{F}X, \mathbb{F}Y) = \mathbb{G}(X, Y)$, $\forall X, Y \in \chi(\widetilde{TM})$.

(2) With respect to an adapted basis, (2.7.3) implies $\theta(X, Y) = \mathbb{G}(\mathbb{F}X, Y)$.

(3) Taking into account Theorem 2.1.1, it follows that θ is a symplectic structure (i.e., $d\theta = 0$).

The space $H^{2n} = (\widetilde{TM}, \mathbb{G}, \mathbb{F})$ is called *the almost Kählerian model* of the Lagrange space L^n. It has a remarkable property, given by the following theorem.

THEOREM 2.7.2. *The canonical metrical connection* D *with coefficients* $C\Gamma(N) = (L^i_{\ jk}, C^i_{\ jk})$ *of the Lagrange space* L^n *is an almost Kählerian connection, i.e.,*

$$D\mathbb{G} = 0, \qquad D\mathbb{F} = 0. \tag{2.7.4}$$

We can use this geometrical model for studying the geometry of the Lagrange space L^n. For instance, the Einstein equations of the (pseudo-)Riemannian space $(\widetilde{TM}, \mathbb{G})$ equipped with the canonical metrical connection $C\Gamma(N)$ are the Einstein equations of the Lagrange space studied in the previous subsection.

G.S. Asanov [16,19,20] showed that the metric $\mathbb{G}$ given by the lift (2.7.2) does not satisfy the principle of the post-Newtonian calculus. This fact holds because the two terms of $\mathbb{G}$ have not the same physical dimensions. This is the reason to introduce a new lift [19,20], which can be used in Gauge theory of physical fields.

Let consider the scalar field

$$\varepsilon = \|y\|^2 = g_{ij}(x, y)y^i y^j, \tag{2.7.5}$$

called the absolute energy of the Lagrange space L^n [19].

We assume $\|y\|^2 > 0$ and consider the following lift of the fundamental tensor g_{ij}:

$$\overset{0}{\mathbb{G}} = g_{ij}\, dx^i \otimes dx^j + \frac{a^2}{\|y\|^2} g_{ij}\, \delta y^i \otimes \delta y^j, \tag{2.7.6}$$

where $a > 0$ is a constant, imposed by applications to Theoretical Physics (in order to preserve the physical dimensions of that two terms of $\tilde{\mathbb{G}}$).

Let consider also the tensor field

$$\overset{0}{\mathbb{F}} = -\frac{\|y\|}{a} \frac{\partial}{\partial y^i} \otimes dx^j + \frac{a}{\|y\|} \frac{\delta}{\delta x^i} \otimes \delta y^i, \tag{2.7.7}$$

on $\widetilde{TM}$ and the 2-form

$$\overset{0}{\theta} = \frac{a}{\|y\|}\theta, \tag{2.7.8}$$

where θ is defined by (2.7.3).

THEOREM 2.7.3.

(1) *The pair* $(\overset{0}{\mathbb{G}}, \overset{0}{\mathbb{F}})$ *is an almost Hermitian structure on the manifold* $\widetilde{TM}$, *depending only on the fundamental function* $L(x, y)$ *of the space* L^n.

(2) *The almost symplectic structure* $\overset{0}{\theta}$ *associated to the structure* $(\overset{0}{\mathbb{G}}, \overset{0}{\mathbb{F}})$ *is given by* (2.7.8).

(3) $\overset{0}{\theta}$ *being conformal to symplectic structure* θ, *the pair* $(\overset{0}{\mathbb{G}}, \overset{0}{\mathbb{F}})$ *is conformal to the almost Kählerian structure* $(\mathbb{G}, \mathbb{F})$.

2.8. *Generalized Lagrange spaces*

A first natural generalization of the notion of Lagrange space is provided by the notion of a generalized Lagrange space. This notion was introduced by the present author in [19,20].

DEFINITION 2.8.1. A generalized Lagrange space is a pair $GL^n = (M, g_{ij}(x,y))$, where $g_{ij}(x,y)$ is a symmetric d-tensor field of type $(0,2)$, having the rank n and constant signature on $\widetilde{TM}$.

We continue to call $g_{ij}(x,y)$ the *fundamental* tensor on GL^n.

One easily seen that any Lagrange space $L^n = (M, L(x,y))$ is a generalized Lagrange space with the fundamental tensor

$$g_{ij}(x,y) = \frac{1}{2}\frac{\partial^2 L(x,y)}{\partial y^i \partial y^j}. \tag{2.8.1}$$

But not any generalized Lagrange space GL^n is a Lagrange space L^n.

Indeed, if $g_{ij}(x,y)$ is given, the system of partial differential equations (2.8.1) does not admit always solutions $L(x,y)$.

PROPOSITION 2.8.1.

(1) *A necessary condition in order that the system* (2.8.1) *admits a solution $L(x,y)$ is that the d-tensor field $\frac{\partial g_{ij}}{\partial y^k} = 2C_{ijk}$ be completely symmetric.*

(2) *If the condition* (1) *is verified and the functions $g_{ij}(x,y)$ are* 0*-homogeneous with respect to y^i, then the function*

$$L(x,y) = g_{ij}(x,y)y^i y^j + A_i(x)y^i + U(x) \tag{2.8.2}$$

is a solution of the system of partial differential equations (2.8.1) *for any arbitrary d-covector field $A_i(x)$ and any arbitrary function $U(x)$ on the base manifold M.*

If the system (2.8.1) does not admit any solution, we say that the generalized Lagrange space $GL^2 = (M, g_{ij(x,y)})$ is not reducible to a Lagrange space.

REMARK 2.8.1. The Lagrange spaces L^n with the fundamental function (2.8.2) gives an important class of Lagrange spaces which includes the Lagrange space of Electrodynamics.

EXAMPLES.

(1) The pair $GL^n = (M, g_{ij})$ with the fundamental tensor field

$$g_{ij}(x,y) = e^{2\sigma(x,y)}\gamma_{ij}(x), \tag{2.8.3}$$

where σ is a function on $\widetilde{TM}$ and $\gamma_{ij}(x)$ is a pseudo-Riemannian metric on the manifold M, is a generalized Lagrange space if the d-covector field $\frac{\partial\sigma}{\partial y^i}$ does not vanish.

It is not reducible to a Lagrange space. R. Miron and R. Tavakol [20] proved that $GL^n = (M, g_{ij}(x,y))$ defined by (2.8.3) satisfies the Ehlers–Pirani–Shield's axioms of General Relativity.

(2) The pair $GL^n = (M, g_{ij}(x, y))$, with

$$g_{ij}(x, y) = \gamma_{ij}(x) + \left(1 - \frac{1}{n^2(x, y)}\right) y_i y_j, \qquad y_i = \gamma_{ij}(x) y^j, \tag{2.8.4}$$

where $\gamma_{ij}(x)$ is a pseudo-Riemannian metric and $n(x, y) > 1$ is a smooth function (n is a refractive index), is a generalized Lagrange space GL^n, which is not reducible to a Lagrange space.

The restriction of the fundamental tensor $g_{ij}(x, y)$ to a section $S_V\colon x^i = x^i$, $y^i = V^i(x)$ (V^i being a vector field) of the bundle $\pi : TM \to M$, is given by $g_{ij}(x, V(x))$. It provides the known Synge's metric tensor of Relativistic Optics [20].

For a generalized Lagrange space $GL^n = (M, g_{ij}(x, y))$, an important problem is to determine a nonlinear connection obtained from the fundamental tensor $g_{ij}(x, y)$. In the particular cases given by the previous two examples, this is possible, but not generally.

We point-out a method to determine a nonlinear connection N, strongly related to the fundamental tensor g_{ij} of the space GL^n, if such a nonlinear connection exists.

Consider the *absolute energy* $\varepsilon(x, y)$ of the space GL^n:

$$\varepsilon(x, y) = g_{ij}(x, y) y^i y^j, \tag{2.8.5}$$

$\varepsilon(x, y)$ is a Lagrangian.

The Euler–Lagrange equations of $\varepsilon(x, y)$ are

$$\frac{\partial \varepsilon}{\partial x^i} - \frac{d}{dt}\frac{\partial \varepsilon}{\partial y^i} = 0, \qquad y^i = \frac{dx^i}{dt}. \tag{2.8.6}$$

Of course, according to the general theory, the energy E_ε of the Lagrangian $\varepsilon(x, y)$ is $E_\varepsilon = y^i \frac{\partial \varepsilon}{\partial y^i} - \varepsilon$ and it is preserved along the integral curves of the differential equations (2.8.6).

If $\varepsilon(x, y)$ is a regular Lagrangian (in this case we say that the space GL^n is weakly regular), it follows that the Euler–Lagrange equations determine a semispray with the coefficients

$$2G^i(x, y) = \overset{\vee}{g}{}^{is}\left(\frac{\partial^2 \varepsilon}{\partial y^s \partial x^j} y^j - \frac{\partial \varepsilon}{\partial x^s}\right) \qquad \left(\overset{\vee}{g}_{ij} = \frac{1}{2}\frac{\partial^2 \varepsilon}{\partial y^i \partial y^j}\right). \tag{2.8.7}$$

Consequently, the nonlinear connection N with the coefficients $N^i_j = \frac{\partial G^i}{\partial y^j}$ is determined only by the fundamental tensor $g_{ij}(x, y)$ of the space GL^n.

In the case when we can not derive a nonlinear connection from the fundamental tensor g_{ij}, we give a priori a nonlinear connection N and study the geometry of the pair (GL^n, N) by the methods of the geometry of Lagrange space L^n.

For instance, using an adapted basis $(\frac{\delta}{\delta x^i}, \frac{\partial}{\partial y^i})$ to the distributions N and V, respectively, and its dual basis $(dx^i, \delta y^i)$, we can lift $g_{ij}(x, y)$ to $\widetilde{TM}$:

$$G(x, y) = g_{ij}(x, y)\, dx^i \otimes dx^j + \frac{a}{\|y\|^2} g_{ij}(x, y)\, \delta y^i \otimes \delta y^j$$

and consider the almost complex structure

$$\mathbb{F} = -\frac{\|y\|}{a}\frac{\partial}{\partial y^i} \otimes dx^i + \frac{a}{\|y\|}\frac{\delta}{\delta x^i} \otimes \delta y^i,$$

with $\varepsilon(x, y) > 0$ and $\|y\| = \varepsilon^{1/2}(x, y)$.

The space $(\widetilde{TM}, \mathbb{F})$ is an almost Hermitian space associated to the pair (GL^n, N).

3. Finsler spaces

An important class of Lagrange spaces is provided by the Finsler spaces.

The notion of Finsler space was introduced by Paul Finsler in 1918 and was developed by remarkable mathematicians, as L. Berwald, E. Cartan, H. Busemann, H. Rund, S.S. Chern, M. Matsumoto [4,6,14,19,24] and many others.

This notion is a generalization of a Riemann space, which gives an important geometrical framework in Theoretical Physics. Therefore, the Finsler spaces are basically in the geometrical theory of physical fields [8,10,20,28,29].

In the last 40 years, some remarkable books on Finsler geometry and its applications were published by H. Rund, M. Matsumoto, R. Miron and M. Anastasiei, A. Bejancu, Abate-Patrizio, D. Bao, S.S. Chern and Z. Shen, P. Antonelli, R. Ingarden and M. Matsumoto, R. Miron, D. Hrimiuc, H. Shimada and S. Sabău, G.S. Asanov, M. Crampin, P.L. Antonelli, S. Vacaru and S. Ikeda.

In the present section, we study the Finsler spaces, considered as Lagrange spaces and applying the mechanical principles. This method simplifies the theory of Finsler spaces. We will study: Finsler metric, Cartan nonlinear connection derived from the canonical spray, Cartan metrical connection and its structure equations. Some examples: Randers spaces, Kropina spaces and some new classes of spaces more general than Finsler spaces: almost Finsler Lagrange spaces and Ingarden spaces.

3.1. *Finsler metrics*

DEFINITION 3.1.1. A *Finsler space* is a pair $F^n = (M, F(x, y))$, where M is a real n-dimensional differentiable manifold and $F : TM \to \mathbb{R}$ is a scalar function which satisfies the following axioms:

(1) F is a differentiable function on $\widetilde{TM}$ and continuous on the null section of the bundle $\pi : TM \to M$.
(2) F is a positive function.
(3) F is positive 1-homogeneous with respect to the variables y^i.
(4) The Hessian of F^2, having the components

$$g_{ij}(x, y) = \frac{1}{2}\frac{\partial F^2}{\partial y^i \partial y^j}, \tag{3.1.1}$$

is positive definite on the manifold $\widetilde{TM}$.

Of course, the axiom (4) is equivalent to the following:

(4′) The pair $(M, F^2(x, y)) = L^n_F$ is a Lagrange space with positive definite fundamental tensor g_{ij}. L^n_F is called the Lagrange space associated to the Finsler space F^n. It follows that all properties of the Finsler space F^n derived from the fundamental function F^2 and the fundamental tensor g_{ij} are the properties of the associated Lagrange space L^n_F.

REMARKS.

(1) Sometimes we ask for g_{ij} to be of constant signature and $\operatorname{rank}(g_{ij}(x, y)) = n$ on $\widetilde{TM}$.

(2) Any Finsler space F^n is a Lagrange space L^n_F, but the converse does not hold.

EXAMPLES.

(1) A Riemannian manifold $(M, \gamma_{ij}(x))$ determines a Finsler space $F^n = (M, F(x, y))$, where

$$F(x, y) = \sqrt{\gamma_{ij}(x) y^i y^j}. \tag{3.1.2}$$

The fundamental tensor is $g_{ij}(x, y) = \gamma_{ij}(x)$.

(2) Let consider in a preferential local system of coordinates, the following function:

$$F(x, y) = \sqrt[4]{\left(y^1\right)^4 + \cdots + \left(y^n\right)^4}. \tag{3.1.3}$$

Then F satisfies the axioms (1)–(4). This example is due to B. Riemann.

(3) Antonelli–Shimada's ecological metric is given, in a preferential local system of coordinates on $\widetilde{TM}$, by

$$F(x, y) = e^{\phi} L, \quad \phi = \alpha_i x^i \quad (\alpha_i \text{ are positive constants}),$$

where

$$L = \left\{\left(y^1\right)^m + \left(y^2\right)^m + \cdots + \left(y^n\right)^m\right\}^{1/m}, \quad m \geqslant 3, \tag{3.1.4}$$

with m an even integer.

(4) Randers metric is defined by

$$F(x, y) = \alpha(x, y) + \beta(x, y), \tag{3.1.5}$$

where $\alpha^2(x, y) := a_{ij}(x) y^i y^j$, $(M, a_{ij}(x))$ is a Riemannian manifold and $\beta(x, y) := b_i(x) y^i$.

The fundamental tensor g_{ij} is expressed by

$$g_{ij} = \frac{\alpha + \beta}{\alpha} h_{ij} + d_i d_j, \qquad h_{ij} := a_{ij} - \overset{0}{l}_i \overset{0}{l}_j,$$

$$d_i = b_i + \overset{0}{l}_i, \qquad \overset{0}{l}_i := \frac{\partial \alpha}{\partial y^i}. \tag{3.1.5'}$$

One can prove that g_{ij} is positive definite under the condition $b^2 = a^{ij} b_i b_j < 1$. In this case, the pair $F^n = (M, \alpha + \beta)$ is a Finsler space.

The first example motivates the following theorems.

THEOREM 3.1.1. *If the base manifold M is paracompact, then there exist functions $F: TM \to \mathbb{R}$ such that the pair (M, F) is a Finsler spaces.*

THEOREM 3.1.2. *The system of axioms of a Finsler space is minimal.*

We state some properties of a Finsler space F^n:

(1) The components of the fundamental tensor $g_{ij}(x, y)$ are 0-homogeneous with respect to y^i.

(2) The components of the 1-form

$$p_i = \frac{1}{2} \frac{\partial F^2}{\partial y^i} \tag{3.1.6}$$

are 1-homogeneous with respect to y^i.

(3) The components of the Cartan tensor

$$C_{ijk} = \frac{1}{4} \frac{\partial^3 F^2}{\partial y^i \partial y^j \partial y^k} = \frac{1}{2} \frac{\partial g_{ij}}{\partial y^k} \tag{3.1.7}$$

are 1-homogeneous with respect to y^i. Consequently we have

$$C_{oij} = y^s C_{sij} = 0. \tag{3.1.8}$$

If X^i and Y^i are d-vector fields, then $\|X\|^2 := g_{ij}(x, y) X^i X^j$ and $\langle X, Y \rangle := g_{ij}(x, y) X^i Y^j$ are scalar fields.

Assuming $\|X\|_u \neq 0$, $\|Y\|_u \neq 0$, the angle $\varphi = \angle(X, Y)$ at a point $u \in \widetilde{TM}$ is given by

$$\cos \varphi = \frac{\langle X, Y \rangle (u)}{\|X\|_u \cdot \|Y\|_u}.$$

The vectors X_u, Y_u are orthogonal if $\langle X, Y \rangle (u) = 0$.

PROPOSITION 3.1.1. *On a Finsler space F^n, the following identities hold:*

(1) $F^2(x, y) = g_{ij}(x, y) y^i y^j$.

(2) $p_i y^i = F^2$.

PROPOSITION 3.1.2.

(1) *The* 1*-form*

$$\omega = p_i \, dx^i \tag{3.1.9}$$

is globally defined on $\widetilde{TM}$.

(2) *The* 2*-form*

$$\theta = d\omega = dp_i \wedge dx^i \tag{3.1.10}$$

is globally defined on $\widetilde{TM}$.

(3) θ *is a symplectic structure on* $\widetilde{TM}$.

DEFINITION 3.1.2. A Finsler space $F^n = (M, F)$ is called *reducible* to a Riemann space if its fundamental tensor $g_{ij}(x, y)$ does not depend on the variables y^i.

PROPOSITION 3.1.3. *A Finsler space* F^n *is reducible to a Riemann space if and only if the tensor* C_{ijk} *vanishes identically on* $\widetilde{TM}$.

3.2. *Geodesics*

On a Finsler space $F^n = (M, F(x, y))$, one defines the notion of arclength of a smooth curve. Let c be a parametrized curve on the manifold M

$$c : t \in [0, 1] \to \left(x^i(t)\right) \in U \subset M, \tag{3.2.1}$$

U being a domain of a local chart on M.

The extension $\tilde{c}$ of c to $\widetilde{TM}$ has the equations

$$x^i = x^i(t), \quad y^i = \frac{dx^i}{dt}(t), \quad t \in [0, 1]. \tag{3.2.1$'$}$$

Thus the restriction of the fundamental function $F(x, y)$ to $\tilde{c}$ is $F(x(t), \frac{dx}{dt}(t))$, $t \in [0, 1]$.

We define the *length* of the curve c with extremities $c(0)$, $c(1)$ by the number

$$\ell(c) = \int_0^1 F\left(x(t), \frac{dx}{dt}(t)\right) dt. \tag{3.2.2}$$

The number $\ell(c)$ does not depend on the local coordinates on $\widetilde{TM}$ and, by means of the 1-homogeneity of function the F, $\ell(c)$ does not depend on the parametrization of the curve c, $\ell(c)$ depends only on the curve c.

We can chose a canonical parameter on c, considering the following function $s = s(t)$, $t \in [0, 1]$,

$$s(t) = \int_{t_0}^t F\left(x(\tau), \frac{dx}{dt}(\tau)\right) d\tau.$$

This function is derivable and its derivative is

$$\frac{ds}{dt} = F\left(x(t), \frac{dx}{dt}(t)\right) > 0, \quad t \in (0, 1).$$

So the function $s = s(t)$, $t \in [0, 1]$, is inversible. Let $t = t(s)$, $s \in [s_0, s_1]$ be its inverse. The change of parameter $t \to s$ has the property

$$F\left(x(s), \frac{dx}{ds}(s)\right) = 1. \tag{3.2.3}$$

Variational problem on the functional ℓ gives the curves on $\widetilde{TM}$ which extremize the arclength. These curves are the geodesics of the Finsler space F^n. Thus, they are the solutions of the Euler–Lagrange equations:

$$\frac{\partial F}{\partial x^i} - \frac{d}{dt}\left(\frac{\partial F}{\partial y^i}\right) = 0, \qquad y^i = \frac{dx^i}{dt}. \tag{3.2.4}$$

The system of differential equations (3.2.4) is equivalent to the following system:

$$\frac{\partial F^2}{\partial x^i} - \frac{d}{dt}\frac{\partial F^2}{\partial y^i} = -2\frac{dF}{dt}\frac{\partial F}{\partial y^i}, \qquad y^i = \frac{dx^i}{dt}.$$

THEOREM 3.2.1. *The geodesics parametrized by arclength of the Finsler space F^n are the solutions of the system of differential equations*

$$E_i(F^2) := \frac{\partial F^2}{\partial x^i} - \frac{d}{ds}\frac{\partial F^2}{\partial y^i} = 0, \qquad y^i = \frac{dx^i}{ds}. \tag{3.2.5}$$

Now, remarking that $F^2 = g_{ij} y^i y^j$, the previous equations can be written in the form:

$$\frac{d^2 x^i}{ds^2} + \gamma^i_{jk}\left(x, \frac{dx}{ds}\right)\frac{dx^j}{ds}\frac{dx^k}{ds} = 0, \qquad y^i = \frac{dx^i}{ds}, \tag{3.2.6}$$

where γ^i_{jk} are the Christoffel symbols of the fundamental tensor g_{ij}:

$$\gamma^i_{jk} = \frac{1}{2} g^{ir}\left(\frac{\partial g_{rk}}{\partial x^j} + \frac{\partial g_{jr}}{\partial x^k} - \frac{\partial g_{jk}}{\partial x^r}\right). \tag{3.2.7}$$

A theorem of existence and uniqueness of the solutions of the differential equations (3.2.6) holds.

3.3. *Cartan nonlinear connection*

Considering the Lagrange space $L^n_F = (M, F^2)$ associated to the Finsler space $F^n = (M, F)$, we can obtain some main geometrical object field of F^n.

Theorem 2.3.1 affirms:

THEOREM 3.3.1. *On a Finsler space F^n, the equations*

$$g^{ij}E_j(F^2) := g^{ij}\left(\frac{\partial F^2}{\partial x^j} - \frac{d}{dt}\frac{\partial F^2}{\partial y^j}\right) = 0, \qquad y^i = \frac{dx^i}{dt},$$

can be written in the form

$$\frac{d^2x^i}{dt^2} + 2G^i\left(x, \frac{dx}{dt}\right) = 0, \qquad y^i = \frac{dx^i}{dt}, \tag{3.3.1}$$

where

$$2G^i(x,y) = \gamma^i_{jk}(x,y)y^j y^k. \tag{3.3.1'}$$

Consequently Eqs. (3.3.1) give the integral curves of the semispray

$$S = y^i\frac{\partial}{\partial x^i} - 2G^i(x,y)\frac{\partial}{\partial y^i}. \tag{3.3.2}$$

Since G^i are 2-homogeneous functions with respect to y^i, it follows that S is a *spray*. S determines a canonical nonlinear connection N with the coefficients

$$N^i_j = \frac{\partial G^i}{\partial y^j} = \frac{1}{2}\frac{\partial}{\partial y_j}\left\{\gamma^i_{rs}(x,y)y^r y^s\right\}. \tag{3.3.3}$$

N is called the Cartan nonlinear connection of the space F^n.

The tangent bundle $T(TM)$, the horizontal distribution N and the vertical distribution V give the direct decomposition of vector spaces

$$T_u(\widetilde{TM}) = N(u) \oplus V(u), \quad \forall u \in \widetilde{TM}. \tag{3.3.4}$$

An adapted basis to N and V is $(\frac{\delta}{\delta x^i}, \frac{\partial}{\partial y^i})$ and its dual adapted basis is $(dx, \delta y^i)$, where

$$\begin{cases} \frac{\delta}{\delta x^i} = \frac{\partial}{\partial x^i} - N^j_i(x,y)\frac{\partial}{\partial y^j}, \\ \delta y^i = dy^i + N^i_j(x,y)\,dx^j. \end{cases} \tag{3.3.4'}$$

THEOREM 3.3.2.
(1) *The horizontal curves in* F^n *are given by*

$$x^i = x^i(t), \qquad \frac{\delta y^i}{dt} = 0.$$

(2) *The autoparallel curves of the Cartan nonlinear connection* N *coincide with the integral curves of the spray* S, *defined by* (3.3.2).

3.4. *Cartan metrical connection*

Let $N(N^i_j)$ be the Cartan nonlinear connection of the Finsler space F^n. According to Section 2.5, one introduces the canonical metrical N-linear connection of the space F^n.

For Finsler spaces, the system of axioms from Theorem 2.5.1, can be written in the Matsumoto's form [14,20].

THEOREM 3.4.1.
(1) *For any Finsler space* $F^n = (M, F)$, *there exists a unique linear connection* D *on the manifold* $\widetilde{TM}$, *with the coefficients* $C\Gamma = (N^i_j, F^i_{jk}, C^i_{jk})$, *which satisfies the following axioms*:
 (A_1) *The deflection tensor field* $D^i_j = y^i_{|j}$ *vanishes*;
 (A_2) $g_{ij|k} = 0$ (*D is h-metrical*);
 (A_3) $g_{ij}|_k = 0$ (*D is v-metrical*);
 (A_4) $T^i_{jk} = 0$ (*D is h-torsion free*);
 (A_5) $S^i_{jk} = 0$ (*D is v-torsion free*).
(2) *The coefficients* $(N^i_j, F^i_{jk}, C^i_{jk})$ *are as follows*:
 (a) N^i_j *are the coefficients of the Cartan nonlinear connection*;
 (b) F^i_{jk}, C^i_{jk} *are expressed by the generalized Christoffel symbols*:

$$F^i_{jk} = \frac{1}{2} g^{is} \left(\frac{\delta g_{sk}}{\delta x^j} + \frac{\delta g_{js}}{\delta x^k} - \frac{\delta g_{jk}}{\delta x^s} \right),$$

$$C^i_{jk} = \frac{1}{2} g^{is} \left(\frac{\partial g_{sk}}{\partial y^j} + \frac{\partial g_{js}}{\partial y^k} - \frac{\partial g_{jk}}{\partial y^s} \right). \tag{3.4.1}$$

(3) *This connection depends only on the fundamental function* F.

The proof can be found in the books [19,20].

The previous connection is called the Cartan metrical connection of the Finsler space F^n.

Now we can develop the geometry of Finsler spaces, exactly as the geometry of the associated Lagrange spaces $L^n_F = (M, F^2)$.

Also, in the case of Finsler spaces, the geometrical model $H^{2n} = (\widetilde{TM}, \mathbb{G}, \mathbb{F})$ is an almost Kählerian space.

A very interesting example is given by the Randers spaces, introduced by R.S. Ingarden.

A Randers space is a Finsler space $F^n = (M, \alpha + \beta)$ equipped with a Cartan nonlinear connection N and denoted by $RF^n = (M, \alpha + \beta, N)$.

The geometry of these spaces are very intensively studied by many geometers. We refer to the monograph of D. Bao, S.S. Chern and Z. Shen [6].

The Randers spaces RF^n can be generalized considering the Finsler spaces $F^n = (M, \alpha + \beta)$, where $\alpha(x, y)$ is the fundamental function of a Finsler space $F'^n = (M, \alpha)$. The Finsler space $F^n = (M, \alpha + \beta)$ equipped with the Cartan nonlinear connection N of the space $F'^n = (M, \alpha)$ is a generalized Randers space [19,20]. Obviously, this notion has some advantages, since we can consider some remarkable Finsler spaces F'^n.

As an application of the previous notions, we define the notion of Ingarden space IF^n [3,1]. This is the Finsler space $F^n = (M, \alpha + \beta)$ equipped with the nonlinear connection $N = \gamma^i_{jk}(x)y^k - F^i_j(x)$, $\gamma^i_{jk}(x)$ being the Christoffel symbols of the Riemannian metric $a_{ij}(x)$, which defines $\alpha^2 = a_{ij}(x)y^i y^j$ and the electromagnetic tensor $F^i_j(x)$ determined by $\beta = b_i(x)y^i$. While the spaces RF^n have not the electromagnetic field $\mathcal{F} = \frac{1}{2}(D_{ij} - D_{ji})$, the Ingarden spaces have such tensor fields and they give the well-known Maxwell equations [19]. Also, the autoparallel curves of the nonlinear connection N are given by the known Lorentz equations.

An example of a special Lagrange space derived from a Finsler one [19] is the following. Let consider the Lagrange space $L^n = (M, L(x, y))$, with the fundamental function

$$L(x, y) = F^2(x, y) + \beta,$$

where F is the fundamental function of a priori given Finsler space $F^n = (M, F)$ and $\beta = b_i(x)y^i$.

These spaces are called *almost Finsler Lagrange spaces* (shortly AFL-spaces) [19,20]. They generalize the Lagrange space from Electrodynamics.

Indeed, the Euler–Lagrange equations of AFL-spaces are exactly the Lorentz equations

$$\frac{d^2x^i}{dt^2} + \gamma^i_{jk}(x, y)\frac{dx^j}{dt}\frac{dx^k}{dt} = \frac{1}{2}F^i_j(x)\frac{dx^j}{dt}.$$

As a conclusion of these three sections, we remark that the class of Riemann spaces $\{\mathcal{R}^n\}$ is a subclass of the class of Finsler spaces $\{F^n\}$, the class $\{F^n\}$ is a subclass of the class of Lagrange spaces $\{L^n\}$ and this is a subclass of the class of generalized Lagrange spaces $\{GL^n\}$. So, we have the following sequence of inclusions:

(I) $\{\mathcal{R}^n\} \subset \{F^n\} \subset \{L^n\} \subset \{GL^n\}$.

Therefore, we can say that the Lagrange geometry is the geometric study of the terms of the sequence of inclusions (I).

4. The geometry of $T^{(k)}M$

The importance of Lagrange geometries consists by the fact that the variational problems for Lagrangians have several applications to various fields, as: Mathematics, Physics, Theory of Dynamical Systems, Optimal Control, Biology, Economy, etc.

All the above mentioned applications have imposed also the introduction of the notion of higher-order Lagrange spaces. The base manifold of such space is the bundle of accelerations of higher order. The methods used in the construction of the geometry of higher-order Lagrange spaces are the natural extensions of those used in the theory of Lagrangian geometries exposed in Sections 1–3.

The concept of higher-order Lagrange space was introduced by the present author [16,17]. The problems raised by the geometrization of Lagrangians of order $k > 1$ have been investigated by many mathematicians: Ch. Ehresmann, P. Libermann, J. Pommaret, J.T. Synge, M. Crampin, P. Saunders, G.S. Asanov, D. Krupka, M. de Léon, H. Rund, W.M. Tulczyjew, A. Kawaguchi, K. Yano, K. Kondo, D. Grigore, R. Miron et al. [12,9,16, 31].

In this section we present, briefly, the following problems:

(1) The geometry of total space of the bundle of higher-order accelerations.
(2) The definition of higher-order Lagrange space, based on nondegenerate Lagrangians of order $k \geqslant 1$.
(3) The problem of prolongation of Riemannian structures given on the base manifold M to Riemannian structures on the total space of the bundle of accelerations of order $k \geqslant 1$; we prove the existence of Lagrange spaces of order $k \geqslant 1$.
(4) The elaboration of the geometrical ground for variational calculus involving Lagrangians which depend on higher-order accelerations.
(5) The introduction of the notion of higher-order energies and proof of the conservation law.
(6) The notions of k-semispray, nonlinear connection, the canonical metrical connection and the structure equations.
(7) The Riemannian $(k-1)n$-almost contact model of the Lagrange space of order k.

For more information, we refer to the books [19,20,22].

4.1. *The bundle of acceleration of order $k \geqslant 1$*

In Analytical Mechanics a real n-dimensional differentiable manifold M is considered as the space of configurations of a physical system. A point $(x^i) \in M$ is called a configuration. A mapping $c: t \in I \to (x^i(t)) \in U \subset M$ is a law of moving (evolution), t is the time, a pair (t, x) is an event and the k-tuple $(\frac{dx^i}{dt}, \ldots, \frac{1}{k!}\frac{d^k x^i}{dt^k})$ gives the velocity and generalized accelerations of order $1, \ldots, k-1$. The factors $\frac{1}{h!}$ $(h = 1, \ldots, k)$ are introduced here for the simplicity of calculus. In this section, we omit the word "generalized" and call $\frac{1}{h!}\frac{d^h x^i}{dt^h}$ shortly, the acceleration of order h. A law of moving $c: t \in I \to c(t) \in U$ is called a parametrized curve by time t.

In order to obtain the differentiable bundle of accelerations of order k, we use the accelerations of order k, by means of the geometrical concept of contact of order k for two curves on the manifold M.

Two curves $\rho, \sigma : I \to M$ in M have at the point $x_0 \in M$, $\rho(0) = \sigma(0) = x_0 \in U$ (U is a domain of a local chart on M) have a *contact* of order k if we have

$$\left.\frac{d^\alpha (f\circ\rho)(t)}{dt^\alpha}\right|_{t=0} = \left.\frac{d^\alpha (f\circ\sigma)(t)}{dt^\alpha}\right|_{t=0} \quad (\alpha = 1, \ldots, n). \tag{4.1.1}$$

It follows that: the curves ρ and σ have at the point $x_0 = \rho(0) = \sigma(0)$ a contact of order k if and only if the accelerations of order $1, 2, \ldots, k$ on the curve ρ at x_0 have the same values as the corresponding accelerations on the curve σ at the point x_0.

The relation "*to have a contact of order k*" is an equivalence. Let $[\rho]_{x_0}$ be a class of equivalence and $T^k_{x_0}M$ the set of equivalence classes. Consider the set

$$T^kM = \bigcup_{x_0\in M} T^k_{x_0}M, \tag{4.1.2}$$

$$\pi^k : [\rho]_{x_0} \in T^kM \to x_0 \in M, \quad \forall[\rho]_{x_0}. \tag{4.1.2'}$$

Thus the triple (T^kM, π^k, M) can be endowed with a natural differentiable structure, exactly as in the cases $k = 1$, when (T^1M, π^1, M) is the tangent bundle.

If $U \subset M$ is a coordinate neighborhood on the manifold M, $x_0 \in U$ and the curve $\rho : I \to U$, $\rho_0 = x_0$ is analytical on U, given by the equations $x^i = x^i(t)$, $t \in I$, then $T^k_{x_0}M$ can be represented by

$$x_0^i = x^i(0), \quad y_0^{(1)i} = \frac{dx^i}{dt}(0), \quad \ldots, \quad y_0^{(k)i} = \frac{1}{k!}\frac{d^kx^i}{dt^k}(0). \tag{4.1.3}$$

Setting

$$\phi : ([\rho]_{x_0}) \in T^kM \to \phi([\rho]_{x_0}) = \left(x_0^i, y_0^{(1)i}, \ldots, y_0^{(k)i}\right) \in R^{(k+1)n}, \tag{4.1.4}$$

it follows that the pair $((\pi^k)^{-1}(U), \phi)$ is a local chart on T^kM induced by the local chart (U, φ) on the manifold M.

So a differentiable atlas of the manifold M determines a differentiable atlas on T^kM and the triple (T^kM, π^k, M) is a differentiable bundle. Of course the mapping $\pi^k : T^kM \to M$ is a submersion.

(T^kM, π^k, M) is called the k-accelerations bundle or tangent bundle of order k or k-osculator bundle [16]. A change of local coordinates $(x^i, y^{(1)i}, \ldots, y^{(k)i}) \to (\tilde{x}^i, \tilde{y}^{(1)i}, \ldots, \tilde{y}^{(k)i})$ on the manifold T^kM, according to (4.1.3), is given by

$$\begin{cases} \tilde{x}^i = \tilde{x}^i(x^1, \ldots, x^n), \quad \operatorname{rank}\left(\frac{\partial \tilde{x}^i}{\partial x^j}\right) = n, \\ \tilde{y}^{(1)i} = \frac{\partial \tilde{x}^i}{\partial x^j} y^{(1)j}, \\ 2\tilde{y}^{(2)i} = \frac{\partial \tilde{y}^{(1)i}}{\partial x^j} y^{(1)j} + 2\frac{\partial \tilde{y}^{(1)i}}{\partial y^{(1)j}} y^{(2)j}, \\ \vdots \\ k\tilde{y}^{(k)i} = \frac{\partial \tilde{y}^{(k-1)i}}{\partial x^j} y^{(1)j} + 2\frac{\partial \tilde{y}^{(k-1)i}}{\partial y^{(1)j}} y^{(2)j} + \cdots + k\frac{\partial \tilde{y}^{(k-1)i}}{\partial y^{(k-1)j}} y^{(k)j}. \end{cases} \tag{4.1.5}$$

We have the following identities:

$$\frac{\partial \tilde{y}^{(\alpha)i}}{\partial x^j} = \frac{\partial \tilde{y}^{(\alpha+1)i}}{\partial y^{(1)j}} = \cdots = \frac{\partial \tilde{y}^{(k)i}}{\partial y^{(k-\alpha)j}} \quad \left(\alpha = 0, \ldots, k-1,\ y^{(0)} = x\right). \qquad (4.1.5')$$

We denote a point $u \in T^kM$ by $u = (x, y^{(1)}, \ldots, y^{(k)})$ and its coordinates by $(x^i, y^{(1)i}, \ldots, y^{(k)i})$.

A section of the bundle (T^kM, π^k, M) is a mapping $S : M \to T^kM$ satisfying $\pi^k \circ S = 1_M$. A local section S has the property $\pi^k \circ S_{|U} = 1_U$.

If $c : I \to M$ is a smooth curve, locally represented by $x^i = x^i(t)$, $t \in I$, then the mapping $\tilde{c} : I \to T^kM$ given by

$$x^i = x^i(t), \quad y^{(1)i} = \frac{1}{1!}\frac{dx^i}{dt}(t), \quad \ldots, \quad y^{(k)i} = \frac{1}{k!}\frac{d^{(k)}x^i}{dt^k}(t), \quad t \in I, \qquad (4.1.6)$$

is *the extension of order k* to T^kM of c. We have $\pi^k \circ \tilde{c} = c$.

If the differentiable manifold M is paracompact, then T^kM is a paracompact manifold.

We shall use the manifold $\widetilde{TM} = T^kM \setminus \{0\}$, where 0 is the null section of π^k.

4.2. *Liouville vector fields*

The natural basis at a point $u \in T^kM$ of $T_u(T^kM)$ is given by

$$\left(\frac{\partial}{\partial x^i}, \frac{\partial}{\partial y^{(1)i}}, \ldots, \frac{\partial}{\partial y^{(k)i}}\right)_u.$$

A change of local coordinates changing (4.1.5) transforms the natural basis by the following rule:

$$\begin{aligned}
\frac{\partial}{\partial x^i} &= \frac{\partial \tilde{x}^j}{\partial x^i}\frac{\partial}{\partial \tilde{x}^j} + \frac{\partial \tilde{y}^{(1)j}}{\partial x^i}\frac{\partial}{\partial \tilde{y}^{(1)j}} + \cdots + \frac{\partial \tilde{y}^{(k)j}}{\partial x^i}\frac{\partial}{\partial \tilde{y}^{(k)j}},\\
\frac{\partial}{\partial y^{(1)i}} &= \frac{\partial \tilde{y}^{(1)j}}{\partial y^{(1)i}}\frac{\partial}{\partial \tilde{y}^{(1)j}} + \cdots + \frac{\partial \tilde{y}^{(k)j}}{\partial y^{(1)i}}\frac{\partial}{\partial \tilde{y}^{(k)j}},\\
&\vdots\\
\frac{\partial}{\partial y^{(k)i}} &= \frac{\partial \tilde{y}^{(k)j}}{\partial y^{(k)i}}\frac{\partial}{\partial \tilde{y}^{(k)j}},
\end{aligned} \qquad (4.2.1)$$

at the point $u \in T^kM$.

The natural cobasis $(dx^i, dy^{(1)i}, \ldots, dy^{(k)i})_u$ is transformed by (4.1.5) as follows:

$$d\tilde{x}^i = \frac{\partial \tilde{x}^i}{\partial x^j}dx^j,$$

$$
\begin{aligned}
d\tilde{y}^{(1)i} &= \frac{\partial \tilde{y}^{(1)i}}{\partial x^j} dx^j + \frac{\partial \tilde{y}^{(1)i}}{\partial y^{(1)j}} dy^{(1)j}, \\
&\vdots \\
d\tilde{y}^{(k)i} &= \frac{\partial \tilde{y}^{(k)i}}{\partial x^j} dx^j + \frac{\partial \tilde{y}^{(k)i}}{\partial y^{(1)j}} dy^{(1)j} + \cdots + \frac{\partial \tilde{y}^{(k)i}}{\partial y^{(k)j}} dy^{(k)j}.
\end{aligned}
\tag{4.2.1'}
$$

The formulae (4.2.1) and (4.2.1′) allow to determine some important geometric object fields on the total space of accelerations bundle T^kM.

The vertical distribution V_1 is locally generated by the vector fields $\{\frac{\partial}{\partial y^{(1)i}}, \ldots, \frac{\partial}{\partial y^{(k)i}}\}$, $i = 1, \ldots, n$. V_1 is integrable and of dimension kn. The distribution V_2 locally generated by $\{\frac{\partial}{\partial y^{(2)i}}, \ldots, \frac{\partial}{\partial y^{(k)i}}\}$ is also integrable, of dimension $(k-1)n$ and it is a subdistribution of V_1. This procedure may be continued.

The distribution V_k locally generated by $\{\frac{\partial}{\partial y^{(k)i}}\}$ is integrable and of dimension n. It is a subdistribution of the distribution V_{k-1}. We have the following sequence:

$$
V_1 \supset V_2 \supset \cdots \supset V_k.
$$

Using again (4.2.1), we deduce:

THEOREM 4.2.1. *The following operators in the algebra of functions $\mathcal{F}(T^kM)$:*

$$
\begin{aligned}
\overset{1}{\Gamma} &= y^{(1)i} \frac{\partial}{\partial y^{(k)i}}, \\
\overset{2}{\Gamma} &= y^{(1)i} \frac{\partial}{\partial y^{(k-1)i}} + 2y^{(2)i} \frac{\partial}{\partial y^{(k)i}}, \\
&\vdots \\
\overset{k}{\Gamma} &= y^{(1)i} \frac{\partial}{\partial y^{(1)i}} + 2y^{(2)i} \frac{\partial}{\partial y^{(2)i}} + \cdots + ky^{(k)i} \frac{\partial}{\partial y^{(k)i}}
\end{aligned}
\tag{4.2.2}
$$

are vector fields on T^kM. They are independent on the manifold $\widetilde{T^kM}$ and $\overset{1}{\Gamma} \subset V_k$, $\overset{2}{\Gamma} \subset V_{k-1}$, ..., $\overset{k}{\Gamma} \subset V_1$.

The vector fields $\overset{1}{\Gamma}, \overset{2}{\Gamma}, \ldots, \overset{k}{\Gamma}$ are called the *Liouville vector fields*.

THEOREM 4.2.2. *For any function $L \in \mathcal{F}(\widetilde{T^kM})$, the following entries are 1-forms on the manifold $\widetilde{T^kM}$:*

$$
d_0 L = \frac{\partial L}{\partial y^{(k)i}} dx^i,
$$

$$d_1 L = \frac{\partial L}{\partial y^{(k-1)i}} dx^i + \frac{\partial L}{\partial y^{(k)i}} dy^{(1)i},$$

$$\vdots$$

$$d_k L = \frac{\partial L}{\partial x^i} dx^i + \frac{\partial L}{\partial y^{(1)i}} dy^{(1)i} + \cdots + \frac{\partial L}{\partial y^{(k)i}} dy^{(k)i}. \tag{4.2.3}$$

Clearly, $d_k L = dL$.

In applications, we also use the nonlinear operator

$$\Gamma = y^{(1)i} \frac{\partial}{\partial x^i} + 2y^{(2)i} \frac{\partial}{\partial y^{(1)i}} + \cdots + k y^{(k)i} \frac{\partial}{\partial y^{(k-1)i}}. \tag{4.2.4}$$

It is not a vector field on $\widetilde{T^k M}$.

DEFINITION 4.2.1. A k-tangent structure J on $T^k M$ is an $\mathcal{F}(T^k M)$-linear mapping $J : \mathcal{X}(T^k M) \to \mathcal{X}(T^k M)$:

$$J\left(\frac{\partial}{\partial x^i}\right) = \frac{\partial}{\partial y^{(1)i}}, \quad J\left(\frac{\partial}{\partial y^{(1)i}}\right) = \frac{\partial}{\partial y^{(2)i}}, \quad \ldots,$$
$$J\left(\frac{\partial}{\partial y^{(k-1)i}}\right) = \frac{\partial}{\partial y^{(k)i}}, \quad J\left(\frac{\partial}{\partial y^{(k)i}}\right) = 0. \tag{4.2.5}$$

It is not difficult to see that J has the properties:

(1) J is globally defined on $T^k M$,

(2) J is an integrable structure,

(3) J is locally expressed by

$$J = \frac{\partial}{\partial y^{(1)i}} \otimes dx^i + \frac{\partial}{\partial y^{(2)i}} \otimes dy^{(1)i} + \cdots + \frac{\partial}{\partial y^{(k)i}} \otimes dy^{(k-1)i}, \tag{4.2.6}$$

(4) $\operatorname{Im} J = \operatorname{Ker} J$, $\operatorname{Ker} J = V_k$,

(5) $\operatorname{rank} J = kn$,

(6) $J \overset{k}{\Gamma} = \overset{k-1}{\Gamma}, \ldots, J \overset{2}{\Gamma} = \overset{1}{\Gamma}, J \overset{1}{\Gamma} = 0$,

(7) $J \circ J \circ \cdots \circ J = 0$ ($k+1$ factors).

In the next subsection, we shall use the functions

$$I^1(L) = \mathcal{L}_{\overset{1}{\Gamma}} L, \quad \ldots, \quad I^k(L) = \mathcal{L}_{\overset{k}{\Gamma}} L, \quad \forall L \in \mathcal{F}(T^k M), \tag{4.2.6}$$

where $\mathcal{L}_{\overset{\alpha}{\Gamma}}$ is the operator of Lie derivation with respect to the Liouville vector field $\overset{\alpha}{\Gamma}$.

The functions $I^1(L), \ldots, I^k(L)$ are called the *main invariants* of the function L. They play an important role in the variational calculus.

4.3. *Variational problem*

DEFINITION 4.3.1. A differentiable Lagrangian of order k is a mapping $L:(x, y^{(1)}, \dots, y^{(k)}) \in T^k M \to L(x, y^{(1)}, y^{(k)}) \in \mathbb{R}$, differentiable on $\widetilde{T^k M}$ and continuous on the null section $0: M \to \widetilde{T^k M}$ of the bundle $(T^k M, \pi^k, M)$.

If $c: t \in [0,1] \to (x^i(t)) \in U \subset M$ is a curve, with extremities $c(0) = (x^i(0))$ and $c(1) = (x^i(1))$ and $\tilde{c}: [0,1] \to \widetilde{T^k M}$ is its extension, then the integral of action of $L \circ \tilde{c}$ is defined by

$$I(c) = \int_0^1 L\left(x(t), \frac{dx}{dt}(t), \dots, \frac{1}{k!}\frac{d^k x}{dt^k}(t)\right) dt. \tag{4.3.1}$$

REMARK. One proves [16] that if $I(c)$ does not depend on the parametrization of the curve c, then the following Zermelo conditions hold:

$$I^1(L) = \dots = I^{k-1}(L) = 0, \qquad I^k(L) = L. \tag{4.3.2}$$

Generally, these conditions are not satisfied.

The variational problem involving the functional $I(c)$ from (4.3.1) will be studied as a natural extension of the theory exposed in Section 2.2.

On the open set U, we consider the curves

$$c_\varepsilon: t \in [0,1] \to \left(x^i(t) + \varepsilon V^i(t)\right) \in M, \tag{4.3.3}$$

where ε is a real number, sufficiently small in absolute value such that $\operatorname{Im} c_\varepsilon \subset U$, $V^i(t) = V^i(x(t))$ being a regular vector field on U, restricted to c. We assume all curves c_ε have the same end points $c(0)$ and $c(1)$ and their osculating spaces of order $1, 2, \dots, k-1$ coincide at the points $c(0), c(1)$. This means:

$$\begin{aligned} &V^i(0) = V^i(1) = 0, \\ &\frac{d^\alpha V^i}{dt^\alpha}(0) = \frac{d^\alpha V^i}{dt^\alpha}(1) = 0 \quad (\alpha = 1, \dots, k-1). \end{aligned} \tag{4.3.3'}$$

The integral of action $I(c_\varepsilon)$ of the Lagrangian L is:

$$I(c_\varepsilon) = \int_0^1 L\left(x + \varepsilon V, \frac{dx}{dt} + \varepsilon\frac{dV}{dt}, \dots, \frac{1}{k!}\left(\frac{d^k x}{dt^k} + \varepsilon\frac{d^k V}{dt^k}\right)\right) dt. \tag{4.3.4}$$

A necessary condition for $I(c)$ to be an extremal value for $I(c_\varepsilon)$ is

$$\left.\frac{dI(c_\varepsilon)}{d\varepsilon}\right|_{\varepsilon=0} = 0. \tag{4.3.5}$$

Thus, we have

$$\frac{dI(c_\varepsilon)}{d\varepsilon} = \int_0^1 \frac{d}{d\varepsilon} L\left(x+\varepsilon V, \frac{dx}{dt}+\varepsilon\frac{dV}{dt}, \dots, \frac{1}{k!}\left(\frac{d^k x}{dt^k}+\varepsilon\frac{d^k V}{dt^k}\right)\right) dt.$$

The Taylor expansion of L at $\varepsilon = 0$, gives

$$\left.\frac{dI(c_\varepsilon)}{d\varepsilon}\right|_{\varepsilon=0} = \int_0^1 \left(\frac{\partial L}{\partial x^i}V^i + \frac{\partial L}{\partial y^{(1)i}}\frac{dV^i}{dt} + \dots + \frac{1}{k!}\frac{\partial L}{\partial y^{(k)i}}\frac{d^k V^i}{dt^k}\right) dt. \qquad (4.3.7)$$

Now, using the notations

$$\overset{\circ}{E}_i(L) := \frac{\partial L}{\partial x^i} - \frac{d}{dt}\frac{\partial L}{\partial y^{(1)i}} + \dots + (-1)^k \frac{1}{k!}\frac{d^k}{dt^k}\frac{\partial L}{\partial y^{(k)i}} \qquad (4.3.8)$$

and

$$I_V^1 L = V^i \frac{\partial L}{\partial y^{(k)i}}, \quad I_V^2(L) = V^i \frac{\partial L}{\partial y^{(k-1)i}} + \frac{dV^i}{dt}\frac{\partial L}{\partial y^{(k)i}}, \quad \dots,$$

$$I_V^k = V^i \frac{\partial L}{\partial y^{(1)i}} + \frac{dV^i}{dt}\frac{\partial L}{\partial y^{(2)i}} + \dots + \frac{1}{(k-1)!}\frac{d^{k-1}V^v}{dt^{k-1}}\frac{\partial L}{\partial y^{(k)i}}, \qquad (4.3.9)$$

we obtain the identity

$$\frac{\partial L}{\partial x^i}V^i + \frac{\partial L}{\partial y^{(1)i}}\frac{dV^i}{dt} + \dots + \frac{1}{k!}\frac{\partial L}{\partial y^{(k)i}}\frac{d^k V^i}{dt^k}$$
$$= \overset{\circ}{E}_i(L) + \frac{d}{dt}\left\{I_V^k(L) - \frac{1}{2!}\frac{d}{dt}I_V^{k-1}(L) + \dots + (-1)^{k-1}\frac{1}{k!}\frac{d^{k-1}}{dt^{k-1}}I_V^1(L)\right\}. \qquad (4.3.10)$$

Applying (4.3.7) and taking account of (4.3.10) and (4.3.3′), with

$$I_V^\alpha(L)\big(c(0)\big) = I_V^\alpha(L)c(1) = 0 \quad (\alpha = 1, 2, \dots, k),$$

we get

$$\left.\frac{dI(c_\varepsilon)}{d\varepsilon}\right|_{\varepsilon=0} = \int_0^1 \overset{0}{E}_i(L)V^i\,dt. \qquad (4.3.11)$$

Since $V^i(t)$ is an arbitrary vector field, Eqs. (4.3.5) and (4.3.11) lead to the following result.

THEOREM 4.3.1. *In order that the integral of action $I(c)$ to be an extremal value for the functionals $I(c_\varepsilon)$, given by* (4.3.4), *is necessary that the following Euler–Lagrange equations hold*:

$$\begin{cases} \overset{0}{E}_i(L) := \frac{\partial L}{\partial x^i} - \frac{d}{dt}\frac{\partial L}{\partial y^{(1)i}} + \cdots + (-1)^k \frac{1}{k!}\frac{d^k}{dt^k}\frac{\partial L}{\partial y^{(k)}} = 0, \\ y^{(1)i} = \frac{dx^i}{dt}, \ldots, y^{(k)i} = \frac{1}{k!}\frac{d^k x^i}{dt^k}. \end{cases} \tag{4.3.12}$$

One proves [16] that $\overset{0}{E}_i(L)$ is a covector field. Consequently the equation $\overset{0}{E}_i(L) = 0$ has a geometrical meaning.

Consider the scalar field

$$\mathcal{E}^k(L) = I^k(L) - \frac{1}{2!}\frac{d}{dt}I^{k-1}(L) + \cdots + (-1)^{k-1}\frac{1}{k!}\frac{d^{k-1}}{dt^{k-1}}I^1(L) - L. \tag{4.3.13}$$

It is called the *energy of order* k of the Lagrangian L.

THEOREM 4.3.2 [16]. *For any Lagrangian $L(x, y^{(1)}, \ldots, y^{(k)})$, the energy of order k, $\mathcal{E}^k(L)$, is conserved along every solution of the Euler–Lagrange equations $\overset{0}{E}_i(L) = 0$, $y^{(1)i} = \frac{dx^i}{dt}, \ldots, y^{(k)i} = \frac{1}{k!}\frac{d^k x^i}{dt^k}$.*

REMARK. Introducing the notion of energy of order $1, 2, \ldots, k-1$, we can prove a Nöther theorem for the Lagrangians of order k.

Now we remark that for any C^∞-function $\phi(t)$ and any differentiable Lagrangian $L(x, y^{(1)}, \ldots, y^{(k)})$, the following equality holds:

$$\overset{0}{E}_i(\phi L) = \phi\, \overset{0}{E}_i(L) + \frac{d\phi}{dt}\, \overset{1}{E}_i(L) + \cdots + \frac{d^k\phi}{dt^k}\, \overset{k}{E}_i(L), \tag{4.3.14}$$

where $\overset{1}{E}_i(L), \ldots, \overset{k}{E}_i(L)$ are d-covector fields (called Graig–Synge covectors [16]). We consider the covector

$$\overset{k-1}{E}_i(L) = (-1)^{k-1}\frac{1}{(k-1)!}\left(\frac{\partial L}{\partial y^{(k-1)i}} - \frac{d}{dt}\frac{\partial L}{\partial y^{(k)i}}\right). \tag{4.3.15}$$

It is important in the theory of k-semisprays on Lagrange spaces of order k.

The Hamilton–Jacobi equations of a space $L^n = (M, L(x, y))$ introduced in Section 2.4 can be extended to higher-order Lagrange spaces by using the Jacobi–Ostrogradski momenta. Indeed, the energy of order k, $\mathcal{E}^k(L)$, is a polynomial function in $\frac{dx^i}{dt}, \ldots, \frac{d^k x^i}{dt^k}$, given by

$$\mathcal{E}^k(L) = p_{(1)i}\frac{dx^i}{dt} + p_{(2)i}\frac{d^2x^i}{dt^2} + \cdots + p_{(k)i}\frac{d^k x^i}{dt^k} - L, \tag{4.3.16}$$

where

$$
\begin{aligned}
p_{(1)i} &= \frac{\partial L}{\partial y^{(1)i}} - \frac{1}{2!}\frac{d}{dt}\frac{\partial L}{\partial y^{(2)i}} + \cdots + (-1)^{k-1}\frac{1}{k!}\frac{d^{k-1}}{dt^{k-1}}\frac{\partial L}{\partial y^{(k)i}},\\
p_{(2)i} &= \frac{1}{2!}\frac{\partial L}{\partial y^{(2)i}} - \frac{1}{3!}\frac{d}{dt}\frac{\partial L}{\partial y^{(3)i}} + \cdots + (-1)^{k-2}\frac{1}{k!}\frac{d^{k-2}}{dt^{k-2}}\frac{\partial L}{\partial y^{(k)i}},\\
&\vdots\\
p_{(k)i} &= \frac{1}{k!}\frac{\partial L}{\partial y^{(k)i}}.
\end{aligned}
\tag{4.3.17}
$$

$p_{(1)i}, \ldots, p_{(k)i}$ are called *the Jacobi–Ostrogradski momenta.*

The following important result has been established by M. de Léon and others [16].

THEOREM 4.3.3. *Along the integral curves of the Euler–Lagrange equations* $\overset{0}{E}_i(L) = 0$, *the following Hamilton–Jacobi–Ostrogradski equations hold*:

$$
\begin{aligned}
&\frac{\partial \mathcal{E}^k(L)}{\partial p_{(\alpha)i}} = \frac{d^\alpha x^i}{dt^\alpha} \quad (\alpha = 1, \ldots, k),\\
&\frac{\partial \mathcal{E}^k(L)}{\partial x^i} = -\frac{dp_{(1)i}}{dt},\\
&\frac{1}{\alpha!}\frac{\partial \mathcal{E}^k(L)}{\partial y^{(\alpha)i}} = -\frac{dp_{(\alpha+1)i}}{dt} \quad (\alpha = 1, \ldots, k-1).
\end{aligned}
$$

REMARK. The Jacobi–Ostrogradski momenta determine the 1-forms:

$$
\begin{aligned}
p_{(1)} &= p_{(1)i}\,dx^i + p_{(2)i}\,dy^{(1)i} + \cdots + p_{(k)i}\,dy^{(k-1)i},\\
p_{(2)} &= p_{(2)i}\,dx^i + p_{(3)i}\,dy^{(1)i} + \cdots + p_{(k)i}\,dy^{(k-2)i},\\
&\vdots\\
p_{(k)} &= p_{(k)i}\,dx^i.
\end{aligned}
$$

4.4. *Semisprays. Nonlinear connections*

A vector field $S \in \chi(T^kM)$ with the property

$$
JS = \overset{k}{\Gamma}
\tag{4.4.1}
$$

is called a *k-semispray* on T^kM. S is uniquely written in the form

$$S = y^{(1)i}\frac{\partial}{\partial x^i} + \cdots + ky^{(k)i}\frac{\partial}{\partial y^{(k-1)i}} - (k+1)G^i\left(x, y^{(1)}, \ldots, y^{(k)}\right)\frac{\partial}{\partial y^{(k)i}}, \tag{4.4.2}$$

or shortly,

$$S = \Gamma - (k+1)G^i\frac{\partial}{\partial y^{(k)i}}. \tag{4.4.2$'$}$$

The set of functions G^i is the set of the *coefficients* of S. With respect to (4.1.5), Section 4.1, G^i are transformed by

$$(k+1)\tilde{G}^i = (k+1)G^j\frac{\partial \tilde{x}^i}{\partial x^j} - \Gamma\tilde{y}^{(k)i}. \tag{4.4.3}$$

A curve $c : I \to M$ is called a *k-path* on M with respect to S if its extension $\tilde{c}$ is an integral curve of S. A k-path is characterized by the $(k+1)$-differential equation:

$$\frac{d^{k+1}x^i}{dt^{k+1}} + (k+1)G^i\left(x, \frac{dx}{dt}, \ldots, \frac{1}{k!}\frac{d^k x}{dt^k}\right) = 0. \tag{4.4.4}$$

We shall show that a k-semispray determines the main geometrical object fields on T^kM, as: the nonlinear connection N, the N-linear connection D and their structure equations. The connections N and D are basic for the geometry of the manifold T^kM.

DEFINITION 4.4.1. A subbundle HT^kM of the tangent bundle $(TT^kM, d\pi^k, T^kM)$ complementary to the vertical subbundle V_1T^kM:

$$TT^kM = HT^kM \oplus V_1T^kM \tag{4.4.5}$$

is called a *nonlinear connection.*

The fibres of HT^kM determine a horizontal distribution

$$N : u \in T^kM \to N_u = H_uT^kM \subset T_uT^kM, \quad \forall u \in T^kM,$$

complementary to the vertical distribution V_1, i.e.,

$$T_uT^kM = N_u \oplus V_{1,u}, \quad \forall u \in T^kM. \tag{4.4.5$'$}$$

If the base manifold M is paracompact, then there exist nonlinear connections on T^kM. The local dimension of N is $n = \dim M$.

Consider a nonlinear connection N and denote by h and v the horizontal and vertical projectors with respect to N and V_1:

$$h + v = I, \quad hv = vh = 0, \quad h^2 = h, \quad v^2 = v.$$

As usual we denote

$$X^H = hX, \quad X^V = vX, \quad \forall X \in \chi(T^k M).$$

A horizontal lift with respect to N is a $\mathcal{F}(M)$-linear mapping $l_h : \mathcal{X}(M) \to \mathcal{X}(T^k M)$ which has the properties

$$v \circ l_h = 0, \qquad d\pi^k \circ l_h = I_d.$$

There exists a unique local basis adapted to the horizontal distribution N. It is given by

$$\frac{\delta}{\delta x^i} = l_h\left(\frac{\partial}{\partial x^i}\right) \quad (i = 1, \dots, n). \tag{4.4.6}$$

The linearly independent vector fields of this basis can be uniquely written in the form:

$$\frac{\delta}{\delta x^i} = \frac{\partial}{\partial x^i} - \underset{(1)}{N}{}^j_i \frac{\partial}{\partial y^{(1)j}} - \dots - \underset{(k)}{N}{}^j_i \frac{\partial}{\partial y^{(k)j}}. \tag{4.4.7}$$

The systems $(\underset{(1)}{N}{}^j_i, \dots, \underset{(k)}{N}{}^j_i)$ of differential functions on $T^k M$ gives the *coefficients* of the nonlinear connection N.

By means of (4.4.6), it follows that:

PROPOSITION 4.4.1. *With respect to a change of local coordinates on the manifold $T^k M$, we have*

$$\frac{\delta}{\delta x^i} = \frac{\partial \tilde{x}^j}{\partial x^i} \frac{\delta}{\delta \tilde{x}^j}, \tag{4.4.7$'$}$$

and

$$\underset{(1)}{\tilde{N}}{}^i_m \frac{\partial \tilde{x}^m}{\partial x^j} = \underset{(1)}{N}{}^m_j \frac{\partial \tilde{x}^i}{\partial x^m} - \frac{\partial \tilde{y}^{(1)i}}{\partial x^j},$$

$$\vdots$$

$$\underset{(k)}{\tilde{N}}{}^i_m \frac{\partial \tilde{x}^m}{\partial x^j} = \underset{(k)}{N}{}^m_j \frac{\partial \tilde{x}^i}{\partial x^m} + \dots + \underset{(1)}{N}{}^m_j \frac{\partial \tilde{y}^{(k-1)i}}{\partial x^m} - \frac{\partial \tilde{y}^{(k)i}}{\partial x^j}. \tag{4.4.8}$$

REMARK. Equations (4.4.8) characterize a nonlinear connection N with the coefficients $\underset{(1)}{N}{}^j_i, \dots, \underset{(k)}{N}{}^j_i$.

These considerations lead to an important result.

THEOREM 4.4.1 (I. Bucătaru [16]). *If S if a k-semispray on T^kM, with the coefficients G^i, then the following system of functions*:

$$\underset{(1)}{N^i_j} = \frac{\partial G^i}{\partial y^{(k)i}}, \quad \underset{(2)}{N^i_j} = \frac{\partial G^i}{\partial y^{(k-1)i}}, \quad \ldots, \quad \underset{(k)}{N^i_j} = \frac{\partial G^i}{\partial y^{(1)j}} \tag{4.4.9}$$

gives the coefficients of a nonlinear connection N.

The k-tangent structure J, defined by (4.2.5), transforms the horizontal distribution $N_0 = N$ into a vertical distribution $N_1 \subset V_1$ of dimension n, complementary to the distribution V_2. Then it transforms the distribution N_1 into the distribution $N_2 \subset V_2$, complementary to the distribution V_3, and so on. Of course we have $\dim N_0 = \dim N_1 = \cdots = \dim N_{k-1} = n$.

Therefore we can write:

$$N_1 = J(N_0), \quad N_2 = J(N_1), \quad \ldots, \quad N_{k-1} = J(N_{k-2}), \tag{4.4.10}$$

and we obtain the direct decomposition:

$$T_u T^k M = N_{0,u} \oplus N_{1,u} \oplus \cdots \oplus N_{k-1,u} \oplus V_{k,u}, \quad \forall u \in T^k M. \tag{4.4.11}$$

An adapted basis to $N_0, N_1, \ldots, N_{k-1}, V_k$ is given by

$$\left\{ \frac{\delta}{\delta x^i}, \frac{\delta}{\delta y^{(1)i}}, \ldots, \frac{\delta}{\delta y^{(k-1)i}}, \frac{\partial}{\partial y^{(k)i}} \right\} \quad (i = 1, \ldots, n), \tag{4.4.12}$$

where

$$\begin{cases} \dfrac{\delta}{\delta y^{(1)i}} = \dfrac{\partial}{\partial y^{(1)i}} - \underset{(1)}{N^j_i} \dfrac{\partial}{\partial y^{(2)i}} - \cdots - \underset{(k-1)}{N^j_i} \dfrac{\partial}{\partial y^{(k)i}}, \\ \vdots \\ \dfrac{\delta}{\delta y^{(k-1)i}} = \dfrac{\partial}{\partial y^{(k-1)i}} - \underset{(1)}{N^j_i} \dfrac{\partial}{\partial y^{(k)j}}. \end{cases} \tag{4.4.13}$$

With respect to (4.1.4), we have

$$\frac{\delta}{\delta \tilde{y}^{(\alpha)i}} = \frac{\partial \tilde{x}^j}{\partial x^i} \frac{\delta}{\delta \tilde{y}^{(\alpha)j}} \quad \left(\alpha = 0, 1, \ldots, k, \ y^{(0)i} = x^i, \ \frac{\delta}{\delta y^{(k)i}} = \frac{\partial}{\partial y^{(k)i}} \right). \tag{4.4.14}$$

Let $h, v_1, \ldots, v_k$ be the projectors determined by (4.4.11):

$$h + \sum_1^k v_\alpha = I, \quad h^2 = h, \quad v_\alpha v_\alpha = v_\alpha, \quad h v_\alpha = 0;$$

$$v_\alpha h = 0, \quad v_\alpha v_\beta = v_\beta v_\alpha = 0 \quad (\alpha \neq \beta).$$

If we denote

$$X^H = hX, \quad X^{V_\alpha} = v_\alpha X, \quad \forall X \in \mathcal{X}(T^k M), \tag{4.4.15}$$

we have

$$X = X^H + X^{V_1} + \cdots + X^{V_k}. \tag{4.4.16}$$

With respect to the adapted basis (4.4.12), one has

$$X^H = X^{(0)i} \frac{\delta}{\delta x^i}, \quad X^{V_\alpha} = X^{(\alpha)i} \frac{\delta}{\delta y^{(\alpha)i}} \quad (\alpha = 1, \ldots, k).$$

THEOREM 4.4.2. *The nonlinear connection N is integrable if and only if*

$$[X^H, Y^H]^{V_\alpha} = 0, \quad \forall X, Y \in \chi(T^k M) \; (\alpha = 1, \ldots, k).$$

4.5. *The dual coefficients of a nonlinear connection*

Consider a nonlinear connection N, with the coefficients $(\underset{(1)}{N^i_j}, \ldots, \underset{(k)}{N^i_j})$. The dual basis of the adapted basis (4.4.12) has the form

$$(\delta x^i, \delta y^{(1)i}, \ldots, \delta y^{(k)i}), \tag{4.5.1}$$

where

$$\begin{cases} \delta x^i = dx^i, \\ \delta y^{(1)i} = dy^{(1)i} + \underset{(1)}{M^i_j}\, dx^j, \\ \vdots \\ \delta y^{(k)i} = dy^{(k)i} + \underset{(1)}{M^i_j}\, dy^{(k-1)j} + \cdots + \underset{(k-1)}{M^i_j}\, dy^{(1)j} + \underset{(k)}{M^i_j}\, dx^j, \end{cases} \tag{4.5.2}$$

and

$$\begin{cases} \underset{(1)}{M^i_j} = \underset{(1)}{N^i_j}, \quad \underset{(2)}{M^i_j} = \underset{(2)}{N^i_j} + \underset{(1)}{N^m_j}\, \underset{(1)}{M^i_m}, \quad \ldots, \\ \underset{(k)}{M^i_j} = \underset{(k)}{N^i_j} + \underset{(k-1)}{N^m_j}\, \underset{(1)}{M^i_m} + \cdots + \underset{(1)}{N^m_j}\, \underset{(k-1)}{M^i_m}. \end{cases} \tag{4.5.3}$$

The system of functions $(\underset{(1)}{M^i_j}, \ldots, \underset{(k)}{M^i_j})$ is called the system of dual coefficients of the nonlinear connection N. If the dual coefficients of N are given, then we uniquely obtain from (4.5.3) the *primal* coefficients $(\underset{(1)}{N^i_j}, \ldots, \underset{(k)}{N^i_j},)$ of N.

With respect to (4.1.4), the dual coefficients of N are transformed by the rule

$$\underset{(1)}{M}{}^m_j \frac{\partial \tilde{x}^i}{\partial x^m} = \underset{(1)}{\tilde{M}}{}^i_m \frac{\partial \tilde{x}^m}{\partial x^j} + \frac{\partial \tilde{y}^{(1)i}}{\partial x^j},$$

$$\vdots$$

$$\underset{(k)}{M}{}^m_j \frac{\partial \tilde{x}^i}{\partial x^m} = \underset{(k)}{\tilde{M}}{}^i_m \frac{\partial \tilde{x}^m}{\partial x^j} + \underset{(k-1)}{\tilde{M}}{}^i_m \frac{\partial \tilde{y}^{(1)m}}{\partial x^j} + \cdots + \underset{(1)}{\tilde{M}}{}^i_m \frac{\partial \tilde{y}^{(k-1)m}}{\partial x^j} + \frac{\partial \tilde{y}^{(k)i}}{\partial x^j}. \tag{4.5.4}$$

These transformations of the dual coefficients characterize the nonlinear connection N.

THEOREM 4.5.1 (R. Miron, Gh. Atanasiu [16]). *For any k-semispray S with the coefficients G^i, the functions*

$$\underset{(1)}{M}{}^i_j = \frac{\partial G^i}{\partial y^{(k)j}}, \quad \underset{(2)}{M}{}^i_j = \frac{1}{2}\Big(S \underset{(1)}{M}{}^i_j + \underset{(1)}{M}{}^i_m \underset{(1)}{M}{}^m_j\Big), \quad \ldots,$$

$$\underset{(k)}{M}{}^i_j = \frac{1}{k}\Big(S \underset{(k-1)}{M}{}^i_j + \underset{(1)}{M}{}^i_m \underset{(k-1)}{M}{}^m_j\Big) \tag{4.5.5}$$

are the dual coefficients of a nonlinear connection, which depends only on the k-semispray S.

As an application we can prove

THEOREM 4.5.2.

(I) *With respect to an adapted basis* (4.4.12), *the Liouville vector fields $\overset{1}{\Gamma}, \ldots, \overset{k}{\Gamma}$ can be expressed by*

$$\overset{1}{\Gamma} = z^{(1)i} \frac{\delta}{\delta y^{(k)i}}, \qquad \overset{2}{\Gamma} = z^{(1)i} \frac{\delta}{\delta y^{(k-1)i}} + 2z^{(2)i} \frac{\delta}{\delta y^{(k)i}},$$

$$\vdots$$

$$\overset{k}{\Gamma} = z^{(1)i} \frac{\delta}{\delta y^{(1)i}} + 2z^{(2)i} \frac{\delta}{\delta y^{(2)i}} + \cdots + kz^{(k)i} \frac{\delta}{\delta y^{(k)i}}, \tag{4.5.6}$$

where

$$\begin{cases} z^{(1)i} = y^{(1)i}, \quad 2z^{(2)i} = 2y^{(2)i} + \underset{(1)}{M}{}^i_m y^{(1)m}, \quad \ldots, \\ kz^{(k)i} = ky^{(k)i} + (k-1) \underset{(1)}{M}{}^i_m y^{(k-1)m} + \cdots + \underset{(k-1)}{M}{}^i_m y^{(1)m}. \end{cases} \tag{4.5.7}$$

(II) *With respect to* (4.1.4), *we have*

$$\tilde{z}^{(\alpha)i} = \frac{\partial \tilde{x}^i}{\partial x^j} z^{(\alpha)j} \quad (\alpha = 1, \ldots, k). \tag{4.5.7'}$$

This is the reason for which we call $z^{(1)i}, \ldots, z^{(k)i}$ *the distinguished Liouville vector fields* (shortly, *d-vector fields*). These vectors are important in the geometry of the manifold $T^k M$.

A 1-form $\omega \in \mathcal{X}^*(T^k M)$ is uniquely written as

$$\omega = \omega^H + \omega^{V_1} + \cdots + \omega^{V_k},$$

where

$$\omega^H = \omega \circ h, \qquad \omega^{V_\alpha} = \omega \circ v_\alpha \quad (\alpha = 1, \ldots, k).$$

For any function $f \in \mathcal{F}(T^k M)$, the 1-form df is

$$df = (df)^H + (df)^{V_1} + \cdots + (df)^{V_k}.$$

With respect to an adapted cobasis, one has

$$(df)^H = \frac{\delta f}{\delta x^i} dx^i, \quad (df)^{V_\alpha} = \frac{\delta f}{\delta y^{(\alpha)i}} \delta y^{(\alpha)i} \quad (\alpha = 1, \ldots, k). \tag{4.5.8}$$

Let $\gamma : I \to T^k M$ be a parametrized curve, locally expressed by

$$x^i = x^i(t), \quad y^{(\alpha)i} = y^{(\alpha)i}(t) \quad (t \in I)\ (\alpha = 1, \ldots, k).$$

The tangent vector field $\frac{d\gamma}{dt}$ is given by

$$\frac{d\gamma}{dt} = \left(\frac{d\gamma}{dt}\right)^H + \left(\frac{d\gamma}{dt}\right)^{V_1} + \cdots + \left(\frac{d\gamma}{dt}\right)^{V_k}$$
$$= \frac{dx^i}{dt}\frac{\delta}{\delta x^i} + \frac{\delta y^{(1)i}}{dt}\frac{\delta}{\delta y^{(1)i}} + \cdots + \frac{\delta y^{(k)i}}{dt}\frac{\delta}{\delta y^{(k)i}}.$$

The curve γ is called horizontal if $\frac{d\gamma}{dt} = (\frac{d\gamma}{dt})^H$. It is characterized by the system of differential equations

$$x^i = x^i(t), \quad \frac{\delta y^{(1)i}}{dt} = 0, \quad \ldots, \quad \frac{\delta y^{(k)i}}{dt} = 0. \tag{4.5.9}$$

A horizontal curve γ is called an *autoparallel* curve of the nonlinear connection if $\gamma = \tilde{c}$, where $\tilde{\tilde{c}}$ is the extension of a curve $c : I \to M$.

The *autoparallel* curves of the nonlinear connection N are characterized by the system of differential equations

$$\frac{\delta y^{(1)i}}{dt}=0,\quad \ldots,\quad \frac{\delta y^{(k)i}}{dt}=0,$$
$$y^{(1)i}=\frac{dx^i}{dt},\quad \ldots,\quad y^{(k)i}=\frac{1}{k!}\frac{d^kx^i}{dt^k}. \tag{4.5.9'}$$

4.6. *Prolongation to the manifold T^kM of the Riemannian structures given on the base manifold M*

Applying the previous theory of nonlinear connections on the total space of the accelerations bundle T^kM, we can solve the classical problem of the prolongation of a Riemann (or pseudo-Riemann) structure g given on the base manifold M. This problem was stated by L. Bianchi and was studied by several remarkable mathematicians, as: E. Bompiani, Ch. Ehresmann, A. Morimoto and S. Kobayashi. But the solution of this problem, as well as the solution of the prolongation to T^kM of the Finsler or Lagrange structures have been recently given by R. Miron and Gh. Atanasiu [16]. We will expose it here with very few proofs.

Let $\mathcal{R}^n=(M,g)$ be a Riemann space, g being a Riemannian metric defined on the base manifold M, having the local coordinates $g_{ij}(x)$, $x\in U\subset M$. We extend g_{ij} to $\pi^{-1}(U)\subset T^kM$, by setting

$$\left(g_{ij}\circ\pi^k\right)(u)=g_{ij}(x),\quad \forall u\in\pi^{-1}(U),\ \pi^k(u)=x.$$

The functions $g_{ij}\circ\pi^k$ will be denoted by g_{ij}.

The problem of prolongation of the Riemannian structure g to T^kM may be formulated as follows.

Let g be a Riemannian structure g on the manifold M. Find a Riemannian structure G on T^kM such that G is determined only by g.

As usually, we denoted by $\gamma^i_{jk}(x)$ the Christoffel symbols of g.

THEOREM 4.6.1. *There exist nonlinear connections N on the manifold $\widetilde{T^kM}$ determined only by the given Riemannian structure $g(x)$. One of them has the following dual coefficients*:

$$\underset{(1)}{M^i_j}=\gamma^i_{jm}(x)y^{(1)m},$$

$$\underset{(2)}{M^i_j}=\frac{1}{2}\left\{\Gamma\underset{(1)}{M^i_j}+\underset{(1)}{M^i_m}\,\underset{(1)}{M^m_j}\right\},$$

$$\vdots$$

$$\underset{(k)}{M^i_j} = \frac{1}{k}\left\{\Gamma \underset{(k-1)}{M^i_j} + \underset{(1)}{M^i_m} \underset{(k-1)}{M^m_j}\right\}, \tag{4.6.1}$$

where Γ is the operator defined by (4.2.4).

REMARK. Γ can be substituted with any k-semispray S, since $\underset{(1)}{M^i_j}, \ldots, \underset{(k-1)}{M^i_j}$ do not depend on the variables y^k.

One proves that N is integrable if and only if the Riemann space $\mathcal{R}^n = (M, g)$ is locally flat.

Let consider the adapted cobasis $(\delta x^i, \delta y^{(1)i}, \ldots, \delta y^{(k)i})$ to the nonlinear connection N and to the vertical distributions $\underset{(1)}{N}, \ldots, \underset{(k-1)}{N}, V_k$. It depends on the dual coefficients (4.6.1), so it depends only on the structure $g(x)$.

Now, consider the following *lift* of $g(x)$ to T^kM:

$$G = g_{ij}(x)\, dx^i \otimes dx^j + g_{ij}(x)\, \delta y^{(1)i} \otimes \delta y^{(1)j} + \cdots + g_{ij}(x)\, \delta y^{(k)i} \otimes \delta y^{(k)j}. \tag{4.6.2}$$

THEOREM 4.6.2. *The pair* $\operatorname{Prol}^k \mathcal{R}^n = (\widetilde{T^kM, G})$ *is a Riemann space of dimension* $(k+1)n$, *whose metric G given by* (4.6.2) *depends only on the a priori given Riemann structure* $g(x)$.

The announced problem is solved.

We point-out some remarks:

(1) The d-Liouville vector fields $z^{(1)i}, \ldots, z^{(k)i}$, from (4.5.7) are constructed only by using the Riemannian structure g:

(2) The following function:

$$L\big(x, y^{(1)}, \ldots, y^{(k)}\big) = g_{ij}(x) z^{(k)i} z^{(k)j} \tag{4.6.3}$$

is a regular Lagrangian which depends only on the Riemann structure $g_{ij}(x)$.

(3) The previous theory holds good for pseudo-Riemannian structures $g_{ij}(x)$ too.

4.7. *N-linear connections on T^kM*

The notion of an N-linear connection on the manifold T^kM can be studied as a natural extension of that of N-linear connection on TM, defined in Section 1.4.

Let N be a nonlinear connection on T^kM having the primal coefficients $\underset{(1)}{N^i_j}, \ldots, \underset{(k)}{N^i_j}$ and the dual coefficients $\underset{(1)}{M^i_j}, \ldots, \underset{(k)}{M^i_j}$.

DEFINITION 4.7.1. A linear connection D on the manifold T^kM is called *distinguished* if D preserves by parallelism the horizontal distribution N. It is an N-linear connection if one has the following property:

$$DJ = 0. \tag{4.7.1}$$

THEOREM 4.7.1. *A linear connection D on T^kM is an N-linear connection if and only if*

$$\begin{cases} (D_X Y^H)^{V_\alpha} = 0 \quad (\alpha = 1, \dots, k), \\ (D_X Y^{V_\alpha})^H = 0, \quad (D_X Y^{V_\alpha})^{V_\beta} = 0 \quad (\alpha \neq \beta); \\ D_X(JY^H) = J D_X Y^H, \quad D_X(JV^\alpha) = J D_X V^\alpha. \end{cases} \tag{4.7.2}$$

Of course, for any N-linear connection D we have

$$Dh = 0, \quad Dv^\alpha = 0 \quad (\alpha = 1, \dots, k).$$

Since

$$D_X Y = D_{X^H} Y + D_{X^{V_1}} Y + \cdots + D_{X^{V_k}} Y,$$

setting

$$D_X^H = D_{X^H}, \quad D_X^{V_\alpha} = D_{X^{V_\alpha}} \quad (\alpha = 1, \dots, k),$$

we can write

$$D_X Y = D_X^H Y + D_X^{V_1} Y + \cdots + D_X^{V_k} Y. \tag{4.7.2}$$

The operators D^H, D^{V_α} are not covariant derivations, but they have similar properties with the covariant derivations. The notion of d-tensor fields can be introduced and studied exactly as in Section 1.3.

With respect to the adapted basis (4.4.12) and its adapted cobasis (4.5.1), we represent a d-tensor field of type (r, s) as

$$T = T^{i_1 \dots i_r}_{j_1 \dots j_s} \frac{\delta}{\delta x^{i_1}} \otimes \cdots \otimes \frac{\delta}{\delta y^{(k)i_r}} \otimes dx^{j_1} \otimes \cdots \otimes \delta y^{(k)j_s}. \tag{4.7.3}$$

A transformation of coordinates (4.1.4), determines the following transformation rules:

$$\tilde{T}^{i_1 \dots i_r}_{j_1 \dots j_s} = \frac{\partial \tilde{x}^{i_1}}{\partial x^{h_1}} \cdots \frac{\partial \tilde{x}^{i_r}}{\partial x^{hr}} \frac{\partial x^{k_1}}{\partial x^{j_1}} \cdots \frac{\partial x^{k_s}}{\partial x^{j_s}} T^{h_1 \dots h_r}_{k_1 \dots k_s}. \tag{4.7.3'}$$

Thus, $\{1, \frac{\delta}{\delta x^i}, \dots, \frac{\delta}{\delta y^{(k)i}}\}$ generate the tensor algebra of d-tensor fields.

The theory of N-linear connections, described in Section 1 for $k = 1$, can be extended step by step to N-linear connections on the manifold T^kM.

With respect to the adapted basis (4.4.12), an N-linear connection D has the following form:

$$D_{\frac{\delta}{\delta x^j}}\frac{\delta}{\delta x^i}=L^m_{ij}\frac{\delta}{\delta x^m},\quad D_{\frac{\delta}{\delta x^j}}\frac{\delta}{\delta y^{(\alpha)i}}=L^m_{ij}\frac{\delta}{\delta y^{(\alpha)m}}\quad (\alpha=1,\ldots,k),$$

$$D_{\frac{\delta}{\delta y^{(\beta)j}}}\frac{\delta}{\delta x^i}=\underset{(\beta)}{C^m_{ij}}\frac{\delta}{\delta x^m},\quad D_{\frac{\delta}{\delta y^{(\beta)j}}}\frac{\delta}{\delta y^{(\alpha)i}}=\underset{(\beta)}{C^m_{ij}}\frac{\delta}{\delta y^{(\alpha)m}}$$

$$(\alpha,\beta=1,\ldots,k). \tag{4.7.4}$$

The system of functions

$$D\Gamma(N)=\Big(L^m_{ij},\underset{(1)}{C^m_{ij}},\ldots,\underset{(k)}{C^m_{ij}}\Big) \tag{4.7.5}$$

represents *the coefficients* of D.

With respect to (4.1.4), L^m_{ij} are transformed by the same rule as the coefficients of a linear connection defined on the base manifold M. Other coefficients $\underset{(\alpha)}{C^h_{ij}}$ $(\alpha=1,\ldots,k)$ are transformed as the d-tensors of type $(1,2)$.

If T is a d-tensor field of type (r,s) given by (4.7.3) and $X=X^H=X^i\frac{\delta}{\delta x^i}$, then, by means of (4.7.4), one has

$$D^H_X T=X^m T^{i_1\ldots i_r}_{j_1\ldots j_s|m}\frac{\delta}{\delta x^{i_1}}\otimes\cdots\otimes\frac{\delta}{\delta y^{(k)i_r}}\otimes dx^{j_1}\otimes\cdots\otimes\delta y^{(k)j_s}, \tag{4.7.6}$$

where

$$T^{i_1\ldots i_r}_{j_1\ldots j_s|m}=\frac{\delta T^{i_1\ldots i_r}_{j_1\ldots j_s}}{\delta x^m}+L^{i_1}_{hm}T^{hi_2\ldots i_r}_{j_1\ldots j_s}+\cdots-L^h_{j_sm}T^{i_1\ldots i_r}_{j_1\ldots h}. \tag{4.7.7}$$

The operator "$|$" is be called *the h-covariant derivative.*

Consider the v_α-covariant derivatives $D^{V_\alpha}_X$, for $X=\overset{(\alpha)}{X^i}\frac{\delta}{\delta y^{(\alpha)i}}$ $(\alpha=1,\ldots,k)$. Then, (4.7.3) and (4.7.4) lead to

$$D^{V_\alpha}_X T=\overset{(\alpha)}{X^m}\,T^{i_1\ldots i_r}_{j_1\ldots j_s}\overset{(\alpha)}{|}_m\frac{\delta}{\delta x^{i_1}}\otimes\cdots\otimes\frac{\delta}{\delta y^{(k)i_r}}\otimes dx^{j_1}\otimes\cdots\otimes dx^j_s, \tag{4.7.8}$$

where

$$T^{i_1\ldots i_r}_{j_1\ldots j_s}\overset{(\alpha)}{|}_m=\frac{\delta T^{i_1\ldots i_r}_{j_1\ldots j_s}}{\delta y^{(\alpha)m}}+\underset{(\alpha)}{C^{i_1}_{hm}}\,T^{hi_2\ldots i_r}_{j_1(\alpha)j_s}+\cdots-\underset{(\alpha)}{C^h_{j_sm}}\,T^{i_1\ldots i_r}_{j_1\ldots j_{s-1}h}\quad(\alpha=1,\ldots,k).$$

The operators "$\overset{(\alpha)}{|}$", $\alpha=1,\ldots,k$, are called *v_α-covariant derivatives.*

Any operators "$_|$" or "$\overset{(\alpha)}{|}$" has the usual properties with respect to the sum and tensor product of d-tensors.

Now, with respect to the adapted basis (4.4.12), we can determine the torsion T and the curvature R of an N-linear connection D, following the same method as in the case $k = 1$.

We have the following torion d-tensors:

$$T^i_{jk} = L^i_{jk} - L^i_{kj}, \quad \underset{(\alpha)}{S^i_{jk}} = \underset{(\alpha)}{C^i_{jk}} - \underset{(\alpha)}{C^i_{kj}} = 0 \quad (\alpha = 1, \dots, k) \tag{4.7.9}$$

and curvature d-tensors

$$R^i_{h\,jm}, \quad \underset{(\alpha)}{P}{}^{\,i}_{h\,jm}, \quad \underset{(\beta\alpha)}{S}{}^{\,i}_{h\,jm} \quad (\alpha, \beta = 1, \dots, k). \tag{4.7.10}$$

The connection 1-forms of the N-linear connection D are

$$\omega^i_j = L^i_{jh}\, dx^h + \underset{(1)}{C^i_{jh}}\, \delta y^{(1)h} + \cdots + \underset{(k)}{C^i_{jh}}\, \delta y^{(k)h}. \tag{4.7.11}$$

THEOREM 4.7.2. *The structure equations of an N-linear connection D on the manifold T^kM are given by*

$$\begin{aligned} &d(dx^i) - dx^m \wedge \omega^i_m = -\overset{(0)}{\Omega}{}^i, \\ &d(\delta y^{(\alpha)i}) - \delta y^{(\alpha)m} \wedge \omega^i_m = -\overset{(\alpha)}{\Omega}{}^i, \\ &d\omega^i_j - \omega^m_j \wedge \omega^i_m = -\Omega^i_j, \end{aligned} \tag{4.7.12}$$

where $\overset{(0)}{\Omega}{}^i$, $\overset{(\alpha)}{\Omega}{}^i$ *are the torsion 2-forms (see* [16]) *and* Ω^i_j *are the curvature 2-forms*

$$\begin{aligned} \Omega^i_j = {}& \frac{1}{2} R^i_{j\,pq}\, dx^p \wedge dx^q \\ & + \sum_{\alpha=1}^{k} \underset{(\alpha)}{P^i_{j\,pq}}\, dx^p \wedge \delta y^{(\alpha)q} + \sum_{\alpha,\beta=1}^{k} \underset{(\alpha\beta)}{S^i_{j\,pq}}\, \delta y^{(\alpha)p} \wedge \delta y^{(\beta)q}. \end{aligned}$$

The Bianchi identities of D can be derived from (4.7.12).

The nonlinear connection N and the N-linear connection D allow to study the geometrical properties of the manifold T^kM equipped with these two geometrical object fields.

5. Lagrange spaces of higher order

The concept of higher-order Lagrange space was introduced and studied by the present author [16,17].

A Lagrange space of order k is defined as a pair $L^{(k)n} = (M, L)$, where $L: T^kM \to \mathbb{R}$ is a differentiable regular Lagrangian having the fundamental tensor of constant signature. Applying the variational problem to the integral of action of L, we determine a canonical k-semispray, a canonical nonlinear connection and a canonical metrical connection. All these notions are basic for the geometry of the space $L^{(k)n}$.

5.1. *The spaces* $L^{(k)n} = (M, L)$

DEFINITION 5.1.1. A Lagrange space of order $k \geqslant 1$ is a pair $L^{(k)m} = (M, L)$, where M is a real n-dimensional manifold M and $L: (x, y^{(1)}, \dots, y^{(k)}) \in T^kM \to L(x, y^{(1)}, \dots, y^{(k)}) \in \mathbb{R}$ a differentiable Lagrangian, such that its Hessian, whose elements are

$$g_{ij} = \frac{1}{2}\frac{\partial^2 L}{\partial y^{(k)i}\partial y^{(k)j}}, \tag{5.1.1}$$

satisfies

$$\operatorname{rank}(g_{ij}) = n \quad \text{on } \widetilde{T^kM} \tag{5.1.2}$$

and the d-tensor g_{ij} has a constant signature.

Of course, we can prove that g_{ij} is a symmetric d-tensor field of type $(0, 2)$. It is called the *fundamental* (or *metric*) tensor of the space $L^{(k)n}$, while L is called its *fundamental function*.

The geometry of the manifold T^kM equipped with $L(x, y^{(1)}, \dots, y^{(k)})$ is called the geometry of the space $L^{(k)n}$. We study this geometry using the theory from the previous section. Consequently, starting from the integral of action $I(c) = \int_0^1 L(x, \frac{dx}{dt}, \dots, \frac{1}{k!}\frac{d^kx}{dt^k})\,dt$, we determine the Euler–Lagrange equations $\overset{0}{E}_i(L) = 0$ and the Craig–Synge covectors $\overset{1}{E}_i(L), \dots, \overset{k}{E}_i(L)$. According to (4.3.15), one gets

THEOREM 5.1.1. *The equations* $g^{ij}\overset{k-1}{E}_i(L) = 0$ *determine a* k*-semispray*

$$S = y^{(1)i}\frac{\partial}{\partial x^i} + 2y^{(2)i}\frac{\partial}{\partial y^{(1)i}} + \dots + ky^{(k)i}\frac{\partial}{\partial y^{(k-1)i}} - (k+1)G^i\frac{\partial}{\partial y^{(k)i}}, \tag{5.1.3}$$

where the coefficients G^i *are given by*

$$(k+1)G^i = \frac{1}{2}g^{ij}\left\{\Gamma\left(\frac{\partial}{\partial y^{(k)i}}\right) - \frac{\partial}{\partial y^{(k-1)i}}\right\}, \tag{5.1.4}$$

and Γ *is the operator defined by* (4.2.4).

The semispray S depends only on the fundamental function L. S is called canonical and is globally defined on the manifold $\widetilde{T^kM}$.

Taking into account Theorem 4.5.1, we have

THEOREM 5.1.2. *The system of functions*

$$\underset{(1)}{M^i_j} = \frac{\partial G^i}{\partial y^{(k)j}}, \quad \underset{(2)}{M^i_j} = \frac{1}{2}\Big(S\underset{(1)}{M^i_j} + \underset{(1)}{M^i_m}\underset{(1)}{M^m_j}\Big), \quad \dots,$$

$$\underset{(k)}{M^i_j} = \frac{1}{k}\Big(S\underset{(k-1)}{M^i_j} + \underset{(1)}{M^i_m}\underset{(k-1)}{M^m_j}\Big) \tag{5.1.5}$$

are the dual coefficients of a nonlinear connection N determined only by the fundamental function L of the space $L^{(k)n}$.

N is the canonical nonlinear connection of $L^{(k)n}$.

The adapted basis $\{\frac{\delta}{\delta x^i}, \frac{\delta}{\delta y^{(1)i}}, \dots, \frac{\delta}{\delta y^{(k)i}}\}$ has its dual $\{\delta x^i, \delta y^{(1)i}, \dots, \delta y^{(k)i}\}$. They are constructed by the canonical nonlinear connection. So, the horizontal curves are characterized by the system of differential equations (4.5.9), and the autoparallel curves of N are given by (4.5.9′).

The condition that N be integrable is expressed by $[\frac{\delta}{\delta x^i}, \frac{\delta}{\delta x^j}]^{V_\alpha} = 0$ ($\alpha = 1, \dots, k$).

5.2. *Examples of spaces $L^{(k)n}$*

(1) Let consider the Lagrangian

$$L\big(x, y^{(1)}, \dots, y^{(k)}\big) = g_{ij}(x) z^{(k)i} z^{(k)j}, \tag{5.2.1}$$

where $g_{ij}(x)$ is a Riemannian (or pseudo-Riemannian) metric on the base manifold M and $z^{(k)i}$ is the Liouville d-vector field

$$k z^{(k)i} = k y^{(k)i} + (k-1)\underset{(1)}{M^i_m} y^{(k-1)m} + \dots + k \underset{(k-1)}{M^i_m} y^{(1)m}, \tag{5.2.2}$$

constructed by means of the dual coefficients of the canonical nonlinear connection N from the problem of prolongation to T^kM of $g_{ij}(x)$ (see (4.6.1)). Then the Lagrangian depends only on $g_{ij}(x)$.

The pair $L^{(k)n} = (M, L)$ is a Lagrange space of order k. Its fundamental tensor is $g_{ij}(x)$, because the d-vector $z^{(k)i}$ is linearly in the variables $y^{(k)}$'s.

(2) Let $\overset{\circ}{L}(x, y^{(1)})$ be the Lagrangian from Electrodynamics

$$\overset{\circ}{L}\big(x, y^{(1)}\big) = mc\gamma_{ij}(x) y^{(1)i} y^{(1)j} + \frac{2e}{m} b_i(x) y^{(1)i}. \tag{5.2.3}$$

$$\mathbb{F}^2(X) = -X + \sum_{a=1}^{n}\sum_{\alpha=1}^{k-1} \overset{\alpha a}{\eta}(X) \underset{\alpha a}{\xi}, \quad \forall X \in \mathcal{X}\big(\widetilde{T^k M}\big), \ \overset{\alpha a}{\eta} \circ \mathbb{F} = 0.$$

Let $N_{\mathbb{F}}$ be the Nijenhuis tensor of the structure $\mathbb{F}$,

$$N_{\mathbb{F}}(X, Y) = [\mathbb{F}X, \mathbb{F}Y] + \mathbb{F}^2[X, Y] - \mathbb{F}[\mathbb{F}X, Y] - \mathbb{F}[X, \mathbb{F}Y].$$

The structure (5.4.1) is said to be normal if

$$N_{\mathbb{F}}(X, Y) + \sum_{a=1}^{n}\sum_{\alpha=1}^{k-1} d\overset{\alpha a}{\eta}(X, Y) = 0, \quad \forall X, Y \in \mathcal{X}\big(T^k M\big).$$

A characterization of the normality of the structure $\mathbb{F}$ is given by the following

THEOREM 5.4.2. *The almost* $(k-1)n$*-contact structure* (5.4.2) *is normal if and only if, for any* $X, Y \in \mathcal{X}(\widetilde{T^k M})$, *we have*

$$N_{\mathbb{F}}(X, Y) + \sum_{a=1}^{n}\sum_{\alpha=1}^{k-1} d\big(\delta y^{(\alpha)a}\big)(X, Y) = 0.$$

The lift of the fundamental tensor g_{ij} of the space $L^{(k)n}$ with respect to N is defined by

$$\mathbb{G} = g_{ij}\, dx^i \otimes dx^j + g_{ij}\, \delta y^{(1)i} \otimes \delta y^{(1)j} + \cdots + g_{ij}\, \delta y^{(k)i} \otimes \delta y^{(k)j}. \tag{5.4.3}$$

Obviously, $\mathbb{G}$ is a pseudo-Riemannian structure on the manifold $\widetilde{T^k M}$, determined only by the space $L^{(k)n}$.

THEOREM 5.4.3. *The pair* $(\mathbb{G}, \mathbb{F})$ *is a Riemannian* $(k-1)n$*-almost contact structure on* $\widetilde{T^k M}$.

In this case, the following condition holds:

$$\mathbb{G}(\mathbb{F}X, Y) = -\mathbb{G}(\mathbb{F}Y, X), \quad \forall X, Y \in \mathcal{X}\big(\widetilde{T^k M}\big).$$

Therefore, the triple $(\widetilde{T^k M}, \mathbb{G}, \mathbb{F})$ is a metrical $(k-1)n$-almost contact space, called the *geometrical model of the Lagrange space* $L^{(k)n}$ of order k.

Using this space, we can study the electromagnetic and gravitational fields on the spaces $L^{(k)n}$ [16].

5.5. *Generalized Lagrange spaces of order k*

The notion of a generalized Lagrange space of higher order is a natural extension of that studied in Section 2.

DEFINITION 5.5.1. A generalized Lagrange space of order k is a pair $GL^{(k)n} = (M, g_{ij})$, where M is a real differentiable n-dimensional manifold and g_{ij} a symmetric d-tensor field of type $(0, 2)$ on $\widetilde{T^kM}$, having the properties:

(a) g_{ij} has a constant signature on $\widetilde{T^kM}$;

(b) $\operatorname{rank}(g_{ij}) = n$ on $\widetilde{T^kM}$.

g_{ij} is called the *fundamental tensor* of $GL^{(k)n}$.

Any Lagrange space of order k, $L^{(k)n} = (M, L)$, determines a space $GL^{(k)n}$ with fundamental tensor

$$g_{ij} = \frac{1}{2} \frac{\partial^2 L}{\partial y^{(k)i} \partial y^{(k)j}}. \tag{5.5.1}$$

The converse statement does not hold. If $g_{ij}(x, y^{(1)}, \ldots, y^{(k)})$ is a priori given, it is possible that the system of partial differential equations (5.5.1) does not admit any solution $L(x, y^{(1)}, \ldots, y^{(k)})$. A necessary condition that the system (5.5.1) admits solutions L is that the d-tensor field

$$\underset{(k)}{C}{}_{ijh} = \frac{1}{2} \frac{\partial g_{ij}}{\partial y^{(k)h}} \tag{5.5.2}$$

be completely symmetric.

If the system (5.5.1) has solutions with respect to L, we say that the space $GL^{(k)n}$ is reducible to a Lagrange space of order k. Otherwise, GL^n is said to be nonreducible to a Lagrange space $L^{(k)n}$.

EXAMPLES.

(1) Let $\mathcal{R} = (M, \gamma_{ij}(x))$ be a Riemannian space and $\sigma \in \mathcal{F}(T^kM)$. Consider the d-tensor field:

$$g_{ij} = e^{2\sigma} \left(\gamma_{ij} \circ \pi^k\right). \tag{5.5.3}$$

If $\frac{\partial \sigma}{\partial y^{(k)h}}$ is a nonvanishing d-covector on the manifold $\widetilde{T^kM}$, then the pair $GL^{(k)n} = (M, g_{ij})$ is a generalized Lagrange space of order k and it is not reducible to a Lagrange space $L^{(k)n}$.

(2) Let $\mathcal{R}^n = (M, \gamma_{ij}(x))$ be a Riemann space and $\operatorname{Prol}^k \mathcal{R}^n$ be its prolongation of order k to $\widetilde{T^kM}$.

Consider the Liouville d-vector field $z^{(k)i}$ of $\operatorname{Prol}^k \mathcal{R}^n$. It is expressed by the formula (4.5.7), Section 4.5. We can introduce the d-covector field $z_i^{(k)} = \gamma_{ij} z^{(k)j}$.

We assume that there exists a function $n(x, y^{(1)}, \ldots, y^{(k)}) \geqslant 1$ on $\widetilde{T^kM}$. Thus

$$g_{ij} = \gamma_{ij} + \left(1 - \frac{1}{n^2}\right) z_i^{(k)} z_j^{(k)} \tag{5.5.4}$$

is the fundamental tensor of a space $GL^{(k)n}$. This space is not reducible to a space $L^{(k)n}$, if the function $n \neq 1$.

These two examples prove the existence of generalized Lagrange spaces of order k.

In the last example, $k = 1$ leads to the metric (2.8.4) of the Relativistic Optics (n being the refractive index).

On a generalized Lagrange space $GL^{(k)n}$ is difficult to find a nonlinear connection N derived only by the fundamental tensor g_{ij}. Therefore, assuming that N is a priori given, we shall study the pair $(N, GL^{(k)n})$. Thus, a theorem of existence and uniqueness of metrical N-linear connections holds:

THEOREM 5.5.1.

(1) *There exists a unique N-linear connection D satisfying*

$$g_{ij|h} = 0, \quad g_{ij} \overset{(\alpha)}{|}_h = 0 \quad (\alpha = 1, \dots, k),$$
$$T^i_{jk} = 0, \quad \underset{(\alpha)}{S^i_{jk}} = 0 \quad (\alpha = 1, \dots, k).$$

(2) *The coefficients of D are given by the generalized Christoffel symbols* (*see* (5.3.2)).

(3) *D depends only on g_{ij} and N.*

Using this theorem, we may study the geometry of Generalized Lagrange spaces.

5.6.

The notion of Finsler spaces of order k, introduced by the present author, was investigated in the book *The Geometry of Higher-Order Finsler Spaces*, Hadronic Press, 1998. It is a natural extension to the manifold $\widetilde{T^k M}$ of the theory of Finsler spaces given in Section 3. A substantial contribution in the geometry of these spaces is due to H. Shimada and S. Sabău [17].

The impact of such spaces in Differential Geometry, Variational Calculus, Analytical Mechanics and Theoretical Physics is decisive. Finsler spaces play an important role in applications to Biology, Engineering, Physics or Optimal Control. Also, the introduction of the notion of Finsler space of order k is required by the solution of the problem of prolongation to $T^k M$ of the Riemannian or Finslerian structures defined on the base manifold M.

In order to introduce the Finsler space of order k we need some considerations on the concept of homogeneity of functions on the manifold $T^k M$ [17].

A function $f : T^k M \to \mathbb{R}$, C^∞-differentiable on $\widetilde{T^k M}$ and continuous on the null section of the bundle $\pi^k : T^k M \to M$, is called homogeneous of degree $r \in \mathbb{Z}$ on the fibres of $T^k M$ (briefly r-homogeneous) if for any $a \in \mathbb{R}^+$ we have

$$f\left(x, ay^{(1)}, a^2 y^{(2)}, \dots, a^k y^{(k)}\right) = a^r f\left(x, y^{(1)}, \dots, y^{(k)}\right).$$

The Euler theorem holds:

A function $f \in \mathcal{F}(T^k M)$*, differentiable on* $\widetilde{T^k M}$ *and continuous on the null section of* π^k *is* r*-homogeneous if and only if*

$$\mathcal{L}_{\overset{k}{\Gamma}} f = rf, \tag{5.6.1}$$

$\mathcal{L}_{\overset{k}{\Gamma}}$ *being the Lie derivative with respect to the Liouville vector field* $\overset{k}{\Gamma}$.

A vector field $X \in \mathcal{X}(T^k M)$ is r-homogeneous if

$$\mathcal{L}_{\overset{k}{\Gamma}} X = (r-1)X. \tag{5.6.1$'$}$$

DEFINITION 5.6.1. A Finsler space of order k, $k \geqslant 1$, is a pair $F^{(k)n} = (M, F)$, where M is a real differentiable manifold of dimension n and $F : T^k M \to \mathbb{R}$ a function having the following properties:

(1) F is differentiable on $\widetilde{T^k M}$ and continuous on the null section of the bundle $(T^k M, \pi^k, M)$.
(2) F is positive.
(3) F is k-homogeneous on the fibres of the bundle $T^k M$.
(4) The Hessian of F^2, having the elements

$$g_{ij} = \frac{1}{2} \frac{\partial^2 F^2}{\partial y^{(k)i} \partial y^{(k)j}}, \tag{5.6.2}$$

is positive definite on $\widetilde{T^k M}$.

It follows that *the fundamental tensor* g_{ij} is nonsingular and 0-homogeneous on the fibres of $T^k M$.

Also, we remark that any Finlser space $F^{(k)n}$ can be considered as a Lagrange space $L^{(k)n} = (M, L)$, whose fundamental function L is F^2.

By means of the solution of the problem of prolongation of a Finsler structure $F(x, y^{(1)})$ to $T^k M$, we can construct some important examples of spaces $F^{(k)n}$.

A Finsler space with the property g_{ij} depend only on the points $x \in M$ is called a Riemann space of order k and denoted by $\mathcal{R}^{(k)n}$.

Consequently, we have the following sequence of inclusions, similar with that from Section 3:

$$\{\mathcal{R}^{(k)n}\} \subset \{F^{(k)n}\} \subset \{L^{(k)n}\} \subset \{GL^{(k)n}\}. \tag{5.6.3}$$

We may say that the Lagrange geometry of order k is the geometrical theory of the sequence (5.6.3).

Of course the geometry of $F^{(k)n}$ can be studied as the geometry of the Lagrange space of order k, $L^{(k)n} = (M, F^2)$. Thus the canonical nonlinear connection N is *the Cartan nonlinear connection of* $F^{(k)n}$ and the metrical N-linear connection D is the Cartan N-metrical connection of the space $F^{(k)n}$ [16,17].

References

[1] P.L. Antonelli, ed., *Finslerian Geometries. A Meeting of Minds*, FTPH, vol. 109, Kluwer Academic (2000).

[2] P.L. Antonelli, R.S. Ingarden and M. Matsumoto, *The Theory of Sprays and Finsler Spaces with Applications in Physics and Biology*, FTPH, vol. 58, Kluwer Academic (1993).

[3] P.L. Antonelli and R. Miron, eds, *Lagrange and Finsler Geometry. Applications to Physics and Biology*, FTPH, vol. 76, Kluwer Academic (1996).

[4] G.S. Asanov, *Finsler Geometry, Relativity and Gauge Theories*, Reidel, Dordrecht (1985).

[5] D. Bao, S.S. Chern and Z. Shen, eds, *Finsler Geometry*, Contemporary Mathematics, vol. 196, Amer. Math. Soc. (1995).

[6] D. Bao, S.S. Chern and Z. Shen, *An Introduction to Riemann–Finsler Geometry*, Graduate Texts in Math., Springer-Verlag (2000).

[7] A. Bejancu, *Finsler Geometry and Applications*, Ellis Harwood (1990).

[8] M. Crampin, *On the inverse problem of the calculus of variations for systems of second-order ordinary differential equations*, Finslerian Geometry, FTPH, vol. 109, Kluwer Academic (2000), 139–151.

[9] M. De Leon and P.R. Rodrigues, *Generalized Classical Mechanics and Field Theory*, North-Holland (1985), 73–124.

[10] S. Ikeda, *Advanced Study in Applied Geometry*, Seizansha, Kanagawa (1995).

[11] J. Kern, *Lagrange geometry*, Arch. Math. **25** (1974), 438–443.

[12] O. Krupkova, *The Geometry of Ordinary Variational Equations*, Lecture Notes in Math., vol. 1678, Springer (1997).

[13] P. Libermann and C.-M. Marle, *Symplectic Geometry and Analytical Mechanics*, Reidel (1987).

[14] M. Matsumoto, *Foundations of Finsler Geometry and Special Finsler Spaces*, Kaisheisha Press, Otsu (1986).

[15] R. Miron, *Lagrange geometry*, Math. Comput. Modelling **20** (4/5) (1994).

[16] R. Miron, *The Geometry of Higher Order Lagrange Spaces. Applications to Mechanics and Physics*, FTPH, vol. 82, Kluwer Academic (1997).

[17] R. Miron, *The Geometry of Higher-Order Finsler Spaces*, Hadronic Press, USA (1998).

[18] R. Miron, *The Geometry of Higher-Order Hamilton Spaces. Applications to Hamiltonian Mechanics*, FTPH, vol. 132, Kluwer Academic (2003).

[19] R. Miron and M. Anastasiei, *The Geometry of Lagrange Spaces: Theory and Applications*, FTPH, vol. 59, Kluwer Academic (1994).

[20] R. Miron and M. Anastasiei, *Vector Bundles and Lagrange Spaces with Applications to Relativity*, vol. 1, Geometry Balkan Press, Bucureşti (1997).

[21] R. Miron, M. Anastasiei and I. Bucătaru, *The geometry of Lagrange spaces*, Handbook of Finsler Geometry, P.L. Antonelli, ed., Kluwer Academic (2003).

[22] R. Miron, D. Hrimiuc, H. Shimada and S.V. Sabău, *The Geometry of Hamilton and Lagrange Spaces*, FTPH, vol. 118, Kluwer Academic (2001).

[23] M. Puta, *Hamiltonian Mechanical Systems and Geometric Quantization*, Mathematics and Its Applications, vol. 260, Kluwer Academic (1993).

[24] H. Rund, *The Differential Geometry of Finsler Spaces*, Grundlehren Math. Wiss., vol. 101, Springer-Verlag (1959).

[25] D.J. Sanders, *The Geometry of Jet Bundles*, Cambridge Univ. Press, New York (1989).

[26] M.R. Santilli, *Foundations of Theoretical Mechanics I: The Inverse Problem in Newtonian Mechanics*, Springer-Verlag (1978).

[27] M.R. Santilli, *Foundations of Theoretical Mechanics II: Birkhoffian Generalization of Hamiltonian Mechanics*, Springer-Verlag (1981).

[28] J.L. Synge, *Relativity: General Theory*, North-Holland (1966).

[29] S. Vacaru, *Interactions, Strings and Isotopies in Higher Order Anisotropic Superspaces*, Hadronic Press, Palm Harbor, USA (1998).

[30] I. Vaisman, *Lectures on the Geometry of Poisson Manifolds*, Birkhäuser, Basel (1994).

[31] K. Yano and S. Ishihara, *Tangent and Cotangent Bundles. Differential Geometry*, Dekker, New York (1973).

CHAPTER 8

Certain Actual Topics on Modern Lorentzian Geometry*

Francisco J. Palomo**

Departamento de Matemática Aplicada, Universidad de Málaga, Complejo Tecnológico, Campus Teatinos, 29071-Málaga, Spain

E-mail: fjpalomo@ctima.uma.es

Alfonso Romero**

Departamento de Geometría y Topología, Universidad de Granada, 18071-Granada, Spain

E-mail: aromero@ugr.es

Contents

*Both authors are partially supported by MEC-FEDER Grant MTM2004-04934-C04-01.

**The authors express their deep thanks to Prof. Franki Dillen, from Katholieke Universiteit Leuven, for giving the opportunity to participate in this volume of the Handbook of Differential Geometry.

HANDBOOK OF DIFFERENTIAL GEOMETRY, VOL. II

Edited by F.J.E. Dillen and L.C.A. Verstraelen

Preface

Since Lorentz–Minkowski spacetime was extended to a curved spacetime by A. Einstein in order to model nonzero gravitational fields, Lorentzian geometry has been the mathematical theory which is used by general relativity. In the beginning, this geometry had essentially a local character, perhaps under the influence of a first great interest for accurate descriptions of our nearest universe. The important contributions to singularity theory of S.W. Hawking, R. Penrose, R.P. Geroch and others, in the 70's, made use of sophisticated global techniques inspired on an extraordinarily developed Riemannian geometry, allowed Lorentzian geometry to pass from language and terminology to the basic tool to predict the behavior of the entire universe in the past and the future. This situation was a great incentive for the development and advance of new techniques in the study of cosmological models more and more adapted to the physical reality. Now, Lorentzian geometry is a very active research area of differential geometry with specific problems of purely geometric nature and interest.

Our main aim with these lecture notes is to give an extensive panoramic view of the research on four relevant topics on Lorentzian geometry. From natural limitations of time and space, we have to make a choice for our contribution. This has not been an easy job, because the number of interesting research areas in Lorentzian geometry is impressive and increases every year. For convenience of the reader, we have dealt with important basic results before introducing several current problems and open questions. We have sometimes compared several Lorentzian results to Riemannian or indefinite (non-Lorentzian) ones, emphasizing on mathematical behaviors which are specific of Lorentzian geometry. Definitively, we have tried to show the reader a part of the wonderful world of research in Lorentzian geometry.

1. Some aspects on the topology of Lorentzian manifolds

A Lorentzian manifold is an $n(\geqslant 2)$-dimensional semi-Riemannian manifold with index one; that is with signature $(+,\dots,+,-)$. An n-dimensional semi-Riemannian manifold of index s such that $0 < s < n$ is said to be indefinite.

1.1. *Lorentzian metrics and line fields*[1]

It should be noted that if a manifold in the terminology of [51] is considered; i.e. without the assumption of having a countable basis in its topology, then it can be shown [57] that

PROPOSITION. *If a manifold (in the previous sense) admits a Lorentzian metric, then it must be paracompact.*

[1]This subsection is based on a talk [22] given by the second author in the Seminar of Geometry of Kyungpook National University, Taegu, Korea, in November, 1998.

More generally, it was shown in [84, Corollary 25] that the same conclusion holds if one assumes the existence of an affine connection (therefore, paracompactness is also achieved if the existence of an indefinite Riemannian metric is assumed). Now consider a paracompact manifold[2] M, it is classical that, by using a partition of the unity, we can always construct a Riemannian metric on M. But, the same procedure does not work in the Lorentzian case. In fact, although we can consider a Lorentzian metric on each coordinate open subset of M, it may be not possible to glue the locally defined Lorentzian metrics, as in the Riemannian case, to produce a Lorentzian metric defined on the whole manifold M. Therefore, it is natural to ask

When does a manifold admit a Lorentzian metric?

The answer is the well-known result:

PROPOSITION. *An $n(\geqslant 2)$-dimensional manifold M admits a Lorentzian metric if and only if it admits a 1-dimensional distribution.*

To sketch the proof, Greub, Halperin and Vanstone [37] consider a Lorentzian metric g on M, and let g_R be an arbitrary Riemannian metric on M. A $(1,1)$-tensor field P on M can be defined by

$$g(u,v) = g_R\big(P(u), v\big)$$

for all $u, v \in T_pM$, $p \in M$. Clearly, P is g_R-selfadjoint and, hence, at any point $p \in M$, there is a g_R-orthonormal basis of T_pM consisting of eigenvectors of P. None of the eigenvalues is zero, $n-1$ are positive and one is negative. Put $\mathcal{D}_p$ the eigenspace associated to the negative eigenvalue of P at p, then $\mathcal{D}$ defines a 1-dimensional distribution (or line field) on M. It should be noted that $\mathcal{D}$ clearly depends on the arbitrary Riemannian metric chosen g_R.

Conversely, if a 1-dimensional distribution $\mathcal{D}$ on M is given, fix an arbitrary Riemannian metric g_R on M. There exist an open covering $\{U_\alpha\}$ of M and vector fields $X_\alpha \in \mathfrak{X}(U_\alpha)$ such that we can write locally,

$$\mathcal{D} = \operatorname{Span}\{X_\alpha\}, \quad \text{with } g_R(X_\alpha, X_\alpha) = 1.$$

By putting

$$g_L(u,v) = g_R(u,v) - 2g_R\big(u, X_\alpha(p)\big)g_R\big(v, X_\alpha(p)\big),$$

for any tangent vectors $u, v \in T_pM$ with $p \in U_\alpha$, it is easily deduced that g_L does not depend on α and provides us with the desired Lorentzian metric on M.

Any noncompact manifold admits a nonvanishing vector field. Thus, as a direct consequence of last result, any $n(\geqslant 2)$-dimensional noncompact manifold admits a Lorentzian

[2] Along this paper a manifold will be assumed to be of class C^∞ and with a countable basis in its topology.

metric. On the other hand, an $n(\geqslant 2)$-dimensional compact manifold M admits a 1-dimensional distribution if and only if its Euler–Poincaré characteristic $\chi(M)$ is zero. Therefore, any $(2n+1)$-dimensional compact orientable manifold admits a Lorentzian metric.

The existence of a 1-dimensional distribution on a manifold is closely related to the existence of a nonvanishing vector field. In fact, it is a standard topological result that

PROPOSITION. *An $n(\geqslant 2)$-dimensional compact manifold M admits a nonvanishing vector field if and only if $\chi(M)=0$.*

On a simply connected manifold (compact or not), every 1-dimensional distribution on M arises from a global nonvanishing vector field $X \in \mathfrak{X}(M)$. But,

> A 1-dimensional distribution cannot be lifted in general to a global nonvanishing vector field,

as the following example [36] shows.

Let $M = \mathbb{S}^1 \times SO(3)$. Since M is parallelizable, every vector field $X \in \mathfrak{X}(M)$ can be contemplated as a map

$$X : M \to \mathbb{R}^4$$

and, by fixing a diffeomorphism $\psi : \mathbb{R}P^3 \to SO(3)$, a 1-dimensional distribution $\mathcal{D}$ as a map

$$\mathcal{D} : M \to SO(3).$$

In particular, the canonical projection on the second factor $\mathcal{D}_2$ defines a natural 1-dimensional distribution on $M = \mathbb{S}^1 \times SO(3)$. If we assume that $\mathcal{D}_2$ lifts to a vector field X without any zero, then, taking into account that $\mathbb{R}^4 - \{0\}$ is simply connected, one easily shows that $SO(3)$ would be also simply connected, which is not true. Hence $\mathcal{D}_2$ cannot be lifted to a global vector field on $\mathbb{S}^1 \times SO(3)$.

1.2. *Structure of globally hyperbolic spacetimes*

Several causality conditions on a spacetime M (i.e. a time oriented Lorentzian manifold) are now related to topological and metric properties of M. Recall that (M, g) is called strongly causal at $p \in M$ if for any neighborhood $\mathcal{O}$ of p there is a neighborhood $\mathcal{U} \subset \mathcal{O}$ of p such that every causal curve segment with endpoints in $\mathcal{U}$ lies entirely in $\mathcal{O}$, (M, g) is strongly causal if it is strongly causal at any point [63, Chapter 14].

As usual, we shall write

$$J^+(p) = \{x \in M \colon\ p \leqslant x\} \quad \text{and} \quad J^-(p) = \{x \in M \colon\ x \leqslant p\}$$

for causal future and causal past of p, respectively. If a spacetime (M, g) is strongly causal and for each $p, q \in M$ the set $J^+(p) \cap J^-(q)$ is compact, (M, g) is called globally hyperbolic [63, Chapter 14].

Global hyperbolicity is one of the most important causality conditions which are imposed to spacetime. Global hyperbolic spacetimes have the relevant property that any pair of causally related points may be joined by a nonspacelike geodesic segment of maximal length.

In a classical result [35], Geroch proved that if (M, g) is globally hyperbolic, then it admits a topological Cauchy hypersurface S, and therefore there is a homeomorphism

$$\Phi : M \to \mathbb{R} \times S.$$

In global Lorentzian geometry, it has been a challenge to improve this topological fact to a smooth result. Although Seifert [83] claimed the existence of smooth Cauchy hypersurfaces, several authors have pointed out that his proof is complicated and seems unclear, see [12] for a detailed discussion of this topic and also the references therein. In fact, most of Lorentzian geometers do not affirm that such a smooth Cauchy hypersurface must exist and in the reference book [10, p. 65] this folklore question appears as open. From the beginning of Causality theory, the question of existence in any globally hyperbolic spacetime of a smooth spacelike Cauchy hypersurface has been widely analyzed.

An elegant and clear proof for the existence of a smooth spacelike Cauchy hypersurface on every globally hyperbolic spacetime (M, g) has been recently given in [12]. And so, a global diffeomorphism between M and $\mathbb{R} \times S$, where S is any Cauchy hypersurface exists for every globally hyperbolic spacetime. The proof of this remarkable result goes, roughly speaking, by following two steps. First, by fixing two Cauchy (topological) hypersurfaces S_1 and S_2 with S_1 in the past of S_2, Bernal and Sánchez construct for each point $p \in S_2$ a smooth function h_p with compact support such that ∇h_p is timelike or 0 in the past of S_2. Secondly, by using paracompactness, a locally finite set of these functions h_p can be summed in such a way that the sum, h, restricted to $J^+(S_1) \cap J^+(S_2)$ admits regular level hypersurfaces which provide us the desired smooth Cauchy hypersurface.

As a natural continuation, the same authors [13] have improved the above result into a metric splitting theorem for globally hyperbolic spacetime (see also [80]). The splitting problem has become an important subject in global Lorentzian geometry (see, for instance, [10, Chapter 14]). Classically, if a Lorentzian manifold (M, g) is under a suitable assumption of prototype splitting Lorentzian results, then an isometry between (M, g) and a Lorentzian product manifold $(\mathbb{R} \times S, -dt^2 + \bar{g})$, where $(S, \bar{g})$ is a Riemannian manifold, is provided.

As far as we know, a new splitting philosophy has been introduced in [13] where the Lorentzian metric g does not appear as a sum of metric tensors. This splitting theorem becomes an important result in global Lorentzian geometry, and it is stated as follows.

THEOREM. *Any globally hyperbolic spacetime (M, g) is isometric to*

$$\left(\mathbb{R} \times S, -\beta\, d\mathcal{T}^2 + \bar{g}\right),$$

where S is a smooth spacelike Cauchy hypersurface, $\mathcal{T} : \mathbb{R} \times S \to \mathbb{R}$ is the natural projection, $\beta : \mathbb{R} \times S \to (0, \infty)$ a smooth function, and $\bar{g}$ a 2-covariant symmetric tensor field on $\mathbb{R} \times S$, satisfying:

1. $\nabla\mathcal{T}$ *is timelike and* $\mathcal{T}$ *is a time function.*
2. *Each hypersurface* $S_{\mathcal{T}}$ *at constant* $\mathcal{T}$ *is a spacelike Cauchy hypersurface.*
3. *The radical of* $\bar{g}$ *at each* $(t,x)\in\mathbb{R}\times S$ *is* $\mathrm{Span}\{\nabla\mathcal{T}(t,x)\}$.

2. Geodesics and completeness

Let (M,g) be a Lorentzian manifold and let $p\in M$. The causal character of a tangent vector $v\in T_pM$ is *timelike* (respectively *null*, *spacelike*) if $g(v,v)<0$ (respectively $g(v,v)=0$ and $v\neq 0$, $g(v,v)>0$ or $v=0$). If γ is a geodesic of (M,g), we will say that it is timelike, null or spacelike according to the causal character of its velocity γ'. An inextendible geodesic is called *complete* if it is defined for all $t\in\mathbb{R}$, and *incomplete* otherwise. The Lorentzian manifold (M,g) is said to be *timelike*, *null* or *spacelike complete*, according to the causal character of its complete geodesics. It is said (geodesically) *complete* if any of its inextendible geodesics is complete.

2.1. *Completeness in the Lorentzian setting*

There is no analogous result to the Hopf–Rinow theorem for Lorentzian manifolds. In fact, there are well-known examples of compact Lorentzian manifolds which are incomplete, cf. [71] and references therein (see also [69]). For an incomplete compact Lorentzian manifold (M,g) we can assert that there is no Riemannian metric on M such that its Levi-Civita connection agrees with the one of the Lorentzian metric g, giving a distinguishing feature of global Lorentzian geometry from the global Riemannian one. Thus, it seems natural to find geometric assumptions which, imposed to a compact Lorentzian manifold, could achieve its completeness.

On the other hand, geodesics in a Lorentzian manifold are separated into three classes according to their causal character, and another natural question on completeness arises in the Lorentzian setting.

Are the three kinds of causal completeness logically independent?

In the noncompact case there are several answers to this question. Kundt in [55] showed the first example of a Lorentzian manifold which is complete in a causal character but incomplete in another one. Geroch [34] and Beem [9] gave the remainder examples to show that there is no logical dependence among these completeness conditions in the noncompact case. Nevertheless, in several particular cases there are relations among them, recall, for example, that Lafuente in [56] proved the complete logical equivalence among the three types of completeness for locally symmetric Lorentzian manifolds, and it is a classical result that any symmetric semi-Riemannian manifold is complete. More general symmetry conditions as semi-symmetry and pseudo-symmetry have shown to be fruitful both in geometry and physics [42]. It is natural to ask the question whether the Lafuente result could be given for these Lorentzian manifolds.

In [69] it was conjectured that an incomplete compact Lorentzian manifold must be null incomplete. A partial answer has been given by Carrière and Rozoy in [20] for Lorentzian metrics on a torus showing that the conjecture is generically true (in measure theory sense). They argued how a counter-example can be found (but they did not construct it). As far as we know, this problem remains open (see [71] for related questions).

The above examples of Lorentzian manifolds, which are complete in one causal character but incomplete in another one, are constructed by taking a Lorentzian manifold and multiplying its metric tensor by a suitable conformal factor. In general,

> Can completeness be gained (or lost) by conformal changes of the metric tensor in the Lorentzian setting?

For globally hyperbolic spacetimes, which are noncompact Lorentzian manifolds, the works of Seifert [82] and Clarke [23] allow us to say that all globally hyperbolic spacetimes are conformally timelike and null complete; that is, a metric which is pointwise conformal to the given one, is timelike and null complete (recall that Nomizu and Ozeki proved [62] that any Riemannian metric on a noncompact manifold is pointwise conformal to a complete metric). But it is not known, even in this case, what occurs for spacelike completeness. On the other hand, for compact Lorentzian manifolds, null completeness is a conformal invariant but the problem is open for spacelike and timelike completeness.

A manifold M is said to be of *Lorentzian type* if it admits a Lorentzian metric. From the results in Section 1.1, we know that M is of Lorentzian type if and only if it is noncompact or it is compact with zero Euler–Poincaré characteristic.

> Does a Lorentzian type manifold admit a complete Lorentzian metric?

Some partial answers will be given, after the introduction of suitable techniques, in the next subsection.

Finally, recall also that, from the Hopf–Rinow theorem, a complete Riemannian manifold is always geodesically connected. In the noncompact Lorentzian case this property does not hold, the de Sitter space $\mathbb{S}_1^n$ being an example of a complete Lorentzian manifold which is not geodesically connected. Moreover, an incomplete Lorentzian manifold may be geodesically connected or geodesically nonconnected [32] (see also [81]).

2.2. *Completeness and symmetries*

It is well known that a homogeneous Riemannian manifold must be complete. But the same assertion is not true in the Lorentzian case; thus, there exist (noncompact) homogeneous Lorentzian manifolds which are incomplete (see [63], for instance). Nevertheless, Marsden [58] proved

THEOREM. *Any compact homogeneous Lorentzian*[3] *manifold M must be complete.*

The proof consists of constructing a partition of the tangent bundle into compact subsets which are invariant under the geodesic flow. So, any inextendible geodesic must be complete (see [71] for details). It should be remarked that an essential fact in the proof is

[3] or indefinite

to assert that for each point $p \in M$ there exist n ($= \dim M$) globally defined Killing vector fields which are independent at p (note that the Killing vector field K which extends $v \in T_pM$ preserves the causal character of v near p, but, of course, it could change far from p). In the same philosophy of the Marsden result, but now making use of a different assumption, the second author and M. Sánchez [73] proved

THEOREM. *A compact Lorentzian manifold (M, g) which admits a timelike conformal vector field K, must be complete.*

The idea of the proof is to see that any geodesic $\gamma : [0, b[\to M$, $0 < b < \infty$, has its velocity γ' contained in a compact subset of TM. According to the assumptions, it is only necessary to check that $g(K \circ \gamma, \gamma')$ is bounded. But we have

$$\frac{d}{dt} g(K \circ \gamma, \gamma') = \frac{1}{2} C\sigma \circ \gamma,$$

where C is the constant $g(\gamma', \gamma')$ and σ is defined by $\mathcal{L}_K g = \sigma g$. Therefore, $\frac{d}{dt} g(K \circ \gamma, \gamma')$ and, as a consequence $g(K \circ \gamma, \gamma')$, is bounded on $[0, b[$.

Kamishima proved in [47], making use of a different technique, that a compact Lorentzian manifold of constant sectional curvature which admits a timelike Killing vector field, must be complete. The previous result does not use any assumption on the curvature and, moreover, it is stated in terms of conformal vector fields.

It should be emphasized that simply the existence of a nontrivial Killing vector field does not imply the completeness of a compact Lorentzian manifold, as the Clifton–Pohl torus shows. Even more, completeness is not obtained if the assumption of the existence of a timelike conformal vector field is extended to the existence of a causal conformal vector field. In fact, the following Lorentzian metric on $\mathbb{R}^2$:

$$g = (1 - \cos 2\pi x)(dx^2 - dy^2) + (1 + \cos 2\pi x)(dx \otimes dy + dy \otimes dx)$$

may be naturally induced on a torus $\mathbb{T}^2$ and the vector field $\partial/\partial y$ defines a causal Killing vector field on $\mathbb{T}^2$, but this Lorentzian torus is incomplete (see [70] for details).

As a practical consequence we can give a partial answer to the question previously stated. In fact, let M be a compact manifold of Lorentzian type. Assume there exists $K \in \mathfrak{X}(M)$ such that $K(p) \neq 0$, for all $p \in M$, and its flow defines an action of $\mathbb{S}^1$ on M. As consequence of [64, Theorem 4.3.1] we have a Riemannian metric g_R on M such that K is Killing for g_R (see [88, Section 3] for the explicit construction). Now

$$g_L = g_R - \frac{2}{g_R(K, K)} \omega \otimes \omega,$$

where $\omega(X) = g_R(K, X)$ for all X, is a Lorentzian metric on M and K is Killing and timelike for g_L. Therefore, g_L is complete.

Finally, it should be noted that a generalization of the previous completeness theorem was given in [72].

2.3. *Completeness and curvature*

Now, another viewpoint is discussed to study the completeness of compact Lorentzian manifolds.

> How can the curvature of a compact Lorentzian manifold influence completeness?

Carrière [19] gave an important answer when he solved a famous conjecture by Markus on affine manifolds. In particular, he proved that a compact flat Lorentzian manifold must be geodesically complete. Later, Klinger [50] dealt with the case of nonzero constant sectional curvature. Carrière and Klinger results give

THEOREM. *Every compact Lorentzian manifold of constant sectional curvature must be geodesically complete.*

Of course, there exist well-known examples of compact Lorentzian manifolds of constant sectional curvature c both in the cases $c = 0$ and $c < 0$. But for the case $c > 0$, one can deduce the following remarkable result.

THEOREM. *There are no compact Lorentzian manifolds (M, g) of constant sectional curvature $c > 0$.*

In the 2-dimensional case, it is an easy consequence of the Gauss–Bonnet theorem for Lorentzian metrics (see [8] or [14], for instance). For the $n(\geqslant 3)$-dimensional case, recall that there is a classical result of Calabi and Markus [18] (see also [63, Proposition 9.16]) which asserts that the fundamental group $\pi_1(M)$ is finite when (M, g) is a complete Lorentzian manifold with $\dim M = n \geqslant 3$ and constant sectional curvature $c > 0$. If M is additionally assumed to be compact, then we have, making use of the Klinger result, that (M, g) should be complete. Therefore, (M, g) will be isometric to a finite quotient of the De Sitter space $\mathbb{S}_1^n(1/\sqrt{c})$ (see [63, Corollary 8.26], for instance). This contradicts the compactness of M.

Nevertheless, Einstein Lorentzian manifolds with positive Ricci curvature were shown to exist in the compact case [89].

As a complement to the Carrière and Klinger results, one could ask if the sectional curvature of a compact Lorentzian manifold, when it is not constant, is related to completeness. The answer is the following surprising fact [70].

THEOREM. *There are two Lorentzian metrics on the 2-dimensional torus with the same Gauss curvature at each point and one is geodesically complete whereas the other one is not.*

3. Curvature of Lorentzian manifolds

Let (M, g) be an $n(\geqslant 2)$-dimensional (connected) Lorentzian manifold. We shall write ∇ for its Levi-Civita connection, R for its Riemannian curvature tensor,[4] Ric for its Ricci

[4] According our convention, $R(X,Y)Z = \nabla_X\nabla_Y Z - \nabla_Y\nabla_X Z - \nabla_{[X,Y]}Z$, is the curvature tensor.

tensor, $\widetilde{\mathrm{Ric}}$ for its associated quadratic form, S for its scalar curvature and $\mathcal{K}$ for its sectional curvature on nondegenerate tangent planes.

3.1. *Sectional curvature*

For a Lorentzian[5] manifold (M, g), the sectional curvature $\mathcal{K}$ is defined, on nondegenerate tangent planes, exactly as in the Riemannian case. Recall that a plane Π of T_pM is said to be nondegenerate (respectively spacelike, Lorentzian) whenever the restriction of g to Π is nondegenerate (respectively positive definite, Lorentzian). Nevertheless, the behavior of $\mathcal{K}$ presents here several remarkable differences to the definite case. The pioneer contribution to the study of the sectional curvature in this setting was done by Wolf [90, Theorems 2.9, 4.1] who showed that among the isotropic manifolds with indefinite metric, only the ones of constant sectional curvature have bounded sectional curvature. Later, Kulkarni proved an amazing result [53], without any assumption of homogeneity.

THEOREM. *If (M, g) is an $n(\geqslant 3)$-dimensional Lorentzian manifold and its sectional curvature is bounded from above (or from below) for all nondegenerate tangent planes, then (M, g) has constant sectional curvature.*[6]

It is interesting to point out that the proof of the Kulkarni theorem works pointwise, that is

PROPOSITION. *For every point p in any $n(\geqslant 3)$-dimensional Lorentzian manifold, (M, g), the following conditions on T_pM are equivalent*:

1. *$\mathcal{K}$ is constant on nondegenerate planes of T_pM.*
2. *There exists a constant $a \in \mathbb{R}$ such that $a \leqslant \mathcal{K}$, or there exists a constant $b \in \mathbb{R}$ such that $\mathcal{K} \leqslant b$, on nondegenerate planes of T_pM.*

Therefore, making use of the classical Schur theorem, (M, g) will have constant sectional curvature if $\mathcal{K}$ is pointwise bounded.

It follows from the Kulkarni theorem that the concepts of an $n(\geqslant 3)$-dimensional Lorentzian (or indefinite) manifold being positively curved or negatively curved based on the sign of $\mathcal{K}$ are vacuous, except in the case of constant curvature; in particular, pinching results on $\mathcal{K}$ have no meaning in the Lorentzian setting. This fact contrasts with the Riemannian case, where the inequalities on sectional curvature has been used to get very strong well-known pinching theorems.

For any $n(\geqslant 3)$-dimensional Lorentzian manifold (M, g), its sectional curvature $\mathcal{K}$ can be considered as a function

$$\mathcal{K}: G_2^*(M) \to \mathbb{R}$$

[5] or indefinite

[6] Of course, the Kulkarni result remains true for any indefinite Riemannian manifold.

on the Grassmannian $G_2^*(M)$ of all nondegenerate tangent planes on M. Of course, $G_2^*(M)$ is a proper open subset of the ordinary Grassmannian $G_2(M)$ of all tangent planes on M, and $\mathcal{K}$ is continuous on $G_2^*(M)$. Note that even if M is assumed to be compact, $G_2^*(M)$ will be not compact because its fiber is not compact, and hence $\mathcal{K}$ has neither maximum nor minimum unless it is constant, as the Kulkarni theorem asserts. This is in contrast with the Riemannian case, where $\mathcal{K}: G_2(M) \to \mathbb{R}$ has a maximum and a minimum values when M is compact [15, Section 9.3].

The Kulkarni result was the starting point for a wide research on the sectional curvature of Lorentzian (or indefinite) metrics. Among them, Dajczer and Nomizu [25] (see also [63, Proposition 8.28]) proved

THEOREM. *If the sectional curvature of an $n(\geqslant 3)$-dimensional Lorentzian*[7] *manifold is bounded in absolute value on all timelike (or spacelike) tangent planes, then the manifold must have constant sectional curvature.*

Moreover, Nomizu found in [61]

THEOREM. *Assume that, for each spacelike tangent vector $v \in T_pM$ of a Lorentzian*[7] *manifold (M, g), there is a number $\delta > 0$ such that the sectional curvature satisfies*

$$|\mathcal{K}(\Pi)| \leqslant \delta$$

for all spacelike (respectively Lorentzian) tangent planes Π containing v (i.e. every pencil of spacelike (respectively Lorentzian) planes determined by v). Then (M, g) has constant sectional curvature.

Beem and Parker [11] studied the value distribution of the sectional curvature of Lorentzian (and indefinite) metrics as follows. For a point p of a Lorentzian manifold (M, g), let I_t (respectively I_s) be the image under $\mathcal{K}$ of all Lorentzian (respectively spacelike) planes at T_pM. Clearly, if $\mathcal{K}$ is constant c at p, then $I_t = I_s = \{c\}$. Otherwise, we have

THEOREM. *Let p be a point of a Lorentzian manifold M of dimension $n \geqslant 3$. If the sectional curvature is not constant at p, then both I_s and I_t are intervals of infinite length.*

More recently, Kishta [49] proved, as a nice application of the Kulkarni result.

THEOREM. *If for every $p \in M$ the subgroup of isometries fixing the point p is transitive on the unit timelike tangent vectors of T_pM, then M has constant sectional curvature.*

On the other hand, in [25] (see also [10, Lemma 2.1]) several boundedness conditions of the Ricci curvature are studied. It is proved that

[7] or indefinite

PROPOSITION. *For every $n(\geqslant 3)$-dimensional Lorentzian*[8] *manifold, each of the following conditions implies that (M,g) is Einstein (i.e.* $\mathrm{Ric} = \lambda g$, *for some* $\lambda \in \mathbb{R}$),

1. *For every null tangent vector v we have* $\widetilde{\mathrm{Ric}}(v) = 0$.
2. *For every unit timelike tangent vector w we have* $|\widetilde{\mathrm{Ric}}(w)| \leqslant \delta$, *where δ is a fixed positive number.*

3.2. *Volume comparison and curvature*

Roughly, bound specifications on the sectional curvature force Lorentzian manifolds to have constant sectional curvature. So, this is an a priori difficulty to think about Lorentzian volume comparison results (of course, a Lorentzian metric has a canonical measure which is analogously defined to the canonical measure of a Riemannian metric). Nevertheless, some comparison results between volumes of Lorentzian manifolds have been obtained by Andersson and Howard in [7] making use of an inequality on the curvature tensor. On the other hand, Ehrlich, Jung and Kim [29] also obtained volume comparison results considering inequalities involving the sectional curvature on tangent planes with a definite causal character.

As noted by Ehrlich and Sánchez [31], there are another two difficulties in order to state volume comparison results in Lorentzian geometry:

1. Let $I^+(p)$ be the chronological future of p [10, p. 5] in a time oriented Lorentzian manifold (M,g) and d the Lorentzian distance function [10, p. 8] of (M,g). The *inner* metric balls $B^+(p,\varepsilon) = \{q \in I^+(p)\colon\ d(p,q) < \varepsilon\}$, $\epsilon > 0$, need not to be open subsets in M.
2. It can be showed that for any pair of Lorentzian space forms $Q(c)$, $\hat{Q}(\hat{c})$, with different sectional curvatures c and $\hat{c}$, and for any point $p \in Q(c)$ there are two normal neighborhoods W, W' of p such that the corresponding transplanted neighborhoods $\hat{W}, \hat{W}'$ of $\hat{p}$ in $\hat{Q}(\hat{c})$ satisfy $\mathrm{Vol}(W) < \mathrm{Vol}(\hat{W})$ and $\mathrm{Vol}(W') > \mathrm{Vol}(\hat{W}')$ (see [31, Corollary 5.3] for details).

Ehrlich and Sánchez introduced in [31] the notion of *standard subset for comparison of Lorentzian volumes at a point p*, in order to avoid such difficulties. They call a subset $U \subset M$ to be standard for comparison of Lorentzian volumes at a point p if there is an open $\tilde{U} \subset T_pM$ satisfying:

1. $\tilde{U}$ is an open subset of $J^+(0_p)$, the causal future of the origin in T_pM.
2. $\tilde{U}$ is starshaped from the origin (i.e. if $v \in \tilde{U}$, then $tv \in \tilde{U}$ for any $t \in (0,1)$), the exponential map at p, $\exp_p$, is defined on all $\tilde{U}$, and the restriction of $\exp_p$ to $\tilde{U}$ is a diffeomorphism onto its image $U = \exp_p(\tilde{U})$.
3. The closure of $\tilde{U}$ is compact.

Let $Q(c)$ be an n-dimensional Lorentzian space form of constant sectional curvature $c \in \mathbb{R}$. Choose $p_0 \in Q(c)$ and $i: T_pM \to T_{p_0}Q(c)$ a linear isometry. We consider

$$F = \exp_{p_0} \circ\, i \circ (\exp_p|_U)^{-1} : U \to Q(c)$$

and put $\tilde{U}_0 = i(\tilde{U})$, $U_0 = \exp_{p_0}(\tilde{U}_0) = F(U)$.

---
[8] or indefinite

A tangent vector v to U is said to be *radial* if it can be written as

$$v = \frac{d}{dt}\bigg|_{t_0} \exp_p(tu_p)$$

for some $u_p \in T_pM$. A tangent plane Π to U is said to be *radially timelike* if it contains a timelike radial tangent vector.

Now, we can state the following Günther–Bishop type theorem for Lorentzian manifolds [31, Theorem 2.1].

THEOREM. *Let (M, g) be a Lorentzian manifold, let U be a standard subset for comparison of Lorentzian volumes at a point $p \in M$, and assume the following two conditions hold*:

1. *For any radially timelike tangent plane Π we have $\mathcal{K}(\Pi) \geqslant c$, and*
2. $\exp_{p_0} : \tilde{U}_0 \to U_0$ *is a diffeomorphism.*

Then

$$\mathrm{Vol}(U) \geqslant \mathrm{Vol}(U_0)$$

and equality holds if and only if $F : U \to U_0$ is an isometry.

A Lorentzian result analogue to the classical Bishop comparison theorem has been also given in [31, Theorem 2.2] and it is stated as follows.

THEOREM. *Let (M, g) be a Lorentzian manifold, let U be a standard subset for comparison of Lorentzian volumes at a point $p \in M$, and assume*

$$\widetilde{\mathrm{Ric}}(v) \geqslant (n-1)cg(v, v),$$

for any timelike radial vector v tangent to U. Then

$$\mathrm{Vol}(U) \leqslant \mathrm{Vol}(U_0)$$

and equality holds if and only if $F : U \to U_0$ is an isometry.

Finally, it should be noted that a Bishop–Gromov type theorem has been also proved in [31, Theorem 2.3].

3.3. *Null sectional curvature*

For any $n(\geqslant 3)$-dimensional Lorentzian manifold (M, g), as mentioned before, the Grassmannian of nondegenerate tangent planes $G_2^*(M)$ on M is a proper open subset of the Grassmannian of all tangent planes $G_2(M)$. Put

$$G_2^o(M) = G_2(M) \setminus G_2^*(M),$$

the Grassmannian of degenerate tangent planes on M. $G_2^o(M)$ is not in the domain of the sectional curvature $\mathcal{K}$. In a natural way it appears

When can $\mathcal{K}$ be continuously extended to $G_2^o(M)$?

Degenerate tangent planes play an important role in the study of the geometry of Lorentzian manifolds. In fact, Harris introduced the notion of null sectional curvature for degenerate tangent planes [43], which has shown to be fruitful to get some comparison theorems [43] and to characterize Robertson–Walker spacetimes [44] (see also [10, Definition A.6]). The null sectional curvature is defined as follows, let $\Pi \subset T_pM$ be a degenerate (or null) tangent plane and $v \in \Pi$ a null tangent vector, then the null sectional curvature of Π with respect to v is given by

$$\mathcal{K}_v(\Pi) = \frac{g(R(x,v)v,x)}{g(x,x)},$$

where x is any nonnull vector in Π. Note that $\mathcal{K}_v(\Pi)$ is independent of the choice of the nonnull vector $x \in \Pi$, but it does depend quadratically on v. Therefore, it is not, in general, a function on the Grassmannian of degenerate tangent planes $G_2^o(M)$. It should be remarked that the null sectional curvature is a Lorentzian notion and cannot be defined for general indefinite metrics.

For a Lorentzian manifold, its null sectional curvature is a useful tool to get global information as the following facts show. First, Harris showed [43, Proposition 2.3]

PROPOSITION. *The null sectional curvature vanishes on any degenerate tangent plane of an $n(\geqslant 3)$-dimensional Lorentzian manifold (M,g) if and only if it has constant sectional curvature.*

This result can be used to give an answer to the previous question reproving a well-known theorem by Thorpe [87]. In fact, it is remarkable that $\mathcal{K}$ may be continuously extended to $G_2^o(M)$ if and only if (M,g) has constant sectional curvature.

Moreover, the Ricci tensor on a null tangent vector v can be easily obtained as a sum of the $n-2$ null sectional curvatures of degenerate planes through v,

$$\widetilde{\mathrm{Ric}}(v) = \sum_{i=3}^{n} \mathcal{K}_v(\Pi_i),$$

where $\Pi_i = \mathrm{Span}\{v, e_i\}$ and $\{e_3, \dots, e_n\}$ are tangent vectors such that $g(v,e_i) = 0$ and $g(e_i,e_j) = \delta_{ij}$.

The null sectional curvature may be also related with conjugate points along null geodesics. Roughly speaking, in these results the null sectional curvature plays an analogous role as the sectional curvature does in the Riemannian case. In fact, the following result [43, Proposition 2.6], may be seen as a null analogue to the classical Bonnet–Myers theorem.

THEOREM. *Let $\gamma : [0, a] \to M$ be a null geodesic in an $n(\geqslant 3)$-dimensional Lorentzian manifold (M, g). Suppose that for some $\delta > 0$ we have*

$$\mathcal{K}_{\gamma'(t)}(\Pi) \geqslant \delta$$

for every degenerate tangent plane containing $\gamma'(t)$. If $a \geqslant \pi/\sqrt{\delta}$, then there exists on γ a conjugate point to $\gamma(0)$.

More recently, a Morse–Schönberg type theorem was stated in [38, Proposition 4.4] as follows.

THEOREM. *Let $\gamma : [0, a] \to M$ be a null geodesic in an $n(\geqslant 3)$-dimensional Lorentzian manifold (M, g). Suppose that for some $\delta > 0$ we have*

$$\mathcal{K}_{\gamma'(t)}(\Pi) \leqslant \delta$$

for every degenerate tangent plane with $\gamma'(t) \in \Pi$, and $\gamma(a)$ is a conjugate point to $\gamma(0)$ along γ, then

$$a \geqslant \pi/\sqrt{\delta}.$$

For a time oriented Lorentzian manifold (M, g), there is a natural way to consider the null sectional curvature as a function on $G_2^o(M)$. In order to do that, we make a choice of a (globally defined) timelike vector field $Z \in \mathfrak{X}(M)$, then define the null congruence on (M, g) relative to Z as the subset of the tangent bundle TM given by

$$C_Z M = \{v \in TM\colon\ g(v, v) = 0 \text{ and } g(v, Z_{\pi(v)}) = 1\},$$

where $\pi : TM \to M$ is the natural projection [44,52]. It is not difficult to see that $C_Z M$ is in fact a codimension two submanifold of TM. Now, an algebraic reasoning (strongly dependent on the Lorentzian character of the metric tensor) says that for each degenerate tangent plane $\Pi \subset T_p M$, there exists a unique null tangent vector in $\Pi \cap C_Z M$. Therefore, we can define a Z-normalized null sectional curvature, $\mathcal{K}^Z$, by putting

$$\mathcal{K}^Z(\Pi) = \mathcal{K}_{(\Pi \cap C_Z M)}(\Pi).$$

Contrary to the well-known behavior of the sectional curvature described by the classical Schur lemma, the Z-normalized null sectional curvature may be a nonconstant point function on the Lorentzian manifold (M, g). In fact, if we normalize Z to get $U = [-g(Z, Z)]^{-1/2} Z$, it is shown in [52] and [44] that

THEOREM. *For any $n(\geqslant 3)$-dimensional time oriented Lorentzian manifold (M, g), the following assertions are equivalent:*

(a) *There exists a (smooth) function $\nu : M \to \mathbb{R}$ such that for each $p \in M$ we have $\nu(p) = \mathcal{K}^U(\Pi)$, for every degenerate tangent plane $\Pi \subset T_p M$.*

(b) *The Riemannian curvature tensor of* (M,g) *satisfies*
1. $R(X,Y)V = k\{g(Y,V)X - g(X,V)Y\}$,
2. $R(X,U)U = \theta X$,

for all $X, Y, V \in U^{\perp}$, *where* $k, \theta \in C^{\infty}(M)$.

It should be noted that if the previous assertion (b) holds, then all tangent planes containing U_p have the same sectional curvature $-\theta(p)$, and all tangent planes contained in $U_p^{\perp}$ have the same sectional curvature $k(p)$, for any $p \in M$. As a direct consequence we get $\nu(p) = \theta(p) + k(p)$, for all $p \in M$. Observe also that the Z-normalized null sectional curvature satisfies $\mathcal{K}^Z = \frac{\nu}{-g(Z,Z)}$.

Previous items (b.1) and (b.2) are shown to be equivalent when $n \geqslant 4$ [52], to the notion of infinitesimal isotropy of the Lorentzian manifold (M,g) relative to U, as introduced by Karcher in [48]. In that paper, looking for a characterization of Friedmann cosmological models, Karcher proved that

THEOREM. *A Lorentzian manifold* (M,g), $\dim M \geqslant 4$, *is infinitesimal isotropic relative to a unit timelike vector field* U, *with everywhere nonzero* U*-normalized null sectional curvature, if and only if* (M,g) *satisfies the following conditions*:
1. *The distribution* $U^{\perp}$ *is integrable.*
2. *The integral manifolds of* $U^{\perp}$ *are totally umbilical and have constant sectional curvature.*
3. *The manifold* (M,g) *is locally conformal to a flat Lorentzian space.*

Although dimension 3 seems to be nonrelevant from a physical point of view, it would be an interesting problem to decide if the Karcher result can be extended to that dimension. In [38] a negative answer was given to that question. In fact, consider the (unit) sphere $\mathbb{S}^3$ endowed with the Lorentzian metric g given by $g = g^0 - 2\omega \otimes \omega$ (compare to g_L in Section 1.1), where g^0 is the canonical Riemannian metric on $\mathbb{S}^3$ and $\omega(v) = g^0(v, U_p)$, for all $v \in T_p\mathbb{S}^3$, $U_p = \mathbf{i}p$, for all $p \in \mathbb{S}^3$. Clearly U is a unit timelike vector field on $\mathbb{S}^3$ and it can deduced that the Lorentzian manifold $(\mathbb{S}^3, g)$ has nonzero constant U-normalized null sectional curvature but the distribution $U^{\perp}$ is not integrable.

We finish this subsection with a question which naturally arises in this setting. The null sectional curvature has been considered as a function on the set $G_2^o(M)$ of all degenerate tangent planes on the time oriented Lorentzian manifold (M,g). So, it is natural to ask

> Is there a (smooth) manifold structure on $G_2^o(M)$ such that the Z-normalized null sectional curvature $\mathcal{K}^Z$ becomes a smooth function on $G_2^o(M)$?

In order to face the question let us consider the Lorentz–Minkowski space $\mathbb{L}^n = (\mathbb{R}^n, g_1 = -dx_1^2 + \sum_{i=2}^n dx_i^2)$, with $n \geqslant 3$, the Grassmannian $G_2(\mathbb{R}^n)$ of all 2-dimensional linear subspaces of $\mathbb{R}^n$, and put

$$\Lambda_n = \{\Pi \in G_2(\mathbb{R}^n)\colon\ g_1|_{\Pi} \text{ is degenerate}\}.$$

It is not difficult to show that Λ_n is a (regular) compact submanifold of $G_2(\mathbb{R}^n)$ with $\dim(\Lambda_n) = 2n - 5$.

Now, for an $n(\geqslant 3)$-dimensional Lorentzian manifold (M,g), $G_2^o(M)$ can be naturally endowed with a manifold structure such that it becomes a submanifold of $G_2(M)$ and such that it is a fiber bundle over M with fiber type Λ_n.

On the other hand, the null congruence associated to a timelike vector field Z is an orientable submanifold of TM and the map $\pi : C_Z M \to M$ is a fiber bundle with fiber type $\mathbb{S}^{n-2}$. The natural projection

$$\mathfrak{p} : G_2^o(M) \to C_Z M, \quad \Pi \mapsto \Pi \cap C_Z M,$$

gives us a fiber bundle with fiber the real projective space $\mathbb{R}P^{n-3}$.

These facts can be summarized in the following diagram:

$$\begin{array}{ccccc} & \Lambda_n & & \mathbb{S}^{n-2} & \\ & \downarrow & & \downarrow & \\ \mathbb{R}P^{n-3} \to & G_2^o(M) & \xrightarrow{\mathfrak{p}} & C_Z M & \\ & \pi \circ \mathfrak{p} \searrow & & \swarrow \pi & \\ & & M & & \end{array}$$

(see [65] for details).

The Z-normalized null sectional curvature is then written as follows:

$$\mathcal{K}^Z(\Pi) = \mathcal{K}_{\mathfrak{p}(\Pi)}(\Pi),$$

and so, an analogous argument as the one in [15, Proposition 9.3.1] permits us to show that $\mathcal{K}^Z$ is in fact a smooth function

$$\mathcal{K}^Z : G_2^o(M) \to \mathbb{R}.$$

In particular, it is continuous. Taking into account that $G_2^o(M)$ is compact whenever M is compact, we can assert

PROPOSITION. *For every time oriented compact $n(\geqslant 3)$-dimensional Lorentzian manifold (M,g) and a timelike vector field Z on M, there exist $a, b \in \mathbb{R}$ such that its Z-normalized null sectional curvature $\mathcal{K}^Z$ satisfies*

$$a \leqslant \mathcal{K}^Z \leqslant b.$$

3.4. *Null conjugate points and curvature*

The null congruence associated to a timelike vector field Z on a time oriented $n(\geqslant 3)$-dimensional Lorentzian manifold (M,g) may be endowed with a natural Lorentzian metric. In order to see that, consider the Sasaki metric on TM, induced from the Lorentzian metric g, which is semi-Riemannian with index 2. Its restriction to $C_Z M$ is a Lorentzian metric $\hat{g}$ and the natural projection $\pi : C_Z M \to M$ then becomes a semi-Riemannian submersion with spacelike fibers [38,39].

This procedure permits us to construct from each time oriented $n(\geqslant 3)$-dimensional Lorentzian manifold (M,g) and a timelike vector field $Z \in \mathfrak{X}(M)$, a new $2(n-1)$-

dimensional Lorentzian manifold. It should be noted that for $n = 3$, $(C_Z(M), \hat{g})$ is a 4-dimensional Lorentzian manifold and the horizontal lift through π of Z gives a timelike vector field on $C_Z(M)$. Thus, the null congruence on a time oriented 3-dimensional Lorentzian manifold gives a 4-dimensional spacetime.

In order to check this construction produces many Lorentzian metrics, recall the following result by Mounoud [59] (see also [60]).

THEOREM. *Let M be a 3-dimensional orientable compact manifold and $\mathcal{L}(M)$ the space of all Lorentzian metrics on M. Then $\mathcal{L}(M)$, for a natural topology, possesses an infinity of connected components.*

Now, consider for instance $M = \mathbb{S}^3$. In this case every Lorentzian metric g on $\mathbb{S}^3$ is time orientable, and so, if Z denotes a timelike vector field of $(\mathbb{S}^3, g)$, then associated to g and Z we have the corresponding 4-dimensional spacetime $(C_Z(\mathbb{S}^3), \hat{g})$.

No extra hypothesis on the timelike vector field Z of (M, g) has been assumed until now. Next, we are going to impose a geometric assumption on Z which becomes a certain type of natural symmetry of g. The concept of symmetry is basic in physics. In general relativity, symmetry is usually based on a local one-parameter group of isometries generated by a Killing or, more generally conformal, vector field. In fact, the main simplification for the search of exact solutions to the Einstein equation is to assume, a priori, the existence of such symmetries [24,28]. We remark that a completely general approach to symmetries in general relativity has been developed in [96]. In the above mentioned references, the causal character of the Killing or conformal vector field is not always prefixed. However, it is natural to assume that this vector field is timelike. This is supported by very well-known examples of exact solutions. At the same time, under this assumption, the integral curves of such vector field provide a privileged class of observers (in the sense of [77]) or test particles in spacetime.

Let us now consider the case that the timelike vector field Z on (M, g) is conformal. Then $C_Z M$ possesses two important properties [38,39]:

(1) $C_Z M$ is invariant by the geodesic flow and so, for every null geodesic γ_v of (M, g), with $v \in C_Z M$, we have $\gamma_v'(t) \in C_Z M$, provided $\gamma_v(t)$ is defined, and the velocity curve γ_v' gives rise to a null geodesic of $(C_Z M, \hat{g})$. Furthermore, each null geodesic β of (M, g) may be reparametrized to obtain a null geodesic α that satisfies $\alpha'(t) \in C_Z M$ for all t where α is defined. In fact, it can be shown that $g(\beta', Z) = a \in \mathbb{R}$, $a \neq 0$ thus, if we put $\alpha(t) = \beta(t/a)$ then, we achieve $g(\alpha', Z) = 1$.

(2) The canonical measure $d\mu_{\hat{g}}$ associated with the Lorentzian metric $\hat{g}$ on $C_Z M$ is preserved by the geodesic flow $\{\Phi_t\}$, which, making use of the completeness result in Section 2.2, gives us the following

THEOREM. *For any $n(\geqslant 3)$-dimensional compact Lorentzian manifold (M, g) which admits a timelike conformal vector field Z, we have*

$$\int_{C_Z M} (f \circ \Phi_t)\, d\mu_{\hat{g}} = \int_{C_Z M} f\, d\mu_{\hat{g}},$$

for every $f \in C^0(M)$ and $t \in \mathbb{R}$.

We end this subsection with several integral inequalities which relate null conjugate points to global geometric properties. They will require the previous quoted properties of $C_Z M$ for the proofs. Moreover, null sectional curvature is used in order to characterize when equalities hold.

THEOREM. *Let (M, g) be an $n(\geqslant 3)$-dimensional compact Lorentzian manifold which admits a timelike conformal vector field Z. If there exists $a \in (0, \infty)$ such that every null geodesic $\gamma_v : [0, a] \to M$, with $v \in C_Z M$, has no conjugate point of $\gamma_v(0)$ in $[0, a)$, then*

$$\mathrm{Vol}(C_Z M, \hat{g}) \geqslant \frac{a^2}{\pi^2(n-2)} \int_{C_Z M} \widetilde{\mathrm{Ric}}\, d\mu_{\hat{g}}. \tag{3.1}$$

Moreover, equality holds if and only if the $(Z/\sqrt{-g(Z,Z)})$-normalized null sectional curvature of (M, g) is the point function $\frac{-\pi^2 g(Z,Z)}{a^2}$.

(See [38, Theorem 3.2] for details.) Under the same assumptions of the above result (compactness and existence of a timelike conformal vector field Z). It is shown in [41, Theorem 4.1] that

THEOREM. *If (M, g) has no null conjugate points (i.e. if any null geodesic has no conjugate points) then,*

$$\int_{C_Z M} \widetilde{\mathrm{Ric}}\, d\mu_{\hat{g}} \leqslant 0, \tag{3.2}$$

and equality holds if and only if (M, g) has constant sectional curvature $c \leqslant 0$.

As showed in [38], the integral $\int_{C_Z M} \widetilde{\mathrm{Ric}}\, d\mu_{\hat{g}}$ may be computed in terms of geometric quantities on (M, g). Concretely,

$$\int_{C_Z M} \widetilde{\mathrm{Ric}}\, d\mu_{\hat{g}} = \frac{\omega_{n-2}}{n-1} \int_M \big[n\widetilde{\mathrm{Ric}}(U) + S\big]\big[-g(Z,Z)\big]^{\frac{-n}{2}}\, d\mu_g, \tag{3.3}$$

where ω_{n-2} is the volume of the $(n-2)$-dimensional standard Riemannian sphere and $U = Z/\sqrt{-g(Z,Z)}$. This permits to write the previous results by means of integral inequalities on M, without explicit mentioning the null congruence $C_Z M$. However, these expressions are formally more complicated that the above ones.

Concerning general facts on null conjugate points in Lorentzian manifolds, there is a simple but surprising result [63, Example 10.11] which asserts that

PROPOSITION. *There are no conjugate points along any null geodesic in a Lorentzian manifold of constant sectional curvature.*

Of course, the converse is not true. The following question arises then in a natural way.

> When does a Lorentzian manifold with no conjugate points along its null geodesics has constant sectional curvature?

Under the assumption of compactness and the existence of a timelike conformal vector field, the integral inequality (3.2) gives an answer to this question.

If the conformal vector field Z is specialized to be Killing, then the above inequalities may be notably improved by using the following result [38, Lemma 3.7].

PROPOSITION. *Let (M, g) be an n-dimensional compact Lorentzian manifold that admits a timelike Killing vector field Z. Then,*

$$\int_M \widetilde{\mathrm{Ric}}(U)\big[-g(Z,Z)\big]^{\frac{-n}{2}}\, d\mu_g \geqslant 0, \tag{3.4}$$

and equality holds if and only if Z is parallel.

Using the previous result, we get [38, Corollary 3.8]

THEOREM. *Let (M, g) be an $n(\geqslant 3)$-dimensional compact Lorentzian manifold that admits a timelike Killing vector field Z. If there is $a \in (0, +\infty)$ such that every null geodesic $\gamma_v : [0, a] \to M$, $v \in C_Z M$, has no conjugate point of $\gamma_v(0) \in [0, a)$, then*

$$\int_M \big[-g(Z,Z)\big]^{\frac{-n+2}{2}}\, d\mu_g \geqslant \frac{a^2}{\pi^2(n-1)(n-2)} \int_M S\big[-g(Z,Z)\big]^{\frac{-n}{2}}\, d\mu_g.$$

Moreover, equality holds if and only if $g(Z, Z)$ is constant and the universal covering of (M, g) is isometric to the semi-Riemannian product

$$\big(\mathbb{R} \times \mathbb{S}^{n-1}(r), -dt^2 + g^0\big),$$

where $r = a/(\sqrt{-g(Z,Z)})\pi$ is the radius of the sphere.

And under the same assumptions of previous result [41, Theorem 4.4]

THEOREM. *If (M, g) has no conjugate points along its null geodesic, then*

$$\int_M S\big[-g(Z,Z)\big]^{\frac{-n}{2}}\, d\mu_g \leqslant 0,$$

and equality holds if and only if (M, g) is isometric to a flat Lorentzian n-torus up to a (finite) covering. In particular, in this case U is parallel, the first Betti number of M is not zero and the Levi-Civita connection of g is Riemannian.

The integral inequalities (3.1) and (3.2) have been used in [38] to show several properties of null geodesics of the natural Lorentzian metric on the odd dimensional sphere $\mathbb{S}^{2n+1}$ (see Section 3.3) and, so, to describe the topology of the null conjugate locus of any point of $\mathbb{S}^{2n+1}$ [40].

We are going to show two applications of these integral inequalities in very different stages.

In Riemannian geometry there is a classical remarkable result of Hopf [46] which asserts that a Riemannian torus with no conjugate points must be flat. In contrast, any Lorentzian surface is free of null conjugate points (see [10, Lemma 10.45], for instance). Moreover, the above explained integral tools work for any dimension $n \geqslant 3$. However, in [41, Theorem 4.8] it has been shown that

THEOREM. *A compact Lorentzian surface admitting a timelike Killing vector field with no conjugate points along its timelike* (*or spacelike*) *geodesics must be flat.*

It should be noted that the assumption is made on timelike (or spacelike) conjugate points, but if (S, g) is a Lorentzian surface free of timelike (or spacelike) conjugate points, then the Lorentzian manifold $(S \times \mathbb{S}^1, g + d\theta^2)$ is free of null conjugate points. This product is compact whenever S is compact and, moreover, it inherits a natural timelike Killing vector field from S. Therefore the integral inequality (3.2) works on $(S \times \mathbb{S}^1, g + d\theta^2)$, concluding the sketch of proof.

Consider now a Lorentzian torus which admits a timelike Killing vector field. Then it must be (globally) conformally flat [78]. Note that an answer to the question

When is a Lorentzian torus which admits a timelike Killing vector field flat?

is given making use of previous theorem.

On the other hand, let $P(B, \mathbb{S}^1)$ be a principal fiber bundle with structure group $\mathbb{S}^1$ and projection $\tau : P \to B$ over an n-dimensional manifold B. From each Riemannian metric g_B on B and each connection form $\omega : TP \to \mathbf{i}\mathbb{R} = \mathfrak{s}^1$, we define a Lorentzian metric g^ω on P as follows:

$$g^\omega(X, Y) = g_B\big(\tau_*(X), \tau_*(Y)\big) + g_{\mathfrak{s}^1}\big(\omega(X), \omega(Y)\big),$$

where $g_{\mathfrak{s}^1}(\mathbf{i}t_1, \mathbf{i}t_2) = -t_1 t_2$. The Lorentzian metric g^ω is called a Kaluza–Klein metric on P (the metric given in Section 3.3 on $\mathbb{S}^3$ may be of course obtained with this procedure on the Hopf fibration).

PROPOSITION. *If B is compact and simply connected and (P, g^ω) has no conjugate points along its causal geodesics, then $P \simeq B \times \mathbb{S}^1$ (i.e. the principal bundle $P(B, \mathbb{S}^1)$ must be trivial*).

In order to sketch the proof, note that the fundamental vector field $\mathbf{i}^* \in \mathfrak{X}(P)$ corresponding to $\mathbf{i} \in \mathfrak{s}^1$ is a unit timelike Killing vector field. For every $k \in \mathbb{Z}$, $k \geqslant 1$, let $(\mathbb{T}^k = \mathbb{S}^1 \times \cdots \times \mathbb{S}^1, h)$ be a k-dimensional Riemannian flat torus. The Lorentzian manifold $(P \times \mathbb{T}^k, g^\omega + h)$ has no conjugate points along its null geodesics. Moreover, $\mathbf{i}^*$ may be seen as a timelike Killing vector field on $P \times \mathbb{T}^k$. Therefore, a straightforward computation from (3.2) and (3.3) gives

$$\int_P \big[(n + k + 1)\widetilde{\mathrm{Ric}}(\mathbf{i}^*) + S\big]\, d\mu_{g^\omega} \leqslant 0.$$

Thus,

$$\int_P \widetilde{\mathrm{Ric}}(\mathbf{i}^*)\, d\mu_{g^\omega} \leqslant 0$$

and from (3.4) we get that $\mathbf{i}^*$ must be parallel, therefore the curvature of ω vanishes identically. The announced result easily follows taking into account that B is simply connected.

We end the subsection with a result related to the previous inequality (3.2). In fact, the no conjugacy assumption on a fixed causal geodesic of a Lorentzian manifold has been used by Ehrlich and Kim in [30] to obtain a generalization of the Hawking–Penrose conjugacy theorem [45] of singularity theory, which is stated as follows.

THEOREM. *Let $\gamma : (-\infty, +\infty) \to M$ be a complete nonspacelike geodesic in a Lorentzian manifold (M, g) (of $\dim M \geqslant 3$ if γ is null) without conjugate points. Then,*

$$\liminf_{s \to +\infty} \int_{-s}^{s} \widetilde{\mathrm{Ric}}(\gamma')\, dt \leqslant 0.$$

Moreover, if γ is timelike then equality holds if and only if

$$\mathcal{K}(\Pi) = 0,$$

for every Lorentzian tangent plane containing $\gamma'(t)$. In the case that γ is null, equality holds if and only if

$$\mathcal{K}_{\gamma'(t)}(\Pi) = 0,$$

for every degenerate tangent plane Π containing $\gamma'(t)$.

4. The Bochner technique on Lorentzian manifolds

4.1. *Focusing aims and difficulties*

Classically, the Bochner technique consists of the vanishing of certain geometric objects of interest on a Riemannian manifold under the assumption of positive or negative definite curvature everywhere. This technique constitutes now a basic topic in Riemannian geometry, and its extension to Lorentzian geometry is not by any means obvious. Although the actual development of the Bochner technique has been impressive [66, Chapter 7], [93] and references therein, [85], we recall two pioneering and important results for testing the serious difficulties of a direct application of the Bochner technique to Lorentzian geometry.

THEOREM. *Let (M, g) be an $n(\geqslant 2)$-dimensional Riemannian manifold such that its Ricci tensor is negative semi-definite everywhere and negative definite at some point p_0. If X is a Killing vector field on (M, g) such that the function $|X|^2$ has a relative maximum, then $X = 0$.*

This result is essentially the famous theorem by Bochner in [16], including the improvement by Wu in [92]. A well-known geometric consequence can be derived from this result on the isometry Lie group $\mathrm{Iso}(M, g)$ of a compact Riemannian manifold (M, g), which is also compact (see, for instance, [51, Chapter VI, Theorem 3.4]). In fact, taking into account that, under the compactness assumption, the Lie algebra of $\mathrm{Iso}(M, g)$ is naturally identified with the Lie algebra of Killing vector fields on (M, g), we have

COROLLARY. *Let (M, g) be a compact Riemannian manifold. If its Ricci tensor is negative semi-definite everywhere and negative definite at some point p_0, then* $\mathrm{Iso}(M, g)$ *is finite.*

The proof of the Bochner theorem follows from the elementary maximum principle for subharmonic functions and the following well-known formula:

$$\Delta\frac{1}{2}|X|^2 = |\nabla X|^2 - \widetilde{\mathrm{Ric}}(X), \tag{4.1}$$

where Δ is the Laplacian on functions, $|X|^2 = g(X, X)$ and $|\nabla X|^2 = g(\nabla X, \nabla X)$ is the square norm of the $(1, 1)$-tensor field ∇X.

Of course, formula (4.1) remains true for a Killing vector field X on any semi-Riemannian manifold (M, g). But:

- Although the D'Alembertian $\Box$ of a Lorentzian metric g is defined formally equal to the Laplacian in the Riemannian case, and formula (4.1) holds if we change the Laplacian Δ to the D'Alembertian $\Box$, it should be noticed that $\Box$ is not an elliptic operator, and so $\Box f = 0$ on a compact Lorentzian manifold does not imply that f is constant.
- The induced metric on $(1, 1)$-tensor fields is indefinite in the Lorentzian case and so $|\nabla X|^2$ has no definite sign, even in the case where X is Killing.
- The negative semi-definite Ricci tensor assumption is not realistic in the Lorentzian case. Note that if $\mathrm{Ric} = \lambda g$ holds and Ric is negative (or positive) semi-definite then $\lambda = 0$.
- Finally, the isometry group $\mathrm{Iso}(M, g)$ of a Lorentzian manifold (M, g) is also a Lie group. However, in general the compactness of M does not imply that $\mathrm{Iso}(M, g)$ is compact [26] (see also [78,79]).

In spite of these difficulties, the Bochner technique has been introduced in Lorentzian manifolds [74–76,68] (see also [86]). In this section, we will summarize strategies and main results with outlined proofs from these papers. Comments relating to the Riemannian case will be also included. This development of the Bochner technique is indeed a part of the analysis on manifolds in the Lorentzian case (compact Lorentzian manifolds are also of interest in physics [95]).

4.2. *Integral approach in the compact case*

Let (M, g) be an arbitrary semi-Riemannian manifold, then for any $X \in \mathfrak{X}(M)$ we have

$$X\,\mathrm{div}(X) = -\widetilde{\mathrm{Ric}}(X) + \mathrm{div}(\nabla_X X) - \mathrm{trace}\big(A_X^2\big), \tag{4.2}$$

where div denotes the divergence on (M, g) and A_X is the linear operator on T_pM defined by $A_X(v) = -\nabla_v X$, for any $v \in T_pM$. Note that it satisfies trace $A_X = -\operatorname{div}(X)$. Furthermore, if X is assumed to be Killing then $\operatorname{div}(X) = 0$, $\nabla\frac{1}{2}|X|^2 = -\nabla_X X$ and $\operatorname{trace}(A_X^2) = -|\nabla X|^2$ and so (4.2) reduces to Eq. (4.1).

Since $\operatorname{div}(\operatorname{div}(X)X) = X\operatorname{div}(X) + (\operatorname{div}(X))^2$, if M is compact, we can deduce from (4.2), making use of the classical divergence theorem,

$$\int_M \{\widetilde{\operatorname{Ric}}(X) + \operatorname{trace}(A_X^2) - (\operatorname{trace} A_X)^2\}\, d\mu_g = 0, \tag{4.3}$$

where $d\mu_g$ is the canonical measure induced from g.

In the Riemannian case the difference $\delta_X = \operatorname{trace}(A_X^2) - (\operatorname{trace} A_X)^2$ is signed in some relevant cases, e.g., $\delta_X \leqslant 0$ if X is conformal (Killing, in particular), and $\delta_X \geqslant 0$ if X is harmonic. However, none of them yield a constant sign for δ_X in the Lorentzian case even assuming that X has a fixed causal character. For instance, if X is Killing, the linear operator A_X is skew-adjoint with respect to the Lorentzian metric g. Hence, trace $A_X = -\operatorname{div}(X) = 0$ and $\delta_X = \operatorname{trace}(A_X^2)$. Contrary to the definite case, the last equality does not give a sign to δ_X.

We are going to introduce a special kind of vector fields on Lorentzian manifolds which allow us to avoid these difficulties. Recall that a *reference frame* on a Lorentzian manifold (M, g) is a vector field Z on M which satisfies $g(Z, Z) = -1$. In general relativity, a reference frame in spacetime is seen as a vector field such that each of its integral curves is an observer (i.e. a particle with unit mass) [77, Definition 2.3.1]. If Z is a reference frame in (M, g) then we can write the following orthogonal decomposition:

$$T_pM = \operatorname{Span}\{Z_p\} \oplus Z_p^{\perp},$$

at any $p \in M$, where $Z_p^{\perp} = (\operatorname{Span}\{Z_p\})^{\perp}$ is the g-orthogonal complement of $\operatorname{Span}\{Z_p\}$. Clearly, $Z_p^{\perp}$ is spacelike and A_Z-invariant. So, we call A'_Z the corresponding linear operator of $Z_p^{\perp}$. Besides the algebraic advantages of A'_Z over A_Z, A'_Z contains almost all the information that A_Z does. In fact, a direct computation yields

$$\operatorname{trace} A'_Z = \operatorname{trace} A_Z \quad \text{and} \quad \operatorname{trace}(A'^2_Z) = \operatorname{trace}(A^2_Z).$$

Consequently, for a reference frame Z the difference δ_Z is expressed as $\delta_Z = \operatorname{trace}(A'^2_Z) - (\operatorname{trace} A'_Z)^2$. Now we can decompose $A'_Z = S'_Z + H'_Z$, at any $p \in M$, where S'_Z (respectively H'_Z) is self-adjoint (respectively skew-adjoint) with respect to the positive definite inner product $g|_{Z_p^{\perp}}$. An easy algebraic argument shows that $\delta_Z = -\sigma'_Z + \operatorname{trace}(H'^2_Z)$, where $\sigma'_Z = (\operatorname{trace} S'_Z)^2 - \operatorname{trace}(S'^2_Z)$. Now we are in a position to state [75]

THEOREM. *Let (M, g) be an n-dimensional compact Lorentzian manifold. For any reference frame Z on (M, g) we have*

$$\int_M \{\widetilde{\operatorname{Ric}}(Z) - \sigma'_Z + \operatorname{trace}(H'^2_Z)\}\, d\mu_g = 0.$$

We will call this integral formula the *Lorentzian Bochner formula*.

If $\dim M = 2$ then $A'_Z = \lambda(p)I$, at any $p \in M$, and therefore $H'_Z = 0$ and $\sigma'_Z = 0$, taking into account that the Gauss curvature of (M, g) satisfies $\mathcal{K} = -\widetilde{\mathrm{Ric}}(Z)$, then the Lorentzian Bochner formula reduces to $\int_M \mathcal{K}\, d\mu_g = 0$, which is the well-known Gauss–Bonnet theorem for time-orientable Lorentzian metrics (and hence for any Lorentzian metric) on M.

Note that $\mathrm{trace}(H'^2_Z) \leqslant 0$ and equality holds at $p \in M$ if and only if $H'_Z = 0$ at this point p, and so we get [75]

THEOREM. *Let (M, g) be an $n(\geqslant 3)$-dimensional compact Lorentzian manifold and let Z be a reference frame on (M, g). If $\sigma'_Z \geqslant 0$, then*

$$\int_M \widetilde{\mathrm{Ric}}(Z)\, d\mu_g \geqslant 0$$

and equality holds if and only if $H'_Z = 0$ and $\sigma'_Z = 0$.

There are remarkable families of reference frames Z of geometric interest which satisfy the assumption $\sigma'_Z \geqslant 0$. In fact, a reference frame Z is said to be *spatially conformal* (respectively *spatially stationary* or *rigid*) if $(\mathcal{L}_Z g)(U, V) = 2\rho g(U, V)$, where $\rho : M \to \mathbb{R}$ (respectively $(\mathcal{L}_Z g)(U, V) = 0$) for all $U, V \perp Z$ [77,33]. It is easy to see that a reference frame Z is spatially conformal (respectively spatially stationary) if and only if ${}^tA'_Z = -A'_Z - 2\rho I$, where ${}^tA'_Z$ denotes the adjoint operator of A'_Z and I the identity transformation of $Z_p^\perp$ (respectively ${}^tA'_Z = -A'_Z$). A Lorentzian manifold which admits a timelike conformal (respectively Killing) vector field is called *conformally stationary* [3] (respectively *stationary*).

The study of conformal vector fields in semi-Riemannian geometry is a subject of interest [54] and the existence of a timelike conformal (or Killing) vector field on a Lorentzian manifold has been shown to be useful in solving several mathematical problems [47,71, 2–5] (see also Section 3.4).

Note that if X is a timelike conformal (respectively Killing) vector field on (M, g), then the reference frame $Z = (1/\sqrt{-g(X, X)})X$ is spatially conformal (respectively spatially stationary). However, there exist spatially conformal reference frames which cannot be obtained in that way. In fact, if $\dim M = 2$, then every reference frame is indeed spatially conformal. But a (time-orientable) incomplete Lorentzian torus does not admit a timelike conformal vector field as showed in Section 2.2.

If Z is a spatially conformal reference frame then $S'_Z = -\rho I$ and $\sigma'_Z = (n-1)(n-2)\rho^2$. Therefore, we get [75]

THEOREM. *Let (M, g) be an $n(\geqslant 3)$-dimensional compact Lorentzian manifold. If (M, g) admits a spatially conformal reference frame Z, then*

$$\int_M \widetilde{\mathrm{Ric}}(Z)\, d\mu_g \geqslant 0$$

and equality holds if and only if $A'_Z = 0$.

The equality case in previous result can be improved by assuming that Z is rigid. Taking into account that when Z is rigid, $A'_Z = 0$ holds if and only if Z is irrotational (i.e. the distribution $Z^\perp$ is integrable), we get [74]

COROLLARY. *Let (M, g) be an $n(\geqslant 3)$-dimensional compact Lorentzian manifold admitting a rigid reference frame Z. Then*

$$\int_M \widetilde{\mathrm{Ric}}(Z)\, d\mu_g \geqslant 0$$

and equality holds if and only if Z is irrotational.

As a direct consequence we have [75]

COROLLARY. *If an $n(\geqslant 3)$-dimensional compact Lorentzian manifold (M, g) is Einstein with* $\mathrm{Ric} = \lambda g$, $\lambda \in \mathbb{R}$, *and admits a spatially conformal reference frame (in particular, a timelike conformal vector field), then $\lambda \leqslant 0$.*

Kamishima proved in [47, Theorem A] that if a compact Lorentzian manifold with constant sectional curvature $c \in \mathbb{R}$ admits a timelike Killing vector field, then $c \leqslant 0$. So the previous corollary is a clear extension to his result. It should be noted that the tools by Kamishima in [47] are strongly depending on the Lie group machinery of Lorentzian space forms, and therefore very different from the ones involved in this technique.

The following classification theorem was given in [74] for the Killing case and later extended in [75] to the conformal case.

THEOREM. *Let (M, g) be a Ricci-flat $n(\geqslant 3)$-dimensional compact Lorentzian manifold. If (M, g) admits a timelike conformal vector field X, then*

X is parallel, the first Betti number of M is not zero and the Levi-Civita connection of g is Riemannian.

Moreover, if one of the following conditions holds:

(1) *(M, g) is homogeneous,*
(2) *(M, g) is flat (in particular if $n = 3$),*
(3) $n = 4$,

then (M, g) is isometric (up to a finite covering in the cases (2) *and* (3)) *to a flat n-dimensional Lorentzian torus.*

That a flat compact Lorentzian manifold admitting a timelike Killing vector field is affinely diffeomorphic to a Riemannian manifold with nonzero first Betti number was proved by Kamishima in [47, Theorem A(1)]. The previous theorem then is a wide extension of his result.

A compact Ricci-flat Lorentzian manifold admitting a timelike conformal vector field is not necessarily either flat or diffeomorphic (up to a covering) to an n-torus. A (nonhomogeneous) counterexample for any dimension $n \geqslant 5$ can be constructed as follows. Let

$N \subset \mathbb{C}P^m$ be a compact complex hypersurface with degree $m+1$ of the complex projective space $\mathbb{C}P^m$ with complex dimension $m \geqslant 3$. For instance, if $m = 3$, N is a K3 complex surface. The first Chern class of N vanishes and therefore, by a well-known theorem by Yau [94], N admits a Ricci-flat but nonflat Riemannian metric g_N. If we put $M = \mathbb{S}^1 \times \overset{(k)}{\cdots} \times \mathbb{S}^1 \times N$, endowed with the Lorentzian metric $-d\theta^2 + g_N$, where $d\theta^2$ is the usual Riemannian metric of $\mathbb{S}^1$, when $k = 1$, and with the Lorentzian metric $-d\theta_1^2 + d\theta_2^2 + \cdots + d\theta_k^2 + g_N$ when $k \geqslant 2$, then we get the desired example.

On the other hand, it is known that there exist flat n-dimensional compact Lorentzian manifolds, $n \geqslant 3$, which cannot be covered by an n-torus (see [47] and references therein). Thus, the previous result can be seen as an obstruction to the existence of timelike conformal vector fields on these Lorentzian manifolds.

Whereas a Ricci-flat homogeneous Riemannian manifold must be flat [1], there exist Ricci-flat homogeneous Lorentzian manifolds which are not flat [17]. In a natural way, the following question arises.

When is a Ricci-flat homogeneous Lorentzian manifold flat?

An answer is then given in the above theorem.

We end this subsection giving another application of the previous result. Making use of the De Rham–Wu decomposition theorem [91] we get the following obstruction [75].

THEOREM. *If an* $n(\geqslant 3)$*-dimensional compact Lorentzian manifold* (M, g) *is simply-connected and satisfies* $\operatorname{Ric}(Y, Y) \leqslant 0$ *for all timelike* Y*, then* (M, g) *admits no timelike conformal vector field.*

4.3. *Hessian approach in the noncompact case*

In the last subsection, the previous strategy for applying the Bochner technique on Lorentzian manifolds will be changed. Compactness is not assumed and projective (in particular affine) vector fields are considered. These vector fields have a rich, wide geometry [27,67,6] which has also shown to be useful in physics to analyze spacetime (see [27, Chapters 7, 8] and references therein); in particular the existence of the symmetry of a spacetime defined by a projective or affine vector field has provided new exact solutions for the Einstein equation.

Recall that a vector field X on a Lorentzian[9] manifold (M, g) is said to be *projective* if its local flows preserve geodesics of (M, g) in a set-theoretic sense. If the fluxes of X preserve geodesics in a mapping sense then it is called *affine*. It is known that X is projective if and only if there exists a 1-form μ on M such that

$$(\mathcal{L}_X \nabla)(U, V) = \mu(U)V + \mu(V)U,$$

for all $U, V \in \mathcal{X}(M)$, in fact $\mu = (1/(n+1))d(\operatorname{div}(X))$ (see [67, Propositions 5.27, 5.28], for instance). Note that X is affine if $\mu = 0$.

[9] or semi-Riemannian

Therefore, if X is projective, we get

$$\nabla_U \nabla_V X - \nabla_{\nabla_U V} X = R(U, X)V + \mu(U)V + \mu(V)U,$$

for all $U, V \in \mathfrak{X}(M)$.

Now, we want to compute the Hessian of the function $\frac{1}{2}|X|^2$, where X is a projective vector field on M. Using the previous formula we get

$$\left(\operatorname{Hess} \frac{1}{2}|X|^2\right)(U, V) = -g\big(R(U, X)X, V\big) + g(\nabla_U X, \nabla_V X) \\ + \mu(U)g(X, V) + \mu(V)g(X, U), \tag{4.4}$$

for all $U, V \in \mathfrak{X}(M)$.

Now, we are in a position to state [76]

THEOREM. *Let (M, g) be an $n(\geqslant 2)$-dimensional Lorentzian manifold and let X be a projective vector field on M. If the function $|X|^2$ attains a relative maximum at some $p_0 \in M$ and X_{p_0} is causal, then*

$$g\big(R(v, X_{p_0})X_{p_0}, v\big) \geqslant 0,$$

for all $v \in T_{p_0}M$ orthogonal to X_{p_0}, and therefore

$$\widetilde{\operatorname{Ric}}(X_{p_0}) \geqslant 0.$$

In particular:

(a) *If X_{p_0} is timelike, then the sectional curvature of any (nondegenerate) plane Π in $T_{p_0}M$ containing X_{p_0} satisfies*

$$K(\Pi) \leqslant 0, \tag{4.5}$$

and if the equality holds for such planes Π, then $\nabla_v X = 0$ for all $v \perp X_{p_0}$.

(b) *If X_{p_0} is null, then $\nabla_{X_{p_0}} X$ is proportional to X_{p_0}. In the case $n \geqslant 3$, the null sectional curvature with respect to X_{p_0} of any degenerate plane Π containing X_{p_0} satisfies*

$$\mathcal{K}_{X_{p_0}}(\Pi) \geqslant 0.$$

If the equality holds for all such planes, then $\nabla_v X_{p_0}$ is proportional to X_{p_0} for all $v \perp X_{p_0}$.

We only give a sketch of the proof for the timelike case (the lightlike case works similarly). As p_0 is a critical point of the function $|X|^2$ then $g(\nabla_w X, X_{p_0}) = 0$, for all $w \in T_{p_0}M$. When X_{p_0} is timelike, the fact $\nabla_w X \perp X_{p_0}$ implies $g(\nabla_w X, \nabla_w X) \geqslant 0$, and equality holds for some vector w if and only if $\nabla_w X = 0$. On the other hand, if p_0 is assumed to be a relative maximum, then $(\text{Hess}\,\frac{1}{2}|X|^2)_{p_0}$ must be negative semi-definite. Therefore, from (4.4) we get

$$g\big(R(v, X_{p_0})X_{p_0}, v\big) \geqslant g(\nabla_v X, \nabla_v X) \geqslant 0, \tag{4.6}$$

for all $v \in T_{p_0}M$ orthogonal to X_{p_0}. Now (4.6) clearly implies (4.5). Moreover, $\nabla_v X$ is spacelike and, therefore, the consequence of the equality in (4.5) follows on from (4.6).

As a consequence previous result yields [76]

COROLLARY. *If an $n(\geqslant 3)$-dimensional compact Lorentzian manifold (M, g) is Einstein with* $\mathrm{Ric} = \lambda g$, $\lambda \in \mathbb{R}$, *and admits a timelike projective vector field, then* $\lambda \leqslant 0$.

Now, the assumption in the previous theorem on the vector field is changed to a stronger one. Suppose that X is affine and leave the remaining assumptions the same. From (4.4) with $\mu = 0$ we get

$$0 \geqslant \left(\text{Hess}\,\frac{1}{2}|X|^2\right)_{p_0}(X_{p_0}, X_{p_0}) = g(\nabla_{X_{p_0}} X, \nabla_{X_{p_0}} X),$$

which implies $\nabla_{X_{p_0}} X = 0$. We can state then the following result [76].

PROPOSITION. *Let (M, g) be an $n(\geqslant 2)$-dimensional Lorentzian manifold and let X be an affine vector field on M. If the function $|X|^2$ attains a relative maximum at some $p_0 \in M$, X_{p_0} is timelike and*

$$\mathcal{K}(\Pi) = 0,$$

for any plane Π in $T_{p_0}M$ containing X_{p_0}, then $(\nabla X)_{p_0} = 0$.

We end this subsection with a result which shows how the Bochner technique can be applied to get nonexistence results of certain types of Lorentzian manifolds.

PROPOSITION. *Let (M, g) be a Lorentzian manifold with dimension $2n$, $n \geqslant 1$, and let X be a Killing vector field on (M, g). Assume the function $|X|^2$ attains a local minimum at $p_0 \in M$ and $|X|^2(p_0) < 0$. Then there is a tangent plane $\Pi \subset T_pM$ such that $X_p \in \Pi$ and*

$$\mathcal{K}(\Pi) \geqslant 0.$$

The proof easily follows if we note that the spacelike subspace $X_{p_0}^{\perp}$ of $T_{p_0}M$ is A_X-invariant and has $\dim X_p^{\perp} = 2n - 1$. Therefore, the corresponding operator A'_X of $X_{p_0}^{\perp}$ must have at least an eigenvalue which is zero because A'_X is skew-adjoint. Hence, there exists $v \in X_{p_0}^{\perp}$, $v \neq 0$, such that $\nabla_{v_0} X = 0$.

On the other hand, $(\text{Hess}\,\frac{1}{2}|X|^2)_{p_0}$ must be positive semi-definite because p_0 is assumed to be a local minimum and therefore (4.4) is claimed to get $\mathcal{K}(\Pi) \geqslant 0$, where $\Pi = \mathrm{Span}\{v, X_{p_0}\}$.

COROLLARY. *There exists no $2n(\geqslant 4)$-dimensional compact Lorentzian manifold which admits a timelike Killing vector field and satisfies*

$$\mathcal{K}(\Pi) < 0,$$

for any Lorentzian tangent plane Π.

Using the Gauss–Bonnet–Chern formula [21], one can deduce that an even dimensional compact semi-Riemannian manifold with constant sectional curvature $k \neq 0$ has nonzero Euler–Poincaré characteristic. Therefore, there exists no even dimensional compact Lorentzian manifold with nonzero constant sectional curvature (recall that, for the case of positive constant sectional curvature, this assertion is contained in the consequence of the Klinger theorem stated in Section 2.3).

Note that only the assumption on the behavior of the sectional curvature on Lorentzian planes, in previous corollary, does not give in general a concrete sign to the integrand function in the Gauss–Bonnet–Chern formula.

References

[1] D.V. Alekseevski and B.N. Kimelfeld, *Structure of homogeneous Riemannian spaces with zero Ricci curvature*, Funct. Anal. Appl. **9** (1975), 97–102.

[2] L.J. Alías, A. Romero and M. Sánchez, *Uniqueness of complete spacelike hypersurfaces of constant mean curvature in generalized Robertson–Walker spacetimes*, Gen. Relativity Gravitation **27** (1995), 71–84.

[3] L.J. Alías, A. Romero and M. Sánchez, *Spacelike hypersurfaces of constant mean curvature in certain spacetimes*, Nonlinear Anal. **30** (1997), 655–661.

[4] L.J. Alías, A. Romero and M. Sánchez, *Spacelike hypersurfaces of constant mean curvature and Calabi–Bernstein type problems*, Tôhoku Math. J. **49** (1997), 337–345.

[5] L.J. Alías, A. Romero and M. Sánchez, *Spacelike hypersurfaces of constant mean curvature in spacetimes with symmetries*, Pub. R.S.M.E. **1** (2000), 1–14.

[6] A.V. Aminova, *Projective transformations of pseudo-Riemannian manifolds*, J. Math. Sci. **113** (2003), 367–470.

[7] L. Andersson and R. Howard, *Comparison and rigidity theorems in semi-Riemannian geometry*, Comm. Anal. Geom. **6** (1998), 819–877.

[8] A. Avez, *Formule de Gauss–Bonnet–Chern en métrique de signature quelconque*, C. R. Acad. Sci. Paris **255** (1962), 2049–2051.

[9] J.K. Beem, *Some examples of incomplete spacetimes*, Gen. Relativity Gravitation **7** (1976), 501–509.

[10] J.K. Beem, P.E. Ehrlich and K.L. Easley, *Global Lorentzian Geometry*, 2nd ed., Pure and Applied Math., vol. 202, Dekker (1996).

[11] J.K. Beem and P.E. Parker, *Values of pseudo-Riemannian sectional curvature*, Comment. Math. Helv. **59** (1984), 319–331.

[12] A.N. Bernal and M. Sánchez, *On smooth Cauchy hypersurfaces and Geroch's splitting theorem*, Commun. Math. Phys. **243** (2003), 461–470.

[13] A.N. Bernal and M. Sánchez, *Smoothness of time functions and metric splitting of globally hyperbolic spacetimes*, Commun. Math. Phys., in press.

[14] G.S. Birman and K. Nomizu, *The Gauss–Bonnet theorem for* 2*-dimensional spacetimes*, Michigan Math. J. **31** (1984), 77–81.

[15] R.L. Bishop and R.J. Crittenden, *Geometry of Manifolds*, Academic Press (2001).

[16] S. Bochner, *Vector fields and Ricci curvature*, Bull. Amer. Math. Soc. **52** (1946), 776–797.

[17] M. Cahen and N. Wallach, *Lorentzian symmetric spaces*, Bull. Amer. Math. Soc. **76** (1970), 585–591.

[18] E. Calabi and L. Markus, *Relativistic space forms*, Ann. of Math. **75** (1962), 63–76.

[19] Y. Carrière, *Autour de la conjeture de L. Markus sur las variétés affines*, Invent. Math. **95** (1989), 615–628.

[20] Y. Carrière and L. Rozoy, *Complétude des métriques Lorentziennes*, Bol. Soc. Brasil. Mat. **25** (1994), 223–235.

[21] S.S. Chern, *Pseudo-Riemannian geometry and the Gauss–Bonnet formula*, An. Acad. Brasil. Ciênc. **35** (1963), 17–26.

[22] Y.S. Choi and Y.J. Suh, *Remarks on the topology of Lorentzian manifolds*, Comm. Korean Math. Soc. **15** (2000), 641–648.

[23] C.J.S. Clarke, *On the geodesic completeness of causal space-times*, Proc. Cambridge Philos. Soc. **69** (1971), 319–324.

[24] A.A. Coley and B.O.J. Tupper, *Special conformal Killing vector spacetimes and symmetry inheritance*, J. Math. Phys. **30** (1989), 2616–2625.

[25] M. Dajczer and K. Nomizu, *On the boundedness of Ricci curvature of an indefinite metric*, Bol. Soc. Brasil. Mat. **11** (1980), 25–30.

[26] G. D'Ambra, *Isometry groups of Lorentz manifolds*, Invent. Math. **92** (1988), 555–565.

[27] K.L. Duggal and R. Sharma, *Symmetries of spacetimes and Riemannian manifolds*, Math. and Its Appl., vol. 487, Kluwer Academic (1999).

[28] D. Earley, J. Isenberg, J. Marsden and V. Moncrief, *Homothetic and conformal symmetries of solutions to Einstein's equations*, Commun. Math. Phys. **106** (1986), 137–158.

[29] P.E. Ehrlich, Y.T. Jung and S.-B. Kim, *Volume comparison theorems in Lorentzian manifolds*, Geom. Dedicata **73** (1998), 39–56.

[30] P.E. Ehrlich and S.-B. Kim, *From the Riccati inequality to the Raychaudhuri equation*, Contemp. Math. **170** (1994), 65–78.

[31] P.E. Ehrlich and M. Sánchez, *Some semi-Riemannian volume comparison theorems*, Tôhoku Math. J. **52** (2000), 331–348.

[32] J.L. Flores and M. Sánchez, *Geodesic connectedness and conjugate points in GRW spacetimes*, J. Geom. Phys. **30** (2000), 285–314.

[33] E. García-Río and D.N. Kupeli, *Singularity versus splitting theorems for stably causal spacetimes*, Ann. Global Anal. Geom. **14** (1996), 301–312.

[34] R.P. Geroch, *What is a singularity in General Relativity*? Ann. Phys. **48** (1968), 526–540.

[35] R.P. Geroch, *Domains of dependence*, J. Math. Phys. **11** (1970), 437–449.

[36] W.H. Greub, *Line fields on Lorentzian manifolds*, Differential Geometric Methods in Math. Phys., Lectures Notes in Physics, vol. 139 (1981), 290–309.

[37] W.H. Greub, S. Halperin, and R. Vanstone, *Connections, Curvature and Cohomology*, vol. 1, Academic Press, New York (1972).

[38] M. Gutiérrez, F.J. Palomo and A. Romero, *A Berger–Green type inequality for compact Lorentzian manifolds*, Trans. Amer. Math. Soc. **354** (2002), 4505–4523; Erratum **355** (2003), 5119–5120.

[39] M. Gutiérrez, F.J. Palomo and A. Romero, *Conjugate points along null geodesics on Lorentzian manifolds with symmetries*, Proc. X Fall Workshop on Geometry and Physics, Miraflores de la Sierra (Madrid) 2001, Pub. R.S.M.E. **4** (2003), 169–182.

[40] M. Gutiérrez, F.J. Palomo and A. Romero, *A new approach to the study of conjugate points along null geodesics on certain compact Lorentzian manifolds*, Lorentzian Geometry Benalmádena 2001, Pub. R.S.M.E. **5** (2003), 171–182.

[41] M. Gutiérrez, F.J. Palomo and A. Romero, *Lorentzian manifolds with no null conjugate points*, Math. Proc. Cambridge Philos. Soc. **137** (2004), 363–375.

[42] S. Haesen, *Inequalities and symmetries of numerical curvatures and their applications in General Relativity*, Ph.D. dissertation, Katholieke Universiteit Leuven (2004).

[43] S.G. Harris, *A triangle comparison theorem for Lorentz manifolds*, Indiana Univ. Math. J. **31** (1982), 289–308.

[44] S.G. Harris, *A characterization of Robertson–Walker spaces by null sectional curvature*, Gen. Relativity Gravitation **17** (1985), 493–498.

[45] S.W. Hawking and R. Penrose, *Singularities of gravitational collapse and cosmology*, Proc. Roy. Soc. London A **314** (1970), 529–548.

[46] E. Hopf, *Closed surfaces without conjugate points*, Proc. Nat. Acad. Sci. USA **34** (1948), 47–51.

[47] Y. Kamishima, *Completeness of Lorentz manifolds of constant curvature admitting Killing vector fields*, J. Differential Geom. **37** (1993), 569–601.

[48] H. Karcher, *Infinitesimale Charakterisierung von Friedmann–Universen*, Arch. Math. **38** (1982), 58–64.

[49] M.A. Kishta, *Sectional curvature results in a Lorentzian manifold*, J. Inst. Math. Sci. Math. Ser. **4** (1991), 249–253.

[50] B. Klinger, *Complétude des variétes lorentziennes à courbure constante*, Math. Ann. **306** (1996), 353–370.

[51] S. Kobayashi and K. Nomizu, *Foundations of Differential Geometry*, vol. 1, Wiley, New York (1963).

[52] L. Koch-Sen, *Infinitesimal null isotropy and Robertson–Walker metrics*, J. Math. Phys. **26** (1985), 407–410.

[53] R. Kulkarni, The values of sectional curvature in indefinite metrics, *Comment. Math. Helv.* **54** (1979), 173–176.

[54] W. Kühnel and H.B. Rademacher, *Conformal vector fields on pseudo-Riemannian spaces*, Differential Geom. Appl. **7** (1997), 237–250.

[55] W. Kundt, *Note on the completeness of spacetimes*, Z. Phys. **172** (1963), 488–499.

[56] J. Lafuente-López, *A geodesic completeness theorem for locally symmetric Lorentz manifolds*, Rev. Mat. Univ. Complut. Madrid **1** (1988), 101–110.

[57] K.B. Marathe, *A condition for paracompactness of a manifold*, J. Differential Geom. **7** (1972), 571–573.

[58] J.E. Marsden, *On completeness of pseudo-Riemannian manifolds*, Indiana Univ. Math. J. **22** (1973), 1065–1066.

[59] P. Mounoud, *Sur l'espace des métriques de Lorentz d'une variété*, Ph.D. dissertation, Université Montpellier II (2001).

[60] P. Mounoud, *Some topological and metric properties of the space of the Lorentz metrics*, Differential Geom. Appl. **15** (2001), 47–57.

[61] K. Nomizu, *Remarks on sectional curvature of an indefinite metric*, Proc. Amer. Math. Soc. **89** (1983), 473–476.

[62] K. Nomizu and H. Ozeki, *The existence of complete Riemannian metrics*, Proc. Amer. Math. Soc. **12** (1961), 889–891.

[63] B. O'Neill, *Semi-Riemannian Geometry with Applications to Relativity*, Academic Press, New York (1983).

[64] R.S. Palais, *On the existence of slices for actions of non-compact Lie groups*, Ann. of Math. **73** (1961), 295–323.

[65] F.J. Palomo, *Puntos conjugados de geodésicas luminosas en variedades de Lorentz*, Ph.D. dissertation, Universidad de Málaga (2003).

[66] P. Petersen, *Riemannian Geometry*, Graduate Texts in Math., vol. 171, Springer-Verlag (1998).

[67] W.A. Poor, *Differential Geometric Structures*, McGraw–Hill (1981).

[68] A. Romero, *The introduction of Bochner's technique on Lorentzian manifolds*, Nonlinear Anal. **47** (2001), 3047–3059.

[69] A. Romero and M. Sánchez, *On completeness of geodesics obtained as a limit*, J. Math. Phys. **34** (1993), 3768–3774.

[70] A. Romero and M. Sánchez, *New properties and examples of incomplete Lorentzian tori*, J. Math. Phys. **35** (1994), 1992–1997.

[71] A. Romero and M. Sánchez, *On completeness of compact Lorentzian manifolds*, Geometry and Topology of Submanifolds VI, World Scientific (1994), 171–182.

[72] A. Romero and M. Sánchez, *On completeness of certain families of semi-Riemannian manifolds*, Geom. Dedicata **53** (1994), 103–117.

[73] A. Romero and M. Sánchez, *Completeness of compact Lorentz manifolds admitting a timelike conformal-Killing vector field*, Proc. Amer. Math. Soc. **123** (1995), 2831–2833.

[74] A. Romero and M. Sánchez, *An integral inequality on compact Lorentz manifolds and its applications*, Bull. London Math. Soc. **28** (1996), 509–513.

[75] A. Romero and M. Sánchez, *Bochner's technique on Lorentz manifolds and infinitesimal conformal symmetries*, Pacific J. Math. **186** (1998), 141–148.

[76] A. Romero and M. Sánchez, *Projective vector fields on Lorentzian manifolds*, Geom. Dedicata **93** (2002), 95–105.

[77] R. Sachs and H. Wu, *General Relativity for Mathematicians*, Graduate Texts in Math., vol. 48, Springer-Verlag (1977).

[78] M. Sánchez, Structure of Lorentzian tori with a Killing vector field, *Trans. Amer. Math. Soc.* **349** (1997), 1063–1080.
[79] M. Sánchez, Lorentzian manifolds admitting a Killing vector field, *Nonlinear Anal.* **30** (1997), 643–654.
[80] M. Sánchez, *Causal hierarchy of spacetimes, temporal functions ans smoothness of Geroch's splitting*, A revision, preprint 2004, arXiv:gr-qc/o411143 v1.
[81] M. Sánchez, *On causality and closed geodesics of compact Lorentzian manifolds and static spacetimes*, Differential Geom. Appl., in press.
[82] H.J. Seifert, *The causal boundary of space-times*, Gen. Relativity Gravitation **1** (1971), 247–259.
[83] H.J. Seifert, *Smoothing and extending cosmic time functions*, Gen. Relativity Gravitation **8** (1977), 815–831.
[84] M. Spivak, *Differential Geometry*, vol. 2, Publish or Perish, Berkeley (1979).
[85] S.E. Stepanov, *The Bochner technique for an m-dimensional compact manifold with $SL(m,\mathbb{R})$-structure*, St. Petersburg Math. J. **10** (1999), 703–714.
[86] S.E. Stepanov, *An analytic method in general relativity*, Theoret. Math. Phys. **122** (2000), 402–414.
[87] J.A. Thorpe, *Curvature and the Petrov canonical forms*, J. Math. Phys. **10** (1969), 1–7.
[88] A.W. Wadsley, *Geodesic foliations by circles*, J. Differential Geometry **10** (1975), 541–549.
[89] M. Wang, Some examples of homogeneous Einstein manifolds in dimension seven, *Duke Math. J.* **49** (1982), 23–28.
[90] J.A. Wolf, *Isotropic manifolds of indefinite metrics*, Comment. Math. Helv. **39** (1964), 21–64.
[91] H. Wu, *On the de Rham decomposition theorem*, Illinois J. Math. **8** (1964), 291–311.
[92] H. Wu, *A remark on the Bochner technique in differential geometry*, Proc. Amer. Math. Soc. **78** (1980), 403–408.
[93] H. Wu, *The Bochner Technique in Differential Geometry*, Mathematical Reports, vol. 3, part 2, Harwood Academic, London (1988).
[94] S.T. Yau, *Calabi's conjecture and some new results in algebraic geometry*, Proc. Nat. Acad. Sci. USA **74** (1997), 1798–1799.
[95] U. Yurtsever, *Test fields on compact space-times*, J. Math. Phys. **31** (1990), 3064–3078.
[96] E. Zafiris, *Irreducible decomposition of Einstein's equations in spacetimes with symmetries*, Ann. Phys. **263** (1998), 155–178.

Author Index

Upright numbers refer to pages on which the author (or his/her work) is mentioned in the text of a chapter. Italic numbers refer to reference list pages. (No distinction is made between first and coauthor(s).)

Subject Index

9780444520524